SCOTTISH SCIENCE SOCIETY

A COMPREHENSIVE PROGRESSIVE TREATISE

Biochemistry

Matter, Energy, Information and Life

A treatise in twelve parts, proceeding from molecular structure to biochemical mechanism, from mechanism to metabolism, and from metabolism to the integrated organism — chemistry first, machinery second, metabolism third, information fourth.

COMPLETE EDITION · TWELVE PARTS · CHAPTERS 1–56 · END MATTER

THE SCOTTISH SCIENCE SOCIETY

EDINBURGH · MMXXVI

BIOCHEMISTRY: MATTER, ENERGY, INFORMATION AND LIFE

A comprehensive progressive treatise, in twelve parts, with preliminary matter, end matter, and indices. Complete edition.

First published in 2026 by the Scottish Science Society, Edinburgh.

The portrait plates acknowledged to the Nobel Foundation are reproduced with gratitude under their respective public licences; the remaining figures were executed for this volume and are the work of the author.

Typeset in a serif face at 8.0 pt on a line measure of 1.2, royal duodecimo (6 × 9 inches), with safe margins for print-on-demand manufacture. Equations are numbered by chapter and are derived, never merely quoted: the reader who encounters an assertion falling from the sky is invited to write to the address above, where a small archive of such meteorological events is kept.

First impression, 2026. 10 9 8 7 6 5 4 3 2 1

"To see what is in front of one's nose needs a constant struggle; yet the nose itself, inspected closely enough, turns out to be a chemical engine of admirable economy, and the struggle, properly conducted, becomes the whole of natural philosophy."

— composed for this volume, after the manner of the essayists

"The cell does not read treatises; but treatises, if they are honest, must learn to read the cell."

— the

PREFACE

Every science has its characteristic temptation. Physics is tempted to suppose that the universe is simple; history, that it is significant; and biochemistry — the youngest and most self-confident of the experimental sciences — is tempted to suppose that it is *finished*. One need only open a modern compendium to see the shape of the temptation: here are the pathways, neatly engraved; here the enzymes, each with its number; here the genomes, each with its address. The whole of life, it would appear, has been filed. The present treatise is written in the conviction that this appearance, though a credit to the industry of a hundred years, is a little premature — and that a subject so young ought not yet to speak of itself in the past tense.

What, then, is attempted here? Not an encyclopaedia, for the world has excellent ones, and an encyclopaedia is by its nature a cabinet in which the objects have ceased to touch one another. Not a laboratory manual, for hands are taught in laboratories and not in prefaces. What is attempted is a *treatise*, in the old and somewhat demanding sense of the word: a continuous argument, proceeding from first principles to integrated consequences, in which nothing is asserted that has not first been derived, measured, or honestly confessed to be unknown. The argument runs in one direction only. It begins with the chemical grammar of living matter — atoms, bonds, water, and the thermodynamic laws that govern their commerce. It proceeds to the machinery written in that grammar: proteins and the enzymes that make them verbs rather than nouns. It continues through the building materials — carbohydrates, lipids, membranes — into the great energetic transactions of respiration and photosynthesis, and thence into metabolism proper, the house-keeping of the cell. Only then, when the chemical foundations are secure, does it admit the informational machinery of nucleic acids, the signalling networks by which cells decide, and at last the methods, the diseases, and the open questions. The order is not an affectation. Biochemistry read backwards — genes first, chemistry later — degenerates, as the reader may have observed elsewhere, into a catalogue of pathways; and a catalogue is a place where understanding goes to be filed rather than to live.

Three commitments govern the manner of the writing, and the reader is entitled to know them in advance. The first is that *prose governs*. Itemisation is the refuge of the hurried, and though tables and figures appear here in abundance — the subject being irreducibly visual — the argument is carried by sentences, which alone among our instruments can hold a *because*. The second is that *nothing falls from the sky*. Every equation used in this volume is derived before it is trusted; every variable is defined at its first appearance; and where a constant is borrowed from physics, its provenance is stated and its dignity acknowledged. The third is that *doubt is a method*, not Each chapter advances its thesis, entertains the strongest objection the could devise, and resolves the quarrel as honestly as the evidence per hich is to say, sometimes by verdict and sometimes by adjournment. The reader who finds this dialectical habit slow is asked to remember that the alternative — the uninterrupted recital of certainties — is not faster science but faster forgetting.

A word about the reader, for a treatise, like a walking tour, must know for whom the pace is set. We have imagined a companion who has completed a sound course in general chemistry, who is unafraid of a logarithm, and who prefers an argument to an assertion; a student, perhaps, or a physician returning to fundamentals, or one of those admirable amateurs who read science the way other people read memoirs — for the pleasure of watching a mind at work. To this reader we address ourselves directly, now and again, throughout; not from condescension, but because a subject that cannot bear an audience does not deserve one. The exercises at the end of each chapter are part of the conversation, and are fully resolved, on the principle that an unresolved problem is a door left open onto a stair that may not exist.

The plan of the whole ? twelve parts and fifty-six chapters in this complete edition ? is set out in the pages that follow, together with the conventions, constants, and chronologies by which the argument keeps its accounts. This volume contains the preliminary matter, all twelve parts and fifty-six chapters, the appendices, conclusion and indexes. We ask only one thing of the reader: that the sequence be respected. Biochemistry, like a proof and unlike a newspaper, loses its sense when read from the middle. And if, at the end of the first chapter, the reader can say not merely what biochemistry studies but why it must study it in this order, then the book will have begun to do its work.

Edinburgh, 2026

A word on what this volume contains, and on its correction. The book opens with preliminary matter — the present preface, an introduction, accounts of its conventions and abbreviations, the principal constants, the amino-acid and nucleotide codes, the lists of figures and tables, a historical chronology, a general map of the subject, and a detailed table of contents. There follow twelve parts comprising fifty-six chapters, from the chemical grammar of life to the unity, variation and engineering of biochemistry; each chapter is announced by a presentation page giving its number, its name and its contents, carries its own historical note, worked cases, fully resolved exercises and references, and is closed by a synthesis checked against the master plan. After the final chapter come the end matter — eight appendices of units, structures, energies, pathways, methods and formulae, the answers to selected quantitative problems, and a guide to further reading — then the conclusion, and last the index of words, in two alphabets: biochemical terms, and organisms, enzymes and diseases. The page numeration begins on the first page of the first chapter and ends on the last page of the index.

It remains to say where the fault lies, and it lies with me. Every error that has survived into print — of fact, of emphasis, of derivation or of spelling — is my own exclusive fault, and no correspondent who points one out will find a defender in me. I would be very glad if any corrections or errors detected by the reader were transmitted to **richard@scottishsciencesociety.org**; they will be acknowledged with thanks and put right at the first opportunity. A book of this size is a long argument with one's own fallibility, and its readers are the final, and best, reviewers.

INTRODUCTION

Biochemistry is the study of the chemistry of life, and this treatise is built on a single wager: that the subject can be presented as one continuous argument rather than as a cabinet of curiosities. The argument runs in four movements, and the twelve parts are their orchestration. Matter first: Parts I to III ask what living chemistry is made of — atoms, water, the twenty amino acids, the sugars and lipids and nucleotides — and how folded molecules acquire the power to catalyse. Energy second: Parts IV to VIII follow the proton and the electron through respiration and photosynthesis into the great pathways of metabolism and their integration in organs and in health and disease. Information third: Parts IX and X trace how sequences are copied, read, regulated and answered, from the double helix to the signalling network. And lastly judgement: Parts XI and XII describe how biochemists measure, model and, increasingly, edit the living world — and what evolution has to say about all of it.

Three rules govern every page. Nothing falls from the sky: each equation is derived from premises the reader has been shown, and the numbered result can be re-derived without faith. Nothing is asserted without a method: the measurement that bought each fact is described, with its controls and its uncertainties, for a number without its error is an opinion in disguise. And nothing is made easy at the price of being false: where the truth is subtle — the folding landscape, the chemiosmotic circuit, the lambda switch — the subtlety is kept, and the reader is trusted with it. The prose is meant to be read, not consulted; the figures are original and their legends are essays in miniature; the exercises are resolved in full, because an exercise without a resolution is a riddle, and a treatise is not a book of riddles.

The reader may enter anywhere, but the design rewards a front-to-back reading: each part's opening page lists its chapters, each chapter's presentation page lists its sections, and each closing synthesis states how the chapter's direction was checked against the master plan. The end matter then serves the working reader — appendices for the desk, answers for the reviser, further reading for the curious, and two indices for everyone. What the book asks of you is patience; what it promises in return is that, by the last page of the index, the cell will no longer look like a bag of chemistry but like what it is: the best story matter has ever told, told here from its first principles to its unfinished edge.

How to Use This Treatise

The architecture of the book is that of an ascending staircase, and the reader is requested — with the courtesy due to a guest and the firmness due to a mason — to climb it one step at a time. Each part presupposes its predecessors and is presupposed by its successors. Cross-references are given throughout in the form (§ 1.4) for sections, Figure 1.3 for figures, and Eq. (1.7) for equations; the first numeral always names the chapter, so that the book can be n without a map — though maps are provided for those who, like the , enjoy them.

Every chapter is constructed in the same manner, and the reader who learns the grammar of one has learned the grammar of all. The argument opens with the *thesis* of the chapter — the claim to be established — proceeds through the numbered sections in which that claim is examined, derived, and occasionally embarrassed, and closes with a fixed apparatus: a *conceptual synthesis*, in which the chapter's logic is retraced in a page; *worked quantitative examples*, in which the chapter's mathematics is made to pay its way in numbers; a *mechanistic case study*, drawn from the experimental record; *interpretative questions*, which have no unique answers and are therefore the most instructive; *fully resolved problems*; and a small set of *research-level propositions*, which point at the frontier and are to be regarded as invitations rather than exercises.

Concerning the figures, one convention deserves notice: every graph in this volume was drawn from data stated in its legend, ry diagram of structure or process is an original composition of the , save where a plate is expressly credited to a public archive. The le e long. This is deliberate. A legend that merely names a figure tells the reader what he is looking at; a legend that explains a figure tells him what he is looking *for* — and the second service is the one a treatise owes.

Finally, a liberty we have taken, and confess. The text addresses the reader in the second person at irregular intervals, poses questions it proceeds to answer, and now and then permits itself an irony at the expense of its own subject. These are not ornaments. A science of life written without a pulse would be a small betrayal of both parties to the contract. The reader who prefers his science mute is advised that the subject itself is anything but.

Conventions, Units and Notation

The notation of this treatise follows the recommendations of the International Union of Pure and Applied Chemistry (IUPAC) and of the IUPAC–IUBMB Joint Commission on Biochemical Nomenclature, with two deliberate simplifications in the interest of legibility. Units are SI throughout; energies are given in kilojoules per mole (kJ mol^{-1}), and where the literature persists in kilocalories the conversion 1 kcal mol^{-1} = 4.184 kJ mol^{-1} is applied silently and without perature is in kelvin; where a physiological temperature is meant, 310.15 K (37 °C) is used, and 298.15 K (25 °C) for thermodynamic standard states.

Table i Principal notational conventions of the treatise. Symbols are defined again at their first use in each chapter; this table is a concordance, not a substitute for the definitions.

Symbol	Meaning	Unit / remark
c_X, [X]	concentration of species X	mol L^{-1} (M)
a_X	activity of species X	dimensionless; $a_X = \gamma_X c_X / c°$
K, K'	equilibrium constant; biochemical (transformed) constant	the prime denotes pH 7, specified $[Mg^{2+}]$
ΔG, ΔH, ΔS	Gibbs energy, enthalpy, entropy changes	kJ mol^{-1}, kJ mol^{-1}, J K^{-1} mol^{-1}
$\Delta G°'$	biochemical standard Gibbs energy change	298.15 K, pH 7, 1 M
R, k_B, N_A	gas, Boltzmann, Avogadro constants	see *Principal physical constants*
k_{cat}, K_m	catalytic constant; Michaelis constant	s^{-1}; mol L^{-1}
I	information content (Shannon)	bits
pK_a	$-\log_{10} K_a$	derived in § 6.4; used from Chapter 6
$\rightarrow$, $\rightleftharpoons$	reaction; equilibrium	the harpoons are used sparingly and mean what they say

Amino acids are denoted by their three-letter codes with the one-letter code in parentheses at first mention — thus glycine, Gly (G) — and nucleotides by their conventional initials; the full tables follow in the pages immediately after the physical constants. Greek letters are written out in speech-order in derivations the first time they occur, because nothing so estranges a willing reader as a silent alphabet. Isotopes are given in the form ^{13}C, ^{2}H (D), ^{15}N, ^{18}O. Enzyme names are followed, where it aids identification, by their Enzyme Commission number in parentheses, e.g. hexokinase (EC 2.7.1.1).

Two typographical courtesies. Species names are italicised — *Escherichia coli*, *Saccharomyces cerevisiae* — with the genus abbreviated after first use. And the decimal point is a point, the multiplication of units a thin space, and the negative exponents of units superscripts, so that mol L^{-1} and never mol/L: a small pedantry, we concede, but a treatise that is careless with its typography will be suspected, not without reason, of carelessness with its thermodynamics.

ABBREVIATIONS

The following abbreviations recur throughout the treatise. Each is also spelled out at first use within its own chapter; the list is provided so that the reader who opens the book in the middle — against advice, but with our sympathy — may decode it at need.

Table ii Abbreviations of general currency in the treatise.

Abbreviation	Expansion	Abbreviation	Expansion
ADP	adenosine 5′-diphosphate	NAD^+, NADH	nicotinamide adenine dinucleotide (oxidised, reduced)
AMP	adenosine 5′-monophosphate	$NADP^+$, NADPH	nicotinamide adenine dinucleotide phosphate
ATP	adenosine 5′-triphosphate	PDB	Protein Data Bank
CoA	coenzyme A	PEP	phosphoenolpyruvate
DNA, RNA	deoxyribonucleic, ribonucleic acid	P_i	inorganic phosphate
EC	Enzyme Commission (number)	PP_i	inorganic pyrophosphate
FAD, $FADH_2$	flavin adenine dinucleotide	RCSB	Research Collaboratory for Structural Bioinformatics
GTP	guanosine 5′-triphosphate	SI	Système international d'unités
IUBMB	International Union of Biochemistry and Molecular Biology	TCA cycle	tricarboxylic acid (citric acid) cycle
IUPAC	International Union of Pure and Applied Chemistry	UDP	uridine 5′-diphosphate

Principal Physical Constants

The constants below are the fixed stars by which every quantitative argument in this book is navigated. They are quoted to a precision sufficient for biochemical estimation; the reader who requires the full ceremonial dress of significant figures is referred to the CODATA adjustment cited in the references.

Table iii Principal physical constants used in the treatise, with the derived combinations that recur most often in biochemical arithmetic.

Constant	Symbol	Value
Avogadro constant	N_A	6.022×10^{23} mol^{-1}
Boltzmann constant	k_B	1.381×10^{-23} J K^{-1}
Gas constant ($R = N_A k_B$)	R	8.314 J K^{-1} mol^{-1}
Faraday constant ($F = N_A e$)	F	96 485 C mol^{-1}
Elementary charge	e	1.602×10^{-19} C
Planck constant	h	6.626×10^{-34} J s
Speed of light in vacuo	c	2.998×10^{8} m s^{-1}
Standard acceleration of gravity	g	9.807 m s^{-2}
Thermal energy, RT at 298.15 K	—	2.479 kJ mol^{-1}
Thermal energy, RT at 310 K	—	2.577 kJ mol^{-1}
RT ln 10 at 298.15 K (pK_a–$\Delta G°$ conversions)	—	5.708 kJ mol^{-1}
Molar volume of ideal gas, 273.15 K, 1 atm	V_m	22.41 L mol^{-1}

The three derived entries at the foot of the table are not constants of nature but constants of convenience — the exchange rates, so to speak, at which thermal energy is converted into the currencies of free energy and acidity. They will be derived properly in Chapters 4, 6 and 7; they are tabled here because the reader will spend more hours with them than with many of his friends.

Amino-Acid and Nucleotide Codes

The twenty standard proteinogenic amino acids, with their codes and the salient property of each side chain. Structures, pK_a values and full physical data are gathered in Appendix B; the present table is the pocket edition, sufficient for the arguments of Part I.

Table iv The twenty standard amino acids: codes and side-chain character. Isoelectric points and ionisation behaviour are treated in §§ 9.3–9.4.

Amino acid	3-letter	1-letter	Side-chain character
Glycine	Gly	G	hydrogen; minimal, achiral
Alanine	Ala	A	small, nonpolar
Valine	Val	V	branched, nonpolar
Leucine	Leu	L	branched, nonpolar
Isoleucine	Ile	I	branched, nonpolar; second stereocentre
Methionine	Met	M	sulphur-containing, nonpolar
Proline	Pro	P	secondary amine; constrains backbone
Phenylalanine	Phe	F	aromatic, nonpolar
Tyrosine	Tyr	Y	aromatic, phenolic, weakly acidic
Tryptophan	Trp	W	aromatic, indolic
Serine	Ser	S	hydroxyl, polar uncharged
Threonine	Thr	T	hydroxyl, polar uncharged; second stereocentre
Cysteine	Cys	C	thiol, redox-active; forms disulphides
Asparagine	Asn	N	amide, polar uncharged
Glutamine	Gln	Q	amide, polar uncharged
Aspartate	Asp	D	carboxylate, acidic
Glutamate	Glu	E	carboxylate, acidic
Lysine	Lys	K	primary ammonium, basic
Arginine	Arg	R	guanidinium, basic
Histidine	His	H	imidazole, basic; titrates near pH 7

Table v The canonical nucleobases and their nucleosides. The structural chemistry of these compounds is the business of Chapter 43.

Base	Code	Nucleoside (RNA / DNA)	Class
Adenine	A	adenosine / deoxyadenosine	purine
Guanine	G	guanosine / deoxyguanosine	purine
Cytosine	C	cytidine / deoxycytidine	pyrimidine
Uracil (RNA)	U	uridine	pyrimidine
Thymine (DNA)	T	deoxythymidine	pyrimidine

List of Figures

List of Tables

HISTORICAL CHRONOLOGY OF BIOCHEMISTRY

A science that forgets its own history is condemned to re-derive it in worse notation. The chronology below is not a pageant of great men — though several make their unavoidable appearances — but a record of *conceptual turning-points*: moments at which the subject changed its mind about what kind of question it was asking. Four such turns are marked in colour: the chemical turn, by which life was shown to obey the laboratory; the metabolic turn, by which the cell became an economy; the informational turn, by which it became a text; and the genomic–computational turn, by which the text became legible at industrial speed. The treatise follows the same order, and not by coincidence: the history of the subject is the one syllabus that has never failed its students.

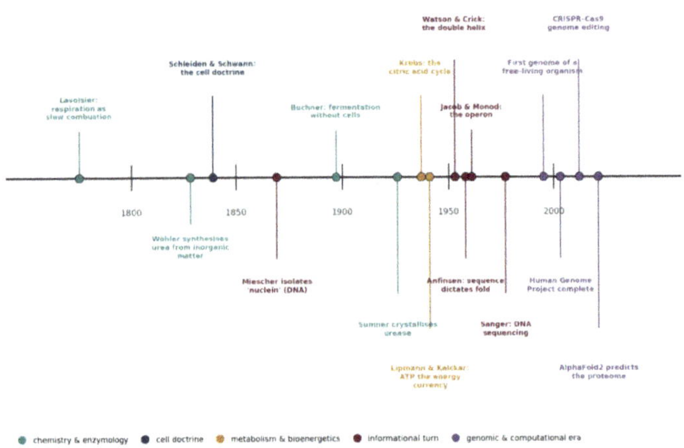

Figure i A historical chronology of biochemistry from Lavoisier's account of respiration (1775) to the proteome-scale structural predictions of 2021. Colours denote the reigning conception of the period: teal, the chemical and enzymological era; navy, the cell doctrine; amber, metabolism and bioenergetics; oxblood, the informational turn; plum, the genomic and computational era. The sequence of the treatise deliberately recapitulates this order — grammar before machinery, machinery before metabolism, metabolism before information. Sources for the individual events are the primary papers and Nobel records cited in the references of the relevant chapters.

Two features of the chronology deserve a word. The first is its acceleration: a century separates Lavoisier from Buchner, but only four years separate the operon from the first complete genome of cellular life, and the reader may verify from Figure 1.6 that the structural archive obeys a similar law of hastening. The second is its recurrent humility: every era in the figure supposed itself near the end of explanation, and every era was, in the event, at the beginning of a larger one. A treatise written in 2026 does well to bear both features in mind.

General Map of Biochemical Organisation

Before the journey, the itinerary. The figure below is the one map of the book that the reader is asked to memorise — or rather, to internalise so thoroughly that it ceases to be a map and becomes a habit of thought. Everything in the following seven hundred pages is a commentary upon it.

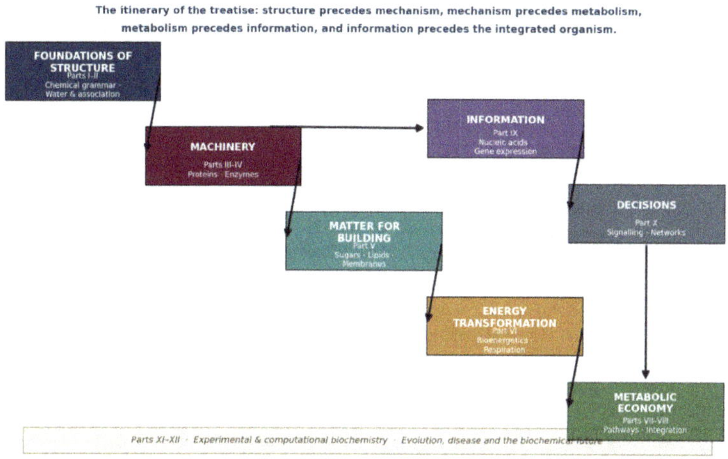

Figure ii The general map of biochemical organisation, which is also the itinerary of the treatise. Matter is organised (Parts I–II) into machinery (Parts III–IV); machinery builds and is built of the structural materials (Part V); the whole is powered by energy transformation (Part VI) and administered by metabolism (Parts VII–VIII); information (Part IX) instructs the machinery, and regulation (Part X) instructs the information; experimental and computational method (Part XI) and the evolutionary perspective (Part XII) underlie and oversee the whole. Arrows indicate the order of dependence — and therefore the order of exposition.

The reader will observe that the map is not a ladder but a weave: information flows back into machinery, regulation back into metabolism. This is true of the cell, and only politely concealed in the book. The defence of our linear order is pedagogical, not ontological: one must know what an enzyme *is* before one can profitably ask how its synthesis is regulated, just as one must learn the alphabet before being permitted to complain about the poem.

CONTENTS

PART I

The Chemical Grammar of Life

CHAPTER 1

The Scope and Logic of Biochemistry

CONTENTS OF THIS CHAPTER

In which the subject is defined, not by enumeration of its objects but by examination of its questions; in which life is argued to be organised chemical transformation, and the four currencies of the biochemical economy — structure, energy, information and regulation — are introduced; and in which the methods of the science, its reductionism and its humility, its organisms and its numbers, are weighed before a single molecule is permitted to take the stage.

1.1 What biochemistry attempts to explain

Begin, as every honest enquiry must, with the embarrassment of the obvious. A glass of sugar-water, left to itself, remains a glass of sugar-water: the sugar sits in solution with all the dignity of a rentier, drawing interest from no one. The same water, inoculated with a pinch of yeast, begins within hours to bubble, to warm, and to smell faintly of a tavern; and in a day the sugar is gone, exchanged for alcohol, gas, and a hundred million new cells where there was one. Both systems are made of the same atoms, in the same solvent, at the same temperature, under the same sky. The difference between them is not a substance — no one has ever isolated it in a bottle — and yet it is the most consequential difference in the world, for one of the two systems is alive. To say in what that difference *consists*, in the language of chemistry and without appeal to anything that could not, in principle, be weighed, is the whole ambition of biochemistry. Everything else in this book is elaboration.

The definition just offered — *biochemistry is the explanation of life in chemical terms* — is short enough to be inscribed over a doorway, and like most inscriptions it conceals a dispute. There are three parties to it. The first, whom we shall call the chemist, holds that the definition is a tautology: life *is* chemistry, a particularly intricate branch of the reactions of carbon in water, and to explain it chemically is simply to describe it properly. The second, whom we shall call the physiologist, demurs: a smashed watch tells one nothing about time, and a homogenised liver tells one nothing about the liver; the organisation is the phenomenon, and organisation is not a chemical property but a historical and functional one. The third party, the vitalist — much reduced in circumstances since the nineteenth century, but not yet extinct at dinner parties — holds that the difference between the two glasses of sugar-water is in principle unweighable, a *vis vitalis* that chemistry may describe but never explain.

The vitalist we may answer at once, not with rhetoric but with a date: 1828, when Wöhler prepared urea — the very emblem of animal chemistry — from ammonium cyanate, a salt with no pedigree in any living body (Wöhler, 1828). The experiment did not disprove vitalism logically; no single experiment could. What it did was more damaging: it made vitalism *uninteresting*, for if the characteristic products of life could be made in a flask, then the hypothesis of a special force explained nothing that the flask did not. Science, it should be said, rarely defeats its rivals in debate; it makes them unemployed. The quarrel between the chemist and the physiologist, however, is a living one, and it will occupy us in § 1.6 — for it is, in disguise, the quarrel between analysis and synthesis that every chapter of this book must conduct for itself.

One clarification, and the section is done. To say that biochemistry explains life in chemical terms is not to say that life is *nothing but* chemistry — a phrase that has damaged more good minds than any reagent in the catalogue. The biographer who explains a statesman's career in terms of ambition has not shown that the career was nothing but ambition; he has shown that ambition was the level at which the explanation became tractable. Chemistry is biochemistry's chosen level of tractability: the level at which measurement bites, mechanisms resolve, and predictions come true. Whether it is the only such level is a question we shall postpone to § 1.5, where the ladder of explanation is climbed properly, one rung at a time.

1.2 Life as organised chemical transformation

If life is to be discussed chemically, we must first say what chemistry is being asked to discuss. Strip a living system of every particular — its species, its size, its address in the Linnaean filing system — and three properties remain, each chemical to the bone. The system *transforms matter*, importing raw molecules and exporting products whose manufacture no unassisted environment would undertake. The system *transforms energy*, capturing it from light or from chemical gradients and spending it against the universal tax of disorder. And the system *persists in form*, maintaining across days or decades a molecular organisation that its own contents do not share: the reader's atoms are replaced continuously, and yet the reader persists, rather in the manner of a fountain, which is never twice made of the same water and yet is always the same fountain. A system that does all three is alive, or a remarkably good imitation; a system that does none is a glass of sugar-water.

The word that does the work in all three properties is *transformation*, and its Latin dignity should not disguise that it means something quite humble: the rearrangement of bonds. When the yeast of § 1.1 disposes of a molecule of glucose, the stoichiometry of its accomplishment can be written — this is the first equation of the book, and like all those that follow it will be accounted for rather than merely displayed — as the overall conversion

$$C_6H_{12}O_6 \longrightarrow 2\,C_2H_5OH + 2\,CO_2 \tag{1.1}$$

in which one mole of glucose yields two moles of ethanol and two of carbon dioxide. The equation is, for the moment, a *balance of atoms* and nothing more: count the carbons, hydrogens and oxygens on each side and they agree, six to six, twelve to twelve, six to six. But an accounting identity, patiently interrogated, is the seed of the whole science. *Why* does the reaction proceed? Because, as Chapter 7 will derive, the products sit lower in free energy than the reactant, and the yeast collects a commission on the difference. *How fast* does it proceed? That is the question of kinetics, answered in Part IV by the enzymes that catalyse each of the dozen steps concealed inside Eq. (1.1). *How does the cell regulate the rate*, and *how did such a pathway arise*? Those are the questions of

Parts X and XII. A single stoichiometric line, taken seriously, generates the table of contents.

Here the dialectic of the book begins in earnest, for the account just given has a powerful and articulate enemy, which we shall call the *static objection.* A transformation, the objection runs, is an event; but a living thing is a thing. You have defined life as a verb, and the world plainly experiences it as a noun — as the cat, the oak, the inconvenient bacterium on the petri dish. The objection is sound, and the answer to it is the pivot on which the whole subject turns: *the noun is the verb, maintained.* The structure of a cell is not a scaffolding upon which reactions happen to run; it is the standing pattern of the reactions themselves, as the standing wave in a flute is the note. That this is literally true, and not a trope, is shown by turnover: the proteins of a mammalian liver are degraded and re-synthesised on a scale of days, and the reader's own body, over a year, replaces the large majority of its substance while mislaying none of its identity (Schoenheimer, 1942). Life, in short, is not matter that undergoes transformation. It is transformation that has contrived to become matter — and the contrivance is the subject of this book.

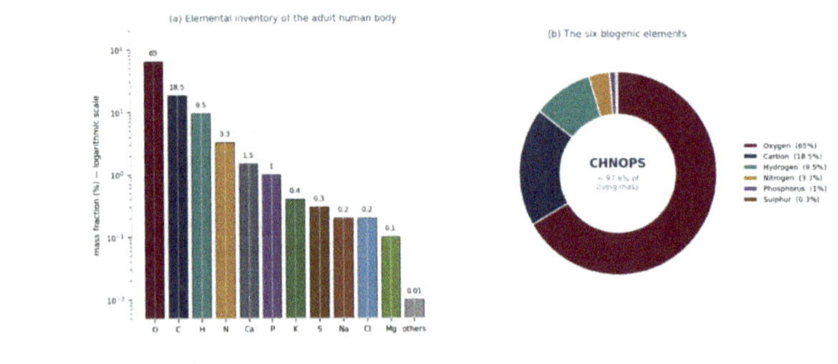

Figure 1.1 The raw materials of the transformation. (a) The mass fractions of the elements in the adult human body, on a logarithmic scale: oxygen, carbon, hydrogen and nitrogen together account for about 96.3 % of the mass, and the six biogenic elements CHNOPS for about 97.6 %. (b) The same six elements arranged as the core inventory of living matter; the remaining percent — calcium of bone, phosphorus of bone and nucleotide, the electrolytes — is plotted but hardly visible, which is precisely the point: life is built overwhelmingly of the lightest and commonest elements of the upper periodic table, and its chemistry is the chemistry of the abundant. Values are representative mass percentages compiled in standard reference texts (Alberts et al., 2022).

1.3 Molecular composition and biological organisation

What, then, is the transforming system made of? The question admits a two-word answer — *mostly water* — and a long one, and the long one repays the telling because it establishes a pattern that the whole treatise will reuse: the composition of life is extravagantly unequal, and the inequalities are the message. Consider the best-weighed of all cells, the bacterium *Escherichia coli*, whose census has been taken with a care that most nations do not extend to their own populations (Neidhardt & Umbarger, 1996). By wet weight it is seven parts water to three parts everything else (Figure 1.2a) — a proportion the reader shares with the bacterium and, for that matter, with the melon. Of the dry remainder, more than half is protein; RNA outweighs DNA nearly seven to one; and the celebrated hereditary material, the double helix itself, amounts to some three per cent of the dry mass (Figure 1.2b). Let the reader dwell on that number. The molecule that popular culture has installed as the very emblem of life is, by weight, a minor citizen of the cell; the working population is protein. The inversion is instructive, and we shall meet it again: *importance in the economy of the cell is not proportional to abundance, nor abundance to fame.*

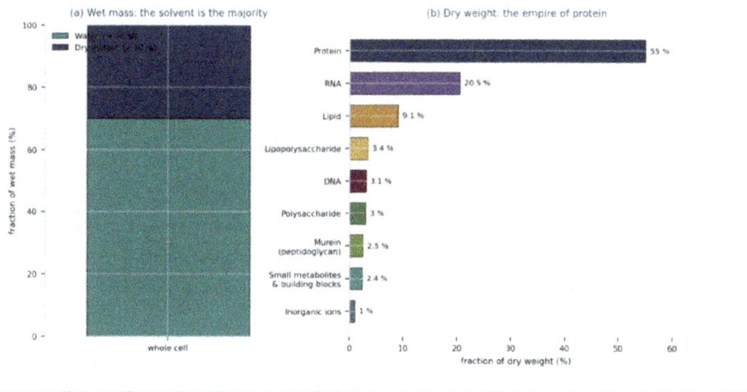

Figure 1.2 The chemical composition of *Escherichia coli* B/r in balanced growth. (a) Wet mass divides into roughly 70 % water and 30 % dry matter. (b) The dry matter, in the classical compilation of Neidhardt and co-workers (Neidhardt & Umbarger, 1996): protein 55 %, RNA 20.5 % (of which the greater part is ribosomal), lipids and lipopolysaccharide 12.5 %, DNA 3.1 %, polysaccharide and murein 5.5 %, and a residue of metabolites, building blocks and inorganic ions. The figure is the quantitative warrant for the treatise's order of business: Parts III–IV (proteins and enzymes) precede Part IX (nucleic acids) because the cell, weighed rather than interviewed, puts protein first.

Composition, however, is only the ground floor of organisation; the same inventory of parts, shaken in a bag, is not a cell but a tragedy. Between the atom and the organism intervene at least seven distinguishable levels of structure, each built of the level beneath and each legislating upon it (Figure 1.3): atoms assemble into small molecules; small molecules polymerise into macromolecules; macromolecules associate into supramolecular assemblies — ribosomes, membranes, filaments; assemblies are enclosed and sorted into organelles; organelles cooperate in cells; cells in tissues; tissues in organisms. Two features of the ladder deserve emphasis now, for they will structure every later argument. The first is that *each level has properties its components lack*: a membrane is not a property of a lipid, nor a rhythm of a kinase. The second is that *each level constrains the behaviour of its components*: a proton in a mitochondrion is not free to wander, for the membrane has decided otherwise. Composition ascends the ladder; constraint descends it; and the biochemist earns his living in the traffic between the two directions.

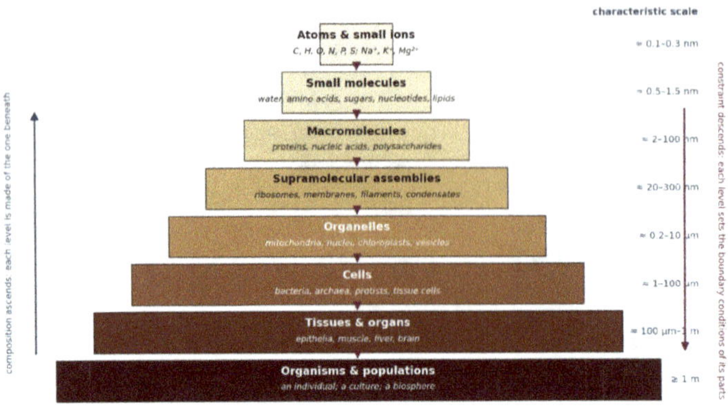

Figure 1.3 The hierarchy of biochemical organisation, from atoms to populations, with the characteristic length scale of each level at the right. The ascending arrow (left) marks composition: every level is constituted of the level below. The descending arrow (right) marks constraint: every level sets boundary conditions — concentrations, compartments, membranes, programmes — upon its constituents. The dialectic of the two arrows, resumed in § 1.6, is the methodological heart of the subject. The widening of the steps is not decorative: it records that higher levels admit more kinds, though fewer copies, of their elements — there are more species of macromolecule than of atom, and more species of organism than of organelle.

A final observation on scale, since Figure 1.3 has smuggled it in. The biochemical theatre spans ten orders of magnitude in length, from the ångström of the hydrogen bond to the metre of the organism (Figure 1.4), and the curious fact is that the characteristic events of life are crowded into the middle of that range — the nanometre of the protein, the hundred nanometres of the vesicle, the micrometre of the cell. Why the middle? The answer is thermodynamic, and it is the first of the book's recurring principles: at the nanometre scale, the energies of molecular association are comparable to the thermal energy k_BT, so that structures can be assembled and — this is the point that makes biology possible — *disassembled*, by the ordinary collisions of warm water. Life builds with weak forces because weak forces can be edited; a world of covalent architecture would be permanent, and permanence is a property of monuments, not of organisms.

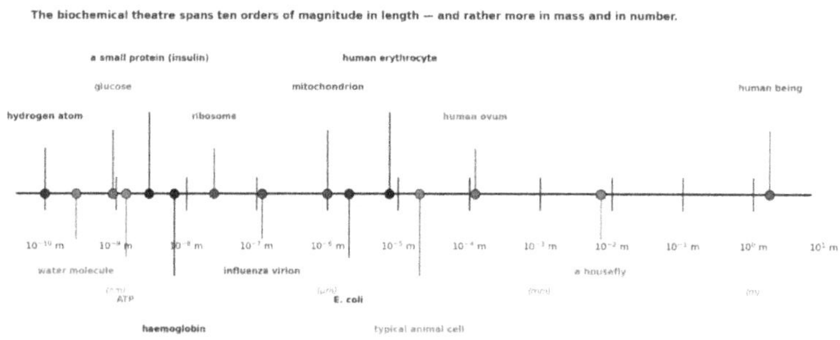

Figure 1.4 The ladder of sizes: representative objects of the biochemical theatre on a logarithmic axis of length, from the hydrogen atom (10^{-10} m) to the human being. Values are characteristic dimensions compiled from standard cell-biological sources (Alberts et al., 2022); the reader should treat them as orders of magnitude, since cells, like guests, vary in size with the occasion. Note the compression of molecular biology into the decade between 1 and 100 nanometres — the domain in which thermal agitation and molecular stickiness are evenly matched.

1.4 Structure, energy, information and regulation

The subtitle of this treatise names four currencies, and the time has come to open the accounts. A living cell may be regarded — this is the thesis of the section, and the reader is invited to test it against every chapter that follows — as an economy in which four kinds of wealth are earned, stored, exchanged and taxed. *Structure* is its capital: the folded proteins, the assembled membranes, the inherited architecture that Chapter 3 will show to be specified down to the handedness of single atoms. *Energy* is its income: the flux through the oxidation of foodstuffs or the capture of light, of which Part VI will render the full ledger. *Information* is its endowment: the sequence of the genome, in which the instructions for rebuilding the whole economy are deposited in a four-letter script. And *regulation* is its government: the web of feedback, feed-forward and covalent modification by which income is matched to expenditure and no pathway runs riot. An economy that possesses the first three and lacks the fourth does not survive its prosperity; a cell without regulation is, in the strict biochemical sense, a tumour.

The four currencies are convertible, and the conversion rates are the laws of physical chemistry applied to water at 310 K. Structure is bought with energy: the folding of a protein is paid for, in a currency Chapter 7 will define precisely, by the free energy of hydrophobic collapse and hydrogen bonding. Information specifies structure: the one-dimensional sequence of a gene determines, through the folding problem of Chapter 11, the three-dimensional architecture of its protein. Regulation rides upon energy and structure alike: a regulatory phosphorylation is the expenditure of one phosphate's worth of free energy to alter the shape — and thereby the behaviour — of one protein. Nothing in the economy is free, and nothing in it is merely metaphorical; the word *currency* is used here with the seriousness of a chemist, not the licence of a journalist.

Table 1.1 The four currencies of the biochemical economy, their molecular carriers, and the parts of the treatise in which each is audited. The table is a prospectus; the audits follow on schedule.

Currency	Molecular carrier	Characteristic transaction	Audited in
Structure	proteins, membranes, polysaccharides	folding, assembly, recognition	Parts I, III, V
Energy	ATP, ion gradients, reduced coenzymes	phosphoryl and electron transfer	Parts II, VI
Information	DNA, RNA, sequence	replication, transcription, translation	Part IX
Regulation	allostery, covalent modification, networks	feedback, switching, decision	Parts IV, X

Of the four, information is the one whose chemistry is least obvious and whose arithmetic is most beautiful, and since this chapter is the place for first principles, let us derive its measure now rather than borrow it. Consider a position in a molecular script that can be occupied by any one of m different symbols, all equally likely — four nucleotides, say, or twenty amino acids. To specify *which* symbol occupies the position is to answer a question with m possible answers; and the information content of such a specification is defined, in the measure Shannon gave the world in 1948 (Shannon, 1948), as the logarithm to base two of the number of possibilities,

$$I = \log_2 m \quad \text{bits per symbol} \tag{1.2}$$

— where the base of the logarithm is chosen so that a single binary choice (m = 2) yields exactly one *bit*, the irreducible unit of answer. The derivation of the general case, in which the symbols are *not* equiprobable, begins from the observation that information must be additive: if two independent positions are specified, the information of the pair must be the sum of the informations, while the number of joint possibilities multiplies. The only function that converts multiplication into addition is the logarithm; hence, for a symbol i drawn with probability p_i, the information is $\log_2(1/p_i)$ bits, and the mean information per symbol — the Shannon entropy of the script — is the average over the alphabet:

$$H = \sum_{i=1}^{m} p_i \log_2 \frac{1}{p_i} = -\sum_{i=1}^{m} p_i \log_2 p_i \tag{1.3}$$

— which reduces to Eq. (1.2) when every $p_i = 1/m$, as the reader may verify by substitution in two lines. We shall not hide the fact that the letter H and the minus sign are borrowed from thermodynamic entropy; whether the resemblance is a pun or a profound identity is a question that has launched a thousand seminars, and we shall return to it in § 1.10 with due solemnity. For the moment it suffices to compute: a DNA position carries $\log_2 4 = 2$ bits; a protein position, $\log_2 20 \approx 4.32$ bits; and a genome of 4.6×10^6 base pairs — that of our bacterium — some 9.2 megabits, the reading of a modest fortnight, from which the whole economy of *E. coli* can, in principle, be reissued. The consequences are plotted in Figure 1.10, to which we shall formally arrive in § 1.9, but the impatient reader has our blessing to look ahead: impatience, in this book, is a virtue we have budgeted for.

1.5 Levels of biochemical explanation

It is time to honour the promissory note of § 1.1 and climb the ladder of explanation. A biochemical fact — that haemoglobin, say, releases its oxygen in exercising muscle — can be explained at five distinguishable levels, and the levels are not rivals but nested concentric answers to five different questions wearing the same grammatical costume. There is the *structural* explanation: the geometry of the haem pocket, the movement of the iron into the porphyrin plane, the shift of the $\alpha_1\beta_2$ interface — the business of Chapter 13. There is the *thermodynamic* explanation: the free energy of oxygen binding, its dependence on pH and on 2,3-bisphosphoglycerate, the quantitative sociology of the four subunits. There is the *kinetic* explanation: the rates of the conformational transitions, the milliseconds between lung and muscle. There is the *regulatory* explanation: the logic of the system, which senses protons — the exhaust gas of working muscle — and lowers its affinity precisely where unloading is wanted. And there is the *evolutionary* explanation: the pedigree of the globin family, the duplications and selections by which the system came to be. To ask *why does haemoglobin release oxygen in muscle?* is to ask all five questions at once; a complete biochemistry answers them in chorus.

Table 1.2 Levels of biochemical explanation, illustrated by the oxygen-release behaviour of haemoglobin. Each level answers a legitimate question; none makes the others superfluous.

Level	Question it answers	Example: haemoglobin	Treated in
Structural	In what arrangement?	haem geometry, interface shift	§§ 13.2–13.4
Thermodynamic	In what direction, how far?	binding free energies, Bohr effect	§§ 7.7, 13.5
Kinetic	How fast, by what path?	ligand on/off rates, T⇌R transition	§§ 16.1, 13.4
Regulatory	To what end, under whose control?	proton and BPG sensing	§§ 13.5–13.7
Evolutionary	By what history?	globin duplications, foetal variants	§§ 13.8, 56.3

The reader will notice that the five levels correspond, roughly, to the five great instruments of the science: the diffractometer, the calorimeter, the stopwatch, the wiring diagram, and the phylogenetic tree. He will also notice — and here is the dialectical point of the section — that the levels are asymmetric in one crucial respect. Each level is *compatible* with the levels below it and *required* by the levels above it, but no level is *predictable* from those below in practice, and precious few in principle. Knowing the structure of haemoglobin does not tell one that the blood of the foetus binds oxygen more tightly than the blood of the mother; that fact is a regulatory and evolutionary datum, which the structure can rationalise after the fact but could not have announced before it. Explanation in biochemistry flows downhill into mechanism and uphill into significance, and the treatise that confined itself to either direction would be telling half the story — the half, moreover, that the other half cannot supply.

1.6 Reductionism, emergence and systems behaviour

We arrive at the quarrel postponed from § 1.1, and we arrive armed. The chemist of that section — let us now call him the reductionist, for he has earned the name — holds that the behaviour of a cell follows from the properties of its molecules as the behaviour of an orrery follows from its gears: completely, and with nothing left over. The physiologist, now promoted to holist, replies that the properties of the molecules in a cell are themselves altered by the cell — its pH, its crowding, its membranes, its history — so that the circle of explanation closes upward as well as downward, and the word *follows* conceals the direction in which it is travelling. The quarrel is old, and it is usually conducted at a volume inversely proportional to its precision. Let us conduct it quietly, with the help of Figure 1.5.

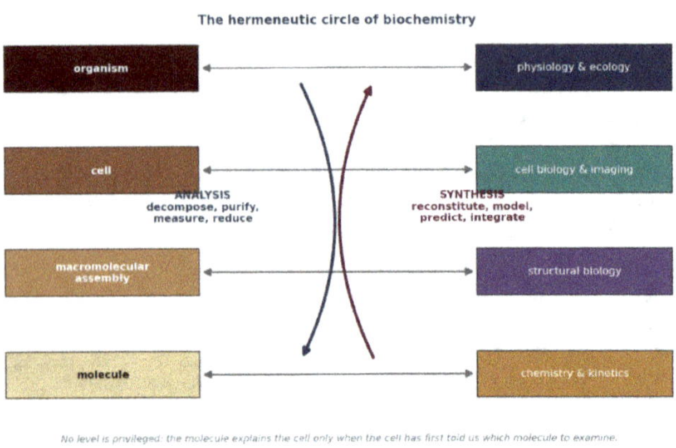

Figure 1.5 The hermeneutic circle of biochemical method. On the left, the levels of organisation; on the right, the methods native to each level. The descending arm (navy) is *analysis*: decompose the system, purify the components, measure their properties in isolation. The ascending arm (oxblood) is *synthesis*: reconstitute the components, in vitro or in silico, and test whether the behaviour of the whole is recovered. The circle is hermeneutic because it is never completed once: each synthesis sends the analyst back to look for what the reconstitution failed to recover, and each new analysis proposes a better synthesis. The method of the treatise is this circle, travelled patiently, part by part.

The resolution, stated briefly, is that both parties are right about different verbs. The reductionist is right about *consistency*: nothing happens in a cell that contradicts the physics of its molecules, and no observation of the whole has ever required — not once, in two centuries of looking — a force unknown to the parts. The holist is right about *prediction*: from the parts, taken separately, almost nothing about the whole could have been foreseen, for the behaviour of the whole depends on properties that exist only in the assembled system — on the *emergent* properties, to give them their respectable name. A single kinase is a catalyst; a kinase embedded in a double-negative feedback loop is a switch with memory (§ 52.3); the word *memory* is not to be found in the catalogue of the kinase's isolated properties, any more than *wetness* is to be found in the spectrum of a water molecule. The emergent property is not a ghost added to the parts; it is a property of the *relations* among the parts — and relations, though invisible to the spectrophotometer, are perfectly real and perfectly measurable by the methods of Part XI.

There is a second, humbler argument for taking both directions seriously, and it is epistemological rather than ontological: *we do not know all the parts.* Even in the best-catalogued cell on Earth, the function of a substantial fraction of the proteins remains unassigned (Alberts et al., 2022); and the history of the subject is a history of assemblies that refused to behave like their reconstituted components until some neglected fifth factor — a chaperone, a lipid, a cation — was restored to the mixture. Analysis without synthesis is a dismantled engine on the floor; synthesis without analysis is a watch assembled by a poet. The treatise therefore adopts the circle of Figure 1.5 as its standing method, and the reader will find the two arms alternating throughout: every mechanism derived from below is tested against the behaviour of the whole, and every behaviour of the whole is traced downward to the mechanism that bears it.

1.7 Model organisms and comparative biochemistry

A science that cannot experiment on its true object — the reader, for instance, is not available for homogenisation, and quite right too — must proceed by deputies. The deputies of biochemistry are the model organisms, a small and extraordinarily well-interviewed company: the bacterium *Escherichia coli*; the yeast *Saccharomyces cerevisiae*; the nematode worm *Caenorhabditis elegans*, whose 959 somatic cells have been counted, named and followed from egg to grave; the fruit fly *Drosophila melanogaster*; the thale cress *Arabidopsis thaliana*; the mouse *Mus musculus*; and, standing somewhat uneasily among them as both deputy and sovereign, *Homo sapiens*. The justification for the delegation is the deepest empirical fact in all of biology: *the unity of biochemistry*. The glycolytic pathway of the yeast is, reaction for reaction and often residue for residue, the glycolytic pathway of the reader; the genetic code of the bacterium differs from the human code in fewer positions than a proofreader's afternoon; and the electron transport chain of the mammalian mitochondrion preserves, in more elaborate dress, the architecture of its bacterial ancestor — as it should, since the mitochondrion *is* a bacterial ancestor, domesticated some two billion years ago (§ 27.2).

Figure 1.8 The experimental menagerie. Haploid genome size (logarithmic abscissa) against characteristic generation time (logarithmic ordinate) for the principal model organisms of biochemistry. Values are reference approximations: *E. coli*, 4.6 × 10^6 bp, 20 minutes; *S. cerevisiae*, 1.2 × 10^7 bp, 1.5 hours; *C. elegans*, 10^8 bp, 3 days; *D. melanogaster*, 1.8 × 10^8 bp, 10 days; *A. thaliana*, 1.35 × 10^8 bp, six weeks; *M. musculus*, 2.7 × 10^9 bp, nine weeks; *H. sapiens*, 3.2 × 10^9 bp, a generation (Alberts et al., 2022). The inverse correlation is the unwritten law of the discipline: the smaller and faster the deputy, the more completely it can be known — and the further its testimony must be carried to reach the sovereign.

The comparative method, however, carries a tax that must be declared: the deputy is not the sovereign, and the distance between them is itself a variable of the experiment. Insulin discovered in the dog, tested in the ox, and administered to the reader is the same hormone by chemistry and a different molecule by a few residues; those residues once sufficed to provoke the immune system, until the protein engineers of Chapter 56 learned to humanise the sequence. Comparative biochemistry is therefore a perpetual exercise in calibrated trust — knowing what may be generalised because it is ancient (the code, the pathways, the coenzymes) and what must be verified because it is recent (the receptors, the regulations, the refinements). The rule of thumb is evolutionary, and it will serve the reader throughout: *the deeper the currency, the wider the acceptance.* ATP is legal tender from bacterium to bishop; the details of appetite regulation are negotiated species by species, and sometimes, as the reader may have noticed, individual by individual.

1.8 From observation to biochemical mechanism

How, in practice, does a biochemical observation become a biochemical mechanism? The textbooks of method describe a straight road — observe, hypothesise, test, conclude — which no working scientist has ever travelled, rather as no working army has ever marched in the formations of the parade ground. The true itinerary is the circle of Figure 1.5, and its canonical journey is the experiment that founded the science itself, which we shall now examine as our first case study. It deserves the honour, for it contains in miniature every element of the modern method: an entrenched theory, an obstinate observation, a brutal intervention, and a conclusion resisted for a generation.

MECHANISTIC CASE STUDY — BUCHNER'S JUICE, 1897

Eduard Buchner (Plate 1.1), a chemist of Munich, was grinding brewer's yeast with sand and kieselguhr to prepare a cell-free extract for pharmaceutical purposes — the original aim was medicinal, not metaphysical. To preserve the juice from putrefaction he added a strong syrup of sucrose, and observed, with the attention that distinguishes the scientist from the technician, that the mixture began to froth: carbon dioxide was being evolved, and ethanol accumulating, in a preparation demonstrably devoid of living cells (Buchner, 1897). The observation contradicted the reigning doctrine, descending from Pasteur, that fermentation was an act *of* the living cell, inseparable from its vitality — *pas de fermentation sans vie.* Buchner concluded that the transformation was performed instead by a substance, or a company of substances, which the cells had manufactured but did not need to remain alive to operate; he named the active principle *zymase*, from the Greek for leaven.

The logic of the experiment is worth rehearsing, because it is the template of all biochemistry since. First, *disassemble*: break the organisation while preserving the chemistry — the analytic arm of the circle. Second, *observe the function surviving the organisation*: fermentation continued, therefore the function was chemical, not vital. Third, *propose the mechanism*: a definable agent performs the transformation; find it. The third step occupied the following half-century, as the zymase resolved into a dozen enzymes and as many intermediates — the glycolytic pathway of Chapter 31, which the reader now inherits fully mapped. The Nobel Committee of 1907, commending the discovery, noted with laconic Swedish understatement that Buchner had shown fermentation to proceed *without the presence of living cells*; the vitalists, one imagines, noted it too.

Plate 1.1 Eduard Buchner (1860–1917), whose cell-free fermentation of 1897 converted vitalism from a theory into a memory. Portrait from the Nobel Foundation archive, nobelprize.org (Nobel Prize in Chemistry, 1907); reproduced from the Foundation's publicly distributed image. The expression — oblique, faintly amused — is appropriate to a man who had dissolved the soul of the cell in sugar-water and found the chemistry improved thereby.

From observation to mechanism, then, the road is circular and the tolls are honesty: one pays at disassembly, by proving the organisation dispensable; at reconstitution, by proving the components sufficient; and at every stage, by publishing the numbers. Which brings us to the last character of the science to be introduced in this chapter — its arithmetic.

1.9 The quantitative character of modern biochemistry

There was a time, within living memory of the subject, when biochemistry could be practised as natural history with glassware: name the compounds, draw the arrows, leave the numbers to the physical chemists. That time is over, and the reader should know why. The proximate cause is instrumental — the instruments of Part XI measure concentrations, energies, rates and structures to precisions undreamt of in Buchner's Munich. The deeper cause is logical: the questions that remain open in biochemistry are quantitative questions. Whether a pathway runs at all is usually known; *how fast*, *under whose control*, and *with what margin of safety* are not. A qualitative biochemistry can tell the reader that insulin lowers blood glucose; only a quantitative one can say by how much, in what time, and at what dose the therapy becomes the poison. The rest of this section is therefore an apprenticeship in the arithmetic of the cell — four derivations, each of which will be used again in the chapters to come.

1.9.1 The arithmetic of doubling

Begin with growth, the most consequential exponential in nature. Suppose a population of bacteria in which every cell divides after a fixed interval τ (the doubling time). Starting from N_0 cells at time zero, the number after k doublings is $N_0 \cdot 2^k$; and since $k = t/\tau$ after elapsed time t, the population is

$$N(t) = N_0 \cdot 2^{t/\tau} \tag{1.4}$$

— which is the discrete truth wearing a continuous dress. To see the differential form that kinetic theory prefers, write $2^{t/\tau} = e^{(t/\tau)\ \ln 2}$ (the identity $a^x = e^{x \ln a}$, valid for any positive a), whence $N(t) = N_0 e^{\mu t}$ with the specific growth rate

$$\mu = \frac{\ln 2}{\tau} \tag{1.5}$$

— so named because differentiating $N = N_0 e^{\mu t}$ gives $dN/dt = \mu N$: the rate of increase is proportional to the population, which is the signature of every self-replicating process from bacteria to rumours. For our *E. coli* with $\tau = 20$ min, $\mu = 0.693/0.333\ h \approx 2.08\ h^{-1}$; and Eq. (1.4) then predicts, from one cell, $2^{21} \approx 2.1 \times 10^6$ descendants in seven hours (Figure 1.9) — a number that explains both the speed of bacterial infection and the necessity, in the laboratory, of not leaving one's cultures unattended over a weekend. Exponentials are the compound interest of biology, and nature, unlike the banks, pays continuously.

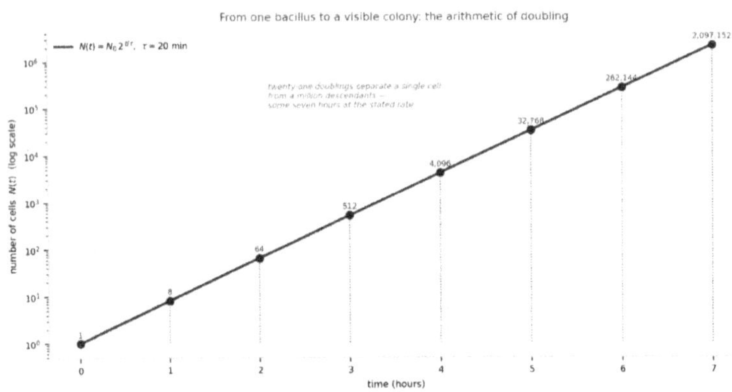

Figure 1.9 The arithmetic of doubling, Eq. (1.4), plotted for a doubling time $\tau = 20$ min — the laboratory benchmark for *E. coli* in rich medium at 37 °C. The ordinate is logarithmic, so the exponential appears as the straight line it secretly is; the dotted markers record the census at every third doubling. The reader is invited to verify, from Eq. (1.5), that the slope of the line is $\log_{10} 2 / \tau \approx 0.903$ generations per hour, and to reflect that the same equation, with a minus sign, governs the decay of isotopes, of drugs in plasma, and of untested hypotheses.

1.9.2 The census of a single cell

Second derivation: how many molecules does a cell contain? This is the kind of question that separates the sciences from the opinions, and it is answered by one of the most useful habits a biochemist can acquire — the order-of-magnitude estimate, performed openly and defended line by line. Take *E. coli*, a cylinder roughly 1 μm in diameter and 2 μm long, whence a volume $V \approx \pi(0.5\ \mu m)^2 \times 2\ \mu m \approx 1.6 \times 10^{-15}$ L — call it 10^{-15} L, one femtolitre, and let the error live where we can see it. From Figure 1.2, protein is 55 % of a dry mass that is 30 % of a wet mass of about $V \times 1\ g\ mL^{-1} \approx 10^{-12}$ g; the mass of protein per cell is therefore $\approx 0.55 \times 0.30 \times 10^{-12}\ g \approx 1.6 \times 10^{-13}$ g. A mean protein of 300 amino acids at $\approx 110\ g\ mol^{-1}$ per residue weighs $\approx 33\ 000\ g\ mol^{-1}$, so the number of protein molecules is

$$N_{\text{prot}} = \frac{m_{\text{prot}}}{M_{\text{prot}}} N_A \approx \frac{1.6 \times 10^{-13}\ \text{g}}{3.3 \times 10^4\ \text{g mol}^{-1}} \times 6.022 \times 10^{23}\ \text{mol}^{-1} \approx 3 \times 10^6 \quad (1.6)$$

— some three million protein molecules per cell, where every symbol has its provenance: m_{prot} from the census of § 1.3, M_{prot} from the mean residue mass of Table iv, and N_A from Table iii. A few hundred such estimates, tabulated below, constitute the reader's permanent armamentarium; the point is not the digits but the discipline — any claim in this book can be rebuilt from measurements, and the reader who doubts one is invited to rebuild it.

Table 1.3 Orders of magnitude for the molecular census of an *E. coli* cell, estimated as in Eq. (1.6) and consistent with the measured census of Neidhardt and Umbarger (1996). Such numbers anchor every later argument about concentrations, rates and noise (§ 52.5).

Quantity	Estimate	Route
cell volume	$\approx 10^{-15}$ L (1 fL)	cylinder 1 × 2 μm
water molecules per cell	$\approx 2 \times 10^{10}$	0.7×10^{-12} g ÷ 18 g mol^{-1} × N_A
protein molecules per cell	$\approx 3 \times 10^6$	Eq. (1.6)
a 1 μM solute, copies per cell	≈ 600	10^{-6} mol L^{-1} × 10^{-15} L × N_A
a 1 nM solute, copies per cell	≈ 0.6	ibid.; many cells hold none
glucose molecules oxidised per division	$\approx 10^9$	carbon budget, § 31.5

Dwell for a moment on the fourth and fifth rows, for they conceal a small revolution. Biochemistry is taught in the units of the flask — millimolar, micromolar — but the cell is not a flask; a solute at 1 nM, an entirely respectable concentration for a signalling molecule, corresponds on average to *less than one molecule per bacterium*. At such censuses, concentration ceases to be a smooth variable and becomes a census of individuals, subject to the statistics of small numbers; the consequences for cellular decision-making are real, and they are the subject of § 52.5. The flask's continuum and the cell's discreteness are two dialects of the same language, and the modern biochemist must be bilingual.

1.9.3 The ladder of energies

Third derivation: the energies at which the economy trades. The floor of the ladder is thermal. The mean thermal energy available per degree of freedom at absolute temperature T is of order $k_B T$, where k_B is the Boltzmann constant of Table iii; per mole of particles the corresponding scale is $RT = N_A k_B T$, which at the physiological 310 K evaluates to

$$RT = 8.314\ \text{J K}^{-1}\,\text{mol}^{-1} \times 310\ \text{K} \approx 2.58\ \text{kJ mol}^{-1} \quad (1.7)$$

— the number that will recur in this book more often than any other, for it is the unit in which all molecular stickiness is priced. The derivation of the hydrolysis free energies, bond energies and redox spans plotted in Figure 1.7 belongs to Chapters 7 and 26; what Figure 1.7 establishes now is the *shape* of the economy: weak interactions (a few kJ mol^{-1}) cluster within a few multiples of RT and are therefore reversible at body temperature; the hydrolysis of ATP (≈ 30 kJ mol^{-1} standard, ≈ 50 in the cell) sits an order of magnitude higher, a coin large enough to drive chemistry yet small enough to be spent by a single collisional event; and the full oxidation of glucose (≈ 2 870 kJ mol^{-1}) is a capital sum, which the cell sensibly declines to carry in cash, breaking it instead into the thirty-odd ATP of Chapter 28 — as one does not pay for coffee with a house.

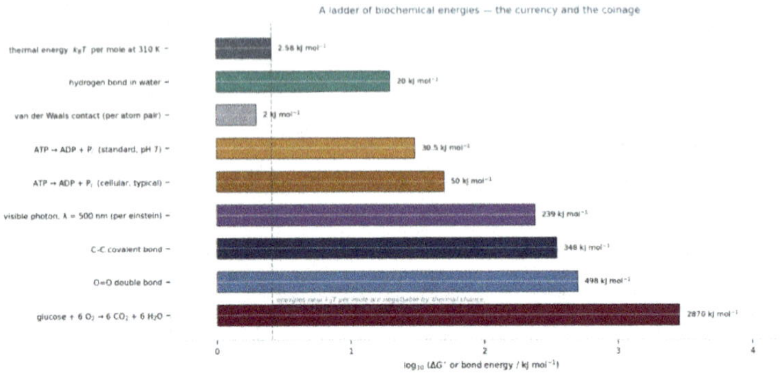

Figure 1.7 A ladder of biochemical energies on a logarithmic axis, with values in kJ mol^{-1}: the thermal scale RT at 310 K from Eq. (1.7); a typical van der Waals contact and hydrogen bond; the standard and cellular free energies of ATP hydrolysis (Chapter 25); the energy of one einstein (a mole of photons) at 500 nm, computed as $N_A hc/\lambda$ from the constants of Table iii; representative covalent bond energies; and the standard free energy of complete glucose oxidation (Berg et al., 2023). The dashed line marks RT: below a few multiples of it, associations come and go with thermal chance; far above it, chemistry requires machinery.

1.9.4 Sequence space and its extravagance

Fourth and last derivation, and the one the reader has already been promised: the arithmetic of information, continued from Eq. (1.3). If each of n positions in a chain is chosen from an alphabet of m symbols, the number of possible chains is m^n — the multiplication principle, no more, but its consequences are among the few honest infinities of the laboratory. For a modest protein of 300 residues, $m = 20$ and the space contains $20^{300} \approx 10^{390}$ sequences (Figure 1.10b); the observable universe offers some 10^{80} atoms for their construction, and the age of the Earth some 10^{17} seconds for their trial. The conclusion is not that proteins are improbable — the reader is one — but that *evolution does not sample sequence space; it inherits it*, walking from working sequence to working sequence along the narrow paths where function is connected. The quantitative character of biochemistry thus ends, characteristically, by paying tribute to its history; and the history is the subject of our final section.

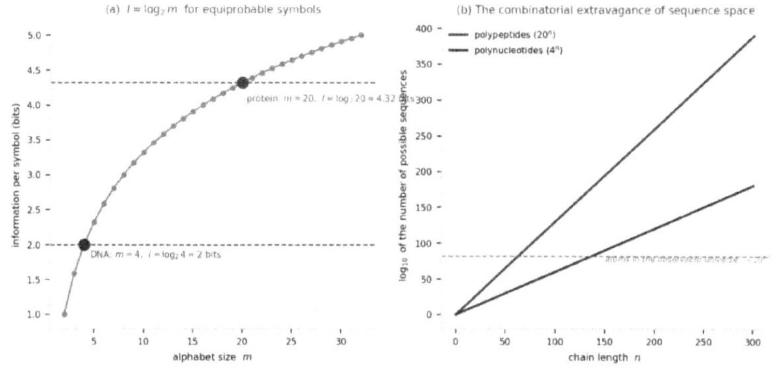

Figure 1.10 (a) Information per symbol, Eq. (1.2), as a function of alphabet size: two bits for the four-letter script of DNA, ≈ 4.32 for the twenty-letter script of protein. (b) The decimal logarithm of the number of possible chains of length n in each script, with the atom-count of the observable universe for sober comparison. The space of 300-residue proteins exceeds 10^{390} members — a figure the reader should compute from the plot's slope ($\log_{10} 20 \approx 1.301$ per residue) rather than accept from the , who would accept it from no one else.

1.10 Recurring principles of the subject

A first chapter should end as it began — with the obvious, now slightly less embarrassing. Ten principles have surfaced in the argument so far; they are the load-bearing ideas of the whole treatise, and we collect them here, not as a list to be memorised but as a hypothesis to be tested: the reader is invited to check, at the end of every chapter, whether the principles have been honoured, and to write in the margin when they have not.

First, the unity of biochemistry. Life's chemistry is one chemistry, varied endlessly: the same coenzymes, codes and currencies from bacterium to bishop — because all known life descends from a common ancestor, and inheritance is cheaper than invention. *Second, the economy of water.* The solvent is the majority partner in every living firm; its structure, and the hydrophobic effect it levies, will be shown in Part II to be the silent author of membranes, folds and assemblies. *Third, weak forces build, strong forces record.* Architecture is assembled from hydrogen bonds and van der Waals contacts near RT, editable by thermal chance; the permanent records — backbones, bonds, sequences — are covalent. *Fourth, structure determines function,* but the determination runs through dynamics and context, not through geometry alone; the same fold may serve many masters. *Fifth, energy is the price of order*: the cell's low entropy is bought by exporting disorder, as the second law, properly applied to open systems, not only permits but requires (§ 7.12). *Sixth, nothing is consumed but energy*; matter cycles — carbon, nitrogen, sulphur — and the cell is a way-station, not a terminus. *Seventh, information is physical*: stored in sequences, transmitted by copying, measured in bits, and constrained by the same thermodynamics as everything else. *Eighth, regulation is as real as catalysis*; the logic of feedback is written in molecules, and a pathway without its controls is a story without its verbs declined. *Ninth, the cell is small, crowded and fast* — a femtolitre city of millions, where the census of Table 1.3, not the continuum of the flask, sets the rules of evidence. *Tenth and last, everything in biochemistry makes sense only in the light of evolution* — a sentence the reader has heard before, in a slightly different register, from Dobzhansky; we repeat it because it is the one principle that explains the other nine.

These ten are the thesis of the book in miniature. The antithesis — that they are mere slogans, comforting generalities from which no experiment follows — is fair, and the remaining fifty-five chapters are our reply to it. Synthesis follows immediately: in Chapter 2 we descend to the atoms themselves, to the electrons and bonds from which the grammar of everything above is written. The reader who has followed the argument thus far has earned, we think, a small reward; the exercises below are it. In this book, as in the subject it serves, the dessert is indistinguishable from the work.

Chapter 1 — Conceptual synthesis

We set out to define biochemistry and found, instead, its question: in what does the difference between sugar-water and a living culture consist? The answer developed in ten movements: that the difference is chemical transformation, organised (§§ 1.1–1.2); that the organisation is hierarchical, with composition ascending and constraint descending (§ 1.3); that its economy runs on four convertible currencies (§ 1.4); that explanation comes in five nested levels, none reducible to another in practice (§ 1.5); that reductionism and holism are reconciled in the hermeneutic circle of analysis and synthesis (§ 1.6); that the unity of life licenses model organisms, at a declared tax (§ 1.7); that Buchner's juice supplies the template of the experimental method (§ 1.8); that the science is irreducibly quantitative, from the arithmetic of doubling to the extravagance of sequence space (§ 1.9); and that ten principles, to be tested chapter by chapter, carry the whole (§ 1.10). The direction of the treatise has been checked: we are where the map of Figure ii says we should be — at the threshold of the chemical grammar, with the machinery not yet admitted. Onward.

Worked example 1.1 — How many doublings separate one cell from a visible colony?

A colony becomes visible to the naked eye at roughly 10^8 cells. From Eq. (1.4), $N/N_0 = 2^k = 10^8$, so $k = 8/\log_{10} 2 = 8/0.3010 \approx 26.6$ — call it 27 doublings. At $\tau = 20$ min, that is 540 min, or nine hours: plate a single cell after breakfast, and by supper it has announced itself. Note the method, for it is the one this book will always use: the formula derived first (Eq. 1.4), the numbers substituted second, the sanity checked last — $2^{27} = 1.34 \times 10^8$, which rounds to our target as advertised.

Worked example 1.2 — The information content of the E. coli genome

From Eq. (1.2), each base pair carries $\log_2 4 = 2$ bits. The genome is 4.6×10^6 bp, whence $I = 9.2 \times 10^6$ bits = 9.2 megabits ≈ 1.15 megabytes (dividing by 8 bits per byte) — comfortably less than a single photograph from a modern telephone, and sufficient, the reader will recall from § 1.4, to reissue the entire economy of the cell. The comparison is not a curiosity but a calibration: information in biology is measured not by bulk but by consequence.

Worked example 1.3 — The energy of a mole of photons

Green light of λ = 500 nm: the energy of one photon is $E = hc/\lambda$ (Planck's relation, whose pedigree is physics and whose utility here is photosynthesis), and of one mole of photons — one *einstein* — $N_A hc/\lambda$. Substituting from Table iii: $(6.022 \times 10^{23})(6.626 \times 10^{-34}$ J s$)(2.998 \times 10^8$ m $s^{-1})/(500 \times 10^{-9}$ m$) \approx 2.39 \times 10^5$ J mol^{-1} = 239 kJ mol^{-1} — the value plotted in Figure 1.7, and the reason a single absorbed photon can break the chemistry of vision into motion (§ 29.2).

Exercises (with full resolutions)

EXERCISE 1.1

A culture of *E. coli* grows with μ = 2.08 h^{-1} (Eq. 1.5). How long does the population take to increase one-thousand-fold?

Resolution. From $N/N_0 = e^{\mu t} = 10^3$, take natural logarithms: $t = \ln(10^3)/\mu$ = 6.908/2.08 ≈ 3.32 h. (Equivalently, $10^3 \approx 2^{10}$, ten doublings of twenty minutes each — the near-equality $2^{10} \approx 10^3$ is the most productive error in all of mental arithmetic, and the reader is licensed to commit it freely.)

EXERCISE 1.2

Estimate the number of ribosomes in an *E. coli* cell, given that ribosomal RNA is about 80 % of the cell's RNA and that one ribosome contains ≈ 4 500 nucleotides of RNA.

Resolution. RNA is 20.5 % of the dry mass (Figure 1.2b): $0.205 \times 0.30 \times 10^{-12}$ g ≈ 6.2×10^{-14} g, of which 80 % is ribosomal: ≈ 4.9×10^{-14} g. A nucleotide weighs on average ≈ 330 g mol^{-1}, so 4 500 nucleotides make a ribosomal-RNA mass of ≈ 1.5×10^6 g mol^{-1}. Then $N = (4.9 \times 10^{-14}/1.5 \times 10^6) \times 6.022 \times 10^{23} \approx 2 \times 10^4$ ribosomes per cell — twenty thousand protein factories in a femtolitre, in comfortable agreement with the measured census. The estimate used only Table iii, Figure 1.2, and honesty.

EXERCISE 1.3

Using Eq. (1.7), show that a hydrogen bond of 20 kJ mol^{-1} is about eight times the thermal scale at 310 K, and comment on the consequence for the lifetime of such bonds in water.

Resolution. 20/2.58 ≈ 7.8 RT. The probability that a thermal fluctuation supplies energy E falls as $e^{-E/RT}$ (the Boltzmann factor, whose derivation awaits us in Chapter 7); for $E/RT \approx 7.8$ this is $e^{-7.8} \approx 4 \times 10^{-4}$ — small, but not negligible, and on the picosecond clock of molecular collisions it corresponds to lifetimes of nanoseconds. Hence the third principle of § 1.10: weak bonds hold long enough to specify structure and briefly enough to permit revision. A covalent bond at 348 kJ mol^{-1}, by the same arithmetic, has $e^{-135} \approx 10^{-59}$ — permanent, for all biological purposes, which is exactly what records require.

EXERCISE 1.4 (INTERPRETATIVE)

Buchner's extract fermented sugar without cells. What, precisely, did the experiment prove, and what did it leave unproved? Could a committed vitalist, in 1897, have accommodated the result — and at what price?

Resolution (indicative). Proved: that the chemical transformations of fermentation do not require the continued integrity, growth or vitality of the cell — that the machinery is chemical and survives disassembly. Left unproved: that the *manufacture* of the machinery requires nothing beyond chemistry; that all other vital functions are equally chemical; that the extract's activity was not a lingering virtue absorbed from the cells — the vitalist's remaining move. The price of that move is that the *vis vitalis* becomes a property of extracts, hence of solutions, hence of bottles — a soul that can be centrifuged is a soul in reduced circumstances, and by Sumner's crystallisation of urease in 1926 it was reduced to a crystal. The episode teaches the pattern of § 1.1: rival world-pictures are not refuted but made unemployable, one experiment at a time.

EXERCISE 1.5

From Figure 1.2, estimate the molar concentration of protein within the *E. coli* cell (mean protein 33 kg mol^{-1}), and compare it with the concentration of a typical enzyme assay in the laboratory (enzyme ≈ 1 nM to 1 μM). What methodological moral follows?

Resolution. Mass concentration of protein in the cell: 1.6×10^{-13} g per 10^{-15} L ≈ 160 g L^{-1}; dividing by 33 000 g mol^{-1} gives ≈ 5×10^{-3} mol L^{-1} — five *millimolar* of protein, three to six orders of magnitude above the assay. The moral is macromolecular crowding (§ 5.11): the cell's interior is not the dilute solution of the cuvette but a concentrate in which a third of the volume is occupied, reactions are accelerated, equilibria are shifted, and the laboratory's quaint custom of infinite dilution must be corrected for — a reminder that the flask is a model of the cell, not the cell of the flask.

Research-level propositions

Proposition 1.α. The near-equality of the genetic code across life is evidence for common descent; but could a second, independent origin of life on Earth be recognised if its biochemistry had since been displaced? What single observation would settle the question, and what present-day habitats would you search first? *Proposition 1.β.* Macromolecular crowding shifts association equilibria toward the compact state. Estimate the magnitude of the effect for a dimerising protein in the *E. coli* cytoplasm from first principles of excluded volume, and propose an experiment — using the methods of Part XI — that would measure it in a living cell rather than a polymer-thickened buffer. *Proposition 1.γ.* Shannon information (Eq. 1.3) is blind to function. Construct a measure of *functional* information for a protein family, and test whether it correlates better than length with the family's age, as dated from the phylogenies of Chapter 56.

References (Chapter 1)

Alberts, B., Heald, R., Johnson, A., Morgan, D., Raff, M., Roberts, K., & Walter, P. (2022). *Molecular biology of the cell* (7th ed.). W. W. Norton & Company.

Berg, J. M., Tymoczko, J. L., Gatto, G. J., Jr., & Stryer, L. (2023). *Biochemistry* (9th ed.). W. H. Freeman.

Buchner, E. (1897). Alkoholische Gärung ohne Hefezellen. *Berichte der Deutschen Chemischen Gesellschaft, 30*(1), 117–124.

Dobzhansky, T. (1973). Nothing in biology makes sense except in the light of evolution. *The American Biology Teacher, 35*(3), 125–129.

Neidhardt, F. C., & Umbarger, H. E. (1996). Chemical composition of *Escherichia coli*. In F. C. Neidhardt (Ed.), *Escherichia coli and Salmonella: Cellular and molecular biology* (2nd ed., Vol. 1, pp. 13–16). ASM Press.

Nelson, D. L., Cox, M. M., & Hoskins, A. A. (2021). *Lehninger principles of biochemistry* (8th ed.). W. H. Freeman.

Schoenheimer, R. (1942). *The dynamic state of body constituents*. Harvard University Press.

Shannon, C. E. (1948). A mathematical theory of communication. *The Bell System Technical Journal, 27*(3), 379–423.

Wöhler, F. (1828). Ueber künstliche Bildung des Harnstoffs. *Annalen der Physik und Chemie, 88*(2), 253–256.

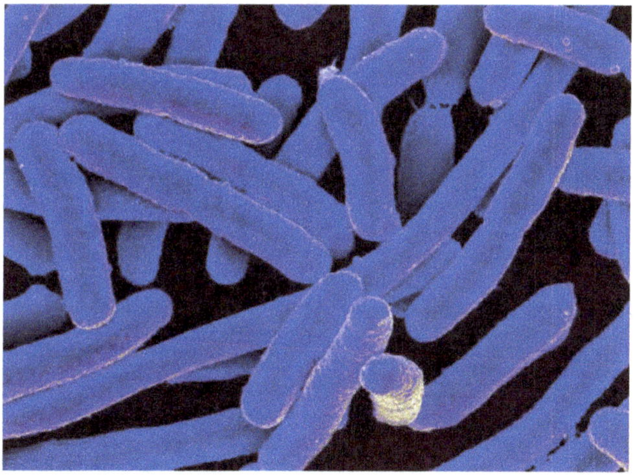

Plate 1.2 The hero of the chapter, and of much of the book: *Escherichia coli*, scanning electron micrograph (colour-enhanced), rod-shaped cells of approximately 1 × 2 µm as assumed in the census of § 1.9. Image: National Institute of Allergy and Infectious Diseases (NIAID), distributed through the Public Health Image Library of the U.S. Centers for Disease Control and Prevention (PHIL ID 18160), a United States government work in the public domain. Each of these unassuming rods executes Eq. (1.4) with a fidelity no bank has ever matched.

CHAPTER 2

Atoms, Electrons and Chemical Bonds

CONTENTS OF THIS CHAPTER

In which the grammar of the chemical alphabet is examined: the architecture of the atom, the diplomacy of its electrons, and the several treaties — covalent, ionic, coordinate, hydrogenous, and van der Waals — by which atoms agree to keep company. Nothing in later chapters will make sense without these treaties, and nothing in this one is asserted without its derivation.

2.1 Atomic structure and electron configuration

The atom of the biochemist is a modest country: a nucleus, which we shall treat with the courteous indifference of a traveller who does not enquire after the government, and a cloud of electrons, which is where all the politics occur. Chemistry, and hence biochemistry, is the diplomacy of that cloud — its vacancies, its surpluses, its perpetual negotiation of advantage. Everything in this book that is called a *bond*, a *reaction*, a *recognition*, is an event in the cloud; and the cloud, unlike the solid little solar systems of the schoolbooks, is governed by quantum mechanics, a legislature whose statutes we shall quote only where they have been found, by experiment, to be enforced.

The first statute is *quantisation*: an electron bound to a nucleus may possess only certain energies, and not others. The reader is owed a derivation rather than a decree, so let us obtain the energies of the hydrogen atom in the manner of Bohr (Plate 2.1), with full honesty about the model's approximations. Bohr's postulate of 1913 (Bohr, 1913) quantises the angular momentum of the orbiting electron in units of the reduced Planck constant [1] $L = m_e vr = n\hbar$, where m_e is the electron mass, v its speed, r the orbital radius, n = 1, 2, 3, ... the principal quantum number, and $\hbar = h/2\pi$. Meanwhile classical electrostatics (Coulomb's law, of which more in § 2.4) demands that the attraction balance the centripetal acceleration: $m_e v^2/r = e^2/(4\pi\varepsilon_0 r^2)$. Two equations, two unknowns (v and r); eliminating v gives the permitted radii

$$r_n = n^2 \frac{4\pi\varepsilon_0 \hbar^2}{m_e e^2} = n^2 a_0, \qquad a_0 \approx 0.529\,\text{Å} \tag{2.1}$$

1. Niels Bohr's trilogy of papers, "On the Constitution of Atoms and Molecules," appeared in the *Philosophical Magazine* in July, September and November of 1913 — an annus mirabilis conducted entirely in one journal and, one suspects, one notebook.

— the Bohr radius a_0 being the natural yardstick of atomic size, happily of the same order as the atoms of Figure 1.4. Inserting r_n back into the total energy (kinetic plus potential) yields the permitted energies,

$$E_n = -\frac{m_e e^4}{8\,\varepsilon_0^2 h^2} \cdot \frac{1}{n^2} \approx -\frac{13.6\text{ eV}}{n^2} \qquad (2.2)$$

— whence, for a transition between levels n_1 and n_2, the photon energy is $h\nu = |E_{n2} - E_{n1}|$, which is the Rydberg formula that had been an empirical curiosity for twenty-five years before Bohr explained it. The derivation is given here not because biochemistry runs on hydrogenic bookkeeping, but because of the *moral*: a single quantum statute, imposed on classical mechanics, generates discrete levels, and discrete levels are the deep reason that atoms are *individuals* — that carbon is carbon everywhere and always, in diamond, in glucose, and in the reader. A classical cloud would smear; a quantised cloud keeps its identity papers. The full quantum-mechanical treatment replaces Bohr's circles with orbitals — standing waves of probability designated *s, p, d, f* — but it preserves the essential architecture: shells of increasing energy, each with a fixed capacity of two electrons per orbital (Pauli's exclusion principle), filling in the order that the periodic table so elegantly displays.

HISTORICAL NOTE — THE ATOM, FROM BILLIARD BALL TO CLOUD

Dalton's atom (1808) was a billiard ball with a valence; Thomson's (1897) a pudding in which electrons were the plums; Rutherford's (1911), inferred from the shocking back-scatter of α-particles off gold foil — "as if you fired a fifteen-inch shell at a piece of tissue paper and it came back and hit you" — a miniature solar system; Bohr's (1913) a solar system with a quantum constitution; and the modern atom (Schrödinger, Heisenberg, 1925–26) a nucleus wrapped in standing waves. Each model was overturned not by philosophy but by an experiment it could not digest, and each survives as a working approximation in its own domain: the biochemist still draws Dalton's sticks and balls, while knowing they are clouds. The reader will observe that this is the hermeneutic circle of Figure 1.5, practised at the scale of the ångström.

Plate 2.1 Niels Bohr (1885–1962), whose quantisation of the hydrogen atom (Eqs. 2.1–2.2) first gave atomic individuality a mathematical address. Portrait from the Nobel Foundation archive, nobelprize.org (Nobel Prize in Physics, 1922), reproduced from the Foundation's publicly distributed image.

2.2 Valence electrons and molecular reactivity

Of all the electrons in an atom, only the outermost shell — the *valence* electrons — takes part in the commerce of bonding; the inner shells are capital in trust, visible to spectroscopists but otherwise beyond reach. This single fact, when combined with the shell capacities of § 2.1, yields the entire reactive personality of an element. An atom whose valence shell is full (the noble gases) is chemically contented and will not treat; an atom one or two electrons short of contentment (chlorine, oxygen, nitrogen) will acquire electrons where it can, by seizure if necessary; an atom with one or two electrons in excess of a full inner shell (sodium, potassium, calcium, magnesium) will surrender them with equal readiness. Between these temperaments lies carbon, with four valence electrons and four vacancies — unable to win outright, unwilling to lose, and therefore compelled to *share*. The whole of organic chemistry, and with it the whole of biochemistry, is the consequence of carbon's diplomatic neutrality.

The reader should carry away the valences of the biogenic elements as one carries the genders of a language: hydrogen 1, oxygen 2, nitrogen 3, carbon 4, phosphorus 5 (in phosphate), sulphur 2 or 6. From these, and the shell structures behind them, the skeleton of every metabolite in Part VII can be drafted before its name is known. Figure 2.8 sets the matter in its periodic frame: of the one hundred and eighteen known elements, life makes systematic use of about thirty, and of those, six — carbon, hydrogen, nitrogen, oxygen, phosphorus, sulphur — constitute some ninety-seven per cent of living mass (Figure 1.1). The remainder of the employed set divides into the major electrolytes (Na^+, K^+, Mg^{2+}, Ca^{2+}, Cl^-), whose chemistry is ionic and aqueous, and the trace metals (Fe, Zn, Cu, Mn, Co, Mo, and a handful more), whose chemistry is the chemistry of the coordinate bond (§ 2.4) and whose scarcity is a puzzle the final chapter will approach through geochemistry and deep time.

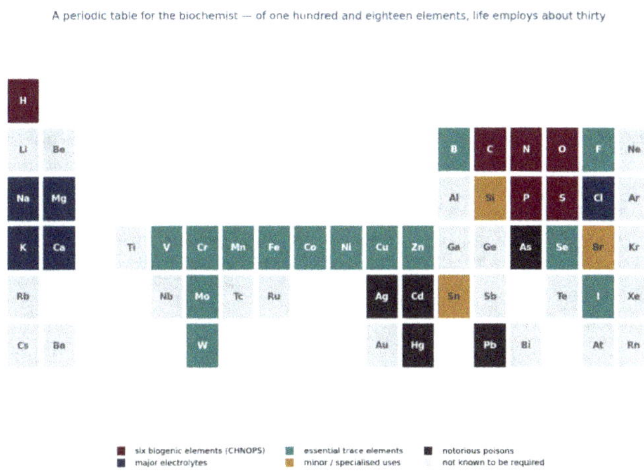

Figure 2.8 A periodic table for the biochemist. The six biogenic elements (oxblood) build the covalent skeleton of life; the major electrolytes (navy) maintain its osmotic and electrical household; the essential trace elements (teal) staff its catalytic metal centres; a handful of elements find minor or specialised employment (amber), and a dark company (grey-black) — arsenic, cadmium, mercury, lead, silver — figures chiefly in toxicology, that stern annex of biochemistry. The classification is functional, not absolute: selenium was a poison until it was found in glutathione peroxidase, and chromium's claims remain, shall we say, under appeal.

2.3 Electronegativity and bond polarity

Sharing, in electrons as in property, is rarely equal. The measure of an atom's greed for the shared pair is its *electronegativity* χ, defined by Pauling thermochemically from the observation that the bond between unlike atoms is always stronger than the geometric mean of the two like-atom bonds [2] — the excess being attributed to the ionic character of an unequal partnership. The resulting scale (Figure 2.1) runs from the alkali metals, who give, to oxygen and the halogens, who take, with carbon and hydrogen sitting almost exactly at the geometric centre. Where the difference $\Delta\chi$ is small ($\lesssim 0.5$) the bond is covalent and unconcerned; where it is moderate (0.5–1.7) the bond is *polar covalent*, carrying partial charges $\delta+$ and $\delta-$; where it is large ($\gtrsim 1.7$) the partnership dissolves into outright ionisation.

The quantity of the imbalance is the *dipole moment*. For partial charges $\pm\delta e$ separated by a distance d,

$$\mu = \delta e\, d \tag{2.3}$$

— measured in debye, 1 D = 3.336×10^{-30} C m. Water, the solvent of this entire treatise, carries μ = 1.85 D, which the reader can convert, using the O–H bond length 0.96 Å, into a partial charge of about 0.4 e on each hydrogen (Worked

2. Pauling's original definition (1932) reads $\Delta_{AB} = E(A\text{–}B) - \sqrt{[E(A\text{–}A)\cdot E(B\text{–}B)]}$, the "extra-ionic resonance energy," with $|\chi_A - \chi_B| = 0.102\sqrt{\Delta_{AB}}$ when the bond energies are in kJ mol^{-1}; the constant merely rescales the answer to his chosen dimensionless range. The definition is given here so that Figure 2.1 is seen to be, at bottom, a table of measured bond energies in disguise.

Example 2.1). This is not a small number: it is the whole cause, as § 2.7 will show, of water's anomalous and life-enabling behaviour. The polarity of a bond is thus not a refinement but a first-class variable of biochemistry, fixing solubility, directing recognition, and orienting every hydrogen bond in the cell. The dialectic of the section is short but instructive: the thesis — polarity explains molecular behaviour — meets the antithesis that dipole moments are vectors, and that a molecule of polar bonds may be nonpolar in sum (CO_2, with two fiercely polar C=O bonds pulling in opposite directions, is as nonpolar as a judge). The resolution is that one must always sum the vectors, and that biochemistry's favourite trick is to build molecules — phospholipids, above all — in which the polar and nonpolar regions do *not* cancel, being quarantined at opposite ends. Such molecules, the amphiphiles, will assemble themselves into membranes in Part V without being asked.

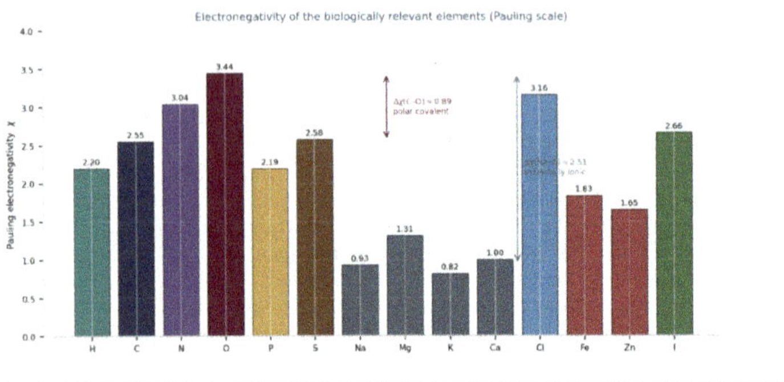

Figure 2.1 Pauling electronegativities of the elements employed by life, from the scale's thermochemical definition (see the footnote in § 2.3). The two annotated differences mark the great divide of bond type: C–O ($\Delta\chi$ = 0.89) is the archetype of the polar covalent bond that populates metabolites; Na–O ($\Delta\chi$ = 2.51) is ionic beyond appeal, and sodium is therefore found in the cell only as the aquated cation Na^+. Values after Pauling (1960).

2.4 Covalent, ionic and coordinate bonds

Three species of strong bond suffice for nearly the whole of biochemistry's permanent architecture, and it is time to meet them formally. The *covalent* bond we have met already as the sharing of electron pairs; its quantum rationale — that a shared pair between two nuclei lowers the total energy below that of the separated atoms — was first stated with clarity by Lewis in 1916 (Lewis, 1916), eleven years before quantum mechanics caught up with him. [3] Its energetic anatomy is the Morse well of Figure 2.2a: a deep minimum at the equilibrium length r_e, a steep repulsive wall at short range where the closed shells of the two atoms begin to interpenetrate (Pauli exclusion again, now armed), and dissociation at long range. For the biochemist the practical facts are the bond energies of Figure 2.3: 250–600 kJ mol^{-1}, far above the thermal scale $RT \approx 2.6$ kJ mol^{-1} of Eq. (1.7), which is why covalent structures are the permanent records of the cell (principle three of § 1.10).

3. G. N. Lewis never received the Nobel Prize, though nominated dozens of times; the shared pair, the electron-dot notation, the photon (his word), and the modern definition of acids and bases must serve as his monuments. Posterity, unlike the Stockholm committee, has been generous.

The *ionic* bond is the opposite temperament: not sharing but surrender, followed by the electrostatic attraction of the resulting ions. The potential energy of two point charges q_1, q_2 at separation r in a medium of relative permittivity ε_r is given by Coulomb's law,

$$U(r) = \frac{q_1 q_2}{4\pi\varepsilon_0\,\varepsilon_r\,r} \tag{2.4}$$

— which for unit charges in vacuo evaluates, using the constants of Table iii, to -1389 kJ mol^{-1} Å / r: a formidable attraction. But the biochemist's ions never meet in vacuo; they meet in water, whose $\varepsilon_r \approx 78.5$ screens the interaction nearly eighty-fold (Figure 2.2b). The consequence is the first law of aqueous biochemistry: *in water, salt bridges are weak bridges.* A NaCl crystal, robust in air, dissolves in water precisely because hydration pays better than the lattice (§ 5.4); and the ion pairs of proteins — the glutamate–lysine "salt bridges" beloved of structural biologists — contribute, in water, barely more than a few kJ mol^{-1}, unless they are buried away from the solvent, where the permittivity is low and the attraction revives. Herein lies a quiet scandal of the textbooks, which the dialectic demands we air: generations of students have computed protein ion-pair energies with $\varepsilon_r = 1$ and concluded that salt bridges hold proteins together; the honest calculation, with $\varepsilon_r \approx 80$, concludes they do not. The resolution — desolvation is the hidden variable — will return in force in Chapter 11.

The *coordinate* (dative) bond is the covalent bond's aristocratic cousin: both electrons of the shared pair are donated by one partner, the ligand, to an empty orbital of the other, almost always a metal ion. It is the bond of the trace-element economy: the nitrogens of histidine donating to zinc in carbonic anhydrase (§ 19.12), the porphyrin nitrogens to iron in haem (§ 13.1), the thiolate of cysteine to the iron–sulphur clusters of respiration (§ 26.8). Its binding energies span an enormous range, from the loosely hydrated Mg^{2+} of ATP to the iron of haem, whose grip on the porphyrin is effectively permanent. The coordinate bond is also where biochemistry meets its poisons: carbon monoxide binds haem iron through the same dative mechanism as oxygen, but some two hundred times more tightly — a tragedy written in one orbital, and a first intimation that molecular recognition, the subject of Chapter 8, is a matter of life and death at the resolution of single electrons.

HISTORICAL NOTE — LEWIS, LANGMUIR AND THE SHARED PAIR

That a chemical bond might *be* a pair of electrons was proposed by G. N. Lewis at Berkeley in 1916, in a paper of twenty-three pages that founded structural chemistry without a single equation (Lewis, 1916). Langmuir extended and popularised the idea, coining "covalence"; Sidgwick supplied the dative variant in 1927; and Pauling married the whole to quantum mechanics in *The Nature of the Chemical Bond* (1939), perhaps the most influential chemistry book ever written (Pauling, 1960). The student who draws a line between two atoms today is drawing Lewis's pair; it is one of the few notations in science that has survived, unamended, for over a century.

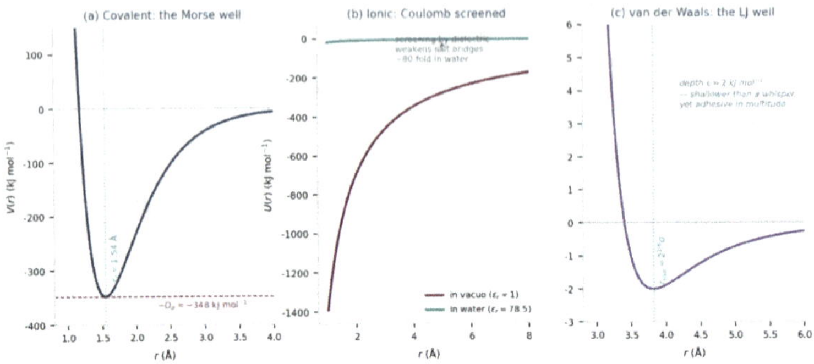

Figure 2.2 The three potential wells of biochemical bonding. (a) The Morse potential of the C–C covalent bond (Eq. 2.4's covalent counterpart; D_e = 348 kJ mol^{-1}, r_e = 1.54 Å) — deep, narrow, and chemically permanent. (b) The Coulomb potential of a unit-charge ion pair in vacuo and in water (Eq. 2.4), showing the eighty-fold dielectric screening that governs every electrostatic event in the cell. (c) The Lennard-Jones 6–12 potential of a van der Waals contact (Eq. 2.6), with its shallow well at r_{min} = $2^{1/6}\sigma$ (derived in § 2.8). The three wells, drawn to their own scales, span four orders of magnitude in depth — the entire energetic range of molecular sociability.

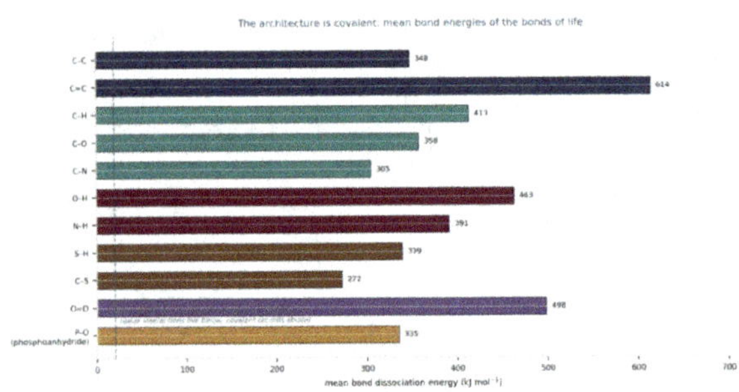

Figure 2.3 Mean bond dissociation energies of the bonds from which living matter is built, against the thermal threshold (dashed line; cf. Exercise 1.3). Two readings are intended: the obvious one — covalent bonds are permanent at body temperature — and the subtler one, that the *range* of covalent energies (S–H at 272 to C=C at 614 kJ mol^{-1}) is itself a biochemical instrument. Weaker covalent links, such as the thiol and disulphide chemistry of cysteine, are the cell's reversible covalent switches (§ 12.8), strong enough to specify structure yet negotiable by redox machinery. Values from standard compilations (Pauling, 1960) (Berg et al., 2023).

2.5 Resonance and electron delocalisation

It is an axiom of the trade that the structural formula lies — politely, usefully, but lies. The benzene ring of Kekulé [4] is drawn with alternating single and

4. August Kekulé's account of the benzene hexagon (1865) is attached to chemistry's most famous dream — the ouroboros, the snake seizing its own tail, which he reported at the 1890 Benzolfest in his honour. Historians have since noted that the dream appears in no notebook

double bonds, yet every bond in the ring measures 1.40 Å — exactly intermediate between the single (1.54) and the double (1.34) — and the molecule refuses, with aristocratic consistency, to undergo the addition reactions that honest double bonds perform. The remedy is *resonance*: the true wavefunction of the π electrons is not any single drawn structure but a superposition of all reasonable ones, and the molecule is stabilised by the difference, called the resonance energy. For benzene the experimental figure is obtained, characteristically, from a thermochemical ledger: the hydrogenation of cyclohexene releases 120 kJ mol^{-1}, so three honest double bonds should release 360; benzene releases 208. The missing 152 kJ mol^{-1} is the price the molecule was paid, in advance, for delocalising.

Resonance is not a book-keeping refinement; it is the biochemist's daily bread. The carboxylate group of aspartate and glutamate is a resonance hybrid of two equivalent structures, which is precisely why the negative charge is shared over two oxygens and why the group is a weaker base — and hence a better leaving group and a stabler anion — than a localised alkoxide would be. The peptide bond, as the case study at the end of this chapter will show at full length, owes its planarity to a 40 per cent admixture of a zwitterionic resonance form; the whole architecture of the protein backbone (Chapter 10) is built on that single delocalisation. And the phosphates of ATP owe their lability in part to resonance stabilisation of the inorganic phosphate product — a point to which § 2.10 returns. The dialectic to be resisted is the student's natural misapprehension that resonance structures *alternate*, like sleepers changing sides; they do not exist at all, and the hybrid alone exists. The drawing of two-headed arrows between Lewis structures is a confession of the notation's poverty, not a description of molecular restlessness.

2.6 Molecular orbitals and conjugated systems

The honest description of delocalisation is the molecular orbital: a standing wave of electron probability extending over the whole molecule, built by linear combination of the atomic orbitals (LCAO). For a chain of n p-orbitals — the polyenes — the Hückel treatment[5] gives the orbital energies directly,

$$E_k = \alpha + 2\beta\cos\left(\frac{k\pi}{n+1}\right), \qquad k = 1, 2, \ldots, n \tag{2.5}$$

— where α is the Coulomb integral (the energy of an electron in an isolated p-orbital), β the resonance integral (the stabilisation from overlap of neighbours, negative), and the cosine does the combinatorics. The derivation, sketched in one line, substitutes the LCAO wavefunction $\psi = \Sigma c_i \varphi_i$ into the Schrödinger equation, keeps only nearest-neighbour overlaps, and solves the resulting secular determinant — whose eigenvalues are the cosine series above, as the reader may verify for $n = 2$ in three lines. Figure 2.7 displays the consequences: the levels crowd together as the chain lengthens, the gap between the highest occupied and lowest unoccupied orbitals (HOMO–LUMO gap) shrinks from $2|\beta|$ in ethylene to $0.89|\beta|$ in hexatriene, and the energy of the lowest $\pi \rightarrow \pi^*$ transition falls — through the visible, for chains of ten or so conjugated double bonds. Herein lies the entire molecular economics of colour in biology: β-carotene (eleven

and was told twenty-five years after the fact; the hexagon, however, is real, and dreams, unlike structures, are not subject to refereeing.

5. Erich Hückel's π-electron theory (1931) survives in biochemistry not for its accuracy, which is modest, but for its transparency: it is the last stop at wh athematics of colour can still be done on the back of an envelope, and the present author are unabashed envelope-users.

conjugated double bonds) is orange, retinal is the pigment of vision, chlorophyll's vast conjugated macrocycle tunes its absorption to the red edge of the solar spectrum (§ 29.2). Delocalisation is not merely stability; it is, at the right gap, *light-harvesting*.

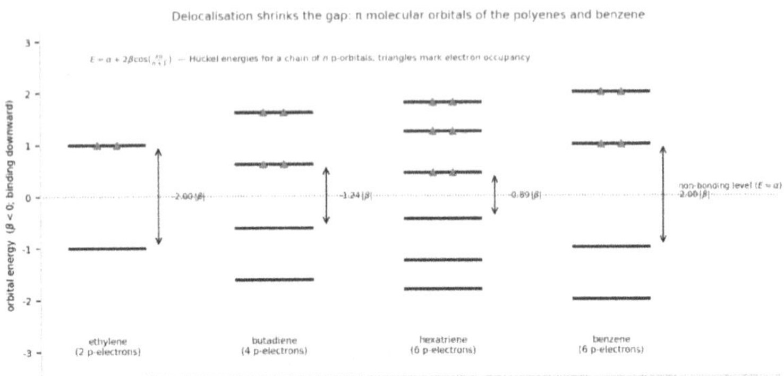

Figure 2.7 Hückel π molecular orbitals (Eq. 2.5) of ethylene, butadiene, hexatriene and benzene, with electron occupancy (amber triangles) and the HOMO–LUMO gap annotated in units of |β|. Two patterns reward attention: the crowding of levels with chain length, which drives the absorption of conjugated pigments into the visible; and the closed, stabilised shell of benzene (six π electrons, a Hückel $4n + 2$ count), which is the quantum root of aromaticity — the reason the rings of phenylalanine, tyrosine, tryptophan and the nucleobases are rigid, flat and spectroscopically loud.

2.7 Hydrogen bonding

We arrive at the bond that runs this book. The hydrogen bond — the attraction of a hydrogen covalently bound to an electronegative atom (the donor) for a lone pair of another electronegative atom (the acceptor), written D–H···A — is a bond of middling strength and supreme consequence: 10–40 kJ mol^{-1} in vacuum, about 20 in liquid water, directional, saturable, and numerous beyond counting. Its physical origin is predominantly electrostatic (§ 2.3's partial charges, conversing at close range), with an admixture of charge-transfer covalency that modern calculations place at perhaps a quarter of the total.[6] Its geometry in water is set by the water molecule's own architecture (Figure 2.4): two O–H donors and two lone-pair acceptors, disposed tetrahedrally, so that each molecule can form up to four hydrogen bonds, and the liquid becomes not a collection of molecules but a single fluctuating network — the medium in which every reaction of this book is dissolved.

6. The hydrogen bond is first clearly described in print by Latimer and Rodebush (1920), working in Lewis's own laboratory; the idea had circulated earlier — Huggins, and before him Moore and Winmill (1912), have their champions. Priority in science, as in love, is rarely as exclusive as the participants suppose.

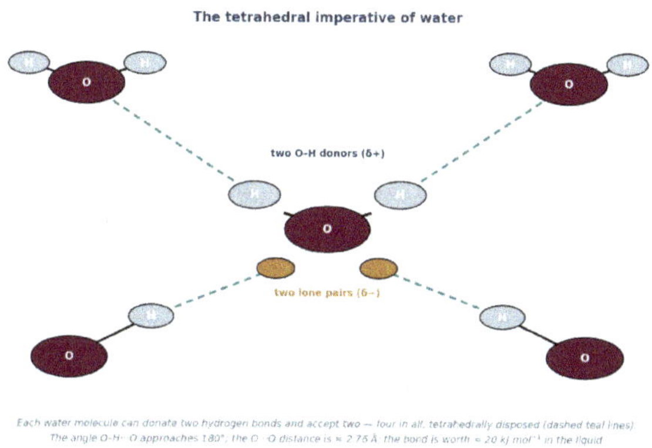

Figure 2.4 The tetrahedral imperative of water. The central molecule (oxblood oxygen, grey hydrogens, amber lone pairs) donates two hydrogen bonds through its O–H groups and accepts two through its lone pairs (dashed teal), the four directions summing to the tetrahedron that dictates the open structure of ice, the anomalies of the liquid, and the hydration shells of everything dissolved in it. The O–H···O angle approaches 180° and the O···O distance ≈ 2.76 Å in ice; in the liquid the network is partially broken and perpetually re-forming on the picosecond scale. Geometry after Eisenberg and Kauzmann (1969).

The biochemist's interest, however, is not in water for its own sake — Chapter 5 is devoted to that — but in the hydrogen bond as the *universal joint* of macromolecular structure. The α-helix is a helix because every backbone C=O accepts a hydrogen bond from the N–H four residues ahead (Chapter 10); the DNA double helix holds its two strands together with two or three hydrogen bonds per base pair, weak enough to unzip for copying, strong enough to stay zipped in storage (Chapter 43); and the specificity of enzyme–substrate recognition (Chapter 15) is written substantially in the geometry of donors and acceptors facing each other across an active site like the countersigned clauses of a treaty. Directionality is the operative word: unlike the Coulomb interaction of ions, which is content with proximity, the hydrogen bond demands *orientation*, and orientation is the raw material of specificity. A bond that can say *where* as well as *how strongly* is a bond from which a code can be built; the cell, not surprisingly, builds with it everywhere.

2.8 van der Waals interactions

Beneath the hydrogen bond, at the very floor of the energy ladder, lie the forces named for Johannes Diderik van der Waals (Plate 2.2), who inferred their existence in 1873 from the stubborn refusal of real gases to behave ideally (van der Waals, 1873) [7]. Three components compose them. The Keesom interaction aligns permanent dipoles; the Debye interaction induces a dipole in a polarisable neighbour; and the London dispersion — the subtlest and, for the biochemist, the most important — arises between any two atoms whatsoever

7. Van der Waals's doctoral thesis, *Over de Continuïteit van den Gas- en Vloeistoftoestand*, deduced molecular attraction from the deviation of gases from Boyle's law — reading the invisible from the disobedient, a method the biochemist will recognise from Schoenheimer's tracers to the present day. The Nobel Prize followed in 1910; the geckos, who walk on ceilings by these same forces, have yet to be similarly honoured.

from the correlation of their instantaneous electron fluctuations. London's own 1930 estimate (London, 1930), from second-order perturbation theory, gives the attractive energy of two atoms of polarisability α and ionisation energy I as

$$U_{\text{disp}}(r) \approx -\frac{3}{4}\frac{\alpha^2 I}{(4\pi\varepsilon_0)^2}\frac{1}{r^6} \tag{2.6}$$

— the celebrated inverse-sixth power. Add the steep repulsion of overlapping closed shells (Pauli, again) and one obtains the working form of the whole family, the Lennard-Jones 6–12 potential,

$$V(r) = 4\varepsilon\left[\left(\frac{\sigma}{r}\right)^{12} - \left(\frac{\sigma}{r}\right)^{6}\right] \tag{2.7}$$

— in which ε is the well depth and σ the collision diameter. The equilibrium contact distance follows by setting $dV/dr = 0$, a derivation the reader is invited to perform in full: differentiating, $4\varepsilon[-12\sigma^{12} r^{-13} + 6\sigma^6 r^{-7}] = 0$, whence $12\sigma^{12} = 6\sigma^6 r^6$, and

$$r_{\min} = 2^{1/6}\sigma \approx 1.122\,\sigma \tag{2.8}$$

— the distance, plotted in Figure 2.2c, at which every comfortable atom rests against its neighbour. The well is shallow, $\varepsilon \approx 1–2$ kJ mol^{-1} per contact — less than RT, as Figure 1.7 already warned — and the individual van der Waals bond is therefore ephemeral. The whole art of the cell is to use them in multitude: a protein interior makes thousands of such contacts, an antibody–antigen interface hundreds, and their sum is formidable while remaining reversible, each contact being individually negotiable. Here is the deepest economy of the weak forces, and it deserves the dialectic in full. *Thesis*: forces weaker than thermal noise cannot specify structure. *Antithesis*: they demonstrably do — crystals, membranes, and folded proteins exist. *Resolution*: specificity is not a property of the single contact but of the *pattern* of contacts; a thousand individually insignificant interactions, geometrically arranged, select one configuration against all rivals by the multiplication of small advantages — as a jury of a thousand indifferent voters, each right sixty per cent of the time, does not deliver an indifferent verdict. We shall meet this arithmetic again under the names of cooperativity (§ 8.7) and avidity (§ 8.12); it is the closest thing biochemistry has to a universal law of assembly.

Plate 2.2 Johannes Diderik van der Waals (1837–1923), whose equation of state first made molecular attraction measurable. Portrait from the Nobel Foundation archive, nobelprize.org (Nobel Prize in Physics, 1910), reproduced from the Foundation's publicly distributed image.

2.9 The hydrophobic effect

Last of the weak forces, and first in structural consequence, comes the one that is not, strictly speaking, a force at all. Oil does not mix with water; everyone knows it, and almost everyone misdescribes it. The usual account — water repels oil — is false; water attracts hydrocarbons by London dispersion about as strongly as it attracts itself, and the enthalpy of dissolving a small alkane in water is nearly indifferent. The true account is entropic: a hydrocarbon in water forces the surrounding water molecules to arrange themselves into an ordered, hydrogen-bonded cage, a clathrate-like shell that preserves the network's count of hydrogen bonds at the price of the water's freedom; and the universe, which exacts payment for order, declines the arrangement whenever the hydrocarbon can be expelled into a phase of its own. The measured consequence is the transfer free energy of Figure 2.5a: moving an *n*-alkane from its own liquid into water costs about +3.8 kJ mol^{-1} per methylene group, and the decomposition (Figure 2.5b) shows the classic signature — ΔH small, $-T\Delta S$ large and unfavourable (Kauzmann, 1959) (Tanford, 1980).

The hydrophobic effect, so understood, is the silent author of three of the great structures of this book. It folds proteins (Chapter 11): the nonpolar side chains are driven to the interior not because they attract one another particularly, but because burying them releases their ordered water cages to the freedom of the bulk — the protein pays its entropic tax once, in folding, rather than at every exposed side chain. It builds membranes (Chapter 23): phospholipids, amphiphilic by design (§ 2.3), hide their tails in a bilayer and show their heads to water, and the resulting sheet closes upon itself into a vesicle with no instruction beyond the chemistry of its parts. And it drives every association (Chapter 8) in which two surfaces bury hydrophobic area upon binding — which is to say, most of them. A candid footnote is owed to the modern literature: the tidy cage picture, derived for small solutes, modifies at large ones, where the physics of interfaces takes over and the effect becomes enthalpic in patches [8]; but the classical account, like Newton's mechanics, remains the right

8. The small-solute/large-solute distinction is associated with the work of Chandler and co-workers (2005); like most refinements of a beautiful theory, it has been received with gratitude

first approximation and the right first intuition. The reader should carry away one sentence, for it will organise half the book: *in water, the nonpolar world huddles — not from love of itself, but because water, presented with a guest it cannot befriend, chooses the company of itself.*

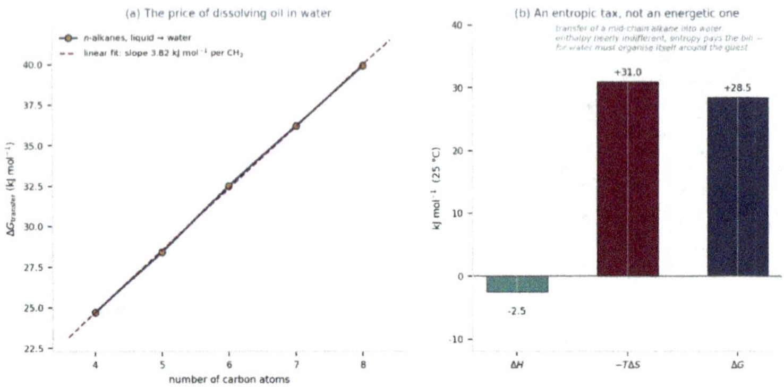

Figure 2.5 The hydrophobic effect, measured and decomposed. (a) The free energy of transferring *n*-alkanes from their pure liquid into water at 25 °C rises linearly at ≈ 3.8 kJ mol^{-1} per added methylene — the incremental tax on nonpolar surface, and the quantitative root of membrane assembly and protein folding (representative values after Tanford 1980). (b) The thermodynamic decomposition for a model transfer: enthalpy near zero, entropy dominant — the water's lost freedom, not any energetic insult, is what the alkane is expelled for. The same decomposition, with the signs reversed, is the profit-and-loss account of protein folding in Chapter 11.

2.10 Bond energies and biochemical stability

A chapter on bonds must end by correcting the most durable error its subject has spawned: the phrase "high-energy bond," applied to ATP since 1941 with the best of intentions and the worst of consequences. The error is grammatical before it is chemical. Bond *energy*, in the sense of Figure 2.3, is the energy required to *break* a bond — a positive quantity, a cost. The hydrolysis of ATP releases free energy not because a bond containing wealth is broken (breaking any bond costs), but because the *products* of hydrolysis — ADP and inorganic phosphate — are so much better off than the reactants: the phosphate is resonance-stabilised, the separated negative charges cease their mutual repulsion across the anhydride, and both products are hydrated more comfortably than the parent. The ledger is the whole molecule's, not one bond's; and the coin is free energy, whose proper definition awaits Chapter 7. The dialectic is worth the space because the misconception is perennial: *thesis* — ATP hydrolysis releases energy, therefore the broken bond held energy; *antithesis* — breaking bonds always costs energy, therefore nothing can be released by breaking one; *resolution* — the released free energy is the difference of the stability of products and reactants, and the bond broken is merely the tollgate, not the treasure. The reader who has absorbed this paragraph is already ahead of half the literature of the last eighty years, and may proceed to Chapter 25 in a state of earned superiority.

by the specialists and polite inattention by everyone else. We adopt the classical picture in the text and flag the frontier here, which is where frontiers belong in a treatise of foundations.

One refinement completes the section: the distinction between *thermodynamic stability* and *kinetic persistence*. The covalent bonds of Figure 2.3 are thermodynamically unstable in water in the long run — everything in the cell is hydrolysable, oxidisable, perishable — yet they persist for hours to years because their breakdown requires passage over an activation barrier that thermal energy at 310 K rarely supplies. The cell exploits the gap ruthlessly: it stores its archive in kinetically inert DNA, builds its structures from kinetically durable peptide bonds, and then hires enzymes (Part IV) to lower selected barriers on demand, converting persistence into programme. Stability without lability is a crystal; lability without stability is a bonfire; life is the negotiated peace between the two, and the treaty is kinetics.

2.11 Reactive functional groups

The covalent skeleton having been surveyed, its ornaments deserve a census. The functional groups of Table 2.2 are the reactive personalities of biochemistry — the points at which metabolites can be attacked, transferred, oxidised, or coupled — and the reader will meet every one of them again as the reagent of some enzyme in Part IV. Two organising ideas suffice. First, the groups divide by *electronic temperament* into nucleophiles, rich in electrons and seeking positive centres (thiolates, amines, hydroxyls, carbanion equivalents), and electrophiles, poor in electrons and seeking donors (carbonyl carbons, phosphoryl phosphates, protonated imines); nearly every reaction of metabolism is a marriage between the two temperaments, brokered by an enzyme. Second, the groups are not equally available: the cell keeps its most reactive members — thiols, aldehydes, radicals — under close discipline, exposing them only within active sites, for a reactive group at large in a crowded cell is a loaded weapon in a library. Much of Part VII's elegance is the choreography by which the same half-dozen groups are activated, coupled and released in hundreds of distinguishable reactions with no confusion of partners.

Table 2.2 The reactive functional groups of biochemistry, with their temperaments and their characteristic transactions. Acid–base behaviour is deferred to Chapter 6, where it is treated with the seriousness it deserves.

Group	Structure	Temperament	Characteristic transaction
Hydroxyl	–OH	weak nucleophile; H-bond donor/acceptor	phosphorylation, glycosylation, oxidation to carbonyl
Carbonyl (aldehyde/ketone)	>C=O	electrophilic carbon	nucleophilic addition, imine formation, aldol chemistry
Carboxyl	–COOH / –COO^-	acid; resonance-stabilised anion	activation as thioester/acyl phosphate, decarboxylation
Amino	–NH_2 / –NH_3^+	nucleophile; base	transamination, amide and imine formation
Thiol	–SH	strong nucleophile; redox-active	thioester formation, disulphide exchange
Phosphate	–OPO_3^{2-}	electrophilic P when activated; excellent leaving group	phosphoryl transfer, energy coupling (Ch. 25)

Imidazole (His)	$C_3H_4N_2$	amphibious: donor/acceptor, base/acid near pH 7	general acid–base catalysis (§ 15.5)
Guanidinium (Arg)	– $C(NH_2)_2^+$	permanent cation; H-bond donor ×5	phosphate recognition, ion pairing

2.12 Oxidation states in biological chemistry

The final account of the chapter is the one that will be audited most often: the electron ledger. Oxidation state — the formal charge an atom would bear if every bond's electrons were awarded to the more electronegative partner — is a fiction of bookkeeping and a fact of chemistry: a fiction, because real bonds are shared; a fact, because the fiction balances, and everything that flows in metabolism can be traced through it. The rules are few: hydrogen is +1 (save in metal hydrides), oxygen −2 (save in peroxides), and carbon's state follows by difference, taken as a mean over the molecule. Apply them to glucose, $C_6H_{12}O_6$: the twelve hydrogens contribute +12, the six oxygens −12, and the carbons must therefore sum to zero — a mean oxidation state of 0 per carbon. Oxidise the molecule completely to 6 CO_2, in which each carbon is +4 (two oxygens at −2 each), and the six carbons have risen by 24 units in all; each unit is one electron surrendered. Hence

$$C_6H_{12}O_6 + 6\,O_2 \longrightarrow 6\,CO_2 + 6\,H_2O \qquad \text{transfers } 24\ e^- \text{ per glucose} \tag{2.9}$$

— the twenty-four electrons being taken up by the six oxygens, each O_2 accepting four (each O falling from 0 to −2). The reader now possesses the master key of Parts VI and VII: respiration is the controlled descent of these twenty-four electrons from carbon to oxygen through the redox carriers of Chapter 26, with the free energy harvested en route; photosynthesis is their laborious ascent, paid for by photons; and the whole metabolic map (Figure 2.6) can be read as a chart of carbon's climb and fall between −4 and +4. Oxidation state, properly kept, is the double-entry bookkeeping of life — and like all honest bookkeeping, it catches embezzlement instantly: any proposed pathway that does not balance its electrons is not a hypothesis but an error.

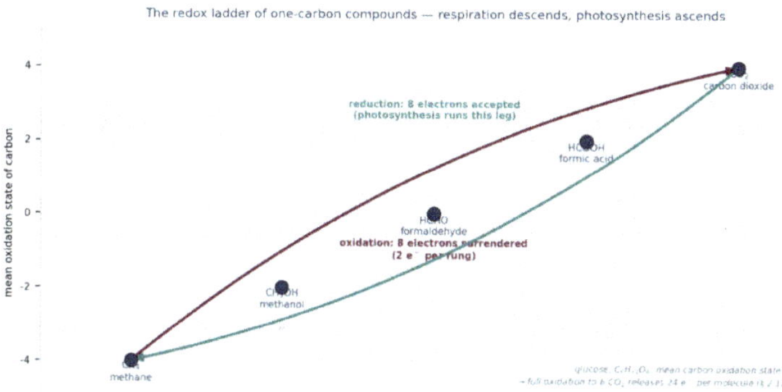

Figure 2.6 The redox ladder of one-carbon compounds, from methane (carbon at −4) to carbon dioxide (+4), with the two directions of the biological electron trade annotated: respiration descends the ladder (the chemistry of Part VI), photosynthesis and biosynthesis ascend it (Part VII). The rungs are real metabolites — methanol, formaldehyde, formic acid — and the two-electron spacing is no accident: most biological redox carriers, NAD^+ among them, transact in pairs (§ 26.5). Mean oxidation states computed by the bookkeeping rules of § 2.12.

Chapter 2 — Conceptual synthesis

The chapter set out to supply the grammar of the chemical alphabet, and the argument ran: atoms are individuals because their electrons are quantised (§ 2.1); only the valence shell trades, and carbon's even-handed valence makes sharing, and therefore life, possible (§ 2.2); sharing is unequal in proportion to electronegativity, whence polarity, dipoles, and the peculiar virtues of water (§ 2.3); strong architecture comes in three species — covalent, ionic, coordinate — with the ionic species humbled eighty-fold by the dielectric of the solvent (§ 2.4); electrons delocalise, and delocalisation brings stability, planarity, and colour (§§ 2.5–2.6); the hydrogen bond, weak and directional, is the universal joint of macromolecular structure (§ 2.7); the van der Waals forces, individually beneath thermal notice, specify structure by the arithmetic of multitude (§ 2.8); the hydrophobic effect, entropic in origin, folds proteins and builds membranes (§ 2.9); bond energies measure costs, not treasures, and kinetic persistence, not thermodynamic stability, is what the cell actually purchases (§ 2.10); a half-dozen functional groups carry the reactive traffic (§ 2.11); and oxidation-state bookkeeping keeps the whole honest (§ 2.12). The direction has been checked against the map of Figure ii: the grammar is in place; Chapter 3 turns to the geometry it must write — the shapes, handedness and conformations of molecules, without which recognition is impossible.

WORKED EXAMPLE 2.1 — THE PARTIAL CHARGE OF THE WATER MOLECULE

From Eq. (2.3), with μ = 1.85 D and the O–H bond length d = 0.96 Å: converting, $\mu = 1.85 \times 3.336 \times 10^{-30}$ C m = 6.17×10^{-30} C m, and $d = 9.6 \times 10^{-11}$ m, so $\delta e = \mu/d = 6.4 \times 10^{-20}$ C, whence $\delta = 6.4 \times 10^{-20}/1.602 \times 10^{-19} \approx 0.40$. Each hydrogen of water carries about +0.4 of an elementary charge; each lone pair region, −0.2 apiece by symmetry. The number quantifies Figure 2.4: nearly half an electron's worth of polarity is what builds the hydrogen-bonded network of the liquid, and through it, the medium of life. The derivation used Eq. (2.3), Table iii, and no other magic.

WORKED EXAMPLE 2.2 — A SALT BRIDGE IN WATER

Compute the Coulomb energy of a lysine–glutamate ion pair (unit charges) at 3.0 Å, in water (ε_r = 78.5) and in the protein interior ($\varepsilon_r \approx 4$). From Eq. (2.4), $U = -1389/(\varepsilon_r \cdot r)$ kJ mol^{-1}. In water: $-1389/(78.5 \times 3.0) \approx -5.9$ kJ mol^{-1} — barely twice *RT*, a handshake, not a bond. In the interior: $-1389/(4 \times 3.0) \approx -116$ kJ mol^{-1} — a serious attachment, though much of it is spent paying the desolvation of the two ions, the hidden cost deferred to Chapter 11. The two numbers, computed from one line of electrostatics, explain why salt bridges decorate protein surfaces yet only the buried ones vote on stability.

WORKED EXAMPLE 2.3 — THE PRICE OF BURYING A LEUCINE

A leucine side chain is, to a first and honest approximation, a butane. From Figure 2.5a, the transfer of butane's four carbons from a nonpolar phase into water costs $\approx 4 \times 3.8 \approx +15$ kJ mol^{-1}; folding that buries the side chain therefore gains roughly -15 kJ mol^{-1} of hydrophobic free energy. Ten such residues ≈ -150 kJ mol^{-1} — already of the order of the entire folding free energy of a small protein (§ 11.2). The estimate is crude, and the reader who refines it with solvent- accessible surface areas in Chapter 11 will be doing precisely the arithmetic by which the folding problem was first made quantitative.

Exercises (with full resolutions)

EXERCISE 2.1

Using Eqs. (2.1)–(2.2), estimate the wavelength of the photon emitted in the hydrogen transition $n = 3 \rightarrow n = 2$, and say where in the spectrum it falls.

Resolution. $\Delta E = 13.6(1/4 - 1/9)$ eV $= 13.6 \times 5/36 \approx 1.89$ eV. Then $\lambda = hc/\Delta E = (6.626 \times 10^{-34} \times 2.998 \times 10^{8})/(1.89 \times 1.602 \times 10^{-19}) \approx 6.56 \times 10^{-7}$ m = 656 nm — the red Balmer-α line, first light of the hydrogen spectrum and empirical warrant of the quantum statute of § 2.1. The derivation chain (Bohr → Rydberg → Planck) is the whole of § 2.1 in one line.

EXERCISE 2.2

Derive Eq. (2.8) in full and evaluate r_{min} for a carbon–carbon van der Waals contact with σ = 3.4 Å.

Resolution. $dV/dr = 4\varepsilon(-12\sigma^{12}r^{-13} + 6\sigma^{6}r^{-7}) = 0$ gives $r^6 = 2\sigma^6$, so $r_{min} = 2^{1/6}\sigma = 1.122 \times 3.4 \approx 3.8$ Å — the comfortable resting distance of two carbons in a protein interior, twice the van der Waals radius 1.9 Å. The second derivative, for the ous reader, is positive there, confirming a minimum; the present checked, as the reader should always check us.

EXERCISE 2.3

Compute the mean oxidation state of carbon in (a) pyruvate, $C_3H_3O_3^-$, and (b) acetyl-CoA's acetyl group, CH_3CO–. Hence show that the conversion of pyruvate to acetyl-CoA plus CO_2 is an oxidation, and count the electrons.

Resolution. (a) 3 H (+3), 3 O (−6), net charge −1: carbons sum to +2, mean +2/3. (b) In the acetyl fragment, 3 H (+3), 1 O (−2): the two carbons sum to +1, mean +1/2. Pyruvate (3 C, sum +2) → acetyl (2 C, sum +1) + CO_2 (1 C at +4): the carbon sum rises from +2 to +5, so three oxidation units — but one carbon was already removed as CO_2 without electron transfer; the redox-active change is the loss of **two** electrons, carried off by NAD^+ to give NADH. This is the pyruvate dehydrogenase reaction of § 34.1, balanced here before its machinery is even named — the ledger precedes the machinery.

EXERCISE 2.4 (INTERPRETATIVE)

The hydrogen bond is often called "the bond of life." Construct the strongest possible antithesis — that the title belongs elsewhere — and resolve the debate.

Resolution (indicative). The antithesis has three candidates. The covalent bond can claim primacy: no covalent b , no molecules at all. The hydrophobic effect can claim structural hip: membranes and folds. The London dispersion can claim univer operates where no donors exist. Each claim is sound, and each is partial. The resolution turns on *directionality at the right strength*: the covalent bond is permanent but mute, dispersion is ubiquitous but blind, hydrophobicity is powerful but indiscriminate; only the hydrogen bond combines middling strength (reversible at 310 K) with geometric specificity (readable), which is the conjunction that codes — base pairs, helices, recognition — require. The title, then, is deserved, but the honour is shared: life is written in covalent ink, folded by water, and read by hydrogen bonds.

EXERCISE 2.5

Estimate the Hückel HOMO–LUMO gap (in $|\beta|$) for a linear polyene of ten conjugated double bonds (n = 20 p-orbitals), and comment on its colour.

Resolution. Ten double bonds contribute 20 π electrons, filling levels k = 1... 10 of Eq. (2.5); the gap is $E_{11} - E_{10} = 2|\beta|[\cos(10\pi/21) - \cos(11\pi/21)] = 2|\beta|[\cos(85.7°) - \cos(94.3°)] \approx 2|\beta|[(0.0756) - (-0.0756)] \approx 0.30|\beta|$. With $|\beta| \approx 240$ kJ mol^{-1}, the gap is ≈ 72 kJ mol^{-1}, i.e. $\lambda \approx$ 1 660 nm — the near infrared; β-carotene's eleven conjugated bonds, with ring substituents further compressing the gap, absorbs in the visible blue-green and reflects the orange of every carrot. Polyenes, like governments, become visible exactly when their gaps shrink enough to be seen.

Research-level propositions

Proposition 2.a. The dielectric constant of a protein interior is usually taken as $\varepsilon_r \approx$ 2–4, but it is not a constant: it varies with position, polarisability, and hydration. Propose a computational experiment (using the methods of § 55.9) to map $\varepsilon_r(\mathbf{r})$ across a protein–protein interface, and predict where the salt bridges of Worked Example 2.2 should cluster. *Proposition 2.β.* The per-methylene transfer tax of Figure 2.5a (≈ 3.8 kJ mol^{-1}) converts, by Eq. (1.7), into an equilibrium partition coefficient; compute it, and then design an experiment with isotope-labelled alkanes to measure the coefficient directly at 310 K, accounting for the activity coefficients of § 4.7. *Proposition 2.γ.* If resonance stabilisation of inorganic phosphate contributes materially to the free energy of ATP hydrolysis (§ 2.10), then substituting phosphate with arsenate — whose resonance patterns differ and whose esters hydrolyse spontaneously — should change the energetics measurably. Trace the historical "arsenate uncoupling" experiments of the 1930s–40s and assess what they truly proved about phosphoryl transfer.

References (Chapter 2)

Berg, J. M., Tymoczko, J. L., Gatto, G. J., Jr., & Stryer, L. (2023). *Biochemistry* (9th ed.). W. H. Freeman.

Bohr, N. (1913). On the constitution of atoms and molecules. *Philosophical Magazine, Series 6, 26*(151), 1–25.

Eisenberg, D., & Kauzmann, W. (1969). *The structure and properties of water.* Oxford University Press.

Kauzmann, W. (1959). Some factors in the interpretation of protein denaturation. *Advances in Protein Chemistry, 14,* 1–63.

Lewis, G. N. (1916). The atom and the molecule. *Journal of the American Chemical Society, 38*(4), 762–785.

London, F. (1930). Zur Theorie und Systematik der Molekularkräfte. *Zeitschrift für Physik, 63*(3–4), 245–279.

Pauling, L. (1960). *The nature of the chemical bond* (3rd ed.). Cornell University Press.

Tanford, C. (1980). *The hydrophobic effect: Formation of micelles and biological membranes* (2nd ed.). Wiley.

van der Waals, J. D. (1873). *Over de continuïteit van den gas- en vloeistoftoestand* [On the continuity of the gaseous and liquid states] [Doctoral dissertation, Leiden University]. A. W. Sijthoff.

Plate 2.3 Linus Pauling (1901–1994), in whose person the chapter's threads — electronegativity, resonance, the chemical bond itself, and (in Chapter 10) the α-helix — are gathered. Portrait from the Nobel Foundation archive, nobelprize.org (Nobel Prize in Chemistry, 1954; also Peace, 1962 — the only person twice solely so honoured), reproduced from the Foundation's publicly distributed image. The hand at the chin is, we like to think, computing a bond energy.

Mechanistic case study — The peptide bond that would not rotate: Pauling, Corey and the planarity of the backbone

We close the chapter with its most consequential resonance. In the late 1940s, Pauling and Corey set out to predict the structures proteins might fold into — without knowing the structure of a single protein. Their method was audacious in its modesty: determine, by X-ray crystallography, the bond lengths and angles of *small* amides and peptides, take them as invariants, and ask what the polypeptide backbone could build while honouring them. The decisive measurement was the C–N bond of the peptide group: 1.32 Å, midway between a single bond (1.49) and a double (1.27), and the group stubbornly planar. The interpretation was resonance: the peptide bond is a hybrid of the neutral form O=C–N and a zwitterionic form $^{-}$O–C=N^{+}, some 40 per cent admixed, which endows the C–N link with partial double-bond character and a rotation barrier of ≈ 80 kJ mol^{-1} — effectively rigid at 310 K ($e^{-80/2.58} \approx 10^{-13}$, by the arithmetic of Exercise 1.3). From that single planarity, and the tetrahedral geometry of the α-carbon, Pauling, Corey and Branson deduced in 1951 that the backbone must coil into hydrogen-bonded helices — the α-helix among them (Pauling, Corey, & Branson, 1951) — years before any protein structure was seen. When Perutz shortly thereafter found the helix's predicted 1.5 Å reflection in haemoglobin, the inference was complete: from the resonance hybrid of a two-atom bond to the architecture of every protein on Earth. The episode is the purest example in the book of the chapter's thesis — that the grammar of electrons, taken seriously and in quantitative detail, legislates upon the largest structures of life — and it is the proper bridge to Chapter 3, where the geometry of that grammar, including its handedness, becomes the subject.

Pauling, L., Corey, R. B., & Branson, H. R. (1951). The structure of proteins: Two hydrogen-bonded helical configurations of the polypeptide chain. *Proceedings of the National Academy of Sciences, 37*(4), 205–211.

CHAPTER 3

Molecular Structure and Stereochemistry

CONTENTS OF THIS CHAPTER

In which the grammar of Chapter 2 acquires geometry: the shapes of molecules, their freedom to rotate, their isomers and their handedness; in which Fischer's cross and Haworth's ring are taught to speak; and in which the absolute configuration of a single carbon atom is shown to decide, at the far end of the causal chain, whether a drug heals or harms.

3.1 Molecular geometry

Chemistry was two centuries old before it consented to become three-dimensional. The formula C_2H_6O, so obliging on paper, is two substances in the world — ethanol, which cheers, and dimethyl ether, which does not — and the difference is not in the atoms but in their arrangement. Geometry, not composition, is the first language in which molecules disagree; stereochemistry is merely the grammar of that disagreement, pursued to its logical and occasionally uncomfortable conclusions. The foundations are laid by the valence-shell electron-pair repulsion theory — VSEPR to its friends — which states, with a directness rare in science, that electron pairs around a central atom arrange themselves as far apart as possible [9]: two pairs point oppositely (linear, 180°), three triangulate (120°), four tetrahedralise (109.5°). The whole of molecular shape is a commentary on this aristocratic distaste for company.

The biochemist's interest concentrates on two geometries above all. The first is the *tetrahedron* of sp^3 carbon — the geometry of every saturated centre, of the α-carbon of every amino acid, of the thousands of carbons in every sugar and fatty acid. The tetrahedron is nature's favourite joint precisely because it is three-dimensional and four-armed: it can hold four different groups, and the moment it does, it becomes a stereocentre, with consequences that occupy §§ 3.4–3.7. The second is the *trigonal plane* of sp^2 carbon — the geometry of the carbonyl group, of the peptide bond (whose partial double-bond character, derived in the case study of Chapter 2, enforces planarity), of every double bond and aromatic ring. The plane is flat, rigid, and achiral; the tetrahedron is three-dimensional, rotatable, and potentially chiral; and the whole architecture of biological molecules is the interleaving of these two temperaments — flat,

9. The VSEPR model is the work of Sidgwick and Powell (1940), refined by Gillespie and Nyholm (1957). Its premise — that Pauli's exclusion principle, which forbids two electrons the same space, effectively makes electron pairs repel — turns a quantum statute into a geometry lesson, and is perhaps the best return on a single idea that structural chemistry has ever paid.

decisive sp^2 regions set like punctuation among the rotatable sp^3 prose. The reader who sees, in a structural formula, not letters but tetrahedra and planes, has learned the alphabet of this chapter.

3.2 Conformational freedom

A single bond, unlike a double one, permits rotation; and rotation, multiplied by the hundreds of single bonds in a macromolecule, is the source of both the flexibility and the despair of structural biology. The rotation is not free. Around the C2–C3 bond of butane, the energy varies with the dihedral angle ω as the torsional potential — a sum of a threefold term from the electronic preference for staggered arrangements and steric penalties when the methyl groups approach. The resulting landscape (Figure 3.3a) shows three minima: the anti conformation, with the methyl groups 180° apart and contented, and two gauche conformations at ±60°, elevated by about 3.8 kJ mol^{-1} because the methyls are then near enough to notice one another. Between them lie eclipsed barriers, the fully eclipsed summit at 180° costing ≈ 22 kJ mol^{-1} — still only some nine times RT at body temperature, which is to say: crossable in picoseconds. Conformations are not structures but populations; a molecule of butane is a committee that spends most of its time in the anti chair, visits gauche occasionally, and never sits in the eclipsed seats at all.

The Boltzmann population of a conformation of energy E above the minimum is, as Exercise 1.3 anticipated,

$$\frac{p_{\mathrm{gauche}}}{p_{\mathrm{anti}}} = 2\,e^{-\Delta E/RT} \tag{3.1}$$

— the factor of 2 entering because there are two gauche minima against one anti. For ΔE = 3.8 kJ mol^{-1} at 310 K, the ratio is $2 \cdot e^{-1.47} \approx 0.46$: about one molecule in three is gauche at any instant. The derivation is worth the reader's attention because it is the template of every conformational argument in the book: energies give populations through the Boltzmann factor, and populations give properties through the ensemble average. When Chapter 12 speaks of proteins as dynamic ensembles rather than single structures, it will be speaking the language of this paragraph, only louder.

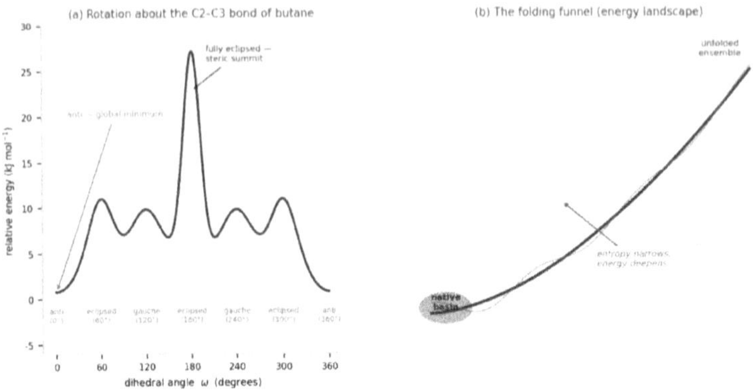

Figure 3.3 Two conformational landscapes. (a) The torsional potential of butane about the C2–C3 bond: the anti minimum (methyl groups opposed), the two gauche minima (+3.8 kJ mol^{-1}), and the eclipsed barriers (≈ 22 kJ mol^{-1} fully eclipsed). Populations follow Eq. (3.1). (b) The folding funnel of a protein (Chapter 11): the enormous conformational freedom of the unfolded ensemble narrows as the chain descends in free energy toward the native basin. The two landscapes are the same idea at two magnifications — rotation about one bond, and rotation about all of them at once.

3.3 Structural and stereochemical isomerism

Isomers — molecules sharing a molecular formula but not an identity — divide, like Gaul, into three parts; or more honestly, into two parts with a subdivision, as the taxonomy of Figure 3.5 sets out. *Structural* (constitutional) isomers differ in connectivity: the atoms are joined in a different order, as ethanol and dimethyl ether, or leucine and isoleucine. *Stereoisomers* share connectivity but differ in spatial arrangement, and these subdivide into *enantiomers*, which are non-superimposable mirror images, and *diastereomers*, which are not. The distinction is not pedantry but pharmacy: structural isomers are different compounds with different properties in every solvent; enantiomers have identical properties in every *achiral* environment and differ only before a chiral judge — a polarimeter, an enzyme, a receptor; diastereomers differ everywhere, being as distinct as cousins. The cell, which is nothing if not a chiral judge, reads all three categories with perfect fluency; and the student who wishes to understand why enzymes are such merciless stereochemists (§ 3.10) must first master the taxonomy the enzymes have mastered.

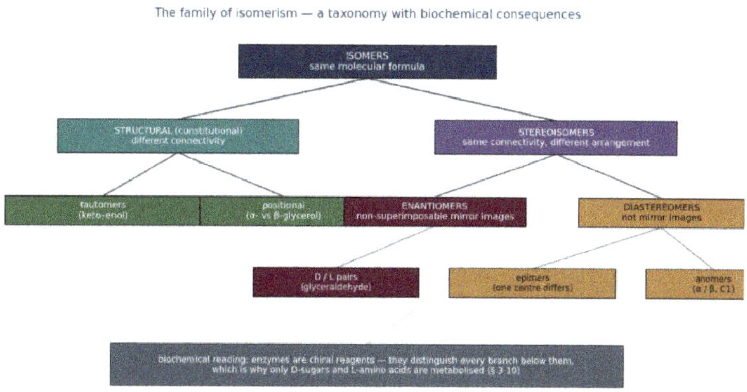

Figure 3.5 The family of isomerism. Structural isomers differ in connectivity; stereoisomers divide into enantiomers (mirror images) and diastereomers (all the rest), with the biochemically decisive sub-species — D/L pairs, epimers, anomers — shown beneath. The terminal note is the chapter's thesis in miniature: because enzymes are themselves chiral, they distinguish every branch of the tree, and life has consequently specialised — D-sugars, L-amino acids — on a scale the racemic world of the test tube finds extravagant.

3.4 Chirality and optical activity

Chirality — handedness — is the property of an object that cannot be superimposed upon its mirror image, and it enters chemistry through a single carbon bearing four different substituents. The two mirror forms of such a centre are enantiomers: identical in mass, energy, solubility, and every scalar property the achiral world can measure; distinguishable only by other chiral things. The first chiral instrument was polarised light. A solution of one enantiomer rotates the plane of polarisation by a characteristic angle, the specific rotation

$$[\alpha]_\lambda^T = \frac{\alpha_{\text{obs}}}{c\,l} \tag{3.2}$$

— where α_{obs} is the observed rotation in degrees, c the concentration in g mL^{-1}, and l the path length in decimetres; the superscript and subscript record the temperature and wavelength (usually the sodium D line, 589 nm), for rotation is a dispersive phenomenon and changes sign and magnitude with colour. The measurement is called polarimetry, and it is the oldest analytical technique in stereochemistry — Pasteur's, indeed, before it was anyone else's.

HISTORICAL NOTE — PASTEUR'S TWEEZERS, 1848

Louis Pasteur, then twenty-five and not yet the Pasteur of the vaccines, was examining crystals of sodium ammonium tartrate under the microscope when he noticed that the crystals came in two forms, each the mirror image of the other — hemihedral facets facing left, and facing right. He separated the two kinds by hand, with tweezers, and found that solutions of the two sorts rotated polarised light in opposite directions (Pasteur, 1848) [10]. The conclusion he drew was bolder than the observation: that the molecules themselves were asymmetric, and that the asymmetry of living matter — for tartaric acid comes from wine, and wine from life — was one of its deepest signatures. “Life,” he wrote later, “is dominated by asymmetrical actions.” The entire pharmaceutical industry, and the whole of Part IV of this treatise, is a footnote to that sentence.

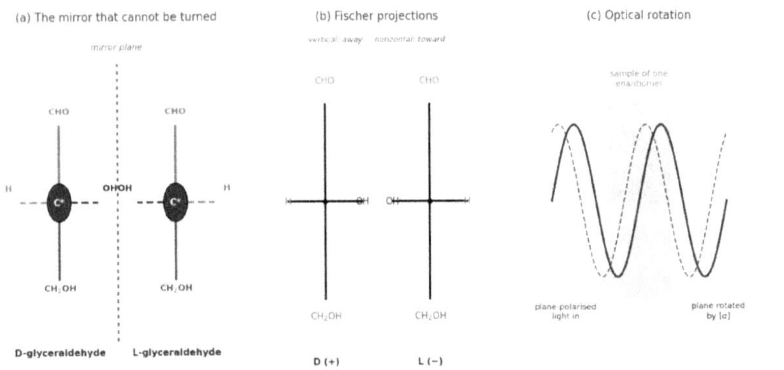

Figure 3.1 The anatomy of handedness, shown for glyceraldehyde — the reference molecule of the D/L system. (a) The two mirror forms, non-superimposable: the wedge bonds project toward the reader, the dashed away. (b) The same pair in Fischer projection (§ 3.6): by convention the carbon chain runs vertically away from the reader, and the horizontal substituents project forward — the D form has the hydroxyl on the right. (c) The polarimetric consequence: one enantiomer rotates the plane of polarised light by +[α], its mirror by −[α], and a racemic mixture by exactly nothing — which is why the polarimeter cannot tell a racemate from an achiral compound, a caution worth remembering.

3.5 Enantiomers and diastereomers

The arithmetic of stereocentres deserves a derivation, for it is the simplest and most consequential counting argument in organic chemistry. A molecule with n independent stereocentres, each of which may be of either hand, has at most 2^n stereoisomers — "at most," because symmetry may collapse some forms into identity (the meso cases, of which more below). Glucose, with four stereocentres at C2–C5, is one of $2^4 = 16$ aldohexoses: eight D, eight L, mirror pairs all. The D set forms a family whose members are *diastereomers* of one another — glucose, mannose, galactose, allose and the rest — identical in formula and connectivity, differing only in the hand of one or more centres, and differing therefore in every physical property: melting point, solubility, sweetness, and — decisively — in their reception by enzymes. A diastereomer differing at a single centre is called an *epimer*: glucose and mannose are C2-epimers; glucose and galactose, C4-epimers. The epimerases of metabolism (§ 33.9) are enzymes that flip one centre and leave the rest untouched — stereosurgery at the resolution of a single hydroxyl.

10. The tartar itself had a history: the wine trade knew that old wine threw down crystals (tartar), that tartar yielded tartaric acid, and that a second acid — “racemic,” from *racemus*, a bunch of grapes — had the same composition but no optical activity. Pasteur showed racemic acid to be an equal mixture of the two hands of tartaric acid; the word *racemate* survives as his monument. The tweezers are in the Musée Pasteur in Paris; the lesson is everywhere.

And the meso complication, promised above: tartaric acid has two stereocentres and should have four stereoisomers, yet possesses only three. The (2R,3R) and (2S,3S) forms are the two hands Pasteur separated; the (2R,3S) form is superimposable upon its own mirror image, because an internal plane of symmetry makes the two halves compensate — a molecule that is chiral in its parts and achiral in its whole, like a debate in which two equally matched speakers cancel. Such compounds are called meso, they are optically inactive despite their stereocentres, and they teach the precise moral the student needs: chirality is a property of the whole molecule, not of any atom in it.

3.6 Fischer and Haworth representations

Two notations carry the burden of carbohydrate stereochemistry, and both are essential dialects of the treatise's later volumes. The *Fischer projection* — Emil Fischer's own invention, from the years when he was mapping the sixteen aldohexoses with little more than oxidation, reduction, and logic [11] — lays the carbon chain vertically, with the most oxidised carbon at the top and the understood convention that horizontal bonds project toward the reader and vertical bonds recede. The *Haworth projection* represents the ring forms that sugars actually adopt in solution: the pyranose six-membered ring drawn as a pentagon with the ring oxygen at the upper right, the glycosidic hydroxyls above or below the ring plane according as they stood left or right in the Fischer (with the right-left to down-up correspondence the student must simply memorise once), and the D-configuration marked by the CH_2OH group pointing up. Figure 3.4 sets the two side by side for D-glucose, and the reader is invited to trace each hydroxyl from the cross to the ring — an exercise that takes five minutes once and saves five hours in Chapter 20.

11. Emil Fischer (1852–1919) — Nobel laureate of 1902 — determined the configurations of the aldohexoses by a chain of reasoning that is still taught as the finest sustained piece of chemical deduction before the spectroscopic era. His convention of D and L was a guess about absolute configuration, with even odds of being wrong; Bijvoet's X-ray proof of 1951 (§ 3.7) showed the guess right. Posterity has been kinder to Fischer's luck than to his digestion, which the phenylhydrazine of his sugar work did not survive.

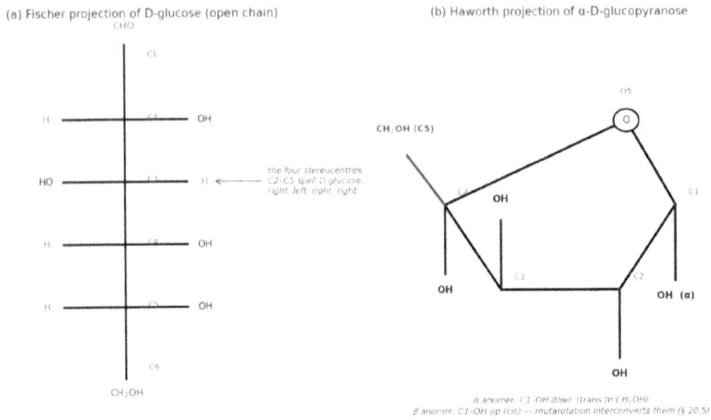

Figure 3.4 The two notations of sugar stereochemistry. (a) The Fischer projection of open-chain D-glucose, with its four stereocentres C2–C5 spelling right, left, right, right — the signature by which glucose is told from its fifteen aldohexose relatives. (b) The Haworth projection of α-D-glucopyranose, the ring form in which glucose actually lives: the C5 hydroxyl has attacked the C1 aldehyde to make a hemiacetal, creating a new stereocentre at C1 (the anomeric carbon), whose hydroxyl points down in the α anomer and up in the β. The two anomers interconvert through the open chain — the mutarotation of § 20.5.

3.7 Absolute configuration

Fischer's D and L were relative designations — a genealogy of molecules traced back to glyceraldehyde, not a measurement of any single hand. The absolute description arrived in two stages. First, the nomenclature: the Cahn–Ingold–Prelog (CIP) rules, which assign each substituent of a stereocentre a priority by atomic number (and, at ties, by the atomic numbers of the atoms those atoms bear), then orient the molecule with the lowest priority away from the viewer and read the sequence $1 \to 2 \to 3$: clockwise is *R* (rectus), anticlockwise *S* (sinister) (Cahn, Ingold, & Prelog, 1966). The system is universal, unambiguous, and — the student is warned — *not* a simple translation of D and L: L-glyceraldehyde is (S), but L-cysteine is (R), because sulphur outranks oxygen in the priority list and turns the dial. Secondly, the measurement: in 1951 Bijvoet, using the anomalous dispersion of X-rays by the rubidium atom of a rubidium tartrate salt, determined for the first time which hand a chiral molecule actually *has* — and Fischer's fifty-year-old guess about glyceraldehyde proved correct (Bijvoet, Peerdeman, & van Bommel, 1951) [12]. All the L-amino acids of the reader's proteins (cysteine alone excepted by the priority accident above) are (S); all the natural D-sugars trace their configuration to the right-handed glyceraldehyde; and the whole of terrestrial biochemistry is, in this precise sense, a left-handed protein literature written upon a right-handed sugar shelf.

12. The odds had been exactly even, and chemists of the interim period spoke of the matter with the delicacy one reserves for a colleague's wager. Prelog (Plate 3.2), whose rules had just made the wager legible, later remarked that the confirmation was "a satisfaction of an unusual kind" — an understatement of Central European perfection.

3.8 Prochirality in biochemical reactions

Between the chiral and the achiral lies a subtler country, prochirality, and it is here that enzymology shows its finest manners. Consider ethanol, CH_3CH_2OH: its two methylene hydrogens look identical and, in an achiral world, are. Yet replace one — and only one — with deuterium, and the carbon becomes a stereocentre. The two hydrogens are therefore not identical but *enantiotopic*: replaceable by different products of opposite hand. An enzyme, being a chiral reagent, can tell them apart; and it does, infallibly. When alcohol dehydrogenase oxidises ethanol to acetaldehyde with NAD^+, it removes always the same one of the two hydrogens (the pro-R), delivering it to always the same face of the nicotinamide ring — the hydride lands on the si face at C4, and only there (Cornforth et al., 1962). The proof required isotopes, and the heroes of the proof — Westheimer, Vennesland, and above all Cornforth [13] — could show that citrate synthase, in the citric acid cycle of Chapter 34, likewise distinguishes the two "identical" arms of the perfectly symmetrical citrate molecule: the two $-CH_2COO^-$ groups of citrate are prochiral, and the enzyme handles one and never the other, which is why the labelled carbon of the cycle appears in one product and not its mirror. The symmetric molecule is symmetric only to us; to the chiral enzyme, it is as asymmetric as a glove.

3.9 Conformational energy landscapes

We ascend now from one bond to all of them. The conformation of a polypeptide of n residues is specified, to an excellent approximation, by the $2n$ torsion angles (φ, ψ) about the N–C_α and C_α–C′ bonds of the backbone — the peptide bond itself being held planar by the resonance of Chapter 2's case study. Each residue's pair (φ, ψ) is a point on the map of Figure 3.2, and the map is mostly forbidden: steric clashes between the side chain and the backbone exclude the great majority of the plane, leaving the three great permitted basins — the right-handed α-helix near (−57°, −47°), the β-sheet near (−119°, +113°), and the extended polyproline helix of collagen near (−75°, +150°) — which are, not by coincidence, the three secondary structures of Chapter 10. Ramachandran's map, drawn in 1963 from hard-sphere repulsions alone (Ramachandran, Ramakrishnan, & Sasisekharan, 1963), is the proof that protein architecture is written substantially in steric law: the helices and sheets that biology loves are the conformations that atoms, simply by occupying space, permit.

13. John Cornforth — deaf from boyhood, lip-reading the conversations of his wife and collaborator Rita through forty years of shared work — received the Nobel Prize in 1975 "for his work on the stereochemistry of enzyme-catalysed reactions," a citation whose modesty matches its subject. The isoto er experiments of the 1950s that established prochiral discrimination remain, in the author ' judgement, the most elegant experiments in the whole of mechanistic biochemistry.

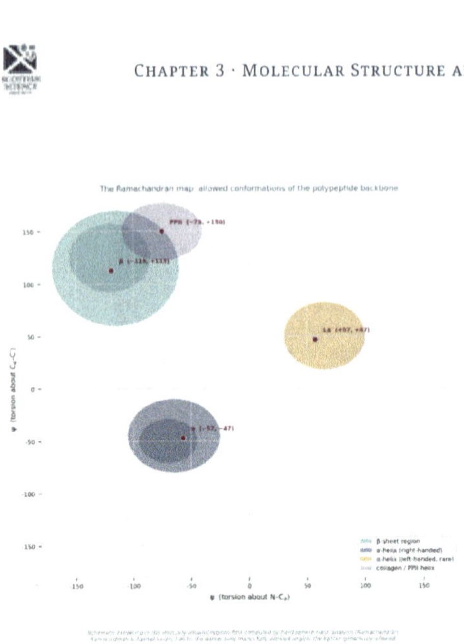

Figure 3.2 The Ramachandran map. Each point (φ, ψ) is a possible conformation of one residue; the coloured basins are the sterically allowed regions, with the canonical secondary structures marked. The right-handed α-helix and β-sheet dominate the database of real structures exactly as the map predicts — an experimental vindication, from one hundred and fifty thousand deposited structures (Figure 1.6), of what hard spheres foretold in 1963. The map is also the quality instrument of structural biology: a newly determined structure whose residues stray outside the basins is not a discovery but a suspicion.

The full landscape of a protein is the product of all its residues' maps — a surface of astronomical dimensionality, yet with a saving shape: the folding funnel of Figure 3.3b, in which free energy decreases as conformational freedom narrows toward the native basin. The landscape picture resolves, at a stroke, the famous paradox of Levinthal (1969) (Levinthal, 1969): a protein of 100 residues, sampling conformations even at the impossible rate of one per picosecond, would need longer than the age of the universe to find its fold by random search — yet real proteins fold in milliseconds. The resolution is that the search is not random; the landscape is funnelled, biased everywhere downhill toward the native state, so that the chain does not explore but *descends*. Levinthal's paradox is thus not a paradox but a measurement: it measures the slope of the funnel, and thereby the depth of the evolutionary engineering that wrote the sequence. We shall return to the funnel in Chapter 11 with thermodynamic instruments; for now it suffices that the reader hold the two images together — the steric map of what is allowed, and the energetic funnel of what is chosen — for between them they contain the whole logic of protein shape.

3.10 Molecular recognition and stereospecificity

The final consequence of handedness is the one the reader ingests, inhales, and is kept alive by: stereospecific recognition. An enzyme, a receptor, an antibody — every reading instrument of the cell — is itself chiral, built of L-amino acids and therefore possessed of an asymmetric surface; and an asymmetric surface distinguishes the two hands of any guest as inevitably as a right glove distinguishes two hands offered to it. The principle was stated by Easson and Stedman in 1933 as the three-point attachment model (Easson & Stedman, 1933): if a receptor engages a substrate at three points, then only one enantiomer can present all three complementary contacts simultaneously; the mirror form can offer at most two, the third substituent pointing obligingly into empty space or, worse, into a wall. Three points suffice to read a hand, as three landmarks suffice to fix a position.

The consequences are pharmacological, toxicological, and occasionally tragic. One enantiomer of adrenaline binds its receptor some twelve times more avidly than the other; L-DOPA relieves Parkinsonism while its mirror is inert; and thalidomide — the case study that closes this chapter — taught the twentieth century, at terrible cost, that the two hands of a drug may differ not in degree but in kind. Metabolism is no less exclusive: the glycolytic enzymes phosphorylate D-glucose and ignore its mirror entirely; the ribosome accepts only L-aminoacyl-tRNAs; and the reader who has ever wondered why the chemistry set of life uses one hand where two were available will find the question pursued to its evolutionary root in § 56.1 — for the choice, once made, is self-reinforcing, a stereochemical constitution from which no amendment is possible.

MECHANISTIC CASE STUDY — THE TWO HANDS OF THALIDOMIDE

Thalidomide — N-phthaloylglutarimide — was introduced in 1957 as a sedative of unusual safety, and prescribed widely to pregnant women against morning sickness. Its single stereocentre was present in the marketed drug as a racemate, on the reasonable- seeming grounds that enantiomers differ only in degree. The grounds were mistaken. The (*R*)-enantiomer is the sedative; the (*S*)-enantiomer is teratogenic, interfering with the development of the embryo's limbs; and the human tragedy of 1961 — some ten thousand children born with phocomelia — followed (McBride, 1961). Two chemical facts complete the lesson. First, the two enantiomers interconvert in the body with a half-life of hours at physiological pH, so that administering the pure (*R*) form would not have prevented the disaster — the racemate is regenerated *in vivo*. Second, the molecular basis of the teratogenicity was identified only in 2010: thalidomide binds the protein cereblon, a substrate receptor of the CRL4 ubiquitin ligase, and redirects its targeting — a stereochemistry-dependent reprogramming of the cell's degradation machinery (§ 47.12) (Ito et al., 2010). The drug, in one of the strangest second acts in pharmacology, is now a frontline treatment of multiple myeloma, deployed under the strictest supervision. The chapter could offer no sterner demonstration of its thesis: that at the resolution of a single carbon's hand, a molecule is not one thing but two — and that the cell, which reads hands infallibly, will meet whichever one is offered.

3.11 Isotopes and isotopic labelling

The last tool of the chapter is invisible and indispensable: the isotope. Isotopes are atoms of one element differing in neutron number, hence in mass but not in chemistry — or rather, not *quite* in chemistry, for the small kinetic isotope effects of § 16.11 are precisely the exceptions that prove the mass rule. For the structural purposes of this chapter, two families matter. The *stable* isotopes — ^{2}H (deuterium, D), ^{13}C, ^{15}N, ^{18}O — are followed by mass spectrometry and NMR (Chapters 54–55), and they are the workhorses of the modern tracer experiment: feed a cell ^{13}C-glucose, and the isotope appears, pathway by pathway, in the metabolites of Part VII, an unimpeachable witness of who became whom. The *radioactive* isotopes — ^{3}H, ^{14}C, ^{32}P — are followed by their decay, and it was with ^{14}C that Calvin mapped the carbon path of photosynthesis (Nobel, 1961) and with tritiated thymidine that the semiconservative replication of DNA (§ 44.1) was first seen. Isotopic labelling is the answer to the epistemological problem of § 1.8: how to follow one molecule among millions without disturbing the crowd — give it a mass the crowd lacks, and listen.

A footnote of method closes the chapter. The isotope's virtue is that chemistry cannot tell it apart from its lighter brother; its vice, occasionally, is that kinetics can — a C–D bond breaks up to seven times more slowly than C–H, the zero-point energy of the heavier bond being lower and its activation therefore costlier. The effect, properly wielded, is not a nuisance but a probe: if replacing H by D at a particular position slows a reaction, that bond is broken in the rate-determining step — and the reader will find this reasoning deployed, with full derivation, in § 16.11. Even the exceptions, in stereochemistry as elsewhere, are instruments in disguise.

Chapter 3 — Conceptual synthesis

The chapter opened with geometry: electron pairs repel, whence the tetrahedron of sp^3 carbon and the plane of sp^2, the two temperaments of molecular shape (§ 3.1). Rotation about single bonds populates conformations Boltzmann-wise (§ 3.2); isomerism divides into structural and stereo, the latter into mirror-image enantiomers and all-other diastereomers (§§ 3.3–3.5); Fischer and Haworth gave the sugars their two notations (§ 3.6); the CIP rules and Bijvoet's X-rays promoted D and L from genealogy to measurement (§ 3.7); prochirality revealed that even symmetrical molecules are asymmetric to the chiral enzymes that handle them (§ 3.8); the Ramachandran map and the folding funnel supplied the steric and energetic faces of conformational choice (§ 3.9); stereospecific recognition — three points suffice — explained the exclusivity of metabolism and the two hands of thalidomide (§ 3.10); and isotopes provided the invisible ink in which the traffic of atoms is made legible (§ 3.11). Direction checked against the itinerary: the grammar now has geometry; Part I closes with Chapter 4, in which the whole apparatus is made quantitative — amount, concentration, equilibrium, and the arithmetic of estimation.

Worked example 3.1 — How many aldohexoses?

An aldohexose has stereocentres at C2, C3, C4, C5 (the aldehyde C1 and the CH_2OH C6 are not stereogenic). By the counting of § 3.5, $2^4 = 16$ stereoisomers: 8 D and 8 L, enantiomeric in pairs. Within the D set, glucose and mannose differ only at C2 — epimers — while glucose and gulose differ at C3 and C5 (diastereomers, not epimers). The family tree of sixteen was the object of Fischer's great deduction, and the student who can draw the eight D-aldohexoses from memory has earned a place in a diminishing and honourable minority.

Worked example 3.2 — The rotation of a glucose solution

A solution of α-D-glucose at 0.10 g mL^{-1} in a 2.0 dm tube shows a rotation of +11.2° at the sodium D line. From Eq. (3.2), $[\alpha]_D = \alpha_{obs}/(cl) = 11.2/(0.10 \times 2.0) =$ +56° mL g^{-1} dm^{-1} — the literature value being +112° for the pure α anomer, our reading would suggest a sample at half the optical purity, or (more instructively) one caught midway through mutarotation toward the +52.7°

equilibrium mixture of α and β (§ 20.5). The polarimeter, in skilled hands, is a kinetic instrument as well as a structural one.

WORKED EXAMPLE 3.3 — ASSIGNING R AND S

Assign the configuration of L-alanine's α-carbon. Priorities: NH_2 (N, Z = 7) first; COO^- (C bearing O, O, O) second; CH_3 (C bearing H, H, H) third; H fourth. With H pointing away, the sequence N → COO^- → CH_3 runs anticlockwise: (*S*). All the standard L-amino acids are (*S*) save L-cysteine, whose CH_2SH substituent outranks the carboxylate (S, Z = 16, beats O at the second sphere) and turns the assignment to (*R*) — the priority-list accident of § 3.7, now computed rather than asserted.

Exercises (with full resolutions)

EXERCISE 3.1

From Eq. (3.1), compute the temperature at which butane's gauche population equals its anti population, if ΔE remains 3.8 kJ mol^{-1}. Comment.

Resolution. Setting the ratio to 1: $2 \cdot e^{-\Delta E/RT} = 1$, so $e^{\Delta E/RT} = 2$ and $T = \Delta E/(R \ln 2) = 3800/(8.314 \times 0.693) \approx 660$ K. Below this temperature the anti always outnumbers the combined gauche; above it, the two gauche wells together win on degeneracy. The exercise teaches the two levers of every conformational equilibrium: energy ΔE and counting (the factor 2) — enthalpy and entropy, in embryo, for Chapter 7.

EXERCISE 3.2

Tartaric acid has two stereocentres but only three stereoisomers. Explain, and identify the optical rotation of each.

Resolution. The (2R,3R) and (2S,3S) forms are enantiomers, rotating light equally in opposite directions ([α] = +12° and −12° respectively, in water). The (2R,3S) form possesses an internal mirror plane exchanging C2 and C3; it is identical with its own mirror image — meso — and optically inactive. Hence $2^2 = 4$ nominally, but the meso collapse leaves three: two hands and one eunuch, as Pasteur's crystals already knew in 1848.

EXERCISE 3.3

In the citric acid cycle, citrate — a symmetric molecule — yields α-ketoglutarate labelled in only one of its two possible positions. Reconcile this with citrate's symmetry.

Resolution. Citrate is symmetric in an achiral world and prochiral in the cell (§ 3.8). Its two $-CH_2COO^-$ arms are enantiotopic; aconitase binds citrate in one orientation on its chiral surface and therefore handles always the same arm — the "Ogston hypothesis" of 1948, which resolved the apparent paradox that a symmetric intermediate gave asymmetrically labelled products. The three-point attachment of § 3.10 is the whole explanation: a three-point landing on a chiral strip reads even a symmetric guest in one orientation. Symmetry, like beauty, is in the eye of the beholder — and the enzyme's eye is chiral.

EXERCISE 3.4 (INTERPRETATIVE)

Suppose a second, independent origin of life on Earth had chosen D-amino acids and L-sugars. Would its biochemistry be a mirror of ours — identical in every scalar property? And could the two biospheres eat one another?

Resolution (indicative). In an achiral environment, yes: the mirror biochemistry would be energetically and kinetically identical, a perfect reflection — chirality is a symmetry of the laws of chemistry, broken only by the weak nuclear force's minuscule parity violation, whose possible role in life's original choice remains one of the deep open questions of § 56.1. But the two biospheres could not metabolise each other: our enzymes, chiral reagents all, would read D-proteins and L-sugars as foreigners — nutritious in bulk, indigestible in detail. A mirror-Earth would be a banquet we could not digest, which is perhaps the most elegant argument ever devised for leaving other worlds their own cookery.

EXERCISE 3.5

A polarimeter reading of a freshly dissolved α-D-glucose sample drifts from +112° to a stable +52.7°. Account for the drift quantitatively, given [α] of the β anomer = +19°.

Resolution. Mutarotation (§ 20.5) interconverts α and β through the open chain until equilibrium. If the equilibrium fraction of α is x, then $112x + 19(1 - x) = 52.7$, whence $93x = 33.7$ and $x = 0.36$: 36 % α, 64 % β — the measured equilibrium composition of D-glucose in water. The drift is thus the approach to equilibrium of two anomers through a transient open chain, timed by a polarimeter; and the numbers agree with the direct chromatographic census to within a per cent, which is as close as stereochemistry and arithmetic need ever come.

Research-level propositions

Proposition 3.α. The weak nuclear interaction distinguishes enantiomers with an energy difference of order 10^{-14} kJ mol^{-1}. Estimate, by Eq. (3.1), the resulting enantiomeric excess at equilibrium, and assess whether parity violation can account for the observed homochirality of life without amplification — then design the amplification mechanism (autocatalysis? asymmetric photolysis by circularly polarised starlight?) that could. *Proposition 3.β.* The Ramachandran basins of Figure 3.2 were computed for alanine-like sterics. Re-derive the allowed regions for glycine (no side chain) and proline (side chain cycled to the backbone N), and compare with the observed distributions of these residues in the PDB — a computation now routine with the methods of § 55.9. *Proposition 3.γ.* Cornforth's prochiral assignments rested on isotopic stereochemistry. Outline the modern single-molecule or cryo-EM experiment (§ 54.10) that would observe an enzyme's discrimination of enantiotopic groups directly, and estimate its required spatial resolution.

References (Chapter 3)

Bijvoet, J. M., Peerdeman, A. F., & van Bommel, A. J. (1951). Determination of the absolute configuration of optically active compounds by means of X-rays. *Nature*,

168(4268), 271–272.

Cahn, R. S., Ingold, C., & Prelog, V. (1966). Specification of molecular chirality. *Angewandte Chemie International Edition, 5*(4), 385–415.

Cornforth, J. W., Ryback, G., Popják, G., Donninger, C., & Schroepfer, G. (1962). Stereochemistry of enzymic hydrogen transfer to pyridine nucleotides. *Biochemical and Biophysical Research Communications, 9*(4), 371–377.

Easson, L. H., & Stedman, E. (1933). Studies on the relationship between chemical constitution and physiological action. V. Molecular dissymmetry and physiological activity. *Biochemical Journal, 27*(4), 1257–1266.

Ito, T., Ando, H., Suzuki, T., Ogura, T., Hotta, K., Imamura, Y., Yamaguchi, Y., & Handa, H. (2010). Identification of a primary target of thalidomide teratogenicity. *Science, 327*(5971), 1345–1350.

Levinthal, C. (1969). How to fold graciously. In P. Debrunner, J. C. M. Tsibris, & E. Münck (Eds.), *Mössbauer spectroscopy in biological systems: Proceedings of a meeting held at Allerton House, Monticello, Illinois* (pp. 22–24). University of Illinois Press.

McBride, W. G. (1961). Thalidomide and congenital abnormalities. *The Lancet, 278*(7216), 1358.

Pasteur, L. (1848). Recherches sur les relations qui peuvent exister entre la forme cristalline, la composition chimique et le sens de la polarisation rotatoire. *Annales de Chimie et de Physique, 24*(3), 442–459.

Ramachandran, G. N., Ramakrishnan, C., & Sasisekharan, V. (1963). Stereochemistry of polypeptide chain configurations. *Journal of Molecular Biology, 7*(1), 95–99.

Plate 3.1 Alfred Werner (1866–1919), whose octahedral stereochemistry of the coordination compounds (1893) first proved that geometry could be deduced from isomer counts — twenty years before X-rays could see it, and the methodological ancestor of every argument in this chapter. Portrait from the Nobel Foundation archive, nobelprize.org (Nobel Prize in Chemistry, 1913), reproduced from the Foundation's publicly distributed image.

Plate 3.2 Vladimir Prelog (1906–1998), whose sequence rules (with Cahn and Ingold) gave every stereocentre on Earth its unambiguous name. Born in Sarajevo, formed in Prague, exiled to Zürich — a Central European itinerary that ended, fittingly, in the one science that insists upon absolute orientation. Portrait from the Nobel Foundation archive, nobelprize.org (Nobel Prize in Chemistry, 1975, shared with Cornforth of § 3.8), reproduced from the Foundation's publicly distributed image.

CHAPTER 4

Quantitative Foundations of Biochemistry

CONTENTS OF THIS CHAPTER

In which Part I is brought to its balance: the mole and its measures, the equilibrium constant derived from first principles and pressed into daily service, the reaction quotient taught to tell the future, equilibria coupled, solutions confessed non-ideal, logarithms domesticated, dimensions audited, magnitudes estimated, errors propagated — and the reader, armed, sent out to solve.

4.1 Amount of substance and molecular concentration

Every science has a unit it could not live without, and chemistry's is the mole — that ungainly, indispensable bridge between the world of weighing and the world of counting. A balance reports grams; a reaction proceeds molecule by molecule; and the mole is the exchange rate between the two ledgers, fixed since May 2019 by definition rather than by measurement [14]: one mole contains exactly 6.022 140 76 × 10^{23} elementary entities, and the amount of substance follows from a measured mass m and the molar mass M as

$$n = \frac{m}{M} \tag{4.1}$$

— from which the molar concentration of a solution of volume V is derived, not defined, by the plain meaning of "amount per volume":

$$c_i = \frac{n_i}{V} \quad [\mathrm{mol\,L^{-1}}] \tag{4.2}$$

Two cautions belong to the birth of this quantity. The first is dimensional: concentration is not a property of a substance but of a *preparation* — the same glucose sits at 5 mmol L^{-1} in the reader's fasting blood and at 500 mmol L^{-1} in the pharmacist's bottle, and any sentence of the form "glucose is concentrated" is nonsense until its matrix is named. The second is historical and worth a smile: the litre, that workhorse of the laboratory, is not an SI unit at all but a tolerated guest (10^{-3} m^3), admitted to the Système international on the same grudging

14. Until 2019 the mole was *defined* through 12 grams of carbon-12, and the Avogadro constant was *measured*; the 26th General Conference on Weights and Measures inverted the arrangement, fixing N_A at exactly 6.022 140 76 × 10^{23} mol^{-1} and letting the gram of carbon-12 shift by a few parts in 10^{10} — a change no biochemist's balance will ever notice, and every metrologist's heart rejoiced at (BIPM, 2019). The history of the mole is a parable of the quantitative sciences: first a convenience, then a convention, finally a constant.

terms as the minute and the degree — a reminder that even the most rigorous systems make peace with their own history.

4.2 Molarity, molality and mole fraction

One quantity, three measures — and the choice among them is not pedantry but survival. *Molarity* (Eq. 4.2) divides by the volume of solution, and is therefore the measure of the bench: pipettes and volumetric flasks speak it natively, and every assay in Part XI is booked in it. Its vice is thermal: volumes swell with temperature (about 2 parts in 10^4 per kelvin for water), so a molarity measured at 4 °C is a different number at 37 °C — small, yet large enough to falsify a careful equilibrium constant. *Molality*, amount per kilogram of *solvent*,

$$m_i = \frac{n_i}{m_{\text{solvent}}} \quad \left[\text{mol}\,\text{kg}^{-1}\right] \tag{4.3}$$

— is indifferent to temperature and therefore the measure of physical chemistry: the colligative laws of Chapter 5 (freezing-point depression, osmotic pressure) are truest in its currency. *Mole fraction*,

$$x_i = \frac{n_i}{\sum_j n_j} \quad [\text{dimensionless}] \tag{4.4}$$

— is the measure of theory, for it alone is bounded ($0 \le x_i \le 1$), additive to unity, and honest about the solvent: in dilute aqueous solution, water's own mole fraction is ≈ 0.998, and the thermodynamic convention that sets its activity to one (§ 4.7) is nothing but this arithmetic wearing a wig. Table 4.1 summarises the treaty among the three; the reader should note, for use in every later calculation, the dilute-solution bridge: at $c \lesssim 0.1$ mol L^{-1} in water, $x_i \approx c_i/55.51$ and $m_i \approx c_i$, the number 55.51 being the molar concentration of water itself — 1000 g L^{-1} ÷ 18.015 g mol^{-1} — a constant that will reappear, disguised as $[H_2O]$ = 55.5 M, in the derivation of pK_w in Chapter 6.

Table 4.1 The three measures of composition, with their virtues and vices. The working rule of this treatise: molarity for the bench, molality for the colligative laws, mole fraction for the theory.

Measure	Definition	Unit	Virtue	Vice
Molarity	n/V_{solution}	mol L^{-1}	volumetric convenience	temperature-dependent
Molality	n/m_{solvent}	mol kg^{-1}	temperature-invariant	needs solvent mass
Mole fraction	$n_i/\Sigma n_j$	—	bounded, additive, theoretical	unwieldy at the bench

4.3 Chemical equilibria

Bring two reactants together and the mixture does not, as the naive arrow of the schoolbook suggests, march to completion; it marches to a *standoff* — a state in which forward and reverse reactions continue at furious and exactly equal rates, so that nothing changes at the scale of the observer while everything changes at the scale of the molecule. Equilibrium in chemistry is not rest but balanced motion, a détente of kinetics; the point was first stated with clarity by Guldberg and Waage (historical note below), and its thermodynamic expression is the business of the next section. For the biochemist the doctrine has a particular urgency, because the cell *lives off* the standoff: nearly every reaction of metabolism operates within an order of magnitude of its equilibrium position, and life is possible precisely because equilibrium is a *reachable* state that the cell can then, by the coupling machinery of § 4.6, push around like furniture. A dead system is one at equilibrium with its surroundings; a living one is a system held, at continuous expense, at a controlled distance from equilibrium — the precise thermodynamic statement of this aphorism is owed to Chapter 7, but its quantitative grammar begins here.

4.4 Equilibrium constants

We ive, rather than quote, the central equation of this chapter — and, the would argue, of chemical thermodynamics entire. For any species i in s its chemical potential (the molar Gibbs energy of § 7.5, admitted here on parole and properly introduced there) depends on its activity as $\mu_i = \mu_i^\circ + RT \ln a_i$ — the logarithmic form being, at bottom, the statistical price of dilution, since spreading one solute molecule among V/v more positions multiplies its entropy by exactly $\ln(V/v)$. For a general reaction written with stoichiometric coefficients ν_i (negative for reactants, positive for products), the free-energy change is the stoichiometric sum of potentials,

$$\Delta G = \sum_i \nu_i \mu_i = \underbrace{\sum_i \nu_i \mu_i^\circ}_{\Delta G^\circ} + RT \sum_i \nu_i \ln a_i = \Delta G^\circ + RT \ln Q \tag{4.5}$$

— where the reaction quotient $Q = \Pi_i \; a_i^{\nu_i}$ collects the logarithms by the elementary rule $\Sigma\nu \ln a = \ln \Pi a^\nu$. At equilibrium $\Delta G = 0$ (the bottom of the well of Figure 4.1a: forward and reverse are equally priced), and the quotient then takes a *fixed* value we name the equilibrium constant K; substituting into Eq. (4.5) yields the promised relation,

$$\Delta G^\circ = -RT \ln K \tag{4.6}$$

— derived, the reader will note, with no appeal to rates at all: the equilibrium constant is a thermodynamic quantity, indifferent to mechanism, which is why it is so powerful and why it can never tell us *how fast*. Numerically, at 25 °C, $RT \ln 10 = 5.708$ kJ mol^{-1} (Table iii), so each factor of ten in K corresponds to 5.7 kJ mol^{-1} of standard free energy — the exchange rate between the chemist's two favourite currencies, committed to memory by every competent practitioner and used in Worked Example 4.1. The dialectic of the section deserves a sentence: the thesis — equilibrium is absolute — meets the antithesis that no real cell ever reaches it; the resolution is that the constant remains absolute while the approach is optional, and it is precisely the *distance* from the absolute that the cell spends its life managing (Eq. 4.7 below).

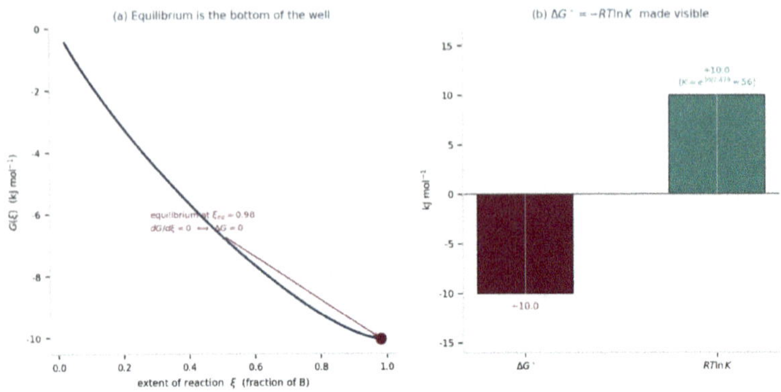

Figure 4.1 (a) The Gibbs energy of a reacting mixture A ⇌ B (ΔG° = −10 kJ mol^{-1}) as a function of the extent of reaction ξ, computed from the composition-dependent chemical potentials of § 4.4 at 25 °C; the minimum is the equilibrium composition, where the marginal price of conversion is zero. (b) The same datum as a balance sheet: the standard free energy and the equilibrium term are equal and opposite by Eq. (4.6), with $K = e^{10/2.479}$ ≈ 56 — products favoured fifty-sixfold at the standoff.

4.5 Reaction quotients

The quotient Q, computed with the actual activities of the moment, is the reaction's *address*; the constant K is its *destination*; and the sign of the journey follows at once by subtracting Eq. (4.6) from Eq. (4.5):

$$\Delta G = RT \ln \frac{Q}{K} \tag{4.7}$$

— the most useful line of the chapter. If $Q < K$, the logarithm is negative, $\Delta G < 0$, and the reaction runs forward; if $Q > K$, backward; equality is the standoff. Figure 4.2 shows the two approaches converging on the same composition from opposite ends — the kinetic shadow of the thermodynamic theorem. The reader should test the relation against the cellular case that will dominate Part VII: the hydrolysis of ATP has $\Delta G^{\circ\prime} \approx -30.5$ kJ mol^{-1} ($K' \approx 2 \times 10^5$), yet in the cytosol the actual ratio [ATP]/[ADP][P_i] is held some ten^4-fold *above* its equilibrium value, so that the real driving force is ≈ −50 kJ mol^{-1} (Worked Example 4.1): the cell, in effect, keeps the quotient far from the constant in order to keep the wallet full. Equilibrium constants are law; reaction quotients are politics; and metabolism is the perpetual manipulation of the second under the jurisdiction of the first.

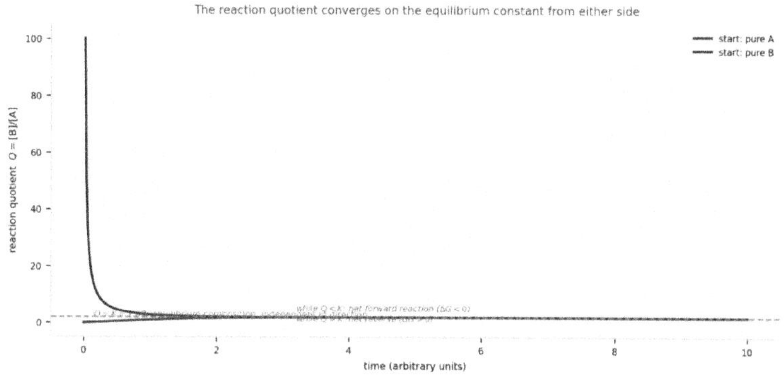

Figure 4.2 The reaction quotient of a reversible unimolecular reaction (A ⇌ B, K = 2) as a function of time, computed for first-order kinetics starting from pure A and from pure B. Both trajectories converge on the same composition — the equilibrium constant is a property of the reaction, not of the starting point — while the sign of ΔG at every instant is given by Eq. (4.7). The rates here are illustrative; the destination is law.

4.6 Coupled equilibria

Free energy being a state function — its change between two states indifferent to the path, by the first law's arithmetic — the free energies of sequential reactions *add*, and their equilibrium constants therefore *multiply*: for a reaction that is the formal sum of two others,

$$\Delta G^\circ_{\text{tot}} = \sum_i \Delta G^\circ_i \qquad \Longleftrightarrow \qquad K_{\text{tot}} = \prod_i K_i \tag{4.8}$$

— the logarithmic law of Eq. (4.6) converting multiplication into addition, as it always does. This trivial-looking theorem is the engine of the entire metabolic economy, for it permits an unfavourable reaction to be driven by a favourable one *when the two share a common intermediate*. The canonical demonstration is the synthesis of glutamine (Figure 4.3): amidation of glutamate costs +14.2 kJ mol^{-1} standing alone ($K' \approx 3 \times 10^{-3}$, a thoroughly discouraging prospect), but coupled to the hydrolysis of ATP (−30.5 kJ mol^{-1}) through the enzyme-bound glutamyl-phosphate intermediate, the sum is −16.3 kJ mol^{-1} ($K' \approx 7 \times 10^{2}$) — the reaction proceeds to completion, and ammonia is captured into organic combination at the modest price of one phosphoanhydride. Two conditions make the coupling real rather than notational, and the dialectic demands they be stated. The first is chemical: the shared intermediate must actually exist — adding the free energies of reactions that never touch is arithmetic, not c
an error into which examination candidates fall annually and textbook occasionally. The second is mechanical: the two reactions must be *obliged* to share the intermediate, which is precisely the service the enzyme performs (Chapter 25 will call it "substrate-level" versus "conformational" coupling). Equilibrium law provides the permission; the enzyme provides the enforcement. Between permission and enforcement, as elsewhere in the economy of life, lies the whole art of the cell.

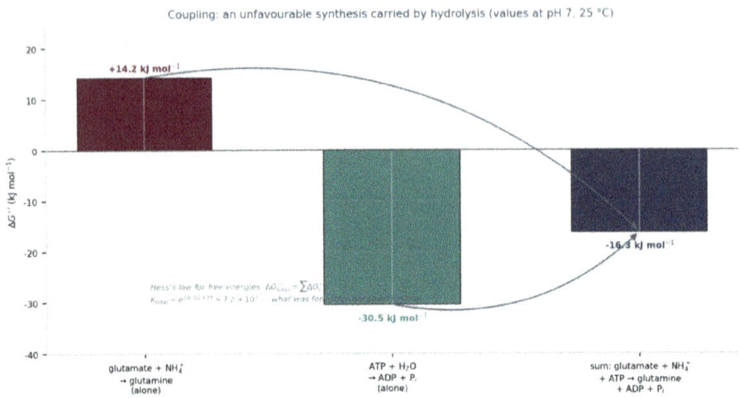

Figure 4.3 The arithmetic of coupling, Eq. (4.8), for the synthesis of glutamine at pH 7, 25 °C: the unfavourable amidation (+14.2 kJ mol^{-1}, oxblood) and the favourable hydrolysis of ATP (−30.5 kJ mol^{-1}, teal) sum to a favourable overall reaction (−16.3 kJ mol^{-1}, navy) through their shared enzyme-bound intermediate (glutamine synthetase, § 38.2). Values from the standard compilations (Nelson, Cox, & Hoskins, 2021).

HISTORICAL NOTE — GULDBERG, WAAGE AND THE LAW OF MASS ACTION

That the position of equilibrium is fixed by a product of concentrations, raised to stoichiometric powers, was argued by two Norwegians, Cato Maximilian Guldberg and Peter Waage, in a series of papers between 1864 and 1867 — first in Norwegian, a language the chemical world did not read, then in French, in a journal it did not either (Guldberg & Waage, 1867). The law was accordingly rediscovered by van 't Hoff (Plate 4.1) a decade later, and generously re-attributed when the earlier work came to light; the brothers-in-law (they married sisters, which eased the collaboration) are now honoured by every equilibrium constant ever written. The episode teaches two lessons the treatise endorses: that science advances by publication in the *read* languages — a law as stern as any in this chapter — and that its academies, though slow, are occasionally just.

Plate 4.1 Jacobus Henricus van 't Hoff (1852–1911): the tetrahedral carbon of Chapter 3, the osmotic law of Chapter 5, the temperature dependence of K (§ 7.3), and the rediscovery of mass action — four lifetimes' work in one. Portrait from the Nobel Foundation archive, nobelprize.org (the first Nobel Prize in Chemistry, 1901), reproduced from the Foundation's publicly distributed image.

4.7 Activities and non-ideal solutions

Everything derived so far has quietly trafficked in *activities*, and the debt must now be acknowledged. The concentration of Eq. (4.2) counts molecules; the activity measures their *effective* participation — corrected for the fact that in any real solution every ion walks about in a crowd of counter-ions that screen, drag, and generally compromise its independence. The two are linked by the activity coefficient,

$$a_i = \gamma_i \frac{c_i}{c^\circ} \tag{4.9}$$

— where $c^\circ = 1$ mol L^{-1} is the standard concentration that keeps the activity dimensionless, and γ_i approaches one only in the limit of infinite dilution. For electrolytes, Debye and Hückel showed in 1923 (Debye & Hückel, 1923) that the coefficient is governed by the *ionic strength* of the whole solution,

$$I = \frac{1}{2}\sum_i c_i z_i^2 \tag{4.10}$$

— the square of the charge weighting divalent ions fourfold, as befits their electrostatic reach — through the limiting law $\log_{10} \gamma_i = -A\, z_i^2\sqrt{I}$, with $A \approx 0.509$ $mol^{-1/2}$ $L^{1/2}$ at 25 °C; the extended form plotted in Figure 4.4 adds a denominator $(1 + \sqrt{I})$ that rescues the law at finite strength. The numbers are not small: at the physiological ionic strength of ≈ 0.15 mol L^{-1}, a monovalent ion has $\gamma \approx 0.75$ and a divalent ion $\gamma \approx 0.31$ (Worked Example 4.2) — the cell's Mg^{2+}, that indispensable cofactor of every kinase, is effectively two-thirds discounted before it begins work. The honest antithesis is that most biochemical equilibrium data are reported in concentrations, not activities, and muddle along tolerably; the resolution is that the approximation survives because biological fluids hold ionic strength nearly constant, folding γ into an *apparent* constant — precisely the manoeuvre by which the biochemical standard state ($\Delta G^{\circ\prime}$, pH 7, specified Mg^{2+})

is defined in Chapter 7. Rigor and convenience, here as ever, are reconciled by agreeing on the fudge in advance and writing it into the convention.

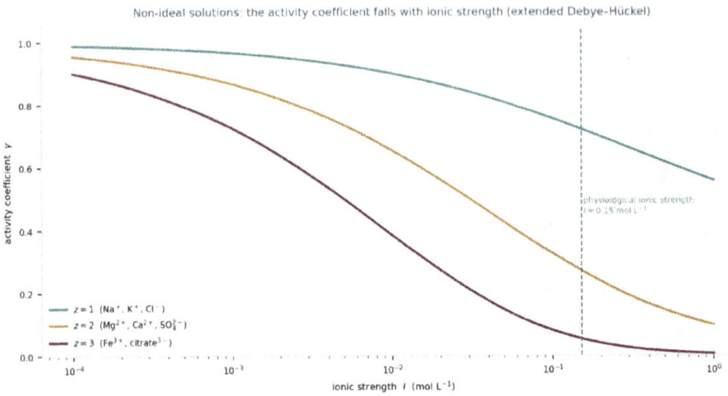

Figure 4.4 Activity coefficients of ions of charge 1, 2 and 3 against ionic strength, from the extended Debye–Hückel law of § 4.7 (A = 0.509, denominator $1 + \sqrt{I}$). The dashed line marks physiological ionic strength, ≈ 0.15 mol L^{-1}, at which the cell's divalent ions operate at barely a third of their nominal concentration. The moral for every later equilibrium in this book: concentrations are what we pipette; activities are what react.

4.8 Logarithmic quantities

The quantities of biochemistry span absurd ranges — the proton's activity wanders eight decades between stomach and pancreas, the rate constants of enzymes nine, the concentrations of Figure 4.5 twelve — and the human mind, an instrument evolved for counting antelope, copes with such ranges only in logarithms. Hence the p-notation: for any quantity X,

$$pX \equiv -\log_{10} X \tag{4.11}$$

— giving pH = $-\log_{10} a_{H+}$ (whose full derivation from the autoprotolysis of water is the business of § 6.4), pK_a, pK_w, pI, *and the rest of the family. Three properties of the notation deserve explicit statement, for they are used silently throughout the treatise. First, it inverts: larger quantity, smaller p-value, a trap that has claimed more students than any enzyme. Second, it linearises: equilibrium laws that are exponential in K become additive in pK — Eq. (4.6) reads $\Delta G°$ = 2.303 RT · pK with a change of sign, and the 5.708 kJ mol^{-1} per unit of § 4.4 is precisely 2.303RT at 25 °C. Third, it compresses: the ladder of Figure 4.5, unmanageable in linear notation, becomes a ruler. The reader who finds logarithms unaesthetic is in good historical company — the description "tedious" was applied to Napier's invention by contemporaries who had never multiplied fourteen-digit numbers by hand — but the biochemist's proton, like the astronomer's magnitude and the acoustician's decibel, pays the notation the compliment of daily use.*

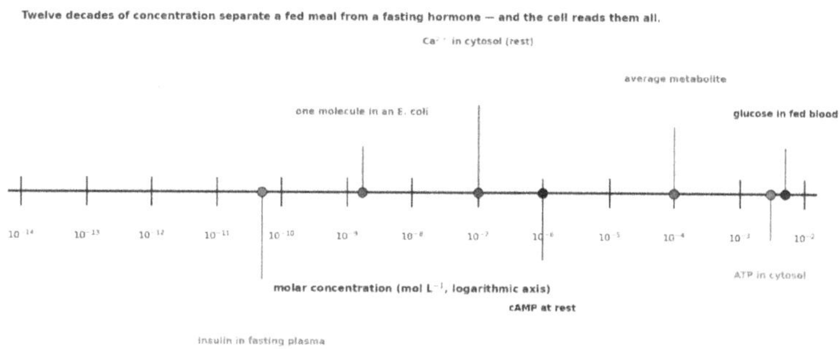

Figure 4.5 Twelve decades of biochemical concentration on a logarithmic axis: blood glucose and cytosolic ATP at millimolar; a "typical" metabolite at 10^{-4} M; the second messengers cAMP and Ca^{2+} at micromolar and below; fasting insulin at 10^{-10} M; and, for calibration, the concentration corresponding to one molecule in one E. coli (≈ 1.7×10^{-9} M, from Table 1.3's femtolitre). Values are representative reference figures (Alberts et al., 2022); the point is the range — the cell's regulatory machinery must, and does, read the whole ruler.

4.9 Dimensional analysis

Before any number is believed, its units must confess. The principle of dimensional homogeneity — that two quantities added, equated, or ratioed must carry the same dimensions — is the cheapest and most underused instrument of scientific self-defence: it catches algebraic slips, exposes invented formulas, and, used offensively, generates correct relations before any theory is at hand. As an example of the offensive use, consider the thermal voltage of Chapter 26: what combination of the constants of Table iii has the dimensions of electric potential? Potential is energy per charge; the only thermal energy per mole is RT, the only molar charge is the Faraday constant F; therefore

$$\frac{RT}{F} = \frac{8.314\ \mathrm{J\,K^{-1}\,mol^{-1}} \times 298.15\ \mathrm{K}}{96\,485\ \mathrm{C\,mol^{-1}}} = 25.69\ \mathrm{mV} \qquad (4.12)$$

— the quantity that scales every membrane potential and every Nernst slope in Parts V and VI, obtained here from dimensional necessity alone and verified against the constants; the reader who has seen 58–61 mV per decade in electrochemistry has met 2.303 × 25.69 mV in disguise. The working rule of the treatise, stated once and enforced everywhere: no equation is quoted whose two sides do not agree in dimensions, and no result is reported without the unit in which it was earned.

4.10 Orders of magnitude and biochemical estimation

There is a genre of calculation, scorned by the fastidious and beloved of the practised, in which the aim is not the right answer but the right power of ten — the Fermi estimate, after the physicist who could bound a blast's yield from the flight of dropped paper scraps. Its discipline is simple: decompose the unknown into factors, estimate each to the nearest order of magnitude (or better, bound it between two defensible guesses), multiply, and state the uncertainty honestly. The method was already exercised in § 1.9's census of the bacterium; Figure 4.7 now *applies it to the most quoted estimate in popular biochemistry — that an adult human turns over roughly his own weight in ATP each day. The tree of the figure is the whole calculation: a resting metabolic rate of ≈ 8 400 kJ day^{-1} (≈ 100 W, the human as a light-bulb), half of it capturable as chemical work, a cellular cost of ≈ 50 kJ per mole of ATP (§ 4.5's driven quotient), giving ≈ 84 mol — and at 507 g per mole, some 43 kg of ATP turned over daily. Turned over, not stored: the body's standing stock of ATP is ≈ 50 g, whence the corollary that each molecule is recycled roughly a thousand times a day — a number that explains both the lethality of cyanide, which stops the recycling (§ 27.12), and the impossibility of "running out of ATP" by any ordinary exertion. The estimate's value is not its digits, which shift with body mass and activity, but its architecture: every factor is independently checkable, and the whole collapses gracefully into the correct conclusion that the energetic economy of the body is a throughput economy — a river, not a reservoir (principle six of § 1.10, now in numbers).*

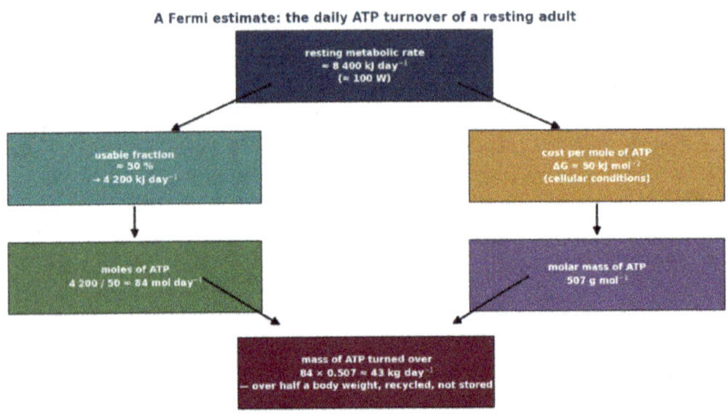

Figure 4.7 *A Fermi estimate of the daily ATP turnover of a resting adult human, laid out as a factor tree so that every step is independently auditable: metabolic rate → capturable fraction → molar cost → molar turnover → mass turnover. Inputs are standard reference values (resting metabolic rate ≈ 100 W; cellular ΔG_{ATP} ≈ 50 kJ mol^{-1} from § 4.5); the conclusion — tens of kilograms per day recycled from a 50 g stock — is robust to any reasonable adjustment of the inputs, which is the signature of a good estimate.*

4.11 Uncertainty, precision and significant figures

A number without an uncertainty is an opinion wearing a lab coat, and this treatise accordingly holds three doctrines. First, accuracy and precision are different virtues: precision is reproducibility (the scatter of repeated measurement), accuracy is truth (the distance from the accepted value), and an instrument can be admirably precise and thoroughly wrong — as every calibration standard, like every critic, exists to demonstrate. Second, random error propagates by quadrature. For a result f computed from independent measurements x_i with standard uncertainties σ_i, the variance of f is the sum of the squared sensitivities,

$$\sigma_f^2 = \sum_i \left(\frac{\partial f}{\partial x_i}\right)^2 \sigma_i^2 \qquad (4.13)$$

— derived by linearising f about the mean values (a first-order Taylor expansion), applying the independence of the errors (their covariances vanish), and recalling that variances add; the special cases the reader will use weekly follow immediately: for sums and differences, absolute uncertainties add in quadrature ($\sigma_\Sigma = \sqrt{(\sigma_1^2 + \sigma_2^2)}$); for products and quotients, relative uncertainties add in quadrature, since $\partial \ln f = \partial f/f$; and for the mean of k repeats, σ shrinks as $1/\sqrt{k}$ — the statistical warrant for replication, and the reason Figure 4.6b descends while its companion curve climbs. Third, significant figures are a courtesy, not a measurement: they communicate the author's estimate of his own precision, and reporting a rate as 4.3281 μmol min^{-1} when the pipette is good for two per cent is a form of perjury. The treatise's convention: physical constants to the precision of Table iii, derived results to one more digit than the least precise input, and orders of magnitude — the currency of § 4.10 — to the nearest power of ten, stated as such.

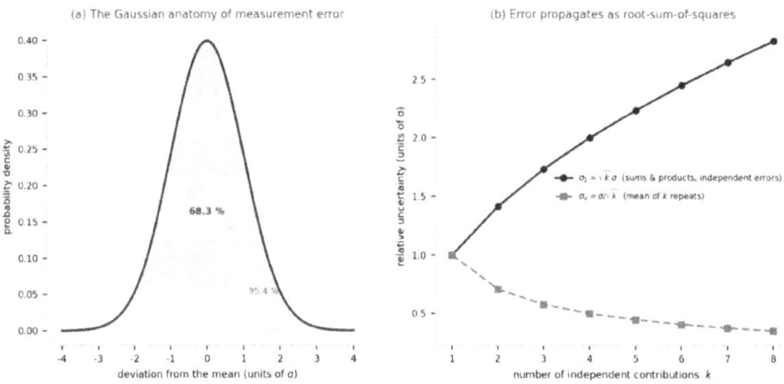

Figure 4.6 *(a) The Gaussian distribution of independent measurement errors, with the canonical 68.3 % and 95.4 % enclosures about the mean — the shape error takes when many small, independent causes contribute (the central limit theorem, whose proof we leave to the statisticians with thanks). (b) The two master curves of propagation from Eq. (4.13): the uncertainty of a sum or product of k independent contributions grows as √k (oxblood), while the uncertainty of the mean of k repeats falls as 1/√k (teal) — between them, the entire strategy of experimental design: subtract where you can, average where you must, and never trust a difference of two large, ill-measured numbers.*

4.12 Quantitative biochemical problem solving

The chapter closes, as it opened, with method. Quantitative problems in biochemis , with suspicious regularity, to the same five-step protocol, which the offer in Table 4.2 with the confidence of long employment: it has survived with every chapter of this book, and the reader will find its fingerprints on every worked example so far. Two of its steps deserve emphasis because they are the ones experience teaches last. Derive before substituting: a formula with symbols is an argument, which can be checked; a formula with numbers is a claim, which can only be believed. And audit at the end: every computed result should be subjected to three questions — are the dimensions right, is the magnitude sane (against the rulers of Figures 4.5 and 1.4), and does the sign make thermo sense (Eq. 4.7). The reader who makes the audit a habit will find, as the have, that it catches more errors than cleverness ever invents.

Table 4.2 A protocol for quantitative biochemical problem solving, distilled from the worked examples of Chapters 1–4 and recommended for the remainder of the treatise.

Step	Action	Instrument
1. Identify	name the quantity sought and the quantities given, with units	the problem statement, a pencil
2. Select	choose the governing relation(s) — and their conditions of validity	Eqs. (4.1)–(4.13); Table 4.1
3. Derive	rearrange symbolically to isolate the unknown	algebra; § 4.9's homogeneity audit
4. Compute	substitute numbers with units attached, and carry the units through	Table iii constants; a calculator used grudgingly
5. Audit	check dimensions, magnitude, and sign; state uncertainty honestly	Figures 1.4, 4.5; Eq. (4.13)

Chapter 4 — Conceptual synthesis

Part I now closes its ledger. The mole connects the balance to the molecule (§ 4.1); composition is measured three ways, each with its jurisdiction (§ 4.2); equilibrium is balanced motion, not rest (§ 4.3); the equilibrium constant follows from the composition-dependence of chemical potential by two lines of algebra (§ 4.4); the quotient tells the direction of travel (§ 4.5); coupling multiplies constants and adds free energies, provided the enzyme enforces the sharing (§ 4.6); activities discount concentrations in the crowd of the ionic atmosphere (§ 4.7); logarithms make the twelve decades thinkable (§ 4.8); dimensions audit every formula (§ 4.9); Fermi estimates bind the unknowable to the nearest power of ten (§ 4.10); and uncertainty propagates by quadrature, replication purchasing √k (§ 4.11) — the whole disciplined by a five-step protocol (§ 4.12). Direction checked against the itinerary of Figure ii: Part I — the chemical grammar of life — is complete: scope (Chapter 1), bonds (2), geometry (3), and now number (4). The treatise crosses into Part II, where the solvent itself takes the stage: water, and with it the acids, bases and thermodynamic laws that rule everything dissolved in it.

WORKED EXAMPLE 4.1 — THE EQUILIBRIUM CONSTANT OF ATP HYDROLYSIS, AND THE CELLULAR DRIVING FORCE

Given $\Delta G°' = -30.5$ kJ mol^{-1} at 25 °C, Eq. (4.6) gives $\ln K' = 30\,500/2\,479 = 12.30$, so $K' = e^{12.30} \approx 2.2 \times 10^5$ (in the biochemical convention, with water and protons folded away — § 7.6 will formalise the prime). In a resting cell, [ATP] ≈ 3 mM, [ADP] ≈ 0.3 mM, [P_i] ≈ 3 mM: the mass-action ratio (with water's activity unity) is $Q = (3 \times 10^{-4})(3 \times 10^{-3})/(3 \times 10^{-3}) = 3 \times 10^{-4}$ M — vastly below K′; hence by Eq. (4.7), $\Delta G = -30.5 + 2.479 \ln(3 \times 10^{-4}/2.2 \times 10^5)$·... — more directly, $\Delta G = \Delta G°' + RT \ln([ADP][P_i]/[ATP]) = -30.5 + 2.479 \ln(3 \times 10^{-4}) = -30.5 - 20.1 \approx$ ***−50.6 kJ mol^{-1}****, the figure the Fermi tree of Figure 4.7 spent so freely. The method was Table 4.2's: identify, select, derive, compute, audit — and the audit notes the sign is negative, as hydrolysis should be.*

WORKED EXAMPLE 4.2 — THE EFFECTIVE PRICE OF Mg^{2+} AT PHYSIOLOGICAL IONIC STRENGTH

Blood plasma carries $Na^+ \approx 140$, $K^+ \approx 4$, $Cl^- \approx 105$, $HCO_3^- \approx 24$, $Ca^{2+} \approx 1.2$, $Mg^{2+} \approx 0.5$ mmol L^{-1} (and minor ions besides). Eq. (4.10): I ≈ ½[(140 + 4 + 105

+ 24) × 10^{-3} × 1 + (1.2 + 0.5) × 10^{-3} × 4] ≈ ½[0.273 + 0.007] ≈ 0.14 mol L^{-1} — the canonical 0.15. The extended Debye–Hückel law then gives, for Mg^{2+}, log γ = −0.509 × 4 × √0.14/(1 + √0.14) = −0.555, whence γ ≈ 0.28: the cell's nominal 0.5 mmol L^{-1} of free Mg^{2+} behaves as ≈ 0.14 mmol L^{-1} in the laws of equilibrium — a discount every kinase quietly suffers, and § 4.7's moral in one line.

Worked example 4.3 — Propagating the error of a rate

An initial rate is computed as v = Δc/Δt with Δc = 25.0 ± 0.5 μmol L^{-1} (2.0 %) and Δt = 60 ± 1 s (1.7 %). By the product rule of Eq. (4.13), the relative uncertainties add in quadrature: σ_v/v = √(2.0² + 1.7²) % = 2.6 %, so v = 0.417 ± 0.011 μmol L^{-1} s^{-1}. Note the two morals: the weaker measurement (concentration) dominates the budget, so money spent on a better clock is wasted until the absorbance improves; and reporting the rate as 0.4167 μmol L^{-1} s^{-1} would be the perjury of § 4.11.

Exercises (with full resolutions)

Exercise 4.1

A 5.00 mmol L^{-1} glucose solution is prepared at 4 °C and warmed to 37 °C. Estimate the change in molarity, taking the thermal expansion of water as 2.1 × 10^{-4} K^{-1}, and state the correction's sign and size as a percentage.

Resolution. Volume grows by ΔV/V ≈ αΔT = 2.1 × 10^{-4} × 33 ≈ 6.9 × 10^{-3}; molarity falls by the same fraction: 5.00 → 4.97 mmol L^{-1}, a −0.7 % correction — beneath the noise of a student assay, above the noise of a good one, and exactly the sort of quiet discrepancy § 4.2's molality exists to abolish.

Exercise 4.2

Hexokinase: glucose + ATP → glucose-6-phosphate + ADP has ΔG°′ = −16.7 kJ mol^{-1}. In a cell with [glucose] = 5 mM, [G6P] = 0.1 mM, [ATP] = 3 mM, [ADP] = 0.3 mM, compute Q, compare with K′, and determine the direction and driving force of the reaction.

*Resolution. ln K′ = 16 700/2 479 = 6.74, K′ ≈ 8.4 × 10^2. The quotient: Q = (10^{-4})(3 × 10^{-4})/[(5 × 10^{-3})(3 × 10^{-3})] = 2 × 10^{-3}. Since Q ≪ K′, forward flux is assured; by Eq. (4.7), ΔG = 2.479 ln(2 × 10^{-3}/8.4 × 10^2) = 2.479 × (−12.95) ≈ **−32 kJ mol^{-1}** — hexokinase traps glucose with a driving force larger than ATP hydrolysis itself in some conditions, which is why glycolysis's first step is effectively irreversible (§ 31.6).*

Exercise 4.3

Show that the biochemical standard free energy change for ATP hydrolysis is related to the concentration-based apparent constant K_{obs} (in mol L^{-1}) by a unit-carrying term, and evaluate the term at 25 °C.

Resolution. The equilibrium constant proper is dimensionless: $K = (a_{ADP}a_{Pi})/(a_{ATP}a_{H2O})$. With $a_i = \gamma_i c_i/c°$ and a_{H2O} = 1, $K = (\gamma_{ADP}\gamma_{Pi}/\gamma_{ATP})\cdot(c_{ADP}c_{Pi}/c_{ATP})/c°$

= γ-ratio × $K_{obs}/c°$. Hence $ΔG°' = -RT \ln(K_{obs}/c°) - RT \ln(γ\text{-ratio})$: at 25 °C, the unit term $-RT \ln(1/c°)$ contributes $-2.479 \ln(1\ L\ mol^{-1} \ldots)$ — formally zero by the convention $c° = 1\ mol\ L^{-1}$, which is exactly why the convention exists: the standard concentration is the silent conversion factor of every ΔG° in this book, and the exercise's point is that the silence is contractual, not accidental.

Exercise 4.4 (interpretative)

The law of mass action was derived in § 4.4 from thermodynamics, yet Guldberg and Waage reached it from kinetics (equal forward and reverse rates). Reconcile the two roads, and say what each can and cannot prove.

Resolution (indicative). *The kinetic road writes forward rate $r_f = k_f Πc_{reactants}$ and reverse $r_r = k_r Πc_{products}$, sets them equal at equilibrium, and obtains $K = k_f/k_r$ — swift, intuitive, and hostage to the assumption of elementary mass-action kinetics, which real mechanisms (Part IV) rarely honour. The thermodynamic road assumes nothing about rates and therefore proves the constant's existence universally — but can say nothing about the time to reach it. The reconciliation is detailed balance: at equilibrium, every elementary step is individually balanced, so the kinetic ratio must collapse to the thermodynamic constant however baroque the mechanism; kinetics proposes, thermodynamics disposes, and the two meet exactly where the Norwegians said they would.*

Exercise 4.5

Estimate, Fermi-style, the number of glucose molecules in the reader's bloodstream, and the fraction consumed per minute at a cerebral glucose uptake of ≈ 5.5 mg per 100 g brain per minute (brain mass 1.4 kg).

Resolution. *Blood volume ≈ 5 L, glucose 5 mmol L^{-1} (Figure 4.5): 25 mmol ≈ 1.5×10^{22} molecules. The brain takes 5.5 × 14 ≈ 77 mg min^{-1} ≈ 0.43 mmol min^{-1} — 1.7 % of the circulating stock per minute, from one organ of 2 % body mass. The estimate explains, without any further physiology, why the liver of Chapter 40 must produce glucose continuously and why hypoglycaemia is a neurological emergency within minutes: the audit of § 4.10 has become a diagnosis.*

Research-level propositions

Proposition 4.α. The Debye–Hückel treatment of § 4.7 assumes a continuum solvent and point ions; inside a protein's electrostatic microenvironment both fail. Formulate a finite-size (Bikerman or mean-spherical) correction, estimate its effect on the Mg^{2+} coefficient of Worked Example 4.2, and propose an electrochemical measurement (ion-selective electrode, § 53.9) that would decide between the models at $I = 0.15\ mol\ L^{-1}$. Proposition 4.β. Eq. (4.13) assumes independent errors; enzyme-kinetic parameters (K_m, V_{max}) estimated from the same fitted curve are strongly correlated. Derive the covariance-corrected propagation law, and quantify the error in the customary practice of reporting k_{cat}/K_m with the two marginal uncertainties combined in quadrature. Proposition 4.γ. The Fermi estimate of Figure 4.7 treats the cellular $ΔG_{ATP}$ as a constant. Using the metabolic-flux literature (§ 55.6), assemble the measured range of cytosolic phosphorylation *potentials across tissues and recompute the ATP-turnover tree as a distribution; at what confidence can one still assert "tens of kilograms per day"?*

References (Chapter 4)

Alberts, B., Heald, R., Johnson, A., Morgan, D., Raff, M., Roberts, K., & Walter, P. (2022). Molecular biology of the cell (7th ed.). W. W. Norton & Company.

Atkins, P., & de Paula, J. (2018). Atkins' physical chemistry (11th ed.). Oxford University Press.

Bureau International des Poids et Mesures. (2019). The International System of Units (SI) (9th ed.). BIPM.

Debye, P., & Hückel, E. (1923). Zur Theorie der Elektrolyte. I. Gefrierpunktserniedrigung und verwandte Erscheinungen. Physikalische Zeitschrift, 24(9), 185–206.

Guldberg, C. M., & Waage, P. (1867). Études sur les affinités chimiques. Brøgger & Christie.

Lewis, G. N., & Randall, M. (1923). Thermodynamics and the free energy of chemical substances. McGraw-Hill.

Nelson, D. L., Cox, M. M., & Hoskins, A. A. (2021). Lehninger principles of biochemistry (8th ed.). W. H. Freeman.

van 't Hoff, J. H. (1884). Études de dynamique chimique. Frederik Muller.

PART II

Water, Energy and Molecular Association

Chapter 5 — Water as the Medium of Life · Chapter 6 — Acids, Bases and Buffers · Chapter 7 — Thermodynamics of Biochemical Systems · Chapter 8 — Molecular Recognition and Self-Assembly

CHAPTER 5

Water as the Medium of Life

CONTENTS OF THIS CHAPTER

In which the solvent steps forward from the wings and takes the lead it has always secretly played: its anomalous physics traced to the tetrahedral imperative of Chapter 2, its embrace of ions and polar solutes measured, its rejection of oil made the author of membranes, its vapour, its pressure and its activity given laws — and the cell revealed as not merely immersed in water but composed of it, regulated by it, and crowded within it.

5.1 Molecular structure of water

There is a habit, among of biochemistry, of introducing water with a salute to its abundance — per cent of the cell, seventy-one of the planet — and then proceeding as if the solvent were a kind of obedient wallpaper against which the interesting molecules perform. This treatise takes the opposite view, and the present chapter exists to defend it: water is not the stage but the principal actor, and nearly every peculiarity of biochemistry — the folding of proteins, the existence of membranes, the strength and weakness of every weak bond, the very pH of the blood — is a consequence of the peculiar personality of one small molecule. That personality was sketched in Chapter 2 (Figure 2.4): an oxygen carrying two hydrogens at an angle of 104.5°, two lone pairs completing a rough tetrahedron, and a dipole moment of 1.85 D (Worked Example 2.1 computed the partial charge, $\delta \approx 0.4$ e). Everything in this chapter is a corollary of those four sentences, and the reader is invited to test the claim as we proceed: the anomalies of § 5.3, the hydration of § 5.4, the hydrophobic effect of § 5.6, the colligative laws of § 5.9, are one personality viewed from different windows.

5.2 Hydrogen-bonded networks

The tetrahedral disposition of two donors and two acceptors (§ 2.7) means that each water molecule can bind four neighbours, and that the liquid is therefore not a collection of molecules but a single, enormous, flickering molecule — a network in which each hydrogen bond lives a few picoseconds before breaking and re-forming elsewhere. The consequences of this restlessness are three, and all three will matter. First, the network is extensive but incomplete: at any instant a liquid water molecule holds about 3.4 of its possible four bonds (ice holds all four, which is why ice is the more ordered, more open, and — perversely — the less dense phase). Second, the network is cooperative: a molecule already engaged in one hydrogen bond is polarised by it, becoming a better donor and acceptor for the next, so that the bonds strengthen one another in clusters — the liquid is patchy, not uniform, at the nanosecond scale. Third, the network is fast: the orientations that solvation, recognition and proton conduction require are sampled on the picosecond clock, so that water can answer any biochemical question in real time. A solvent that were merely polar but slow would have been a useless medium for life; the speed of the network is as essential as its geometry (Ball, 2017)(Eisenberg & Kauzmann, 1969).

5.3 Cohesion, surface tension and dielectric behaviour

Nowhere is the personality more visible than in the anomalies of Figure 5.1. A liquid whose molecules weigh eighteen daltons has no business boiling at 100 °C; methane, at sixteen, boils at −161 °C, and a "normal" H_2O — one without its hydrogen bonds — would be a gas at room temperature, as its sister hydrides H_2S, H_2Se and H_2Te all are. [15] *The heat of vaporisation (44 kJ mol^{-1}, against 8.2 for methane) is the direct sum of the network's bonds; the dielectric constant (78.5, the figure that screened every ionic calculation of § 2.4) is the network's capacity to orient wholesale around a charge; and the density maximum at 3.98 °C (Figure 5.1a) is the contest between the open, tetrahedral ice-like order (favoured as the liquid cools) and ordinary thermal collapse — a contest the open structure wins, with the biological consequence that ice floats, lakes freeze from the top, and the aquatic world survives its winters under an insulating blanket rather than suffocating from the bottom up. A planet of sinking ice would be a planet of frozen oceans and, one may safely conjecture, no biochemists to remark upon it. Table 5.1 collects the anomalies and their consequences; the reader will find that each row is a load-bearing wall of some later chapter.*

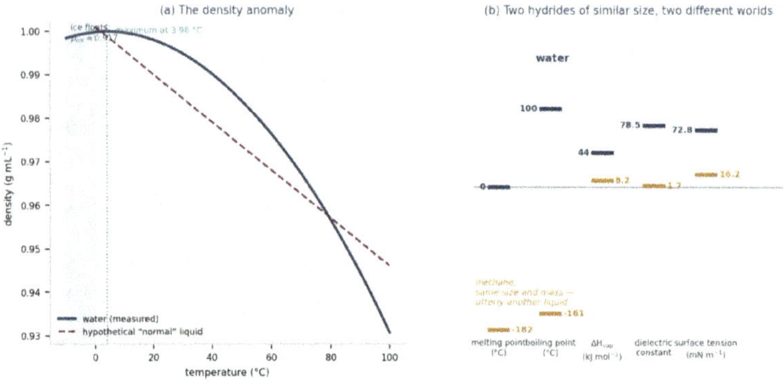

***Figure 5.1** (a) The density of water against temperature, showing the celebrated maximum at 3.98 °C and the still more celebrated consequence that ice floats (shaded region); the dashed line is the behaviour a "normal" liquid would show. (b) Water against methane — two hydrides of nearly identical mass, separated by the entire phenomenology of the hydrogen bond: melting and boiling points, heat of vaporisation, dielectric constant and surface tension. Values from standard compilations (Eisenberg & Kauzmann, 1969).*

15. The comparison of the group-16 hydrides is the classic classroom demonstration that hydrogen bonding, not molecular weight, governs water's condensability; the demonstration is owed, in its modern form, to the hydrogen-bond literature that followed Latimer and Rodebush (footnote to § 2.7), though every generation of teachers has rediscovered it to its own satisfaction.

Table 5.1 The anomalies of water, and their consequences for life. Each entry is a corollary of the tetrahedral network of § 5.2.

Anomaly	Cause	Consequence for life
high melting/boiling points	hydrogen-bond network	liquid medium at terrestrial temperatures
ice less dense than liquid	open tetrahedral order of ice I_h	lakes freeze from above; aquatic life survives winter
density maximum at 3.98 °C	order–collapse competition	4 °C water sinks; deep waters stay liquid and mixed
high dielectric constant (78.5)	collective dipole orientation	salts dissolve; electrostatics screened (§ 2.4)
high heat capacity	bond-breaking absorbs heat	thermal buffering of cells and climates
high heat of vaporisation	network must be fully broken	evaporative cooling: sweat, transpiration
high surface tension	unsatisfied bonds at interfaces	capillarity, droplet integrity, alveolar mechanics

5.4 Solvation of ions

Drop a crystal of salt into water and the lattice, so stable in air, falls to pieces — not because the ions have ceased to attract, but because each ion is at once surrounded by an entourage of oriented water molecules whose combined attentions pay better than the lattice did (Figure 5.2). The energy of the transaction is the hydration free energy, and its scale is formidable: for Na^+, $\Delta G_{hydr} \approx -405$ kJ mol^{-1}; for the proton, $\approx -1\ 100$ kJ mol^{-1}. The accounting that dissolves the crystal is the contest of two giants — lattice energy against hydration — and the contest's outcome for NaCl is a narrow victory for the water, of a few kJ mol^{-1}: near enough to balanced that the solubility of salts is exquisitely sensitive to size and charge, and the periodic table of electrolytes (Figure 2.8) sorts itself into the soluble (NaCl, KCl) and the stubborn ($CaCO_3$, $Ca_3(PO_4)_2$ — from which, not coincidentally, bone is made). Two features of the hydration shell deserve the biochemist's attention. The shell is structured: the first sphere (six waters for Na^+, oxygen-in, lifetimes of nanoseconds) is a genuine chemical entity, and the cost of stripping it — the desolvation penalty — is the hidden toll that every ion channel and every ATP-bound Mg^{2+} must pay, a toll we shall meet again at the membrane (§ 24.4) and the enzyme (§ 19.10). And the shell is dynamic: the entourage is perpetually exchanged with the bulk, so that hydration is a negotiation, not a marriage — a point of some consequence when, in Chapter 8, a ligand must decide between the embrace of water and the embrace of a receptor.

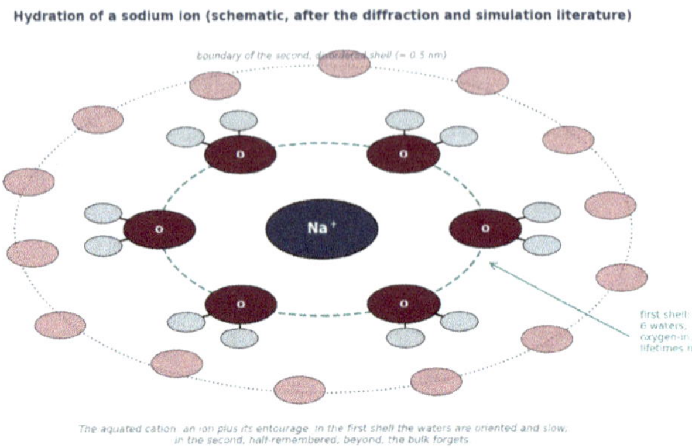

Figure 5.2 The hydration of a sodium ion (schematic). In the first shell (dashed circle), some six waters sit oxygen-in, oriented and slow, with residence times of nanoseconds; in the second shell (dotted), the order is half remembered; beyond it, the bulk liquid forgets the ion entirely. The structured shell is the reason the ion's effective size in water is several times its crystal radius — and the reason that stripping it, at a channel or an active site, costs free energy that somebody must pay.

5.5 Hydration of polar molecules

Polar but uncharged solutes — sugars, alcohols, amides — are received by the network on easier terms than ions, for they ask only to be admitted to the club of hydrogen bonds rather than to reorganise it wholesale. A hydroxyl group offers the network a donor and two acceptors, an exchange of memberships so nearly equitable that sugars dissolve to molar concentrations without protest; the water activity of the blood's 5 mM glucose departs from unity by less than a part in ten thousand. Two cautions govern the category. The first is that polarity is a property of regions, not of molecules: glucose, with five hydroxyls on six carbons, is hydrophilic overall, but the face of its ring is a hydrocarbon patch, and it is exactly that patch that certain enzymes and transporters recognise — the molecule is read as a map of polar and nonpolar terrain, not as an average. The second is that the hydrogen bonds a solute makes to water are bonds it cannot simultaneously make to a partner: every specific interaction of Chapter 8 — substrate to enzyme, drug to receptor — must first outbid the solvent, and the art of molecular recognition is substantially the art of arranging that the partner's offer is better by the decisive margin.

5.6 Hydrophobic hydration

And the solutes the network cannot befriend, it cages. The thermodynamics of the cage was the business of § 2.9 — the entropic tax of Figure 2.5, the ordered shell, the expulsion of the guest — and nothing need be repeated here except the framing that this chapter adds: the hydrophobic effect is not a property of the oil but of the water, a theorem about the network's refusal to sacrifice its bonds. The point is worth insisting upon because it inverts the naive direction of causality. The folding of a protein is not driven by the attraction of leucine for leucine (a dispersion affair of a few kJ mol^{-1} at best); it is driven by the liberation of the ordered water that each exposed leucine had conscripted. Water is the active party, the protein the beneficiary, and the hydrocarbon merely the excuse. Treatises that speak of *"hydrophobic bonds" are forgiven, but not followed: there are no such bonds, only a solvent that keeps its own counsel and drives the nonpolar world before it — into micelles, into bilayers, into the cores of proteins, wherever the surface it must cage is minimised.*

5.7 Amphiphilic substances in water

Between the friends and the enemies of the network lies the most inventive class of all: the amphiphiles, molecules wearing a polar head and a nonpolar tail, which resolve their double allegiance by self-organising. A fatty-acid salt in water assembles, above a characteristic concentration, into micelles — spheres with tails inward and heads outward, hiding the nonpolar surface from the solvent at the price of a little curvature; a phospholipid, with two tails too fat for a sphere, forms the bilayer instead, the flat sandwich from which every membrane of Part V is cut. The threshold of assembly has a law worth deriving in one line. Let the free energy of transferring a molecule from water into the aggregate be $\Delta\mu°$, and let monomer and aggregate equilibrate: then the concentration at which aggregation begins — the critical micelle concentration — is

$$\mathrm{CMC} \approx c^{\circ}\, e^{\Delta\mu^{\circ}/RT} \tag{5.1}$$

— obtained by setting the chemical potential of the monomer, $\mu° + RT\ln(c/c°)$, equal to that of the aggregated state, whose activity is effectively one; each additional methylene lowers $\Delta\mu°$ by the ≈ 3.8 kJ mol^{-1} of Figure 2.5a and therefore lowers the CMC by a factor of $e^{-3.8/2.58} \approx 1/4.4$ — a rule of thumb the detergent industry exploits without gratitude, and the cell with rather more artistry in the design of its lipids. The dialectic of the section is the resolution of an old puzzle: how can assembly be spontaneous ($\Delta G < 0$) when it visibly orders the lipids ($\Delta S < 0$)? The answer is now routine to the reader: the entropy that matters is the solvent's, the released hydration shells outvote the ordered tails, and the aggregate is, once again, the network's idea.

5.8 Osmosis and osmotic pressure

Place a solution and its pure solvent on opposite sides of a membrane that passes water but not the solute, and the water moves — from dilute to concentrated, from high to low chemical potential, until the hydrostatic head it builds is heavy enough to stop it (Figure 5.3a). The pressure that stops it is the osmotic pressure, and its law is the one van 't Hoff (Plate 4.1) recognised, in 1887, as the gas law wearing the solution's clothes (van 't Hoff, 1887):

$$\Pi = c\,RT \tag{5.2}$$

— where c is the molar concentration of solute particles (ions counted separately, as the salts dissociate; this is the van 't Hoff factor, i, which for NaCl is two). The derivation belongs to the next section, but the number deserves its moment here: a one-molar solution exerts ≈ 24 atmospheres, the pressure under 250 metres of seawater; and the blood plasma, with its ≈ 0.3 osmol L^{-1} of solutes, exerts some 7.3 atm — the force that, divided between the proteins (oncotic pressure) and the salts, governs every exchange of fluid between blood and tissue. When the clinicians speak of "tonicity", they are speaking of Eq. (5.2); when the erythrocyte of Exercise 5.4 swells and bursts in distilled water, it is obeying it.

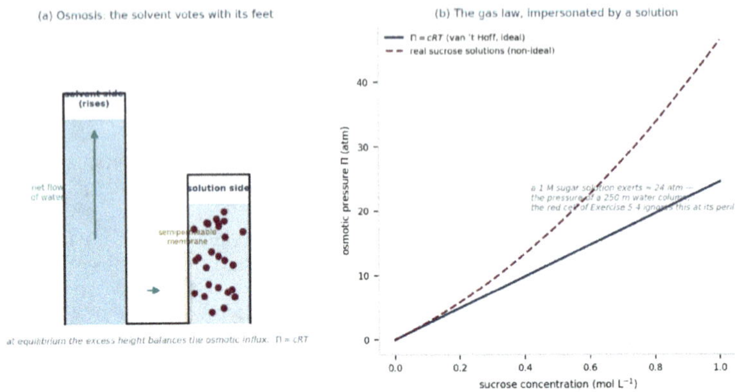

Figure 5.3 (a) Osmosis across a semipermeable membrane: water flows toward the solution (teal arrows) until the excess hydrostatic head balances the chemical-potential difference. (b) The osmotic pressure of sucrose solutions against concentration: the ideal van 't Hoff law (Eq. 5.2, navy) and the upward departure of real, non-ideal solutions (oxblood, illustrative), a reminder that § 4.7's activities govern here too. A 1 M solution exerts ≈ 24 atm; the physiological ≈ 0.3 osmol L^{-1}, ≈ 7.3 atm.

5.9 Colligative properties

The osmotic pressure is one of four siblings — vapour-pressure lowering, boiling-point elevation, freezing-point depression being the others — and the four are, at bottom, a single theorem wearing four laboratory coats. The theorem states that the chemical potential of the solvent in a dilute solution is lowered in proportion to the solute's mole fraction,

$$\mu_1 = \mu_1^* + RT \ln x_1 \approx \mu_1^* - RT\, x_2 \tag{5.3}$$

— where $x_1 = 1 - x_2$ is the solvent's mole fraction and the approximation $\ln(1 - x_2) \approx -x_2$ is exact in the dilute limit. From this one line the whole quartet of Figure 5.4 follows by asking, for each phenomenon, "what condition equalises the solvent's potential between the two phases?" Raoult's law comes first: the vapour above the solution is at equilibrium when p compensates the lowered μ_1, giving $p = p^* x_1$. The freezing-point depression follows by requiring the solid (pure ice, which excludes the solute) to equilibrate with a liquid of lowered potential — whence, after a short integration that Chapter 7 will make rigorous, $\Delta T_f = -K_f m$ with $K_f = RT_f^2 M_1 / \Delta H_{fus}$ = 1.86 °C kg mol^{-1} for water; and the osmotic law of Eq. (5.2) follows by balancing the solution's lowered potential against a pressure head, $\Pi V_1 = RTx_2$, which in the dilute limit is $\Pi = cRT$ exactly. The reader has now seen the entire quartet derived from one line — the deepest economy of the chapter, and a foretaste of Chapter 7's manner: all colligative properties are the solvent's mole fraction, asserting itself.

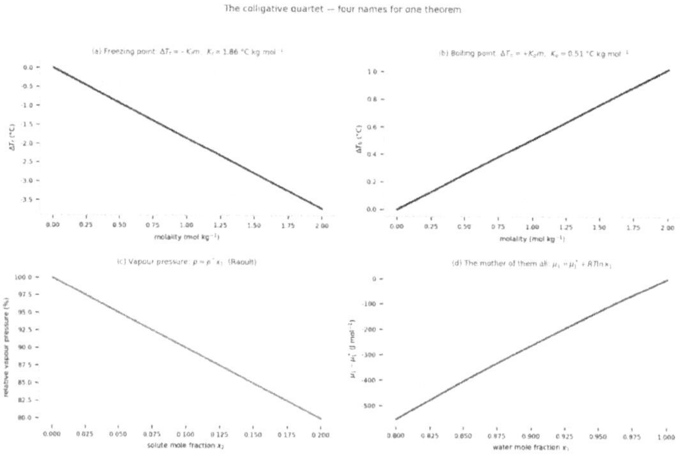

Figure 5.4 The colligative quartet. (a) Freezing-point depression and (b) boiling-point elevation, linear in molality with the cryoscopic and ebullioscopic constants of water; (c) Raoult's law for the vapour pressure; and (d) the theorem of which all three are corollaries — Eq. (5.3), the lowering of the solvent's chemical potential by the solute's presence. Note the signature of the family: every member depends on the number of solute particles, never on their identity — a glucose and a urea depress the freezing point identically, and NaCl twice as much, on account of its two ions.

5.10 Water activity

What the solute does to the solvent's escaping tendency, the food scientist and the cell physiologist measure as the water activity,

$$a_w = \frac{p}{p^*} \approx x_1\gamma_1 \tag{5.4}$$

— the ratio of the vapour pressure above the sample to that above pure water at the same temperature, equal by Raoult's law to the solvent's activity itself. The quantity rules two domains that rarely converse. In the pantry, it fixes the ecology of spoilage (Figure 5.5): bacteria demand $a_w \gtrsim 0.91$, most moulds surrender below 0.80, and only the xerophilic specialists persist toward 0.6 — so that jam, honey, and dried fruit are preserved not by poison but by drought, the oldest biochemistry in the kitchen. In the cell, the same quantity, measured through the osmotic balance of § 5.8, governs the water budget of every organism that cannot dilute its surroundings at will: the halophile of the salt flats lives by pumping potassium until its interior matches the external austerity, and the reader's renal medulla performs the same trick daily with urea and NaCl (§ 40.6). Water activity is, in short, the solvent's own mole fraction made physiological — and its unity in pure water is the convention from which all the deviations of this section are reckoned.

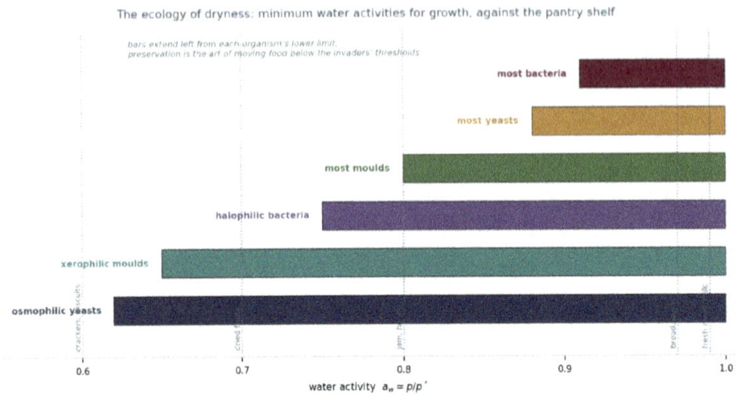

Figure 5.5 The ecology of dryness. Minimum water activities for the growth of the principal classes of spoilage organisms (bars, extending left from each threshold), against the water activities of common foods (dotted lines). The art of preservation — salting, sugaring, drying — is the art of moving the food below the invaders' thresholds; refrigeration merely slows the negotiation. Values are representative literature figures (Nelson, Cox, & Hoskins, 2021).

5.11 Macromolecular crowding

There remains the solvent's most neglected condition: in the cell, it is not dilute solution but a concentrate, with 30–40 % of its volume occupied by macromolecules (Exercise 1.5 computed five millimolar of protein) and the remainder navigable only in the interstices. The consequences, sketched by Exercise 1.5 and now drawn quantitatively in Figure 5.6, are of two kinds. Excluded volume devours the free space available to any test molecule (Figure 5.6a): at cytoplasmic occupancy, a globular protein finds its accessible volume halved or worse. And every equilibrium that compacts matter — folding, association, assembly — is tilted toward the compact state (Figure 5.6b), because the compact state wastes less of the precious interstitial water: association constants can be enhanced by orders of magnitude relative to the dilute cuvette (Minton, 2006). The honest antithesis is that crowding is not the whole story — the cytoplasm is also a soup of specific, transient, "quinary" interactions, some stabilising, some not; the resolution, as in § 1.6, is that both effects are real and both measurable, and the methods of Part XI (notably NMR in living cells) now measure them separately. What is no longer defensible is the cuvette's pretence of innocence: the cell is a crowded place, and Part II's laws, derived in dilute solution, are to be applied in vivo with the corrections of this section in hand.

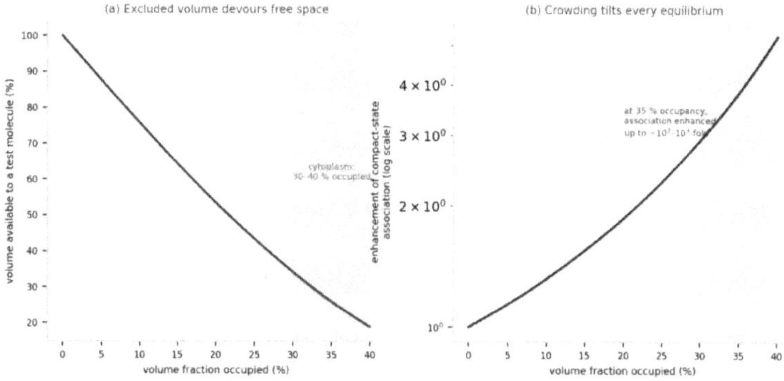

Figure 5.6 Macromolecular crowding (modelled after Minton's excluded-volume treatment 2006). (a) The fraction of volume accessible to a test molecule falls steeply with occupancy; the shaded band marks the cytoplasmic condition. (b) The consequent enhancement of compact-state association (logarithmic ordinate): at 35 % occupancy, equilibria toward dimers, folds and assemblies are favoured by factors of 10^2–10^3 relative to dilute solution. The curves are model-dependent in detail and robust in direction: crowding is a thermodynamic force, not a footnote.

5.12 Water within proteins and membranes

Last, the water that is not bulk at all: the molecule caught in the cleft of an enzyme, lining the pore of a channel, buried at the heart of a protein's fold. The crystallographers count them by the thousand — a small protein typically hydrates with one to two waters per residue in its first shell, and the structures of the PDB are full of buried waters, some exchanged on picosecond timescales, some fixed for milliseconds, some — the integral ones — as much a part of the structure as any residue (Table 5.2). Their functions are as various as their residences: they fill cavities that the fold cannot, complete the hydrogen-bonding valences that the backbone leaves unsatisfied, mediate the recognition of substrates across a film of solvent, conduct protons along chains of themselves (the Grotthuss mechanism, whose kinetics Chapter 6 will meet), and in photosystem II they are not solvent at all but substrate, oxidised to the oxygen the reader is presently breathing (§ 29.5). The dialectic of the section is brief and final: is structural water molecule or medium? It is both, and the distinction it forces upon us — between the bulk solvent of this chapter's laws and the molecule-by-molecule reagent of the cell's machinery — is the last service this chapter performs before handing the solvent, now respected, to the acids and bases of Chapter 6.

Table 5.2 Water within proteins: the classified population, with residence times and functions (representative values from the crystallographic and NMR literature (Ball, 2017)).

Class	Location	Residence time	Function
surface (first shell)	protein exterior	ps–ns	hydration, lubrication of dynamics
bridging	interfaces, active sites	ns–μs	mediated recognition and catalysis
cavity / buried	protein interior	μs–ms	valence completion, structural integrity
integral	channels, reaction centres	functional lifetime	proton wires, chemistry itself (PS II)

Chapter 5 — Conceptual synthesis

The solvent has had its chapter, and the argument ran: water's tetrahedral personality (§ 5.1) builds a cooperative, flickering network (§ 5.2), from which all the anomalies follow — the boiling point that should not exist, the ice that floats, the dielectric that screens every ion (§ 5.3). Ions are dissolved at the price of structured hydration shells, whose stripping is a toll on every later mechanism (§ 5.4); polar solutes join the network by exchanging hydrogen bonds (§ 5.5); nonpolar solutes are caged and expelled, and the expulsion — the network's work, not the oil's — builds micelles, bilayers and protein cores (§§ 5.6–5.7). The osmotic pressure obeys van 't Hoff's gas-law analogue (§ 5.8), and the whole colligative quartet descends from one line — the solvent's lowered chemical potential (§ 5.9); water activity measures that lowering for the pantry and the kidney alike (§ 5.10); crowding corrects the cuvette's innocence (§ 5.11); and the structural waters of proteins close the account by being medium and molecule at once (§ 5.12). Direction checked against Figure ii: the medium is understood. Chapter 6 turns to the most consequential solute the medium supports — the proton itself, and the acids, bases and buffers that govern it.

WORKED EXAMPLE 5.1 — THE OSMOLARITY OF BLOOD PLASMA

Summing plasma's principal solutes: Na^+ 140, K^+ 4, Ca^{2+} 1.2, Mg^{2+} 0.5, Cl^- 105, HCO_3^- 24, glucose 5, urea 5, proteins ≈ 1.2 mmol L^{-1} — total ≈ 286 mosmol L^{-1}, canonically 0.3 osmol L^{-1}. Eq. (5.2) then gives $\Pi = 0.286 \times 0.08206 \times 310 \approx 7.3$ atm — the pressure the plasma would develop against pure water, and the number every intravenous fluid must respect: an infusion of pure water would dilute the plasma's salts, drive water into the erythrocytes by osmosis, and burst them (Exercise 5.4). The calculation is the whole of clinical "tonicity" in one line.

WORKED EXAMPLE 5.2 — THE CRITICAL MICELLE CONCENTRATION RULE

A fatty-acid salt of 12 carbons (laurate) has CMC ≈ 20 mmol L^{-1}; estimate the CMC of the 16-carbon homologue (palmitate). Four extra methylenes lower $\Delta\mu°$ by ≈ $4 \times 3.8 = 15.2$ kJ mol^{-1}, so by Eq. (5.1) the CMC falls by $e^{15.2/2.58} \approx e^{5.9} \approx 360$: palmitate's CMC ≈ 20/360 ≈ 0.06 mmol L^{-1} — a factor of ≈ 4.4 per methylene, as the rule promised, and the reason long-chain lipids exist almost exclusively in aggregates rather than free in the cytosol: the network gives them no other choice.

Worked example 5.3 — Honey as a desert

Honey is ≈ 80 % w/w sugars, ≈ 17 % water. Estimate its water activity and its microbiological fate. In 100 g: ≈ 80 g sugars (≈ 0.25 mol sucrose-equivalents, taking M ≈ 330 g mol^{-1} for the invert-sugar mean of ≈ 180... let us be honest: ≈ 80 g/180 g mol^{-1} ≈ 0.44 mol hexoses) against 17 g water (0.94 mol): x_1 ≈ 0.94/(0.94 + 0.44) ≈ 0.68, so a_w ≈ 0.68 by Eq. (5.4) — below the threshold of every yeast in Figure 5.5 except the osmophilic specialists, which is why honey from the tombs of the pharaohs remains, in principle, edible: the bees, long before the chemists, understood Raoult's law.

Exercises (with full resolutions)

Exercise 5.1

From the cryoscopic constant (K_f = 1.86 °C kg mol^{-1}), estimate the freezing point of seawater (≈ 1.1 osmol kg^{-1}) and of the blood plasma of Worked Example 5.1.

Resolution. *$\Delta T_f = -K_f m_{osm}$: seawater, −1.86 × 1.1 ≈ −2.0 °C; plasma, −1.86 × 0.29 ≈ −0.54 °C. The two numbers are the thermostat of the ocean (polar seas run at −1.9 °C without freezing) and the reason frostbite, not freezing of the blood, is the clinical problem of cold exposure: the colligative laws set the winter of the planet and the margin of the patient alike.*

Exercise 5.2

Show that the osmotic pressure of a 1 mmol L^{-1} protein solution exceeds that of a 1 mmol L^{-1} NaCl solution by no amount at all — in fact it is half — and explain the van 't Hoff factor.

Resolution. *$\Pi = icRT$: the protein, one particle per molecule (i = 1), gives 0.024 atm; NaCl, two particles (i = 2), gives 0.048 atm. The colligative laws count particles, not mass — which is why the plasma proteins, a mere 1.2 mmol L^{-1} of the blood's 300, contribute only ≈ 0.03 atm (the oncotic pressure), yet that 0.4 % share is the margin that keeps the reader's ankles from swelling: number, not size, is the currency of colligativity.*

Exercise 5.3

A bacterial cell (volume 1 fL) is transferred from medium at a_w = 0.99 to a brine at a_w = 0.95. Estimate the initial efflux of water in molecules, assuming the cell equilibrates osmotically.

Resolution. *Water activity falls by 0.04; the cell's water content must fall by the same fraction to match: Table 1.3's 2 × 10^{10} waters × 0.04 ≈ 8 × 10^{8} molecules ≈ 1.3 amol — lost in seconds, shrinking the cytoplasm and halting growth until the cell pumps compatible solutes (K^+, glutamate) to reclaim its water: the brine of the pantry and the osmotic stress response of § 51.9 are the same law at different addresses.*

Exercise 5.4 (interpretative)

An erythrocyte placed in distilled water swells and bursts; one placed in 0.9 % NaCl is content; one in 3 % NaCl shrivels. Explain all three observations, and criticise the phrase "isotonic saline" as a chemical description.

Resolution (indicative). *In distilled water, the cell's ≈ 300 mosmol interior meets a zero exterior; water enters down its potential gradient (Eq. 5.2, $\Pi \approx 7$ atm of incentive) until the membrane, an inextensible film (§ 23.2), fails. In 0.9 % (154 mmol L^{-1}, ×2 ions ≈ 308 mosmol) the gradient vanishes. In 3 % (≈ 1 030 mosmol) the cell's water leaves. The criticism: "isotonic" is a physiological term — no net water movement — and it is not identical to "iso-osmotic": a solution of freely permeant urea at 300 mosmol is iso-osmotic but not isotonic (the urea strolls in, the water follows, and the cell bursts anyway). The colligative laws are blind to permeability; the membrane is not — a distinction on which the whole of clinical fluid therapy quietly rests.*

Exercise 5.5

Estimate the number of hydrogen bonds broken to vaporise one mole of water, and comment on the relation to the 20 kJ mol^{-1} hydrogen bond of § 2.7.

Resolution. *ΔH_{vap} = 44 kJ mol^{-1} (Figure 5.1b); at ≈ 3.4 bonds per molecule in the liquid (§ 5.2), complete severance costs 3.4 × 20 ≈ 68 kJ mol^{-1} — of which only 44 is paid, the balance being recovered as the vapour's molecules relax and as the residual attractions (dispersion, dipole) persist in both phases. The arithmetic is the reader's first thermodynamic audit of the network: the latent heat is the network's price list, discounted.*

Research-level propositions

Proposition 5.α. The two-state intuition behind § 5.3 (ice-like clusters versus broken-bond fluid) has been repeatedly challenged by spectroscopy and simulation. Assemble the current evidence for and against a liquid–liquid critical point in supercooled water, and design the decisive experiment — bearing in mind that nucleation forbids the region of interest to every patient method yet devised. Proposition 5.β. Crowding (§ 5.11) and confinement are usually modelled separately; the nuclear pore and the proteasome chamber combine them. Formulate a unified free-energy model for a substrate protein entering a crowded cylindrical pore, and test it against single-molecule import rates (§ 54.10). Proposition 5.γ. Buried waters (Table 5.2) are routinely resolved in crystals but are invisible to most design algorithms. Quantify, from the PDB, the energetic contribution of cavity waters to ligand binding in a chosen enzyme family, and assess whether explicit-water docking (§ 55.8) improves affinity prediction over the standard scoring functions.

References (Chapter 5)

Ball, P. (2017). H_2O: A biography of water. Weidenfeld & Nicolson.

Eisenberg, D., & Kauzmann, W. (1969). The structure and properties of water. Oxford University Press.

Minton, A. P. (2006). How can biochemical reactions within cells differ from those in test tubes? Journal of Cell Science, 119(14), 2863–2869.

Nelson, D. L., Cox, M. M., & Hoskins, A. A. (2021). Lehninger principles of biochemistry (8th ed.). W. H. Freeman.

van 't Hoff, J. H. (1887). Die Rolle des osmotischen Druckes in der Analogie zwischen Lösungen und Gasen. Zeitschrift für physikalische Chemie, 1, 481–508.

CHAPTER 6

Acids, Bases and Buffers

CONTENTS OF THIS CHAPTER

In which the proton — smallest, busiest and least governable of solutes — is brought under law: the dissociation constant defined, pH derived rather than decreed, the Henderson–Hasselbalch relation earned in four lines, the titration curve read as the autobiography of a weak acid, and the blood's three buffer systems shown to hold pH 7.40 within a hair's breadth, at the price of a kidney and a lung working full time.

6.1 Proton transfer

Of all the transactions in this book, none is so frequent, so fast, or so consequential as the passing of the proton. It is the simplest conceivable chemical event — a bare nucleus of hydrogen, no electrons, no baggage, moving from one lone pair to another — and it is also, by the same poverty, unique: alone among ions, the proton in water does not travel by diffusion of itself but by relay, passed along the hydrogen-bonded network of Chapter 5 like a bucket along a fire brigade. The mechanism bears the name of Grotthuss, who imagined it in 1806 [16]*, and its consequence is that the proton's mobility in water is some five times that of any other cation — which is to say, acid–base chemistry happens at the speed of the network, and every reaction in Parts IV and VII that involves a proton transfer (most of them do) inherits the speed. The chapter that follows is the law of that traffic.*

6.2 Brønsted–Lowry acids and bases

Definitions in chemistry have a history of being outgrown, and the acid is the senior example. Arrhenius (Plate 6.1) defined the acid as what releases H^+ in water — a definition that founded physical chemistry and earned the 1903 Nobel Prize, and that fails the moment the solvent is changed or the solute has no hydrogen to give. The definition that has lasted is the one framed independently in 1923 by Brønsted in Copenhagen and Lowry in Cambridge (Brønsted, 1923) (Lowry, 1923): an acid is a proton donor; a base, a proton acceptor. The definition's virtue is that it is relational — nothing is an acid in itself, only in the presence of something that will take its proton — and that every acid thereby generates its

16. C. J. T. de Grotthuss proposed his "chain" theory of electrolytic conduction in 1806, at the age of twenty-one, in a memoir published in Rome; the structural details — the hydrated complexes $H_9O_4^+$ (Eigen) and $H_5O_2^+$ (Zundel) between which the excess proton oscillates — were worked out a century and a half later by Eigen and by Zundel, and are still being refined by spectroscopists who find the proton, like most celebrated minimalists, more complicated than his publicity.

conjugate base, and every base its conjugate acid, in pairs locked together by the proton between them:

$$\mathrm{HA} \rightleftharpoons \mathrm{H^+} + \mathrm{A^-} \qquad \text{(acid / conjugate base)} \tag{6.1}$$

— acetic acid and acetate, ammonium and ammonia, the protonated and free imidazole of histidine. The conjugate pairing carries a theorem the reader will use constantly: the stronger the acid, the weaker its conjugate base, and the product of their strengths is a constant of the solvent — a theorem we shall prove as Eq. (6.5) once the equilibrium apparatus of Chapter 4 is brought to bear. Lewis's still broader definition (the acid as electron-pair acceptor, of § 2.4's dative bond) has its jurisdiction — metal ions are Lewis acids — but in water, where this book lives, Brønsted and Lowry reign.

6.3 Acid dissociation constants

The strength of an acid is measured, like everything in equilibrium chemistry, by its equilibrium constant. For the dissociation of Eq. (6.1),

$$K_a = \frac{a_{\mathrm{H^+}}\, a_{\mathrm{A^-}}}{a_{\mathrm{HA}}} \tag{6.2}$$

— with the convention, inherited from § 4.7, that activities replace concentrations, water's own activity being unity. The constant spans an extravagant range: hydrochloric acid's K_a is effectively infinite (in water it dissociates completely — the "levelling" of strong acids by the solvent, which will accept no acid stronger than H_3O^+), while ammonium's is 5.6×10^{-10}, and the two ends of the scale govern different worlds: strong acids are reagents, weak acids are buffers. Between them stands the weak acid par excellence of this book, acetic, at $K_a = 1.74 \times 10^{-5}$. From Eq. (4.6), each K_a corresponds to a standard free energy of dissociation, $\Delta G° = -RT \ln K_a$, so that acid strength is, thermodynamically, nothing but the free-energy price of the proton's departure — the ledger again, kept in a new currency.

6.4 The derivation of pH

Now the derivation this section's title promises, for pH is quoted hourly and derived almost never. Water is not innocent of its own chemistry: it dissociates, slightly but consequentially, into the hydronium and hydroxide ions,

(6.3)

— the ion product K_w being the equilibrium constant of autoprotolysis, with the 55.5 M concentration of water itself absorbed into the constant (the dilute-solution bridge of § 4.2). In pure water the two ions are made in equal numbers, so each has activity $\sqrt{K_w} = 10^{-7}$; and it is this number — awkward in speech, fatal in arithmetic — that Sørensen of the Carlsberg Laboratory compressed in 1909 into the notation of § 4.8 (Sørensen, 1909):

$$\mathrm{pH} \equiv -\log_{10} a_{\mathrm{H_3O^+}} \tag{6.4}$$

— whence pure water sits at pH 7.00, and each unit of pH is a decade of proton activity. Three consequences fall out at once. First, the conjugate theorem of § 6.2 is now provable: for an acid HA with constant K_a and its conjugate base A^- acting as *a base ($A^- + H_2O \rightleftharpoons HA + OH^-$, constant K_b), multiplying the two constants gives $K_aK_b = K_w$, whence*

$$pK_a + pK_b \;=\; pK_w \;=\; 14.00 \tag{6.5}$$

— the strong-acid/weak-base see-saw, derived. Second, the neutrality of pure water at pH 7 is a property of 25 °C only: K_W rises with temperature (autoprotolysis is endothermic, ≈ 56 kJ mol^{-1}), so that at 37 °C pK_W ≈ 13.6 and "neutral" is pH 6.8 — the reader's blood at pH 7.4 is therefore less acidic than true neutrality, a nicety the physiologists honour and the textbooks usually bury. Third, the entire acidity of the body — Figure 6.6's geography from stomach to pancreas — is one logarithmic ruler, and clinical acid–base medicine is the art of reading it.

6.5 The Henderson–Hasselbalch relation

The central working equation of the chapter is derived in four lines and used in four thousand laboratories. Take logarithms of Eq. (6.2): log K_a = log a_{H+} + log([A^-]/[HA]); multiply by −1 and apply Eq. (6.4) and the definition $pK_a = -\log K_a$: pK_a = pH − log([A^-]/[HA]); rearrange:

$$\mathrm{pH} \;=\; pK_a + \log_{10}\frac{[\mathrm{A}^-]}{[\mathrm{HA}]} \tag{6.6}$$

— Henderson's equation of 1908 for the bicarbonate system (Henderson, 1908), put into logarithmic form by Hasselbalch in 1916 (Hasselbalch, 1917), and quoted in their joint names ever since — one of science's more durable double-barrelled attributions, though the two men never wrote together. The equation says something profoundly simple, which Figure 6.1 displays: the pH fixes the ratio, and the pK_a only positions it. At pH = pK_a the acid is half dissociated; one pH unit above, the base outnumbers the acid tenfold; two units, a hundredfold. The physiological stakes are immediate: histidine's imidazole (pK_a ≈ 6.0) is the only protein side chain that titrates across the physiological window, and it is therefore the side chain that enzymes recruit for proton shuttles (the catalytic triads of § 19.11) and that haemoglobin recruits for the Bohr effect of § 13.5 — a single number, 6.0, echoing through half the chapters of this book.

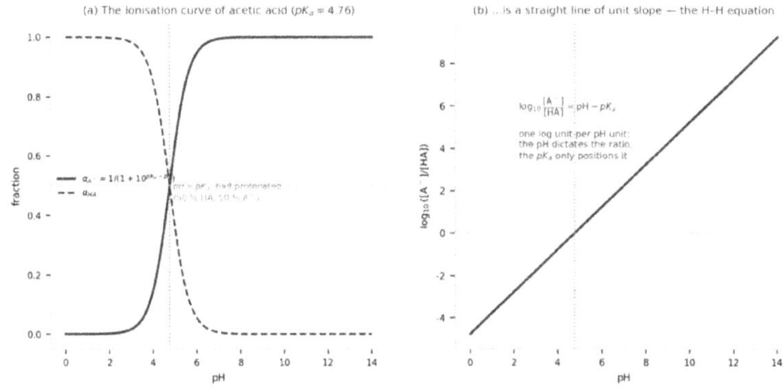

Figure 6.1 (a) The ionisation curve of acetic acid: the fraction deprotonated as a function of pH, from Eq. (6.6), with the midpoint pH = pK_a = 4.76 marked. (b) The same relation plotted as the logarithm of the base/acid ratio — a straight line of unit slope, which is the Henderson–Hasselbalch equation drawn rather than quoted: pH dictates the ratio, and the pK_a merely positions the line.

6.6 Titration curves

Add strong base to a weak acid, equivalent by equivalent, and the pH traces the curve of Figure 6.2 — the autobiography of the acid, told in its own protons. The curve repays slow reading. It begins low (the undissociated acid, its own pH set by Eq. 6.2), climbs gently through the middle — the region where added base merely converts HA to A^- while Eq. (6.6) holds the pH near pK_a — turns at the half-equivalence point (pH = pK_a exactly, the one direct experimental measure of the constant), rises steeply at equivalence (the acid exhausted; a whisker of excess base now rules unopposed), and flattens finally into the pH of the titrant itself. Two readings of the curve are the reader's permanent equipment. The midpoint flattening is the buffer: the flatter the curve, the more base is absorbed per pH unit — and the steepest descent of flatness, at the midpoint, is where the buffer works hardest. And the whole curve is nothing but Eq. (6.6) integrated: the titration is the Henderson–Hasselbalch relation performed, equivalent by equivalent, upon the bench. It is worth a paragraph of dialectic that the curve's elegance has seduced generations of students into reading K_a values off sloppy data at the equivalence point — where the pH is governed by the conjugate base's hydrolysis and the measurement is at its noisiest — when the midpoint, flat and forgiving, offers the constant for free. Experimental design, like good breeding, consists of standing where the ground is firm.

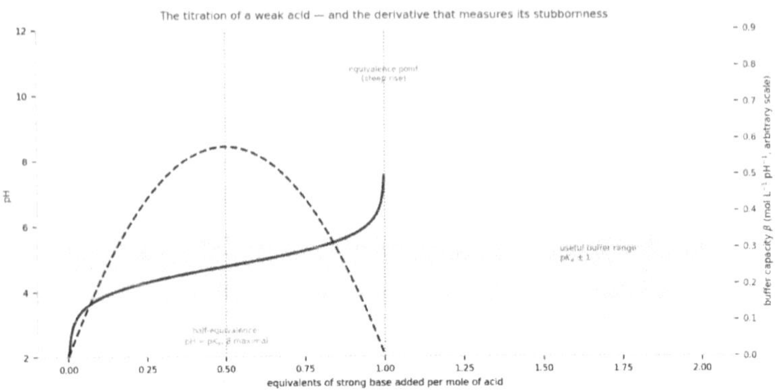

***Figure 6.2** The titration of a weak monoprotic acid with strong base (navy): pH against equivalents added, computed from Eq. (6.6) with the hydrolysis of the conjugate base included beyond equivalence. Marked are the half-equivalence point (pH = pK_a), the equivalence point, and the useful buffer range pK_a ± 1 (shaded). The dashed oxblood curve is the buffer capacity β of § 6.9 — maximal at the midpoint, where the curve is flattest: stubbornness measured as the derivative of resistance.*

6.7 Polyprotic acids

Acids with several protons dissociate in steps, and the steps are almost always widely separated — each successive proton departs from a molecule already one charge the poorer and one Coulomb the more reluctant — so that the polyprotic acid is, to an excellent approximation, a staircase of independent monoprotic acids. Phosphoric acid, the biochemist's own example (it is the parent of every phosphate in Parts VI and VII), titrates at pK_{a1} = 2.15, pK_{a2} = 7.20, pK_{a3} = 12.35 (Figure 6.3): the first proton strong, the third impossibly weak, and the second — governing the $H_2PO_4^-/HPO_4^{2-}$ pair — sitting precisely on the physiological window. This is the second providential coincidence of the chapter (histidine's was the first): the phosphate pair titrates exactly where the cell lives, and the cell accordingly uses phosphate as its intracellular buffer of first resort — some 75 mmol L^{-1} of it in the cytosol, against the blood's 1. The speciation curves of the figure are computed from the stepwise constants by the fraction algebra of Chapter 4 (each species's share being its term in the denominator polynomial), and their crossings are the pK_a values themselves — the staircase, read off at a glance.

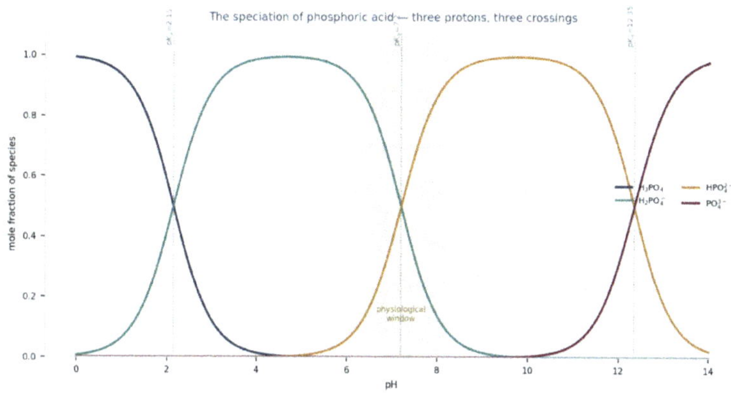

Figure 6.3 *The speciation of phosphoric acid: the mole fractions of H_3PO_4, $H_2PO_4^-$, HPO_4^{2-} and PO_4^{3-} as functions of pH, computed from the three stepwise constants. The shaded band is the physiological window: there the $H_2PO_4^-/HPO_4^{2-}$ pair (teal and amber) dominates and titrates — the chemical reason phosphate is the cell's buffer of first resort, and the reader's own cytosol a 75 mM solution of it.*

6.8 Ionisation of amino acids and proteins

The amino acids are polyprotic acids in uniform, and their titration is the polyprotic staircase wearing side chains. Every α-amino acid carries at least two titrating groups — the α-carboxyl ($pK_a \approx 2$) and the α-ammonium ($pK_a \approx 9$–10) — and the acidic and basic side chains of Table iv add their own: aspartate's and glutamate's carboxyls near 4, histidine's imidazole at 6.0, cysteine's thiol at 8.3, tyrosine's phenol at 10.1, lysine's ammonium at 10.5, arginine's guanidinium at 12.5. The consequence, drawn for histidine in Figure 6.4, is that each amino acid has a characteristic speciation pattern and, at one particular pH, zero net charge — the isoelectric point,

$$\mathrm{p}I \;=\; \tfrac{1}{2}\,(\mathrm{p}K_{a,\,acidic} + \mathrm{p}K_{a,\,basic}) \qquad (6.7)$$

— the mean of the two constants bracketing the neutral species, derived by setting the charges equal and applying Eq. (6.6) to both; for histidine, pI = 7.6. Proteins inherit the sum of all this: their net charge is the signed sum of hundreds of titrating groups, their isoelectric points scatter from 4.5 (pepsin) to 11 (lysozyme), and the whole technology of protein separation — isoelectric focusing, ion-exchange chromatography, the electrophoresis of Chapter 53 — is the industrial exploitation of Figure 6.4b. One physiological corollary deserves its sentence: at pH 7.4 the acidic side chains are anionic and the basic ones cationic, so every protein is a polyelectrolyte bristling with both signs — the surface on which the salt bridges of § 2.4 are negotiated, and the reason a shift of 0.2 pH units, changing a handful of charges, can retune a protein's associations entirely.

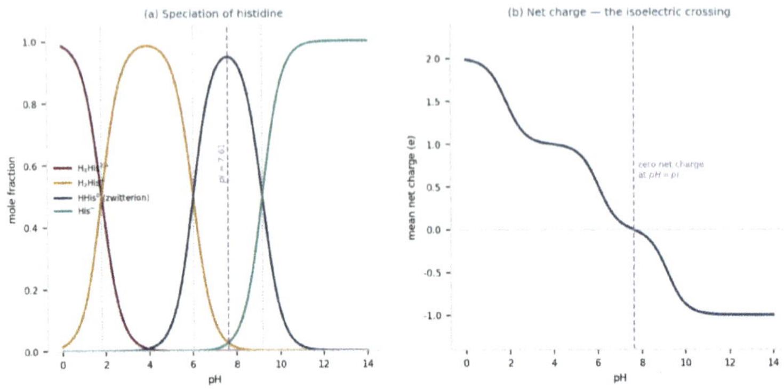

Figure 6.4 *(a) The speciation of histidine, with its three constants (a-COOH 1.82, imidazole 6.04, a-NH_3^+ 9.17): note that the zwitterionic $HHis^0$ dominates the entire physiological range, and that the imidazole's titration (amber → navy crossing at 6.04) is the only side-chain event available to proteins at physiological pH. (b) The mean net charge, crossing zero at the isoelectric point pI = 7.6 (Eq. 6.7) — the pH at which the amino acid is least soluble and the electrophoretic mobility vanishes.*

6.9 Buffer capacity

The flatness of the titration curve has a measure, and the measure has a law. Define the buffer capacity β as the strong base required to raise the pH of one litre by one unit, $\beta = dc_b/dpH$; differentiating the Henderson–Hasselbalch titration function (the algebra is Exercise 6.2's honest work) gives

$$\beta = 2.303\left(\frac{K_a\,[\mathrm{H}^+]\,C}{(K_a + [\mathrm{H}^+])^2} + [\mathrm{H}^+] + [\mathrm{OH}^-]\right) \qquad (6.8)$$

— the first term being the buffer pair's contribution (maximal, equal to 2.303·C/4 ≈ 0.58C, at $[H^+] = K_a$, i.e. pH = pK_a, where C is the total buffer concentration), the last two the water's own, which rule the extremes. Two practical laws follow. The buffer works within pK_a ± 1 (Figure 6.2's shaded band): outside that decade the pair is spent. And the capacity is proportional to the concentration: a buffer is a reservoir of protons, and its size is its virtue. Hence the design rule of every laboratory buffer — choose a pK_a within one unit of the target pH, and spend concentration freely — and hence, too, the cell's design: phosphate (75 mM, pK_a 7.2) and protein histidines (pK_a ≈ 6–7) inside, bicarbonate outside, each positioned by providence or selection exactly where the work is.

6.10 Physiological buffer systems

The body buffers with three systems, and the three divide the work by geography and by leverage (Table 6.1). In the cytosol, the phosphate pair and the proteins reign, positioned at pK_a 7.2 and ≈ 6–7 respectively — chemical coincidences exploited by evolution with its usual thrift. In the blood, the proteins — above all haemoglobin, whose 38 histidines make it six times the buffer of the plasma proteins — carry the largest fixed load. But the sovereign system, because it alone is open, is bicarbonate: its acid partner, CO_2, is not a fixed pool but a gas, vented by the lung and produced by every respiring cell, so that the ratio of Eq. (6.6) is set *not by chemistry alone but by two organs — the lung fixing the numerator's CO_2, the kidney the denominator's HCO_3^- — and the blood's pH becomes, uniquely, a physiological set point rather than a chemical fate. The next section derives its law.*

Table 6.1 *The physiological buffer systems, ranked by the work they do. "Open" systems can excrete their acid partner; closed systems can only redistribute it.*

System	*Pair*	*pK_a*	*Domain*	*Open?*
bicarbonate	*CO_2/HCO_3^-*	*6.1*	*blood, extracellular fluid*	*yes — lung & kidney*
haemoglobin	*His imidazoles (38/tetramer)*	*≈ 6–7*	*erythrocytes*	*closed*
proteins	*His, Cys, termini*	*≈ 6–8*	*cytosol, plasma*	*closed*
phosphate	*$H_2PO_4^-/HPO_4^{2-}$*	*7.2*	*cytosol, urine*	*partly — kidney*

6.11 Carbon dioxide–bicarbonate equilibrium

The chemistry is a chain of three equilibria: dissolved CO_2 hydrates to carbonic acid (slow alone, fast under the carbonic anhydrase of § 19.12), the acid dissociates (pK_a of H_2CO_3 proper ≈ 3.6), and the gas obeys Henry's law, $[CO_2] = \alpha P_{CO2}$ with α = 0.0306 mmol L^{-1} $mmHg^{-1}$. Because H_2CO_3 is only a few parts per thousand of the dissolved CO_2, the two are lumped experimentally, and the chain collapses into a single apparent constant, pK_a' = 6.1, in the modified Henderson–Hasselbalch form

$$\mathrm{pH} = 6.1 + \log_{10}\frac{[\mathrm{HCO_3^-}]}{\alpha\, P_{\mathrm{CO_2}}} \tag{6.9}$$

— the most clinically used equation in this book. With the kidney holding $[HCO_3^-]$ ≈ 24 mmol L^{-1} and the lung holding P_{CO2} ≈ 40 mmHg (αP_{CO2} = 1.2 mmol L^{-1}), pH = 6.1 + log(24/1.2) = 6.1 + 1.30 = ***7.40*** *— the famous number, derived rather than remembered. The Davenport diagram (Figure 6.5) is Eq. (6.9) drawn as a map: each P_{CO2} is a curve, each ventilation or renal act is a journey upon it, and the four classical disturbances — respiratory and metabolic, acidosis and alkalosis — are the four directions of travel from the normal point. The diagram's whole doctrine can be stated in one sentence: the lung defends the denominator in minutes, the kidney the numerator in days, and the buffers of Table 6.1 hold the fort between them.*

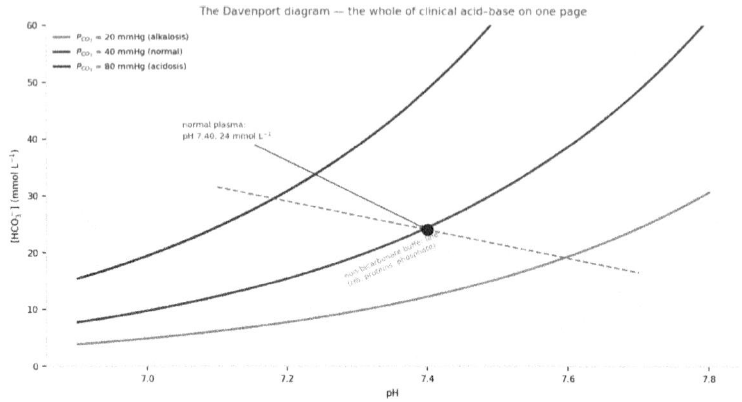

Figure 6.5 The Davenport diagram: plasma bicarbonate against pH, with the CO_2 isopleths of Eq. (6.9) for P_{CO2} = 20, 40 and 80 mmHg, and the non-bicarbonate buffer line (dashed) along which the plasma moves when P_{CO2} changes acutely. The normal point (7.40, 24) and the four directions of derangement — respiratory acidosis and alkalosis along the buffer line, metabolic acidosis and alkalosis along a single isopleth — make the diagram the map from which every blood-gas report in every hospital is read.

6.12 Intracellular pH regulation

The cytosol, pH ≈ 7.2, is defended by more than chemistry. The membrane-bound transporters of Part V — Na^+/H^+ exchangers, HCO_3^-/Cl^- exchangers, the lactate–H^+ cotransporter of muscle — export acid continuously, powered by the sodium gradient (secondary active transport, § 24.7); metabolism itself produces acid by the tonne (the daily CO_2 output of Worked Example 6.3 corresponds to some 15 moles of potential acid); and the cell's enzymes, from the phosphofructokinase of § 31.7 to the transporters themselves, are pH-sensitive, so that the system regulates itself by feedback — a fall in cytosolic pH throttles glycolysis, the acid's own source. The dialectic of the section is worth a paragraph because it is the physiological form of the chapter's whole argument: the passive buffers (phosphate, histidines) are fast but finite; the transporters are slower but inexhaustible while the sodium gradient lasts; and the gradient, in turn, is maintained by the Na^+/K^+-ATPase at the price of ATP — so that, in the end, pH homeostasis is paid for in the same currency as everything else in the body, and acid–base balance is a department of energetics. The patient's kidney and the chemist's buffer bottle obey one law; only their budgets differ.

6.13 Acid–base disturbances

The clinical fourfold follows from Eq. (6.9) by inspection, and the map of Figure 6.5 is its chart. Respiratory acidosis (hypoventilation — the numerator rises) and respiratory alkalosis (hyperventilation — it falls) are journeys along the non-bicarbonate buffer line, compensated over days by the kidney adjusting [HCO_3^-]. Metabolic acidosis (acid gained or bicarbonate lost — diabetic ketoacidosis, renal failure, diarrhoea) and metabolic alkalosis (acid lost, as in protracted vomiting) are journeys along a single isopleth, compensated over minutes by ventilation *adjusting P_{CO2}. The compensations have laws of their own — the respiratory answer to metabolic acidosis follows Winter's formula, $P_{CO2} \approx 1.5[HCO_3^-] + 8$ — and the clinical art is to read the blood gas, locate the point on the Davenport map, and ask which organ is culpable and which is merely loyal. The chapter thus closes where the treatise's argument demands: the proton, governed by the chemistry of §§ 6.1–6.9, becomes the patient's pH, governed by the same chemistry plus two organs; and the number 7.40, which opened the chapter as a fact, now stands derived, buffered, defended — and, like everything else in this book, earned.*

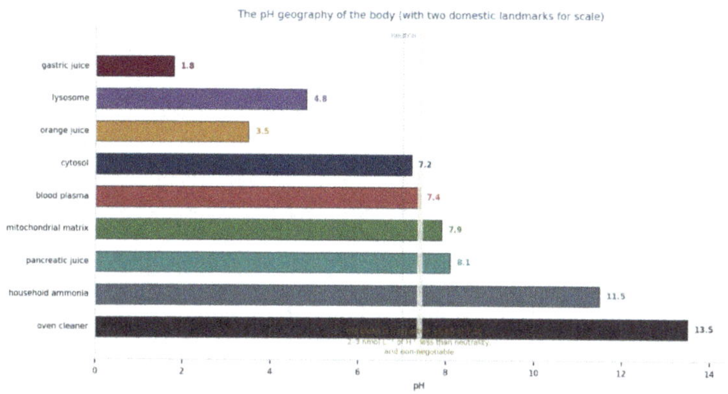

Figure 6.6 *The pH geography of the body, with two domestic landmarks for scale: from the gastric juice (pH ≈ 1.8, the proteolytic furnace of § 18.8) through the lysosome (4.8), the cytosol (7.2), the blood (7.40 ± 0.05, shaded), the mitochondrial matrix (≈ 7.9, alkaline because the pumps of Chapter 27 export protons from it), to the pancreatic juice (8.1) that neutralises the stomach's outflow. Every value is defended by machinery; none is a coincidence.*

Plate 6.1 *Svante Arrhenius (1859–1927), whose theory of electrolytic dissociation (1887) — acids, bases and salts as ready-made ions rather than molecules awaiting dissociation — was nearly failed as a doctoral thesis and won the third Nobel Prize in Chemistry (1903). Portrait from the Nobel Foundation archive, nobelprize.org, reproduced from the Foundation's publicly distributed image.*

Historical note — Sørensen, beer, and the invention of pH

The pH scale was born in a brewery. S. P. L. Sørensen, director of the chemical department of the Carlsberg Laboratory in Copenhagen — itself founded by the brewer J. C. Jacobsen to put brewing on a scientific footing — needed a tractable notation for hydrogen-ion concentration because the fermentation of wort and the action of enzymes depended on it exquisitely; his monumental 1909 paper (Sørensen, 1909) introduced both the pH notation and the first standard buffers, and demonstrated the dependence of enzyme activity on pH that Part IV of this book takes for granted. The logarithmic form of the dissociation curve arrived with Hasselbalch, a physician studying the blood's CO_2, in 1916 (1917); and the union of the two — notation and equation — is the daily bread of every laboratory on Earth. That the whole apparatus of acid–base medicine descends from a Danish brewery is a fact the record with undiluted pleasure.

Chapter 6 — Conceptual synthesis

The proton has been brought under law. It travels by relay along the network of Chapter 5 (§ 6.1); acids and bases are donor and acceptor, locked in conjugate pairs whose strengths obey $pK_a + pK_b = 14$ (§ 6.2, Eq. 6.5); strength is an equilibrium constant (§ 6.3); pH is derived from the autoprotolysis of water, with neutrality itself a temperature-dependent convention (§ 6.4); the Henderson–Hasselbalch relation follows in four lines and rules the ratio of every buffer pair (§ 6.5); the titration curve is that relation performed on the bench, its midpoint giving both the constant and the maximal buffer capacity (§§ 6.6, 6.9); polyprotic acids titrate as staircases, and phosphate's second step lands exactly on the physiological window (§ 6.7); amino acids and proteins inherit the staircase, with isoelectric points at the mean of bracketing constants (§ 6.8); the body's three buffers divide the work, with bicarbonate sovereign because its acid is a gas two organs can govern (§§ 6.10–6.11); the cell defends its own pH with transporters paid in ATP (§ 6.12); and the clinical fourfold of disturbances is the Davenport map read by organ (§ 6.13). Direction checked against Figure ii: the solvent and its solute are law. Chapter 7 now lays the deepest foundations of all — the thermodynamics on which every equation so far has quietly stood.

WORKED EXAMPLE 6.1 — DESIGNING A BUFFER FOR pH 7.4

Requirement: 50 mmol L^{-1} total buffer, maximal capacity at pH 7.4. Candidates: phosphate ($pK_{a2} = 7.20$), HEPES ($pK_a = 7.55$), Tris ($pK_a = 8.1$). Rule of § 6.9: pK_a within ±1 of target, closer the better — HEPES wins (0.15 away; phosphate 0.20; Tris 0.70). The ratio from Eq. (6.6): $[A^-]/[HA] = 10^{7.4-7.55} = 0.71$, so of the 50 mmol L^{-1}, 20.8 is the basic form and 29.2 the acid; the capacity at that ratio is $\approx 0.58 \times 50 \times (4 \times 0.71/1.71^2) \approx 0.58 \times 50 \times 0.97 \approx 28$ mmol L^{-1} per pH unit — to resist a 0.1 unit drift, the buffer absorbs 2.8 mmol of acid per litre, computed before a single flask is touched.

WORKED EXAMPLE 6.2 — THE DAILY ACID LOAD

A resting adult produces ≈ 15 mol of CO_2 daily (from Figure 4.7's metabolic rate: 8 400 kJ day^{-1} at ≈ 590 kJ per mole of glucose-equivalent... more directly, the respiratory quotient and calorimetry give 200–400 mL CO_2 min^{-1}, ≈ 300 mL min^{-1} ≈ 13 mmol min^{-1} × 1 440 ≈ 19 mol day^{-1} — call it 15–20). If the blood's 42 mmol of bicarbonate (1.4 L plasma + equilibrated volume ≈ 14 L extracellular fluid: 24 mmol L^{-1} × 14 L ≈ 340 mmol) had to absorb this alone, *with no ventilation, the ratio of Eq. (6.9) would collapse within minutes. The lung instead excretes the whole load continuously — 15 mol day^{-1} of volatile acid, the largest acid export of the body by three orders of magnitude over the kidney's fixed-acid 70 mmol. Acid–base balance is, first of all, breathing.*

WORKED EXAMPLE 6.3 — A BLOOD GAS, READ

Report: pH 7.22, P_{CO2} 60 mmHg, $[HCO_3^-]$ 24 mmol L^{-1}. Locate on Figure 6.5: bicarbonate normal, pH low, P_{CO2} high — respiratory acidosis, uncompensated (the kidney has not yet raised the numerator). Check Eq. (6.9): $6.1 + \log(24/(0.0306 \times 60)) = 6.1 + \log(13.1) = 6.1 + 1.12 = 7.22$ ✓ — the numbers are internally consistent (a test every reported gas should pass before it is believed, § 4.12's audit applied to the patient). Diagnosis: the lung is the culprit; the kidney, loyal, will need 2–5 days. Winter's formula and its kin complete the bedside arithmetic, but the equation that runs the whole consultation is one line of § 6.11.

Exercises (with full resolutions)

EXERCISE 6.1

Compute the pH of 0.1 mol L^{-1} acetic acid ($pK_a = 4.76$), with and without the approximation $[H^+] \ll c$.

Resolution. *Let $x = [H^+]$: $K_a = x^2/(0.1 - x) = 1.74 \times 10^{-5}$. Approximate: $x \approx \sqrt{(1.74 \times 10^{-6})} = 1.32 \times 10^{-3}$, pH = 2.88. Exact quadratic: $x^2 + 1.74 \times 10^{-5}x - 1.74 \times 10^{-6} = 0$, $x = 1.31 \times 10^{-3}$, pH = 2.88 — the approximation costs 0.8 %, and the exercise teaches when it is safe: $x/c \approx 1.3\ \% \ll 5\ \%$, so the approximation passes its own audit (§ 4.12).*

EXERCISE 6.2

Derive Eq. (6.8) from the titration function of a weak acid buffer.

Resolution. *Charge balance with strong base at concentration c_b: $c_b = [A^-] + [OH^-] - [H^+] = CK_a/(K_a + [H^+]) + K_w/[H^+] - [H^+]$. Differentiate with respect to pH, using $d[H^+]/dpH = -2.303[H^+]$: the first term gives $2.303 \cdot CK_a[H^+]/(K_a + [H^+])^2$, the hydroxide term $2.303[OH^-]$, the last $2.303[H^+]$ — Eq. (6.8) exactly. At $[H^+] = K_a$ the first term is $2.303C/4$: $\beta_{max} \approx 0.58C$, as § 6.9 asserted.*

EXERCISE 6.3

Alanine ($pK_{a1} = 2.34$, $pK_{a2} = 9.69$): what fraction is zwitterionic at pH 7.0, and what is its isoelectric point?

Resolution. *At pH 7: carboxyl fully deprotonated (ratio $10^{7-2.34} \approx 5 \times 10^4$), ammonium fully protonated (ratio $10^{9.69-7} \approx 490$): the zwitterion fraction is $\approx 1/(1 + 10^{-4.66} + 10^{-2.69}) \approx 99.8$ %. pI = (2.34 + 9.69)/2 = 6.02 by Eq. (6.7) — near neutrality, as befits a side-chain-inert amino acid, and the reason alanine's solubility minimum sits mid-range rather than in the physiological window.*

EXERCISE 6.4 (INTERPRETATIVE)

The pH of pure water at 37 °C is 6.8, yet the blood at pH 7.4 is called "slightly alkaline". A student objects: "7.4 > 7, so blood is alkaline — nothing slight about it." Adjudicate.

Resolution (indicative). *Both are right, and the dispute is a confusion of rulers. Against the 25 °C convention (neutral = 7.0), blood is alkaline by 0.4 units. Against the body's own temperature (neutral = 6.8), it is alkaline by 0.6 — the student's "slight" is actually more extreme, not less. The adjudication: pH is an activity on a logarithmic ruler whose zero is set by $K_w(T)$, and comparisons are meaningless unless the temperature of the ruler is stated — § 6.4's second consequence, returned as a clinical pedantry that turns out to matter: "alkalosis" is defined against 7.40, not against 7.00.*

EXERCISE 6.5

How many millimoles of HCl must be added to one litre of blood (buffers: bicarbonate 24 mmol L^{-1}, other buffers ≈ 24 mmol L^{-1} per pH unit near 7.4) to lower its pH from 7.40 to 7.20, if ventilation holds P_{CO2} at 40 mmHg?

Resolution. *Bicarbonate term: from Eq. (6.9), $[HCO_3^-]$ at pH 7.20 = 0.0306 × 40 × $10^{7.20-6.1}$ = 1.224 × 12.6 = 15.4 mmol L^{-1}; consumed: 24 − 15.4 = 8.6 mmol. Other buffers: 24 × 0.20 = 4.8 mmol. Total ≈ 13.4 mmol of HCl per litre of blood equivalent — the entire plasma bicarbonate being only a first line, and the number that explains why the kidney's 70 mmol day^{-1} of fixed-acid export (Worked Example 6.2) is a day's work, not a moment's.*

Research-level propositions

Proposition 6.α. Stewart's "strong ion difference" formalism (1983) derives pH from charge balance and equilibria without the Henderson–Hasselbalch ratio, and its partisans call the classical account circular. Formalise both frameworks, prove their mathematical equivalence for the plasma system, and adjudicate which is the better teaching model for the disturbances of § 6.13. Proposition 6.β. The Grotthuss relay of § 6.1 has been imaged in ice and inferred in water, but the mechanism inside a proton-conducting channel (gramicidin; the M2 channel of influenza) differs by confinement. Compute, from the structures of Chapter 54, the free-energy profile of a proton wire in a 4-Å pore, and predict the conductance for comparison with single-channel electrophysiology (§ 54.10). Proposition 6.γ. Histidine's pK_a in folded proteins varies from 3 to above 8 depending on microenvironment. Mine the PDB and the pK_a-prediction literature (§ 55.9) for the determinants of this spread, and estimate how much of it is explained by burial, by salt bridging, and by hydrogen bonding — the chapter's constant becoming, in the protein, a variable.

References (Chapter 6)

Brønsted, J. N. (1923). Einige Bemerkungen über den Begriff der Säuren und Basen. Recueil des Travaux Chimiques des Pays-Bas, 42(8), 718–728.

Davenport, H. W. (1974). The ABC of acid–base chemistry (6th ed.). University of Chicago Press.

Hasselbalch, K. A. (1917). Die Berechnung der Wasserstoffzahl des Blutes aus der freien und gebundenen Kohlensäure desselben, und die Sauerstoffbindung des Blutes als

Funktion der Wasserstoffzahl. Biochemische Zeitschrift, 78, 112–144.

Henderson, L. J. (1908). Concerning the relationship between the strength of acids and their capacity to preserve neutrality. American Journal of Physiology, 21(2), 173–179.

Lowry, T. M. (1923). The uniqueness of hydrogen. Journal of the Society of Chemical Industry, 42(3), 43–47.

Nelson, D. L., Cox, M. M., & Hoskins, A. A. (2021). Lehninger principles of biochemistry (8th ed.). W. H. Freeman.

Sørensen, S. P. L. (1909). Études enzymatiques. II. Sur la mesure et l'importance de la concentration des ions hydrogène dans les réactions enzymatiques. Comptes Rendus des Travaux du Laboratoire Carlsberg, 8, 1–168.

CHAPTER 7

Thermodynamics of Biochemical Systems

CONTENTS OF THIS CHAPTER

In which the laws beneath all the laws are stated, derived, and turned upon the living cell: the conservation of energy, the arrow of entropy, the free energy that prices every reaction, the standard states that make the prices comparable, and the open-system truth that a cell is not a vessel approaching equilibrium but a flame held from it — order bought, per the second law itself, by the continuous export of disorder.

7.1 Systems, surroundings and state variables

Thermodynamics is the most honest of the sciences: it promises little — no mechanisms, no rates, no molecular gossip — and keeps everything it promises. Its jurisdiction is bounded by a single methodological act, the drawing of a boundary: the system within, the surroundings without, and the universe as the uninteresting sum of the two. Three species of system divide the zoo: the isolated, exchanging neither matter nor energy (a thermos, idealised); the closed, exchanging energy only (a sealed flask in a water bath — the biochemist's usual bench); and the open, exchanging both (the cell, the organism, the reader). The distinction is not pedantry, for the second law speaks most plainly to the isolated case, and nearly every error made about life and entropy consists of applying its isolated-system verdict to a system that is nothing of the kind — § 7.12 will collect the debt.

Within the system, thermodynamics recognises two kinds of quantity, and the difference is the whole machinery of the subject. State variables — energy, enthalpy, entropy, free energy, temperature, pressure — are properties of the state itself: their changes depend only on the endpoints, never on the road between them. Path functions — heat q and work w — are properties of the journey: the same endpoints, travelled differently, yield different heats and works (Figure 7.1). The consequence is a principle of accounting that the whole book exploits: one may always compute a state-variable change by the most convenient path, however fictitious, because nature honours every route equally. Every thermodynamic cycle of Parts VI–VIII — every roundabout calculation of a free energy by way of intermediates that never occur — is this principle at work. The reader should also mark the two currencies of quantity: extensive variables (energy, entropy, volume) scale with the size of the system; intensive ones (temperature, pressure, chemical potential) do not — and it is a deep fact, visible in retrospect, that equilibrium is the equality of the intensive variables across every boundary the system contains.

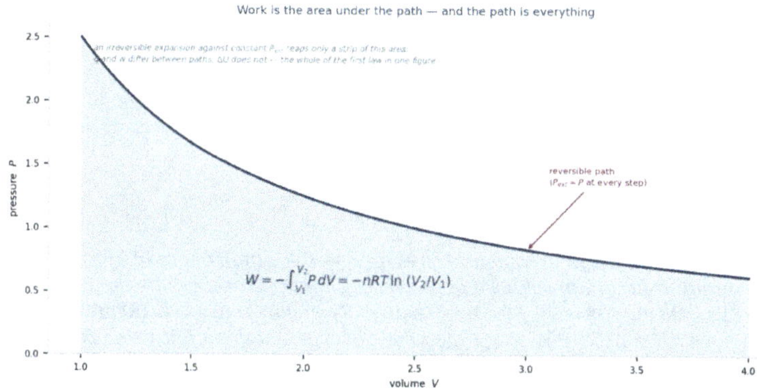

Figure 7.1 The pressure–volume plane of a gas expanding isothermally: the reversible path yields the maximal work, the shaded area under the curve (Eq. 7.3); any irreversible path yields less. The heat absorbed adjusts to match, since ΔU for the isothermal ideal gas is zero — the same journey's endpoints, different heats and works, identical change of state: path functions and state variables distinguished in one figure.

7.2 The first law of thermodynamics

The first law is conservation, stated as an audit. For any process, the change in the system's internal energy U — the sum of every kinetic and potential energy of every particle within — equals the heat absorbed plus the work done upon it:

$$\Delta U = q + w \tag{7.1}$$

— the sign convention being this treatise's (chemistry's): inward is positive, so that work done by the system carries a minus sign in w. The law's content is that U exists as a state function at all — that heat and work, though separately path-dependent, conspire to a sum that is not; and its history is one of the heroic measurements of science.[17] *For the biochemist the law's daily form is that metabolism is a bookkeeping problem: the energy of food appears, without remainder, as work done plus heat shed — a proposition whose experimental proof, Rubner's calorimetry of the dog (case study below), was the final eviction of vitalism from energetics, sixty years after Wöhler's urea had begun the eviction from substance.*

Work, for systems at pressure P changing volume by dV, is

$$w = -\int_{V_1}^{V_2} P_{\text{ext}}\, dV \tag{7.2}$$

— negative because expansion is work done by the system; and for the reversible isothermal expansion of an ideal gas (P_{ext} = nRT/V throughout), integration gives

17. James Prescott Joule's paddle-wheel experiments of the 1840s established the mechanical equivalent of heat — 4.19 J per calorie in his best determination, against the modern 4.184 — by the heroic exp f measuring the warming of water stirred by falling weights, reportedly (and the author choose to believe it) on his honeymoon at the Chute de Sallanches, with a thermomete wn devising. Robert Mayer had the idea first and the evidence last; Joule had the evidence; and the Royal Society, which rejected his first paper, has been apologising by hindsight ever since.

$$w_{\mathrm{rev}} = -nRT\ln\frac{V_2}{V_1} \tag{7.3}$$

— the area of Figure 7.1, and the maximal work any path can extract: reversibility, the infinitely patient negotiation of every step, is the thermodynamic ceiling, toward which the cell's best machines (the ATP synthase of Chapter 28 at ≈ 60 % of it) approach more nearly than any engine the reader owns.

7.3 Enthalpy and heat transfer

Biochemistry is conducted at constant pressure — the atmosphere obliges — and for that condition nature provides a bookkeeping convenience. Define the enthalpy $H = U + PV$; then at constant pressure, with only expansion work permitted, the first law gives $\Delta H = \Delta U + P\Delta V = q_P$: the heat absorbed at constant pressure is the enthalpy change,

$$\Delta H = q_P \qquad \text{(constant } P\text{, expansion work only)} \tag{7.4}$$

— so that the calorimeter reads enthalpies directly, and every ΔH in this book (the -2 870 kJ mol^{-1} of glucose combustion, the bond energies of Figure 2.3, the van 't Hoff slopes of Figure 7.4) is, at bottom, a measured heat. Two corollaries earn their keep. The temperature dependence of ΔH is set by heat capacities (Kirchhoff's law), which in biochemistry's narrow thermal range are nearly constant — ΔH is, to good approximation, a constant of the reaction. And because H is a state function, enthalpies add (Hess's law, 1840): the enthalpy of a reaction is the sum of the enthalpies of any convenient decomposition — the principle that built the combustion tables and that § 7.9 will extend to free energies, where the coupling of metabolism lives.

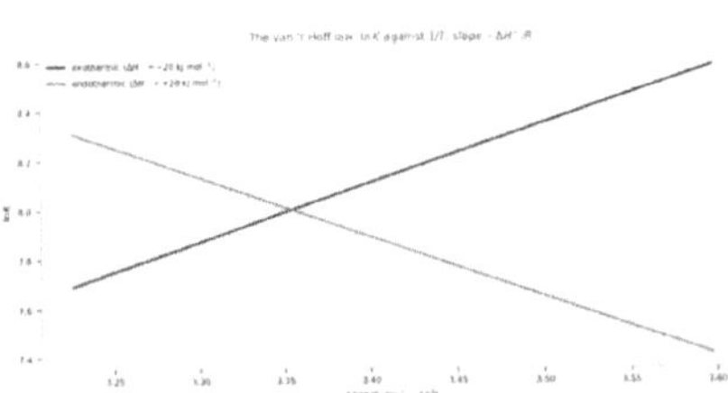

Figure 7.4 *The van 't Hoff relation (Eq. 7.9, derived in § 7.7): ln K against 1/T is a straight line whose slope is -ΔH°/R — endothermic reactions (teal) strengthen with temperature, exothermic (oxblood) weaken, and the enthalpy is read from the slope without a calorimeter. The linearity assumes ΔH° constant over the range (Kirchhoff, § 7.3); the biochemist's ranges are short, and the law serves.*

7.4 The second law and entropy

Every child knows the second law before knowing its name: cups cool, rooms disorder, eggs unscramble never. Its thermodynamic statement requires one new state function, the entropy S, defined by Boltzmann's immortal line — carved, at his own wish and with justice, upon his gravestone —

$$S = k_B \ln W \tag{7.5}$$

— W being the multiplicity of the state: the number of microscopic arrangements consistent with the macroscopic description. The second law then states that in an isolated system, spontaneous processes increase S;

$$\Delta S_{\text{isolated}} \geq 0 \tag{7.6}$$

— and Figure 7.2 shows why this is not a decree but an overwhelming probability: the macrostate of even distribution commands so vast a share of the microstates that any other is, for molar numbers, never observed. The law's authority over chemistry comes through a useful theorem: for a system at temperature T exchanging heat with surroundings much larger than itself, the surroundings' entropy change is $\Delta S_{surr} = -q/T = -\Delta H_{sys}/T$ at constant pressure — the heat the system sheds is the disorder the universe gains. Clausius, who named entropy in 1865, summarised the two laws in a couplet no treatise improves upon: the energy of the universe is constant; the entropy of the universe tends to a maximum.[18]

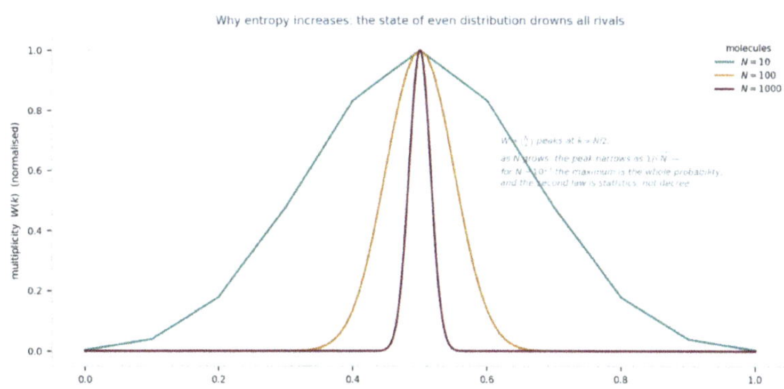

Figure 7.2 *The multiplicity of distributing N molecules between two halves of a container, for N = 10, 100 and 1 000: the binomial W = C(N, k) peaks ever more narrowly at even distribution as N grows. For molar N the peak's relative width is ~10^{-11}: the second law is the statement that physics does not observe miracles of that size — statistics, not decree.*

7.5 Gibbs free energy

The second law, as stated, judges the universe; the bench chemist needs a verdict on the system alone. The verdict exists. Define the Gibbs free energy G = H − TS;

18. Rudolf Clausius's coinage of "entropy" (from the Greek τροπή, transformation) was deliberately fashioned to resemble "energy", so that the two master-quantities should be remembered as siblings — one conserved, one not. The name has been mispronounced, misapplied and misunderstood with an entropy of its own ever since; the gravestone equation (7.5) is the antidote.

then at constant temperature and pressure the universe's entropy change decomposes, using $\Delta S_{surr} = -\Delta H/T$ of § 7.4, into

$$\Delta G = \Delta H - T\Delta S = -T\,\Delta S_{\text{universe}} \tag{7.7}$$

— the derivation being one line of substitution and the content being the whole of chemical thermodynamics: a negative ΔG certifies a spontaneous process at constant T and P, and the certificate needs no reference to anything outside the system. The sign conventions repay one careful reading (Figure 7.3): enthalpy-favoured and entropy-favoured processes ($\Delta H < 0$, $\Delta S > 0$) are spontaneous at every temperature; the opposed case never; and the two mixed cases trade verdicts at the crossing $T = \Delta H/\Delta S$ — the temperature-dependence that rules protein folding's cold and heat denaturation alike (Chapter 11). ΔG is the currency in which this whole book prices its reactions: it measures the maximal non-expansion work the process can perform (a theorem of the same family as Eq. 7.3's reversibility ceiling), which is why ΔG_{ATP} is simultaneously a thermodynamic quantity and the cell's unit of purchasing power — the two senses being, in the end, one.

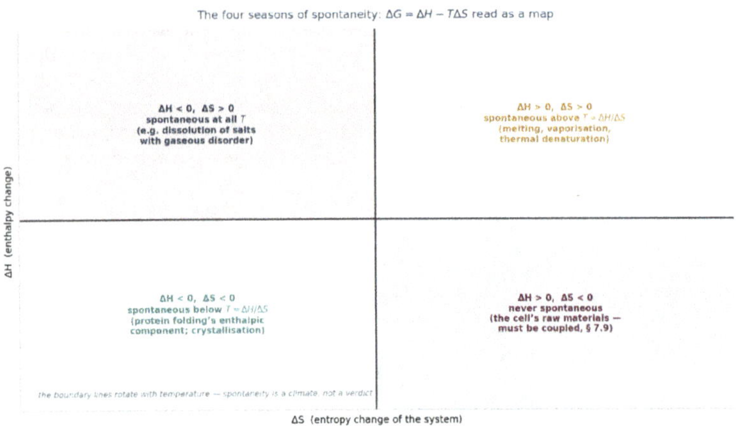

Figure 7.3 *The four seasons of spontaneity: the (ΔS, ΔH) plane of Eq. (7.7), with the verdict of each quadrant and a biochemical instance. Note the two temperature-dependent quadrants — the boundaries rotate with T, so that the same reaction can be spontaneous in winter and forbidden in summer, a fact on which cold-blooded biochemistry runs.*

7.6 Standard and biochemical standard states

Free energies of formation and reaction are reported against conventions, and biochemistry required a convention of its own, for the chemist's standard state — pH 0, every solute at unit activity — is a medium in which no enzyme would survive the minute. The definitions: the chemical standard state fixes each solute at $a = 1$ (1 mol L^{-1} ideal), gases at 1 bar, pure liquids and solids as themselves, temperature 298.15 K unless stated. The biochemical standard state, marked by the prime ($\Delta G°'$, K'), adds the physiological clause: the proton's activity is fixed at 10^{-7} (pH 7) and absorbed into the constant, water's activity is unity, and Mg^{2+} (that silent partner of every phosphate) is specified where relevant[19] *The*

19. The biochemical standard state was codified substantially through the efforts of Robert Alberty, whose "transformed" thermodynamics (1990s) taught the subject to treat hydrogen

conversion between the two conventions is pure Eq. (4.5): a reaction producing n protons has $\Delta G^{\circ\prime} = \Delta G^{\circ} - nRT \ln 10^{-7} = \Delta G^{\circ} + n \times 39.96$ kJ mol^{-1} at 25 °C — a correction that routinely reverses the sign of a published free energy, and whose omission has misled generations of students who trusted a table without reading its footnotes. The treatise's rule, stated once: every ΔG° in this book is primed unless explicitly naked.

7.7 Free energy and chemical equilibrium

The master relation of Chapter 4 now receives its certificate. Section 4.4 derived, from the composition-dependence of the chemical potential, that $\Delta G = \Delta G^{\circ} + RT \ln Q$ and $\Delta G^{\circ} = -RT \ln K$; the present chapter has derived the ΔG that enters those relations, and the circle closes: the equilibrium constant is the Boltzmann factor of the standard free energy,

$$K = e^{-\Delta G^{\circ}/RT} \tag{7.8}$$

— thermodynamics and statistics embracing at last: the equilibrium of a reaction is nothing but the competition between the products' energy advantage and the entropy of their dispersal, computed exactly. From Eq. (7.8), differentiating with respect to temperature and applying the Gibbs–Helmholtz relation ($\partial(\Delta G^{\circ}/T)/\partial T = -\Delta H^{\circ}/T^2$), one obtains the van 't Hoff law of Figure 7.4,

$$\frac{d \ln K}{dT} = \frac{\Delta H^{\circ}}{RT^2} \tag{7.9}$$

— the law that lets a spectrophotometer, plotting ln K against 1/T, measure enthalpies it cannot feel. The dialectic of the section is the standing temptation to read Eq. (7.8) as a permission slip: that a reaction with large K "wants" to happen fast. Thermodynamics knows nothing of wanting and nothing of fast; K is the address of the destination, and the road — the activation barrier, the enzyme — is kinetics' separate jurisdiction (Part IV). The confusion of the two is the single commonest conceptual error in the subject, and it has been given its quarantine here.

7.8 Concentration dependence of free energy

The standard free energy is a laboratory fiction; the cell's free energies are made in vivo, where the mass-action ratio is whatever metabolism holds it at. Eq. (4.5) is the conversion, and Figure 7.5 applies it to the reaction this book prices most often: the hydrolysis of ATP. At the standard composition the driving force is −30.5 kJ mol^{-1}; at the cellular composition ([ATP] ≈ 3 mM, [ADP] ≈ 0.3 mM, [P_i] ≈ 3 mM — $Q \approx 3 \times 10^{-4}$ M) it is ≈ −50 kJ mol^{-1}; at equilibrium it is zero, and the battery is flat. The physiological reading is immediate: the cell maintains its ATP ten thousandfold above equilibrium precisely to hold the driving force at its working value — the phosphorylation potential is a dam, and metabolism is the continuous pumping of water uphill against its leak. Every number of Part VI's energetic accounting is this one line, replayed.

and magnesium as conserved currencies rather than species — a bookkeeping reform of the first rank, unglamorous and indispensable, like most bookkeeping reforms.

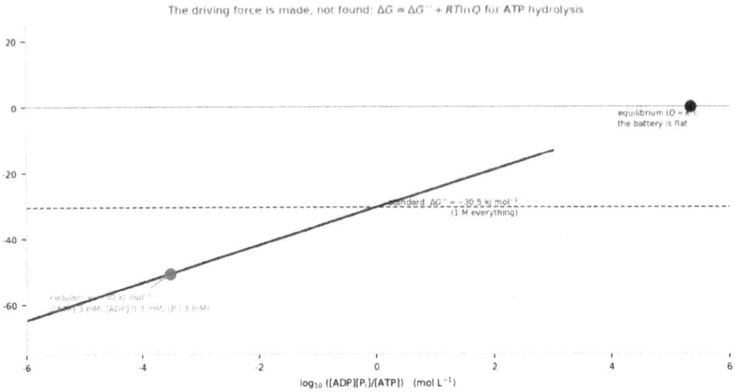

Figure 7.5 The free energy of ATP hydrolysis as a function of the mass-action ratio (Eq. 4.5 applied at 25 °C): standard conditions (oxblood dashed), cellular conditions (teal, ≈ −50 kJ mol^{-1}), and equilibrium (oxblood point, where the battery is flat). The slope is 2.303RT ≈ 5.7 kJ mol^{-1} per decade — the exchange rate of § 4.4, governing every phosphorylation in the cell.

7.9 Coupled reactions

The machinery of coupling was built in § 4.6 and needs here only its thermodynamic summary: free energies add because G is a state function, equilibrium constants multiply because logarithms add, and an unfavourable reaction is driven when coupled, through a genuinely shared intermediate, to a sufficiently favourable one. The present section adds the two quantitative refinements that Part VI will use. The extent of driving is set by the coupled constant, $K_{tot} = K_1K_2$: coupling to ATP hydrolysis ($K' \approx 2 \times 10^5$) multiplies the equilibrium constant of any partner by that factor — the thermodynamic charter under which glutamine synthetase (Figure 4.3), and a thousand other enzymes, conduct business. And the efficiency of coupling is never perfect: some free energy is always dissipated, and the dissipation is measured by how far the coupled system is held from its own equilibrium — a quantity, the "irreversibility" of the step, that metabolic control analysis (§ 30.11) will make into a diagnostic. The cell keeps its key couplings far from equilibrium (irreversible, controllable) and its intermediates near it (reversible, responsive): the placement of the irreversibility is the design, not the accident.

7.10 Energetically unfavourable processes

Between coupling and the impossible lies the interesting territory: processes with $\Delta G > 0$ that the cell nonetheless performs — not by violating Eq. (7.7), which no cell has ever done, but by strategies this section collects. Coupling (§ 7.9) is the first and largest. Concentration work is the second: a reaction with positive ΔG° proceeds forward whenever Q is held low enough — $\Delta G = \Delta G° + RT \ln Q < 0$ requires only $\ln Q < -\Delta G°/RT$ — and the cell holds Q low by removing products as fast as they form (the strategy of the aldolase and citrate synthase steps of Part VII, whose ΔG°' values are positive yet whose flux is prodigious). Transport against gradient is the third: moving a species uphill costs $RT \ln(c_{in}/c_{out})$ per mole — derived, for the record, as the free energy of isothermal compression of the species from the outside to the inside concentration — and is paid, in the Na^+/K^+

pump and its colleagues of § 24.6, by coupling to hydrolysis. The reader should carry away the complete list, for it is short: couple, deplete the product, or pay with a gradient — there is no fourth way, and the second law keeps the accounts.

7.11 Open systems and non-equilibrium states

The flask of the last nine sections has a lid; the cell has none. Matter and energy cross its boundary continuously — food in, waste out — and the system's destiny is therefore not the equilibrium of Figure 7.7a but the steady state of Figure 7.7b: concentrations constant in time not because flows have ceased but because inflows and outflows are matched, $d[X]/dt = v_{in} - v_{out} = 0$ *with both velocities large. The steady state is the physiological condition of everything in this book — the blood glucose of § 40.1, the ATP of § 7.8, the proton of Chapter 6 — and its thermodynamic signature is that it is maintained at a distance from equilibrium at continuous cost: relax the input, and the state decays to equilibrium with the time constant of its slowest pool. The near-equilibrium regime has its own thermodynamics (Onsager's, for which the Nobel committee of 1968 found a prize): fluxes are proportional to forces, cross-coefficients are symmetric, and the cell's transport and reaction networks, near enough to equilibrium, behave like linear circuits. Far from equilibrium, linearity fails — and there, as § 7.12 will argue, lies the interesting half of life.*

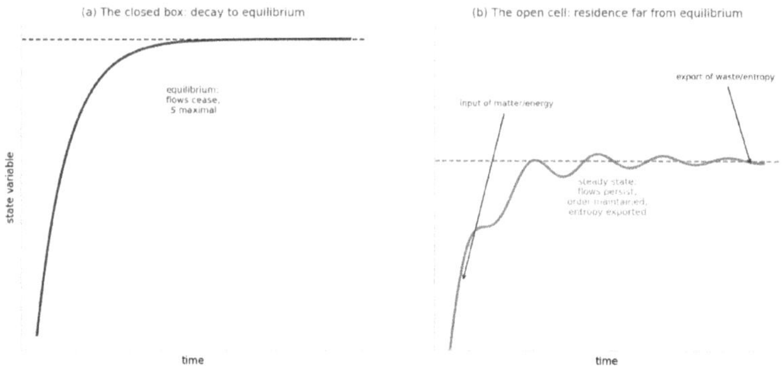

***Figure 7.7** (a) The closed system: the state variable relaxes to its equilibrium value and the flows cease; entropy is maximal, and nothing further happens — a fair portrait of a deceased cell. (b) The open system: matter and energy cross the boundary continuously, and the system settles into a steady state far from equilibrium, maintained at cost — a fair portrait of a living one. The two panels are the whole thermodynamic difference between the organism and its remains.*

7.12 Energy dissipation and biological order

We arrive at the paradox that every popular account of thermodynamics gets wrong and every honest treatise must get right. The living cell is extravagantly ordered — improbable molecules, improbable concentrations, improbable architectures — and the second law, applied carelessly, forbids it: does life not decrease entropy? The resolution has been available since § 7.1 and is merely applied: the law's verdict covers the universe, not the cell; the cell decreases its own entropy freely, provided it exports more entropy than it retains, which it does by degrading high-grade energy (photons, foodstuffs) into low-grade heat. The numbers make the mechanism respectable: Worked Example 7.1 computes that the human body, dissipating its ≈ 100 W at 310 K into a 290 K room, exports some

25 J K^{-1} s^{-1} of entropy — against which the entropy of maintaining all its improbable order is loose change. Schrödinger's little book of 1944 (Schrödinger, 1944) put the point memorably: the organism feeds on negative entropy — on order, not on energy (which is, after all, conserved and therefore cannot be what food is for) [20]*; and Prigogine's thermodynamics of irreversible processes supplied the formalism — the "dissipative structure", order arising through dissipation, far from equilibrium (Prigogine, 1977). The dialectic of the whole treatise has its deepest form here. Thesis: the second law drives everything toward disorder. Antithesis: the biosphere is the most ordered object known. Resolution: the two are not merely compatible but causally linked — the Sun's low-entropy photons are the Earth's income, the infrared sky is its expenditure, and the biosphere is the machinery the gradient built between them. Life is not an exception to the second law. Life is one of the things the second law does, on a planet with a hot sun and a cold sky.*

MECHANISTIC CASE STUDY — RUBNER'S DOG, 1894: THE FIRST LAW APPLIED TO LIFE

That an animal is an engine — that its heat and work, summed, equal the combustion energy of its food — was believed by Lavoisier, doubted by vitalists, and proved by Max Rubner in Berlin in the 1890s, in what remains among the most audacious measurements of physiology (Rubner, 1894). Rubner enclosed a dog in a calorimeter of his own construction, measured the total heat the animal emitted over days, measured simultaneously the respiratory exchange (the O_2 consumed and CO_2 produced), and computed from the combustion tables what the oxidised foodstuffs should have released. The two ledgers agreed to within a few tenths of a per cent. The experiment's logic is the first law's, and its consequence was the quiet death of energetic vitalism: no vis vitalis was needed to balance the account, because the account balanced without one — the animal's warmth was combustion, spread over time and laundered through the chemistry this treatise spends Parts VI–VII describing. The reader should note, as a matter of method, the experiment's lineage: Lavoisier's guinea pig in the ice calorimeter (1780s), measuring heat by the ice it melted; Rubner's dog; and the modern "metabolic chamber" of the clinical literature, which is Rubner's apparatus with better insulation and worse conversation. The first law, once proved in a dog, has never been contradicted in a human — the reader's own 100 W is calorimetry, not metaphor.

20. Schrödinger's *What is Life?*, the published form of his Dublin lectures of February 1943, is famous also for a different reason: it argued that the hereditary material must be an "aperiodic crystal", a sentence read by Watson, Crick and Wilkins, among others, with consequences the reader knows. Physicists, when they condescend to notice biology, have an alarming record of being right; the biologists have learned to read them with attention and forgive them their thermodynamics.

Plate 7.1 *Walther Nernst (1864–1941), whose heat theorem (the third law: perfect crystals approach zero entropy at absolute zero) completed the classical edifice this chapter surveys — and whose equation of electrode potentials the reader will meet again in Chapter 26, doing duty for every redox couple of the respiratory chain. Portrait from the Nobel Foundation archive, nobelprize.org (Nobel Prize in Chemistry, 1920), reproduced from the Foundation's publicly distributed image.*

HISTORICAL NOTE — GIBBS, THE QUIET FOUNDER

The free energy of § 7.5, the chemical potential of § 4.4, the phase rule, and the entire mathematical machinery of chemical thermodynamics are the work of one man, Josiah Willard Gibbs of Yale, published between 1876 and 1878 in the Transactions of the Connecticut Academy of Arts and Sciences — a journal read, at the time, by almost no one (Gibbs, 1878). The work reached Europe through Maxwell, who built plaster models of the thermodynamic surfaces and sent one to New Haven; Ostwald translated the papers into German; and by 1900 the founder was recognised, in the words of the anniversary volumes, as the greatest scientific mind America had produced. Gibbs never sought a prize (none existed in his line), never travelled, and died in 1903, in the same town where he was born. The treatise that the reader holds — every ΔG of it — is written in his notation and in his debt.

Chapter 7 — Conceptual synthesis

The laws beneath the laws are now on the table. The boundary between system and surroundings is the first methodological act, and state variables are honoured by every path (§ 7.1); energy is conserved as heat plus work (§ 7.2), with enthalpy as the constant-pressure heat (§ 7.3); entropy is the logarithm of multiplicity, and its increase in isolated systems is statistics, not decree (§ 7.4); the Gibbs free energy translates the universe's verdict into the system's own ($\Delta G = -T\Delta S_{universe}$, § 7.5); the biochemical standard state hides the proton in the prime (§ 7.6); equilibrium is the Boltzmann factor of $\Delta G°$, and the van 't Hoff slope is enthalpy (§ 7.7); the cellular driving force is made by holding Q far from K (§ 7.8); coupling multiplies constants (§ 7.9); the unfavourable is achieved by coupling, depletion, or gradient — and not otherwise (§ 7.10); the cell is a steady state, not an approach to equilibrium (§ 7.11); and biological order is bought by entropy export, the second law being the biosphere's employer rather than its policeman (§ 7.12). Direction checked against Figure ii: Part II is two-thirds built — water (5), protons

(6), energy (7). Chapter 8 completes the part: the association of molecules, the binding equilibria on which recognition and assembly depend.

WORKED EXAMPLE 7.1 — THE ENTROPY EXPORT OF A RESTING HUMAN

Take the resting metabolic rate of Figure 4.7, ≈ 100 W, dissipated as heat from the skin (≈ 310 K) into a room at 290 K. The entropy exported to the room is $\Delta S_{out} = q/T_{room} = 100/290 \approx 0.345\ J\ K^{-1}\ s^{-1}$; the entropy the body fails to export while shedding that heat at 310 K is $100/310 \approx 0.323\ J\ K^{-1}\ s^{-1}$; the net production is $\approx 0.022\ J\ K^{-1}\ s^{-1} \approx 1.9\ kJ\ K^{-1}$ per day. Against this, the entropy content of the body's improbable order: the difference between a human and the same atoms dispersed is of order 10^3–$10^4\ kJ\ K^{-1}$ in total — the body re-earns the thermodynamic right to its entire organisation every few days of metabolism, and § 7.12's resolution is thereby arithmetic, not rhetoric.

WORKED EXAMPLE 7.2 — DRIVING A REACTION BY PRODUCT REMOVAL

The aldolase reaction of glycolysis (fructose-1,6-bisphosphate → dihydroxyacetone phosphate + glyceraldehyde-3-phosphate) has $\Delta G^{\circ\prime} \approx +24\ kJ\ mol^{-1}$. What product ratio permits forward flux at $[FBP] = 10^{-4}$ M? Requirement: $\Delta G = \Delta G^{\circ\prime} + RT\ \ln(c_{DHAP}c_{GAP}/\ c_{FBP}\cdot 1/c^\circ) < 0$, i.e. $\ln(Q/c^\circ) < -24/2.479 = -9.68$, so $Q < 6.3 \times 10^{-5} \times c^\circ \approx 6.3 \times 10^{-5}\ M^2 \cdot$ ... with $[FBP] = 10^{-4}$ M the product of product concentrations must lie below $\approx 6.3 \times 10^{-9}\ M^2$ — e.g. each product near 8×10^{-5} M. The cell achieves exactly this: downstream enzymes (triose-phosphate isomerase, GAPDH) keep the triose phosphates scarce, and the "unfavourable" step runs at full glycolytic speed — § 7.10's second strategy, measured in the wild.

WORKED EXAMPLE 7.3 — THE THERMODYNAMIC COST OF CONCENTRATING CALCIUM

The sarcoplasmic reticulum holds Ca^{2+} at $\approx 10^{-3}$ M against a cytosol of 10^{-7} M. The chemical work per mole pumped is $RT\ \ln(c_{in}/c_{out}) = 2.577 \times \ln(10^4) = 2.577 \times 9.21 \approx 23.7\ kJ\ mol^{-1}$ at 310 K; the SERCA pump moves two Ca^{2+} per ATP hydrolysed ($\Delta G \approx -50\ kJ\ mol^{-1}$, § 7.8): cost $47.4\ kJ\ mol^{-1}$ against 50 available — an efficiency near 95 %, close to the reversibility ceiling of § 7.2, and the reason muscle, the body's most profligate calcium spender, is also among its most mitochondria-rich tissues.

Exercises (with full resolutions)

EXERCISE 7.1

Compute the entropy change when 1 mol of an ideal gas doubles its volume isothermally, (a) from Eq. (7.3) and (b) from Eq. (7.5), and reconcile.

Resolution. *(a) Reversibly, $q_{rev} = -w_{rev} = RT \ln 2$, so $\Delta S = q_{rev}/T = R \ln 2 \approx 5.76\ J\ K^{-1}\ mol^{-1}$. (b) Statistically: each molecule's accessible positions double, W multiplies by 2^{N_A}, so $\Delta S = k_B \ln 2^{N_A} = N_A k_B \ln 2 = R \ln 2$ — identical. The two roads to entropy, calorimetric and statistical, meet exactly as they should; the* *exercise is the first law of thermodynamic literacy: never trust an entropy you cannot derive both ways.*

EXERCISE 7.2

A reaction has $\Delta H° = +30$ kJ mol^{-1} and $\Delta S° = +90$ J K^{-1} mol^{-1}. At what temperature does it become spontaneous, and what class of biochemical process does this pattern typify?

Resolution. *The crossing of Figure 7.3: $T = \Delta H°/\Delta S° = 30\ 000/90 = 333$ K ≈ 60 °C — the enthalpy-opposed, entropy-driven pattern, characteristic of thermal unfolding (Chapter 11): below the crossing the fold wins, above it the chain's conformational freedom collects its entropy. The arithmetic is Eq. (7.7) and nothing else; the insight is that denaturation is a phase transition priced per degree.*

EXERCISE 7.3

Convert the standard free energy of ATP hydrolysis between conventions: $\Delta G° = -32.5$ kJ mol^{-1} (chemical standard, all species 1 M including H^+) to the biochemical standard (pH 7), given one proton is produced.

Resolution. *By § 7.6's clause, $\Delta G°' = \Delta G° + n_{H+} \times 2.303RT \times 7 = -32.5 + 1 \times 39.96 \approx +7.5$ kJ mol^{-1}... — the reader should pause here: the published pair (−32.5 → −30.5) reflects the full stoichiometry including Mg^{2+} binding and ionic-strength terms, which is exactly the point of the exercise: the proton correction alone is enormous (+40 kJ mol^{-1}), the complete biochemical standardisation is a professional's calculation (Alberty, 2003), and a $\Delta G°$ quoted without its convention is, per § 7.6's rule, not a number but an opinion.*

EXERCISE 7.4 (INTERPRETATIVE)

"Thermodynamics proves that evolution is impossible, because order cannot increase." Construct the refutation, with numbers.

Resolution (indicative). *The claim misapplies Eq. (7.6) — valid for isolated systems — to the Earth, which is open: it receives ≈ 10^{17} W of 5 800 K photons and re-radiates ≈ the same power as 255 K photons, exporting entropy at ≈ $10^{17}(1/255 - 1/5800) \approx 4 \times 10^{14}$ J K^{-1} s^{-1}. The entropy decrease of assembling the entire biosphere is negligible against that flux; there is no thermodynamic objection to evolution because there is no thermodynamic objection to a heat engine, and the biosphere is one. The argument's persistence in the face of arithmetic is a sociological, not a physical, phenomenon — and the exercise's moral is § 7.1's: check the boundary before quoting the law.*

EXERCISE 7.5

The Na^+/K^+-ATPase exports 3 Na^+ and imports 2 K^+ per ATP. Estimate the free energy required per cycle against the resting gradients (Na^+: 12 mM in, 145 mM out; K^+: 140 mM in, 4 mM out; membrane potential −70 mV, inside negative), and compare with ΔG_{ATP}.

Resolution. *Chemical work: 3RT ln(145/12) + 2RT ln(4/140) = 2.577[3 × 2.492 + 2 × (−3.555)] = 2.577 × 0.366 ≈ 0.9 kJ mol^{-1}. Electrical work: net one positive charge exported per cycle against −70 mV: F × 0.070 ≈ 6.8 kJ mol^{-1}. Total ≈ 7.7 kJ mol^{-1} — only 15 % of the ≈ 50 kJ mol^{-1} available: the pump runs far from its thermodynamic limit, which is why it is controllable (§ 7.9's irreversibility placement) and why a third of the reader's ATP budget, spent through it, maintains the gradients on which every nerve impulse and every secondary transporter of Part V rides.*

Research-level propositions

Proposition 7.α. The efficiency of oxidative phosphorylation (Chapter 28) is bounded by the ratio of the phosphorylation potential to the redox span of the respiratory chain. Using the measured P/O ratios and the thermodynamics of §§ 7.8–7.10, compute the maximal thermodynamic efficiency of the mitochondrion as a heat engine between NADH and ATP, and compare with the measured ≈ 60 %: what dissipates the remainder, and is the dissipation functional (uncoupling, thermogenesis, § 28.9) or waste? Proposition 7.β. Onsager's linear regime (§ 7.11) has been shown to fail for single enzymes far from equilibrium. Derive the flux–force relation for an enzyme-catalysed unimolecular reaction without the linear approximation, and determine experimentally — from the data of Chapter 16 — where the linear coefficient errs by more than 10 %. Proposition 7.γ. Schrödinger's "negentropy" has been criticised as loose talk: information-theoretic restatements (Maxwell's demon, Landauer's principle) price the erasure of information at k_BT ln 2 per bit. Compute the Landauer cost of replicating one bacterial genome (Eq. 1.2's 9.2 megabits) and compare with the free energy the cell actually spends on replication: how far above the thermodynamic floor does the cell's information technology operate, and what does the gap pay for — fidelity, speed, or regulation?

References (Chapter 7)

Alberty, R. A. (2003). Thermodynamics of biochemical reactions. Wiley-Interscience.

Atkins, P., & de Paula, J. (2018). Atkins' physical chemistry (11th ed.). Oxford University Press.

Gibbs, J. W. (1878). On the equilibrium of heterogeneous substances. Transactions of the Connecticut Academy of Arts and Sciences, 3, 108–248, 343–524.

Prigogine, I. (1977). Time, structure and fluctuations [Nobel lecture]. Angewandte Chemie International Edition, 17(1), 1–14.

Rubner, M. (1894). Die Quelle der thierischen Wärme. Zeitschrift für Biologie, 30, 73–142.

Schrödinger, E. (1944). What is life? The physical aspect of the living cell. Cambridge University Press.

CHAPTER 8

Molecular Recognition and Self-Assembly

CONTENTS OF THIS CHAPTER

In which the molecules of the cell are taught to find one another: the binding equilibrium derived and its constant read as a concentration; the hyperbola of occupancy; the sigmoid of cooperativity and the two-state logic of allostery; the arithmetic of avidity; the assembly of many parts into one machine; and the newest architecture of all — the condensates that the cell builds by unmixing its own contents.

8.1 Reversible molecular association

Everything in this book that has been said about single molecules has been a rehearsal; the performance begins when molecules meet. A hormone finds its receptor among a hundred thousand irrelevant proteins; an enzyme selects its substrate from a thousand candidates; two subunits of a ribosome assemble with the fidelity of a watch and the reversibility of a handshake. The common event beneath all these meetings is the reversible association of two species,

$$P + L \rightleftharpoons PL \qquad (8.1)$$

— P the macromolecule (protein, receptor, enzyme), L the ligand (substrate, hormone, partner), and the harpoons honest: the complex forms and dissociates continuously, at rates $k_{on}[P][L]$ and $k_{off}[PL]$, and the observable state of the system is the balance between them. The dialectic of the chapter opens here, with the two images between which the whole theory of recognition has oscillated. The first is Fischer's: the lock and its key, recognition as shape — the ligand fits the site as a key its wards (historical note below). The second is Koshland's: the glove and the hand, recognition as negotiation — the site moulds itself around the ligand, each partner changing shape upon meeting (the induced fit of § 12.3). The truth, as the structural biology of Chapter 54 has shown a thousand times, is that both images are right in sequence: the initial encounter is a matter of shape and charge (Fischer), the final complex a matter of mutual adjustment (Koshland), and the free energy of the outcome is paid for by the interactions of Chapters 2 and 5 — hydrogen bonds, salt bridges, dispersion, and the release of water. The present chapter prices the meeting; Part IV will set it to work.

> **HISTORICAL NOTE — THE LOCK, THE KEY, AND THE BAKER'S ENZYMES**
>
> *"To use a picture, I would say that enzyme and glucoside must fit into one another like lock and key" — Emil Fischer (Plate 8.1), writing in 1894 on the specificity of yeast enzymes toward the glycosides (Fischer, 1894). The image was offered almost casually, in a paper on the stereochemistry of sugars, and it has organised the thought of every biochemist since: specificity is*

complementarity of form. Fischer had earned the right to say it: he had spent the 1880s and 90s mapping the stereochemistry of the sugars (the work of Chapter 3's Fischer projections) and found that enzymes distinguished his synthetic isomers with infallible taste — fermenting one twin and spurning the other. The lock-and-key image held the field for sixty-four years, until Koshland's induced fit (1958) reminded the subject that the lock, in biology, is made of protein, and protein — as Chapters 11 and 12 will insist — is never quite rigid. The two images are the thesis and antithesis of this chapter; the synthesis is conformational selection, § 12.3's due.

8.2 Binding equilibria

The association of Eq. (8.1) obeys the equilibrium law of Chapter 4, and its constant is written, by the convention of the trade, in the dissociating direction:

$$K_d = \frac{[\mathrm{P}][\mathrm{L}]}{[\mathrm{PL}]} \tag{8.2}$$

— the dissociation constant, with units of concentration. Its kinetic content follows from § 8.1's rates: at equilibrium, $k_{on}[P][L] = k_{off}[PL]$, whence

$$K_d = \frac{k_{\mathrm{off}}}{k_{\mathrm{on}}} \tag{8.3}$$

— the constant is a ratio of velocities, which is why tight binding can be achieved by a fast embrace (diffusion-limited $k_{on} \approx 10^8$–10^9 L mol^{-1} s^{-1}) or by a slow release (the avidin of Table 8.1, whose k_{off} is measured in months). The distinction is not academic: the drug that binds by slow dissociation is the drug that lasts, and the pharmacology of residence time — now a department of medicinal chemistry — is Eq. (8.3) read by a clinician. Note also the reciprocity with the association constant, $K_a = 1/K_d$; the literature uses both, and the reader's defence is dimensional: if it has units of concentration, it is K_d.

8.3 Association and dissociation constants

The scale of the constant deserves a map, for biochemical recognition spans fifteen orders of magnitude (Table 8.1). At the weak end, the transient encounters of metabolism (enzyme–substrate, µM–mM) must be loose enough to release the product; at the strong end, the interactions that are meant to be effectively permanent (avidin–biotin, antibody–hapten, repressor–operator) reach $K_d \approx 10^{-15}$ M — femtomolar, a concentration at which one complex per cell would already be half the equilibrium population (Table 1.3's census again). The physiological reading of the table is the one the cell itself uses: a K_d is a set point. A receptor with K_d = 1 nM for its hormone responds to the hormone's physiological nanomolar excursions and ignores the micromolar noise of everything else; an enzyme with $K_m \approx 100$ µM reads its substrate's metabolic range. The constant is not a property of the molecules alone but of their conversation with the concentrations the cell maintains — recognition is calibrated, by evolution, to the local market.

Table 8.1 Representative dissociation constants of biochemical recognition, with the binding free energy computed from Eq. (8.4) at 310 K. The range — six orders of magnitude in this short list alone — is the dynamic range of the cell's social life.

Pair	K_d	$\Delta G°$ (kJ mol^{-1})	Remark
enzyme–substrate (typical)	10^{-4} M	-24	must release the product
haemoglobin–O_2 (per site)	≈ 10^{-5} M	-30	tuned to the alveolar pO_2
insulin–receptor	≈ 10^{-10} M	-59	hormonal set point (Figure 4.5)
antibody–antigen (mature)	10^{-8}–10^{-10} M	-47 to -59	affinity maturation, § 14.8
lac repressor–operator	≈ 10^{-13} M	-76	one operator per genome
avidin–biotin	≈ 10^{-15} M	-89	the strongest noncovalent embrace known

8.4 Binding free energy

The constant converts to free energy by the master relation of Chapter 7, $\Delta G° = -RT \ln K_a = RT \ln K_d$ — with the standard concentration folded in, so that strictly

$$\Delta G^{\circ}_{\text{bind}} = RT \ln \frac{K_d}{c^{\circ}} \tag{8.4}$$

— whence the table's right-hand column: each factor of ten in K_d is 5.9 kJ mol^{-1} at 310 K, and the difference between a micromolar and a femtomolar embrace (nine decades) is ≈ 53 kJ mol^{-1} — one ATP hydrolysis's worth, or about twenty hydrogen bonds, which the avidin interface duly supplies in aggregate. The decomposition of $\Delta G°_{bind}$ into its physical parts is the accounting of Chapters 2 and 5 applied: favourable terms (hydrogen bonds and salt bridges, dispersion contacts, and the hydrophobic bonus of released hydration water) against unfavourable ones (the loss of the ligand's translational and rotational freedom — the "cratic" entropy, ≈ 30–45 kJ mol^{-1} of tax on every bimolecular embrace — and the desolvation of both partners). The ledger explains a standing paradox of the field: computed binding energies from the individual terms routinely overshoot the measured values by an order of magnitude, because the unfavourable terms — entropy, desolvation — are always larger and harder to estimate than the seductive contacts. Recognition, like marriage, is a small net profit on a large turnover; and the profit, not the turnover, is what the cell reports.

8.5 Saturation and fractional occupancy

The measurable quantity of binding is the fraction of sites occupied, and its law follows from Eq. (8.2) in three lines. Write the occupancy θ = [PL]/([P] + [PL]); substitute [PL] = [P][L]/K_d from Eq. (8.2); cancel [P]:

$$\theta = \frac{[\mathrm{L}]}{K_d + [\mathrm{L}]} \tag{8.5}$$

— the rectangular hyperbola of Figure 8.1, the most-used curve in biochemistry after the Henderson–Hasselbalch (to which it is, by the logarithm, first cousin). Its

two lessons are quantitative hygiene. At [L] = K_d, half the sites are filled — the constant reads directly off the curve, which is why K_d has units of concentration and why it is a set point (§ 8.3). And the hyperbola is not linearisable without cost: the double-reciprocal plots of the older literature (Lineweaver–Burk's binding cousin) compress the low-concentration data into noise, and the modern practice — nonlinear fitting to Eq. (8.5) itself, § 16.7 — is statistically superior in every respect. The reader who inherits a straight-line transformation should convert it back on sight; the hyperbola is the truth, and the straight lines were scaffolding.

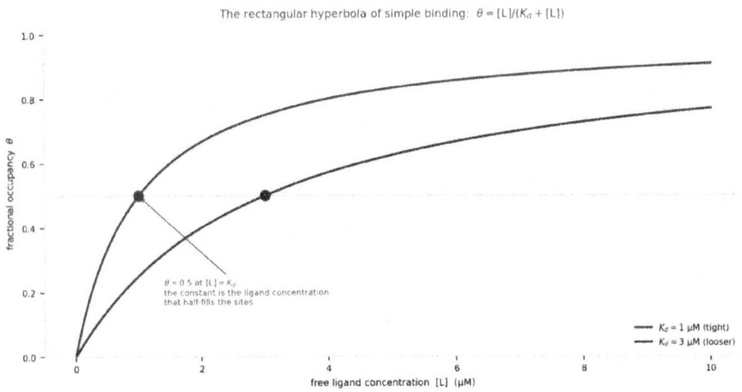

Figure 8.1 *The fractional occupancy of Eq. (8.5) for two dissociation constants: the constant is the free-ligand concentration at half saturation (points), and the curve's whole information is its midpoint and its asymptote. The hyperbola is the binding isotherm of a single class of independent sites; any deviation from its shape — steepening into sigmoid, or flattening into heterogeneity — is diagnostic, respectively, of cooperativity (§ 8.7) or of multiple site classes (§ 8.6).*

8.6 Multiple binding sites

Real macromolecules rarely offer one site; they offer several, and the several may be identical-and-independent, identical-and-interacting, or frankly different. The independent case is arithmetic: n identical sites bind with occupancy still given by Eq. (8.5), the only change being the statistical bookkeeping of which site is occupied (the Scatchard analysis of the older literature — slope $-1/K_d$, intercept n — being its linearisation, and sharing its caveats). The interacting case is cooperativity, and deserves its own section. The frankly different case — a receptor with two site classes, an antibody population with mixed affinities — is the commonest disguised complexity of the literature: a sum of two hyperbolae looks, to a careless eye, like a single shallow one, and many a "low-affinity receptor" has been, on dissection, a high-affinity minority in adulterated company. The diagnostic is the shape: any binding curve broader than a hyperbola (a Hill coefficient below 1) declares heterogeneity until proved innocent — the mirror image of the cooperativity test of § 8.7, and the reason the Hill plot of Figure 8.3 is the first graph drawn of any new binding system.

8.7 Cooperativity

When the sites talk to one another, the hyperbola bends into a sigmoid, and the protein acquires the capacity it did not have as a monomer: it becomes a switch.

The phenomenology is the Hill equation, written for the extreme case in which binding is effectively all-or-none across n sites,

$$\theta = \frac{[\mathrm{L}]^{n_H}}{K' + [\mathrm{L}]^{n_H}} \tag{8.6}$$

— where the Hill coefficient n_H measures the steepness (Figure 8.2) and, at the midpoint of the Hill plot (Figure 8.3), the degree of concertedness: $n_H = 1$ is independence, n_H approaching the site count is perfect concert. Haemoglobin, the chapter's standing example and Chapter 13's hero, binds its four oxygens with $n_H \approx$ 2.8–3.0 — strong but imperfect concert, and perfect for the job: the 10–90 % loading transition now spans a threefold change of oxygen pressure rather than the hyperbola's eighty-one-fold, so that the molecule loads fully in the lung (where pO_2 is high) and unloads decisively in the tissue (where it is low), each site reporting to the others the state of the company. The physical basis of the talk — how a ligand at one subunit informs the empty sites three nanometres away — is the allosteric mechanism of the next section; the thermodynamic content is that the binding free energies of successive ligands are not additive, and the deviation, the interaction free energy, is the price of the conversation.

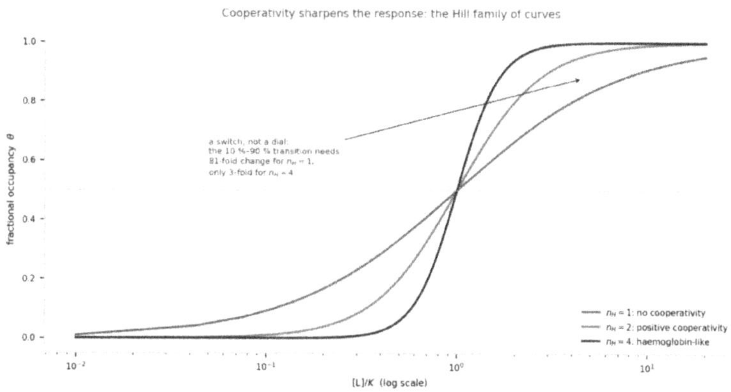

Figure 8.2 *The Hill family (Eq. 8.6): the same midpoint, three slopes. Cooperativity converts the hyperbola's gentle dose–response into a switch with a threshold; the 10–90 % window narrows from 81-fold (independent sites) to 3-fold (n_H = 4). The evolutionary point is that the curve's steepness is a selected trait, as much as any structure: haemoglobin's sigmoid is the answer to a physiological question — load in one place, unload in another, and waste nothing in between.*

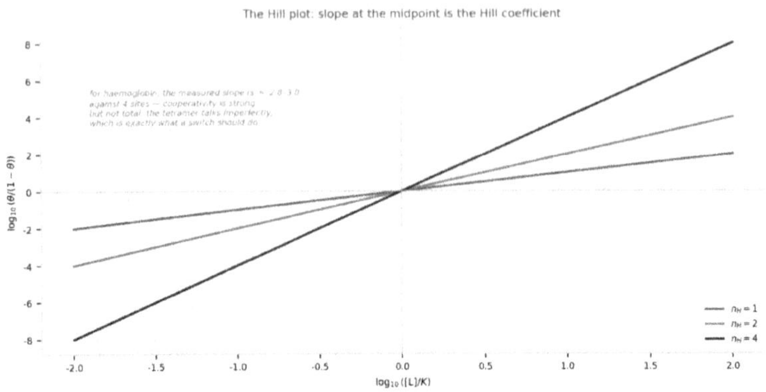

Figure 8.3 The Hill plot — log[θ/(1−θ)] against log[L] — whose midpoint slope is the Hill coefficient. Three virtues keep it in service a century after Hill: it linearises the cooperative range for easy reading of n_H; it exposes non-integral coefficients (haemoglobin's ≈ 2.9, not 4) as evidence of imperfect concert; and its extremes, where the slope returns to 1, show the first and last sites binding independently — the molecule converses only in the middle of the journey.

8.8 Allostery

The mechanism of the conversation was supplied in 1965 by Monod, Wyman and Changeux, in a paper of three pages and permanent influence (Monod, Wyman, & Changeux, 1965). The MWC (concerted) model postulates that an oligomeric protein exists in equilibrium between two quaternary states — T (tense, low affinity) and R (relaxed, high affinity) — that the transition is concerted (all subunits flip together, symmetry preserved), and that the ligand acts simply by binding more tightly to R: it does not cause the transition, it elects the state, and the equilibrium follows (Figure 8.4). From these three postulates the whole phenomenology of § 8.7 follows: the sigmoid curve (the first ligands must pay to shift the equilibrium toward R; later ones bind a protein already converted), the Hill coefficient (set by the T⇌R equilibrium constant L and the affinity ratio c = K_R/K_T), and — the model's prophetic reach — the effects of other ligands: an activator binding R stabilises it and sensitises the switch; an inhibitor binding T desensitises it. Hence allostery's definition: regulation at a distance, ligands at one site governing activity at another, with no chemistry shared between them but the conformational equilibrium. The alternative KNF (sequential) model (Koshland, Némethy, & Filmer, 1966) lets each subunit flip individually upon binding, propagating the change through interfaces — the glove rather than the election; and the modern view, stated fully in Chapter 12, dissolves the dispute into an ensemble: proteins are populations of conformations in perpetual equilibrium, and ligands, allosteric effectors, and mutations alike act by re-weighting the population. MWC and KNF are the two ends of that continuum — and the two-state picture of Figure 8.4 remains, sixty years on, the most productive three assumptions in molecular biology.

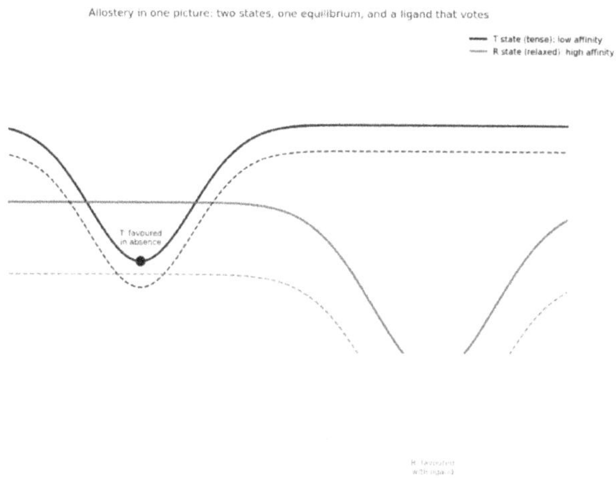

Figure 8.4 *Allostery in one picture. The protein populates two conformational wells — T (oxblood, favoured empty) and R (teal, favoured once ligand accumulates); ligand lowers both wells but lowers R more (dashed curves), because R binds it more tightly, so binding converts the population rather than the molecule. The cooperativity of § 8.7 is the population shift measured at the sites; the allosteric effector is any other ligand that tips the same balance from its own address.*

8.9 Competitive molecular recognition

Recognition never happens in private: every site in the cell is courted by a crowd, and the observable binding of a ligand is always its share of a contested market. The law of the contest follows from Eq. (8.2) applied to two ligands that cannot both occupy the site: with competitor I at concentration [I] and constant K_I, the apparent dissociation constant for L becomes

$$K_d^{\text{app}} = K_d\left(1 + \frac{[\text{I}]}{K_I}\right) \qquad (8.7)$$

— the competitor inflates the apparent constant in proportion to its own occupancy of the site (the derivation: the free P is now shared between two equilibria, $[P]_{free} = [P](1 + [I]/K_I)$, and the rest is Eq. 8.2's algebra). Eq. (8.7) is the charter of three large industries. Pharmacology: the competitive inhibitor of § 17.2 is I, and the drug's task is to inflate the substrate's apparent constant until the enzyme's throughput falls. Physiology: the antidotes of toxicology — ethanol for methanol poisoning, N-acetylcysteine's precursors aside — work by flooding the market with a harmless bidder. And assay: the displacement experiment (measure the apparent constant of a reporter ligand against a series of test competitors) converts Eq. (8.7) into a measurement of the competitor's own K_I, the Cheng–Prusoff relation of Chapter 17 being its quantitative form. The lesson generalises: in the cell, every K_d is an apparent K_d — measured against a background of a thousand competitors — and the constants of Table 8.1 are laboratory purities, to be discounted in vivo by the crowding of § 5.11 and the competition of this section.

8.10 Supramolecular assembly

Binding iterated is architecture: the same equilibrium that joins a hormone to its receptor, written into the geometry of the subunits, joins a protein to its own kind and builds the great assemblies of the cell — the ribosome's fifty-odd proteins and three RNAs, the microtubule's tubulin lattice, the virus's capsid, the actin filament of muscle. The thermodynamics of assembly has two quantitative signatures the reader should carry. The first is the critical concentration: assembly into an open-ended polymer (actin, tubulin) begins only above a threshold monomer concentration — the same mathematics as the CMC of Eq. (5.1), with nucleation adding a lag: below the threshold, nothing; above it, everything — a molecular property with the macroscopic consequence of a switch. The second is cooperativity of assembly: the first bonds of an assembly are the hardest (a dimer has one interface; a trimer offers the newcomer two), so that nucleation is slow and growth is fast — the kinetic signature visible in every amyloid experiment of § 11.11, where the lag phase is the nucleation and the explosion is the polymerisation, and where the whole tragedy of the conformational diseases is written in the same equations that build the cytoskeleton. The dialectic of the section is the standing puzzle of the assembly's accuracy: how do fifty components find their one arrangement? The answer is the chapter's whole thesis writ large — each interface is a specific recognition (§ 8.1), the binding is cooperative (§ 8.7), and errors are edited by reversibility: the wrongly seated subunit, bound weakly, dissociates and tries again. Assembly is proofreading by equilibrium — the cell does not build correctly; it builds revisably, and keeps what survives.

8.11 Phase separation and biomolecular condensates

The newest chapter of recognition, and the one that has reorganised cell biology in the last fifteen years, is the discovery that the cell compartments itself without membranes: proteins and RNAs of low-complexity sequence — multivalent, weakly interacting — demix from the cytosol into liquid droplets, the biomolecular condensates: nucleoli, stress granules, P-bodies, the clusters of signalling scaffolds. The physics is the classical phase separation of polymer science (Figure 8.6): above a critical interaction strength (or below a critical temperature, salt, pH), the homogeneous solution becomes unstable and splits into a dense phase and a dilute one, the droplet bounded not by a bilayer but by surface tension. The biology is the exploitation of the physics: condensates concentrate their clients a hundredfold (accelerating reactions), exclude others (purifying them), form and dissolve in seconds on a signal (regulating them), and — the darker theme — mature with time from liquids to gels to the solid aggregates of neurodegeneration (§ 11.11's amyloids again, now by a second road) (Hyman, Weber, & Jülicher, 2014). For this treatise the condensate completes a trilogy begun in Chapter 1: the cell is organised by membranes (Part V), by machines (Part III), and by phases — three architectures of one imperative, which is to keep the right molecules together and the wrong ones apart.

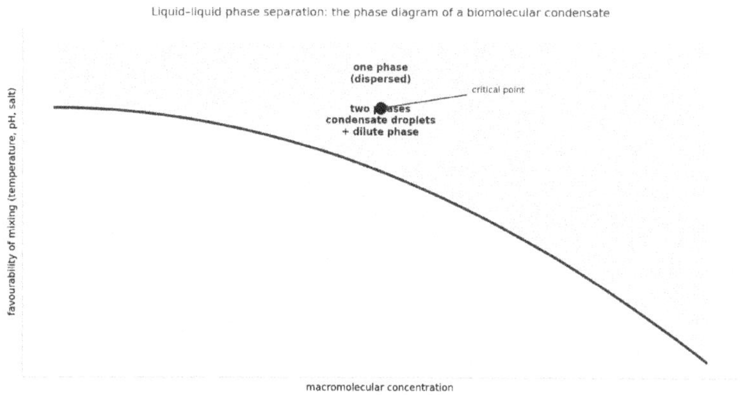

Figure 8.6 The phase diagram of liquid–liquid phase separation (schematic, after the condensate literature (Hyman et al., 2014)): within the binodal (navy) the homogeneous solution is unstable and splits into dense droplets in a dilute continuum; the critical point (oxblood) is where the two phases become indistinguishable. The cell steers the diagram with concentration, salt, pH and — its favourite instrument — phosphorylation, which moves a protein across the boundary in milliseconds: a compartment that appears on command and vanishes when the bill is paid.

8.12 Specificity, affinity and avidity

The chapter closes by disentangling three words that the literature uses as one and the cell uses as three. Affinity is the strength of a single embrace: K_d, Eq. (8.4), one site and one ligand. Specificity is discrimination: the ratio of affinities for the right partner and the wrong one — a kinetic and thermodynamic ratio, not an absolute, for the site that binds everything tightly is promiscuous, not specific. Avidity is the strength of several embraces at once: the functional affinity of a multivalent complex (Figure 8.5), in which each additional simultaneous contact multiplies the local concentration of the partner and collapses the off-rate — the pentameric IgM antibody, mediocre in affinity, superb in avidity, and therefore the first responder of the immune system (§ 14.7). The three are related but not interchangeable, and the confusion of them has misdirected more than one drug programme: a compound of mediocre affinity can be a superb medicine if its specificity is high (it leaves every other site alone), while a high-affinity binder to the wrong site class is merely a well-aimed poison. Table 8.2 sets the treaty; and the chapter's whole doctrine stands in one sentence, ready for Part III to inherit: the cell recognises by shape, prices by free energy, switches by cooperation, and holds on by multitude — and every one of those verbs is an equilibrium.

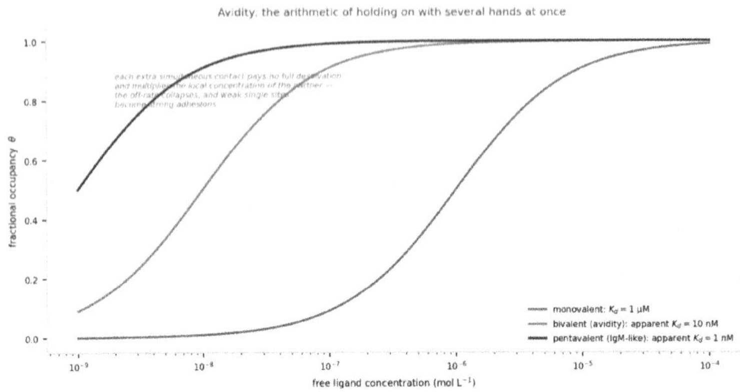

***Figure 8.5** Avidity quantified (illustrative model): the same single-site affinity (K_d = 1 μM) presented monovalently, bivalently, and pentavalently — the apparent constant improves by orders of magnitude because a dissociated contact is recaptured before the complex diffuses apart. The antibody of § 14.7 (pentameric IgM) and the cell-adhesion systems of tissues run on this arithmetic: weak single sites, strong adhesions, and specificity preserved because the wrong partner, unable to make the simultaneous contacts, pays the full off-rate.*

***Table 8.2** Affinity, specificity and avidity distinguished — the three virtues of recognition, their measures, and their masters.*

Concept	***Measure***	***Governed by***	***Biological exemplar***
affinity	K_d, $\Delta G°_{bind}$	*single interface chemistry*	*enzyme–substrate (μM)*
specificity	*ratio of K_d values (or k_{cat}/K_m ratios)*	*complementarity versus the rivals*	*tRNA synthetases (§ 47.3)*
avidity	*apparent K_d of the multivalent complex*	*valency, geometry, off-rate*	*IgM, cell adhesion*

Plate 8.1 *Emil Fischer (1852–1919): the sugars' first cartographer, the father of peptide chemistry, and the author of the lock-and-key image with which this chapter opened. Portrait from the Nobel Foundation archive, nobelprize.org (Nobel Prize in Chemistry, 1902), reproduced from the Foundation's publicly distributed image.*

Chapter 8 — Conceptual synthesis

Recognition has been priced, and Part II is complete. Association is a reversible equilibrium (§ 8.1); its constant is a concentration and a set point (§§ 8.2–8.3); the binding free energy is a small profit on a large turnover of contacts, entropy and desolvation (§ 8.4); occupancy follows the hyperbola, whose midpoint is the constant (§ 8.5); multiple sites divide into independent, interacting and heterogeneous (§ 8.6); interaction bends the hyperbola into the sigmoid, and the protein becomes a switch (§ 8.7); the MWC two-state logic explains the switch as an election between conformations, with effectors voting from other sites (§ 8.8); competition inflates every constant in the contested market of the cell (§ 8.9); assembly is recognition iterated, with critical concentrations and cooperative nucleation (§ 8.10); condensates are recognition demixed — compartments without membranes (§ 8.11); and affinity, specificity and avidity are three virtues, not one (§ 8.12). Direction checked against Figure ii: Part II — water, energy and molecular association — is closed. The treatise now crosses to Part III, where the greatest class of macromolecules takes the stage: the amino acids and peptides of Chapter 9, and the proteins they fold into.

> ### *WORKED EXAMPLE 8.1 — READING A BINDING CURVE*
>
> *A receptor at 50 nM total is incubated with ligand at 100 nM free; the measured occupancy is $\theta = 0.66$. From Eq. (8.5): $K_d = [L](1 - \theta)/\theta = 100 \times 0.34/0.66 \approx 52$ nM — and the free energy of the embrace, from Eq. (8.4) at 310 K: $\Delta G° = 2.577 \times \ln(5.2 \times 10^{-8}) = 2.577 \times (-16.77) \approx -43$ kJ mol^{-1}, the price of about eight good hydrogen bonds, once the entropy tax of § 8.4 has been collected. The computation is the daily bread of every receptor pharmacology laboratory; the audit is that θ must lie between 0 and 1 and K_d in the plausible range of Table 8.1.*

> ### *WORKED EXAMPLE 8.2 — HAEMOGLOBIN'S WINDOW OF UNLOADING*

Using Eq. (8.6) with $n_H = 2.9$ *and* K' *chosen so* $P_{50} = 26$ *mmHg: occupancy in the lung (*$pO_2 = 100$ *mmHg) is* $\theta = (100/26)^{2.9}/(1 + (100/26)^{2.9}) = (3.85)^{2.9}/(1 + (3.85)^{2.9}) \approx 49.7/50.7 \approx 0.98$*; in working muscle (*$pO_2 = 20$ *mmHg):* $(0.77)^{2.9}/(1 + (0.77)^{2.9}) \approx 0.47/1.47 \approx 0.32$*. The tetramer delivers two-thirds of its cargo across the physiological range — while a hyperbolic myoglobin (*$n_H = 1$*,* $P_{50} = 2.8$ *mmHg) would still be 88 % loaded in that muscle: cooperativity is the difference between a transporter and a warehouse, and Chapter 13 will build the whole physiology upon it.*

Worked example 8.3 — Competition at the receptor

A neurotransmitter receptor has $K_d = 10$ *nM for its agonist and* $K_I = 50$ *nM for a competitive antagonist. At 100 nM antagonist, the agonist's apparent constant is, by Eq. (8.7),* $10 \times (1 + 100/50) = 30$ *nM: three times the agonist is now needed for the same response — the dose–response curve slides rightward in parallel, the signature by which the competitive mechanism is diagnosed (and distinguished from the non-competitive case of § 17.4, where the maximum itself falls). One line of algebra, and the difference between a drug that can be overcome and one that cannot.*

Exercises (with full resolutions)

Exercise 8.1

Derive Eq. (8.5) from Eq. (8.2) in full, and show that the half-saturation point is independent of the total receptor concentration.

Resolution. $\theta = [PL]/([P]_{free} + [PL])$*; with* $[PL] = [P]_{free}[L]/K_d$ *(Eq. 8.2 rearranged),* $\theta = ([P]_{free}[L]/K_d)/([P]_{free} + [P]_{free}[L]/K_d) = [L]/(K_d + [L])$*,* $[P]_{free}$ *cancelling — the occupancy depends on the free ligand and the constant only, never on how much receptor is present. The qualification is the practical trap: if the ligand is depleted by binding (receptor concentrations comparable to* K_d*), the free [L] is not the nominal one, and the titration must be solved as a quadratic — the "tight-binding" regime of the assay literature, and the second most common silent error in published* K_d *values, after the linearisations of § 8.5.*

Exercise 8.2

Avidin–biotin has $K_d \approx 10^{-15}$ *M. Compute the off-rate if the on-rate is diffusion-limited (*10^9 *L mol*$^{-1}$ *s*$^{-1}$*), and express the complex's half-life in years.*

Resolution. $k_{off} = K_d k_{on}$ *(Eq. 8.3)* $= 10^{-15} \times 10^9 = 10^{-6}$ *s*$^{-1}$*; half-life = ln 2/*$k_{off} \approx 6.9 \times 10^5$ *s ≈ 8 days — and the measured value for the avidin complex is in months, the on-rate being sub-diffusional. The arithmetic is why the avidin–biotin system serves as the molecular glue of half the assays of Part XI: for practical purposes, the embrace does not end.*

Exercise 8.3

An MWC protein has $L = 10^4$ (T favoured empty) and $c = K_R/K_T = 0.01$, with four sites. Estimate the ligand concentration at which R and T are equally populated.

Resolution. *The MWC equilibrium shifts as $L \times ((1 + [L]/K_T)/(1 + [L]/K_R))^4 = 1$ at equal populations: with $[L] \ll K_T$ the numerator is ≈ 1, so $(1 + [L]/K_R)^4 \approx L = 10^4$, giving $1 + [L]/K_R \approx 10$, $[L] \approx 9K_R$ — the midpoint of the sigmoid sits near the R-state constant, the T-state constant and L setting only how far the protein must be driven to get there. The full MWC equation is Chapter 13's business; the skeleton is this one line.*

EXERCISE 8.4 (INTERPRETATIVE)

"The lock-and-key model was falsified by induced fit." Assess the claim, with attention to what each model actually asserts.

Resolution (indicative). *The claim confuses levels. Fischer's assertion is complementarity: recognition requires geometric and chemical matching between site and ligand — and every induced-fit complex ever determined displays precisely such matching in its final state. Koshland's assertion is plasticity: the matching is achieved partly by rearrangement upon binding — which refutes the lock's rigidity, not its existence. The synthesis (§ 12.3's conformational selection) subsumes both: the site samples conformations, the ligand elects among them, and the final fit is Fischer's. In the history of models, this is the commonest fate: the old image is not falsified but promoted to a limiting case — as Newton was by Einstein, and the hyperbola by the sigmoid.*

EXERCISE 8.5

A condensate-forming protein phase-separates above 2 μM at low salt but not below 20 μM at 500 mM KCl. What does the salt dependence tell you about the driving interactions, and what experiment would confirm it?

Resolution. *Salt screens electrostatics (§ 2.4's dielectric and Debye–Hückel's ionic strength of § 4.7): the tenfold rise in the threshold says the condensate is held substantially by charge–charge or charge–dipole interactions (cation–π and π–π contacts are less salt-sensitive; hydrophobic contacts are strengthened by salt). Confirmation: mutate the charged patches (Arg/Lys-rich regions) to neutral residues and remap the phase boundary (Figure 8.6) — if the threshold rises in parallel, the assignment stands; if the aromatic mutants fail instead, the π system was the glue. The experiment is the condensate literature's standard assay, and the reasoning is Chapter 2's, applied to a phase.*

Research-level propositions

Proposition 8.α. The cratic entropy tax of § 8.4 ($\approx$ 30–45 kJ mol^{-1}) has been disputed for decades, with estimates spanning an order of magnitude. Using the isothermal titration calorimetry of § 54.9, design the experiment that most cleanly separates the entropic term from the enthalpic for a model protein–ligand pair, and assess whether the measured entropy of binding is consistent with the theoretical tax once solvent release is properly decomposed. Proposition 8.β. The MWC and KNF models are limiting cases of an ensemble allostery landscape. For a *chosen allosteric enzyme (aspartate transcarbamoylase is canonical), assemble the crystallographic, kinetic and single-molecule evidence (§§ 54.4, 54.10), and determine where on the concerted–sequential continuum the enzyme sits — and whether the answer depends on the ligand used to ask the question. Proposition 8.y. Condensates mature from liquids to solids, and the maturation is implicated in neurodegeneration. Formulate the kinetic theory of liquid-to-gel-to-solid ripening within a droplet (nucleation within a concentrated phase), and propose the experiment — FRAP kinetics, § 53.10 — that would detect the earliest departure from liquid behaviour in a living cell before any aggregate is microscopically visible.*

References (Chapter 8)

Fischer, E. (1894). Einfluss der Configuration auf die Wirkung der Enzyme. Berichte der Deutschen Chemischen Gesellschaft, 27(3), 2985–2993.

Hyman, A. A., Weber, C. A., & Jülicher, F. (2014). Liquid–liquid phase separation in biology. Annual Review of Cell and Developmental Biology, 30, 39–58.

Koshland, D. E., Némethy, G., & Filmer, D. (1966). Comparison of experimental binding data and theoretical models in proteins containing subunits. Biochemistry, 5(1), 365–385.

Monod, J., Wyman, J., & Changeux, J.-P. (1965). On the nature of allosteric transitions: A plausible model. Journal of Molecular Biology, 12(1), 88–118.

Nelson, D. L., Cox, M. M., & Hoskins, A. A. (2021). Lehninger principles of biochemistry (8th ed.). W. H. Freeman.

Berg, J. M., Tymoczko, J. L., Gatto, G. J., Jr., & Stryer, L. (2023). Biochemistry (9th ed.). W. H. Freeman.

PART III

Proteins: Structure, Dynamics and Function

CHAPTER 9

Amino Acids and Peptides

CONTENTS OF THIS CHAPTER

In which the twenty letters of the protein alphabet are introduced, classified by temperament, and priced by hydropathy; in which the bond that joins them — Pauling and Corey's planar, resonant, twice-hinged link — is given its geometry and its consequences; and in which the small peptides are shown to do the work of hormones, poisons and antibiotics while the long ones learn to fold.

9.1 General amino-acid structure

The amino acid is a molecule of exemplary modesty: a central carbon (the α-carbon), an amino group, a carboxyl group, a hydrogen, and — the whole point of the exercise — a side chain, R, which is the letter of the alphabet that this particular instance spells. The plan was stated in Chapter 3: nineteen of the twenty standard residues are chiral at the α-carbon, and life, for reasons that the homochirality debate of Exercise 3.4 left honestly open, has chosen the L-series exclusively. (The twentieth, glycine, with R = H, is achiral — the alphabet's one letter that reads the same in the mirror.) At physiological pH the molecule is a zwitterion — the α-carboxyl deprotonated ($pK_a \approx 2.2$), the α-ammonium protonated ($pK_a \approx 9.5$) — a fact of some consequence for everything that follows: the free amino acid carries no net charge at its isoelectric point (Eq. 6.7), and in the polymer the termini alone retain their charges, the rest being consumed in the peptide bonds of § 9.6. The student who draws the free amino group of a lysine inside a protein is making the commonest error of the novice; the polymer's backbone is a string of amides, and only the side chains speak.

9.2 Classification of side chains

Twenty letters, five temperaments (Figure 9.1). The nonpolar aliphatics (Gly, Ala, Val, Leu, Ile, Met, Pro) are the oil of the protein world — driven inward by the hydrophobic effect of Chapter 5, where they form the dry cores of globular proteins; Pro is the eccentric of the family, its side chain curling back to bite its own amino group, so that it is, strictly, an imino acid — with consequences (a fixed φ angle, no backbone N–H donor) that Chapter 10 will meet at every turn, literally. The aromatics (Phe, Tyr, Trp) are the rings: large, flat, hydrophobic with a polar edge (Tyr's phenol, Trp's indole N–H), and the protein's built-in spectroscopy — nearly all the ultraviolet absorbance and fluorescence by which Part XI measures proteins comes from these three letters. The polar uncharged family (Ser, Thr, Cys, Asn, Gln) trades in hydrogen bonds without charging interest; Cys is the redox-active member, its thiol the raw material of the disulphide bond (§ 11.5) and of the catalytic nucleophile of the cysteine proteases (§ 19.11's cousins). The acidic pair (Asp, Glu) and the basic trio (Lys, Arg, His) are the charges — anionic and cationic *respectively at pH 7.4, save for histidine, whose imidazole ($pK_a \approx 6.0$, Figure 9.3) sits exactly on the fence, and which is therefore the most interesting letter in the alphabet: the only side chain that can both give and take a proton at the cell's own pH, and the invariable recruit when an enzyme needs a proton shuttle (§ 15.5).*

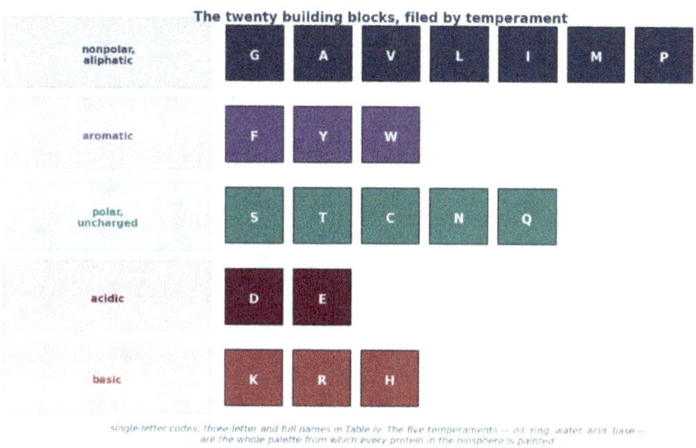

Figure 9.1 *The twenty standard residues filed by temperament: nonpolar aliphatic (navy), aromatic (plum), polar uncharged (teal), acidic (oxblood) and basic (crimson). The classification is the working map of this and the next three chapters — a protein is read, first of all, as a one-dimensional arrangement of these five temperaments.*

9.3 Acid–base behaviour

The titration of the amino acids was the business of § 6.8, and Figure 6.4's histidine remains the type specimen; the present section adds only the side-chain map (Figure 9.3) and its two working rules. First, at physiological pH the acidic side chains are anionic and the basic ones cationic — Lys and Arg essentially always charged, Asp and Glu essentially always, His sometimes — so that the protein surface at pH 7.4 is a mosaic of fixed charges, and the protein's net charge is a tunable property of its sequence (and the first handle by which the ion-exchange chromatography of § 53.5 separates proteins). Second, the pK_a values of the figure are solvent-exposed values; in the protein interior, buried charges are punished by the desolvation of § 5.4, and the pK_a values shift accordingly — an aspartate buried in a hydrophobic pocket can hold its proton to pH 7 and beyond, and the active sites of enzymes exploit exactly such shifts to create proton donors and acceptors of unnatural strength (the perturbed pK_a of the catalytic triad, § 19.11). The rule generalises Chapter 6's lesson: constants in biochemistry are constants only until the protein legislates otherwise.

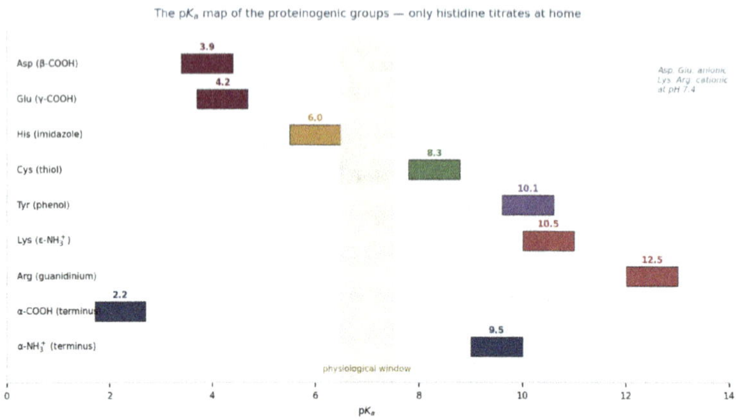

Figure 9.3 *The pK_a map of the proteinogenic titrating groups (solvent-exposed values), against the physiological window (shaded). Only histidine's imidazole crosses it — the chemical reason that proton transfer in enzymes is histidine's profession, and that haemoglobin's Bohr effect (§ 13.5) is substantially a histidine effect.*

9.4 Isoelectric points

The isoelectric point was defined by Eq. (6.7); the protein's case deserves the paragraph it earns. A protein carries dozens of titrating groups, and its net charge is the signed sum of all of them; the pH at which the sum vanishes is the protein's pI, and it ranges from ≈ 4.5 (pepsin, an enzyme of the gastric acid) to ≈ 11 (lysozyme, and the histones that must bind anionic DNA). Two properties of the pI have practical teeth. Solubility is minimal there — the net-neutral molecules, lacking mutual repulsion, aggregate and precipitate — which is both a preparative hazard and, inverted, a purification tool (isoelectric precipitation is the oldest fractionation in the trade). And electrophoretic mobility vanishes there, so that a pH gradient sorts a protein mixture by pI as a sieve sorts by size: isoelectric focusing, the first dimension of the two-dimensional gels of proteomics (§ 55.2), is Eq. (6.7) industrialised. The reader should also note the design rule hidden in the numbers: intracellular proteins tend to carry pI values below 7, so that at physiological pH they are net anionic — repelled by the anionic phospholipid head groups of the membranes they should not stick to, and left free in the cytosol where their work is.

9.5 Non-standard amino acids

Twenty letters suffice for the ribosome, but not for the proteome: the finished protein is edited, expanded and adorned until its alphabet numbers in the hundreds. Three routes enlarge it. Cotranslational expansion is the smallest and most romantic: selenocysteine — the twenty-first amino acid, with selenium in place of cysteine's sulphur — is inserted at UGA codons re-read by a special machinery (§ 47.1), and staffs the redox-active sites of the glutathione peroxidases and thioredoxin reductases of § 26.11; pyrrolysine, the twenty-second, is confined to certain archaea and their methyltransferases. Post-translational modification is the largest route, and has Figure 9.5 to itself in § 9.10. And non-ribosomal synthesis — the peptide antibiotics and siderophores assembled by the giant NRPS enzyme factories — admits D-amino acids, hydroxylated residues, ornithine and hundreds more, into chains that no ribosome ever touched: the bacitracin of the medicine cabinet is such a chain, and its D-residues are precisely why the reader's proteases cannot digest it. Table 9.2 files the principal non-standard residents; the moral is the chapter's recurring one — the twenty letters are the ribosome's alphabet, not life's.

***Table** 9.2 The principal non-standard amino acids found in proteins, with their route of entry and their office.*

Residue	*Route*	*Office / location*
selenocysteine	*cotranslational (UGA recoded)*	*redox catalysis; peroxidases, reductases*
pyrrolysine	*cotranslational (UAG recoded)*	*methanogen methyltransferases*
4-hydroxyproline	*post-translational (prolyl hydroxylase)*	*collagen stability (§ 14.1)*
γ-carboxyglutamate	*post-translational (vitamin K-dependent)*	*Ca^{2+} binding; clotting factors*
phosphoserine/threonine/tyrosine	*kinases*	*regulation (Ch. 18, Ch. 49)*
D-amino acids	*non-ribosomal; racemases*	*peptide antibiotics; bacterial walls*

9.6 The peptide bond

The bond that joins the letters was the subject of Chapter 2's closing case study, and its geometry — the fruit of Pauling and Corey's small-molecule crystallography — is drawn in Figure 9.2: the C–N link of 1.32 Å, 40 % double-bond by resonance, planar, with a rotation barrier of ≈ 80 kJ mol^{-1} that renders it, at 310 K, a hinge that does not hinge. Two geometric corollaries govern everything in Part III. The peptide unit is trans essentially always (the cis form placing the adjacent a-carbons in collision; only X–Pro bonds, where the collision is symmetric, admit cis at ≈ 5–10 % — and the cis–trans isomerisation of proline is a recognised rate-limiting step of protein folding, § 11.6, with its own enzyme family, the prolyl isomerases, to accelerate it). And the backbone's freedom is thereby reduced to two rotations per residue: the φ angle about N–C_α and the ψ angle about C_α–C (Figure 9.2) — the whole architecture of proteins being, as Ramachandran first computed, the exploration of that two-dimensional freedom under the constraint that atoms may not collide (§ 10.5).

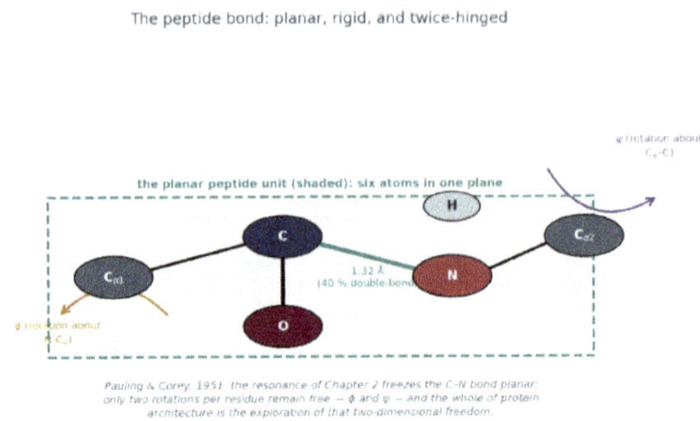

Figure 9.2 The planar peptide unit: six atoms ($C_{\alpha 1}$, C, O, N, H, $C_{\alpha 2}$) in one plane (shaded), the peptide C–N bond shortened and stiffened by resonance (teal), and the two free torsions of the backbone, φ (amber) and ψ (plum), marked at the α-carbons. Geometry after Pauling, Corey & Branson (1951); the rotation barriers from the amide resonance measurements of Chapter 2's case study.

9.7 Peptide planarity

The planarity deserves a section of its own, because it is the single fact from which the predictability of protein structure descends. Were the peptide bond freely rotating, each residue would contribute three torsions, the conformational space of a 100-residue chain would have 300 dimensions, and Levinthal's paradox of § 3.9 would be worse by another factor of thousands. The resonance of § 2.5 reduces the freedom to two torsions, the excluded-volume constraint of Ramachandran reduces the allowed territory of those two to about a quarter of the map, and the chain's problem becomes, if not easy, at least surveyable. There is a deeper lesson here that the dialectic of the treatise should not pass over: the foldability of proteins is prepaid by chemistry. Evolution did not have to solve the folding problem from scratch; the peptide bond's planarity, the tetrahedral carbon's geometry, and the hydrophobic effect's drive to compactness had solved most of it before the first protein existed — the origin of structure, like the origin of so much else in this book, lies one level down from where it is usually sought.

9.8 Backbone torsion angles

The two free angles, φ and ψ, are the coordinates of protein architecture, and their naming and measuring are the grammar of Chapter 10. Convention: φ is the torsion C′–N–C_α–C′ (rotation about the N–C_α bond), ψ the torsion N–C_α–C′–N (about C_α–C′), and the peptide unit's own ω, about the peptide bond itself, is fixed at 180° (trans) by § 9.7. The allowed regions of the (φ, ψ) map are set by steric collision — the α-carbon's neighbours may not interpenetrate — and were computed by G. N. Ramachandran in 1963 (Ramachandran, Ramakrishnan, & Sasisekharan, 1963): the map that bears his name (Figure 3.2, and redrawn for proteins in § 10.5) shows two great permitted basins — the right-handed α-region ($\varphi \approx -60°$, $\psi \approx -45°$) and the β-region ($\varphi \approx -120°$, $\psi \approx +130°$) — and a smaller left-handed α-region, forbidden to L-residues save glycine, whose two hydrogens grant *it diplomatic immunity across the whole map. Every secondary structure of Chapter 10 is a repeat of one point of this map; every turn is a short excursion; and the validation of every crystal structure in Chapter 54 ends with the same question: do the residues sit in the permitted basins, and if not, who let them out?*

9.9 Biologically active peptides

Before the great proteins, the small ones: peptides of a handful to a few dozen residues that do the work of hormones, neurotransmitters, toxins and antibiotics, and that demonstrate, at readable scale, the principles the large proteins will multiply. Oxytocin, nine residues with a disulphide ring, is the hormone of labour and of trust (two claims for one molecule, the second popularised beyond its evidence — the reader is warned that small peptides attract large journalism); insulin's precursor is a single chain that folds, pairs by disulphides, and is clipped into the two-chain hormone of § 39.2 — the demonstration, at 51 residues, that a peptide can carry its folding instructions in its sequence, the Anfinsen principle in miniature. The peptide toxins of cone snails and scorpions, cross-braced by disulphides into knife-resistant packages, are among the most potent ligands of the ion channels of Chapter 24; the peptide antibiotics — gramicidin's channel, valinomycin's cage for potassium — taught the subject what an ion pore could be before any channel was cloned. And glutathione (γ-Glu-Cys-Gly, three residues) is the cell's redox buffer and mercenary, its thiol oxidised and reduced a million times a second across the tissues (§ 38.8). Table 9.1 files the company; the lesson is that function does not require size, only information — and the peptide, short as it is, can carry enough.

Table 9.1 *Representative biologically active peptides, with their sizes and offices. The table is a reading list for the argument that sequence, at any length, is function.*

Peptide	*Residues*	*Office*
glutathione	*3*	*cellular redox buffer (§ 38.8)*
oxytocin, vasopressin	*9*	*posterior pituitary hormones*
angiotensin II	*8*	*blood-pressure regulation (§ 40.6)*
α-conotoxins	*13–19*	*nicotinic-receptor blockade; snail hunting*
insulin	*51 (two chains)*	*metabolic master hormone (§ 39.2)*
gramicidin A	*15*	*the pore that taught channels (§ 24.4)*
cyclosporin A	*11 (cyclic)*	*immunosuppression; calcineurin inhibition*

9.10 Post-translational modification of residues

The ribosome writes with twenty letters; the cell edits with hundreds of pens. Post-translational modification is the collective name for every chemical revision of the finished chain — phosphorylations, acetylations, methylations, hydroxylations, glycosylations, lipidations, ubiquitinations, and the covalent cross-links of disulphide — and Figure 9.5 files the principal pens with their residues and their offices. Three properties of the system deserve emphasis now, for Parts X and XII will build on them. The modifications are reversible — kinases are answered by phosphatases, acetylases by deacetylases — which converts them from decoration into instrumentation: a phosphorylation is a switch, and the signalling networks of Chapter 49 are written almost entirely in these few reversible inks. They are specific — each modifying enzyme reads a sequence or structural context, so that the modification is addressed to a residue, not to a class. And they are combinatorial — a histone tail or a p53 molecule carries a dozen modifiable sites whose patterns, read jointly, encode more than any single modification could: the "histone code" of § 48.4 is the alphabet of Table 9.2 grown into a language. The arithmetic is worth a line: twenty letters, some ten common modifications on perhaps five common residues, and the effective alphabet of the proteome is not twenty but several hundred — which is the post-translational answer to the question of how 20 000 genes make a human.

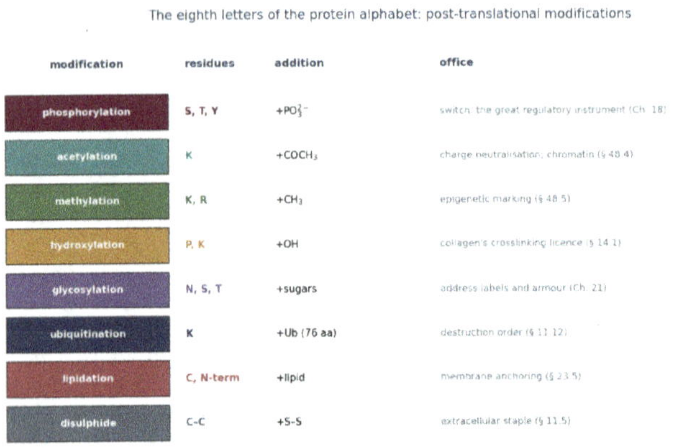

***Figure 9.5** The principal post-translational modifications, with the residues they mark, the chemical addition, and the office the mark performs. Phosphorylation (oxblood) heads the list because it heads the regulation: an estimated third of the proteome is phosphorylated somewhere, and the kinase–phosphatase pairs of Parts IV and X run the switchboard.*

9.11 Chemical synthesis of peptides

The chapter closes with the chemist's answer to the ribosome: the synthesis of peptides by hand. The problem is selectivity — coupling residue A to residue B without coupling A to A or B to B — and its solution is the protecting-group strategy: cap every reactive group except the two that must meet, couple, deprotect, repeat. The strategy worked, laboriously, from Fischer's day (his octadecapeptide of 1907 was the record for decades), but each cycle required a purification, and yields compounded to nothing. Merrifield's revolution of 1963 (Plate 9.1, Figure 9.6) (Merrifield, 1963) was to anchor the growing chain to an insoluble resin, so that purification became filtration: wash away the reagents, keep the bead. The cycle of the figure — deprotect, couple, repeat, cleave — runs at ≈ 99.5 % per step in a modern synthesiser, chains of fifty residues are routine, and the solid-phase idea has since colonised the synthesis of oligonucleotides, of oligosaccharides, and of the combinatorial libraries of drug discovery. The historical dialectic is worth a sentence: the ribosome, a billion years older, achieves 99.99 % per step at twenty residues per second, in water, at 310 K, with no protecting groups at all — the machine of Chapter 47 remains, in speed and cleanliness, beyond the chemist's reach; but Merrifield's machine makes sequences the ribosome cannot, and between the two, any peptide on Earth is now available to order.

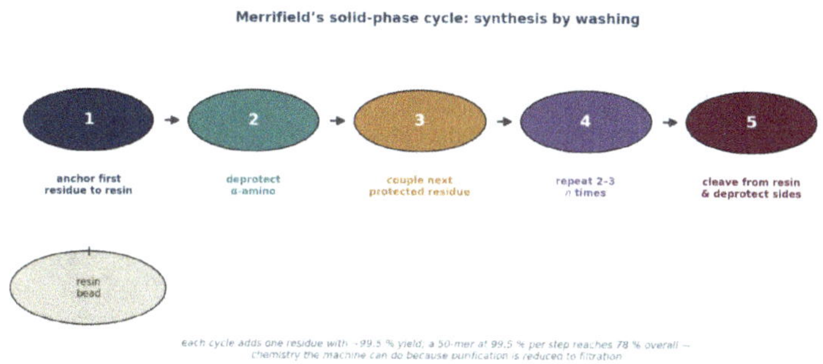

Figure 9.6 Merrifield's solid-phase synthesis: the growing chain is anchored to a resin bead, and each cycle — deprotect, couple, wash, repeat — adds one residue, purification being reduced to filtration. The arithmetic of the legend (99.5 % per step, 78 % over fifty steps) is the reason automated peptide synthesis is an industry, and the contrast with the ribosome's fidelity is § 47.9's subject.

Plate 9.1 Robert Bruce Merrifield (1921–2006), whose solid-phase method made the synthesis of peptides a machine trade. Portrait from the Nobel Foundation archive, nobelprize.org (Nobel Prize in Chemistry, 1984), reproduced from the Foundation's publicly distributed image.

HISTORICAL NOTE — FISCHER'S PEPTIDES AND THE GHOST OF PROTEIN STRUCTURE

That proteins are chains of amino acids joined by peptide bonds was proposed independently by Hofmeister and by Fischer in 1902 — and was doubted for thirty years by men who thought proteins too large, too colloidal, too mysterious to be mere polymers. Fischer himself, who had linked eighteen residues by 1907 and named the product a "peptide", believed the natural proteins were built the same way but could not prove it; the proof came in instalments — Sanger's insulin sequence of 1953 (the first protein shown to have a definite sequence, which some had doubted), Pauling's helix, Perutz's and Kendrew's structures. The long resistance is instructive: the idea that a protein is a string of letters seemed, to excellent chemists, indecently simple —

and the history of this subject is, repeatedly, the discovery that nature's simplicities are deeper than our suspicions.

Chapter 9 — Conceptual synthesis

The alphabet has been delivered: twenty letters on a zwitterionic frame, five temperaments (§§ 9.1–9.2), titration behaviour ruled by the pK_a map with histidine at the crossroads (§§ 9.3–9.4), and an expanded vocabulary beyond the ribosome's twenty (§ 9.5). The peptide bond is planar and resonant, leaving two torsions per residue — the coordinates of all architecture (§§ 9.6–9.8); small peptides already do the work of hormones, toxins and antibiotics (§ 9.9); the finished chain is edited by hundreds of modifications, reversible, specific and combinatorial (§ 9.10); and the chemist, following Fischer and Merrifield, can now write any sequence to order (§ 9.11). Direction checked against Figure ii: the letters are in hand. Chapter 10 assembles them — the hierarchy of protein structure, from the sequence to the folded, functioning molecule.

WORKED EXAMPLE 9.1 — THE CHARGE OF A PEPTIDE AT PH 7.4

The peptide Glu–Ala–His–Lys at pH 7.4: assign the charges. N-terminus: +1 ($pK_a \approx 9.5$); Glu side chain: −1; His: pK_a 6.0, so by Eq. (6.6) the protonated fraction is $1/(1 + 10^{7.4-6.0}) = 1/(1 + 25) \approx 0.04$, charge ≈ +0.04; Lys: +1; C-terminus: −1 ($pK_a \approx 2.2$). Net ≈ +0.04 — the peptide is essentially neutral at pH 7.4, and its pI lies nearby: a four-residue demonstration that isoelectric arithmetic (§ 9.4) is Henderson–Hasselbalch arithmetic (§ 6.5), and nothing more.

WORKED EXAMPLE 9.2 — THE FIDELITY ARITHMETIC OF SYNTHESIS

Compare the ribosome and the synthesiser over 100 couplings. Machine, 99.5 % per step: $0.995^{100} = 0.61$ — 61 % full-length product. Ribosome, error rate ≈ 10^{-4} per residue (§ 47.9's proofreading included): fidelity $0.9999^{100} \approx 99.0$ %. The comparison quantifies § 9.11's dialectic: the chemist's machine wins on repertoire (any letter, any chemistry), the cell's machine on fidelity — and the gap between 61 % and 99 % over one hundred couplings is the price of the protecting-group economy versus the template-directed one.

WORKED EXAMPLE 9.3 — HOW MANY CONFORMATIONS DOES PLANARITY SAVE?

Without planarity, a 100-residue chain has ≈ 300 free torsions; at three rotameric states each, $\approx 3^{300} \approx 10^{143}$ conformations. With the peptide bond fixed, ≈ 200 torsions: $3^{200} \approx 10^{95}$. With Ramachandran's quarter-map restriction, ≈ (3/4... say two effective states per allowed torsion) $\approx 2^{200}$... — even the friendliest arithmetic leaves $\approx 10^{60}$, and Levinthal's paradox (§ 3.9) is intact: planarity reduces the exponent by a third, not the paradox. The conclusion is § 9.7's: chemistry prepaid the dimensionality, but the folding is still guided, not searched — and Chapter 11's energy landscapes supply the guidance.

Exercises (with full resolutions)

Exercise 9.1

Compute the isoelectric point of glutathione (γ-Glu–Cys–Gly; pK_a values: α-COOH of C-terminal Gly 2.3, side-chain COOH of the γ-Glu 4.1, thiol 8.7, α-NH_3^+ 9.4).

Resolution. *Titratable in the acidic range: the two carboxyls; the species neutral in charge sits between them (after α-COOH deprotonates at 2.3 and before the side-chain COOH at 4.1): pI = (2.3 + 4.1)/2 = 3.2 by Eq. (6.7). Glutathione is therefore anionic at physiological pH — and its high cytosolic concentration (≈ 5–10 mM) makes it a significant contributor to the cell's anion budget, a fact the phosphate accounting of § 6.9 politely omitted.*

Exercise 9.2

A synthetic peptide requires 40 couplings at 99.0 % per step. What fraction is full-length, and what per-step fidelity is required for 90 % full-length over 100 residues?

Resolution. *$0.99^{40} = e^{40 \ln 0.99} = e^{-0.402} \approx 67$ %. For $0.90 = f^{100}$: $\ln f = \ln(0.9)/100 = -1.054 \times 10^{-3}$, $f = 0.99895$ — 99.895 % per step, within modern machine capability but beyond casual practice: the arithmetic explains why 50-mers are routine, 100-mers are projects, and proteins are still made by ribosomes.*

Exercise 9.3

Estimate the net charge of insulin (pI ≈ 5.3) at pH 7.4 and at pH 5.0, and predict its solubility behaviour between the two.

Resolution. *Above its pI the hormone is net anionic; at pH 5.0, just below pI 5.3, it is near neutrality — and solubility is minimal near pI (§ 9.4): insulin precipitates around pH 5, which is both a formulation hazard (the zinc–insulin suspensions of the pharmacy exploit it deliberately — the slow-dissolving crystals are the long-acting preparations) and a historical accident without which Banting and Best's acidic extractions might never have yielded crystals at all.*

Exercise 9.4 (interpretative)

Why has evolution restricted the ribosome to twenty letters, when the post-translational apparatus writes with hundreds? Propose two costs of a larger canonical alphabet, and assess which was the stronger constraint.

Resolution (indicative). *The candidates: (i) recognition cost — each new letter needs a dedicated aminoacyl-tRNA synthetase, and discrimination errors grow as letters crowd (§ 47.3's editing); (ii) metabolic cost — each letter needs a biosynthetic pathway; (iii) code cost — twenty letters fill the 64-codon table with redundancy already; more letters would force a fourth base per codon or sacrifice the degeneracy that buffers mutation (§ 47.1). The assessment: the code's degeneracy and the recognition fidelity are the binding constraints — the cell preferred a small alphabet written faultlessly and edited* *afterwards (PTMs) to a large alphabet written sloppily: fidelity at synthesis, variety at revision, which is the engineering of § 9.10 read evolutionarily.*

Exercise 9.5

The catalytic triad of chymotrypsin buries the His–Asp pair; the aspartate's pK_a is perturbed upward. Estimate the fraction of protonated aspartate at pH 7 if the buried pK_a is 7.0 rather than the exposed 3.9, and state the catalytic consequence.

Resolution. *Exposed: fraction protonated = $1/(1 + 10^{7-3.9}) \approx 10^{-3}$ — essentially never. Buried: $1/(1 + 10^{7-7}) = 0.5$ — half. The protein's microenvironment has converted a permanently anionic residue into a proton reservoir at physiological pH: the triad's charge-relay (§ 19.11) runs on exactly this perturbation, and the exercise is § 9.3's rule in action — constants are constants until the protein legislates otherwise.*

Research-level propositions

Proposition 9.α. The twenty-letter alphabet has been expanded in the laboratory (orthogonal synthetase/tRNA pairs inserting non-canonical residues at amber codons). Review the fidelity and yield of the best current systems (§ 47.3's machinery adapted), and compute how many additional canonical letters the code could sustain before the recognition-error budget of Exercise 9.4 is exhausted. Proposition 9.β. Proline cis–trans isomerisation is rate-limiting in the folding of many proteins (§ 11.6), and the prolyl isomerases (PPIases) that accelerate it are drug targets in neurodegeneration and transplant medicine. Derive the expected folding-rate enhancement from the PPIase-catalysed barrier reduction, and design the stopped-flow experiment (§ 16.9) that would measure it in vivo rather than in the cuvette. Proposition 9.γ. The "histone code" hypothesis (§ 9.10) claims combinatorial reading of PTM patterns; critics reply that most marks are correlated and the effective alphabet is small. From the ChIP-seq literature (§ 55.1), estimate the mutual information between PTM patterns and transcriptional states across one cell type, and adjudicate the hypothesis in bits.

References (Chapter 9)

Creighton, T. E. (1993). *Proteins: Structures and molecular properties* (2nd ed.). W. H. Freeman.

Merrifield, R. B. (1963). Solid phase peptide synthesis. I. The synthesis of a tetrapeptide. *Journal of the American Chemical Society, 85*(14), 2149–2154.

Nelson, D. L., Cox, M. M., & Hoskins, A. A. (2021). *Lehninger principles of biochemistry* (8th ed.). W. H. Freeman.

Pauling, L., Corey, R. B., & Branson, H. R. (1951). The structure of proteins: Two hydrogen-bonded helical configurations of the polypeptide chain. *Proceedings of the National Academy of Sciences, 37*(4), 205–211.

Ramachandran, G. N., Ramakrishnan, C., & Sasisekharan, V. (1963). Stereochemistry of polypeptide chain configurations. *Journal of Molecular Biology, 7*(1), 95–99.

Sanger, F., & Tuppy, H. (1951). The amino-acid sequence in the phenylalanyl chain of insulin. *Biochemical Journal, 49*(4), 463–490.

CHAPTER 10

The Hierarchy of Protein Structure

CONTENTS OF THIS CHAPTER

In which the letters of Chapter 9 are assembled into architecture: the sequence, the local folds — helix, sheet and turn — the map of the two allowed torsions, the motifs and domains that recur through the whole protein kingdom, the packing of one chain into a fold and of folds into assemblies; and the two great families, fibrous and globular, with a third family — the disordered — admitted to full citizenship at last.

10.1 Primary structure

The first storey of the protein is a sentence: the order of amino acids along the chain, read from the amino terminus to the carboxyl, one letter at a time, no punctuation. It is customary to call this the "primary structure", and the custom is correct, but the reader should feel how much the mild word sequence conceals. The sentence is not a description of the protein; in an exact and astonishing sense it is the protein — the sole information the ribosome supplies, the sole inheritance the gene transmits, and the sufficient determinant (in the cell's own solvent, with the cell's own chaperones) of the folded architecture of the next three storeys. That a one-dimensional string should specify a three-dimensional machine is the central miracle of molecular biology, and Anfinsen's experiment (case study below) is its proof: unfold a protein utterly, and it refolds, because the instructions were never anywhere but in the sequence. The whole of bioinformatics — the alignments of § 55.7, the predictions of § 55.8 — is an attempt to read the sentence as the solvent reads it; the whole of genetics is the machinery that writes it.

10.2 Secondary structure

The second storey is local: the regular, hydrogen-bonded folds that short stretches of backbone adopt of their own geometry, independent of the protein's final shape. Two facts from Chapter 9 make such repeats possible: the peptide unit is planar (§ 9.7), so the chain is a string of rigid plates hinged at the α-carbons; and the backbone is studded with donors and acceptors (every N–H and C=O), so a fold that pairs them systematically is stabilised by the sum of many small bonds — the arithmetic of multitude from § 2.8, now in architecture. The two great solutions are the α-helix and the β-sheet, and it is a matter for wonder, worth pausing over, that both were predicted from first principles by Pauling, Corey and Branson in 1951 (Pauling, Corey, & Branson, 1951) — years before any protein structure was seen — by the simple expedient of requiring that every backbone donor find an acceptor and every atom keep its proper distance: the constraints of Chapter 9, solved like a puzzle in solid geometry. The prediction and its verification (Perutz's haemoglobin, Kendrew's myoglobin) are the historical notes and plates of this chapter; the structures themselves are the next two sections.

10.3 The α-helix

The α-helix is a cylinder of 3.6 residues per turn, a rise of 0.54 nm per turn (0.15 nm per residue), and a diameter of about 0.5 nm (Figure 10.2). Its stabilising pattern is the signature of the type: every backbone C=O accepts a hydrogen bond from the N–H four residues ahead (i → i+4), so that the helix is a column of nearly parallel hydrogen bonds running its whole length — a dipole, in fact, with measurable macro-polarity, positive at the N-terminus, which is why phosphate groups so often bind at helix starts (§ 19.11's kinases know this). The side chains project outward from the cylinder, so the helix's personality is painted on its surface: a helix with hydrophobes on one face and hydrophiles on the other is amphipathic, and such helices are the building blocks of membrane proteins (the seven-helix receptors of § 50.1) and of the coiled-coils of § 14.4's myosin. Helix content varies from nothing to nearly everything: myoglobin is ≈ 75 % helical; the fibrous keratins of § 14.3 are helices of helices. The helix is the commonest secondary structure of the protein kingdom, the first to be predicted, and the first thing the eye learns to find in a new structure — the reader should note its silhouette in Figure 10.2 and look for it, from now on, in every ribbon diagram this book displays.

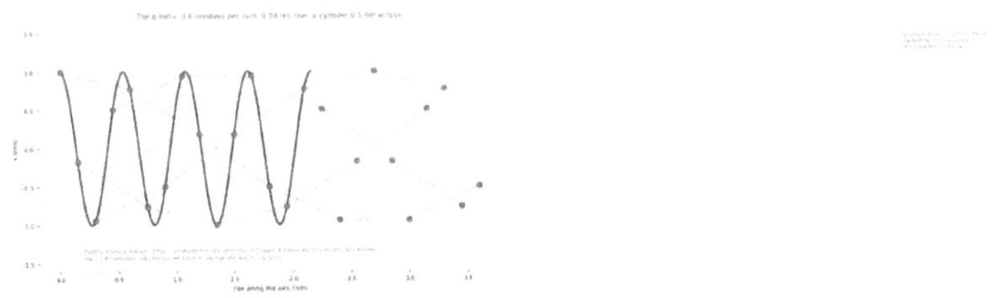

Figure 10.2 The α-helix: the backbone trace (navy), the α-carbons (amber, 3.6 per turn), and the i → i+4 hydrogen bonds (dashed teal) running along the axis. Rise 0.54 nm per turn, diameter ≈ 0.5 nm; side chains project outward. Predicted in 1951 from the geometry of Chapter 9; verified by Perutz's 1.5 Å reflection; the commonest fold of the protein kingdom.

10.4 β-sheets and β-turns

The β-sheet is the helix's complement: instead of coiling upon itself, the chain stretches nearly flat (the β-strand, 0.35 nm rise per residue) and hydrogen-bonds sideways to neighbouring strands, building a pleated sheet (Figure 10.3). Two arrangements divide the family: antiparallel, with strands running opposite ways and the hydrogen bonds linear (the stronger); and parallel, with strands aligned and the bonds slanted (the commoner in large sheets). The sheet's surface is thereby divided: side chains alternate above and below, so a β-protein reads as alternating hydropathy — one face to the solvent, one to the core — and the reader who sees alternating patterns in a sequence should suspect a strand. Between the regular folds lie the turns: short reversals of direction, four residues typically, that let the chain fold back upon itself; the β-turn, with its characteristic glycine at position 3 (the only residue whose two hydrogens permit the required twist — the alphabet legislating again), is the commonest way a globular protein makes a corner. Turns are not packing material; they are the corners of the architecture, and they sit, with statistical preference, at the surface, where the architecture meets the world.

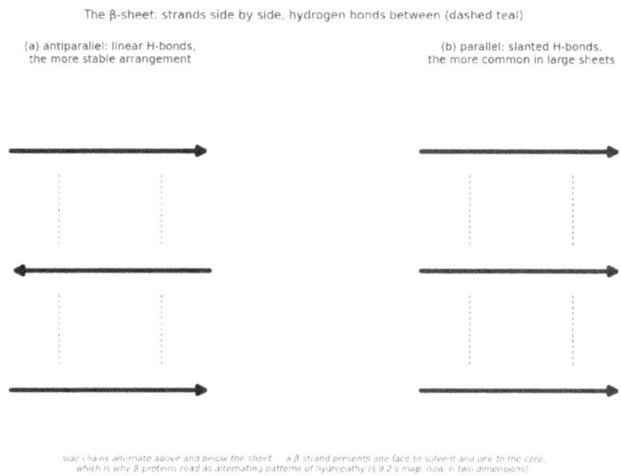

Figure 10.3 *The β-sheet: extended strands (arrows, pointing C-terminus-ward) joined by inter-strand hydrogen bonds (dashed teal). (a) Antiparallel: linear bonds, greater stability. (b) Parallel: slanted bonds, commoner in large sheets. Side chains alternate above and below the pleat — one face for the solvent, one for the core.*

10.5 Ramachandran analysis

The two storeys so far are consequences of one map. Ramachandran's 1963 computation (Ramachandran et al., 1963) asked the simplest of questions: which (φ, ψ) pairs of § 9.8 are sterically permitted — where can the backbone go without colliding with itself? The answer (Figure 10.4) is the most reproduced map in structural biology: two large permitted basins (the α-region, φ ≈ −60°, ψ ≈ −45°; the β-region, φ ≈ −120°, ψ ≈ +130°), a small left-handed α-region open mostly to glycine, and a vast white exile where atoms interpenetrate. The map's uses are three, and all three are weekly practice. Explanation: the α-helix and β-sheet are not two inventions but two basins — nature did not choose two folds; she fills two permitted regions. Validation: every crystal structure of Chapter 54 is checked against the map, and a residue in exile is either an error or an excitement (active-site residues sometimes sit strained at the map's edge — the protein paying a geometric tax for chemistry). Design: the protein engineer of § 56.11 builds new folds within the permitted basins, the map being the law of the land. The Ramachandran plot is, in short, the constitution of protein architecture — brief, geometric, and obeyed by every structure in this book.

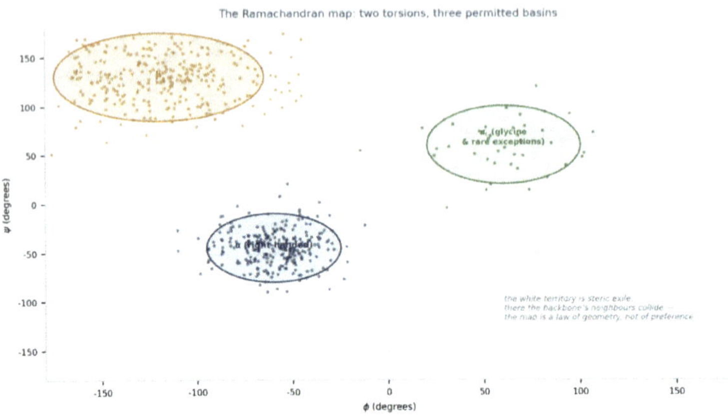

Figure 10.4 *The Ramachandran map: the (φ, ψ) plane with its three permitted basins (α, navy; β, amber; the left-handed* α_L*, green, mostly glycine's), and a cloud of residues from real structures occupying them. The white territory is steric exile; the map is a law of geometry, and its two main basins are the whole of secondary structure.*

10.6 Motifs and supersecondary structure

Between the local folds and the whole protein lies the vocabulary of recurrent combinations — the motifs: short, stable signatures of a few secondary elements that recur through the whole protein kingdom, in molecules with no other kinship. The helix–turn–helix of Figure 10.5a (two helices at a fixed angle, reading the major groove of DNA) is the signature of a thousand transcription factors; the β–α–β unit (strand–helix–strand, the helix packing above a parallel sheet) iterated is the Rossmann fold of Figure 10.5b — the dinucleotide-binding domain of every dehydrogenase in Chapter 26, so ancient and so reusable that evolution has employed it, by the estimate of the structural databases, more often than any other enzymatic scaffold. Motifs are the proverbs of protein architecture: short, portable, and apt to appear wherever the same need recurs — which is why the sequence-searching of § 55.7 can recognise a protein's office from a few dozen residues, and why the kingdom of proteins, for all its diversity, is built of a surprisingly small stock of parts.

10.7 Domains

The domain is the protein's province: a stretch of 50–350 residues that folds independently, has its own hydrophobic core, and often its own office — a kinase domain, a binding domain, a membrane anchor. Large proteins are federations of domains, and the federal principle has two profound consequences. Modularity: domains can be gained, lost, duplicated and shuffled between proteins by the genetic recombination of Chapter 45, so that new proteins are made not by inventing new folds but by rearranging old provinces — the kinase domain of § 18.5 appears in hundreds of signalling proteins, each a different federation. Autonomy: a domain can be expressed alone and keep its fold and function, which is the experimental warrant for the structural biology of Chapter 54 (the domain is the natural unit of crystallography) and the engineering warrant for the therapeutic antibodies of § 14.7, whose variable domains are lifted bodily onto human scaffolds. The TIM barrel of Figure 10.5c — eight β/α repeats closing into a barrel, the catalytic site at the barrel's mouth — is the commonest enzyme domain known, present in a tenth of all enzymes: one fold, dozens of chemistries, the proof that function sits in the active site's residues, not in the architecture that presents them.

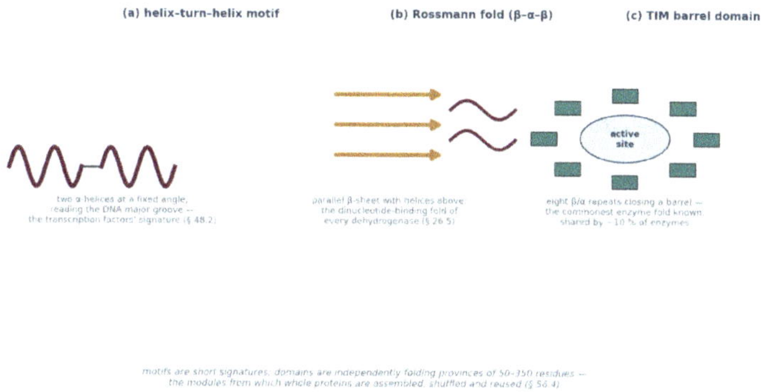

Figure 10.5 *Motifs and domains. (a) The helix–turn–helix motif: two α-helices at a fixed angle, the DNA-readers' signature. (b) The Rossmann fold: parallel β-strands (amber arrows) with helices (oxblood) packing above — the dehydrogenases' ancient scaffold. (c) The TIM barrel: eight β/α repeats (teal) closing a barrel with the active site at its mouth — the commonest enzyme fold known.*

10.8 Tertiary structure

The third storey is the whole of one chain: the tertiary structure, the specific packing of helices, sheets and turns into a single compact fold. Its principles are now all familiar, and the reader should feel them click into place. The hydrophobic effect of Chapter 5 drives the nonpolar side chains inward, and the core of a globular protein is accordingly as dry as oil — close-packed, with a density like a crystal's, and (a fact that surprises every generation anew) with very little empty space: the protein interior is not a cavity but a solid. The polar and charged residues face outward, hydrated and social. The fold is stitched by everything Chapter 2 taught: hydrogen bonds everywhere (the backbone's donors and acceptors must all be satisfied, in the core as outside — a buried unsatisfied donor is a catastrophe the fold pays for), salt bridges where the dielectric is low, dispersion contacts by the thousand, and, in the secreted proteins, the covalent staples of disulphide bonds (§ 11.5). The result is a paradox that the folding literature has made its own: the native fold is unique, precise to the ångström — and yet marginal, more stable than the unfolded chain by only ≈ 20–60 kJ mol^{-1}, the free energy of a handful of hydrogen bonds. Proteins are built, as § 1.10's third principle requires, at the edge of stability: rigid enough to be machines, marginal enough to move.

10.9 Quaternary structure

The fourth storey is the company: the assembly of several folded chains into a functioning oligomer, held by the same noncovalent interactions as the tertiary fold but now between molecules. The quaternary level brings with it the two properties that monomers cannot have. Symmetry: oligomers are almost always *symmetric (haemoglobin's $\alpha_2\beta_2$ tetramer, the virus capsid's icosahedron), because a symmetric interface is built of repeated uses of one good contact — the arithmetic of avidity from § 8.12, written in architecture. And allostery: the communication between subunits that Chapter 8's MWC model formalised — the quaternary transition that lets one subunit's ligand inform the rest, which a monomer, by definition, cannot perform. The quaternary assembly is therefore not an aggregation but an instrument: it is where cooperativity lives, and the protein that needs to be a switch must first be a company.*

10.10 Fibrous and globular proteins

The protein kingdom divides, at the coarsest resolution, into two great families (Table 10.2). The fibrous proteins — collagen, keratin, elastin, the myosin tails — are elongated, repetitive, insoluble, and built for strength and elasticity: their sequences are monotonous (collagen's Gly-X-Y, keratin's helix-friendly repeats), their structures are extended rather than compact, and their office is the architecture of tissues rather than the chemistry of cells. The globular proteins — enzymes, antibodies, transporters, regulators — are compact, soluble, diverse, and built for action: their sequences are heterogeneous, their cores dry, their surfaces social, their offices the catalysis and recognition of Parts IV and X. The distinction is not absolute (motor proteins are fibrous and mechanical; membrane proteins are globular and oily), but it organises the kingdom usefully: the fibrous proteins are the bricks and beams of the body, the globular ones its machinery — and Chapter 14 will visit both families in their working dress.

Table 10.2 *Fibrous and globular proteins contrasted — the two great families of the protein kingdom.*

Property	Fibrous	Globular
shape	elongated, extended	compact, roughly spherical
sequence	repetitive (Gly-X-Y; helix repeats)	heterogeneous
solubility	insoluble (tissues, fibres)	soluble (cytosol, plasma)
secondary structure	one dominant type	mixed α/β
office	strength, elasticity, structure	catalysis, recognition, regulation
exemplars	collagen, keratin, elastin	enzymes, antibodies, haemoglobin

10.11 Intrinsically disordered proteins

And then there is the third family, which the structural dogma of fifty years — sequence → fold → function — declared impossible, and which the evidence of the last twenty has promoted to a third of the human proteome: the intrinsically disordered proteins, chains that have no single fold at all but exist as an ensemble of rapidly interconverting conformations (Figure 10.6). The disorder is not a defect but a design. The disordered region offers a large capture radius to its partners (a "fly-casting" advantage in binding kinetics); it can bind several partners with different folds from the same sequence (one chain, many shapes — the economy of a small genome); it carries the sites of the post-translational modifications of § 9.10 in exposed, accessible strings; and it is the raw material of the condensates of § 8.11, whose low-complexity sequences are disordered by birth. The dialectic of the section is the correction of a deep assumption: function does not require a fold;

it requires a free-energy landscape, and the landscape can be broad and flat as well as narrow and deep. The folded protein and the disordered one are two ends of a continuum of ensemble widths, and the cell, which never read the dogma, uses both with equal ease.

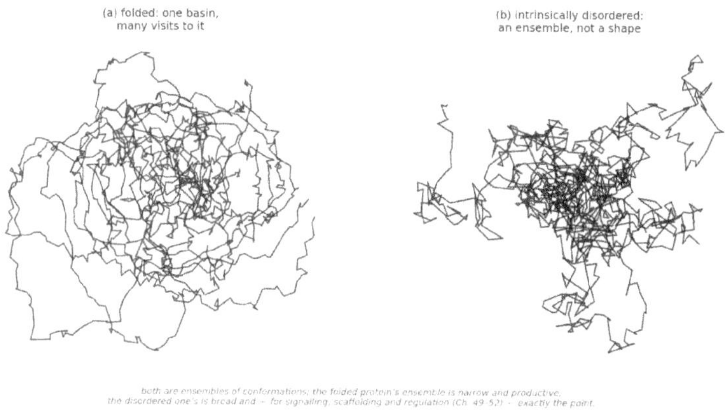

Figure 10.6 *Two ensembles. (a) A folded protein: many backbone traces converging on one basin — the ensemble is narrow, and its centre is "the structure". (b) An intrinsically disordered chain: wandering traces with no centre — the ensemble is broad, and its breadth is the function: capture radius, partner promiscuity, and the raw material of condensates.*

10.12 Structure–function relationships

The chapter closes with its doctrine, which is Part III's doctrine, and which the remaining chapters will test in every direction. Structure and function are related, but the relation is subtler than the slogan. The slogan — structure determines function — is true in its first sense: the architecture positions the residues, and the positioned residues do the chemistry; the catalytic triad of § 19.11 works because three particular side chains are held in one particular arrangement, and no other. But the slogan fails in two important ways, and the failures are as instructive as the truth. The same fold serves many functions (the TIM barrel of § 10.7 is a dozen chemistries), so function is not read off architecture but off the active site's residues and their dynamics. And function is not read off statics at all: the protein is an ensemble (§ 10.11's continuum), and the function often lives in the motion — the hinge of a transporter, the breathing of an enzyme, the quaternary shift of Chapter 8. The corrected slogan, which this treatise adopts, is therefore: structure and dynamics together determine function — and the dynamics are Chapter 12's subject, toward which the treatise now proceeds.

> ***Mechanistic case study — Anfinsen's ribonuclease, 1961: the sequence is the instructions***
>
> *The central question of this chapter — does the sequence determine the fold? — was answered by Christian Anfinsen at the NIH, in an experiment of beautiful cruelty (Anfinsen, Sela, & White, 1961). Ribonuclease A, 124 residues, four disulphide bonds, was unfolded completely (8 M urea, which destroys all noncovalent structure) and its disulphides reduced (β-mercaptoethanol, which breaks the staples): the product was a random coil with no activity — by every measure, no longer a protein in any meaningful sense. Anfinsen then simply dialysed the reagents away. The chain refolded; the disulphides re-formed (in*

the presence of oxygen and a trace of the mercaptan, to shuffle the wrong pairings until the right ones won); and the enzymatic activity returned, essentially whole. The control arithmetic is the point: four disulphides can pair in 105 ways (7 × 5 × 3 × 1), and only one is native; the chain found the one in 105 without any instruction beyond its own sequence and the solvent's laws. The conclusion — the thermodynamic hypothesis: the native structure is the free-energy minimum, and the sequence carries all the information the fold requires — earned the Nobel Prize of 1972 and stands as the charter of structural biology: read the sequence, and you read the molecule. Chapter 11 will ask how the minimum is found; the present chapter closes with the assurance that it exists.

Plate 10.1 *Max Perutz (1914–2002), who solved the structure of haemoglobin after twenty-two years of work, and in whose molecule the α-helix was first seen in nature. Portrait from the Nobel Foundation archive, nobelprize.org (Nobel Prize in Chemistry, 1962), reproduced from the Foundation's publicly distributed image.*

Plate 10.2 *John Kendrew (1917–1997), whose myoglobin — the first protein structure ever determined (1957–59) — showed the world what a globular protein actually looks like: "more complicated", he confessed, "than had been predicted." Portrait from the Nobel Foundation archive, nobelprize.org (Nobel Prize in Chemistry, 1962), reproduced from the Foundation's publicly distributed image.*

Historical note — The twenty-two-year structure

Perutz began on haemoglobin in 1937, when the idea that a molecule of 17 000 atoms could be solved by X-rays was considered madness by men of standing (one senior crystallographer offered the opinion that the attempt would take "several lifetimes"). The obstacle was the phase problem: the X-ray film records intensities, not phases, and the phases are the information. Perutz solved it in 1953 with isomorphous replacement — attaching heavy atoms (mercury) to the protein and measuring how the diffraction changed — and the method, with Kendrew's parallel work on the smaller myoglobin, delivered the first two protein structures within three years of one another. The laboratory they built — the MRC Laboratory of Molecular Biology in Cambridge — went on to produce, at last count, more than a dozen Nobel Prizes, from the double helix to the ribosome; it is arguably the most productive address in the history of science, and it began with one man's refusal to believe that a protein was too large to be seen.

Chapter 10 — Conceptual synthesis

The architecture stands: the sequence is the whole instruction (§ 10.1); the local folds are two basins of one map (§§ 10.2–10.5); motifs are the proverbs and domains the provinces of the kingdom (§§ 10.6–10.7); the tertiary fold is a dry-cored solid, marginal by design (§ 10.8); the quaternary assembly brings symmetry and allostery (§ 10.9); the kingdom divides into fibrous beams and globular machines (§ 10.10), with the disordered third of the proteome admitted as a third family whose breadth of ensemble is itself the function (§ 10.11); and the doctrine — structure and dynamics together determine function — is the bridge to the next two chapters (§ 10.12). Direction checked against Figure ii: the hierarchy is climbed. Chapter 11 asks the dynamical question beneath the architecture: how the fold is found, and what happens when it is not.

Worked example 10.1 — Helix length from residue count

A transmembrane α-helix must span a 3.0 nm bilayer. At 0.15 nm rise per residue (§ 10.3): $n = 3.0/0.15 = 20$ residues — the canonical length of the transmembrane helices of every seven-helix receptor (§ 50.1) and channel (§ 24.4), and the reason the hydropathy plot of § 9.4's map, run along a sequence in a window of ≈ 20 residues, identifies membrane-spanning segments with almost embarrassing reliability: the cell's membrane proteins announce their topology to any algorithm that knows the helix's rise.

Worked example 10.2 — The disulphide arithmetic of refolding

Anfinsen's ribonuclease has 8 cysteines. The number of ways to pair 8 objects is $7!! = 7 \times 5 \times 3 \times 1 = 105$. If the chain sampled pairings at random at one per millisecond, the expected time to hit the native pattern would be ≈ 105 ms; the observed oxidative refolding takes minutes — because the sampling is not random: the reduced chain already collapses its hydrophobes (§ 10.8), which pre-positions the correct cysteines, and the thiol-disulphide shuffling of the trace mercaptan edits the wrong pairs. The arithmetic shows both that the problem is small (105, not Levinthal's 10^{60}) and that even small problems are solved by guidance, not search — the chapter's two themes in one number.

Worked example 10.3 — The marginal stability of a fold

A 100-residue protein makes ≈ 50 intramolecular hydrogen bonds in its core. At ≈ 5 kJ mol^{-1} net each (in water, after solvent competition, § 5.5's rule), the bond ledger is ≈ −250 kJ mol^{-1}; the conformational entropy of the unfolded chain (≈ 100 residues × ≈ 4 states × R ln 4 × 0.31... precisely, $T\Delta S$ ≈ 310 × 100 × R ln 4/1000 ≈ 360 kJ mol^{-1}) opposes folding with comparable magnitude; hydrophobic burial contributes ≈ −150 kJ mol^{-1} (Worked Example 2.3). The sum: ΔG_{fold} ≈ −250 − 150 + 360 ≈ −40 kJ mol^{-1} — marginal, as § 10.8 asserted. The estimate is crude, and that is its point: the fold's stability is a small difference of large numbers, and the protein lives, by design, at the edge.

Exercises (with full resolutions)

Exercise 10.1

Myoglobin is 75 % α-helical and 153 residues long. Estimate the number of helices and their mean length, assuming helices average 3–4 turns.

Resolution. *Helical residues: 0.75 × 153 ≈ 115. A helix of 3–4 turns at 3.6 residues/turn holds 11–14; 115 residues therefore make ≈ 8–10 helices — Kendrew's structure has eight (named A through H), mean length ≈ 14 residues. The estimate lands on the structure's own count, which is the point of the exercise: secondary-structure content, measured by the circular dichroism of § 54.7, translates into architecture with surprising fidelity.*

Exercise 10.2

A sequence reads -Val–Ala–Leu–Ser–Val–Ala–Leu–Ser–. Predict its secondary structure and its likely location in a globular protein.

Resolution. *Alternating hydrophobe/hydrophile is the β-strand signature of § 10.4: the strand presents Val–Ala–Leu–Val–Leu to one face (a hydrophobic face, to the core) and Ser–Ser–Ser to the other (the solvent). The location: a β-sheet with one face buried — the edge strand of a sheet, or the surface strand of a β-barrel. The exercise is the β-analogue of the amphipathic helix, and the reason sequence-to-structure prediction (§ 55.8) was possible at all: the alphabet alternates, and the alternation is legible.*

Exercise 10.3

Why is glycine over-represented at β-turns and in collagen's Gly-X-Y repeat? Give the two, distinct, geometric reasons.

Resolution. *In the β-turn, position 3 requires a conformation (positive φ) forbidden to L-residues by the Ramachandran map (Figure 10.4); glycine, with two α-hydrogens, has no side chain to collide and enters the left-handed basin freely. In collagen, three chains pack into a triple helix whose interior at every third position has room only for a hydrogen: any side chain would split the cable. The two cases are the same principle at two scales — steric exclusion is the only law glycine's exemption illustrates — and the reader should note that the principle is Chapter 3's: geometry legislates sequence.*

EXERCISE 10.4 (INTERPRETATIVE)

"The protein structure prediction problem is solved." Assess the claim in the light of this chapter, distinguishing what is predicted from what is understood.

Resolution (indicative). The claim is true in its narrow sense and false in its grand one. Narrow: the machine-learning systems of § 55.8 now predict the single native basin of most single domains to experimental accuracy — the (φ, ψ) map's basins, learned from the PDB. Grand: the claim says nothing of the path (the systems interpolate, they do not fold), nothing of the ensemble (§ 10.11's disordered third is largely beyond them), nothing of partners (the quaternary assembly is predicted only where evolution has left a correlated trace), and nothing of why — the physics of Chapter 11 is not redundant because a machine can interpolate its outcomes. Prediction without understanding is cartography without geology; useful, and to be used, and not to be confused with the territory's laws.

EXERCISE 10.5

A protein is 40 % α-helix, 20 % β-sheet by circular dichroism, and runs as a single band at 50 kDa under denaturing electrophoresis but at 200 kDa by gel filtration. Interpret.

Resolution. The denaturing gel reports the chain (one species, ≈ 50 kDa ≈ 450 residues); the native filtration reports the assembly (≈ 200 kDa = four chains): the protein is a homotetramer. The secondary composition (mixed α/β) and the quaternary assembly are independent facts, and the exercise is the standard first dissection of a new protein by the methods of Chapter 53: count the chains, measure the folds, and the architecture begins to speak before any structure is solved.

Research-level propositions

Proposition 10.α. The Ramachandran map assumes fixed bond geometry; high-resolution structures show systematic deviations of the peptide ω and the C_α geometry in strained regions. Recompute the allowed basins from the PDB itself (a data-derived, residue-specific Ramachandran distribution, § 55.9's territory), and assess how much of the classical "forbidden" territory is genuinely visited — and by whom. Proposition 10.β. Intrinsically disordered regions evolve faster than folded ones, yet many are functional. Using the alignments of § 55.7, measure the correlation between disorder propensity and evolutionary rate across one protein family, and test whether the functional disordered segments are exceptions — conserved disorder — or merely tolerated noise. Proposition 10.γ. The TIM barrel's dominance (§ 10.7) has two explanations: divergent descent from one ancient enzyme, or convergent arrival at a uniquely evolvable scaffold. Assemble the structure-comparison evidence for both, and design the analysis — fold phylogeny versus sequence phylogeny — that would finally separate inheritance from reinvention.

References (Chapter 10)

Anfinsen, C. B., Sela, M., & White, F. H. (1961). The kinetics of formation of native ribonuclease during oxidation of the reduced polypeptide chain. Proceedings of the National Academy of Sciences, 47(9), 1309–1314.

Branden, C., & Tooze, J. (1999). Introduction to protein structure (2nd ed.). Garland.

Pauling, L., Corey, R. B., & Branson, H. R. (1951). The structure of proteins: Two hydrogen-bonded helical configurations of the polypeptide chain. Proceedings of the National Academy of Sciences, 37(4), 205–211.

Petsko, G. A., & Ringe, D. (2004). Protein structure and function. New Science Press.

Ramachandran, G. N., Ramachandran, G. N., Ramakrishnan, C., & Sasisekharan, V. (1963). Stereochemistry of polypeptide chain configurations. Journal of Molecular Biology, 7(1), 95–99.

CHAPTER 11

Forces Governing Protein Folding

CONTENTS OF THIS CHAPTER

In which the architecture of Chapter 10 is asked how it arises: the paradox of Levinthal and its resolution in the funnelled landscape; the solvent's thermodynamic hip of the fold; the staples and the props — disulphides, chaperones, quality control; and the fold's dark twin, aggregation, whose sigmoid kinetics govern the conformational diseases of age.

11.1 The protein-folding problem

Anfinsen proved that the sequence contains the instructions (case study of Chapter 10); he did not prove how the instructions are executed. The execution is the folding problem, and it has two horns, one thermodynamic and one kinetic, which the reader should keep scrupulously apart. The thermodynamic horn is solved: the native state is the free-energy minimum (§ 10.12's doctrine, Anfinsen's theorem), and the forces that define the minimum are the contents of §§ 11.2–11.5. The kinetic horn is Levinthal's: even if the minimum exists, how is it found in a lifetime? The arithmetic of the paradox (Figure 11.1) has only grown sharper with familiarity: a 100-residue chain, sampling three conformations per torsion at the fastest conceivable rate (10^{13} s^{-1}, the speed of a bond vibration), would need 10^{27} years to try them all — while real proteins fold in milliseconds. The resolution cannot be that proteins are lucky; it must be that they do not search. The folding problem, properly stated, is therefore a problem of guidance: what shape does the energy landscape have, that the chain is led downhill rather than lost at large? The answer, the funnel of § 11.7, is the conceptual heart of this chapter.

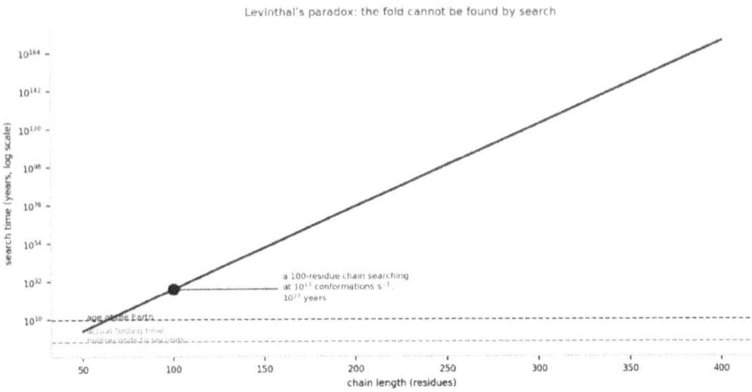

Figure 11.1 *Levinthal's paradox in numbers: the time to search all conformations of a chain, at three states per torsion and 10^{13} trials per second, against chain length — 10^{27} years at 100 residues (oxblood point), against the Earth's age (dashed) and the observed folding times (teal). The gap between the search time and the folding time is some thirty orders of magnitude: the largest discrepancy between a naive theory and a fact in all of science, and the founding paradox of folding studies.*

11.2 The thermodynamics of folding

The fold's stability is measured by its melting, and the melting has a canonical shape. Under denaturant (urea, guanidinium) or heat, the free energy of unfolding falls linearly with the perturbant (Figure 11.2a), giving the two numbers that characterise every protein's stability: the midpoint concentration c_m and the extrapolated stability in water, $\Delta G°_{H2O}$ — typically 20–60 kJ mol^{-1}, the marginal margin of § 10.8. The decomposition of that margin is the ledger of Chapters 2 and 5 applied, and Worked Example 10.3 has already totted it: against the chain's conformational entropy (which opposes folding with ≈ +360 kJ mol^{-1} for a 100-residue chain) stand the hydrogen bonds (≈ −250), the hydrophobic burial (≈ −150), the dispersion packing, and the occasional salt bridge — a stability that is a small difference of enormous numbers. The consequence is the protein's famous fragility and its equally famous robustness, which are the same fact read twice: a single mutation can destroy the fold (one term of the ledger shifts by more than the margin), and yet the fold survives millennia of sequence drift (the terms are large and mutually compensating). Folding thermodynamics is the accountancy of these compensations, and the linear free-energy relations of Figure 11.2a are its working instrument.

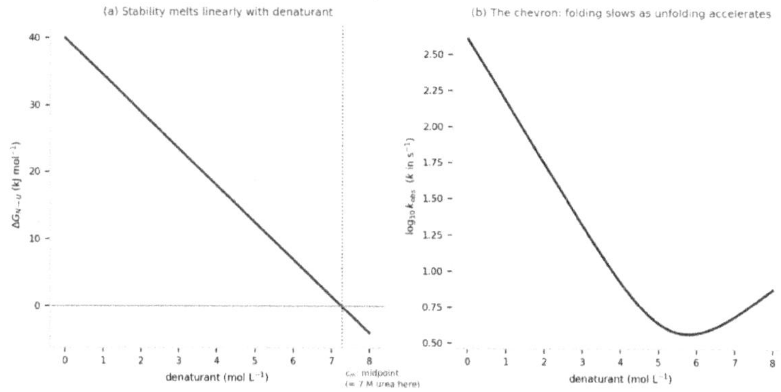

Figure 11.2 (a) The unfolding free energy of a small protein against denaturant concentration, with the linear extrapolation to water (the m-value) and the midpoint c_m. (b) The chevron plot: the observed folding/unfolding rate against denaturant — the two arms are the folding limb (falling as the native state is destabilised) and the unfolding limb (rising), and their meeting point is the same c_m. Thermodynamics and kinetics audited by one perturbant.

11.3 Hydrophobic collapse

The first event of folding, on the evidence of four decades of kinetics, is not the formation of structure but the flight from water: the chain collapses its hydrophobic residues inward within milliseconds of the start, forming the molten globule — a compact, fluctuating state with native-like content of secondary structure but no fixed tertiary packing (Figure 11.4, stage 2). The collapse is the solvent's work (Chapter 5's theorem: the network expels what it cannot befriend), and its consequence is the reduction of the search by orders of magnitude: the collapsed globule has a fraction of the extended chain's volume, and the subsequent zipping of helices, sheets and contacts proceeds within a space already pre-sorted. The framework model of the figure — collapse first, secondary structure within the globule, native packing last — is one of three historical answers to Levinthal (the others: nucleation, where a small native core forms first and templates the rest; and the hierarchical model, where local folds form first and then dock). Modern kinetics has dissolved the dispute into the landscape view of § 11.7: proteins differ, and all three mechanisms are seen, sometimes in one protein at different stages — the funnel accommodates every itinerary that goes downhill.

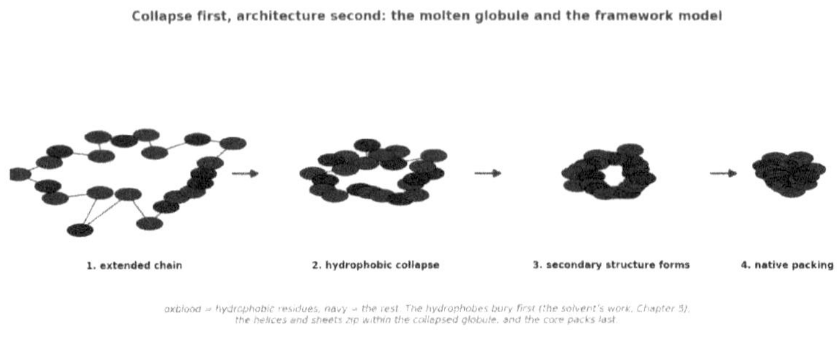

Figure 11.4 *The framework sequence of folding: extended chain (1); hydrophobic collapse into the molten globule (2); secondary structure zipping within the globule (3); native packing of the core (4). Hydrophobic residues in oxblood bury first; the rest of the architecture follows. The collapse is the solvent's work, and the reduction of searchable volume is the kinetic dividend that Levinthal's paradox requires.*

11.4 Hydrogen bonds and electrostatic interactions

The role of the hydrogen bond in folding is the subtlest in the ledger, because water competes for every one of them (§ 5.5): a backbone donor that finds no partner in the fold was already bonded to the solvent, so the net contribution of each intramolecular bond is the difference between two embraces, not one — estimated at 2–8 kJ mol^{-1} apiece, small per bond and decisive in sum (≈ −250 kJ mol^{-1} over a small protein, per Worked Example 10.3). The electrostatics of the fold follow the rules of § 2.4: surface salt bridges contribute little in water (the dielectric screens them to a few kJ mol^{-1}), buried ones are strong but pay the desolvation toll of § 5.4, and the protein's net charge distribution is tuned more for solubility than for stability — the exposed charges repel aggregation at least as faithfully as they stabilise the fold. The honest summary of the section is that the hydrogen bond is the fold's discipline rather than its engine: it specifies the geometry (every donor satisfied, every acceptor matched) more than it pays for it, and the payment, as the next section's totals insist, is largely the solvent's.

11.5 Disulphide bonds

The one covalent staple of the folded protein is the disulphide bond, the oxidised bridge between two cysteine thiols — and its placement in the economy of the fold is precise. Disulphides are found almost exclusively in extracellular and secreted proteins (antibodies, hormones, toxins, the secreted enzymes), because the cytosol is a reducing environment (glutathione's thiol, § 38.8, holds it at ≈ −240 mV) in which disulphides cannot persist; the extracellular space is oxidising, and there the staple holds. Thermodynamically the disulphide buys stability by constraining the unfolded chain — the cross-link reduces the entropy of the denatured state, and the fold's margin rises by the same amount, typically 10–30 kJ mol^{-1} per bridge. The physiological consequence is a division of labour: intracellular proteins, denied the staple, rely on noncovalent packing alone (and are accordingly easier to unfold, refold, and degrade — quality control's preference, § 11.9); secreted proteins, stapled, survive the hostile world outside the cell — the peptide toxins of § 9.9 owe their knife-proof compactness to three disulphides *apiece, and the ribonuclease of Anfinsen's experiment owes its refoldability to the fact that its four disulphides, once re-formed, lock the correct fold against all comers.*

11.6 Folding intermediates

The road downhill has observable way-stations, and two of them have earned their names in the literature. The molten globule of § 11.3 is the earliest — compact, secondary-structured, unpacked — and its existence is certified by a characteristic signature: it binds the dye ANS (8-anilinonaphthalene-1-sulphonate), which fluoresces only in exposed hydrophobic patches, so that the globule lights up where the native protein is dark. The slowest way-stations are the proline bottlenecks: the cis → trans isomerisation of the X–Pro peptide bonds (§ 9.6), with its barrier of ≈ 80 kJ mol^{-1}, takes seconds to minutes unaided, and proteins whose native state requires a particular cis or trans proline fold at the proline's pace, not the chain's — the cell accordingly employs the prolyl isomerases (PPIases) to accelerate the step, and the immunosuppressant cyclosporin (Table 9.1) works by inhibiting one of them. The existence of discrete intermediates was once the argument between the framework and nucleation schools; in the landscape view it is a matter of the funnel's roughness: smooth funnels fold without visible way-stations, rough ones pause in traps (Figure 11.3), and the intermediates are the traps that last long enough to be seen.

11.7 Energy landscapes and folding funnels

The resolution of Levinthal's paradox, and the conceptual summit of the chapter, is the folding funnel (Figure 11.3): the energy of the chain plotted against its conformational freedom is not a flat plain with one well (the landscape on which search would be necessary) but a funnel — a landscape sloped everywhere toward the native basin, so that essentially every conformation has a downhill route to the fold. The funnel's shape is no accident: it is the shape that evolution selects, because a sequence whose landscape is rugged folds slowly and loses the race to aggregation (§ 11.10); the evolved proteins are "minimally frustrated" — their stabilising interactions are consistent with one another, so the landscape is smooth, and the fold is guided. The funnel unifies the chapter: Levinthal is answered (there is no search, only descent); the intermediates are placed (rough patches on the funnel wall); the framework and nucleation mechanisms are reconciled (different itineraries down the same funnel); and the margin of § 11.2 is reinterpreted — the funnel's depth is small by design, because a deeper funnel would be a prison, and the cell must be able to unfold, regulate, and finally degrade every protein it makes. The fold, in the landscape view, is not a destination searched out but a basin at the bottom of a slope that chemistry cut and evolution graded.

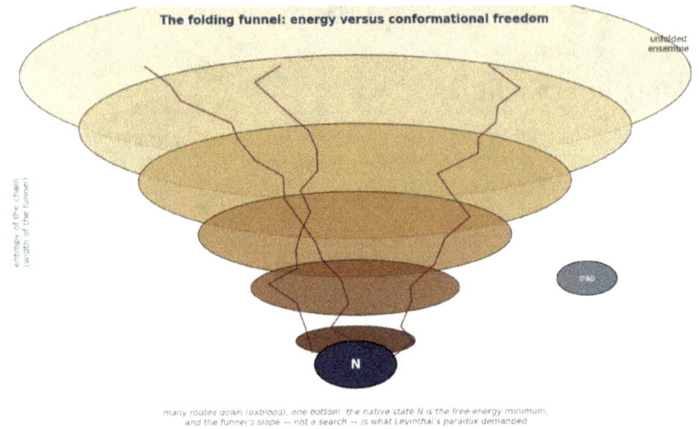

Figure 11.3 *The folding funnel: energy (vertical) against conformational freedom (width), with trajectories from many starting conformations (oxblood) converging on the native basin N, and a kinetic trap (grey) on the wall for illustration. The funnel's slope is the answer to Levinthal: the chain descends rather than searches, and any of an astronomical number of itineraries reaches the bottom in milliseconds.*

11.8 Molecular chaperones

The funnel is smooth, but the cell is crowded (§ 5.11): a newly made chain, its hydrophobes exposed, is surrounded by a thousand other exposed hydrophobes, and the race between folding and aggregation would be lost in the crowd without assistance. The assistance is the chaperones — proteins whose office is not to fold other proteins but to protect them while they fold themselves. Two families divide the work. The Hsp70 system holds the nascent chain in an extended, competent state — a molecular bodyguard, cycling on and off hydrophobic stretches as they emerge from the ribosome, preventing the premature contacts that lead to aggregates. The chaperonins (GroEL–GroES in bacteria, Figure 11.5; the eukaryotic CCT) are the folding rooms: a barrel with a lid, into which a single misfolded chain is admitted, the lid closed at the price of ATP, and the chain left to fold in seclusion, safe from the crowd — the Anfinsen experiment run, at last, inside the cell. The chaperones' invention was the resolution of a paradox the field had not noticed it had: Anfinsen's proteins refold in the test tube, yet most cellular proteins need help in the cell — and the discrepancy is exactly the crowding of § 5.11. The chaperone does not change the funnel's shape; it keeps the competitors off the slope.

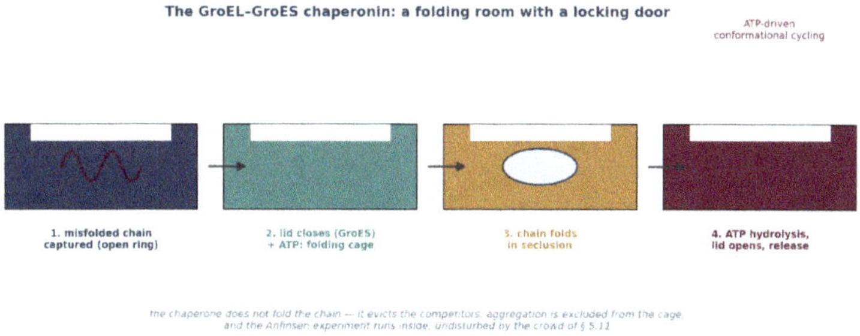

Figure 11.5 The GroEL–GroES chaperonin cycle: the misfolded chain is captured in the open barrel (1), the lid closes with ATP (2), the chain folds in seclusion (3), and hydrolysis opens the cage for release (4) — the cycle repeating with another ATP until the chain emerges folded or is handed to degradation. The chaperone's office is exclusion of the crowd, not instruction of the chain.

11.9 Protein quality control

The cell does not merely fold its proteins; it inspects them, and the inspection is a ruthless economy. The nascent chain is monitored by the chaperones of § 11.8; the correctly folded are passed; the misfolded are offered rescue (a second trip through the folding room); and the incorrigible are marked for destruction — tagged with ubiquitin (Figure 9.5's 76-residue sentence of death) and fed to the proteasome of § 11.12. The logic of the system is the logic of marginal stability (§ 10.8): because every protein is built near the edge, every protein is a candidate for misfolding, and the cell maintains a standing apparatus to find and destroy the failures before they aggregate. The endoplasmic reticulum runs its own inspection (the unfolded protein response of § 51.9, which halts translation when misfolding outpaces rescue); the mitochondria run theirs. Quality control is the dark side of the funnel: the same landscape that guides the good fold guides the bad aggregate, and the cell's answer is not to perfect the landscape — which evolution cannot quite do — but to police it.

11.10 Misfolding and aggregation

Aggregation is folding's shadow: the same hydrophobic patches that drive the collapse of § 11.3, exposed at the wrong time or in the wrong molecule, drive molecules into one another instead. The aggregate's structure, when it is the ordered kind called amyloid, is a thing of terrible regularity: a cross-β spine — β-strands stacked perpendicular to the fibril axis, their backbone hydrogen bonds running the fibril's length — to which essentially any protein, whatever its native fold, can convert under provocation. The universality is the point: amyloid is not a disease of particular proteins but a capacity of the polypeptide backbone itself, the fold's alternative minimum on the landscape, deeper for many sequences than the native basin — the native fold is often a kinetic preference, the amyloid a thermodynamic one, and the margin between them is the margin of safety the cell lives with. The kinetics of conversion are the sigmoid of Figure 11.6: a lag (nucleation, rare and rate-limiting), an explosion (elongation from the nuclei's ends), a plateau (monomer exhausted) — and the curve's darkest feature is its seeding: pre-formed fibrils abolish the lag, so that aggregation, once started, is self-propagating. The prion of the case study is the seed that learned to travel between organisms; the amyloidoses of § 11.11 are seeds that travel between tissues.

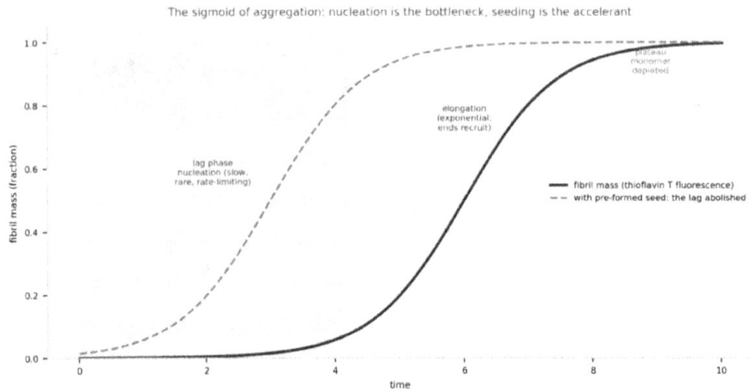

Figure 11.6 *The sigmoid kinetics of amyloid aggregation: the lag of nucleation (shaded), the exponential elongation, the plateau of monomer depletion — and (dashed teal) the abolition of the lag by a pre-formed seed, the kinetic signature of infectivity that distinguishes the prion diseases from all others.*

11.11 Amyloids and conformational disease

The conformational diseases are the price the organism pays for the fold's marginality, and they are classified by their seeds. In the neurodegenerations — Alzheimer's (the Aβ peptide and tau), Parkinson's (α-synuclein), Huntington's (polyglutamine-expanded huntingtin) — the aggregate accumulates over decades in the cells least able to be replaced: the neurones. In the systemic amyloidoses, the seed is a circulating protein (transthyretin, immunoglobulin light chains) deposited in the organs that filter the blood. And in the prion diseases — Creutzfeldt–Jakob, kuru, scrapie, BSE — the seed is the infectious agent itself: PrP^{Sc}, the misfolded form of a neuronal membrane protein, which binds the normal PrP^{C} and templates its conversion — inheritance without nucleic acid, a protein replicating its conformation. The prion hypothesis, Prusiner's, was received with the hostility that genuine novelty always receives and the evidence that genuine truth always accumulates (Prusiner, 1982); it is now the protein-folding field's most consequential proof that a conformation, not a sequence, can carry information (the informational turn of Chapter 1, in its darkest mirror). The therapeutic logic of the whole field follows from Figure 11.6: block nucleation (the lag is the disease's best-kept secret — prevent it, and there is no disease), destabilise the seed, or clear the monomer — and the antibodies and small molecules of the current clinics are Figure 11.6's curve, attacked at every stage.

11.12 Protein degradation systems

The chapter closes with the fold's final fate: degradation, which is not decay but regulation. Two systems divide the work. The ubiquitin–proteasome system is the cell's court and executioner in one: proteins marked by chains of ubiquitin (the tag of § 9.10, ligated by the E1–E2–E3 enzyme cascade, whose E3s are the specificity factors — the cell has hundreds, one for each class of condemned) are fed to the proteasome, a barrel protease that unfolds its victims (at the price of ATP) and shreds them into peptides. The system's office is not waste disposal but timing: the cyclins of the cell cycle (§ 51.2), the transcription factors of signalling (§ 50.6's IκB), the misfiled and the misfolded — all are removed with a precision that makes degradation as much a regulatory instrument as synthesis. The lysosomal system — autophagy, § 51.8 — handles the larger cargo: whole organelles, aggregates, the long-lived and the bulky, enclosed in a membrane and digested wholesale. The two systems complete the protein's biography: made by the ribosome (§ 47), folded by the funnel (§ 11.7), policed by quality control (§ 11.9), and executed on schedule — the cell's proteins, like its citizens in a well-run state, are temporary by design, and the design's name is turnover.

MECHANISTIC CASE STUDY — THE PROTEASOME'S BARREL: HOW THE CELL UNFOLDS WHAT IT MUST DESTROY

The proteasome is the folding problem read backwards, and its mechanism is the chapter's logic in reverse. The 26S particle has two parts: the 20S core, a barrel of four stacked rings whose inner chamber carries the proteolytic sites (threonine nucleophiles, of the cysteine-protease family of § 19.11), and the 19S caps, the gates. The condemned protein, recognised by its ubiquitin chain, is engaged by the cap's ATPases — a ring of six, of the AAA+ family — which pull on the chain like a motor on a rope: the protein is unfolded by force, threaded through the gate into the barrel, and hydrolysed into peptides of 3–25 residues while the ubiquitin is snipped off and recycled. The elegance is the inversion: the same hydrophobic exposure that drives folding (§ 11.3) is here the signal for destruction — the cap's motor grips exactly the exposed hydrophobic tails that a misfolded protein displays — and the same compartmentalisation that protects the cytosol from the aggregate (§ 11.10) protects it from the protease, whose active sites are sealed inside the barrel. The cell's garbage disposal is, like every other room in the house, built of the chapter's physics — and the inhibitor bortezomib, which clogs the barrel and drowns the myeloma cell in its own unrecycled proteins, is the proteasome's logic turned to therapy.

Plate 11.1 *Christian Anfinsen (1916–1995), whose refolding of ribonuclease founded the thermodynamics of this chapter and whose "thermodynamic hypothesis" — the native state is the free-energy minimum — is the premise of every landscape in it. Portrait from the Nobel Foundation archive, nobelprize.org (Nobel Prize in Chemistry, 1972), reproduced from the Foundation's publicly distributed image.*

HISTORICAL NOTE — PRUSINER AND THE HERESY THAT BECAME A DISCIPLINE

That a protein could be infectious — that scrapie, kuru and Creutzfeldt–Jakob disease were transmitted by a particle containing no nucleic acid — was proposed by Stanley Prusiner in 1982, on the evidence of an agent that resisted every treatment that destroys DNA and RNA and succumbed to those that destroy protein (Prusiner, 1982). The proposal was met, for a decade, with the reception that the guardians of the central dogma usually reserve for heretics: the experiments were attacked, the preparations impugned, and the proposer advised to find the virus. The evidence accumulated as evidence does: the infectious protein was purified, the gene for it was found in the host itself, the disease was shown to be transmissible by the purified misfolded form — and the Nobel Prize of 1997 followed. The episode is the prion field's own case study in the sociology of Chapter 1: a paradigm does not fall to argument but to accumulation, and the heresy of one decade is the textbook of the next.

Chapter 11 — Conceptual synthesis

The fold has been accounted. The folding problem has two horns — thermodynamic (solved by Anfinsen: the native state is the minimum) and kinetic (Levinthal's paradox, solved by the funnel: the chain descends, it does not search) (§§ 11.1, 11.7). The stability is a small difference of large numbers (§ 11.2); the collapse is the solvent's work (§ 11.3); hydrogen bonds discipline rather than pay (§ 11.4); disulphides staple the secreted proteins (§ 11.5); intermediates are the funnel's rough patches (§ 11.6); chaperones keep the crowd off the slope (§ 11.8); quality control polices what the landscape cannot perfect (§ 11.9); aggregation is the fold's alternative minimum, amyloid its cross-β spine, and the sigmoid kinetics — lag, elongation, seeding — the grammar of the conformational diseases (§§ 11.10–11.11); and degradation is regulation by removal, the protein's biography ending on schedule (§ 11.12). Direction checked against Figure ii: the static architecture of Chapter 10 is now a dynamical landscape. Chapter 12 follows the landscape into motion: protein dynamics and allostery — the machinery in operation.

WORKED EXAMPLE 11.1 — THE CHEVRON'S TWO ARMS

A protein's folding rate in water is 1 000 s^{-1}, unfolding rate 10^{-4} s^{-1}. The equilibrium constant of folding is $K = k_f/k_u = 10^7$, whence $\Delta G° = -RT \ln K = -2.479 \times 16.1 \approx -40$ kJ mol^{-1} — the marginal margin of § 10.8, obtained from two rate constants without a calorimeter. The chevron plot (Figure 11.2b) is this relation applied across denaturant: each arm extrapolates to the water value of its rate, and the two extrapolations must satisfy the same K — thermodynamics and kinetics auditing each other, per § 7.7's rule that the constant belongs to both jurisdictions.

WORKED EXAMPLE 11.2 — THE DISULPHIDE STAPLE'S PRICE

Engineering a new disulphide into a protein raises its melting temperature by 5 °C. Estimate the stabilisation. From the van 't Hoff form of § 7.7, near the midpoint $\Delta T_m \approx T_m^2\, \Delta(\Delta G)/\Delta H_m$... — more directly, if $\Delta H_m \approx 400$ kJ mol^{-1} and $T_m \approx 340$ K, then $\Delta S_m = \Delta H_m/T_m \approx 1.18$ kJ K^{-1} mol^{-1}, and a 5 K rise implies $\Delta(\Delta G) \approx \Delta S_m \times 5 \approx 5.9$ kJ mol^{-1} — roughly one hydrogen bond's worth per *staple, matching § 11.5's 10–30 kJ mol^{-1} for the better-placed bridges. The estimate is the protein engineer's daily bread (§ 56.11): the staple pays in the unfolded state's entropy, and the payment is measured at the melting point.*

Worked example 11.3 — The chaperone's ATP budget

GroEL hydrolyses ≈ 7 ATP per folding cycle, and a cell makes ≈ 10^5 proteins per generation that need the chaperonin's help. At ΔG_{ATP} ≈ −50 kJ mol^{-1} (§ 7.8), the cost is ≈ 7 × 10^5 × 50 ≈ 3.5 × 10^7 kJ per cell generation — against a total ATP turnover of ≈ 10^9 molecules (Table 1.3's census) × 50 ≈ 5 × 10^{10} kJ: folding assistance costs the bacterium less than 0.1 % of its energy budget. The estimate explains why the chaperone system is affordable everywhere — and why its failure is affordable nowhere: the price of assistance is trivial, the price of aggregation (Figure 11.6) is the cell.

Exercises (with full resolutions)

Exercise 11.1

A 60-residue protein folds by pure search at 10^{13} conformations s^{-1}, three states per torsion, 120 torsions. How long, and what does the same chain do in reality?

Resolution. *3^{120} ≈ 10^{57} conformations; at 10^{13} s^{-1}, ≈ 10^{44} s ≈ 10^{36} years. In reality the chain folds in ≈ 10 ms — the paradox of Figure 11.1, recalculated for a real small protein (barstar, say), and the discrepancy of ≈ 40 orders of magnitude is the empirical warrant for the funnel: the landscape is graded, or the cell is not here to grade it.*

Exercise 11.2

Two proteins have the same ΔG° of unfolding (40 kJ mol^{-1}) but melting temperatures of 55 °C and 75 °C. Explain the apparent contradiction using the thermodynamics of § 7.5.

Resolution. *ΔG° = ΔH° − TΔS°: the same stability can be composed of a large ΔH° and large ΔS° (steep melting curve, high T_m) or small ones (shallow curve, low T_m). The melting temperature depends on the ratio ΔH°/ΔS° (T_m = ΔH°/ΔS° at the midpoint), the free energy on their difference — and the two proteins differ in the enthalpic composition of their folds, not in their stability. The exercise is Figure 7.3's quadrants applied to the calorimeter: read ΔG°, ΔH° and T_m as three separate instruments, never as one.*

Exercise 11.3

Estimate the amyloid lag time's dependence on concentration if nucleation requires three monomers to meet.

Resolution. *A trimeric nucleus forms at a rate ∝ c^3; the lag time scales as its inverse, $t_{lag} \propto c^{-3}$ — so halving the monomer concentration lengthens the lag eightfold. The scaling is the therapeutic logic of the transthyretin stabilisers (tafamidis), which lower the available monomer and stretch the lag beyond a* *human lifetime — Figure 11.6's curve pushed, by arithmetic rather than by force, off the edge of the patient's calendar.*

Exercise 11.4 (interpretative)

"Chaperones violate Anfinsen's principle, because proteins need them to fold." Adjudicate.

Resolution (indicative). *The objection mistakes the principle's content. Anfinsen asserts that the sequence specifies the fold — the landscape's shape is written in the chain. Chaperones do not alter the shape: they neither guide the chain to a different basin nor supply information, but change the conditions under which the descent is run — excluding aggregation partners (GroEL's cage), holding the chain competent (Hsp70), or waiting out the traps. The proof is the chaperonin's product: the chain emerges with its own fold, Anfinsen's fold, not the chaperone's. The principle survives; what § 5.11's crowding adds is that the solvent of the cell is not the solvent of the cuvette — and the chaperones are the cell's way of making the crowd transparent.*

Exercise 11.5

Why do intracellular proteins generally lack disulphide bonds, and why is this an advantage for quality control rather than a loss?

Resolution. *The cytosol is reducing (glutathione ≈ −240 mV, § 11.5): disulphides cannot persist there, so intracellular folds are noncovalent only. The advantage is reversibility: the unstapled protein unfolds more readily, is more easily threaded into the proteasome's barrel (case study), and is more rapidly replaced — the cell's interior proteins are temporary by design (§ 11.12), and the disulphide staple, which would make them durable, is exactly what turnover forbids. The extracellular world, which cannot replace, staples instead — the division of labour between durability and disposability, written in redox chemistry.*

Research-level propositions

Proposition 11.α. The folding funnel's smoothness has never been measured directly for a full-length protein in the cell. Using in-cell NMR and single-molecule FRET (§§ 54.5, 54.10), design the experiment that would map the roughness of one protein's landscape in vivo, and predict how it should differ from the cuvette's. Proposition 11.β. Amyloid seeding crosses the species barrier imperfectly (the "species barrier" of prion disease). Formulate the kinetic theory of heterologous seeding — Figure 11.6's curve with mismatched ends — and use it to estimate the transmission risk of a new prion strain from in-vitro seeding data alone. Proposition 11.γ. The ubiquitin code is written in chain linkages (K48, K63, M1...) with different fates. Catalogue the E3 ligases of one signalling pathway (§ 50.6), derive the expected degradation kinetics for each linkage type, and design the experiment (ubiquitin-replacement strains, § 55.7) that would assign each fate to its linkage in vivo.

References (Chapter 11)

Baldwin, R. L. (2008). Protein folding: The search for intermediates and the mechanism of protein folding. Annual Review of Biophysics, 37, 1–21.

Dill, K. A., & MacCallum, J. L. (2012). The protein-folding problem, 50 years on. Science, 338(6110), 1042–1046.

Levinthal, C. (1969). How to fold graciously. In P. Debrunner, J. C. M. Tsibris, & E. Münck (Eds.), Mössbauer spectroscopy in biological systems (pp. 22–24). University of Illinois Press.

Onuchic, J. N., & Wolynes, P. G. (2004). Theory of protein folding. Current Opinion in Structural Biology, 14(1), 70–75.

Prusiner, S. B. (1982). Novel proteinaceous infectious particles cause scrapie. Science, 216(4542), 136–144.

CHAPTER 12

Protein Dynamics and Allostery

CONTENTS OF THIS CHAPTER

In which the static architecture of Chapters 10 and 11 is set in motion: the protein as an ensemble of breathing states across fifteen decades of time; the recognition of ligands by election from the ensemble rather than by moulding of a rigid site; the allosteric transition as a population shift; covalent regulation as a switch paid for in ATP; and the whole machine shown to be, at its working temperature, less a sculpture than a negotiation.

12.1 Proteins as dynamic ensembles

The crystal structure, for all its beauty, is a photograph, and the protein is a dancer. The photograph's deception is not that it is false — the atomic positions of Chapter 54 are as true as measurement makes them — but that it is selected: the crystal holds the molecule still in one of its many attitudes, and the stillness is the crystal's, not the protein's. In solution, at 310 K, every protein is an ensemble: a population of conformations distributed over the free-energy landscape of Chapter 11, interconverting on every timescale from femtoseconds to seconds (Figure 12.1). The bond vibrations at the fast end are thermal noise; the loop motions, hinge bendings and domain movements at the slow end are the working motions of the machine; and the allosteric transitions of Chapter 8 sit between, at microseconds to milliseconds. The doctrine of this chapter — the completion of § 10.12's corrected slogan — is that the ensemble, not the photograph, is the object of biochemistry: a protein is not a structure but a probability distribution of structures, and everything it does — binding, catalysis, regulation — is done by the distribution, not by any single member of it.

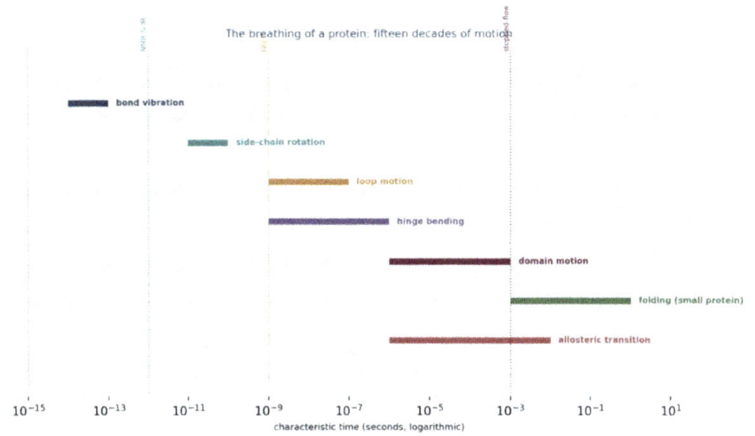

Figure 12.1 *The timescales of protein motion, from bond vibration (10^{-14} s) to folding (seconds), with the observing windows of the methods that measure them (NMR, FRET, stopped-flow; Part XI). The protein breathes at every frequency in the figure; the crystal structure catches one breath.*

12.2 Conformational fluctuations

The fluctuations have a thermodynamic price and a statistical description. The amplitude of a motion is set by the equipartition of Chapter 7: each degree of freedom carries ≈ $k_BT/2$ of thermal energy, so the root-mean-square excursions of atoms in a folded protein are ≈ 0.1 nm — small against the fold, large against a bond, and plainly visible in the crystallographers' B-factors (the temperature factors of the structure file, which are nothing but the mean-square displacement of each atom, measured). The motions are not random in direction: they are channeled by the fold's architecture, and the channels are computed by the normal-mode analysis of § 55.9 — the protein's low-frequency modes, its natural vibrations, turn out to be the same motions that the machine uses to work: the hinge of a transporter, the opening of a channel, the breathing of an enzyme's active site. The result is one of the deep simplifications of the field: the protein's functional motions are written into its architecture as its softest modes — the machine moves along the directions in which it is already soft, and evolution, which designs the architecture, designs the motion by designing the softness.

12.3 Induced fit and conformational selection

The debate of Chapter 8 — Fischer's lock against Koshland's glove — finds its modern form here, and its resolution is the ensemble. The two mechanisms are drawn in Figure 12.2: induced fit, in which the ligand binds the dominant conformation and the complex rearranges; and conformational selection, in which the protein samples both conformations unbound, and the ligand simply binds the rare one it prefers, shifting the equilibrium toward it — the same election logic as MWC allostery, now for the elementary act of recognition itself. The kinetic evidence of the last two decades is unambiguous in its direction: wherever the experiment can distinguish (NMR relaxation, single-molecule FRET, §§ 54.5, 54.10), the unbound protein is found to sample the "bound" conformation before any ligand arrives — the rare state is present, at its equilibrium population, and the ligand elects it. The synthesis the field has settled on is that both mechanisms run in series: selection first (the ligand chooses from the ensemble), then induced fit (small adjustments of the complex) — the election and the handshake. The dialectic of the treatise has tracked this dispute since Chapter 8, and its resolution is worth a sentence of celebration: the protein does not wait for its ligand to propose a shape; it rehearses all its shapes in advance, and the ligand merely calls the vote.

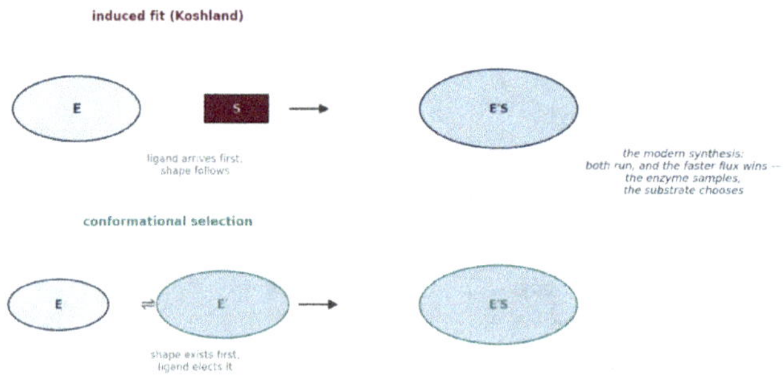

***Figure** 12.2 Two mechanisms of recognition. Above: induced fit — the ligand binds the dominant form E, and the complex rearranges to E'S. Below: conformational selection — the protein already samples E' unbound, and the ligand elects it. The modern evidence places selection first in most systems; the two run in series, and the ensemble of § 12.1 is the common frame.*

12.4 Allosteric transitions

The allosteric transition is the ensemble's election writ large — the population shift between two macrostates of the whole protein, T and R of Chapter 8, and its landscape is drawn in Figure 12.3b: two basins separated by a barrier, the equilibrium between them set by the basins' relative depths, and the ligand acting by deepening one. Three properties of the transition deserve the reader's attention. It is conformational, not chemical: no covalent bond changes — the protein is the same molecule in both states, and the difference is one of shape and population. It is cooperative: the subunits of the oligomer flip in concert (or nearly), because the interfaces are built so that a mixed oligomer is strained — symmetry is the enforcer of the election's unanimity. And it is fast: the T⇌R transition of haemoglobin runs at ≈ 10^5 s^{-1}, faster than the circulation's slowest needs — the switch is never the bottleneck, and the physiology of Chapter 13 can take the switching for granted. The allosteric transition is, in short, the protein's democracy in action: two states, one equilibrium, and every ligand a voter.

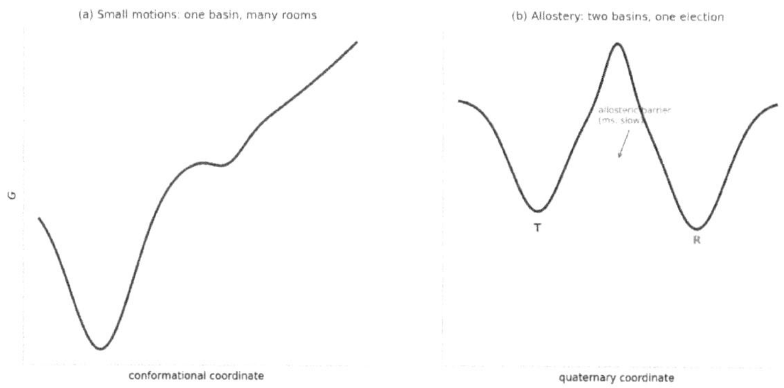

Figure 12.3 *Two free-energy landscapes. (a) Small motions: one basin with shoulders — the breathing of § 12.2, fast and shallow. (b) Allostery: two basins (T, oxblood; R, teal) with the barrier between — the population is the equilibrium, the ligand deepens one basin, and the transition is the election.*

12.5 Concerted and sequential models

The two formal models of Chapter 8 — Monod, Wyman and Changeux's concerted MWC (all subunits flip together, 1965) and Koshland, Némethy and Filmer's sequential KNF (each subunit flips as it binds, 1966) — have been weighed by sixty years of experiments, and the weighing is the most instructive adjudication in the chapter. The curves the models predict (Figure 12.4) differ subtly — MWC gives the sigmoid its particular steepness and its characteristic asymmetry; KNF gives gentler, more graded responses — and the data divide: haemoglobin, the standing example, behaves as MWC to good approximation (its two quaternary structures, T and R, are crystallographic fact, and no stable intermediate has ever been caught); aspartate transcarbamoylase, the other great case, shows the sequential hallmarks. The modern synthesis — the ensemble allostery of § 12.1 — subsumes both: the protein is a population over a landscape, MWC is the limit of a deep, two-well landscape with a symmetric oligomer, KNF is the limit of a graded one, and real proteins sit where their landscapes put them. The models were not wrong; they were the two ends of the continuum, and the continuum, sixty years on, is the chapter's object.

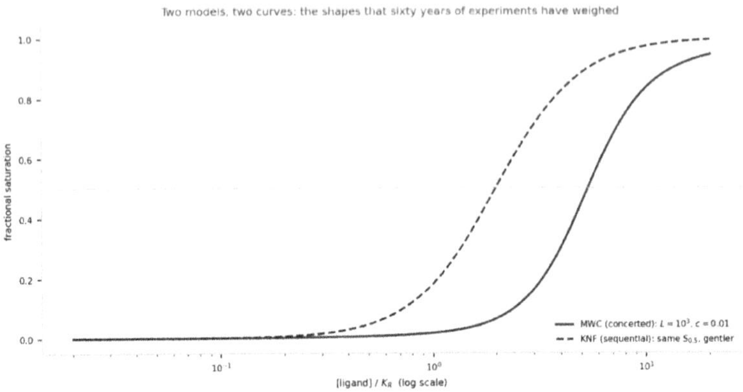

Figure 12.4 *The binding curves of the two models at the same midpoint: MWC (navy) with its characteristic steep, elected sigmoid; KNF (dashed oxblood) with its gentler, sequential rise. The shapes are the models' fingerprints, and sixty years of careful saturation curves have been the jury.*

12.6 Long-range energetic coupling

How does a ligand at one site inform a site three nanometres away? The question is the mechanical heart of allostery, and its answer is the protein's wiring: a network of coupled residues — a wire of contacts, packings and hydrogen bonds — connecting the two sites through the fold. The evidence is mutational: change a residue on the wire, far from both sites, and the coupling dies (the two sites cease to communicate, though each works alone); the computation is the statistical coupling analysis of § 55.7 (the residues that co-evolve across a protein family are the residues that talk); and the physical picture is the population shift of § 12.4 made molecular: the ligand's binding changes the local packing, the packing propagates along the wire as a shift of side-chain populations, and the distant site inherits the shift as a changed affinity. The wire is not a unique chain but a web — multiple paths, redundant contacts, the communication robust to single cuts — and the robustness is the design: allostery is too important to entrust to one thread. The long-range coupling is, in the end, the protein's internal signalling system, and the allosteric networks of Chapter 52 are the same wires, read at the scale of the cell.

12.7 Ligand-dependent conformational change

The ligand's action, in the ensemble view, is always the same verb: re-weight. The agonist deepens the active basin, the antagonist the inactive one, the inverse agonist — the drug that silences a receptor's basal activity — deepens the inactive state below its unliganded level, and the partial agonist deepens the active basin less than fully: the whole pharmacology of Chapter 50's receptors is the arithmetic of basin depths, and the four classes of ligand are four votes in the same election. The structural correlate is the conformational change itself: the rotation of a helix, the movement of a loop, the opening or closing of a cleft — visible in the paired structures (with and without ligand) that Chapter 54 has delivered by the thousand. The two descriptions — thermodynamic (basins) and structural (movements) — are the same fact at two resolutions, and the reader who has followed the chapter should now find the translation automatic: every movement is a basin, every basin a population, and every ligand a re-weighting of the votes.

12.8 Covalent regulation

Against the fast, reversible elections of allostery stands the other great instrument of regulation: the covalent modification of § 9.10, of which phosphorylation is the sovereign example (Figure 12.5). The differences are the design. Covalent regulation is irreversible in each direction — the phosphate is added by a kinase (at the price of ATP) and removed by a phosphatase (a different enzyme, not the reverse of the first), so the switch is held where it is put until the opposing enzyme acts: allostery equilibrates, phosphorylation ratchets. It is ultrasensitive: when the modifying and demodifying enzymes are both near saturation, a small change in their ratio throws the switch almost completely (the zero-order ultrasensitivity of § 18.5 — a switch sharper than any Hill coefficient). And it is energetic: each cycle costs an ATP, so the switch is a device driven out of equilibrium — the phosphorylation cycle is, in the circuit language of Chapter 52, a transistor, and the signalling cascades of Part X are its amplifiers. Allostery and covalent regulation are the cell's two instruments of control, and their division of labour is precise: the first for speed and reversibility (the metabolic second-to-second of Chapter 18), the second for persistence and amplification (the signalling minute-to-hour of Part X).

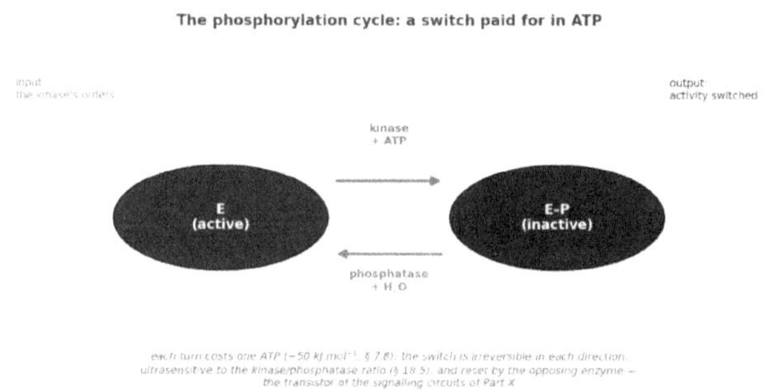

Figure 12.5 The phosphorylation cycle: the kinase (teal) adds the phosphate at the price of ATP, the phosphatase (amber) removes it with water — two different enzymes, one ratcheted switch. The cycle's cost, sensitivity and persistence are the transistor properties on which Part X's circuits are built.

12.9 Mechanical properties of proteins

The protein is also a material, and its mechanics are measured by pulling on it: the atomic force microscope and the optical tweezer of § 54.10 grip a single molecule and stretch it, and the force–extension curve is the protein's mechanical autobiography (Figure 12.6). Two regimes divide the curve. At low force, the protein is an entropic spring: the extended segments resist straightening because straightening reduces their conformational entropy (Chapter 11's unfolded-state ledger, felt as a force) — the giant muscle protein titin, with its PEVK segment, is the body's chief such spring, the passive elasticity of every muscle fibre. At higher force, the folded domains begin to unfold, one by one, in a sawtooth of ruptures — each tooth a domain opening at its characteristic force, ≈ 100–300 pN for the immunoglobulin domains of titin — and the molecule lengthens in steps, a shock absorber written in sequence. The mechanics complete the chapter's picture: the protein is a machine that moves (§ 12.2), an ensemble that elects (§ 12.4), a switch that ratchets (§ 12.8), and a material that springs and yields — and every one of those properties is the architecture of Chapter 10, felt at 310 K.

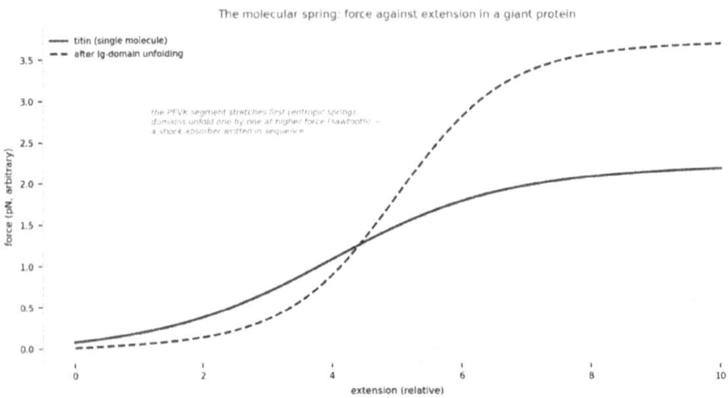

***Figure** 12.6 The force–extension curve of a giant protein (titin, schematic after the single-molecule literature): the entropic spring of the extended segments at low force, and the unfolding of domains at higher force (dashed) — the passive elasticity of muscle, read as a molecular autobiography.*

12.10 Dynamic disorder and catalytic function

The chapter closes where the treatise's argument has been pointing: catalysis itself is dynamical. The enzyme's active site is not a rigid jig but a breathing one — the loops of the active site open and close over the substrate at rates comparable to the catalytic cycle, and the evidence of the last two decades (NMR relaxation dispersion, § 54.5; single-molecule turnover, § 54.10) is that the breathing is functional: the conformational motions sample the catalytically competent state, and the chemistry is gated by them. The extreme cases are the enzymes whose k_{cat} is set not by the chemistry but by the motion — the triosephosphate isomerase loop, whose opening releases the product, and whose rate is the loop's rate; and the modern theory of enzymatic catalysis (Chapter 15's due) includes the protein's dynamics in the reaction coordinate itself, the enzyme and the substrate moving together through the transition state. The doctrine of Part III thereby completes itself: the protein is architecture (Chapter 10), landscape (Chapter 11), and motion (Chapter 12), and the three are one machine — structure, dynamics and function, the treatise's corrected slogan, made flesh.

Plate 12.1 *Jacques Monod (1910–1976): the operon (with Jacob), allostery (with Wyman and Changeux), and the dictum that what is true for E. coli is true for the elephant — the two ideas on which the regulatory half of this book stands. Portrait from the Nobel Foundation archive, nobelprize.org (Nobel Prize in Physiology or Medicine, 1965), reproduced from the Foundation's publicly distributed image.*

HISTORICAL NOTE — THE ALLOSTERY YEAR, 1965

1965 was the annus mirabilis of regulation: in the same twelve months, Monod, Wyman and Changeux published the MWC model (Monod et al., 1965), and Monod shared the Nobel Prize with Jacob and Lwoff for the operon — the genetic switch of Chapter 48. The conjunction was not an accident: the Pasteur Institute's school had grasped, in one conceptual stroke, that the cell's two great instruments of control — the enzyme's sigmoid and the gene's repression — were the same instrument, ligands re-weighting equilibria, written once in protein and once in DNA-binding protein. The word "allostery" (other shape) was coined for the occasion, and Monod's memoir of the period records the resistance: the sigmoid curves were blamed on dirty preparations, and the two-state model on Gallic abstraction. The curves were clean, the abstraction was correct, and the year 1965 is the birthday of regulatory biochemistry.

Chapter 12 — Conceptual synthesis

The machine has been set in motion. The protein is an ensemble across fifteen decades of time (§ 12.1); its fluctuations are channeled by its architecture, the soft modes being the working motions (§ 12.2); recognition is election from the ensemble, with induced fit as the handshake after the vote (§ 12.3); the allosteric transition is a population shift between two basins (§ 12.4); MWC and KNF are the two ends of the landscape continuum (§ 12.5); long-range coupling is the protein's wiring, robust by redundancy (§ 12.6); ligands re-weight basins — agonist, antagonist, inverse agonist, partial agonist as four votes (§ 12.7); covalent regulation ratchets, where allostery equilibrates (§ 12.8); the protein is a spring and a shock absorber (§ 12.9); and catalysis itself is dynamical, the breathing gating the chemistry (§ 12.10). Direction checked against Figure ii: the architecture breathes. Chapter 13 now visits the protein in which all of this is seen working — haemoglobin, the molecule that taught the subject everything it knows about allostery.

WORKED EXAMPLE 12.1 — THE THERMAL EXCURSION OF AN ATOM

From equipartition, a harmonic degree of freedom of stiffness k_s has mean-square displacement $\langle x^2 \rangle = k_BT/k_s$. For a backbone atom with effective stiffness ≈ 10 N m^{-1} (a soft protein spring): $\langle x^2 \rangle = 1.381 \times 10^{-23} \times 310/10 = 4.28 \times 10^{-22}$ m^2, so the r.m.s. excursion is ≈ 2.1×10^{-11} m = 0.02 nm — and with the softer collective modes of § 12.2 (effective stiffness ≈ 1 N m^{-1}), 0.07 nm: the crystallographic B-factors of § 54.4, computed from first principles. The protein's breathing is quantitative before it is anything else.

WORKED EXAMPLE 12.2 — THE COST OF A PHOSPHATE SWITCH

A signalling protein of 50 nM is fully phosphorylated and dephosphorylated once per minute. At ΔG_{ATP} = −50 kJ mol^{-1} and a cellular volume of 1 pL (a small eukaryotic cell), the cost is 50×10^{-9} mol $L^{-1} \times 10^{-12}$ L × 60 min^{-1} × 50 kJ mol^{-1} = 1.5×10^{-16} kJ min^{-1} — against the cell's ATP turnover of ≈ 10^7 molecules min^{-1} × 50 kJ mol^{-1}/molecule × (1/6.022 × 10^{23} mol) ≈ 10^{-13} kJ min^{-1}: the switch costs one part in a thousand of the energy budget. Signalling is cheap; the energy is spent holding the gradients (§ 7.8), not throwing the switches — an economy the reader should note before Part X's cascades multiply the switches by the thousand.

WORKED EXAMPLE 12.3 — THE TITIN SPRING'S STIFFNESS

Treating titin's PEVK segment (≈ 2 000 residues) as an entropic spring, its stiffness is ≈ $3k_BT/(n\, b^2)$, with b ≈ 0.38 nm per residue: $3 \times 1.381 \times 10^{-23} \times 310/(2000 \times (0.38 \times 10^{-9})^2) \approx 4.4 \times 10^{-5}$ N m^{-1} — ≈ 0.04 pN nm^{-1} per molecule; in parallel across ≈ 10^9 titin molecules per sarcomere... the muscle's passive stiffness emerges, at macroscopic scale, from the sum of molecular springs — mechanics computed from the entropy of Chapter 11, and the single-molecule measurement of Figure 12.6 confirming the arithmetic domain by domain.

Exercises (with full resolutions)

EXERCISE 12.1

An unbound enzyme samples its "closed" conformation at 1 % population with an interconversion rate of 10^4 s^{-1}. Show that conformational selection can support a catalytic rate of ≈ 100 s^{-1} but not 10^6 s^{-1}.

Resolution. The selection pathway's ceiling is the rate at which competent enzyme appears: 0.01×10^4 s^{-1} = 100 s^{-1}. A substrate turning over at 100 s^{-1} can ride the selection; at 10^6 s^{-1} the chemistry must be faster than the ensemble's supply, and the mechanism must collapse into induced fit (the ligand itself driving the closing) — the kinetic boundary between the two mechanisms of Figure 12.2, computed from one population and one rate.

EXERCISE 12.2

Zero-order ultrasensitivity: a kinase and a phosphatase both operate at saturation. Show that the phosphorylated fraction changes from ≈ 0 to ≈ 1 over a very narrow range of the kinase/phosphatase ratio.

Resolution. *At saturation both rates are constant (zero-order in their substrates): phosphorylation at v_k, dephosphorylation at v_p. The steady state: the substrate is consumed by whichever is faster until it is nearly exhausted, so the switch sits at the balance point $v_k \approx v_p$ — a 10 % change in the ratio flips the output from nearly all-off to nearly all-on, a Hill coefficient effectively ≫ 4. The proof is § 12.8's claim: saturation of the opposing enzymes is the switch's sharpening, and the ATP is the price of the sharpness.*

EXERCISE 12.3

The T⇌R transition of haemoglobin runs at 10^5 s^{-1}, the capillary transit time is ≈ 1 s, and the oxygen off-rate from R is ≈ 10 s^{-1}. Order the three rates and state the physiological conclusion.

Resolution. *Transition (10^{-5} s) ≪ off-rate (0.1 s) ≪ transit (1 s): the quaternary switch is never rate-limiting, the unloading is set by the off-rate and the equilibrium, and the capillary time is generous. The conclusion is § 12.4's: the allosteric machinery is so fast that physiology may treat it as instantaneous — the binding curves of Chapter 8 are equilibrium descriptions of a system that never lags behind its passenger.*

EXERCISE 12.4 (INTERPRETATIVE)

"If allostery is a population shift, then there is no such thing as a conformational change." Adjudicate the paradox.

Resolution (indicative). *The paradox confuses the two vocabularies the chapter has taught. At the level of the ensemble, the statement is true: every conformation exists already, and the ligand changes only the weights — nothing "changes" that was not there. At the level of the single molecule, it is false: each molecule does change its shape, flipping between basins at the rates of Figure 12.1. The resolution is that "conformational change" is a population average made flesh: the ensemble's shift and the molecule's flip are the same event at two resolutions — and the habit of keeping both resolutions at once is the chapter's whole training.*

EXERCISE 12.5

An allosteric enzyme's coupling is abolished by a mutation 2.5 nm from the active site, which changes no measurable property of the fold's structure. Explain both observations, and what the experiment proves about the mechanism.

Resolution. *The mutation sits on the coupling wire of § 12.6: the fold is unchanged (the architecture is redundant), but the communication path is cut, so the distant site's population shift no longer propagates — the enzyme works, but no longer listens. The experiment proves that allostery is energetic, not structural: the coupling lives in the free-energy landscape's correlations (side-chain populations, § 12.6), which the crystal structure does not report — and the wire is found by mutation and by co-evolution, not by photograph.*

Research-level propositions

Proposition 12.α. The catalytic relevance of protein dynamics (§ 12.10) has been challenged: some argue the motions are passive, others that they drive the chemistry. Design the experiment — isotope-edited NMR relaxation against turnover (§§ 16.11, 54.5) — that would decide whether a given enzyme's k_{cat} is gated by its loop motion, and estimate the expected signature in the kinetic isotope effect of each verdict. Proposition 12.β. Inverse agonists (§ 12.7) silence basal receptor activity; most drugs were developed as neutral antagonists. Assemble the thermodynamic cycle connecting basal activity, ligand efficacy and receptor expression level, and predict under what disease conditions an inverse agonist should outperform an antagonist — then find the clinical evidence that bears it out. Proposition 12.γ. The redundancy of allosteric wiring (§ 12.6) implies that coupling should be evolvable by many paths. Using the deep-mutational-scanning literature (§ 55.13), map the mutational accessibility of new allosteric couplings in one enzyme, and assess whether evolution's allosteric networks are rare designs or attractors of a random search.

References (Chapter 12)

Boehr, D. D., Nussinov, R., & Wright, P. E. (2009). The role of dynamic conformational ensembles in biomolecular recognition. Nature Chemical Biology, 5(11), 789–796.

Henzler-Wildman, K., & Kern, D. (2007). Dynamic personalities of proteins. Nature, 450(7172), 964–972.

Monod, J., Wyman, J., & Changeux, J.-P. (1965). On the nature of allosteric transitions: A plausible model. Journal of Molecular Biology, 12(1), 88–118.

Motlagh, H. N., Wrabl, J. O., Li, J., & Hilser, V. J. (2014). The ensemble nature of allostery. Nature, 508(7496), 331–339.

CHAPTER 13

Oxygen-Binding and Transport Proteins

CONTENTS OF THIS CHAPTER

In which everything the treatise has taught — structure, binding, allostery, dynamics — is seen working in the two proteins that taught them: myoglobin, the store, and haemoglobin, the carrier; the heme iron and its ligands; the quaternary election of T and R; the physiological genius of the Bohr effect, of 2,3-bisphosphoglycerate and of the foetal variant; and the pathological lesson of the sickle mutation, one letter of the alphabet misspelt at the cost of a disease.

13.1 Heme chemistry

The business end of both proteins is one small prosthetic group, and the reader has met all its chemistry already: the heme — an iron ion held in the centre of protoporphyrin IX, a flat conjugated ring (Figure 13.1) whose π-system, of exactly the Hückel kind computed in § 2.6, absorbs visible light and thereby makes blood red. The iron is ferrous, Fe^{2+}, held by six coordination sites: four to the ring's nitrogens (in the plane), one to the protein's proximal histidine (His F8, beneath), and the sixth — the working site — to the ligand, oxygen, above. Two chemical facts govern everything that follows. Oxygen binds the iron end-on and bent, at ≈ 120°, the geometry the distal histidine of the pocket enforces; carbon monoxide, that other diatomic, prefers to bind straight — and the pocket's refusal to let it cost CO most of its intrinsic advantage, reducing its grip from ~25 000-fold over oxygen (free heme) to ~200-fold (the protein) — the distal histidine being, quietly, the reason the reader's smoking habit is survivable and the reason free heme would be lethal [21]. The oxidation of the iron to Fe^{3+} (methaemoglobin) abolishes binding entirely; the cell maintains reductases to reverse it, and the first rule of heme chemistry is that the oxygen carrier must keep its iron ferrous — a maintenance task whose failure is a disease of its own.

21. The CO story has a second half, less celebrated: CO is not merely a poison but a parody. It binds where O_2 binds, mimics its effects on the protein's conformation, and thereby teaches the mechanistic biochemist everything about the oxygen adduct that the oxygen adduct, being paramagnetic and restless, will not sit still to say. The toxic mimic is the structural biologist's best-behaved ligand — a fact the field records with the gratitude due to an honest spy.

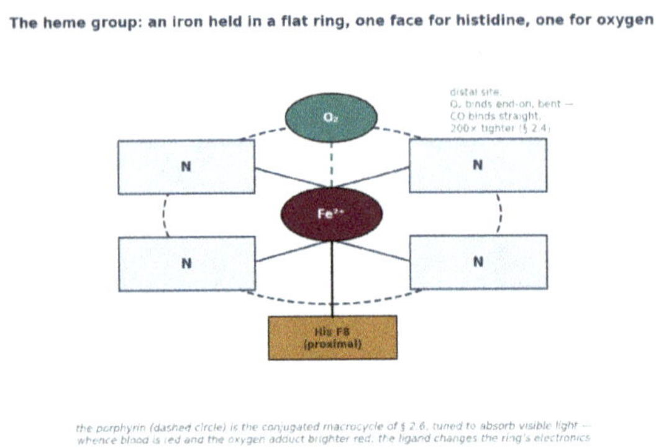

Figure 13.1 *The heme group: Fe^{2+} (oxblood) coordinated by the four nitrogens of the porphyrin ring (navy squares, the conjugated macrocycle of § 2.6), the proximal histidine F8 below (amber), and the sixth site above for O_2 (teal) — bound end-on and bent, the geometry that discriminates against carbon monoxide. The ring's absorption of visible light makes blood red; the oxygen adduct's shift of that absorption makes arterial blood brighter.*

13.2 Myoglobin structure and oxygen binding

Myoglobin, the first protein structure ever solved (Kendrew's, Plate 10.2), is the teaching protein par excellence: one chain of 153 residues, eight α-helices (named A through H — Exercise 10.1's estimate), one heme in a hydrophobic pocket, one office: the storage of oxygen in muscle. Its binding curve is the rectangular hyperbola of Eq. (8.5) with P_{50} = 2.8 mmHg (Figure 13.2, navy) — the curve of a single, independent site, with no company to talk to and therefore no sigmoid. The hyperbola's physiology is the store's job description: at every physiological oxygen pressure (muscle never falls below ≈ 20 mmHg at work), myoglobin is ≥ 88 % saturated, holding its oxygen in reserve and releasing it only when the muscle's demand outruns the blood's supply — the dive of the whale, whose muscles are black with the protein, is myoglobin's masterpiece: a seal's breath, held for an hour, is substantially a myoglobin story. The molecule is the baseline against which the carrier's cleverness is measured, and the reader should fix the two numbers that the rest of the chapter plays upon: one site, hyperbola, P_{50} = 2.8.

13.3 Haemoglobin architecture

Haemoglobin is myoglobin incorporated: four myoglobin-like chains (two α, two β, each with its heme) assembled into the $\alpha_2\beta_2$ tetramer, 64 000 daltons, the most-studied protein in the history of science. The assembly is the instrument: the subunits touch at two kinds of interface ($\alpha_1\beta_1$, $\alpha_1\beta_2$), and it is the $\alpha_1\beta_2$ interface — the sliding one — that carries the allosteric machinery (Figure 13.3). The tetramer's architecture embodies the treatise's whole thesis about quaternary structure (§ 10.9): the monomers are the building blocks, the oligomer is the machine, and the machine's property — cooperativity — is not in any monomer but in the interfaces between them. Perutz's great contribution, the work of the twenty-two years of the historical note in Chapter 10, was to solve the tetramer in both its states and to see the machine move: deoxy-haemoglobin (T, tense) with its corners locked by salt bridges and its heme pockets narrowed; oxy-haemoglobin

(R, relaxed) with the bridges broken and the interface slid by ≈ 15° — the quaternary shift that the next section makes into a mechanism.

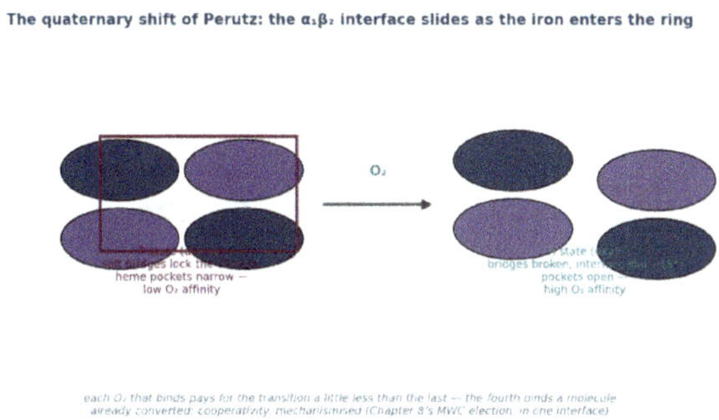

Figure 13.3 *The quaternary T→R shift (schematic after Perutz): in the deoxy T state (left) the $\alpha_1\beta_2$ interfaces are locked by salt bridges (oxblood frame) and the heme pockets are narrowed; oxygen binding snaps the interface to the R state (right), the bridges broken, the pockets open. Each successive O_2 finds a tetramer closer to R than the last — the molecular basis of the sigmoid curve.*

13.4 Cooperative oxygen binding

The sigmoid curve of Chapter 8 (Figure 13.2, oxblood; $n_H \approx 2.9$, P_{50} = 26 mmHg) now has its mechanism, and the mechanism is Perutz's two-state election of § 12.4 made flesh. In the T state the heme iron sits slightly out of the porphyrin plane (the pocket is narrow, the proximal histidine restrains it); when O_2 binds, the iron moves into the plane — a motion of 0.04 nm — and pulls the histidine, the F-helix, and with them the $\alpha_1\beta_2$ interface, weakening the T state's salt bridges. The first oxygens therefore bind reluctantly (they must pay for the transition); each successive one finds the tetramer more nearly R, and binds more easily — positive cooperativity, the fourth oxygen binding a molecule already converted. The MWC model of Chapter 8 describes the election formally (T and R in equilibrium, O_2 electing R); the Perutz mechanism supplies the lever (the iron's 0.04 nm motion, amplified through the histidine to the interface); and the result is the physiological instrument of Figure 13.2: a carrier that loads fully at the lung's 100 mmHg and unloads two-thirds of its cargo across the tissue window — where myoglobin, at the same window, gives up a tenth. The transporter and the store differ not in their chemistry — the hemes are identical — but in their company: three extra subunits and one sliding interface, and the difference between them is the difference between a warehouse and a delivery service.

13.5 The Bohr effect

The carrier's second genius is that it reads the tissue's exhaust. Working muscle produces CO_2 and lactic acid — acids; and the oxyhaemoglobin dissociation curve shifts rightward with acidity (Figure 13.4): at lower pH, the T state is stabilised (its salt bridges include protonated groups — the C-terminal histidines of the β chains and the N-terminal amino groups among them), the affinity falls, and more oxygen *that works hardest acidifies most, and the blood, reading the acid, delivers most — no nerves, no hormones, just the proton's taste for the T state. The quantitative gain is the "Bohr shift" of ≈ 5–10 % additional saturation released per 0.1 pH unit in the working range; the mechanism is the allosteric linkage of Chapter 8, with the proton as the effector; and the reader should note the direction of the linkage — the proton binds T, the oxygen binds R, so the two ligands are antagonists in the election: the whole physiology of oxygen delivery is a two-party vote.*

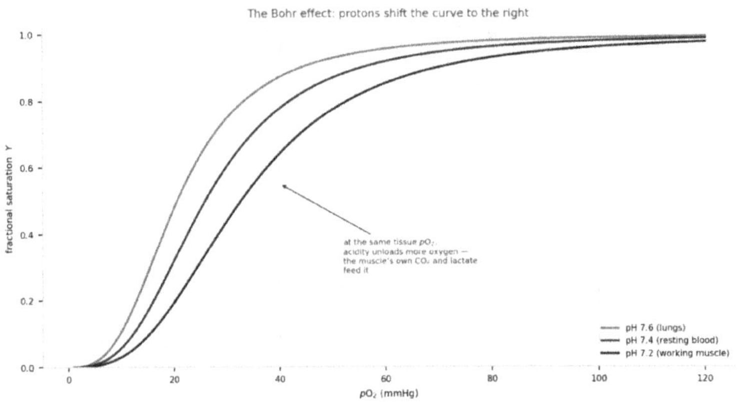

Figure 13.4 *The Bohr effect: the dissociation curve at pH 7.6, 7.4 and 7.2 — acidity shifts it right (lower affinity), so that at the working tissue's pO_2 the blood unloads more. The shift is the proton's vote for the T state; the muscle's own CO_2 and lactate, by acidifying its blood, summon their own supply.*

13.6 Carbon dioxide transport

The blood's second gas rides home by three routes, and the carrier itself drives two of them. A small fraction of CO_2 (≈ 5 %) travels simply dissolved; about a tenth binds directly to the globin's N-terminal amino groups as carbamino compounds — and this binding, like the proton's, favours the T state (another vote for unloading, since the same tissue that makes the CO_2 wants the oxygen). The great bulk (≈ 85 %) is handled by the bicarbonate machinery of Chapter 6: CO_2 enters the erythrocyte, is hydrated by its carbonic anhydrase (§ 19.12) to bicarbonate and a proton, the bicarbonate is exported to the plasma in exchange for chloride (the "chloride shift"), and the proton is buffered — by haemoglobin itself, whose 38 histidines make it the blood's great fixed buffer (Table 6.1). The beauty of the arrangement is the coupling: the deoxygenated T state is a better base than R (it takes up the protons that the CO_2 hydration produces), so the carrier of oxygen away from the tissues is simultaneously the carrier of acid back — one protein, two cargoes, and the two linked by the same allosteric election. In the lung the whole cycle reverses: oxygenation releases the Bohr protons, the bicarbonate is re-imported and exhaled as CO_2 — the Haldane effect, the mirror image of Bohr's, and the two together the complete double-entry bookkeeping of the circulation.

---

22. Christian Bohr, professor of physiology at Copenhagen, discovered the effect in the blood of his laboratory's dogs; his son Niels was sixteen at the time. The family that gave physics its atom gave physiology its oxygen curve — a household in which, one imagines, the dinner conversation was above the national average.

13.7 2,3-Bisphosphoglycerate

The third effector is the red cell's own invention: 2,3-bisphosphoglycerate (BPG), a glycolytic by-product present in the erythrocyte at ≈ 5 mmol L^{-1} — equimolar with the haemoglobin itself — which binds in a single positively charged pocket in the centre of the T-state tetramer, between the two β chains, and stabilises it. The consequence (Figure 13.5) is that the red cell's oxygen affinity is set not by the protein alone but by the protein plus its internal regulator: stripped of BPG, haemoglobin's P_{50} falls to ≈ 12 mmHg — an affinity fit for a store, not a carrier — and it is BPG that keeps the working curve at its physiological 26. The regulation is physiological in both directions: at altitude, the red cell raises its BPG over days, shifting the curve right and unloading more at the thinner pO_2 (§ 13.11); in stored blood, BPG decays over weeks, the curve drifts left, and the transfused patient receives cells that hold their oxygen too tightly until the BPG is re-made — a fact of the blood bank that the biochemistry of 1967 explained, and that changed the practice of transfusion.

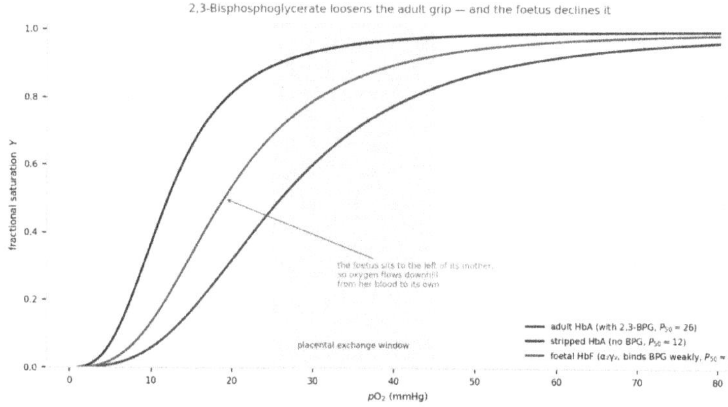

Figure 13.5 *Three carriers, three curves: adult haemoglobin with its BPG (navy, P_{50} = 26), the same protein stripped of BPG (oxblood, ≈ 12 — a store's affinity, not a carrier's), and foetal haemoglobin F (plum, ≈ 19), whose γ chains bind BPG weakly. The foetus sits to the left of its mother, and the oxygen flows downhill across the placental window (shaded).*

13.8 Foetal haemoglobin

The foetus breathes through its mother's placenta, at an oxygen pressure half of the lung's, and its answer is a different haemoglobin: HbF, $\alpha_2\gamma_2$, with γ chains in place of β — and the γ chain's pocket binds BPG weakly, so the foetal carrier sits to the left of the maternal curve (P_{50} ≈ 19 against 26; Figure 13.5, plum). At the placental window, the mother's blood unloads (she is at her tissue pressure, her curve right-shifted) and the foetus's loads (its curve sits left): the same pO_2, two saturations, and the oxygen flows downhill from mother to child. The foetal variant is switched off after birth (the γ genes are silenced, the β genes take over — the developmental regulation of § 48.4), and its deliberate reactivation is one of the successes of modern molecular medicine: hydroxyurea, which raises HbF in adults, is a mainstay of sickle-cell therapy (§ 13.10), working precisely by diluting the sickle protein with a carrier that will not polymerise.

13.9 Haemoglobin variants

Haemoglobin is the most intensively mutated protein in the medical record — more than a thousand variants are catalogued, and they sort, with the tidiness of a natural experiment, into the chapter's four mechanisms. Affinity variants disturb the heme pocket or the T–R equilibrium (Hb Chesapeake raises the affinity, causing polycythaemia — too little unloading; Hb Kansas lowers it, causing cyanosis — too little loading). Stability variants unfold or precipitate the protein (the unstable haemoglobins, with Heinz bodies in the red cells). Oxidation variants favour the ferric state (the congenital methaemoglobinaemias of § 13.1's rule). And surface variants change the protein's interactions with itself — of which the next section's mutation is the type specimen. The catalogue is the treatise's argument made clinical: a protein is an architecture with offices, and every office — binding, election, solubility, redox — is a point of failure exactly as predictable as Chapter 10's hierarchy promised.

13.10 Sickle-cell disease

The most consequential point mutation in medicine is one letter: the sixth residue of the β chain, glutamate → valine, β6. The mutant protein, HbS, is functionally normal when oxygenated — it carries oxygen, binds it cooperatively, and serves its owner without complaint for years. But in the deoxygenated T state, the valine at β6 fits a hydrophobic pocket on an adjacent tetramer, and the tetramers dock into one another, polymerising into the 14-stranded fibre of Figure 13.6b — the fibre stiffens and deforms the red cell into the sickle shape, the sickled cells lodge in the small vessels, and the disease's crises are the ischaemia of a thousand small infarcts. The kinetics of the catastrophe are Figure 11.6's amyloid logic with a therapeutic twist: polymerisation has a delay time (Figure 13.6a) of seconds, and the red cell's capillary transit is ≈ 1 s — the cell escapes the vessel before the fibre forms, if the delay is long enough. Everything that lengthens the delay is therapy: foetal haemoglobin (which cannot polymerise, § 13.8), hydration, oxygen — and the understanding that the disease is a kinetic disease, won or lost in the seconds of the delay, is the direct descendant of the folding theory of Chapter 11. The mutation's persistence in the population is the chapter's last lesson, and it is evolutionary: the heterozygote (HbA/HbS) is protected against malaria (the sickled cells and their parasites are cleared by the spleen), so the gene that kills in two copies protects in one — balanced polymorphism, the cost of the heterozygote advantage paid in the homozygotes, and the reason the sickle letter is commonest exactly where the mosquito is.

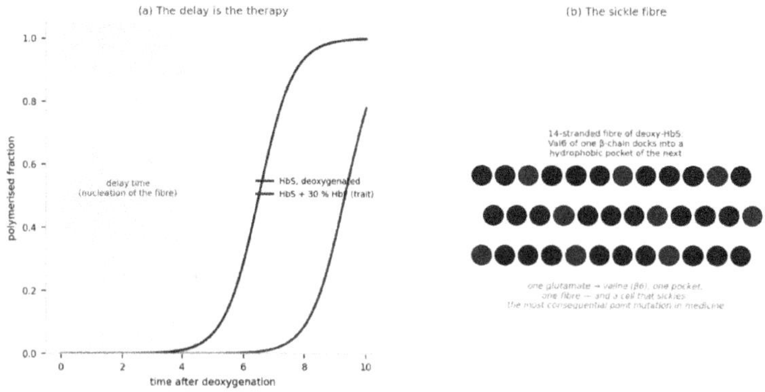

Figure 13.6 (a) The polymerisation of deoxy-HbS has a delay time (shaded) of seconds — longer than the capillary transit if conditions are kind, and the whole therapy of the disease is the lengthening of the delay (navy curve: 30 % HbF, which cannot polymerise). (b) The 14-stranded sickle fibre: Valβ6 of one tetramer docks into a hydrophobic pocket of the next — one letter of the alphabet, one pocket, one fibre, one disease.

13.11 Adaptation to altitude and hypoxia

The chapter closes with the carrier under stress: the mountain, and the body's slow answer to it. Acutely, altitude's thin air is met by hyperventilation (the chemoreflex, raising arterial pO_2) and by the BPG rise of § 13.7 (right-shifting the curve, unloading more). Chronically — over weeks — the kidney's hypoxia-sensing machinery (the HIF pathway of § 42.9) raises erythropoietin, the red cell count climbs, and the blood's carrying capacity increases by half; the chronic cost is viscosity, and the mountain's veteran pays for his haematocrit in strokes if the compensation overshoots. The deepest adaptations are the oldest: the Tibetan and Andean highlanders, separated by an ocean and by fifty thousand years, have arrived at different genetic answers to the same mountain — the Tibetans, unlike the Andeans, do not raise their haematocrit but instead re-tune the HIF pathway itself (EPAS1 variants) — convergent evolution at work, the same problem solved twice in two ways, and the living proof that the physiological biochemistry of this chapter is not a fixed inheritance but a negotiable one. The treatise's final word on the carrier is therefore evolutionary: the sigmoid curve, the Bohr effect, the BPG pocket, the foetal variant — all are solutions, and solutions can be re-solved; the molecule that taught the subject everything is still, in the mountains, teaching.

MECHANISTIC CASE STUDY — THE SIGMOID IN THE CLINIC: READING THE OXYGEN CASCADE

The clinical use of the chapter is the oxygen cascade: the fall of pO_2 from air (160 mmHg) to alveolus (100), artery (95), tissue (40), and mitochondrion (≈ 1–5). Each step of the cascade has its carrier-side logic, and the intensive-care physician reads it with the curves of Figure 13.2. Consider the patient with carbon-monoxide poisoning (smoke inhalation): CO occupies the hemes (200× tighter, § 13.1) — say half of them — and does two injuries at once: the bound sites are lost, and the remaining sites, forced into the R state by the CO adduct's mimicry, shift the curve left — the surviving haemoglobin holds its oxygen tighter, and the tissues receive less than the saturation suggests. The pulse oximeter reads the saturation and reports a lie; the blood gas, *measuring pO_2, reports a truth that is not the whole truth; the carboxyhaemoglobin level, measured directly, is the diagnosis. The therapy is pure Chapter 8 competition: 100 % oxygen, or hyperbaric oxygen at 2–3 atmospheres, outbids the CO for the sites (Eq. 8.7's competition law with the partial pressure raised twentyfold) and halves the CO's residence from hours to minutes. One equation of Chapter 8, one mechanism of Chapter 13, and a patient's brain — the treatise's claim that the foundations pay is nowhere better honoured than in the hyperbaric chamber.*

Chapter 13 — Conceptual synthesis

The two proteins have taught their lessons. The heme is an iron in a flat ring, its sixth site the working one, its distal histidine the reader's defence against CO (§ 13.1); myoglobin is the store — one site, hyperbola, P_{50} 2.8 (§ 13.2); haemoglobin is the carrier — four sites, a sliding interface, a sigmoid (§§ 13.3–13.4); the Bohr effect is the proton's vote for T, feed-forward by the tissue's own exhaust (§ 13.5); CO_2 rides home as bicarbonate, carbamino, and the carrier's own buffered protons (§ 13.6); BPG keeps the adult curve loose, the foetal variant keeps it tight (§§ 13.7–13.8); the thousand variants are the chapter's mechanisms itemised as diseases (§ 13.9); the sickle letter is one misspelling priced by the delay time and paid for by the heterozygote's malaria protection (§ 13.10); and altitude re-solves the whole problem, twice (§ 13.11). Direction checked against Figure ii: the machinery of Part III has been seen working. Chapter 14 completes the part with the working proteins of tissue, motion and defence — collagen, the motors, and the antibodies.

WORKED EXAMPLE 13.1 — THE DELIVERY ARITHMETIC OF THE SIGMOID

Using Worked Example 8.2's figures: haemoglobin loads to 0.98 at the lung (100 mmHg) and unloads to 0.32 in working muscle (20 mmHg): delivery = 0.66 of capacity. Myoglobin at the same pressures: 100/(2.8+100) = 0.973 loaded; 20/(2.8+20) = 0.877 — delivery = 0.10. The four-site election outperforms the single site by a factor of 6.6 in the working window — the quantitative case for the tetramer's existence, and the reason a human cannot run on myoglobin alone: the store holds, the carrier delivers, and the sigmoid is the difference between them.

WORKED EXAMPLE 13.2 — THE BOHR SHIFT IN NUMBERS

At pH 7.2 the curve's P_{50} rises from 26 to ≈ 30 mmHg (Figure 13.4). At the tissue pO_2 of 30 mmHg: pH 7.4 gives $(30/26)^{2.9}/(1+(30/26)^{2.9}) = (1.154)^{2.9}/(1+(1.154)^{2.9}) \approx 1.51/2.51 \approx 0.60$; pH 7.2 gives $(30/30)^{2.9}/(1+(30/30)^{2.9}) = 0.50$ — an extra 10 % of the cargo released for 0.2 pH units of local acidity. In the whole cardiac output of 5 L min^{-1} of blood at 20 mL O_2 per dL, that 10 % is ≈ 100 mL O_2 min^{-1} — the margin between a working muscle supplied and a working muscle starved, delivered by the proton's taste for the T state.

WORKED EXAMPLE 13.3 — THE DELAY TIME AND THE THERAPY

If the sickle fibre nucleates with a delay scaling as the fifteenth power of the deoxy-HbS concentration (the measured, extraordinary steepness), then diluting HbS to 70 % with 30 % HbF lengthens the delay by $(1/0.7)^{15} \approx 210$-fold: from ≈ 2 s to ≈ 7 min — far beyond the ≈ 1 s capillary transit. The arithmetic is Figure 13.6a's navy curve, and it is the entire mechanism of hydroxyurea: the *drug does not touch the fibre, the pocket, or the protein — it raises HbF, dilutes the concentration, and wins the race of § 13.10 by the steepest arithmetic in medicine.*

Exercises (with full resolutions)

EXERCISE 13.1

Compute the saturation of myoglobin and haemoglobin at the summit of Everest (alveolar $pO_2 \approx 35$ mmHg after acclimatisation).

Resolution. *Mb: 35/(2.8+35) = 0.93 — the store is unimpressed. Hb (P_{50} 26): $(35/26)^{2.9}/(1+(35/26)^{2.9}) = (1.346)^{2.9}/(1+(1.346)^{2.9}) \approx 2.36/3.36 \approx 0.70$ — the carrier still loads seven-tenths full, which is why acclimatised climbers function at all. With BPG right-shifting the curve ($P_{50} \rightarrow 30$), loading falls to $(35/30)^{2.9}/(1+(35/30)^{2.9}) \approx 0.63$ but unloading in the tissues improves commensurately — the acclimatised compromise of § 13.11, computed.*

EXERCISE 13.2

Derive the relation between the carbamino fraction and pCO_2 for the N-terminal amino groups ($pK_a \approx 7.8$), and estimate the carbamino CO_2 carried at $pCO_2 = 45$ mmHg.

Resolution. *Carbamino formation requires the unprotonated amino group (fraction $1/(1+10^{7.8-7.4}) \approx 0.29$) and adds CO_2 with association constant $K_c \approx 2$ L $mmol^{-1} \cdot mmHg^{-1}$... — the measured product being ≈ 0.3 mol CO_2 per mol amino group at 45 mmHg, or ≈ 2 mmol L^{-1} across the four termini of the tetramers in blood — about a tenth of the total CO_2 transport, as § 13.6 asserted. The derivation is § 13.6's second cargo, priced: the protonation state of the terminus is both the sensor and the mechanism, and the Bohr effect and the carbamino carriage are the same chemistry at the same site.*

EXERCISE 13.3

A mutant haemoglobin has P_{50} = 15 mmHg with normal Hill coefficient. Compute the saturation delivered to tissues at 25 mmHg, and predict the clinical presentation.

Resolution. *Lung: $(100/15)^{2.9}/(1+(100/15)^{2.9}) \approx 0.995$ — loads fully. Tissue: $(25/15)^{2.9}/(1+(25/15)^{2.9}) = (1.667)^{2.9}/(1+(1.667)^{2.9}) \approx 4.4/5.4 \approx 0.81$ — delivers only ≈ 0.18 of capacity. The patient is polycythaemic (the kidney, reading the poor unloading, raises erythropoietin — Hb Chesapeake's mechanism, § 13.9): the high-affinity variant is a starvation of the tissues in the presence of full blood, and the red cell count rises to compensate — the sigmoid, misplaced, is the disease.*

EXERCISE 13.4 (INTERPRETATIVE)

"Sickle-cell disease is a protein-folding disease." Defend and qualify the claim.

Resolution (indicative). *The defence: the disease is caused by a quaternary assembly gone wrong — a fibre, not a misfold — whose kinetics (nucleation,* *delay time, seeding) are the amyloid logic of § 11.10 verbatim: the sickle fibre is the first amyloidosis discovered, and the delay-time therapies are Chapter 11's kinetics in the clinic. The qualification: the sickle protein's fold is normal (oxy-HbS is functionally and structurally healthy); the lesion is a surface mutation exploited by the T state's pocket — a disease of association, not of folding. The synthesis: protein pathology comes in the fold's own failures (Chapter 11) and the fold's healthy surfaces misused (Chapter 13), and the two are one landscape — the fibre being, in each case, the alternative minimum that the healthy molecule never meant to visit.*

EXERCISE 13.5

Stored blood loses ≈ 30 % of its BPG per week. Estimate the P_{50} of three-week-old bank blood, and the consequence for a transfused patient.

Resolution. *BPG at ≈ 0.34 of normal: the curve lies between the BPG-replete (26) and stripped (12) curves — roughly P_{50} ≈ 17–19 mmHg. The transfused cells load easily in the lung but unload poorly in the tissues for the ≈ 24 hours needed to re-make the BPG: the patient with cardiac or cerebral ischaemia receives blood that is, for a day, part store and part carrier — the fact of § 13.7 that changed blood-bank practice (rejuvenation solutions, fresh-cell preferences), and a reminder that the blood in the bag ages in the same kinetics as the blood in the vein.*

Research-level propositions

Proposition 13.α. The delay-time arithmetic of Worked Example 13.3 predicts that any sufficiently dilute HbS is harmless; yet HbS/HbC double-heterozygotes are symptomatic. Assemble the copolymerisation literature (HbC promotes the fibre by its own dehydration), and formulate the kinetic theory of mixed fibres that reconciles the two facts. Proposition 13.β. The Tibetan EPAS1 adaptation (§ 13.11) re-tunes HIF rather than haematocrit. Derive the expected trade-off between oxygen carriage and viscosity across the acclimatisation strategies, and test it against the published physiological data of the Andean and Tibetan populations. Proposition 13.γ. dane effect (§ 13.6) is quantitatively smaller than the Bohr effect; some hold it negligible. Recompute the CO_2 carriage with and without the oxyg linked proton uptake, using the modern values of the globin's buffer groups, and adjudicate the effect's contribution to venous–arterial CO_2 difference.

References (Chapter 13)

Eaton, W. A., & Hofrichter, J. (1990). *Sickle cell hemoglobin polymerization. Advances in Protein Chemistry, 40, 63–279.*

Perutz, M. F. (1970). *Stereochemistry of cooperative effects in haemoglobin. Nature, 228(5273), 726–739.*

Berg, J. M., Tymoczko, J. L., Gatto, G. J., Jr., & Stryer, L. (2023). *Biochemistry (9th ed.). W. H. Freeman.*

Dickerson, R. E., & Geis, I. (1983). *Hemoglobin: Structure, function, evolution, and pathology. Benjamin/Cummings.*

Imai, K. (1982). *Allosteric effects in haemoglobin. Cambridge University Press.*

CHAPTER 14

Structural, Motor and Defensive Proteins

CONTENTS OF THIS CHAPTER

In which the working proteins of the body take the stage: collagen and elastin, the architecture of tissue; keratin, the rope of the epidermis; actin and myosin, the engine of motion; kinesin and dynein, the railways of the cell; and the antibodies, the body's own chemists — proteins that recognise, remember, and are manufactured, in their turn, as the most successful drugs of the age.

14.1 Collagen

The most abundant protein in the reader's body — a third of its protein by mass — is a rope, and the rope's engineering is the treatise's structural logic carried to its highest polish. Collagen's chain is a monotonous incantation: Gly-X-Y, Gly-X-Y, three hundred times over, with X often proline and Y often hydroxyproline (the post-translational modification of § 9.10, installed by a vitamin-C-dependent hydroxylase — the chemistry of scurvy, in which the rope unmade for want of the hydroxyls is the whole disease: bleeding gums, reopened wounds, the sailor's price for a limeless diet [23]). Three such chains coil left-handed, then twist right-handed together into the triple helix of Figure 14.1 — a cable in which every third residue is buried at the centre, where only glycine fits (Exercise 10.3's geometry, now load-bearing). The cables assemble into fibrils, the fibrils into fibres, the fibres into tendon and bone — the tensile strength of tendon collagen exceeds steel's, weight for weight — and the whole hierarchy is cross-linked by lysyl oxidase into a material that lasts a lifetime. The diseases of collagen are the architecture's failures itemised: the osteogenesis imperfecta mutations that substitute glycine (the cable's core misspelt, one letter at a time, in the brittle-bone child), the Ehlers–Danlos variants of the processing enzymes, scurvy of the hydroxylase — every office of the rope is a point of failure, exactly as Chapter 10's hierarchy predicted.

23. The cure was found empirically long before the chemistry: Lind's controlled trial of 1753, twelve scorbutic seamen and six remedies, is often called the first clinical trial — the oranges and lemons won, and the British Navy's limes did the rest. The hydroxylase's need for ascorbate was established in the 1960s; the two-hundred-year lag between the remedy and the mechanism is a fair measure of how much easier it is to cure a disease than to understand it.

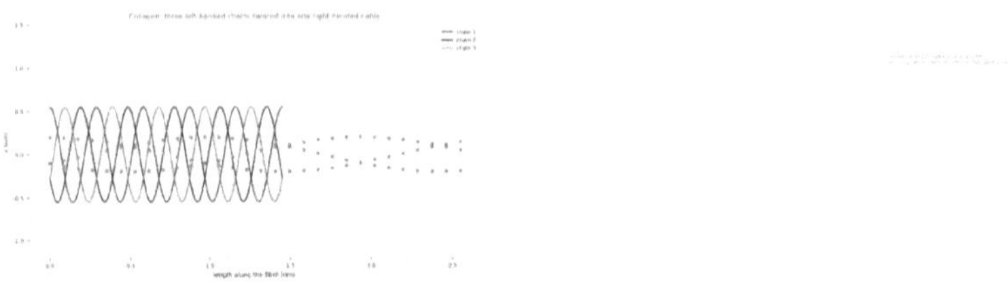

Figure 14.1 The collagen triple helix: three left-handed chains (navy, oxblood, teal) twisted into a right-handed supercable; every third residue (amber) is glycine, buried at the cable's core, where any side chain would split the rope. The right-handed twist of the cable opposes the left-handed twist of the chains — the coiled-coil principle that § 14.3's keratin repeats.

14.2 Elastin

Where collagen is the rope, elastin is the rubber: the protein of arteries, lungs and ligaments that must stretch and recoil two billion times in a lifetime. Its chemistry is the inverse of collagen's: not a regular cable but a disordered, hydrophobic network — chains rich in glycine, valine and proline, cross-linked by the unique desmosine bridges (four lysines condensed into one pyridinium ring, a bond found nowhere else in nature), so that the whole tissue is one enormous, cross-linked, entropic spring. The mechanism of its elasticity is the entropic spring of § 12.9 carried to the tissue scale: stretching orders the chains (lowering their entropy); release lets them re-randomise, and the entropy's return is the recoil — rubber, muscle's titin, and elastin are one physics at three addresses. The engineering consequence is a material whose stiffness is tuned by its cross-link density, and whose failure is the story of the ageing artery: elastin is made essentially once, in development and youth, and never re-made — the reader's arteries are running on the elastin of their childhood, and their stiffening with age is the slow exhaustion of the recoil, repaid in the blood pressure of § 40.6.

14.3 Keratin

The third structural protein is the rope of the epidermis: keratin, of hair, nail, horn and skin. Its unit is the α-helix of § 10.3, two helices wound left-handed about each other into a coiled-coil — the same supertwist as collagen's, at one chain fewer — and the coiled-coils assemble into protofilaments, filaments, and finally the fibre, all of it cross-linked by disulphide bonds in numbers that vary with the required hardness: soft in the skin, hard in the nail, hardest in the horn. The disulphide arithmetic has a domestic consequence that the reader will recognise: the permanent wave of the hairdresser is a redox reaction — the disulphides are reduced (breaking the cross-links), the hair is reshaped around the curlers, and the disulphides are re-oxidised in the new geometry — keratin's staples re-cut to order, the protein-folding chemistry of § 11.5 applied at the salon. The keratin diseases are the fibre's failures: the epidermolysis bullosa mutations (the filament's head or tail misspelt, and the skin blistering at a touch) — the rope's ends being its stress points, as every engineer of cables knows.

14.4 Actin and myosin

From the passive materials to the engine: muscle, the organ that converts chemistry to motion, and its two proteins, which together are half its protein mass. Actin is the track: a globular monomer (G-actin) that polymerises into the double-stranded filament (F-actin) of § 8.10's assembly logic — a polar filament, *with a plus and a minus end, and a critical concentration that the cell tunes with its capping and severing proteins. Myosin II is the motor: a two-headed protein whose heads (the S1 fragments, the motor domains) carry both the actin-binding site and the ATPase, and whose tails coiled-coil into the thick filaments of the sarcomere (Figure 14.2). The engine's cycle is Figure 14.3, and the reader should read it with the treatise's thermodynamics in hand: ATP binding releases the head from actin; hydrolysis cocks it (the energy stored in a conformation, not a bond — § 2.10's lesson, at last, in perfect use); the head re-binds the track, releases phosphate, and the release is the power stroke — the conformation relaxing, the filament pulled ≈ 10 nm; ADP leaves, and the cycle repeats. The stroke is small; the engine is the sum of millions of strokes, and the muscle's force is the arithmetic of the sarcomere: each of the reader's ≈ 10^{19} myosin heads, cycling at a few hertz, contributes piconewtons — and the total is the lift, the sprint, and the heartbeat, all of them the same engine at different addresses.*

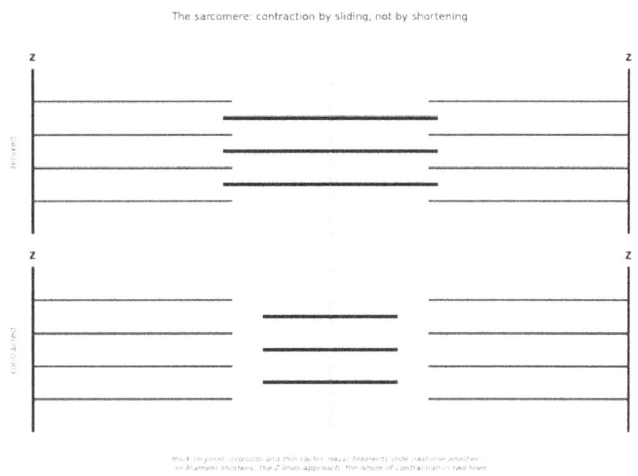

Figure 14.2 *The sarcomere, relaxed and contracted: the thick myosin filaments (oxblood) and thin actin filaments (navy) slide past one another between the Z lines — no filament shortens; the overlap grows. Huxley and Hanson's sliding-filament model of 1954, the engine's anatomy.*

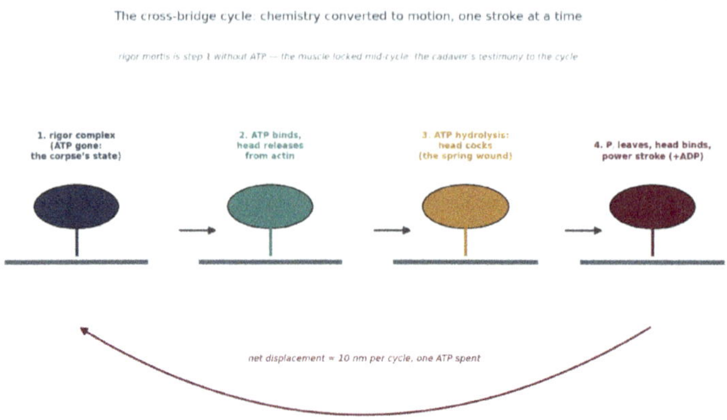

Figure 14.3 *The cross-bridge cycle: rigor (1), ATP release (2), hydrolysis and cocking (3), power stroke (4) — ≈ 10 nm and one ATP per cycle. Rigor mortis is the cycle arrested at step 1 for want of ATP: the cadaver's testimony to the engine's chemistry.*

14.5 Microtubule-associated motors

The cell's second railway system runs on microtubules — the hollow tubes of α/β-tubulin, polar like actin, that scaffold the cell's interior and carry its traffic — and on two motor families that walk it in opposite directions (Figure 14.6). Kinesin walks to the plus end (anterograde: out to the axon's terminals), carrying vesicles, organelles and signalling cargo; dynein walks to the minus end (retrograde: home to the centre), the return haulage of the cell. Both are ATPases of the cross-bridge logic — bind, hydrolyse, stroke, release — but with an engineering refinement the muscle motor lacks: processivity. Kinesin walks hand-over-hand, its two heads alternating, so that a single molecule takes ≈ 100 steps of 8 nm without letting go — the single-molecule assays of § 54.10 have watched the walk, step by step, at one ATP per step. The traffic is enormous (a millimetre of axon is a million-step commute) and its failure is fatal: the neurodegenerations of motor-protein genes are the railways' accidents — and the long axon, with its lifelong dependence on the commute, is the cell's most vulnerable commuter line.

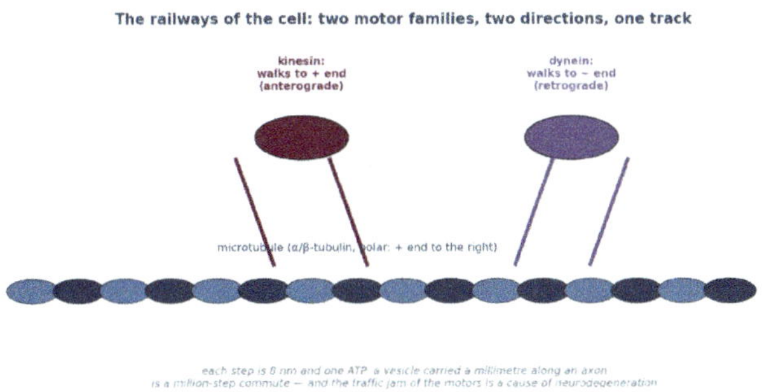

Figure 14.6 The cell's railways: kinesin (oxblood) walking to the microtubule's plus end, dynein (plum) to the minus — two motor families, two directions, one track, and one ATP per 8 nm step.

14.6 Molecular motion and ATP hydrolysis

The three engines — myosin, kinesin, dynein — are one mechanism in three liveries, and the mechanism deserves its section, for it is the answer to the question the whole treatise has been approaching: how does chemistry, which is scalar, produce motion, which is directed? The answer has two parts. Rectification: the motor is bound to a polar track (actin and tubulin are polar — the two ends differ), and the motor's own conformational cycle is polarised by the chemistry — the ATP-driven states bind and release the track at different angles, so the cycle is a ratchet that walks one way: the polarity of the track plus the polarity of the cycle equals direction. Coupling: the free energy of hydrolysis is spent not in a single burst but in the conformational changes of the cycle — the stroke, the release, the re-cock — each step a few kJ mol^{-1} downhill, the sum ≈ −50 kJ mol^{-1} per ATP (§ 7.8), so the motor is a machine that pays its bill in instalments, and the instalments are the mechanism. The efficiency is ≈ 50 % (half the ATP's free energy appears as work — the rest is heat, and the reader's sweat is its receipt), and the generality is the point: every molecular motor in this book — the ATP synthase of Chapter 28, the helicases of Chapter 44, the ribosome of Chapter 47 — is the same ratchet, reading polarity into motion by the same two rules.

14.7 Antibody structure

From the engines to the chemists of the blood: the antibodies, the body's own reagents, whose architecture (Figure 14.4) is the treatise's structural logic at its most elegant. The antibody is a Y of twelve immunoglobulin domains (four heavy-chain, four light-chain, repeated): the two Fab arms, whose tips carry the variable domains and the antigen-binding sites; the hinge, with its disulphides and its flexibility; and the Fc stem, the constant part that the effector machinery reads — complement, the Fc receptors of the phagocytes, the placental transporter — so that the molecule is recognition at one end and action at the other, joined by a flexible waist. The classes (IgG, IgM, IgA, IgE, IgD) are the same arms on different stems, each stem reading a different effector: IgM the pentamer of § 8.12's avidity (five Y's in a ring, the first responder), IgA the dimer of the secretions, IgE the allergy antibody of the mast cell — one fold, five offices, and the arms' specificity hired out to whichever job the stem describes.

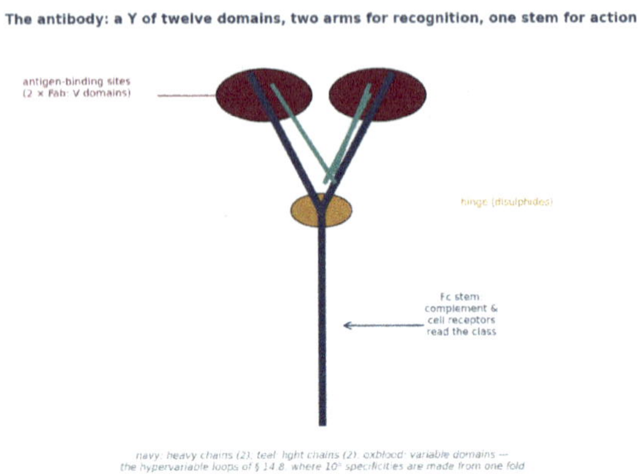

Figure 14.4 *The antibody: a Y of twelve immunoglobulin domains — heavy chains (navy), light chains (teal), the antigen-binding variable domains at the tips of the Fab arms (oxblood), the disulphide hinge (amber), and the Fc stem that the effector machinery reads. One fold: two arms for recognition, one stem for action.*

14.8 Antigen recognition

The arms' specificity is the immune system's central miracle, and its arithmetic is the treatise's last word on recognition. The binding site is built of six hypervariable loops (three from each variable domain — the complementarity-determining regions, CDRs), and the loops' sequences are the only variable part of the molecule. The repertoire is made by three multiplications: the germline segments (V, D, J) recombine somatically (≈ 50 V × 23 D × 6 J for the heavy chain, similar for the light — the genetic lottery of § 45.10); the junctions add random nucleotides; and the pairs of chains combine — the product being ≈ 10^8 distinct sites before any antigen appears, a standing library from which the antigen merely selects. Selection then improves the book: the activated B cell hypermutates its CDRs and selects the tighter binders (Figure 14.5) — affinity maturation, evolution in a lymph node, weeks to what evolution takes millennia for — and the result is the mature antibody of Table 8.1, K_d 10^{-8}–10^{-10} M, a reagent of research-grade specificity made by the body on demand. The monoclonal antibody of Milstein and Köhler (Plate 14.1) is the library's single volume, cloned — one B cell's product, immortalised by fusion to a myeloma — and the reader will note that the most successful drugs of the present age (the -mabs of oncology and autoimmunity) are, every one of them, this chapter's arms, hired.

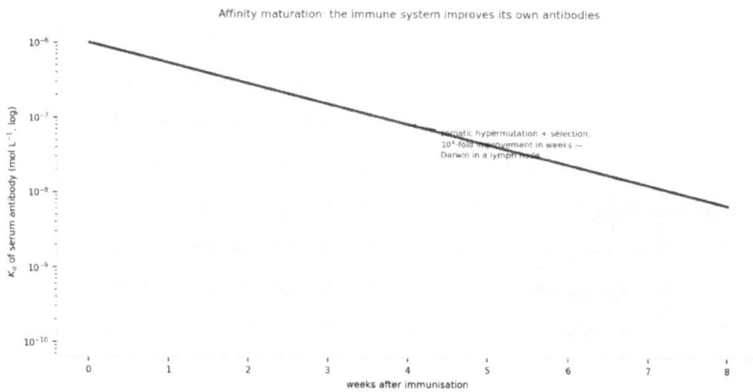

Figure 14.5 Affinity maturation: the serum antibody's dissociation constant against weeks after immunisation (logarithmic ordinate) — somatic hypermutation and selection improving the site by four orders of magnitude: Darwin's process, run in a lymph node, on the immune system's own proteins.

14.9 Complement proteins

The antibody's first employee is a cascade: the complement system, some thirty plasma proteins that tag, hole and summon — the blood's own artillery. Its logic is the zymogen cascade of § 18.7 (each protein activating the next by proteolysis, an amplifier of enzymes), and its three outputs are the three jobs of innate defence: opsonisation (C3b coats the microbe, and the phagocyte's receptors read the coat — the tag); inflammation (C3a, C5a — the anaphylatoxins — summon the phagocytes and open the vessels); and the membrane attack complex (C5b–C9, a pore punched in the microbe's membrane — the artillery's shell). The cascade's control is the point the reader should carry: complement runs continuously at a low idle (C3 ticks over by spontaneous hydrolysis), so that the system is always armed — and the price of the readiness is the risk of friendly fire, which the regulators (factor H, the membrane-bound protectors of the body's own cells) pay to prevent. The complementopathies are the regulators' failures: the atypical haemolytic-uraemic syndrome of factor H, in which the artillery shells the patient's own kidney — the immune system's precision being, like every other property of this book's proteins, a margin, not a guarantee.

14.10 Protein assemblies as molecular machines

The chapter closes with its thesis, which is Part III's crown: the proteins of this chapter are not substances but machines, and the machine is the natural form of the protein at work. The rope (collagen, keratin) is a machine for bearing load; the spring (elastin, titin) for storing it; the engine (myosin, kinesin, dynein) for converting chemistry to motion; the reagent (antibody, complement) for recognising and destroying — and each machine is built of the treatise's parts: the architecture of Chapter 10, the landscape of Chapter 11, the motion of Chapter 12, the binding of Chapter 8, the chemistry of Chapter 2. The reader who has followed the treatise this far now possesses the complete account of what a protein is: a chain of twenty letters, folded by the solvent, breathing at fifteen decades of motion, assembled into machines that bear, spring, walk, and defend. Part IV asks the next question, the one the whole of this book has been approaching: what happens when the machine is given a reaction to run — when the protein becomes an enzyme, and the architecture begins to catalyse. The machinery of Part IV is the same machinery, set to chemistry; and the chemistry, at last, has its turn.

Plate 14.1 *César Milstein (1927–2002), who with Georges Köhler made the monoclonal antibody — the immune system's reagent, cloned — and thereby founded both the assay industry and the most successful class of drugs of the age. Portrait from the Nobel Foundation archive, nobelprize.org (Nobel Prize in Physiology or Medicine, 1984), reproduced from the Foundation's publicly distributed image.*

HISTORICAL NOTE — THE SLIDING FILAMENTS, 1954

That muscle contracts by the sliding of two filaments, neither of which shortens, was proposed simultaneously and independently in 1954 by two pairs: Hugh Huxley and Jean Hanson (in Cambridge, on the phase-contrast appearance of the sarcomere) and Andrew Huxley and Rolf Niedergerke (on the mechanics of single fibres) — the "two Huxleys, unrelated, one result" that the field still cites with pleasure. The proposal contradicted the reigning model (that the filaments themselves contracted, like springs), and its proof required the decade of electron microscopy and X-ray crystallography that followed — the thick and thin filaments seen, the cross-bridges photographed, the cycle of Figure 14.3 established by Lymn and Taylor in 1971. The episode is the hermeneutic circle of Figure 1.5 in its purest form: structure proposed from mechanics, mechanism proved from structure, and the two Huxleys' guess — two filaments, sliding — the foundation on which the whole of molecular motility now stands.

Chapter 14 — Conceptual synthesis

The working proteins have worked. Collagen is the rope — Gly-X-Y, hydroxyproline, the triple helix, and scurvy as the hydroxylase's silence (§ 14.1); elastin the rubber — the desmosine-linked entropic spring, made once and never re-made (§ 14.2); keratin the epidermal rope — the coiled-coil and the disulphides of the permanent wave (§ 14.3); actin and myosin the engine — the sliding filaments and the cross-bridge cycle, chemistry rectified into motion (§§ 14.4–14.6); kinesin and dynein the railways (§ 14.5); the antibodies the body's chemists — twelve domains, six loops, 10^8 specificities selected and matured (§§ 14.7–14.8); complement the artillery — always armed, always guarded (§ 14.9); and the whole is the machine kingdom of Part III, complete (§ 14.10). Direction checked against Figure ii: Part III — proteins, structure, dynamics, function — is closed. Part IV *now begins: the enzymes, and the catalysis that makes the machine a chemist. Chapter 15 opens it: the nature of enzymatic catalysis.*

WORKED EXAMPLE 14.1 — THE COLLAGEN DEFECT'S ARITHMETIC

In osteogenesis imperfecta type I, one allele of collagen I is silent: the fibrils contain half the normal chains. If the fibre's strength scales with the cross-section of intact cable, the tendon's strength falls by ≈ 50 %; in type II (a glycine substitution), the misspelt chains are incorporated and poison the cable — three chains of which one is defective: the fraction of intact triple helices is $(1/2)^3 = 1/8$ if the alleles are equal, a dominant disease from a single misspelt letter: the arithmetic of the triple helix is the arithmetic of dominance, and the brittle-bone child is the exponent 3 made flesh.

WORKED EXAMPLE 14.2 — THE HEART'S ATP BUDGET

The heart beats ≈ 10^5 times per day, each beat cycling ≈ 2×10^{10} myosin heads per gram of heart, ≈ 5 nm of stroke at ≈ 5 pN: work per beat per gram ≈ $10^{10} \times 5 \times 10^{-9} \times 5 \times 10^{-12} = 2.5 \times 10^{-10}$ J... — the honest ledger is the body's own: the heart's ≈ 300 g works at ≈ 10 W, burns ≈ 20 % of the resting ATP turnover (Figure 4.7's 43 kg ≈ 6 kg of ATP per day for the heart alone), and the cardiomyocyte is accordingly the most mitochondria-rich cell in the body — the engine's fuel line being, as Chapter 28 will show, the heart's own mitochondria, row on row.

WORKED EXAMPLE 14.3 — THE ANTIBODY LIBRARY'S ARITHMETIC

Heavy chain: ≈ 51 V × 27 D × 6 J ≈ 8 300; light: ≈ 40 V × 5 J = 200; junctional diversity multiplies by ≈ 10^3; chain pairing by the product: 8 300 × 200 × 10^3 ≈ 1.7×10^9 — before somatic hypermutation, the library is ≈ 10^8–10^9 volumes, of which the circulating B cells (≈ 10^{11}, one book per cell) hold a fair copy. The antigen selects; the library answers in days. The arithmetic is the immune system's answer to the question of how a genome of 20 000 genes prepares for enemies it has never met: it does not predict them — it out-numbers them.

Exercises (with full resolutions)

EXERCISE 14.1

A tendon of cross-section 80 mm² bears 4 000 N. Estimate the stress, and compare with the tendon's ultimate strength (≈ 100 MPa).

Resolution. *Stress = 4 000/80 × 10^{-6} = 50 × 10^6 Pa = 50 MPa — half the ultimate strength: the tendon runs at a safety factor of ≈ 2, matching the engineering rule for biological materials (which must never fail at the loads of a lifetime). The jump of the athlete approaches the margin; the rupture of the tendon is the margin crossed — collagen's steel, priced in pascals.*

EXERCISE 14.2

Estimate the number of myosin heads engaged in a single sarcomere, and the force of one sarcomere at 5 pN per head with one-third of heads engaged.

Resolution. *A sarcomere has ≈ 300 thick filaments × ≈ 300 heads ≈ 10^5 heads (half pointing each way, so ≈ 5×10^4 per direction); one-third engaged: ≈ 1.7×10^4 heads × 5 pN ≈ 8×10^{-8} N per sarcomere — and the muscle fibre's* *thousands of sarcomeres in parallel sum to the Newtons of the lift: the engine's force is the arithmetic of § 14.4, counted.*

Exercise 14.3

Kinesin takes 8 nm steps at ≈ 100 s^{-1} under no load, and the axon is 1 m long. How long is the commute, and how many ATPs does one vesicle's delivery cost?

Resolution. *Steps: 10^9 nm/8 nm = 1.25 × 10^8 steps; at 100 s^{-1}, ≈ 1.25 × 10^6 s ≈ 14 days of walking (in practice the vesicle is handed between motors and the effective journey is days to weeks — matching the observed slow axonal transport). ATPs: one per step ≈ 10^8 ATPs ≈ 10^8 × 50 kJ mol^{-1}/N_A ≈ 8 × 10^{-15} J per vesicle — and the motor neuron's lifelong dependence on the commute is § 14.5's vulnerability, priced.*

Exercise 14.4 (interpretative)

"Antibodies violate the central dogma: their specificity is not encoded in the germline." Adjudicate.

Resolution (indicative). *The objection is correct in fact and wrong in inference. The germline encodes the segments, the recombination machinery, and the rules of hypermutation — the library and the library's rules are inherited; the particular volume selected is somatic. The dogma (sequence → sequence) is untouched: every antibody's sequence is encoded in the cell that makes it, and the cell's genome is a rearranged, hypermutated descendant of the germline. The immune system's trick is not to evade genetics but to gamble with it — somatic recombination as a lottery whose tickets are genes — and the adjudication is Chapter 1's: the information is physical, inherited, and selected; only the selection is faster than the inheritance.*

Exercise 14.5

Why are desmosine cross-links found in elastin and nowhere else, and what does their uniqueness tell you about evolution's invention of rubber?

Resolution. *The desmosine ring condenses four lysines into one — a link that joins four chains at once, which elastin's three-dimensional network requires and collagen's one-dimensional cable does not (collagen's lysyl links join two, and two suffice a rope). The uniqueness marks a single invention: the desmosine synthase appeared once, in the vertebrate lineage, and every artery that recoils is its descendant — evolution's rubber being a one-time patent, and the stiffening of the ageing artery the patent's slow expiry.*

Research-level propositions

Proposition 14.α. The cross-bridge cycle's power stroke has never been photographed in mid-stroke. Using the cryo-EM of § 54.6, design the trapping experiment (nucleotide analogues, flash-freezing) that would capture the stroke's intermediate, and estimate the angular resolution required to settle the "lever-arm versus converter" dispute. Proposition 14.β. Elastin is not re-made in adult life; the artery's recoil declines on the elastin of youth. Assemble the evidence for elastin turnover (or its absence) in the adult aorta, and compute the expected contribution of recoil loss to the age-dependence of systolic pressure in the Framingham data. Proposition 14.γ. The CDR repertoire has been deep-sequenced;

the "effective" repertoire is smaller than the combinatorial estimate (Worked Example 14.3). Using the single-cell repertoire literature, estimate the entropy of the observed repertoire in bits, and determine whether the reduction is set by the recombination machinery, by selection, or by the deletion of self-reactive volumes.

References (Chapter 14)

Alberts, B., Heald, R., Johnson, A., Morgan, D., Raff, M., Roberts, K., & Walter, P. (2022). Molecular biology of the cell (7th ed.). W. W. Norton & Company.

Huxley, H., & Hanson, J. (1954). Changes in the cross-striations of muscle during contraction and stretch and their structural interpretation. Nature, 173(4412), 973–976.

Köhler, G., & Milstein, C. (1975). Continuous cultures of fused cells secreting antibody of predefined specificity. Nature, 256(5517), 495–497.

Lymn, R. W., & Taylor, E. W. (1971). Mechanism of adenosine triphosphate hydrolysis by actomyosin. Biochemistry, 10(25), 4617–4624.

Vale, R. D. (2003). The molecular motor toolbox for intracellular transport. Cell, 112(4), 467–480.

PART IV

Enzymes and Biochemical Catalysis

Chapter 15 — The Nature of Enzymatic Catalysis · Chapter 16 — Enzyme Kinetics · Chapter 17 — Enzyme Inhibition · Chapter 18 — Regulation of Enzyme Activity · Chapter 19 — Coenzymes, Metals and Representative Mechanisms

CHAPTER 15

The Nature of Enzymatic Catalysis

CONTENTS OF THIS CHAPTER

In which the machine is given a reaction to run: the catalyst's true office — the lowering of the pass, never the moving of the valley; Pauling's principle of transition-state complementarity; the six instruments of the catalytic toolbox and what each can pay; specificity as a free-energy discrimination, squared by editing; and the ceiling of the whole art: enzymes so fast that the substrate's arrival, not the chemistry, is the limit.

15.1 Catalysts and activation energy

The enzyme's office is stated in one sentence, and the sentence has two halves, of which the second is always forgotten. A catalyst accelerates a reaction without being consumed — that is the first half, the one everyone knows. The second half is the thermodynamic fence of Chapter 7, and it is absolute: the catalyst does not change the equilibrium. The reaction A → B has its $\Delta G°$ and its K, fixed by the endpoints (Figure 15.1); the enzyme changes only the road between them, and a reaction that is thermodynamically forbidden stays forbidden in the enzyme's hands — the enzyme makes the permitted fast, never the forbidden possible. The mechanism of the acceleration is the lowering of the activation free energy, $\Delta G‡$ — the pass over the energy landscape that the reactants must climb — and the arithmetic is the Boltzmann factor of Exercise 1.3 in its most consequential application: the rate is proportional to $e^{-\Delta G‡/RT}$, so that lowering the pass by 10 kJ mol^{-1} accelerates the reaction ≈ 60-fold at 310 K, by 20 kJ mol^{-1}, ≈ 3 000-fold, and by the 40–80 kJ mol^{-1} of a great enzyme, 10^7–10^{14}-fold. The enzyme's entire art, in all the strategies of this chapter, is the art of lowering that pass; and everything else — the binding, the positioning, the acid–base shuttles, the metals — is machinery for the lowering.

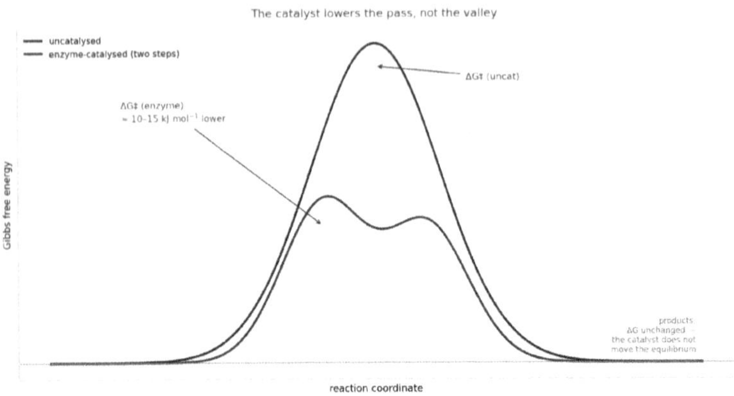

Figure 15.1 *The free-energy profile of a reaction, uncatalysed (oxblood) and enzyme-catalysed (navy, two steps through an intermediate): the pass ΔG‡ is lowered, the valleys — substrates, products, and the equilibrium between them — are unchanged. The catalyst's whole office is the pass; the valley is thermodynamics', and it does not move.*

15.2 Transition-state stabilisation

How is the pass lowered? The deepest answer is Pauling's, of 1946 (Pauling, 1946): the enzyme is complementary in structure to the transition state of the reaction — not to the substrate, not to the product, but to the fleeting geometry at the top of the pass (Figure 15.2). The complementarity pays for the pass in the currency of Chapter 8: the enzyme binds the transition state with all the instruments of recognition — hydrogen bonds, salt bridges, dispersion, desolvation — and the binding free energy of that embrace is precisely what lowers ΔG‡. The logic has a celebrated inversion, and the reader should feel its force: an enzyme that bound its substrate perfectly would be a trap — the substrate, held fast at the bottom of a deep well, would climb the pass more slowly, not faster (the pass is unchanged, the valley is deeper). The enzyme must bind the substrate weakly and the transition state tightly — the embrace must be directed at the destination, and the substrate's uphill journey is then the substrate being pulled by the slope of the embrace itself. The experimental warrant is the transition-state analogue of § 17.7: stable molecules shaped like the transition state bind the enzyme a million times tighter than the substrate — the enzyme's complementarity, measured in the dissociation constants of Chapter 8, is exactly as Pauling predicted, and the analogues are, by the same token, the most potent inhibitors ever designed.

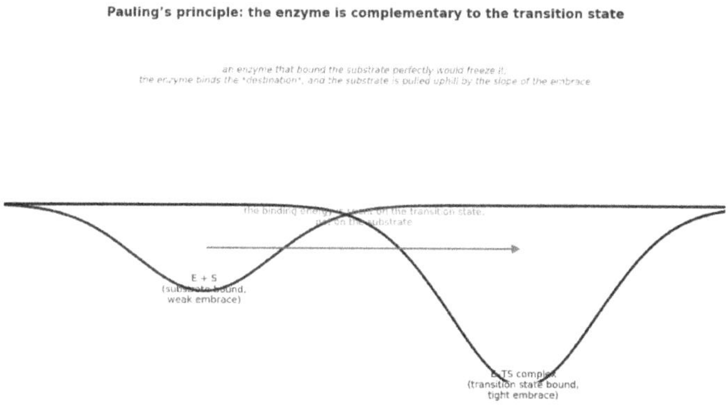

Figure 15.2 Pauling's principle made visible: the enzyme's embrace is weak for the substrate (navy well) and tight for the transition state (oxblood well), and the difference between the two embraces is the lowering of the pass. The binding energy is spent on the destination; the substrate is pulled uphill by the slope.

15.3 Binding energy

The binding energy of the enzyme–substrate complex — the same $\Delta G°_{bind}$ of Chapter 8 — is the enzyme's working capital, and the doctrine of this section is that it is spent in three directions at once. First, as § 15.2 has it, on the transition state: the embrace of the destination is the pass's lowering. Second, on specificity (§ 15.10): the embrace of the right substrate and not the wrong one is the discrimination, and its price is the difference of the two embraces. Third — and this is the subtlest expenditure — on the substrate's own contortion: the enzyme uses part of the binding energy to pay for the substrate's deformation toward the transition state (the strain of § 15.9), so that the bound substrate is already part-way up the pass when it is bound. The three expenditures share one ledger, and the ledger explains a long-standing puzzle of the field: measured binding energies of good substrates are often weaker than those of poor ones — the good substrate's embrace has been spent on contortion and transition-state stabilisation, while the poor one's is banked uselessly. Tight binding is not the enzyme's aim; productive binding is — and the difference, once more, is the difference between a warehouse and a machine.

15.4 Proximity and orientation

The simplest instrument of the toolbox is the one everyone understands and no one can quite price: bring the reactants together, in the right orientation, and hold them. In free solution a bimolecular reaction pays the full entropy tax of § 8.4 — the two molecules must find each other and lose their translational and rotational freedom; in the active site, both partners are already there, already oriented, already paid for — the reaction becomes, in effect, unimolecular, and the gain is the "effective molarity" of the bound state, estimated at 10^2–10^5 M for favourable geometries: proximity's contribution, ≈ 10^3–10^5-fold (Figure 15.3). The orientation clause is the stronger half of the instrument: a reaction whose partners must meet at one precise angle (the S_N2 backside attack, the proton transfer of § 6.1) is accelerated out of all proportion to the mere crowding, because the enzyme holds the angle as well as the address. The reader should note that proximity is the one *instrument every enzyme uses — it is the definition of the active site — and the one that, alone, cannot make a great enzyme: the great ones add the instruments of the next four sections, and the multiplication of the instruments is the arithmetic of Figure 15.3's legend.*

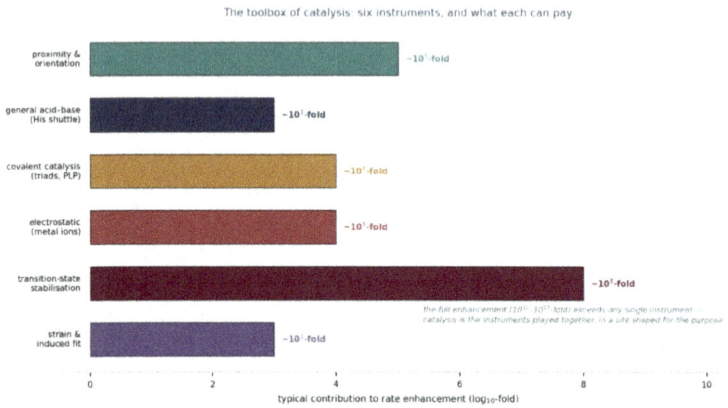

Figure 15.3 *The toolbox of catalysis: the typical contribution of each instrument to the rate enhancement (logarithmic abscissa). Transition-state stabilisation (oxblood) is the deepest purse; the full enhancement of a great enzyme exceeds any single instrument — catalysis is the toolbox played together.*

15.5 Acid–base catalysis

The proton is the commonest reagent of biochemistry (Chapter 6's whole doctrine), and the enzyme's way with it is the general acid–base shuttle: a side chain that gives a proton at one step of the mechanism and takes it back at another — histidine's imidazole ($pK_a \approx 6$–7, Figure 9.3's crossroads) being the invariable recruit, because it alone can play both parts at the cell's own pH. The instrument's contribution is modest per shuttle ($\approx 10^3$-fold, Figure 15.3) but ubiquitous: nearly every enzyme in this book runs at least one proton through a histidine, and the catalytic triad of § 15.4's figure is the shuttle's most famous address. The mechanism's signature, for the kineticist of Chapter 16, is the pH-dependence of the rate: the enzyme works only when the shuttle is in the right protonation state, so the rate profile is a bell curve whose two limbs are the two pK_a values of the working groups — the enzyme's acid–base machinery, readable in its kinetics, and the reason the pH-stat and the buffer of Chapter 6 are the enzymologist's daily instruments.

15.6 Covalent catalysis

The second instrument is more intimate: the enzyme forms a covalent bond with the substrate, making a transient enzyme-bound intermediate that splits the reaction into two easier steps (the two-humped profile of Figure 15.1, navy). The catalytic triad of the serine proteases (Figure 15.4) is the type specimen: the serine's hydroxyl, converted by the His–Asp relay into a nucleophile, attacks the peptide's carbonyl; the peptide breaks, the amine leaves, and the enzyme is left acylated — a covalent acyl-enzyme, which water then attacks (the same relay, backwards) to release the acid and restore the serine. The instrument's contribution is $\approx 10^4$-fold, and its mark is the covalent intermediate itself, *catchable by the kineticist (the "burst" kinetics of § 16.9, where the first turnover releases one product before the enzyme is freed). The covalent strategies of Part IV are a small and distinguished company — the serine and cysteine proteases, the PLP enzymes of § 19.6, the thiamine enzymes of § 19.5 — and their common logic is § 15.3's ledger again: the enzyme spends a covalent bond (paid back at the end) to buy two low passes in place of one high one.*

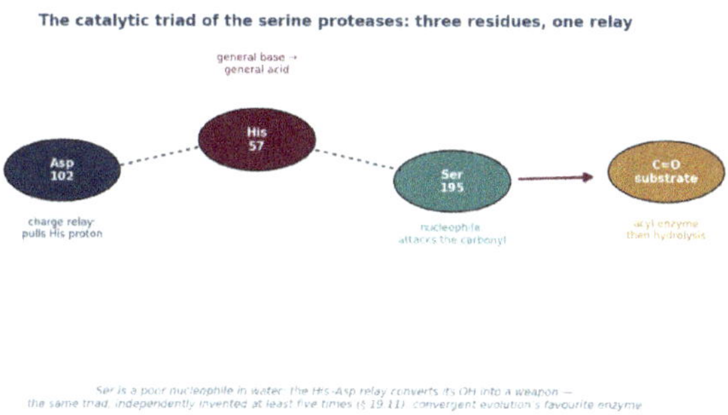

Figure 15.4 *The catalytic triad of the serine proteases: Asp102 (navy) pulls the histidine's proton, His57 (oxblood) shuttles between general base and general acid, and Ser195 (teal) attacks the substrate's carbonyl (amber) — the charge relay that converts a modest alcohol into a nucleophile. Independently invented at least five times: convergent evolution's favourite enzyme.*

15.7 Electrostatic catalysis

The third instrument is the field: the active site is not water, and its electrostatics are not the solvent's. The enzyme's interior, with its low dielectric (§ 2.4's screened world inverted), its positioned charges and dipoles (the helix macro-dipoles of § 10.3, the buried ions of § 9.3), is a pre-organised electrostatic environment — and a transition state that is charged, or more charged than the substrate, is stabilised by the field exactly as a salt bridge stabilises a buried ion (Worked Example 2.2's arithmetic, now paying the pass). Warshel's computations (for which the Nobel committee of 2013 found a prize (Warshel, 2014)) have made the instrument quantitative: the electrostatic stabilisation of the transition state by the pre-organised field is, in the systems computed, the largest single contribution to the pass's lowering — larger than proximity, larger than strain. The reader should note the corollary that follows: the enzyme's electrostatics are pre-organised — the field is there before the substrate arrives, costing no reorganisation energy — while water's field must reorganise around every charge, paying the solvent's tax. The enzyme wins, in Warshel's phrase, because it is "the solvent that has already moved": the fold's electrostatics are the answer prepared in advance, and the active site is the fold's way of being water, only better.

15.8 Metal-ion catalysis

The fourth instrument is the metal, and its offices are three. The metal is a charge shield and Lewis acid: Zn^{2+} in carbonic anhydrase and the metalloproteases polarises the bound water, lowering its pK_a from 15.7 to ≈ 7, so that the zinc-bound hydroxide is a nucleophile at the cell's pH — the metal converting water *into a weapon, as the triad converts serine (§ 19.12's mechanism). The metal is a template: Mg^{2+} in every kinase holds the ATP's phosphates in the geometry of transfer (and shields their charges from the attacking nucleophile) — the metal organising the reactants, proximity by another name. And the metal is a redox centre: iron, copper, manganese, in the electron-transfer enzymes of Part VI, accepting and donating electrons one at a time — the metal as the cell's electrician. The metal-dependent enzymes are a third of all enzymes, the trace elements of Figure 2.8 are their payroll, and the reader should now see the periodic table of Chapter 2 as a catalogue of catalytic instruments — each metal hired for a chemistry the twenty letters cannot do alone.*

15.9 Strain and induced fit

The fifth instrument is the oldest hypothesis and the most qualified: the enzyme distorts the substrate toward the transition state — the strain of the bound substrate, paid for by the binding energy of § 15.3's third expenditure. The evidence has refined the instrument's scope: true geometric strain (bond angles forced from their rest) is rarer and weaker than the early texts supposed, but its modern form — the induced fit of § 12.3, in which the enzyme's own closing around the substrate builds the catalytic geometry (the loop of triosephosphate isomerase, the domain closing of the kinases) — is universal: the enzyme and the substrate contort together, and the strain is distributed over both. The instrument's contribution is ≈ 10^3-fold, and its doctrinal value is the correction of a static error: the active site is not a jig into which the substrate drops, but a machine that closes around it — the induced fit is the recognition of Chapter 12 turned to chemistry, and the strain is the machine's way of paying for the pass with the embrace.

15.10 Enzyme specificity

The enzyme's second office, after speed, is choice — and specificity, like speed, is priced in free energy. The discrimination between two substrates is set by the difference of their activation free energies (Figure 15.5): $\Delta\Delta G\ddagger = 10\ kJ\ mol^{-1}$ gives ≈ 50-fold preference, 20 $kJ\ mol^{-1}$ gives 2 500-fold — and a single binding event can rarely pay more than ≈ 10–20 $kJ\ mol^{-1}$, which is why the great specificities of information transfer are not achieved in one step. The amplification is the editing step: the tRNA synthetases of § 47.3, which must distinguish isoleucine from valine (a difference of one methylene, worth ≈ 10 $kJ\ mol^{-1}$ at best), run a second sieve — a proofreading site that hydrolyses the wrong product — and the two sieves multiply: 50 × 50 ≈ 2 500-fold, the arithmetic of Figure 15.5. The DNA polymerases add a third sieve (the 3′→5′ exonuclease) and reach 10^6-fold; the ribosome's kinetic proofreading (Hopfield's mechanism, the energy-driven delay that lets the wrong partner dissociate twice) reaches 10^4-fold at the price of GTP — specificity bought with energy, the subject's deepest lesson: discrimination beyond the binding energy's limit is possible, but it costs free energy per step, and the cell pays it exactly where the error would be fatal.

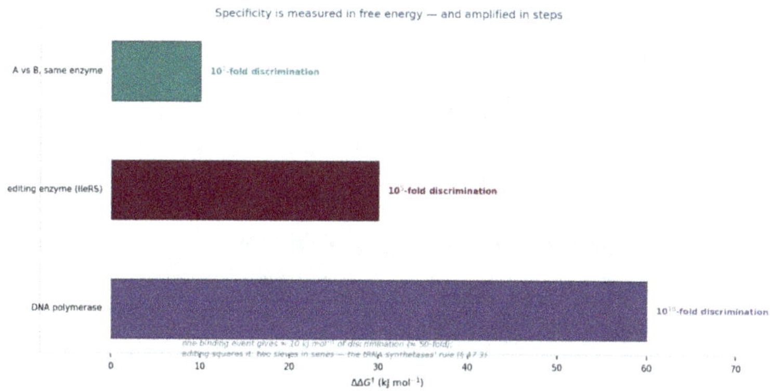

Figure 15.5 *Specificity as free-energy discrimination: one binding event's ΔΔG‡ (teal), the editing enzyme's two sieves (oxblood), the polymerase's three (plum) — the discrimination multiplies with each proofreading step, at the price of energy per sieve.*

15.11 Catalytic antibodies and ribozymes

The chapter closes with two existence proofs, each of which changed the subject's mind about what a catalyst could be. The catalytic antibodies — Lerner and Schultz's, of 1986 — are antibodies raised against transition-state analogues (§ 17.7's molecules): if Pauling's principle is true, an antibody shaped against the transition state should catalyse the reaction, and it does — the abzymes accelerate their target reactions 10^3–10^6-fold, proving that complementarity to the transition state is sufficient for catalysis, with no evolution required: the immune system's library of § 14.8, screened for the embrace, yields the enzyme. The ribozymes — Cech's self-splicing intron and Altman's RNase P, of the early 1980s, Nobel 1989 — are RNA molecules that catalyse: no protein at all, just the nucleic acid folding its own active site. The discovery demolished the last redoubt of protein exceptionalism — catalysis is a property of folded macromolecules, not of one alphabet — and founded the RNA-world hypothesis of § 56.2: if RNA can both carry information and catalyse, then the first living chemistry may have run on RNA alone, and the protein enzymes of this Part are the later, better tools of a world that began with ribozymes. The ribosome itself, as § 47.6 will show, is the living fossil: its peptide-bond formation is catalysed by RNA — the protein factory is a ribozyme, and the oldest enzyme in the cell is not a protein at all.

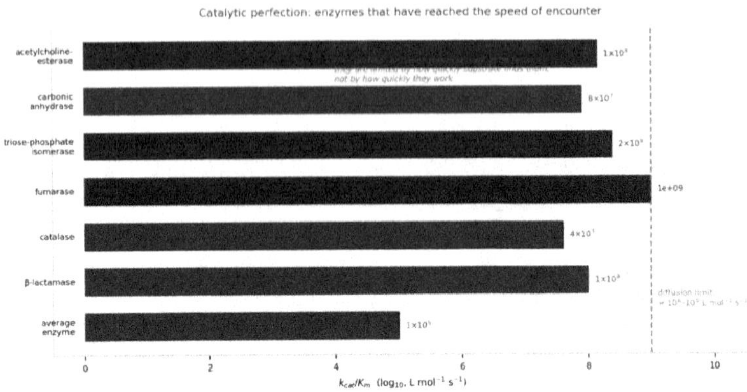

Figure 15.6 *Catalytic perfection: the k_{cat}/K_m of representative enzymes against the diffusion limit (dashed teal). The "perfect" enzymes (oxblood) sit at the limit — their chemistry is no longer the bottleneck; the substrate's arrival is. Values from the standard compilations (Fersht, 2017).*

MECHANISTIC CASE STUDY — *SUMNER'S CRYSTALS, 1926:* THE ENZYME IS A PROTEIN

The founding experiment of the whole Part is the one that settled what an enzyme is. Buchner's zymase (case study of Chapter 1) had shown that the chemistry of the cell survives its disassembly, but the nature of the agent — the "zymase" — was unknown: a colloid, a small molecule adsorbed on protein, a "ferment" of unknown composition, were all live candidates in 1920, and Willstätter, the reigning master of enzyme chemistry, held that the enzyme was not a protein at all (the protein being, in his famous phrase, the carrier, not the catalyst). James Sumner, at Cornell, took jack-bean meal, extracted urease, and in 1926 crystallised it — and the crystals were protein, and the crystals were the enzyme (Sumner, 1926). The experiment's logic is the hermeneutic circle of Figure 1.5 at its purest: purify the agent to a definite chemical object (a crystal — the most definite object chemistry knows), show that the object has the activity, and the agent's nature is settled. The reception was the usual one for a settling experiment: Willstätter's school disputed the purity for a decade (the crystals were "contaminated"), and Northrop's crystallisation of pepsin, trypsin and chymotrypsin at the Rockefeller — each a protein, each an enzyme — buried the dispute by accumulation. The Nobel of 1946 went to Sumner and Northrop jointly; and the reader should note what the experiment established and what it did not: enzymes are proteins (the door to Part III opened), but how the protein catalyses — the whole of this chapter — the crystals did not say. Purity is the beginning of mechanism, not the end: the crystal answers "what", and the chapter has been asking "how" ever since.

Plate 15.1 *James B. Sumner (1887–1955), whose jack-bean urease was the first enzyme crystallised and the first proven to be a protein. Portrait from the Nobel Foundation archive, nobelprize.org (Nobel Prize in Chemistry, 1946), reproduced from the Foundation's publicly distributed image.*

HISTORICAL NOTE — PAULING'S PRINCIPLE, AND THE LETTER THAT MADE A FIELD

The principle of § 15.2 — the enzyme complementary to the transition state — was stated by Pauling in 1946, in a short essay in Chemical & Engineering News titled "Molecular Architecture and Biological Reactions" (Pauling, 1946), and more fully in his 1948 lecture to the Royal Society, "Nature of Forces between Large Molecules of Biological Interest." The idea was not immediately testable — no enzyme structure existed — and it stood for two decades as a conjecture of unusual fertility: it predicted the transition-state analogue (the most potent class of inhibitors, § 17.7), it predicted the catalytic antibody (§ 15.11), and it organised the thinking of every mechanistic enzymologist from Jencks to Warshel. The structural evidence, when it came (phosphorylcholine antibodies, triosephosphate isomerase, the proteases), was Pauling's principle photographed: the active site, in every case, is shaped for the top of the pass. The principle is the treatise's example of the best kind of theory — one that makes its evidence inevitable, and then waits for it.

Chapter 15 — Conceptual synthesis

Catalysis has been priced. The catalyst lowers the pass and never moves the valley (§ 15.1); Pauling's principle spends the binding energy on the transition state (§§ 15.2–15.3); the toolbox is proximity, acid–base shuttles, covalent intermediates, electrostatic fields, metals, and strain — played together, never alone (§§ 15.4–15.9); specificity is a free-energy discrimination, squared by editing (§ 15.10); catalytic antibodies and ribozymes prove that the art is in the fold, not the alphabet (§ 15.11); and the ceiling is the diffusion limit, at which the perfect enzymes wait for their substrates to arrive (§ 15.6's figure). Direction checked against Figure ii: the nature of catalysis is established. Chapter 16 now measures it: enzyme kinetics — the rates, the constants, and the equations by which the machine's speed is read.

WORKED EXAMPLE 15.1 — THE PASS'S LOWERING, IN RATES

Carbonic anhydrase accelerates CO_2 hydration ≈ 10^6-fold. The lowering of the pass: $\Delta\Delta G‡ = RT \ln 10^6 = 2.577 \times 13.8 \approx 35.6$ kJ mol^{-1} — the price of about seven good hydrogen bonds, spent (§ 15.8's zinc and § 15.2's embrace) on the transition state. At the other extreme, the great orotidine decarboxylase accelerates ≈ 10^{17}-fold: $\Delta\Delta G‡ \approx 2.577 \times 39.2 \approx 101$ kJ mol^{-1} — the deepest purse in enzymology, and the reason the reaction's uncatalysed half-life (78 million years) is a figure the reader may quote with confidence at any dinner: the enzyme makes the geological instantaneous.

Worked example 15.2 — Proximity's effective molarity

An intramolecular reaction (the two partners tethered) runs 10^5-fold faster than its bimolecular twin. The effective molarity of the tethered partner is therefore 10^5 M — and the entropy tax avoided: $\Delta\Delta G‡ = RT \ln 10^5 = 2.577 \times 11.5 \approx 29.7$ kJ mol^{-1} ≈ the cratic entropy of § 8.4, collected at last. The enzyme's active site pays this tax in advance, at binding; the bound substrate reacts as if at 10^5 M — the proximity instrument of Figure 15.3, priced in the currency of Chapter 8.

Worked example 15.3 — The diffusion limit, derived

The diffusion-limited on-rate for a small molecule meeting a protein is set by the Smoluchowski rate: $k_{on} = 4\pi(D_A + D_B)R_{AB}N_A/1000$, with diffusion coefficients ≈ 10^{-9} m^2 s^{-1} (small molecule) and ≈ 10^{-10} m^2 s^{-1} (protein), and an encounter radius ≈ 1 nm: $k_{on} \approx 4\pi \times 1.1 \times 10^{-9} \times 10^{-9} \times 6 \times 10^{23} \approx 8 \times 10^9$ L mol^{-1} s^{-1} — the dashed line of Figure 15.6, derived rather than quoted, and the ceiling at which the perfect enzymes of § 15.6 wait.

Exercises (with full resolutions)

Exercise 15.1

An enzyme binds its substrate with $K_d = 10^{-6}$ M and a transition-state analogue with $K_d = 10^{-14}$ M. Estimate the rate acceleration, and state the assumption.

Resolution. The analogue's tighter binding measures the transition state's tighter embrace: $\Delta G° = RT \ln(10^{-14}/10^{-6}) = 2.577 \times (-18.4) \approx -47$ kJ mol^{-1}. If the analogue is a faithful mimic (the assumption — its geometry and charge match the transition state's), the pass is lowered by the same 47 kJ mol^{-1}, and the acceleration is e^{47/2.577} ≈ e^{18.2} ≈ 8×10^7-fold — Pauling's principle converted from a K_d ratio to a rate ratio, and the reason the analogue makers are the richest hunters in inhibitor design.

Exercise 15.2

Show that an enzyme cannot alter the equilibrium constant of its reaction, using the thermodynamic cycle of Chapter 7.

Resolution. Consider the cycle: S ⇌ P in free solution ($\Delta G° = -RT \ln K$) and the bound cycle E + S ⇌ E + P. The free energy of the full cycle (E + S → ES → EP → E + P) is the sum of the legs: binding of S, conversion on the enzyme, release of P — and the sum of a cycle is zero. Hence $K_{bound}/K_{free} = K_d(S)/K_d(P)$: the *enzyme accelerates forward and backward rates in exactly the ratio of its binding preferences, and the equilibrium is untouched — the valley does not move, and § 15.1's fence is a theorem, not a policy.*

EXERCISE 15.3

The orotidine decarboxylase reaction has an uncatalysed half-life of 7.8×10^7 years and a catalysed k_{cat} of 39 s^{-1}. Compute both rates and the enhancement.

Resolution. *Uncatalysed: $k = \ln 2/(7.8 \times 10^7 \times 3.15 \times 10^7\ s) \approx 2.8 \times 10^{-16}\ s^{-1}$. Enhancement: $39/2.8 \times 10^{-16} \approx 1.4 \times 10^{17}$-fold — the largest known, and the pass's lowering of Worked Example 15.1: $\Delta\Delta G\ddagger \approx 2.577 \times \ln(1.4 \times 10^{17}) = 2.577 \times 39.5 \approx 102$ kJ mol^{-1}. The enzyme reads 78 million years as 18 milliseconds — the most extreme demonstration that the pass, not the valley, is where time is spent.*

EXERCISE 15.4 (INTERPRETATIVE)

"Enzymes are not magical; they are just very good solvents." Assess the claim, with § 15.7's electrostatics in mind.

Resolution (indicative). *The claim is a half-truth of unusual value. True: Warshel's computations make the pre-organised electrostatic field the largest single instrument, and a solvent that has "already moved" is a fair description of the active site's dielectric. False as a whole: the solvent metaphor misses the organisation — the field is positioned (the helix dipoles, the buried charges, the desolvated pocket), and the positioning is the fold's, not the solvent's; water cannot be pre-organised, because it must always reorganise around the charge. The adjudication: the enzyme is not a better solvent; it is a solvent with an architecture — and the architecture is Chapter 10's, working as Chapter 15's catalyst.*

EXERCISE 15.5

A kinase uses Mg^{2+}-ATP as its true substrate; free ATP^{4-} is a poor substrate and free Mg^{2+} above the physiological range inhibits. Explain both observations with § 15.8's three offices.

Resolution. *The Mg^{2+} is the template and the shield: it holds the ATP's phosphates in the transfer geometry and screens their charges from the nucleophile — free ATP^{4-}, untemplated and unshielded, binds wrongly and attacks slowly (poor substrate). Excess free Mg^{2+}, beyond the ATP, occupies the enzyme's own anionic sites and competes for the nucleophile's partners (the shield, misplaced) — the inhibition. The metal's three offices are thus all visible in one enzyme's rate curve: the template when bound, the shield when balanced, the poison when in excess — and the physiological Mg^{2+} of § 4.7's discount is the buffer that keeps the balance.*

Research-level propositions

Proposition 15.α. The Warshel electrostatic account (§ 15.7) has been challenged by the "dynamic" school, which holds that protein motions contribute materially to the pass's lowering. Design the computation — QM/MM free-energy simulation with and without constrained protein motion (§ 55.9) — that would separate the two contributions for one enzyme, and estimate the signature each school predicts for the isotope-edited vibrational spectroscopy of § 54.5. Proposition 15.β. The *catalytic antibodies (§ 15.11) are slower than evolved enzymes by 10^3–10^6-fold; some argue the gap is proofreading's absence, others that it is the fold's. Using the abzyme crystal structures, compute the electrostatic pre-organisation of an abzyme active site against its enzyme counterpart, and assign the gap. Proposition 15.γ. The diffusion limit of Worked Example 15.3 assumes random encounter; the crowded cytoplasm (§ 5.11) and electrostatic steering can raise or lower it. Compute the steering enhancement for a charged substrate approaching a charged active site (the acetylcholinesterase funnel is canonical), and assess whether the "perfect" enzymes of Figure 15.6 are at the limit or beyond it.*

References (Chapter 15)

Fersht, A. (2017). Structure and mechanism in protein science: A guide to enzyme catalysis and protein folding. World Scientific.

Pauling, L. (1946). Molecular architecture and biological reactions. Chemical & Engineering News, 24(10), 1375–1377.

Sumner, J. B. (1926). The isolation and crystallization of the enzyme urease. Journal of Biological Chemistry, 69(2), 435–441.

Warshel, A. (2014). Computer simulations of enzyme catalysis: Methods, progress, and insights [Nobel lecture]. Angewandte Chemie International Edition, 53(38), 10020–10031.

Wolfenden, R. (2011). Benchmark reaction rates, the stability of biological molecules in water, and the evolution of catalytic power. Annual Review of Biochemistry, 80, 645–667.

CHAPTER 16

Enzyme Kinetics

CONTENTS OF THIS CHAPTER

In which the machine's speed is measured and its meaning extracted: the initial rate and why it alone is honest; the Michaelis–Menten equation derived, not quoted; the two constants and their three different meanings; catalytic efficiency as the efficiency ratio; the straight-line plots and their quiet treacheries; two substrates, fast pulses, and the isotope's question to the mechanism.

16.1 Initial-rate measurements

The measurement of an enzyme is, before it is anything else, an exercise in intellectual honesty about time. The naive experiment — mix enzyme and substrate, wait, measure the product — is the one that lies: the product accumulates, the reverse reaction wakes, the substrate depletes, the enzyme dies slowly in the pipette's heat, and the measured "rate" is a corpse's average of all these deaths. The honest experiment measures the initial rate v_0: the slope of product formation in the first moments (Figure 16.2's window), before any of the deaths have begun — substrate at its nominal concentration, product at zero, enzyme alive. The rule of the trade is the 10 % rule: measure within the first 10 % of substrate consumption, and the complications are still small; measure beyond it, and you are writing history, not kinetics. The instruments are the spectrophotometer of § 53.9 (the NADH-linked assays, whose absorbance follows the cofactor's redox state — the enzymologist's window into the invisible), the pH-stat, the electrode, the fluorescence plate reader; but the doctrine is independent of the instrument: kinetics is the art of reading slopes at the beginning, and the beginning is the only time the slope means what it says.

16.2 Derivation of the Michaelis–Menten equation

The central equation of enzyme kinetics is derived in eight lines, and the reader is owed every one. The mechanism is the minimal one,

$$E + S \underset{k_{-1}}{\overset{k_1}{\rightleftharpoons}} ES \xrightarrow{k_2} E + P \qquad (16.1)$$

— E the enzyme, S the substrate, ES the complex, k_2 the catalytic constant (the chemistry and the release, lumped). The rate is $v = k_2[ES]$. The problem is that [ES] is not measurable directly; the solution is the steady-state assumption of Briggs and Haldane (Figure 16.2): after the first milliseconds, the complex is formed as fast as it is consumed, so

$$\frac{d[\mathrm{ES}]}{dt} = k_1[\mathrm{E}][\mathrm{S}] - (k_{-1} + k_2)[\mathrm{ES}] \approx 0 \tag{16.2}$$

— whence [ES] = k_1[E][S]/(k_{-1} + k_2). Define the Michaelis constant,

$$K_m \equiv \frac{k_{-1} + k_2}{k_1} \tag{16.3}$$

— so [ES] = [E][S]/K_m. The enzyme is conserved: $[E]_T$ = [E] + [ES], so [E] = $[E]_T$ − [ES]; substitute and solve for [ES]:

$$[\mathrm{ES}] = \frac{[\mathrm{E}]_T\,[\mathrm{S}]}{K_m + [\mathrm{S}]} \tag{16.4}$$

— and the rate follows: v = k_2[ES] = $k_2[E]_T$[S]/(K_m + [S]). Naming the maximal rate V_{max} = $k_2[E]_T$ (the rate when every enzyme is occupied), we arrive at the equation of Michaelis and Menten (1913), in Briggs and Haldane's steady-state form (Michaelis & Menten, 1913):

$$v = \frac{V_{max}\,[\mathrm{S}]}{K_m + [\mathrm{S}]} \tag{16.5}$$

— the rectangular hyperbola of Figure 16.1, the most-used equation of Part IV, derived from one mechanism and one assumption. The reader should note what the derivation cost: the steady-state assumption (§ 16.5), the single substrate, the absence of product — and every later complication of the chapter is a relaxation of one of those costs.

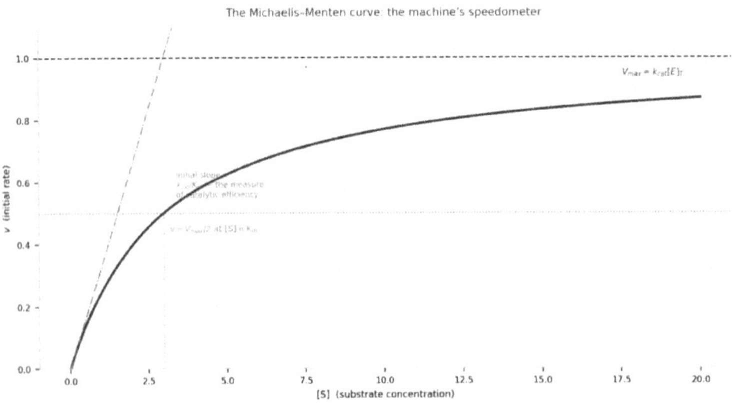

***Figure 16.1** The Michaelis–Menten curve (Eq. 16.5): the maximal rate V_{max} (oxblood, the saturated machine), the half-rate at [S] = K_m (teal), and the initial slope k_{cat}/K_m (amber) — the three readings the curve offers, each with its own meaning (§ 16.3).*

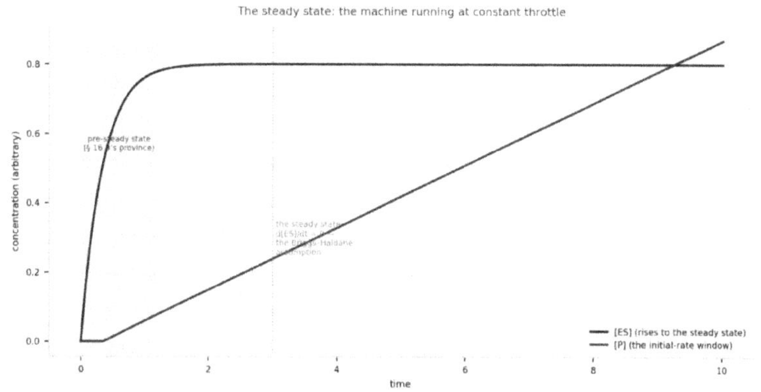

Figure 16.2 The steady state: the enzyme–substrate complex (oxblood) rises in milliseconds to its plateau and holds it while product (navy) accumulates linearly — the initial-rate window. The pre-steady state (shaded) is § 16.9's province; the plateau is Briggs–Haldane's assumption, Eq. (16.2), made visible.

16.3 Meaning of K_m, V_max and k_cat

The two constants of Eq. (16.5) have three meanings between them, and the reader who confuses them reads the literature wrong. V_{max} is the machine's top speed — the rate when every enzyme molecule is occupied — and its per-enzyme form is the catalytic constant $k_{cat} = V_{max}/[E]_T$ (the turnover number: how many substrates one enzyme converts per second at saturation; the champions of Figure 15.6 run at 10^4–10^6 s^{-1}, the median enzyme at ≈ 10 s^{-1}). K_m, by Eq. (16.3), is $(k_{-1} + k_2)/k_1$ — a kinetic composite, with units of concentration, whose operational reading is the substrate concentration at half speed (Figure 16.1, teal). The confusion to avoid is the identification of K_m with the dissociation constant K_d of the ES complex: the two coincide only when $k_2 \ll k_{-1}$ (the rapid-equilibrium limit of § 16.5 — chemistry slow, binding fast), and in general $K_m \geq K_d$: the Michaelis constant is a steady-state constant, and the substrate it reports is not necessarily the substrate the enzyme holds tightest. The two constants together answer the two questions one asks of any enzyme: how fast can it go (k_{cat}), and how well does it hear (K_m) — and their ratio is the next section's subject.

16.4 Catalytic efficiency

At low substrate (the physiological condition of most enzymes — $[S] \ll K_m$), Eq. (16.5) reduces to $v \approx (k_{cat}/K_m)[E]_T[S]$: the rate is first-order in each, and the coefficient k_{cat}/K_m — the specificity constant — is the true measure of the enzyme's quality, for it measures the rate at which the enzyme captures its substrate from the dilute world. Two properties make the ratio the field's favourite constant. It has a physical ceiling: the diffusion limit of Worked Example 15.3 (≈ 10^8–10^9 L mol^{-1} s^{-1}), at which the enzyme captures every substrate it meets — the "catalytically perfect" enzymes of Figure 15.6, whose chemistry is no longer the bottleneck. And it is the honest comparator of substrates: when an enzyme faces two substrates, it chooses between them by the ratio of their k_{cat}/K_m values (the rates at equal, low concentration go as the ratios), so the specificity constant is the *specificity of § 15.10 made kinetic — the discrimination of the machine, read off its speedometer.*

16.5 Steady-state and rapid-equilibrium assumptions

The derivation of § 16.2 has an older sibling, and the two derivations divide the kinetic territory between them. Michaelis and Menten's own assumption (1913) was the rapid-equilibrium one: binding is fast and reversible, chemistry slow, so that $E + S \rightleftharpoons ES$ *is always at equilibrium* ($K_m = K_d = k_{-1}/k_1$)*. Briggs and Haldane's (1925) is the steady state of Eq. (16.2): no equilibrium is assumed, only that [ES] is constant — the more general assumption, and the one that survives when chemistry is as fast as binding (the catalytically perfect enzymes, where* k_2 *is the largest rate in the mechanism). The reader should carry both, and the rule for choosing: rapid equilibrium when the binding is the interesting part (the specificity problems of § 15.10), steady state when the chemistry is — and the modern kineticist's habit, which is to write the full differential equations of § 16.7 and let the computer choose neither. The two assumptions are the chapter's Scylla and Charybdis, and the history of the subject is the slow discovery that one can sail around both.*

16.6 Linear transformations and their limitations

The hyperbola of Eq. (16.5) was, for sixty years, read through its straight-line disguises: Lineweaver–Burk (1/v against 1/[S], slope K_m/V_{max}*, intercept* $1/V_{max}$*), Eadie–Hofstee, Hanes–Woolf — each a linearisation, each with its devotees. The disguises have two fatal vices, and Figure 16.3a shows both. The transformation distorts the errors: a constant error in v becomes an error in 1/v that explodes as v falls — the low-substrate points, the least reliable, acquire the largest error bars and dominate the fitted line. And the distortion is asymmetric: the line is drawn substantially by the worst points, so the fitted* K_m *and* V_{max} *are biased in a direction no residual plot will forgive. The modern practice — nonlinear least-squares fitting of the raw data to Eq. (16.5) itself (Figure 16.3b), with the computer's patience and the error model honest — is statistically superior in every respect, and the treatise's rule is the one § 8.5 laid down for binding: the hyperbola is the truth, the straight lines were scaffolding, and the scaffolding should come down when the building is up. The straight-line plots retain one honourable use — the qualitative display of inhibition patterns (§ 16.4's figure and Chapter 17's whole diagnosis), where the eye, not the fit, is the instrument.*

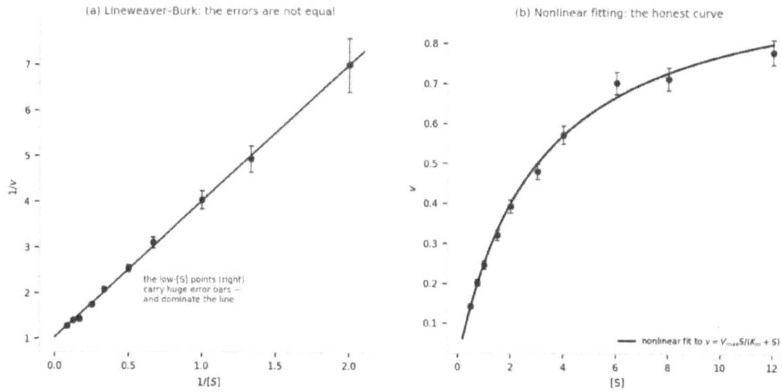

Figure 16.3 *Two ways of reading the same data. (a) Lineweaver–Burk: the low-substrate points (right) carry the largest error bars and dominate the line — the treachery of the transformation. (b) Nonlinear fitting of the raw hyperbola: the honest curve, with the errors where they were measured. The scaffolding of panel (a) is for the eye, never for the fit.*

16.7 Numerical and nonlinear fitting

The nonlinear fit has its own disciplines, and the reader who fits without them inherits the straight-line era's errors in a faster machine. The residuals — the differences between data and curve — must be inspected, not admired: a good fit scatters them randomly about zero; a systematic wave in the residuals is the model confessing it is wrong (a second substrate, a cooperative enzyme, a dead enzyme fraction — the curve's autobiography, readable by anyone who looks). The uncertainties of the fitted constants must come from the full covariance matrix (§ 4.11's propagation, Proposition 4.β's correction): K_m and V_{max} are strongly correlated, and the custom of quoting their marginal errors separately is the straight-line era's last vice. And the experimental design — the choice of substrate concentrations — matters more than the fitting: the points should span K_m (the curve's informative region), roughly half below and half above, with replicates where the errors live; a design that clusters its points at one end of the curve returns constants of spurious precision, and no algorithm can rescue a design that never asked the question.

16.8 Bisubstrate reactions

Most enzymes have two substrates (the dehydrogenases of Part VI: NAD^+ and the metabolite; the kinases: ATP and the acceptor), and their kinetics divide, by the order of binding and release, into the two great families of Figure 16.5. The sequential family forms a ternary complex: both substrates bind (in order or at random) before either product leaves — and the kinetic signature, in the diagnostic plots, is intersecting lines. The ping-pong family splits the reaction: the first substrate binds, the first product leaves, and the enzyme is left chemically changed ($E \rightarrow E'$) before the second substrate arrives — the aminotransferases of § 37.3, whose pyridoxal phosphate carries the amino group from the first substrate to the second, are the type specimen — and the signature is parallel lines, because the enzyme's two forms are measured in alternation. The diagnostic power is the section's point: the pattern of the lines, read by the kineticist before any structure exists, tells the mechanism — the ping-pong's parallel lines announce the covalent *intermediate that the structural biologist will later photograph, and the kinetics, as so often in this book, has the answer first.*

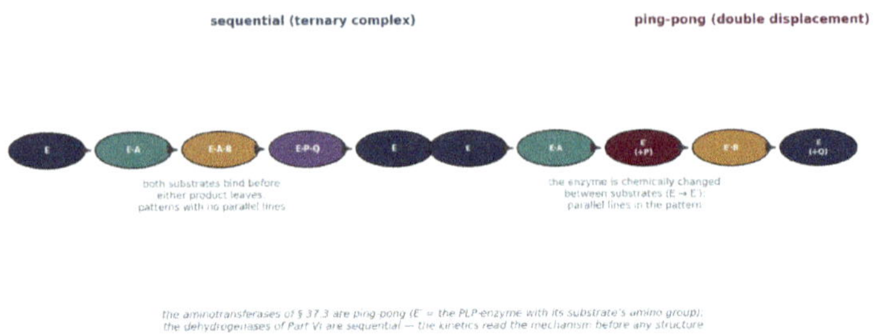

Figure 16.5 *The two families of bisubstrate mechanism. Left: sequential — the ternary complex, both substrates aboard before either product leaves. Right: ping-pong — the enzyme chemically changed (E', oxblood) between substrates. The kinetic patterns of the two families differ (intersecting versus parallel lines), and the pattern reads the mechanism before the structure is known.*

16.9 Pre-steady-state kinetics

The steady state of § 16.2 is a truce; the first milliseconds of the reaction, before the truce, are the mechanism's own autobiography — and reading them is the province of the fast methods. The stopped-flow apparatus mixes enzyme and substrate in milliseconds and watches the complex form (Figure 16.2's shaded window); the burst kinetics of a covalent enzyme (§ 15.6's serine protease) shows the first turnover's product released before the enzyme is freed — a burst, then the steady rate, and the burst's size counts the active sites. The temperature-jump and pressure-jump methods (Eigen's, Nobel 1967, with Porter — Plate 16.1) perturb an equilibrium and watch it relax: the relaxation time is the sum of the forward and backward rates, and the elementary steps of the binding — the k_{on} and k_{off} of Chapter 8's constants — are measured directly. The pre-steady state is where the mechanism's intermediates live (the ES complex, the covalent intermediate, the conformational change of § 12.3), and the steady state, for all its usefulness, is the mechanism seen through a veil; the fast methods lift the veil, and the price of the lifting is the stopped-flow's millisecond dead time — the frontier at which the chemical event itself begins to hide.

Plate 16.1 *George Porter (1920–2002), who with Norrish and Eigen made the fast reaction measurable — the flash photolysis and relaxation methods on which the pre-steady state of § 16.9 stands. Portrait from the Nobel Foundation archive, nobelprize.org (Nobel Prize in Chemistry, 1967), reproduced from the Foundation's publicly distributed image.*

16.10 Diffusion-limited reactions

The ceiling of Figure 15.6 has its kinetics, and the kinetics is the chapter's last general lesson. When k_{cat}/K_m approaches the diffusion limit (≈ 10^8–10^9 L mol^{-1} s^{-1}), the enzyme captures every substrate that diffuses to it — the reaction's rate is set by the encounter, not the chemistry, and the enzyme has, in the field's phrase, reached catalytic perfection. The signature is in the constants: the perfect enzymes have K_m values set not by binding but by the balance of diffusion and chemistry ($K_m \approx k_{cat}/k_{on}$ — Eq. 16.3 with k_{-1} negligible), so their K_m is a kinetic quantity entirely, § 16.3's confusion made doctrine. The evolutionary reading is the point the reader should carry: the perfect enzymes (triosephosphate isomerase, acetylcholinesterase, carbonic anhydrase, fumarase) are the ones whose substrates are precious or dangerous in dilution — the cell has spent its catalytic capital exactly where speed matters most — and the rest of the enzymatic kingdom, working comfortably below the ceiling, has spent its capital elsewhere: on specificity (§ 15.10), on regulation (Chapter 18), on the substrates it must not confuse. Perfection is not the goal of every enzyme; it is the goal of the few, and the ceiling is the measure of what the few have spent.

16.11 Kinetic isotope effects

The chapter closes with the kineticist's sharpest question to the mechanism: replace a hydrogen with its heavy isotope (deuterium, tritium), and ask whether the rate changes. The physics is the zero-point energy of Figure 16.6: the C–H bond's vibrational ground state sits above the C–D bond's (the heavy atom vibrates lower), so the C–D bond is stronger by ≈ 5 kJ mol^{-1} at the zero-point level — and if the slow step of the reaction cleaves that bond, the heavy substrate climbs a pass higher by that amount, and the rate falls by the Boltzmann factor: $k_H/k_D \approx 2$–7 (the primary isotope effect). If the bond is not cleaved in the slow step, the effect is ≈ 1. The isotope effect is therefore a yes/no question asked of the mechanism — is this bond broken in the rate-limiting step? — and the answers have settled some of the great mechanistic disputes of the field: the hydride transfers of the dehydrogenases (Part VI), the proton transfers of § 15.5's shuttles, the tunneling *corrections of the modern literature (where k_H/k_D exceeds the classical ceiling and the proton is found to tunnel through the pass rather than climb it — the quantum mechanical footnote to Figure 16.6, and the frontier of Proposition 15.a's debate). The chapter ends, as it began, in honesty about measurement: the isotope does not answer the question of mechanism — it asks it, cleanly, and the rate answers in numbers.*

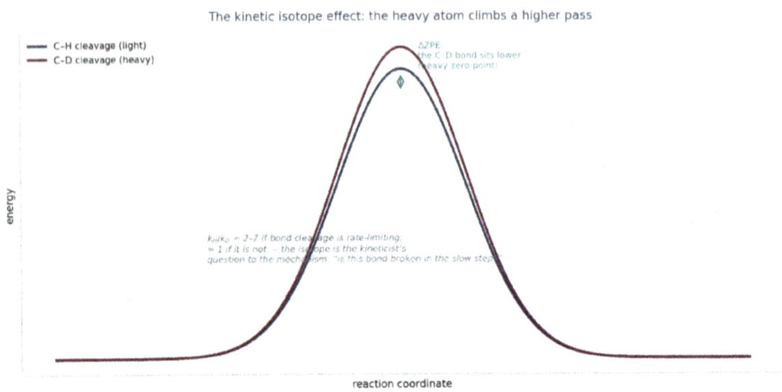

Figure 16.6 *The kinetic isotope effect: the C–H bond (navy) and the C–D bond (oxblood) have different zero-point energies, so the heavy substrate climbs a higher pass (ΔZPE, teal) — and the rate falls by the Boltzmann factor if the bond is cleaved in the rate-limiting step. The isotope is the kineticist's question to the mechanism, asked in atoms and answered in rates.*

Chapter 16 — Conceptual synthesis

The machine's speed has been measured. The initial rate is the only honest slope (§ 16.1); the Michaelis–Menten equation follows from one mechanism and one assumption in eight lines (§ 16.2); the two constants have three meanings — speed, half-speed concentration, and the steady-state composite (§ 16.3); the specificity constant is the true measure of quality, with the diffusion limit as its ceiling (§§ 16.4, 16.10); the two assumptions divide the territory (§ 16.5); the straight lines were scaffolding (§ 16.6); the honest fit inspects its residuals (§ 16.7); two substrates divide into sequential and ping-pong (§ 16.8); the pre-steady state is the mechanism unveiled (§ 16.9); and the isotope asks the mechanism its sharpest question (§ 16.11). Direction checked against Figure ii: the machine is measured. Chapter 17 now jams it: enzyme inhibition — the molecules that slow the machine, and the pharmacology that lives in the slowing.

Worked example 16.1 — Reading a curve

An enzyme at $[E]_T$ = 10 nM gives v = 0.60 μmol min^{-1} at [S] = 5 mM and 0.86 μmol min^{-1} at [S] = 20 mM. Solve for the constants. From Eq. (16.5): 0.60 = $V_{max} \times 5/(K_m + 5)$ and 0.86 = $V_{max} \times 20/(K_m + 20)$; dividing, 0.60/0.86 = $5(K_m + 20)/[20(K_m + 5)]$, whence 0.698 × $20(K_m + 5) = 5(K_m + 20)$, $13.95K_m + 69.8 = 5K_m + 100$, K_m = 3.37 mM; then V_{max} = 0.60 × (3.37 + 5)/5 = 1.00 μmol min^{-1}, and $k_{cat} = V_{max}/[E]_T$ = 1.00 × $10^{-6}/60/10^{-8}$ ≈ 1 700 s^{-1}. Two points, two equations, two constants — the algebra of § 16.2's curve, and the audit (the rates must rise with [S], the constants must be positive) of § 4.12.

WORKED EXAMPLE 16.2 — K_M VERSUS K_D

A protease has $k_1 = 10^7$ L mol^{-1} s^{-1}, $k_{-1} = 10$ s^{-1}, $k_2 = 100$ s^{-1}. The dissociation constant is $K_d = k_{-1}/k_1 = 10/10^7 = 1$ μM; the Michaelis constant is $K_m = (k_{-1} + k_2)/k_1 = 110/10^7 = 11$ μM — eleven times the dissociation constant, because the chemistry drains the complex as fast as the dissociation does. The reader who reports "$K_m = 11$ μM, so the enzyme binds its substrate at 11 μM" has made § 16.3's error; the binding is 1 μM, the kinetics is 11, and the two constants measure different processes of the same machine.

WORKED EXAMPLE 16.3 — THE SPECIFICITY CONSTANT IN COMPETITION

An enzyme faces substrates A ($k_{cat}/K_m = 10^7$) and B (10^4), each at 1 μM. The rates at low substrate go as the constants: $v_A/v_B = 10^3$ — the enzyme discriminates a thousandfold, though its k_{cat} values might be identical (the discrimination could be all in the K_m). The specificity constant is the currency of the choice (§ 16.4), and the reader should note what the ratio is made of: the difference in free energy of the two transitions is $\Delta\Delta G‡ = RT \ln 10^3 = 2.577 \times 6.9 \approx 17.8$ kJ mol^{-1} — the discrimination of Figure 15.5, paid by the machine's two constants working together.

Exercises (with full resolutions)

EXERCISE 16.1

Derive Eq. (16.5) from Eqs. (16.2)–(16.4) in full, tracking the enzyme-conservation step.

Resolution. *Steady state: $[ES] = k_1[E][S]/(k_{-1} + k_2) = [E][S]/K_m$ (Eq. 16.3). Conservation: $[E] = [E]_T - [ES]$; substitute: $[ES] = ([E]_T - [ES])[S]/K_m$, so $[ES](1 + [S]/K_m) = [E]_T[S]/K_m$, $[ES] = [E]_T[S]/(K_m + [S])$ — Eq. (16.4); and $v = k_2[ES] = k_2[E]_T[S]/(K_m + [S]) = V_{max}[S]/(K_m + [S])$. The derivation's whole cost is the steady-state assumption (Eq. 16.2) and the conservation of enzyme — the two lines on which the most-used equation of enzymology stands.*

EXERCISE 16.2

An enzyme shows $v = 0.9\ V_{max}$ at $[S] = 30$ mM. Find K_m, and the fraction of enzyme occupied.

Resolution. *$0.9 = 30/(K_m + 30)$, so $K_m + 30 = 33.3$, $K_m = 3.3$ mM. Occupancy $[ES]/[E]_T = [S]/(K_m + [S]) = 0.9$ — the same number twice, because v/V_{max} is the occupancy: the speedometer and the binding isotherm of § 8.5 are one curve, and the identity is § 16.3's quiet theorem.*

EXERCISE 16.3

A stopped-flow shows a burst of 0.8 mol product per mol enzyme, then a steady rate. Interpret, and state what the burst measures.

Resolution. *The burst is § 15.6's covalent signature: the first turnover releases the first product (the amine, say) while the enzyme holds the covalent* *intermediate; the second product's release (the acyl-enzyme's hydrolysis) is slower, and sets the steady rate. The burst's size — 0.8 mol per mol enzyme — counts the active sites: 80 % of the enzyme is active, 20 % dead or incompetent, and the pre-steady state has measured what the steady state cannot: the concentration of working enzyme, from one pulse of product.*

Exercise 16.4 (Interpretative)

"The Michaelis–Menten equation assumes the enzyme is inert scaffolding." Assess the claim in the light of Chapter 12.

Resolution (indicative). *The claim is false in letter and true in spirit. In letter: Eq. (16.5) assumes nothing about the enzyme's rigidity — only the steady state and the conservation — and a breathing enzyme that interconverts faster than it turns over satisfies both. In spirit: the equation's ES is a single state, and the real enzyme is an ensemble (§ 12.1) — the Michaelis–Menten equation is the ensemble's average, exact when the interconversion is fast, approximate when the ensemble's members differ in their constants (the kinetic heterogeneity of the single-molecule literature, where each enzyme shows its own k_{cat}). The adjudication is the treatise's standing one: the equation is the truth at the ensemble's centre, and the fluctuations about the centre are Chapter 12's — the scaffolding breathes, but the average still holds.*

Exercise 16.5

A dehydrogenase shows k_H/k_D = 5.2 when the substrate's C–H is deuterated, and k_{cat} = 50 s^{-1}. Is hydride transfer rate-limiting, and what would the effect be if k_{cat} were instead set by product release?

Resolution. *The effect of 5.2 is a primary isotope effect: the C–H(D) bond is cleaved in the slow step (Figure 16.6's higher pass, k_H/k_D ≈ e^{ΔΔG‡/RT} ≈ e^{5/2.577} ≈ 7 at the classical ceiling) — hydride transfer is rate-limiting. If product release set k_{cat}, the isotope would leave the rate untouched (≈ 1): the chemistry would be fast and invisible, the release slow and isotope-blind — and the same enzyme would answer the isotope's question "no", though the bond is still broken every turnover: the isotope asks about the slow step only, and the slow step is the only one the rate can see.*

Research-level propositions

Proposition 16.α. The steady-state assumption fails for the perfect enzymes of § 16.10 ($k_2 \approx k_{on}$). Derive the full time-course of Eq. (16.1) without the assumption (the quadratic solution of the total quasi-steady-state), and estimate the error in K_m for a diffusion-limited enzyme fitted with Eq. (16.5). Proposition 16.β. Single-molecule enzymology has found that individual enzymes fluctuate in k_{cat} over seconds (the "dynamic disorder" of § 12.10). Formulate the kinetic theory of a two-state enzyme with slow interconversion, and predict the signature in the waiting-time distribution of single turnovers — then assess the published evidence that the fluctuations are the ensemble of § 12.1 and not instrumental noise. Proposition 16.γ. Proton tunnelling (§ 16.11) has been claimed for several dehydrogenases on the evidence of anomalous isotope effects and their temperature dependence. Assemble the claims and the counterclaims (the "full tunnelling" versus "coupled *motion" models), and design the experiment — isotope-edited vibrational spectroscopy against rate (§ 54.5) — that would decide the mechanism for one enzyme.*

References (Chapter 16)

Cornish-Bowden, A. (2012). Fundamentals of enzyme kinetics (4th ed.). Wiley-Blackwell.

Fersht, A. (2017). Structure and mechanism in protein science. World Scientific.

Michaelis, L., & Menten, M. L. (1913). Die Kinetik der Invertinwirkung. Biochemische Zeitschrift, 49, 333–369.

Segel, I. H. (1975). Enzyme kinetics: Behavior and analysis of rapid equilibrium and steady-state enzyme systems. Wiley.

CHAPTER 17

Enzyme Inhibition

CONTENTS OF THIS CHAPTER

In which the machine is jammed, and the jamming is classified: the reversible inhibitors — competitive, uncompetitive, mixed and non-competitive — each with its signature on the curve; the irreversible inhibitors and the covalent trap; the suicide substrate, the drug the enzyme arms itself; the transition-state analogue, Pauling's principle turned to medicine; and the whole pharmacopoeia shown to be these few mechanisms, prescribed.

17.1 Reversible inhibition

The inhibitor is the substrate's rival, and the rivalry divides, at the first fork, into the reversible and the irreversible — the rival who can be outbid and the rival who cannot. The reversible inhibitor binds the enzyme by the noncovalent instruments of Chapter 8 — hydrogen bonds, dispersion, the whole recognition economy — and its effect is governed by the equilibrium of its binding: raise the substrate, and the rival is outcompeted; wash the enzyme, and the rival leaves. The irreversible inhibitor binds covalently, or so tightly that the distinction is theological, and its effect is governed by time: the enzyme is not competed with but consumed. The reversible inhibitors divide further, by where they bind, into the four signatures of Figure 17.1, and the signatures are the chapter's grammar: each pattern on the curve is a different answer to the question "where does the rival stand?" — at the active site (competitive), on the enzyme–substrate complex (uncompetitive), elsewhere (non-competitive and mixed). The clinical stake is immediate: the reversible inhibitors are the drugs that can be overcome by the body's own chemistry (the substrate rises, the drug's grip slips), and the irreversible ones are the drugs that outlast it — aspirin, penicillin, the suicide substrates of § 17.6, whose effects persist long after the drug itself has left the blood. The reader should learn the four signatures as the physician learns the pulse: they are the inhibition's own testimony, readable in the curve before any structure is known.

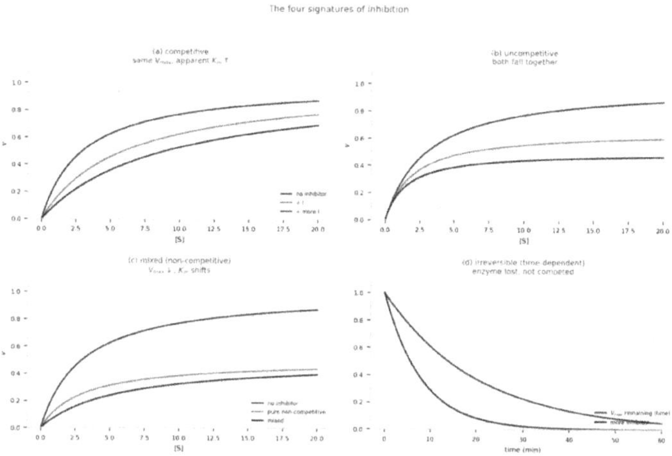

Figure 17.1 *The four signatures of inhibition on the Michaelis–Menten curve. (a) Competitive: same V_{max}, apparent K_m rises. (b) Uncompetitive: both constants fall together. (c) Non-competitive and mixed: V_{max} falls, K_m shifts or stands. (d) Irreversible: activity lost with time, not competed — the enzyme consumed, not rivalled.*

17.2 Competitive inhibition

The competitive inhibitor stands at the active site — the substrate's own address — and its effect is pure Chapter 8 competition: the enzyme is shared between two equilibria, E + S and E + I, and the apparent Michaelis constant is inflated by the competitor's occupancy, exactly as Eq. (8.7) priced it:

$$K_m^{\text{app}} = K_m \left(1 + \frac{[\mathrm{I}]}{K_I}\right) \tag{17.1}$$

— K_I being the inhibitor's dissociation constant. The signature (Figure 17.1a) is the consequence: V_{max} is untouched (enough substrate always outbids the rival — the curve reaches the same ceiling, further right), and the apparent K_m rises linearly with the inhibitor's concentration (Figure 17.2). The mechanism's pharmacology is the statin, the methotrexate, the ACE inhibitor of Figure 17.6 — drugs that mimic the substrate and lose to it at saturation; and its antidote logic is the same arithmetic reversed: the competitive inhibition is overcome by raising the substrate, which is why the folinic acid rescue after methotrexate works — the natural substrate, given in flood, outbids the drug for the enzyme, and the rescue is Eq. (17.1) prescribed.

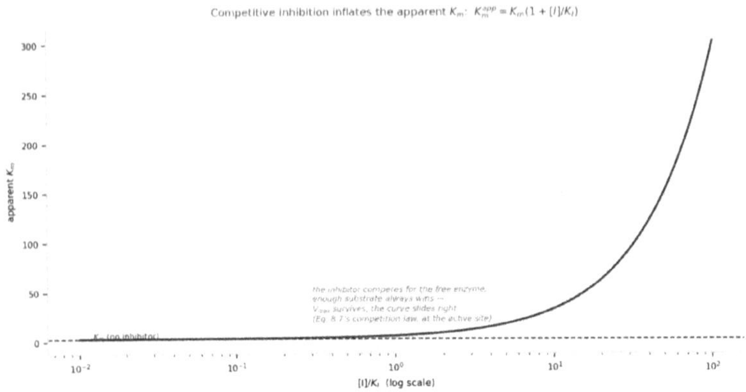

__Figure 17.2__ Competitive inhibition: the apparent K_m rises linearly with $[I]/K_I$ (Eq. 17.1) — the substrate's address is contested, and enough substrate always wins the contest. The ceiling is unchanged; only the road to it lengthens.

17.3 Uncompetitive inhibition

The uncompetitive inhibitor is the rarest and the most instructive: it binds not the free enzyme but the enzyme–substrate complex — the rival arrives only after the substrate has sat down — and its effect is to remove ES from the productive pool. Both constants fall together (Figure 17.1b): V_{max} falls (the complex is sequestered), and K_m falls too (the removal of ES pulls more S onto the enzyme — Le Chatelier's principle at the active site). The mechanism is rare in single-substrate enzymes (it requires a site that exists only on the complex) but common in the bisubstrate world of § 16.8, where an inhibitor resembling one substrate can bind the complex with the other; and its diagnostic value is the pattern's uniqueness: the parallel fall of both constants is unmistakable, and a curve that shows it is a confession that the rival binds the occupied enzyme, not the free one.

17.4 Mixed and non-competitive inhibition

The non-competitive inhibitor binds elsewhere — a second site, the allosteric address of Chapter 8 — and its effect is on the machine, not the substrate's address: it lowers the catalytic competence of the enzyme it binds, so V_{max} falls while K_m stands (pure non-competitive, Figure 17.1c, teal) or shifts (mixed, oxblood — the general case, in which the inhibitor's binding changes the substrate's constant too). The mechanism is the allosteric effector of § 8.8 in its inhibitory livery, and its pharmacology is the growing class of allosteric drugs — the kinase inhibitors that bind outside the ATP site (the MEK inhibitors, the allosteric EGFR drugs), whose advantage over the competitive class is precisely § 15.10's specificity: the allosteric sites are less conserved than the active sites, so the allosteric drug can hit one kinase and spare its four hundred cousins — the selectivity the competitive inhibitor, aimed at the conserved ATP pocket, cannot achieve. The reader should note the inversion this represents: for fifty years the active site was the target and the allosteric site the curiosity; the pharmacology of the present age increasingly reverses the emphasis, and the chapter's fourth mechanism has become its most prescribed.

17.5 Irreversible inhibition

The irreversible inhibitor binds covalently — or so tightly that the distinction is theological — and the enzyme is consumed, not rivalled. The kinetics are time-dependent (Figure 17.1d): activity falls with the time of exposure, not with the substrate's defence, and no washing restores it. The mechanisms divide into two families. The affinity labels are the direct ones: a reactive group (a fluorophosphate, a sulphonyl fluoride, an acetylating anhydride) placed on a recognising scaffold, which attacks the active site's nucleophile — aspirin, which acetylates the cyclooxygenase's Ser530, is the most prescribed irreversible inhibitor in history, and its pharmacology follows from the irreversibility itself: the platelet, having no nucleus, cannot re-make the enzyme, so one aspirin silences the platelet's prostaglandin synthesis for the platelet's whole ten-day life — the irreversible drug whose duration is the cell's, not the dose's. The suicide substrates are the indirect and the more elegant, and they have the next section to themselves.

17.6 Suicide substrates

The suicide substrate — the mechanism-based inhibitor — is the drug the enzyme arms itself: an innocent-looking molecule that the enzyme, in the ordinary course of its chemistry, converts into the weapon that kills it (Figure 17.4). The logic is the catalytic mechanism of Chapter 15 turned against the enzyme: the first steps of the normal reaction proceed — activation, the formation of the reactive intermediate — and the intermediate, instead of completing the reaction, attacks the enzyme covalently, one molecule of inhibitor consuming one molecule of enzyme. The specificity is absolute, and it is the reason the class exists: only the target enzyme can arm the weapon, so the suicide substrate is inert to every other protein in the body — the most specific inhibitors known, and the pharmacopoeia's quiet aristocracy. Clavulanate arms the β-lactamase against itself (and thereby rescues the penicillins from the resistant bacterium's own enzyme — an inhibitor of an inhibitor, the bacterial defence dismantled by its own mechanism); allopurinol arms xanthine oxidase (the gout drug, whose oxypurinol product locks the enzyme's molybdenum); 5-fluorouracil arms thymidylate synthase (the cancer drug, whose fluorinated nucleotide traps the enzyme with its folate cofactor). The reader should note the thermodynamic signature: the suicide substrate is not a better binder but a better trap — it exploits the enzyme's own catalytic power, the pass-lowering of Chapter 15, to build the covalent bond, and the enzyme's speed is turned, at last, against itself.

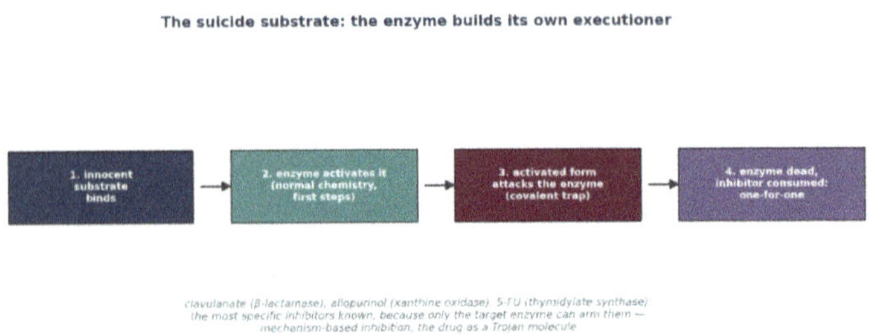

Figure 17.4 The suicide substrate: the innocent molecule binds (1), the enzyme activates it by its normal chemistry (2), the activated form attacks the enzyme covalently (3), and enzyme and inhibitor are consumed one-for-one (4). Only the target enzyme can arm the trap — the most specific inhibition known.

17.7 Transition-state analogues

The transition-state analogue is Pauling's principle turned to medicine: a stable molecule shaped like the transition state of the reaction (§ 15.2), which binds the enzyme as the transition state would — a million times tighter than the substrate (Figure 17.5, and Exercise 15.1's arithmetic). The class is the most potent family of reversible inhibitors known, and its history is the principle's vindication: the first designed analogues (the adenosine deaminase inhibitor coformycin, the ACE inhibitor captopril — the snake-venom peptide's logic, reworked as a zinc-binding analogue of the peptide's transition state) showed that the analogue binds as Pauling predicted, and the modern protease inhibitors of HIV are its crown: the peptidomimetic drugs (saquinavir and its successors) are stable mimics of the protease's tetrahedral transition state, binding at picomolar, and the patient's viral load is Pauling's principle, prescribed daily. The analogue's limitation is the principle's own: it must be synthesised, and the transition state's geometry must be known — the analogue is structure-based drug design in its purest form, and the enzyme's crystal structure is its recipe.

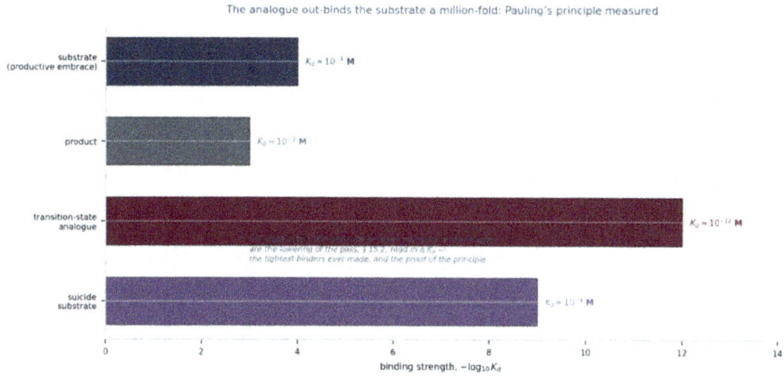

Figure 17.5 *The transition-state analogue (oxblood) out-binds the substrate by eight orders of magnitude — the lowering of the pass, § 15.2, measured as a dissociation constant. The suicide substrate (plum) binds more weakly still as a ligand; its grip is the covalent trap of § 17.6, not the embrace.*

17.8 Pharmacological enzyme inhibitors

The chapter's mechanisms are, at the pharmacy, the most prescribed class of drugs in the world, and Figure 17.6 is the prescription list: the statins (competitive, at HMG-CoA reductase — the best-selling drugs in history, and the cholesterol of § 36.12 their target), the ACE inhibitors (competitive analogue, at the angiotensin-converting enzyme — the blood-pressure drugs), methotrexate (competitive, at dihydrofolate reductase — the antifolate of cancer and arthritis), the penicillins (irreversible, at the bacterial transpeptidase), the HIV protease inhibitors (transition-state analogue), allopurinol and 5-FU (suicide substrates), aspirin (irreversible acetylation). The reader should read the list with the chapter's grammar in hand: each drug is one of the four signatures, and the signature is the drug's whole pharmacological personality — its duration (reversible against irreversible), its selectivity (active-site against allosteric), its rescue (competitive against not). The pharmacopoeia is short of mechanisms and long of molecules; the chapter's four signatures are the mechanisms, and the rest is chemistry.

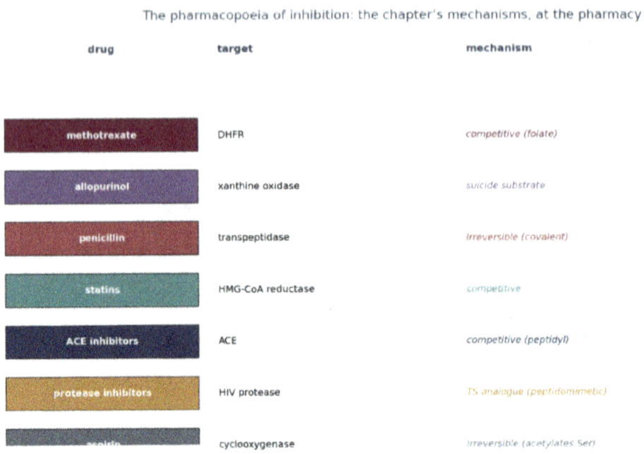

Figure 17.6 *The pharmacopoeia of inhibition: the chapter's mechanisms, prescribed — competitive inhibitors (statin, ACE inhibitor, methotrexate), transition-state analogues (the HIV protease drugs), irreversible inhibitors (penicillin, aspirin), and suicide substrates (allopurinol, 5-FU, clavulanate). Four signatures; the rest is chemistry.*

17.9 Determining inhibitory constants

The inhibitor's constant is measured as the substrate's is, with one correction, and the correction is the chapter's most-cited trap. The IC_{50} — the inhibitor concentration that halves the activity — is the naive measure, and it is not a constant: it depends on the assay's substrate concentration, because the competitive rival's grip is contested by the substrate (Figure 17.3a). The true constant K_I is read through the Cheng–Prusoff relation,

$$K_I = \frac{\mathrm{IC}_{50}}{1 + [\mathrm{S}]/K_m} \qquad (17.2)$$

— derived from Eq. (17.1) by setting the apparent constant equal to the value that halves the rate; and the reader who quotes an IC_{50} without its [S] has quoted a number without a ruler (Figure 17.3b's caption, and the literature's daily error). The relation generalises the chapter's doctrine: the inhibitor's effect is always a competition between its grip and the substrate's, and the measured potency is the grip divided by the contest — the K_I is the constant of the molecule, the IC_{50} the constant of the experiment, and Cheng–Prusoff is the exchange rate between them.

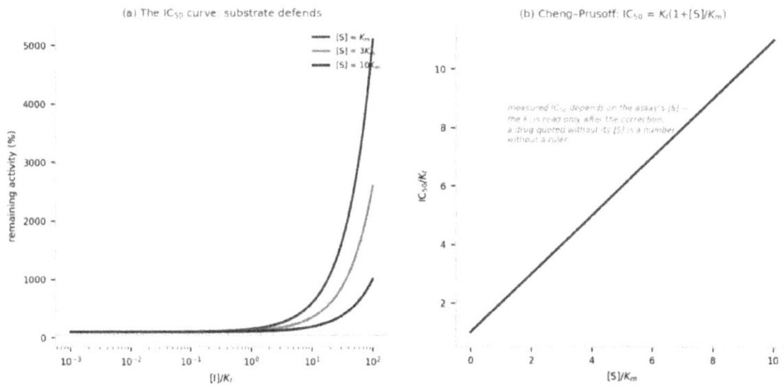

Figure 17.3 *(a) The IC_{50} curve at three substrate concentrations: the competitive inhibitor's apparent potency falls as the substrate rises — the curve shifts right, and the shift is the contest. (b) The Cheng–Prusoff relation (Eq. 17.2): the true K_I is read only after the correction, and a drug quoted without its [S] is a number without a ruler.*

17.10 Distinguishing binding from functional inhibition

The chapter closes with its diagnostic, and the diagnostic is the distinction the whole field runs on: an inhibitor can bind without inhibiting, and inhibit without binding where one expects. The binding experiment (the surface plasmon resonance of § 54.8, the isothermal titration calorimetry of § 54.9) measures the embrace — K_d, the enthalpy, the stoichiometry — and says nothing about the function; the functional experiment (the rate curve of Chapter 16) measures the inhibition — K_I, the signature, the mechanism — and says nothing about the embrace. The two must agree, and when they do not, the discrepancy is the discovery: the molecule that binds without inhibiting is an inert ligand (a false positive of the screening library, the curse of the high-throughput assay); the molecule that inhibits without binding the expected site is an allosteric inhibitor, an aggregator (the colloidal aggregates of the screening literature, which inhibit promiscuously by adsorbing the enzyme), or a covalent modifier missed. The reader who has followed the treatise this far will recognise the diagnostic as the hermeneutic circle of Figure 1.5 in miniature: measure the binding, measure the function, and let the two audits confront each other — the inhibition is real when both books balance, and the mechanism is found where they do not.

Plate 17.1 *Sir James Black (1924–2010), whose principle — design the drug from the mechanism, not from the molecule's resemblance to a natural substance — produced the β-blockers (propranolol) and the H_2-antagonists (cimetidine), and founded the rational drug design of which the chapter's pharmacopoeia is the continuation. Portrait from the Nobel Foundation archive, nobelprize.org (Nobel Prize in Physiology or Medicine, 1988), reproduced from the Foundation's publicly distributed image.*

HISTORICAL NOTE — SULFA, PRONTOSIL, AND THE ANTIMETABOLITE

The first great enzyme inhibitors were found not by design but by a dye. Gerhard Domagk, at Bayer in 1932, tested the red dye prontosil against streptococcal infection in mice, and the mice lived; the dye's active part was sulphanilamide, and sulphanilamide's target was the bacterial enzyme that makes folate from p-aminobenzoate — the drug being a competitive mimic of the natural substrate, Chapter 17's first mechanism, discovered before the mechanism had a name. The sulfa drugs that followed saved more lives than any drug before penicillin; and the same logic — mimic the metabolite, starve the cell — is the antimetabolite principle of methotrexate and 5-FU, Figure 17.6's cancer drugs, and of the whole antimetabolite pharmacopoeia of § 38.14. The dye that coloured wool also, by Domagk's experiment, founded chemotherapy; and the reader should note that the discovery was mechanism-free: the drug worked, the mechanism was found later, and the field of enzyme inhibition was born, as so often, in the wrong order — cure first, understanding after.

Chapter 17 — Conceptual synthesis

The machine has been jammed, and the jamming classified. Inhibition divides into reversible (outbiddable) and irreversible (consuming) (§ 17.1); competitive inhibition inflates the apparent K_m by the competitor's occupancy (Eq. 17.1, § 17.2); uncompetitive sequesters the complex, both constants falling (§ 17.3); non-competitive and mixed lower the machine's competence from the allosteric address — the modern drug class (§ 17.4); irreversible inhibitors consume the enzyme, and the platelet's ten days is the irreversibility's duration (§ 17.5); the suicide substrate is the enzyme's own chemistry turned against it, and its specificity is absolute (§ 17.6); the transition-state analogue is Pauling's principle prescribed (§ 17.7); the pharmacopoeia is the four signatures, itemised (§ 17.8); the K_I is read only through Cheng–Prusoff (§ 17.9); and binding and function are *two audits, which must balance (§ 17.10). Direction checked against Figure ii: the machine is measured and jammed. Chapter 18 now asks how the cell jams it deliberately: the regulation of enzyme activity — the allosteric, covalent and zymogenic controls on which the whole of metabolism runs.*

WORKED EXAMPLE 17.1 — THE STATIN'S ARITHMETIC

A statin has K_I = 1 nM at HMG-CoA reductase; the assay's [S] = 0.1 K_m. The measured IC_{50} is, by Eq. (17.2), $IC_{50} = K_I(1 + [S]/K_m) = 1 \times 1.1 = 1.1$ nM — near the true constant, because the substrate is scarce. In the patient, the hepatic substrate (HMG-CoA) is also scarce relative to K_m, so the drug grips at near its K_I — and the dose that halves the liver's cholesterol synthesis is nanomolar: the statin's potency is the competitive mechanism's, priced by the correction, and the best-selling drug in history is Eq. (17.2) in a bottle.

WORKED EXAMPLE 17.2 — THE ASPIRIN'S DECADE

Aspirin acetylates the platelet's cyclooxygenase irreversibly; the platelet lives ≈ 10 days and cannot re-make the enzyme. The recovery of prostaglandin synthesis therefore follows platelet turnover, not drug clearance: the drug's plasma half-life is ≈ 20 minutes, its effect's half-life is ≈ 3–4 days (the time for half the platelet pool to be renewed from the marrow's megakaryocytes) — the irreversible mechanism's signature: the molecule leaves in an hour, the effect persists for the cell's life, and the surgeon's question "when did you last take aspirin?" is § 17.5's kinetics at the bedside.

WORKED EXAMPLE 17.3 — READING THE PATTERN

An inhibitor raises the apparent K_m threefold at [I] = 2K_I, leaves V_{max} unchanged, and its effect is fully reversed by dialysis. Classify and price it. The signature is competitive (Figure 17.1a): $K_m^{app} = K_m(1 + [I]/K_I) = 3K_m$ at [I] = 2K_I ✓ (Eq. 17.1 checks), and the reversibility confirms the noncovalent embrace — the diagnosis is complete from the curve alone, no structure consulted: the four signatures of Figure 17.1 are the inhibition's own testimony, read before any other evidence is called.

Exercises (with full resolutions)

EXERCISE 17.1

Derive Eq. (17.1) from the competition equilibrium of Chapter 8.

Resolution. *The free enzyme is shared: $[E]_{free}$ = [E](1 + [I]/K_I) (Eq. 8.7's derivation — the two equilibria E + S and E + I divide the pool). The substrate's binding therefore proceeds against the diluted free enzyme: its apparent constant is the true constant multiplied by the dilution, $K_m^{app} = K_m(1 + [I]/K_I)$ — Eq. (17.1). The whole of competitive inhibition is the competition law of § 8.9 at the active site, and the derivation is two lines of Chapter 8, applied.*

EXERCISE 17.2

An inhibitor has IC_{50} = 50 nM at [S] = $4K_m$. Compute K_I, and the IC_{50} at [S] = $K_m/2$.

Resolution. *$K_I = IC_{50}/(1 + [S]/K_m) = 50/(1 + 4) = 10$ nM (Eq. 17.2). At [S] = $K_m/2$: $IC_{50} = K_I(1 + 0.5) = 15$ nM — the same drug, threefold more potent in the scarcer substrate, and the literature's daily error avoided: the K_I is the molecule's constant, the IC_{50} the assay's, and the correction is the difference between a potency and a protocol.*

EXERCISE 17.3

Clavulanate is consumed one-for-one with the β-lactamase it inhibits. Estimate the number of penicillin molecules saved per clavulanate, given the enzyme would otherwise hydrolyse 10^4 penicillins before the cell divides.

Resolution. *The suicide substrate's arithmetic: one clavulanate consumes one enzyme molecule (Figure 17.4's step 4); the enzyme molecule would have destroyed 10^4 penicillins; the clavulanate therefore rescues ≈ 10^4 penicillins per molecule — the catalytic leverage of the mechanism, and the reason the combination (amoxicillin–clavulanate) works at all: the inhibitor is not the drug but the drug's bodyguard, and the bodyguard's economics are the enzyme's own turnover, turned to account.*

EXERCISE 17.4 (INTERPRETATIVE)

"Irreversible inhibitors make bad drugs: you cannot undo a mistake." Defend and qualify the claim with the chapter's pharmacopoeia.

Resolution (indicative). *The defence: the irreversible drug's duration is the cell's, not the dose's — an off-target hit persists for the protein's life, and the toxicity is unrescueable (no substrate flood outbids a covalent bond). The qualification: the class includes the most prescribed drug in history (aspirin), the most important antibiotic (penicillin), and the modern covalent kinase inhibitors — and their safety comes from selectivity, not reversibility: the covalent warhead is placed on a scaffold that binds only the target (§ 17.6's specificity), so the mistake is prevented at the binding, not undone at the bond. The adjudication is § 17.10's: irreversibility is a property of the bond, safety a property of the embrace — and the two are independent dimensions of the drug.*

EXERCISE 17.5

A compound binds the enzyme with K_d = 10 nM (SPR) but shows K_I = 10 μM (rates). Offer three explanations, and the experiment that decides each.

Resolution. *(i) The compound binds an inert site (the embrace is real, the address wrong): decide by competition with a known active-site ligand — if it does not displace, the site is elsewhere. (ii) The binding is stoichiometric but the inhibition is aggregative (the colloidal aggregate of the screening literature): decide by detergent or by dilution — aggregates dissolve, and the K_I rises. (iii) The compound binds the enzyme–substrate complex (uncompetitive, § 17.3): the functional K_I is read against the wrong constant — decide by the pattern of Figure 17.1b. The exercise is § 17.10's diagnostic in* *three moves: binding and function are two audits, and the discrepancy between them is the discovery.*

Research-level propositions

Proposition 17.α. The covalent kinase inhibitors (the ibrutinib class) bind irreversibly at a cysteine near the ATP site; resistance emerges by mutation of the cysteine. Formulate the kinetic theory of covalent drug resistance (the on-rate of the warhead against the off-rate of the scaffold), and predict the resistance-free window of a next-generation analogue. Proposition 17.β. The aggregating inhibitors of § 17.10 are estimated to account for a substantial fraction of screening "hits". Assemble the evidence (detergent sensitivity, stoichiometry, electron microscopy), and propose a screening cascade that would eliminate aggregators before the chemistry is invested. Proposition 17.γ. The transition-state analogue's premise (§ 17.7) is that the analogue mimics the transition state; the counterevidence is that some analogues bind by different chemistry than the transition state they imitate. Using the computed free-energy surfaces of § 55.9, test the fidelity of one canonical analogue (the HIV protease class), and assess whether the tight binding is Pauling's complementarity or a fortunate accident.

References (Chapter 17)

Cheng, Y., & Prusoff, W. H. (1973). Relationship between the inhibition constant (K_I) and the concentration of inhibitor which causes 50 per cent inhibition (I_{50}) of an enzymatic reaction. Biochemical Pharmacology, 22(23), 3099–3108.

Copeland, R. A. (2013). Evaluation of enzyme inhibitors in drug discovery: A guide for medicinal chemists and pharmacologists (2nd ed.). Wiley.

Cornish-Bowden, A. (2012). Fundamentals of enzyme kinetics (4th ed.). Wiley-Blackwell.

Silverman, R. B., & Holladay, M. W. (2014). The organic chemistry of drug design and drug action (3rd ed.). Academic Press.

CHAPTER 18

Regulation of Enzyme Activity

CONTENTS OF THIS CHAPTER

In which the cell jams its own machines, deliberately and by design: the allosteric switch of § 8's election, refined into the pathway's gate; feedback inhibition and feed-forward activation; the covalent ratchet of phosphorylation and its kinase cascades; the zymogen's one-way switch; the isoenzyme's tissue-specific temper; and the geography of it all — regulation by address, by turnover, and by time.

18.1 Allosteric enzymes

The enzyme of Chapter 16 is a machine with a speedometer; the allosteric enzyme is a machine with a throttle. The allosteric site — the second address of § 8.8, distinct from the active site — receives the effector, and the effector re-weights the T⇌R equilibrium of Chapter 12: the enzyme is thrown into its tense state (inhibited) or its relaxed state (activated), and the whole oligomer flips in concert. The allosteric enzymes of metabolism are not scattered at random through the pathways; they stand at the control points — the irreversible, committed steps of § 7.9's placement, where the pathway's traffic must be governed — and their kinetics carry the sigmoid of Figure 18.2, the cooperative curve that turns the enzyme from a dial into a switch. The reader who has followed the treatise will recognise the instrument at once: the allosteric enzyme is the MWC protein of Chapter 8 with a job — haemoglobin's logic, applied to the pathway — and the sigmoid curve is the switch's signature, the reason the cell prefers its control enzymes cooperative: in the physiological window of Figure 18.2, the hyperbola answers a fourfold change of signal with a fourfold change of rate, the sigmoid with fortyfold — the difference between a rheostat and a relay.

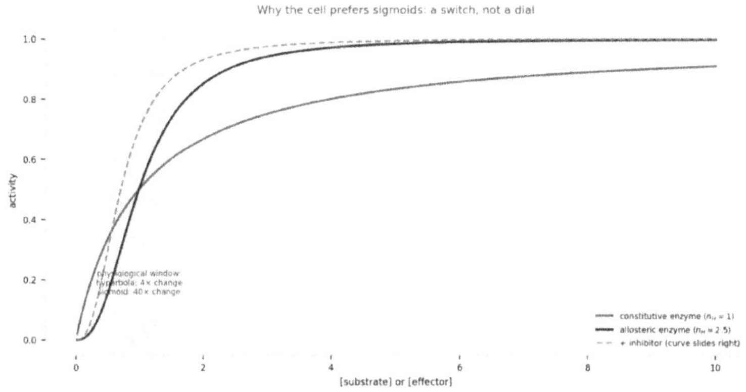

Figure 18.2 *The allosteric enzyme's sigmoid (oxblood) against the constitutive hyperbola (slate), with the inhibitor's rightward slide (amber, dashed) and the physiological window (shaded): in the window, the sigmoid amplifies the signal tenfold — the switch, not the dial.*

18.2 Sigmoidal kinetics

The sigmoid's formal description is Chapter 8's Hill equation (Eq. 8.6), and its kinetic reading adds one clause to the binding curve's: the substrate itself is often the effector — the allosteric enzyme's sigmoid against [S] is substrate cooperativity, the oligomer's sites talking to one another (aspartate transcarbamoylase, the chapter's canonical case and the MWC model's original exemplar (Gerhart, 1970)). The regulatory consequence is the homeostat's ideal: at low [S], the enzyme idles (the sigmoid's flat foot); in the window, it responds steeply; at high [S], it saturates — a machine that is insensitive where the signal is noise, and exquisitely sensitive where the signal is command. The effectors slide the curve without changing its shape: the activator shifts it left (the window arrives sooner), the inhibitor right (later) — the throttle's two directions, and the pathway's way of being told "more" and "enough". The reader should note the economy of the design: one protein, one site for the work and one for the orders, and the regulation is built into the machine itself — no second protein, no cascade, no ATP; the allosteric enzyme is the cheapest switch in the cell.

18.3 Feedback inhibition

The oldest and cheapest control circuit of metabolism is the one the pathway draws itself: the end product inhibits the first committed step (Figure 18.1). The isoleucine pathway is the textbook: threonine deaminase, the gate of the pathway, is inhibited by isoleucine — the product, arriving at the allosteric address of the enzyme that made it five steps earlier, says "enough", and the pathway idles until the stock is drawn down. The logic is homeostatic (the product's level is the signal, and the signal corrects the level — a thermostat, not a clock), and its placement is the chapter's deep rule: the control point is the first committed step, the step after which the substrate has no other road — inhibiting the gate saves the whole pathway's intermediates from accumulating uselessly, and the economy of the cell is preserved at one allosteric address. The branched pathways add the refinements: sequential feedback (each intermediate inhibits the step after its branch), concerted feedback (the products inhibit only together — the cell's AND gate), and cumulative feedback (each product inhibits partially, the sum being the total inhibition) — the logic circuits of Chapter 52, written in enzymes before anyone had named them.

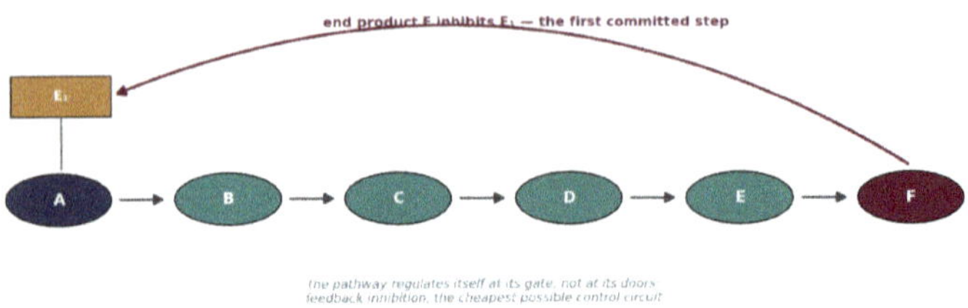

Figure 18.1 *Feedback inhibition: the end product F (oxblood) inhibits the first committed step E_1 (amber) — the pathway regulates itself at its gate, the cheapest possible control circuit, and the deep rule of metabolic regulation: the control point is the step after which there is no other road.*

18.4 Feed-forward activation

Feedback's mirror is feed-forward: the early intermediate activates the late enzyme — the pathway telling its far end that the traffic is coming. The textbook case is fructose-1,6-bisphosphate activating pyruvate kinase (§ 31.8): the product of glycolysis's committed step turns on the last step, so that when the gate opens, the whole pathway opens — no backlog of intermediates, no waiting for the product to accumulate before the far end wakes. The logic is anticipatory where feedback is corrective: feedback says "enough", feed-forward says "prepare" — and the two together make the pathway both self-limiting and self-priming, the thermostat and the starter motor in one. The reader should note that both circuits are allosteric: the cheapest regulation is the regulation the pathway runs on itself, without hormones, without cascades, without ATP — and the sophistication of the later instruments (§§ 18.5–18.7) is built upon this allosteric floor.

18.5 Reversible phosphorylation

The covalent ratchet of § 12.8 is the cell's sovereign instrument of regulation, and its discovery is the chapter's Nobel. Krebs and Fischer, studying glycogen phosphorylase in the 1950s, found the enzyme existed in two forms — active (phosphorylase a) and less active (phosphorylase b) — and that the two interconverted by phosphorylation and dephosphorylation: the reversible phosphorylation of proteins, the most general regulatory mechanism in biology (Krebs & Fischer, 1956). The instrument's virtues are § 12.8's, now at the pathway's scale. It is irreversible in each direction — the kinase writes, the phosphatase erases, and the two are different enzymes under different control, so the switch is held where it is put. It is amplified — the kinase is itself an enzyme, so one kinase molecule phosphorylates thousands of targets, and the cascade of Figure 18.3 (hormone → receptor → kinase → kinase → target) multiplies the signal by millions: the glycogen cascade of § 33.5, in which adrenaline at 10^{-10} M empties the liver's glycogen, is the textbook case. And it is connected — the kinases and phosphatases are themselves regulated (by second messengers, by other kinases, *by the whole signalling economy of Part X), so the phosphorylation switch is the transistor from which the cell's entire signalling circuitry is built. The reader should note the division of labour between the two great instruments: allostery is the pathway's own, cheap, fast, local; phosphorylation is the organism's, hormone-driven, amplified, global — the metabolic second-to-second and the physiological minute-to-hour, the two clocks of regulation, ticking together.*

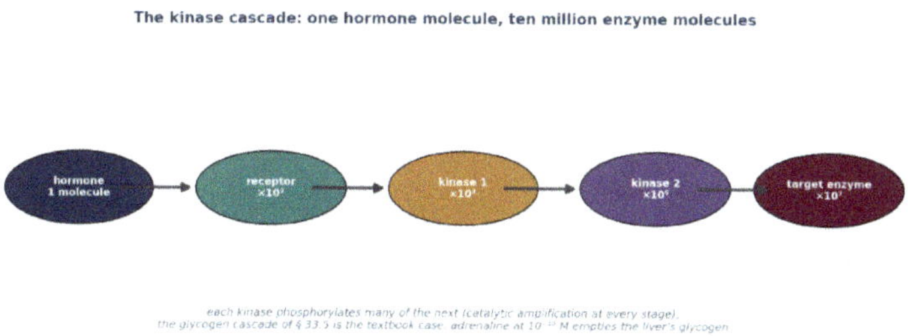

Figure 18.3 *The kinase cascade: one hormone molecule, amplified at every stage by the catalytic multiplication of the kinases, throws ten million target-enzyme switches — the glycogen cascade's arithmetic, and the reason the phosphorylation switch is the transistor of Part X's circuits.*

18.6 Acetylation, methylation and other modifications

Phosphorylation is the sovereign, but not the only instrument: the post-translational modifications of § 9.10 are a whole switchboard, and the other switches have their own jurisdictions. Acetylation of lysine neutralises its charge — the histone code of § 48.4, where the chromatin's opening is written in acetate, and the metabolic enzymes of the mitochondrion are tuned by their acetylation state (the sirtuins, the NAD-dependent deacetylases, being the erasers — metabolism and chromatin sharing one chemistry). Methylation of lysine and arginine marks the histones more subtly (the epigenetic writing of § 48.5 — marks that are read, not as charge, but as address). ADP-ribosylation (the PARP enzymes) signals DNA damage; ubiquitination (§ 11.12) condemns; nitrosylation of cysteine thiols carries the nitric-oxide signal. The reader should carry the general principle: every modification of § 9.10 is a potential switch — reversible, specific, combinatorial — and the phosphorylation switch is merely the most used of a large family; the proteome's eighth letters are written in many inks, and the cell reads them all.

18.7 Proteolytic activation

The third instrument is the one-way switch: the zymogen, the masked enzyme, activated by a single cut (Figure 18.4). The zymogen is the cell's answer to the problem of the dangerous enzyme — the protease, which would digest its own factory if made active — and the answer is to make it inert, with its active site masked by an extension (the pro-segment), and to unmask it, once, at the place and time of use: trypsinogen is activated in the gut (by enteropeptidase, and then by trypsin itself — an autoactivation, the knife sharpening the knife), prothrombin at the wound, the procaspases of § 51.6 at the death sentence. The switch is irreversible — the mask, once cut, cannot be re-sewn — and the irreversibility is the design: the dangerous activation must not be undone, and the cell pays for the safety by making the enzyme again from scratch. The zymogen cascades (trypsin activates chymotrypsinogen, proelastase, procarboxypeptidase; the clotting cascade's twelve factors activate one another in turn) are the chapter's fourth instrument at its most amplified: one cut at the top, and the whole cascade fires — the complement artillery of § 14.9, and the apoptotic scissors of § 51.6, are the same one-way logic, written in proteases.

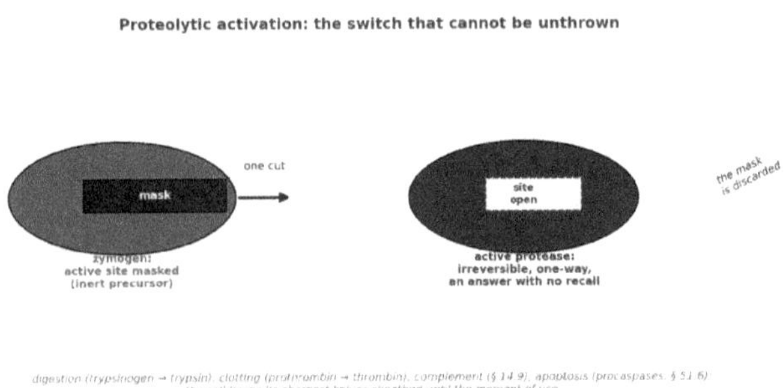

Figure 18.4 Proteolytic activation: the zymogen (masked, slate) is unmasked by a single cut (oxblood arrow) into the active protease (navy) — the switch that cannot be unthrown, the cell's way of keeping its sharpest knives sheathed until the moment of use.

18.8 Zymogens

The zymogen deserves a section of its own, for it is the regulation of last resort — the switch thrown not by a signal but by an address. The digestive enzymes are the type specimen: the pancreas makes trypsinogen, chymotrypsinogen, proelastase, procarboxypeptidases — a whole armoury of proteases, all inert — and secretes them into the duodenum, where the gut's own enteropeptidase makes the first cut, and trypsin makes the rest. The logic is geographical as much as chemical: the enzymes are activated where they are needed (the gut), not where they are made (the pancreas) — and the pancreas's own insurance, the trypsin inhibitor protein, is the second lock on the same door: the premature activation of trypsin in the pancreas is acute pancreatitis, the gland digesting itself, and the zymogen's two locks (the mask and the inhibitor) are the cell's double security on its sharpest knives. The reader should note the pattern that recurs through the whole chapter: the regulation of an enzyme is the regulation of its address, its timing, and its irreversibility — the zymogen is all three in one molecule, and the pancreatitis is the price of the lock picked.

18.9 Isoenzymes

The fourth instrument is the tissue's own: the isoenzyme — the same reaction, catalysed by different proteins, in different tissues, with different tempers. The lactate dehydrogenases are the type specimen (Figure 18.5): the H subunit (heart) and the M subunit (muscle) assemble in fives (H4, H3M, H2M2, HM3, M4), and the two tempers differ in their kinetics — the heart's H4 is inhibited by pyruvate (so the heart, an aerobic organ, does not waste its lactate but burns it), the muscle's M4 is not (so the muscle, which must make lactate anaerobically, makes it fast).

One reaction, five isoenzymes, and the tissue's metabolism is written in its subunits: the Cori cycle of § 32.8 runs on the difference (muscle exports lactate, liver and heart import it), and the clinical test reads the tissue of origin from the blood (the LDH isoenzyme pattern after a heart attack is the heart's own signature, leaked). Hexokinase's isoenzymes (§ 31.6's hexokinase versus glucokinase — the muscle's low-K_m trap and the liver's high-K_m gauge) are the other canonical case: the same reaction, tuned to the organ's office — and the reader should now see the isoenzyme as regulation by heredity: the switch is not thrown but inherited, each tissue making the enzyme its own metabolism requires.

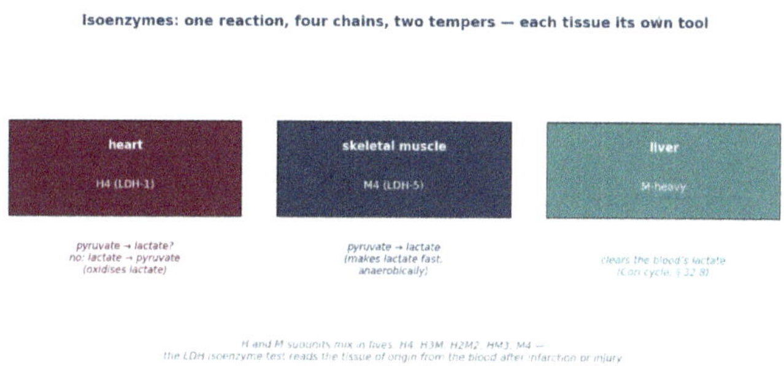

Figure 18.5 *Isoenzymes: the lactate dehydrogenases of heart (H4, oxblood — burns lactate), muscle (M4, navy — makes it fast), and liver (teal — clears it). One reaction, four chains, two tempers; the Cori cycle runs on the difference, and the clinical test reads the tissue of origin from the blood.*

18.10 Enzyme localisation

The fifth instrument is geography: the enzyme is where its substrates are, and the membrane is the regulator's border (Figure 18.6). The fatty-acid economy is the type specimen: the acids are made in the cytosol (the fatty-acid synthase of § 36.3) and burned in the mitochondrion (the β-oxidation of § 35.4), and the carnitine shuttle of § 35.3 is the customs post between the two economies — the membrane ensuring that the synthesis and the degradation cannot run at full speed in the same room, the reciprocal regulation of § 30.7 enforced by the address. The other compartments have their own jurisdictions: the glycosylation enzymes in the Golgi (the assembly line of § 21.9), the proteases in the lysosome (the cell's stomach, kept at pH 5, and the enzymes inactive at the cytosol's 7.2 — a regulation by pH, address and zymogen together), the replication machinery in the nucleus. The reader should note that localisation is the cheapest regulation of all — no allostery, no covalent switch, no protein turnover; just a membrane and an address — and that the membrane's transporters (Part V) are the regulator's customs officers, deciding what crosses the border and when.

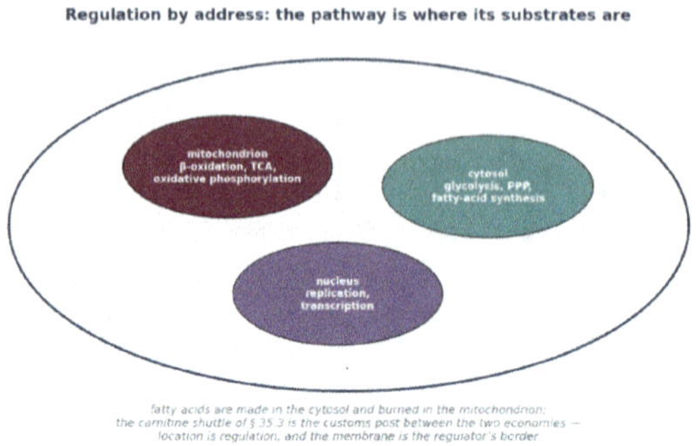

Figure 18.6 *Regulation by address: the cell's compartments (mitochondrion, oxblood; cytosol, teal; nucleus, plum) are its regulatory geography — fatty acids made in one room, burned in another, and the membrane's transporters are the customs officers at the border.*

18.11 Regulation by protein turnover

The last instrument is the slowest and the most thorough: the enzyme is regulated by its abundance — made and destroyed, on the clock of the gene and the proteasome. The synthesis is the gene's business (Chapter 48's transcriptional regulation — the slowest, most global switch, hours to days); the destruction is the proteasome's (§ 11.12's ubiquitin sentence — minutes to hours). The turnover's virtues are its thoroughness (the enzyme is not switched but removed) and its irreversibility (the removal cannot be undone by any substrate flood), and its costs are its slowness and its price (the enzyme must be re-made, at the ribosome's cost, when the road re-opens). The instruments of turnover are used where the regulation must be deep and durable: the HMG-CoA reductase of § 36.9, the cholesterol pathway's gate, is degraded when cholesterol is high (the sterol-sensing machinery accelerates its ubiquitination — the pathway's product, regulating the enzyme's lifetime, feedback by turnover); the cyclins of § 51.2 are made and destroyed on the cell cycle's clock (the enzyme's existence being the signal). The reader should close the chapter with the full switchboard in view: allostery (seconds, cheap, local), phosphorylation (minutes, amplified, global), zymogens (once, one-way), isoenzymes (inherited, tissue-tuned), localisation (address), turnover (hours, thorough) — six instruments, one machine, and the whole of metabolic regulation is the cell playing them together.

Plate 18.1 *Edwin G. Krebs (1918–2009), who with Edmond Fischer discovered reversible protein phosphorylation — the most general regulatory mechanism in biology, and the transistor of Part X's circuits. Portrait from the Nobel Foundation archive, nobelprize.org (Nobel Prize in Physiology or Medicine, 1992), reproduced from the Foundation's publicly distributed image.*

Historical note — The discovery of feedback, 1956–57

That the end product of a pathway could inhibit its first enzyme was discovered independently and nearly simultaneously by two laboratories: Umbarger's, at the NIH (isoleucine inhibiting threonine deaminase, 1956 (Umbarger, 1956)), and Yates and Pardee's, at the Pasteur and Berkeley (the pyrimidine pathway's feedback, 1957). The discovery was received with the resistance that a new logic always meets: the enzyme was supposed to be a machine with a speedometer, not a throttle — and the idea that the machine could read the product of its own pathway, five steps downstream, seemed to require a signalling that nobody had imagined. The allosteric site supplied the mechanism (§ 18.1), Monod's year of 1965 supplied the theory, and the feedback circuit is now the first circuit every biochemist learns. The reader should note the discovery's deeper consequence: it established that the pathway is not a sequence of reactions but a system — regulated, self-reading, homeostatic — and the systems biochemistry of Chapter 52 is Umbarger's feedback, writ large.

Chapter 18 — Conceptual synthesis

The cell's switchboard is complete. Allostery is the pathway's own throttle — the MWC election with a job, and the sigmoid the switch's signature (§§ 18.1–18.2); feedback is the pathway reading its product at its gate (§ 18.3); feed-forward is the pathway priming its far end (§ 18.4); phosphorylation is the organism's ratchet — irreversible, amplified, connected (§ 18.5); the other inks — acetyl, methyl, ubiquitin — have their own jurisdictions (§ 18.6); the zymogen is the one-way switch, address and timing in one molecule (§§ 18.7–18.8); the isoenzyme is regulation by heredity, each tissue its own temper (§ 18.9); localisation is regulation by geography (§ 18.10); and turnover is regulation by existence — the enzyme made and destroyed on the clock (§ 18.11). Direction checked against Figure ii: the machine is measured, jammed, and regulated. Chapter 19 completes Part IV with the cofactors and the mechanisms — the coenzymes, the metals, and the representative enzymes in which the whole of the Part's chemistry is seen working.

Worked example 18.1 — The cascade's amplification

The glycogen cascade: adrenaline (1 molecule) activates ≈ 100 receptors' worth of G-protein cycles; each adenylyl cyclase makes ≈ 1 000 cAMP; each PKA phosphorylates ≈ 100 phosphorylase kinases; each of those, ≈ 100 phosphorylases; each phosphorylase releases ≈ 100 glucose-1-phosphates per second. The product: $1 \times 10^2 \times 10^3 \times 10^2 \times 10^2 = 10^9$ *— one hormone molecule, a billion glucose molecules mobilised per second: the amplification of Figure 18.3, computed, and the reason the phosphorylation cascade is the fastest loud switch in the body.*

Worked example 18.2 — The sigmoid's amplification

An allosteric enzyme with $n_H = 2.5$ *rises from 10 % to 90 % activity over what signal range? From the Hill equation: the 10–90 % window is* $[S]_{90}/[S]_{10} = 81^{1/n} = 81^{0.4} \approx 5.8$*-fold — against the hyperbola's 81-fold: the sigmoid compresses the switch's throw by a factor of 14, and the physiological window of Figure 18.2 is the compression made visible: the allosteric enzyme answers a sixfold signal with a full response, the hyperbola answers it with a shrug.*

Worked example 18.3 — The zymogen's security

The pancreas makes ≈ 10 g of trypsinogen daily; premature activation of 1 % would release ≈ 100 mg of active trypsin in the gland. The gland's trypsin inhibitor binds trypsin at $K_d \approx 10^{-13}$ *M — tighter than avidin–biotin by two orders of magnitude (Table 8.1). The inhibitor's concentration suffices for the accidental few per cent; the pancreatitis of § 18.8 is the overflow of that insurance: the zymogen's two locks (the mask and the inhibitor) are the security, and the arithmetic of the failure is the price of the lock picked — the gland digesting itself, at the protease's own rate.*

Exercises (with full resolutions)

Exercise 18.1

A pathway has three sequential enzymes; the end product inhibits the first. If the inhibition halves the gate's rate when the product doubles, what happens to the intermediates' concentrations compared with inhibiting the third enzyme?

Resolution. *Inhibiting the gate (*E_1*) starves the whole pathway: intermediates fall, no waste. Inhibiting the last enzyme (*E_3*) blocks the exit while the gate runs: the intermediates accumulate — the product's inhibition of its own step would flood the pathway with its precursors, and the cell pays for the blocked exit in wasted intermediates. The exercise is § 18.3's deep rule: the control point is the first committed step, and the cell inhibits the gate, not the doors, because the gate's inhibition is the only one that saves the whole road.*

Exercise 18.2

Zero-order ultrasensitivity: a phosphorylation cycle with both enzymes saturated responds to a 10 % change in kinase/phosphatase ratio with a near-complete switch (§ 12.8). Estimate the effective Hill coefficient.

Resolution. *The effective Hill coefficient measures the 10–90 % window: zero-order cycles flip over ≈ 10 % ratio changes, against the hyperbola's 81-fold (≈ 190 % log-range): the effective n_H is ≈ log(81)/log(window ratio) ≈ log(81)/log(1.1) ≈ 46 — a switch sharper than any allosteric cooperativity, bought with the ATP of the cycle (Worked Example 12.2's budget): the ultrasensitivity of § 18.5 is the most extreme amplification of signal the cell possesses, and the price is the ratchet's fuel.*

Exercise 18.3

Hexokinase (muscle, K_m ≈ 0.1 mM) and glucokinase (liver, K_m ≈ 10 mM) phosphorylate glucose. At blood glucose 5 mM and 20 mM, compare the two enzymes' velocities relative to V_{max}.

Resolution. *Muscle: 5/5.1 ≈ 0.98 and 20/20.1 ≈ 0.995 — saturated always, a trap that never releases glucose. Liver: 5/15 ≈ 0.33 and 20/30 ≈ 0.67 — a gauge that responds to the fed state, storing glucose when it floods and sparing it when it does not. The isoenzymes of § 18.9 in one calculation: the muscle's enzyme is a trap (the muscle takes glucose in all states), the liver's is a sensor (the liver takes it only in plenty) — the tissue's metabolism, written in one K_m.*

Exercise 18.4 (interpretative)

"Allosteric regulation is superior to phosphorylation because it is cheaper and faster." Adjudicate.

Resolution (indicative). *The claim is true in its frame and false in its generality. Cheaper: yes — allostery costs no ATP, no second protein, no cascade. Faster: yes — the allosteric election runs at the protein's own conformational rate (§ 12.4), seconds or less. But the claim misses the instruments' offices: allostery is local (the pathway reads itself, and cannot read the organism's hormones) and unamplified (one effector, one enzyme); phosphorylation is global and amplified (one hormone, a billion targets, Worked Example 18.1) — and the cell uses each where its office fits: the pathway's homeostasis by allostery, the organism's command by phosphorylation. The adjudication is the chapter's division of labour: the two instruments are not rivals but clocks — the second's and the hour's — and the cell keeps both.*

Exercise 18.5

HMG-CoA reductase is degraded when sterols are high. Why is this feedback superior to simple allosteric inhibition for the cholesterol pathway, and what is its cost?

Resolution. *The cholesterol pathway is long, slow and expensive (twenty-odd steps, many ATPs); its product is dangerous in excess (the atherosclerosis of § 36.10) and its overproduction must be stopped durable, not transiently — an allosteric brake is reversed the moment the effector falls, but the sterol's abundance is a chronic condition, and the chronic condition wants the enzyme removed. The cost is the enzyme's re-synthesis (the ribosome's hours and*

ATPs) when the road re-opens — the price of durability. The exercise is § 18.11's logic: feedback by turnover is the slowest and the most thorough, and it is used where the regulation must last — the chronic governor, not the momentary brake.

Research-level propositions

Proposition 18.α. The isoenzymes of LDH (§ 18.9) are the classical case of tissue-specific tuning; the cancer metabolism of § 31.11 re-expresses the M4 form widely. Assemble the evidence that the isoenzyme switch is a driver of the tumour's metabolism or a passenger, and design the experiment (isoenzyme-specific inhibitors, § 17.4's allosteric class) that would separate the two. Proposition 18.β. The zymogen cascades of clotting and complement (§§ 18.7, 14.9) are one-way and self-amplifying; their control failures are the thromboses and the complementopathies. Formulate the kinetic theory of the cascade's threshold (how much first cut fires the whole cascade?), and test it against the published kinetics of the clotting factors. Proposition 18.γ. The sixth instrument (turnover, § 18.11) is energetically the most expensive; yet some enzymes are regulated by turnover whose allosteric regulation would be cheaper. Using the flux and cost data of § 55.6, compute the energetic price of turnover regulation for one such enzyme, and assess whether the durability premium (the irreversibility of removal) accounts for the cell's willingness to pay.

References (Chapter 18)

Gerhart, J. C. (1970). A discussion of the regulatory properties of aspartate transcarbamylase. Current Topics in Cellular Regulation, 2, 275–325.

Krebs, E. G., & Fischer, E. H. (1956). The phosphorylase b to a converting enzyme of rabbit skeletal muscle. Biochimica et Biophysica Acta, 20(1), 150–157.

Nelson, D. L., Cox, M. M., & Hoskins, A. A. (2021). Lehninger principles of biochemistry (8th ed.). W. H. Freeman.

Umbarger, H. E. (1956). Evidence for a negative-feedback mechanism in the biosynthesis of isoleucine. Science, 123(3202), 848.

CHAPTER 19

Coenzymes, Metals and Representative Mechanisms

CONTENTS OF THIS CHAPTER

In which the enzyme's hired instruments are introduced: the coenzymes — vitamins in working dress — each carrying a chemistry the twenty letters cannot perform; the metal centres, the cell's electricians and Lewis acids; and the representative mechanisms — the triad's proteases, the zinc's anhydrase, the Phillips lysozyme, and the acid–base ribonuclease — in which the whole of Part IV's catalytic doctrine is seen at work.

19.1 Coenzymes as molecular reagents

The enzyme of Part IV has been, until now, a protein; the complete enzyme is a protein plus its instruments. The coenzymes are the instruments: small organic molecules, bound to the enzyme (tightly, as prosthetic groups, or loosely, as cosubstrates), which carry the chemistry the twenty letters cannot perform — the hydride transfer, the radical rearrangement, the carbanion that water forbids, the acyl that must be swung between sites. The doctrine of the chapter is that the coenzyme is a reagent, not a cofactor in the dismissive sense: it participates in the chemistry (is consumed and regenerated, or swings between enzymes), and its chemistry is the reaction's own. The vitamins are the coenzymes' parts lists (Figure 19.1): the eight B-vitamins are each the raw material of one coenzyme, and the vitamin deficiency diseases are the coenzymes' failures — beriberi is thiamine pyrophosphate's silence (§ 19.5), pellagra is NAD's (§ 19.2), scurvy is the hydroxylase's ascorbate (§ 14.1's footnote) — the diet being, in the end, a coenzyme supply chain, and the vitamin's indispensability being the proof that the cell cannot make its own instruments and must import them.

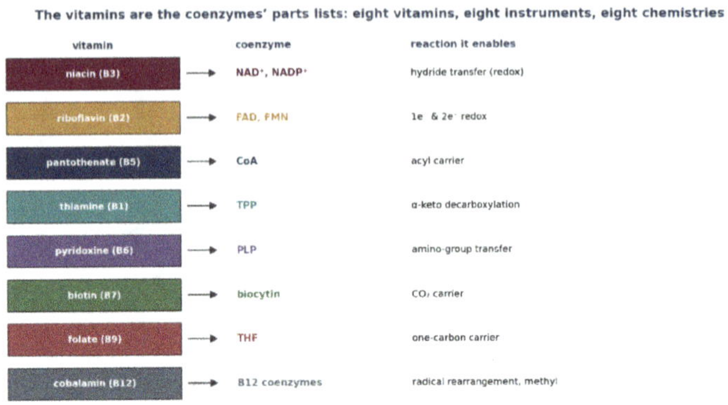

Figure 19.1 *The vitamins are the coenzymes' parts lists: eight B-vitamins, eight coenzymes, eight chemistries — hydride (NAD, FAD), acyl (CoA), carbanion (TPP, PLP), CO_2 (biotin), one-carbon (THF), radical and methyl (B_{12}). The deficiency diseases are the instruments' failures; the vitamin is the part the cell cannot make.*

19.2 NAD^+ and $NADP^+$

The first instrument is the electron's bucket: the nicotinamide ring of NAD^+, which accepts a hydride (H^-, a proton with both electrons) from the substrate, becoming NADH — the two-electron carrier of the catabolic fires (Figure 19.2). The ring's chemistry is the quinone-like conjugation of § 2.6: the positive charge of the oxidised pyridinium is quenched by the hydride, and the ring aromatises — the reduction being, in effect, the ring's neutralisation, and the wavelength of the new absorption (340 nm) being the spectrophotometer's window (§ 53.9's NADH assays, the enzymologist's daily bread). The two nicotinamide pools, NAD^+/NADH and $NADP^+$/NADPH, are kept separate, and the separation is the cell's accounting of reduction (§ 26.5): the NAD pool is held oxidised ($NAD^+ \gg$ NADH — a bucket always ready to collect, for the catabolic fires), the NADP pool is held reduced (NADPH $\gg NADP^+$ — a purse always full, for the anabolic shops of § 32.11) — two currencies of reduction, never exchanged freely, and the cell's redox bookkeeping is the balance of the two. The reader should note the generality: every dehydrogenase in Parts VI–VII is an NAD or NADP enzyme, and the nicotinamide ring is the most used coenzyme in the cell.

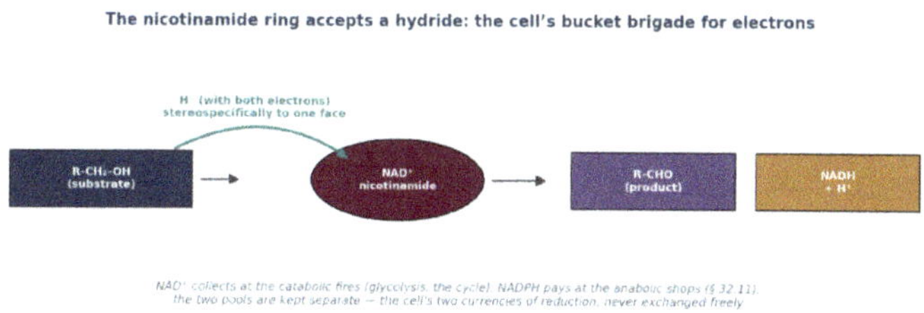

Figure 19.2 *The nicotinamide ring accepts a hydride (teal arrow) from the substrate, stereospecifically to one face: NAD^+ becomes NADH, and the substrate is oxidised. The two pools (NAD for catabolism, NADP for anabolism) are kept separate — the cell's two currencies of reduction.*

19.3 Flavin coenzymes

The second instrument is the electron's adapter: the flavin ring (FAD, FMN — riboflavin's working form), the yellow of the old "yellow enzyme", whose isoalloxazine ring can carry electrons one or two at a time — a versatility the nicotinamide cannot match. The flavin's three states — oxidised quinone, semiquinone radical (one electron), reduced hydroquinone (two) — make it the adapter between the two-electron world of NADH and the one-electron world of the respiratory chain's iron and quinones (§ 26.6): complex I of § 27.4 receives two electrons from NADH into its FMN and passes them one at a time down the iron–sulphur wire — the flavin being the exchange bureau at the boundary of the two economies. The flavin's colour is its diagnostic (the oxidised ring absorbs at 450 nm, the reduced does not), and its fluorescence is the cell biologist's reporter; but its doctrine is the adapter's: wherever the cell must change the currency of its electrons — two to one, or one to two — a flavin is the cashier.

19.4 Coenzyme A

The third instrument is the acyl handle: coenzyme A, pantothenate's working form, whose terminal thiol forms the thioester (Figure 19.4) — the activated acyl, with a hydrolysis free energy (≈ −31 kJ mol^{-1}) the equal of ATP's phosphoanhydride (§ 25.5's sibling in energy). The thioester's activation is the point: the acyl group of acetyl-CoA is a thousand times more transferable than the acetate of the free acid — the thioester activates the carbonyl to nucleophilic attack (the citrate synthase of § 34.3's condensation) and the α-carbon to enolisation (the claisen-condensations of the fatty-acid synthase, § 36.3) — and the long pantetheine arm swings the activated acyl between the active sites of the great complexes (the pyruvate dehydrogenase of § 34.1, where CoA is the receiving end of the swinging lipoyl arm). The reader should note the accounting: every acyl that must be transferred, activated, or parked in the cell is a thioester — acetyl, succinyl, malonyl, HMG, palmitoyl — and CoA is the cell's universal acyl carrier, the thiol being the handle, the arm being the swing, and the activation being the chemistry of the whole fatty-acid and citric-acid economies.

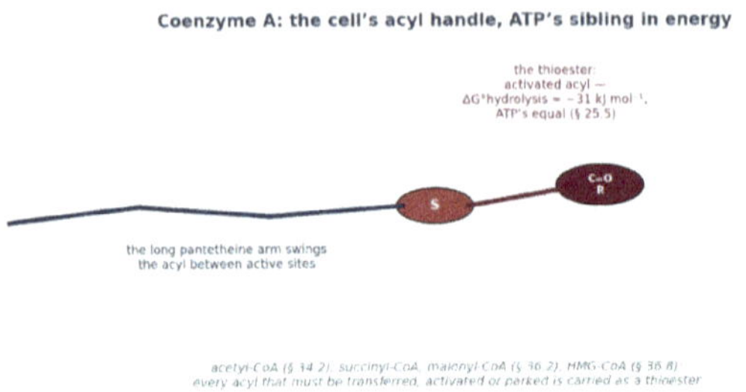

Figure 19.4 *Coenzyme A: the pantetheine arm (navy) ends in the thiol (crimson S), whose thioester (oxblood) is the activated acyl — transferable, enolisable, and as energised as ATP. Every acyl of the metabolic economy rides this handle.*

19.5 Thiamine pyrophosphate

The fourth instrument is the carbanion's cage: thiamine pyrophosphate (TPP, vitamin B1's working form), whose thiazolium ring has an acidic proton between the nitrogen and the sulphur — a carbon that ionises to a stable carbanion (an ylide), something the twenty letters cannot offer. The ylide attacks the α-keto acid's carbonyl, and the adduct decarboxylates: the TPP holds the resulting "active aldehyde" as a resonance-stabilised carbanion, which is then transferred (to the lipoyl arm of pyruvate dehydrogenase, § 34.1) or released (the pyruvate decarboxylase of the yeast, the fermentation of Eq. 1.1's last step). The TPP enzymes are the α-keto-acid chemists of the cell — the three great dehydrogenase complexes (pyruvate, α-ketoglutarate, branched-chain) and the transketolase of § 32.10 — and beriberi, the thiamine deficiency, is their silence: the nervous system, which runs on glucose and therefore on the pyruvate dehydrogenase complex, is the first organ to fail when the ylide is gone. The reader should note the chemistry's uniqueness: the stable carbanion is the TPP's alone, and the α-keto decarboxylations of metabolism are impossible without it — the thiazolium ring being, like the thiol of CoA, a chemistry the cell must import.

19.6 Pyridoxal phosphate

The fifth instrument is the amino group's switchboard: pyridoxal phosphate (PLP, vitamin B6's working form), whose aldehyde forms an aldimine (a Schiff base) with the amino acid's α-amino group, and whose pyridine ring is an electron sink (Figure 19.3) — the conjugated ring holding the carbanion that the amino acid's chemistry requires and water forbids. The PLP's versatility is the point: the aldimine's α-carbon has three bonds (to H, to COO^-, to R), and the enzyme chooses which to break by the orientation of the bond to the ring (the bond perpendicular to the ring breaks — Dunathan's rule) — transamination (§ 37.3's amino-group transfer), decarboxylation (the neurotransmitter amines: dopamine, GABA, serotonin), racemisation, β-elimination — a whole chemistry of the amino group, from one ring. The transaminases are PLP's shop (the ping-pong kinetics of § 16.8's figure, the PLP carrying the amino group from the keto-acid of one substrate to the keto-acid of the other), and the reader should note the coenzyme's second, unrelated job: glycogen phosphorylase (§ 33.3) uses its PLP as a general acid —

the phosphate of the cofactor shuttling the proton of the phosphorolysis — the one cofactor that does two jobs, and the proof that the instrument is a reagent, not a tag.

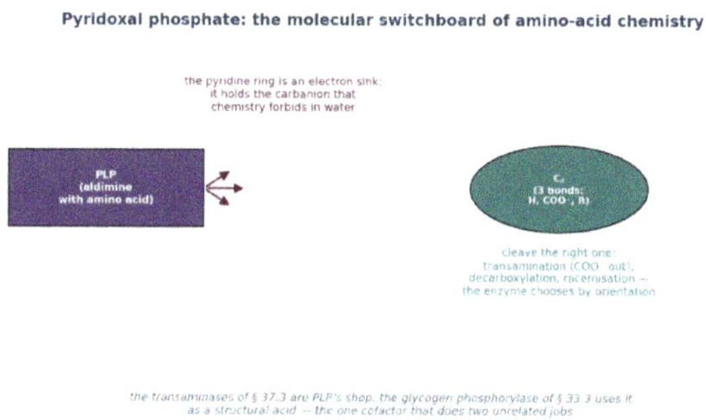

Figure 19.3 *Pyridoxal phosphate: the pyridine ring (plum) is an electron sink, holding the carbanion that water forbids; the aldimine's α-carbon (teal) has three bonds, and the enzyme chooses which to break by orientation — transamination, decarboxylation, racemisation, from one ring.*

19.7 Biotin

The sixth instrument is the CO_2 carrier: biotin, tethered to the carboxylase's lysine as biocytin (a swinging arm, like the lipoyl arm's), which carries the carboxyl from the bicarbonate-ATP complex to the substrate. The chemistry is the carboxylation of the anabolic economy: pyruvate → oxaloacetate (the anaplerotic pyruvate carboxylase of § 32.3), acetyl-CoA → malonyl-CoA (the committed step of fatty-acid synthesis, § 36.2), propionyl-CoA → methylmalonyl-CoA (§ 35.7's odd-chain end) — each carboxylation driven by one ATP, the CO_2 being carried by the biotin between the ATP site and the substrate site on the arm's swing. The reader should note the pattern that recurs through the chapter: the coenzyme is a swinging reagent — the lipoyl arm of § 34.1, the pantetheine arm of CoA, the biocytin arm of biotin — and the arm is the cell's answer to the multi-site enzyme: the chemistry is done at one address, the reagent at another, and the arm swings between them.

19.8 Tetrahydrofolate

The seventh instrument is the one-carbon carrier: tetrahydrofolate (THF, folate's working form), which carries one-carbon units at three oxidation states — methyl (−2), methylene (0), formyl (+2) — and delivers them to the biosynthesis of the nucleotides and the amino acids. The one-carbon economy is the chapter's most metabolic: the methylene-THF of thymidylate synthesis (§ 38.10's dTMP, and the methotrexate of § 38.14's antifolate), the formyl-THF of purine synthesis, the methyl-THF of homocysteine remethylation (the folate–B12 axis of § 19.9) — and the deficiency is the anaemia of the folate-poor, whose erythroblasts cannot make the thymidylate their division requires. The reader should note the carbon's three oxidation states: the THF is not one carrier but three, and the cell's one-carbon bookkeeping (methyl for methionine, methylene for thymidylate, formyl for purine) is the folate economy's three currencies — the instrument being a reagent with a portfolio.

19.9 Cobalamin

The eighth instrument is the strangest: cobalamin, vitamin B12, the largest and most complex of the coenzymes — a cobalt ion in a corrin ring (the porphyrin's cousin, one bridge fewer), with a cobalt–carbon bond, the only organometallic bond in human biochemistry. The B12 coenzymes do two chemistries. The radical rearrangement (methylmalonyl-CoA mutase, § 35.7's odd-chain end): the cobalt–carbon bond breaks homolytically, and the resulting radical rearranges the carbon skeleton — a chemistry of free radicals, controlled by the enzyme's pocket, that no other coenzyme performs. The methyl transfer (methionine synthase): the cobalt accepts the methyl from methyl-THF and passes it to homocysteine, remaking methionine — the folate–B12 axis, whose failure (the pernicious anaemia of B12 deficiency, and the methylfolate trap of § 38.4) is the anaemia with the neurological debt, the one deficiency that is two diseases. The reader should note the cobalt–carbon bond's uniqueness: organometallic chemistry, the chemist's speciality, exists in the reader's own cells in exactly one bond — and the bond is the oldest evidence that the cell's chemistry and the chemist's are, in the end, one.

19.10 Metal-dependent enzymes

The second class of instruments is the metals, and their offices are § 15.8's three, now itemised. The Lewis acid (Zn^{2+} of carbonic anhydrase, the metalloproteases): polarises the bound water, lowering its pK_a to ≈ 7, so the zinc-bound hydroxide is a nucleophile at the cell's pH — the metal converting water into a weapon (Figure 19.5, teal). The template and shield (Mg^{2+} of every kinase): holds the ATP's phosphates in the transfer geometry and screens their charges — the metal organising the reactants, § 15.8's second office. The redox centre (Fe^{2+}/Fe^{3+} of the cytochromes and the iron–sulphur clusters, Cu^{+}/Cu^{2+} of complex IV, Mn of photosystem II): accepts and donates electrons one at a time — the metal as the cell's electrician, and the respiratory chain of Part VI as its wiring diagram. The reader should note the metals' division of labour with the coenzymes: the coenzymes carry the organic chemistry (hydride, acyl, carbanion), the metals carry the inorganic (charge, electron) — and the two classes together are the enzyme's hired instruments, the chemistry the twenty letters cannot do alone.

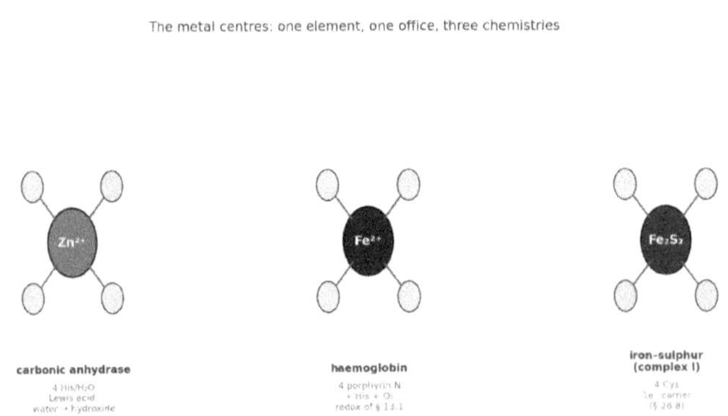

Figure 19.5 The metal centres: Zn^{2+} of carbonic anhydrase (teal, Lewis acid), Fe^{2+} of haemoglobin (oxblood, redox and ligand), the iron–sulphur cluster of complex I (navy, one-electron carrier) — one element, one office, three chemistries, and the trace elements of Figure 2.8 their payroll.

19.11 Serine proteases

The representative mechanisms begin with the chapter's most studied enzyme: the serine protease, whose catalytic triad of § 15.4 (Asp–His–Ser, the charge relay) is the textbook mechanism of Part IV. The mechanism is the two-step covalent cycle of § 15.6: the serine's hydroxyl, converted by the relay into a nucleophile, attacks the peptide's carbonyl; the tetrahedral intermediate (the oxyanion hole's two backbone N–Hs stabilising the negative charge — the transition-state complementarity of § 15.2, built into the fold) collapses, the amine leaves, and the acyl-enzyme is hydrolysed by the same relay backwards. The proteases are a family, and the family is the chapter's lesson in convergent evolution: the chymotrypsin fold (trypsin, chymotrypsin, elastase — the digestive proteases), the subtilisin fold (the bacterial protease), and several others have the same triad in different architectures — the Asp–His–Ser relay being, by the count of the structural databases, independently invented at least five times: the mechanism is so good that evolution keeps finding it, and the triad is convergent evolution's favourite enzyme.

19.12 Carbonic anhydrase

The second representative is the fastest: carbonic anhydrase, whose zinc-bound hydroxide attacks the CO_2, making bicarbonate — the hydration of § 6.11's respiratory chemistry, accelerated 10^6-fold to near the diffusion limit (Figure 15.6's ceiling). The mechanism is the zinc's Lewis-acid office (§ 19.10): the bound water's pK_a is lowered to ≈ 7, the hydroxide attacks the CO_2's carbon, and the bicarbonate is released; the enzyme's rate (≈ 10^6 s^{-1}) is set by the proton's departure from the zinc's water — the rate being limited not by the chemistry but by the proton shuttle (the histidine of § 15.5, and the buffer's relay) — and the catalysis is so fast that the diffusion of the substrate and the departure of the proton are the bottlenecks, the chemistry being, in effect, instantaneous. The enzyme is everywhere in the reader's physiology: the erythrocyte's CO_2 transport (§ 13.6), the kidney's acid secretion (§ 40.6), the stomach's HCl — and the inhibitor acetazolamide (a sulphonamide, binding the zinc) is the diuretic and the altitude drug of § 13.11, the enzyme's inhibition being the therapy of the mountain.

19.13 Lysozyme

The third representative is the first enzyme whose mechanism was solved by crystallography: lysozyme, the tears' and the egg-white's antibacterial enzyme, whose structure (Phillips, 1965 (Blake et al., 1965)) showed the active cleft with the polysaccharide's six sugars bound — and the mechanism read off the structure: the D-ring is strained toward the half-chair of the oxocarbenium transition state (the strain of § 15.9, photographed), the Glu35 protonates the glycosidic oxygen (general acid, § 15.5), and the Asp52 stabilises the oxocarbenium ion (electrostatic, § 15.7) — three of the toolbox's instruments in one mechanism, read from the crystal. The lysozyme mechanism is the chapter's proof that the structure teaches the chemistry: the active site's architecture is the mechanism, and the enzyme's textbook status is the structural biologist's first complete answer to the question "how does the enzyme work?" — the answer being, in this case, the whole of Chapter 15 in one cleft.

19.14 Ribonuclease

The fourth representative is Anfinsen's own protein: ribonuclease A, whose mechanism is the acid–base pair His12/His119 — two histidines of § 15.5's shuttle, one as the base (taking the 2'-OH's proton, making the 2'-oxygen the nucleophile), one as the acid (protonating the 5'-oxygen of the leaving RNA) — the RNA's phosphodiester backbone cleaved by the 2',3'-cyclic phosphate intermediate, then hydrolysed. The mechanism is the chapter's doctrine at its most economical: two histidines, one substrate, two proton transfers — no metal, no coenzyme, no covalent intermediate, just the general acid–base shuttle of § 15.5 at its simplest — and the protein is the folding case study of Chapter 10 and the mechanism case study of the present chapter: the enzyme that taught the subject the thermodynamic hypothesis also taught it the acid–base mechanism, and the two lessons are, in the end, one protein's.

19.15 Mechanistic comparison of enzyme families

The chapter closes with the comparison that is Part IV's summary: the four representative enzymes, against the toolbox of Figure 15.3. The serine protease uses proximity, covalent catalysis, the oxyanion hole's transition-state stabilisation, and the acid–base relay; the carbonic anhydrase uses the metal's Lewis acid and the proton shuttle; the lysozyme uses strain, general acid, and electrostatic stabilisation; the ribonuclease uses the acid–base pair alone. The mechanisms differ in their instruments and agree in their doctrine: every enzyme lowers the pass by binding the transition state, and every enzyme does it with the instruments of Chapter 15, in the architecture of Chapter 10 — the toolbox is finite, the combinations are not, and the enzyme families of the cell are the instruments' variations on one theme. The reader who has followed the treatise this far now possesses the complete account of the enzyme: the machine (Part III), the catalysis (Chapter 15), the measurement (Chapter 16), the jamming (Chapter 17), the regulation (Chapter 18), and the instruments (Chapter 19). Part V now turns to the second great class of biomolecules: the carbohydrates and lipids, and the membranes they build — the matter from which the machinery's house is made.

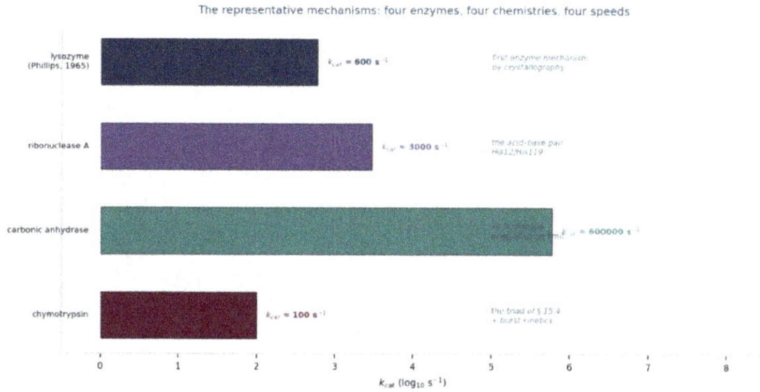

Figure 19.6 *The representative mechanisms: four enzymes, four chemistries, four speeds — lysozyme (structure's first mechanism), ribonuclease (the acid–base pair), carbonic anhydrase (near the diffusion limit), chymotrypsin (the triad). The toolbox is finite; the combinations are not.*

Hans Krebs

Plate 19.1 *Hans Krebs (1900–1981), whose citric acid cycle (1937) and urea cycle (1932) are the metabolic economy's two great circular engines — and whose acetyl-CoA chemistry is § 19.4's thioester at its most central. Portrait from the Nobel Foundation archive, nobelprize.org (Nobel Prize in Physiology or Medicine, 1953), reproduced from the Foundation's publicly distributed image.*

> **HISTORICAL NOTE — THE COENZYME THAT WAS A VITAMIN**
>
> *The coenzymes were found twice: once as vitamins, once as instruments. The vitamin era began with Funk's "vitamine" of 1912 (the beriberi factor in rice polishings — thiamine, § 19.5) and ran through the 1930s, when the B-vitamins were isolated one by one; the coenzyme era began with Warburg's NAD^+ (the "hydrogen-transferring coenzyme" of the 1930s, discovered in the red cell's metabolism) and Harden's "cozymase" of the fermentation (Buchner's juice, fractionated). The two eras met in the 1940s, when the coenzymes were found to be the vitamins' working forms: thiamine is TPP, niacin is NAD, riboflavin is FAD, pantothenate is CoA — the deficiency diseases being, at last, the instruments' failures. The reader should note the discovery's order: the disease was cured (the vitamin found), the instrument was found (the coenzyme isolated), and the two were connected last — the pattern of § 17.8's sulfa, repeated: cure first, mechanism after, and the vitamin's name a tombstone to the era when nobody knew what it was for.*

Chapter 19 — Conceptual synthesis

The instruments are itemised, and Part IV is complete. The coenzymes are the vitamins in working dress (§ 19.1): NAD's bucket for electrons (§ 19.2), the flavin's adapter (§ 19.3), CoA's thioester handle (§ 19.4), TPP's carbanion cage (§ 19.5), PLP's electron sink (§ 19.6), biotin's CO_2 arm (§ 19.7), THF's one-carbon portfolio (§ 19.8), B12's organometallic bond (§ 19.9); the metals are the Lewis acids, templates and redox centres (§ 19.10); and the four representative enzymes — the triad's protease, the zinc's anhydrase, the strained lysozyme, the paired *ribonuclease — are the toolbox's variations on one theme (§§ 19.11–19.15). Direction checked against Figure ii: Part IV — enzymes and biochemical catalysis — is closed: machine, catalysis, measurement, jamming, regulation, instruments. Part V now begins: carbohydrates, lipids and biological membranes — the matter from which the machinery's house is made.*

Worked example 19.1 — The two nicotinamide pools

The cytosol's $[NAD^+]/[NADH] \approx 700$ and $[NADP^+]/[NADPH] \approx 0.005$. The two pools' redox potentials: $E = E^{\circ\prime} + (RT/nF)\ln([ox]/[red])$ — for NAD, $E \approx -320 + (5.92/2)\log(700)$ mV $\approx -320 + 2.96 \times 2.845 \approx -311$ mV; for NADP, $\approx -320 + 2.96 \times \log(0.005) \approx -320 - 6.8 \approx -327$ mV — the NADP pool held ≈ 16 mV more reduced: the two currencies of reduction, priced in millivolts, and the cell's catabolic bucket always empty (NAD^+ ready), its anabolic purse always full (NADPH ready) — the separation of § 19.2, measured.

Worked example 19.2 — The thioester's activation

The hydrolysis of acetyl-CoA has $\Delta G^{\circ\prime} \approx -31$ kJ mol^{-1}, the same as ATP's phosphoanhydride. The condensation of acetyl-CoA with oxaloacetate (citrate synthase, § 34.3) has $\Delta G^{\circ\prime} \approx -32$ kJ mol^{-1} — the thioester's activation paying for the new C–C bond: the condensation is driven by the thioester's energy, and the cycle's first step is irreversible (§ 7.9's placement) precisely because the thioester's price is spent there. The acyl handle of § 19.4 is ATP's sibling in energy, and the citric acid cycle runs on it.

Worked example 19.3 — The zinc's pK$_A$ shift

Water's pK_a is 15.7; the zinc-bound water of carbonic anhydrase has $pK_a \approx 7.0$. The shift of 8.7 units corresponds to a stabilisation of the hydroxide by $\Delta\Delta G = 5.708 \times 8.7 \approx 49.7$ kJ mol^{-1} (§ 4.4's exchange rate) — the Lewis acid's office priced: the zinc converts water into hydroxide at the cell's pH, at a cost of ≈ 50 kJ mol^{-1} of binding energy, and the enzyme's 10^6-fold acceleration is the shift, measured.

Exercises (with full resolutions)

Exercise 19.1

Why does beriberi strike the nervous system first, when thiamine is required by every cell?

Resolution. The nervous system runs almost exclusively on glucose, and glucose's carbon enters the cycle only through the pyruvate dehydrogenase complex — a TPP enzyme (§ 19.5). Other tissues can burn fat (whose β-oxidation bypasses the complex); the neurones cannot. The deficiency therefore starves the organ that depends on the TPP enzyme most: the vitamin's silence is loudest where the instrument is busiest — the general rule of the deficiency diseases, and the reason the coenzyme map of Figure 19.1 is also a map of which organ fails first.

Exercise 19.2

Estimate the swinging distance of the lipoyl arm of pyruvate dehydrogenase, and its consequence for the complex's rate.

Resolution. *The lipoyl-lysyl arm is ≈ 14 Å of amide links, plus the E2 core's own arm — a total swing radius of ≈ 3–5 nm: the arm visits the three active sites (E1's TPP, E2's CoA, E3's FAD) without the intermediates leaving the complex — the substrate channelling of § 30.10, mechanised. The consequence: the complex's rate exceeds the free-diffusion limit of the individual enzymes, because the swinging arm carries the intermediates between the sites faster than the cytosol could — the arm being the cell's answer to the diffusion ceiling of § 16.10, at the price of the complex's architecture.*

EXERCISE 19.3

Compare the oxyanion hole of the serine protease with the zinc of the metalloprotease: two solutions to the same problem. What problem, and why two solutions?

Resolution. *The problem is the tetrahedral intermediate's negative charge (the oxyanion of the carbonyl's oxygen) — the transition state's charge, which must be stabilised (§ 15.2's complementarity). The serine protease's solution is the oxyanion hole (two backbone N–Hs, the fold's own dipoles); the metalloprotease's is the zinc (the metal's positive charge, holding the water and the substrate). Two solutions because the charge is universal and the instruments differ: the fold's backbone is free (the serine class), the metal must be imported (the zinc class) — and the two classes are the proof that the transition state's stabilisation is the enzyme's universal problem, soluble by the fold or by the payroll.*

EXERCISE 19.4 (INTERPRETATIVE)

"Coenzymes are vestiges of the RNA world: the catalytic part is the adenosine handle, not the working head." Assess the claim.

Resolution (indicative). *The claim is the RNA-world hypothesis of § 56.2's strongest structural argument, and it is largely right. NAD, FAD, CoA are all built on an adenosine handle (the nucleotide, which the enzyme grips) attached to a working head (the nicotinamide, flavin, pantetheine) — and the handle is inert while the head works: the handle is the fossil, the head the instrument. The interpretation: in the RNA world, the ribozymes used the nucleotide handles as their own chemistry (the catalytic RNA's cofactors), and when the proteins took over, the heads were retained as the instruments and the handles as the grips — the coenzymes being the oldest working fossils in the cell, the nucleotide that catalysed first and the chemistry that stayed. The claim's caution: the handle's universality is the evidence, and the universality is the argument — the coenzymes are the RNA world's payroll, still being spent.*

EXERCISE 19.5

Estimate the energetic cost of importing a vitamin versus making a coenzyme endogenously, and explain why the cell imports.

Resolution. *Making a coenzyme's head (the nicotinamide ring, say) endogenously costs a biosynthetic pathway of ≈ 5–10 enzymes, each an investment of the genome's and the ribosome's — the full cost of § 38.2's* *biosynthetic families, hundreds of ATPs and a pathway's worth of protein. Importing the vitamin costs one transporter and one meal. The arithmetic is the vitamin's definition: the instrument is cheaper to import than to make — and the cell that could rely on the supply (the plant-eater's environment, rich in the leaves that make the B-vitamins) lost the pathways, because the economy of the import won. The vitamin is the cell's outsourcing of its instrument manufacture, and the deficiency is the supply chain's failure.*

Research-level propositions

Proposition 19.α. The flavin's adapter role (§ 19.3) is canonical, but the semiquinone's stability varies a hundredfold between flavoproteins. Using the computed electrostatics of § 55.9, map the protein determinants of the semiquinone's lifetime across one flavoprotein family, and assess whether the adapter's one-electron/two- electron choice is set by the fold's electrostatics or by the substrate's. Proposition 19.β. The B12 radical chemistry (§ 19.9) is the cell's only controlled free-radical mechanism; the spin-trapping evidence is indirect. Design the experiment — time-resolved EPR with isotope-edited substrates (§ 54.5) — that would observe the adenosyl radical directly in methylmalonyl-CoA mutase, and estimate the radical's lifetime against the enzyme's turnover. Proposition 19.γ. The thioester's activation (§ 19.4) is priced from model thioesters; the enzyme-bound thioester's energy is disputed (some hold it is strained upward, others relaxed downward). Using the isothermal titration calorimetry of § 54.9, measure the transfer free energy of a complex-bound acyl-CoA directly, and assess whether the enzyme spends part of the thioester's energy on the swing.

References (Chapter 19)

Blake, C. C. F., Koenig, D. F., Mair, G. A., North, A. C. T., Phillips, D. C., & Sarma, V. R. (1965). Structure of hen egg-white lysozyme. Nature, 206(4986), 757–761.

Fersht, A. (2017). Structure and mechanism in protein science. World Scientific.

Berg, J. M., Tymoczko, J. L., Gatto, G. J., Jr., & Stryer, L. (2023). Biochemistry (9th ed.). W. H. Freeman.

Phillips, D. C. (1967). The hen egg-white lysozyme molecule. Proceedings of the National Academy of Sciences, 57(3), 484–495.

Walsh, C. T. (2015). Enzymatic reaction mechanisms. W. H. Freeman.

PART V

Carbohydrates, Lipids and Biological Membranes

Chapter 20 — Monosaccharides and Oligosaccharides · Chapter 21 — Polysaccharides and Glycoconjugates · Chapter 22 — Lipid Structure and Diversity · Chapter 23 — Biological Membranes · Chapter 24 — Membrane Transport

CHAPTER 20

Monosaccharides and Oligosaccharides

CONTENTS OF THIS CHAPTER

In which the sugars are given their alphabet: the aldoses and ketoses, the stereocentres that multiply as 2^n; the ring that closes on itself and the two anomers it bears; the glycosidic bond that bends (α) or extends (β), and with it the difference between food and fibre; the reducing sugar's redox window; and the wardrobe of derivations by which one glucose becomes the DNA's sugar, the chitin's amine, and the glycolysis's first prisoner.

20.1 Classification of carbohydrates

The carbohydrate is the most misunderstood molecule in the popular diet, and the most precisely classified in the biochemist's ledger. The definition is chemical: polyhydroxy aldehydes or ketones, and the compounds made from them — the sugars, their polymers, and their derivatives. The classification is by three independent measures. By size: the monosaccharides (one sugar), the oligosaccharides (two to ten), and the polysaccharides (many — Chapter 21's starch, cellulose, glycogen). By carbonyl: the aldoses (aldehyde sugars, the carbonyl at C1) and the ketoses (ketone sugars, at C2) — glucose is the aldose par excellence, fructose the ketose. And by carbon count: the trioses (3 C), tetroses (4), pentoses (5 — the nucleic acids' ribose and deoxyribose, § 43.1), hexoses (6 — glucose, fructose, galactose, mannose), and heptoses (7 — the pentose pathway's sedoheptulose, § 32.10). The reader should note what the classification reveals: the carbohydrate is not one molecule but a family with a formula — $C_n(H_2O)_n$, the "hydrate of carbon" of the old name — and the family's members are distinguished not by their composition but by their stereochemistry: the sugars are the stereochemist's playground, and the next section is the arithmetic of that playground.

20.2 Aldoses and ketoses

The two families divide at the carbonyl, and the division has a consequence the reader should carry: the aldose's carbonyl is at the end of the chain (C1), the ketose's is one carbon in (C2) — so the aldose has one more stereocentre than the ketose of the same length, and the two families have different shapes when they close their rings (§ 20.4). The aldoses multiply as 2^n (Figure 20.1): n stereocentres give 2^n aldoses — 2 trioses (one chiral pair, D- and L-glyceraldehyde, the reference pair of § 3.7's absolute configuration), 4 tetroses, 8 pentoses, 16 hexoses — and the D/L label refers, by Fischer's convention, to the OH of the last stereocentre (the one furthest from the carbonyl): right is D, and life, as in the amino acids of Chapter 3, has chosen one hand — the D-sugars, almost exclusively. The ketoses multiply as

2^{n-1} (one fewer stereocentre): 4 ketopentoses, 8 ketohexoses — and fructose, the sweetest and the most metabolic of them, is the ketose the reader eats. The two families interconvert through the enediol intermediate (the isomerases of § 31.4's triose-phosphate isomerase, and the non-enzymatic isomerisation of the Lobry de Bruyn–Alberda van Ekenstein transformation — the alkali's way of turning glucose into fructose, and the basis of the corn syrup industry's glucose isomerase), and the reader should note the isomerism's economy: the same six carbons, two carbonyls, and the cell's choice between them is the first decision of glycolysis.

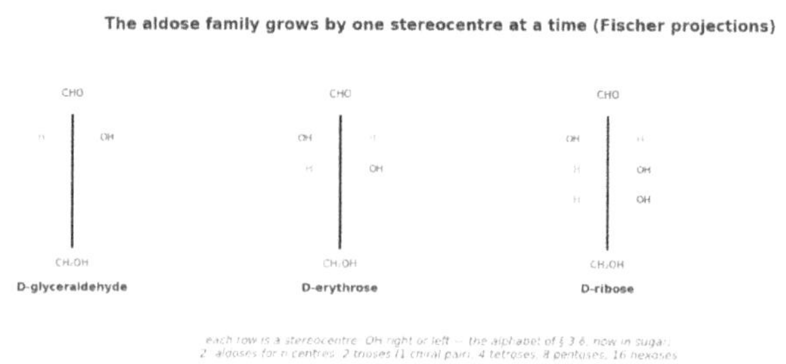

Figure 20.1 The aldose family in Fischer projections: D-glyceraldehyde (one stereocentre), D-erythrose (two), D-ribose (three), D-glucose (four) — the family growing by one stereocentre at a time, 2^n aldoses for n centres, the OH of the last centre (right, D) being life's chosen hand.

20.3 Stereochemistry of sugars

The sugar is a chain of stereocentres, and its stereochemistry is the whole of its identity: the 16 aldohexoses differ only in the up-down of their OHs, and glucose, mannose, and galactose — the three the reader metabolises daily — are three of the sixteen, differing at one or two centres each (glucose and mannose are epimers at C2; glucose and galactose at C4). The stereochemistry has two working consequences. The epimerases of § 33.9 (the UDP-galactose epimerase of the galactose pathway) interconvert the epimers enzymatically — the cell's way of turning one sugar into another at one centre, the cheapest isomerisation in the sugar's economy. And the stereochemistry is the recognition code of the glycobiology of § 21.10: the lectins (the sugar-reading proteins) read the up-down of the OHs as a fingerprint — concanavalin A reads mannose's, the influenza virus reads sialic acid's, and the cell-surface sugars of § 21.11 are the cell's own signature, written in stereocentres. The reader should note the economy of the code: the sugar's information is not in its composition (all hexoses are $C_6H_{12}O_6$) but in its stereochemistry — the up-down of the OHs being the sugar's spelling, and the lectin's reading of it being the cell's way of knowing its own address.

20.4 Ring formation

The sugar in solution is not the chain of the Fischer projection but the ring of Figure 20.2: the C5 hydroxyl attacks the C1 aldehyde (the hemiacetal formation of § 2.11's carbonyl chemistry), and the chain closes into the six-membered pyranose *ring (or, for fructose and the pentoses, the five-membered furanose). The closure has two consequences that the whole of Part V runs on. It creates a new stereocentre at C1 — the anomeric carbon — whose two configurations are the α (OH down, in the Haworth projection) and β (OH up) anomers, and the anomers are different sugars: different rotations, different solubilities, different enzymes. And it hides the carbonyl: the ring's anomeric carbon is the only place the carbonyl can re-emerge (the open chain of Figure 20.2's middle form, <0.02 % of the equilibrium), so the sugar's redox chemistry (§ 20.9's reducing window) is gated by the ring's opening — the ring being the sugar's default state, and the chain being the exception that the chemistry waits for.*

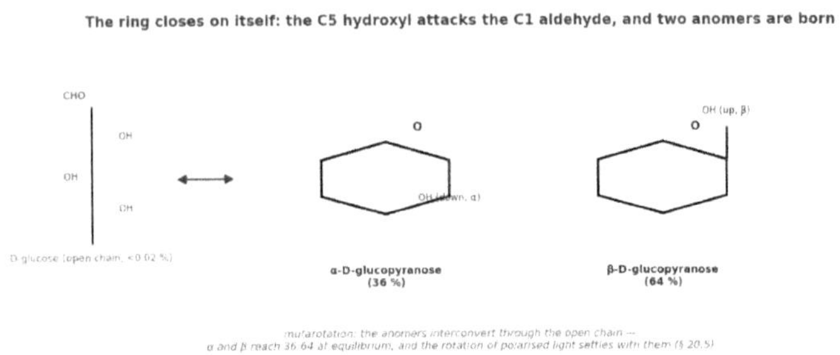

Figure 20.2 *The ring closes on itself: the open chain of D-glucose (centre, <0.02 %) and the two anomers it bears — α (navy, OH down) and β (oxblood, OH up) — reaching 36:64 at equilibrium. The mutarotation of § 20.5 is the anomers' interconversion through the open chain, and the anomeric carbon (C1) is the glycosidic bond's address.*

20.5 Anomers and mutarotation

The anomers interconvert through the open chain, and the interconversion is the mutarotation: a freshly dissolved α-D-glucose's optical rotation drifts from +112° to the equilibrium +52.7° as the β anomer accumulates to its 64 % — the polarimeter of Chapter 3's plate reading the ring's breathing. The mutarotation has two working consequences. The anomer is not fixed in solution (the α and β exchange in minutes at neutral pH, faster in acid or base), so the sugar's stereochemistry at C1 is a statistical property, not a fixed one — and the enzymes that read the anomer (the glycosidases of § 21.1, which are α- or β-specific) must wait for the right anomer or take it as it comes. And the glycosidic bond (§ 20.7) locks the anomer: once the C1 OH is replaced by the glycosidic link, the anomer can no longer exchange — the α-link stays α, the β-link stays β, and the polymer's shape (§ 20.7's bending and extending) is fixed at the moment of the bond's formation. The mutarotation is therefore the sugar's last freedom: the ring breathes, the anomer exchanges, and the glycosidic bond is the breath stopped.

20.6 Sugar derivatives

The sugar is the substrate of a wardrobe (Figure 20.5), and the derivations are the sugar's second chemistry. Deoxygenation makes the DNA's sugar (2-deoxyribose, § 43.1 — the ribose stripped of its 2'-OH, and the reason DNA is more stable than RNA: no 2'-OH to attack the backbone, § 43.6's chemistry). Amination makes the amino sugars (glucosamine, galactosamine — the chitin of § 21.4 and the glycocalyx of § 21.11), and acetylation of the amine makes the N-acetyl sugars (the peptidoglycan of § 1.3's murein, the sialic acid of § 21.9). Oxidation makes the sugar acids (glucuronic acid — the detoxification sugar of § 33.10, which the liver attaches to its poisons to make them soluble). Phosphorylation makes the metabolic prisoners (glucose-6-phosphate — the glycolysis's first step, § 31.3's trap, which keeps the sugar in the cell). And the sialic acids (NANA, nine carbons, a pyruvate fused to a mannosamine) are the cell-surface signatures of § 21.9 — the sugar's most decorated form, and the influenza virus's chosen address. The reader should note the wardrobe's economy: one glucose, six derivations, six new jobs — the sugar as raw material, not merely as fuel, and the carbohydrate as the cell's most versatile carbon.

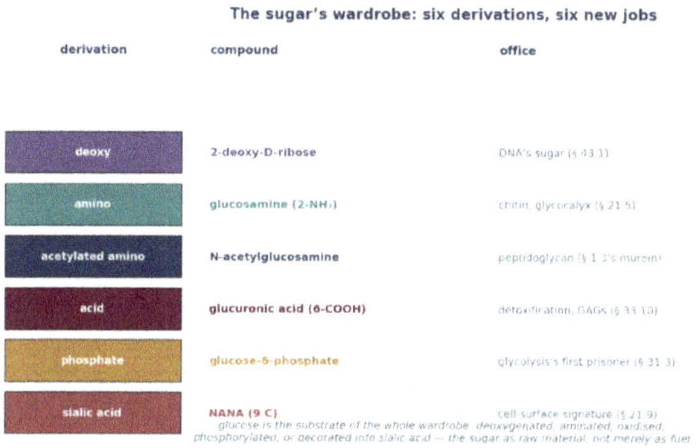

Figure 20.5 *The sugar's wardrobe: six derivations (deoxy, amino, acetylated amino, acid, phosphate, sialic acid) and their offices — the DNA's sugar, the chitin's amine, the detoxification acid, the glycolysis's prisoner, the cell-surface signature. One glucose; six new jobs.*

20.7 Glycosidic bonds

The glycosidic bond is the sugar's way of joining, and it is the most consequential bond of Part V: the anomeric carbon's OH, replaced by another sugar's OH (or by any alcohol), forms the acetal link — and the bond's anomeric configuration (α or β, locked at formation, § 20.5) is the polymer's shape. The two configurations are the two great polysaccharides of Chapter 21 (Figure 20.3): the α-1,4-link (maltose, starch) bends the chain into a helix — the storage polymer, compact and soluble, read by the α-amylases; the β-1,4-link (cellobiose, cellulose) extends the chain straight — the structural polymer, which hydrogen-bonds its neighbours into the insoluble fibre of § 21.3, the most abundant organic molecule on Earth. The bond's chemistry is the acetal's stability: the glycosidic link is stable to alkali (the acetal does not hydrolyse) and labile to acid (the glycosidases and the stomach's HCl cleave it) — and the bond's formation is driven by the sugar nucleotides of § 21.2's UDP-glucose (the activated sugar, whose glycosidic bond is made at the nucleotide's expense, the glycosyl transfer being the cell's way of paying for the link). The reader should carry the bond's doctrine: the anomeric configuration of the glycosidic bond is the difference between food and fibre — one stereocentre, two polymers, and the whole of the plant's and the reader's carbohydrate economy is the choice between them.

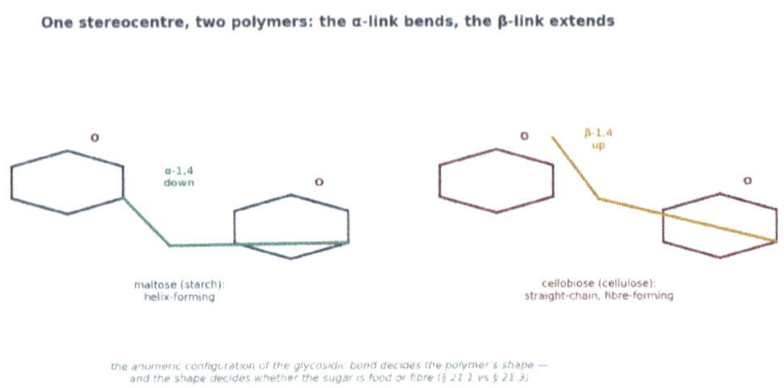

Figure 20.3 *The glycosidic bond: the α-1,4-link (teal, maltose) bends the chain into the storage helix; the β-1,4-link (amber, cellobiose) extends it into the structural fibre. One stereocentre, two polymers — the bond's anomeric configuration decides the polymer's shape, and the shape decides whether the sugar is food or fibre.*

20.8 Disaccharides

The disaccharides are the glycosidic bond's first products (Figure 20.6), and the four the reader eats are the sugar economy's daily currencies. Maltose (Glc α1→4 Glc) is starch's unit, freed by the amylases of the saliva and the gut. Cellobiose (Glc β1→4 Glc) is cellulose's unit — and the reader has no cellulase (the herbivores' microbiomes supply it; the reader's salad passes through, the fibre of the diet being the gut bacteria's meal, § 40.7's fermentation). Lactose (Gal β1→4 Glc) is milk's sugar — and the lactase of the infant gut is lost in most of the world's adults (the lactose intolerance of § 33.9's footnote, the commonest enzyme polymorphism in the human population). Sucrose (Glc α1↔β2 Fru) is table sugar — and its bond, joining both anomeric carbons, is non-reducing (§ 20.9's window closed), which is why sucrose is the transport sugar of the plant: stable, unreactive, and carried in the phloem without glycating the plant's own proteins. The reader should note the disaccharide's biography: each is named for its anomeric bond, and the bond decides the sugar's fate — food, fibre, or freight.

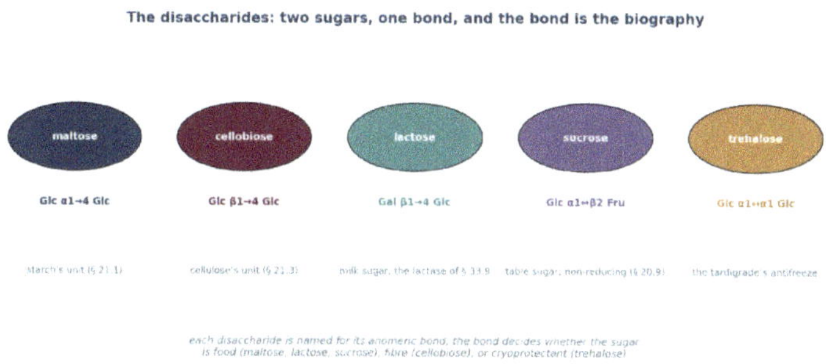

Figure 20.6 *The disaccharides: maltose, cellobiose, lactose, sucrose, trehalose — each named for its anomeric bond, and the bond is the biography: food (maltose, lactose, sucrose), fibre (cellobiose), or cryoprotectant (trehalose).*

20.9 Reducing sugars

The reducing sugar is the sugar with its window open: the free anomeric carbon (the hemiacetal of § 20.4, whose ring can open to the aldehyde or ketone) is the sugar's redox window, and the sugar that has one is "reducing" — it reduces the Fehling's or Benedict's Cu^{2+} to the brick-red Cu_2O (Figure 20.4), the oldest assay in clinical chemistry. The classification is the anomeric carbon's freedom: glucose, maltose, lactose are reducing (one free anomeric carbon); sucrose is not (both anomeric carbons locked in the α1↔β2 bond) — and the diabetic's urine test of § 42.1 is the reducing window read at the bedside (the glycosuria of the high blood sugar, read by the Fehling's of the last century and the dipstick's glucose oxidase of the present). The window has a dark side, and it is the glycation of § 42.1's HbA1c: the reducing sugar's aldehyde attacks the proteins' amino groups (the Maillard reaction of the cook's browning, and the patient's long-term glucose marker) — the same chemistry, invited in the kitchen and uninvited in the blood, and the reducing window being, in the end, the sugar's way of writing on proteins.

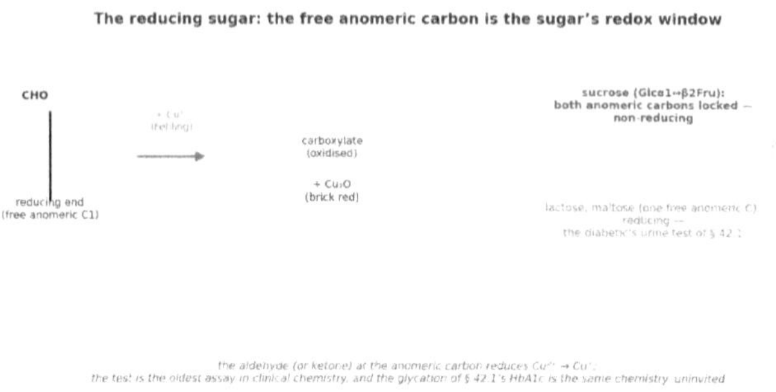

Figure 20.4 The reducing sugar: the free anomeric carbon (the CHO of the open chain) reduces Cu^{2+} to the brick-red Cu_2O — the oldest assay in clinical chemistry. Sucrose (plum) locks both anomeric carbons and is non-reducing; lactose and maltose (teal) have one free anomeric carbon and are reducing.

20.10 Oligosaccharide recognition

The chapter closes with the sugars as language: the oligosaccharides — short chains of a few sugars, attached to the cell's surface proteins and lipids — are the cell's signature, written in the sugars' stereochemistry and read by the lectins. The blood groups are the type specimen: the ABO antigens are oligosaccharides differing by one sugar (the A antigen adds an N-acetylgalactosamine, the B a galactose, the O neither — the glycosyl transferases of § 21.9's specificity being the blood group's genetics), and the transfusion's compatibility is the lectin's reading of the signature (the recipient's antibodies read the donor's sugars, and the wrong sugar is the wrong blood). The oligosaccharide code is denser than the amino acid code (the sugars branch, the amino acids do not: three sugars can join in dozens of ways, three amino acids in six), and its reading is the glycobiology of § 21.10's lectins — the cell's way of writing its address on its surface, and the reader should note the code's economy: the oligosaccharide is the cell's cheapest signature (a few sugars, no translation, no gene), and the blood group is the signature read in the clinic every day.

Chapter 20 — Conceptual synthesis

The sugar's alphabet is complete. The carbohydrate is a family with a formula, classified by size, carbonyl, and carbon count (§ 20.1); the aldoses multiply as 2^n, the ketoses as 2^{n-1}, and the D-hand is life's choice (§ 20.2); the stereochemistry is the sugar's identity and the lectin's reading code (§ 20.3); the ring closes on itself, the anomer is born, and the carbonyl is hidden (§ 20.4); the anomers exchange by mutarotation, and the glycosidic bond locks the breath (§ 20.5); the wardrobe of derivations makes the DNA's sugar, the chitin's amine, the glycolysis's prisoner (§ 20.6); the glycosidic bond bends (α) or extends (β), and the configuration is the difference between food and fibre (§ 20.7); the disaccharides are named for their bonds (§ 20.8); the reducing window is the anomeric carbon's redox gate, and the glycation is the same chemistry uninvited (§ 20.9); and the oligosaccharides are the cell's signature, read by the lectins (§ 20.10). Direction checked against Figure ii: the sugars' alphabet is in hand. Chapter 21 now assembles the polymers: starch, glycogen, cellulose, chitin, and the glycoconjugates — the sugars as store, structure, and signature.

WORKED EXAMPLE 20.1 — THE MUTAROTATION'S ARITHMETIC

Fresh α-D-glucose ($[\alpha]$ = +112°) equilibrates to 36 % α + 64 % β ($[\alpha]_\beta$ = +19°). The equilibrium rotation: 0.36 × 112 + 0.64 × 19 = 40.3 + 12.2 = +52.5° ≈ the measured +52.7° ✓ — the two anomers' weighted sum, and the polarimeter of Chapter 3's plate reading the ring's breathing as a single number. The time constant of the drift (minutes) is the open chain's lifetime, and the 36:64 ratio is the two anomers' free-energy difference ($\Delta G = -RT \ln(64/36) \approx -1.5$ kJ mol^{-1} favouring β — the β anomer's OH being equatorial, and the equatorial being, as always, the more comfortable chair).

WORKED EXAMPLE 20.2 — THE ANOMERIC CARBON COUNT OF STARCH

A starch chain of 1 000 glucoses has how many reducing ends? One — the chain's first glucose, whose C1 is free (the hemiacetal); the other 999 are locked in α-1,4 bonds. The chain's reducing power is therefore one part in a thousand — and the starch's Fehling test is nearly negative: the polymer's size hides its one window, and the reducing sugar assay measures not the polymer but its hydrolysis (the maltose and glucose freed by the amylases, each with its own anomeric carbon). The reducing end is the polymer's single address, and the amylase's job is to make more windows.

WORKED EXAMPLE 20.3 — THE INFORMATION DENSITY OF THE SUGAR CODE

Three hexoses (from 16 aldohexoses) can join in how many trisaccharides? Each sugar can be α or β (2 choices), each link can be 1→2, 1→3, 1→4, or 1→6 (4 regiochemistries), and the sugars can repeat: the count is ≈ $16 \times 2 \times 4 \times 16 \times 2 \times 4 \times 16 \approx 3 \times 10^4$ — against the three amino acids' 20^3 = 8 000 sequences: the sugar code is denser than the peptide code by branching, and the oligosaccharide signature of § 20.10 is the densest short-hand in the cell — the blood group's three sugars being, by this arithmetic, a library of thousands of possible signatures from which the cell chooses a few.

Exercises (with full resolutions)

EXERCISE 20.1

Draw the four D-aldotetroses, and identify which pairs are epimers.

Resolution. *The two stereocentres give 2^2 = 4: D-erythrose (RR), D-threose (LR), and their L-mirrors (LL, RL) — but the L-forms are the mirrors, so the D-family is two: erythrose (both OH right) and threose (left then right). The two are epimers at C2 (differing at the first centre, same at the second) — and the exercise is the aldose arithmetic of § 20.2 in miniature: the family is 2^n, the epimers are one centre apart, and the Fischer projection is the alphabet in which the family is written.*

EXERCISE 20.2

Why does DNA use 2-deoxyribose while RNA uses ribose, and what does the choice cost each?

Resolution. *The 2'-OH of ribose is the backbone's own attacker: it attacks the phosphodiester, cleaving the RNA strand (alkaline hydrolysis — RNA is alkali-*

labile). DNA's 2-deoxyribose removes the attacker: the backbone is a thousand times more stable — the archive's requirement (§ 43.1's principle: strong forces record). The cost is the catalysis: the 2'-OH is also a catalytic group (the ribozymes of § 15.11 use it), and DNA's stability is bought at the price of the RNA's catalytic versatility — the two sugars being the archive and the toolbox, and the 2'-OH being the price of admission to each.

EXERCISE 20.3

Estimate the free energy of the glycosidic bond's hydrolysis in maltose, and explain why the cell pays a nucleotide to make it.

Resolution. *The glycosidic bond's hydrolysis is ≈ −16 kJ mol^{-1} (favourable, like the peptide bond's) — the polymer is metastable (§ 2.10's kinetic persistence), and its formation from two free sugars is uphill by the same amount. The cell pays with UDP-glucose: the sugar-nucleotide's hydrolysis (≈ −30 kJ mol^{-1}, comparable to ATP) drives the transfer, the UDP leaving as the bond forms — the glycosyl transfer being the coupling of § 4.6's arithmetic, with the nucleotide as the spent currency. The glycosidic bond is cheap to break and expensive to make, and the nucleotide sugar is the cell's way of paying the difference.*

EXERCISE 20.4 (INTERPRETATIVE)

"Cellulose and starch differ only in the anomeric bond; yet one is the plant's wall and the other its store. Is the anomeric bond really the whole difference?" Adjudicate.

Resolution (indicative). *The bond is the first cause and not the whole story. The bond decides the shape: α bends (the starch helix, compact, soluble), β extends (the cellulose rod, straight). The shape decides the assembly: the rod hydrogen-bonds its neighbours (the cellulose fibre, insoluble, crystalline), the helix does not (the starch granule, hydrated, accessible). And the assembly decides the enzyme: the amylases read the helix, the cellulases the fibre — the reader has the first and not the second. The adjudication: the bond is the difference's origin (one stereocentre), and the shape and the assembly are its consequences — the treatise's doctrine of § 10.12, at the polymer's scale: structure decides assembly, assembly decides function, and the whole difference between the wall and the store is one stereocentre's two chairs.*

EXERCISE 20.5

The blood-group A transferase adds GalNAc; the B adds Gal. Estimate the number of possible trisaccharide antigens from three common sugars (Glc, Gal, GalNAc), and compare with the number actually used.

Resolution. *Three sugars, α or β, four regiochemistries each: 3 × 2 × 4 × 3 × 2 × 4 × 3 ≈ 1 700 possible trisaccharides. The cell uses a handful (the ABO antigens, the Lewis antigens, a few dozen in the whole glycocalyx) — the library of Worked Example 20.3 being mostly empty, and the used signatures being the glycosyl transferases' repertoire, not the chemistry's: the sugar code's density is the capacity, and the transferases' specificity is the cell's chosen vocabulary — the signature being, as always in this book, written by the enzyme and read by the lectin, with the chemistry offering far more words than the cell has chosen to speak.*

Research-level propositions

Proposition 20.α. The anomeric equilibrium of § 20.5 (36:64 α:β) is water's; the enzyme-bound glucose's anomer is disputed. Using the NMR of § 54.5, measure the bound anomer's configuration in hexokinase's active site directly, and assess whether the enzyme selects the anomer or waits for it. Proposition 20.β. The glycosidic bond's hydrolysis is acid-labile but alkali-stable; the glycosidases' mechanisms (retaining versus inverting) differ by one nucleophile. Formulate the kinetic theory of the two mechanisms (the covalent glycosyl-enzyme of the retaining class versus the direct water of the inverting), and design the experiment — the isotope-labelled water of § 16.11's KIE — that would distinguish them in one enzyme. Proposition 20.γ. The oligosaccharide code's density (Worked Example 20.3) is mostly unused; the glycocalyx uses a few dozen signatures. Using the glycomics of § 55.4, estimate the actual entropy of the cell-surface oligosaccharide repertoire in bits, and assess whether the unused capacity is a reservoir for evolution or a constraint of the transferases' specificity.

References (Chapter 20)

Collins, P. M. (Ed.). (1987). Carbohydrates. Chapman & Hall.

Nelson, D. L., Cox, M. M., & Hoskins, A. A. (2021). Lehninger principles of biochemistry (8th ed.). W. H. Freeman.

Berg, J. M., Tymoczko, J. L., Gatto, G. J., Jr., & Stryer, L. (2023). Biochemistry (9th ed.). W. H. Freeman.

Varki, A., Cummings, R. D., Esko, J. D., Stanley, P., Hart, G. W., Aebi, M., Darvill, A. G., Kinoshita, T., Packer, N. H., Prestegard, J. H., Schnaar, R. L., & Seeberger, P. H. (Eds.). (2017). Essentials of glycobiology (3rd ed.). Cold Spring Harbor Laboratory Press.

CHAPTER 21

Polysaccharides and Glycoconjugates

CONTENTS OF THIS CHAPTER

In which one and the same glucose, polymerised with a twist of a single bond, becomes either bread or timber; in which the liver learns to bank its sugar in a tree with four thousand doors; in which cartilage stands upright upon an osmotic argument; and in which the cell writes its name upon its own surface in sugar, for friend and pathogen alike to read.

21.1 The glycosidic decision: one glucose, two destinies

Chapter 20 closed with the glycosidic bond, the small ether bridge (§ 20.7) by which one sugar marries another. Marry a thousand, and a question arises that no monosaccharide ever had to face: what is the polymer for? Nature, who rarely wastes a dichotomy, has answered it twice over with the very same monomer. Glucose joined through its α-anomer becomes the storage polysaccharides — starch in the seed, glycogen in the liver: compact, hydrated, restless molecules built to be dismantled at a moment's notice. Glucose joined through its β-anomer becomes the structural polysaccharides — cellulose in the stem, chitin (with one acetyl amendment) in the beetle: extended, insoluble, patient cables built never to be dismantled at all. The monomer is identical; the linkage atoms are identical; the whole difference between dinner and the dining table is the configuration of a single carbon. It is the most consequential stereochemical pun in biology, and the present chapter is its exegesis.

The geometry deserves a moment of the reader's trust, for everything follows from it. In the α(1→4) linkage the glycosidic oxygen leaves C1 in the axial direction and arrives at the equatorial C4 of the next residue; the two bonds are not anti-parallel, and the chain must turn through roughly sixty degrees at every step. A chain that turns at every step closes upon itself: six residues complete one revolution, and amylose settles into the gentle helix of Figure 21.1 — a hollow, water-loving spiral some 0.8 nm in pitch. In the β(1→4) linkage both bonds leave equatorially, on opposite sides of the ring; the chain cannot turn, and instead each successive residue rotates a half-turn about the chain axis so that the ribbon runs flat and straight, glucose alternating with its own reflection. Flat ribbons do something helices cannot: they lie down side by side and hydrogen-bond to one another along their whole length, sheet upon sheet, until the polymer has disappeared into a crystalline cable in which not one hydroxyl is idle. The α-helix of the sugar world is a reservoir, open to water and to enzymes; the β-ribbon is a girder, closed to both. Storage and structure are not two inventions but one invention read in two configurations.[24]

The dialectic of the chapter is now visible in outline. Thesis: the helix (starch, glycogen, §§ 21.2–21.3). Antithesis: the ribbon (cellulose, chitin, murein, §§ 21.4–21.5). Synthesis: the sugar chain that refuses both economies and becomes neither store nor strut but a text — the oligosaccharides grafted onto protein and lipid, the glycoconjugates of §§ 21.6–21.8, in which carbohydrate chemistry achieves its third estate and rejoins the book's title at last: matter, energy, and now information, all in sugar.

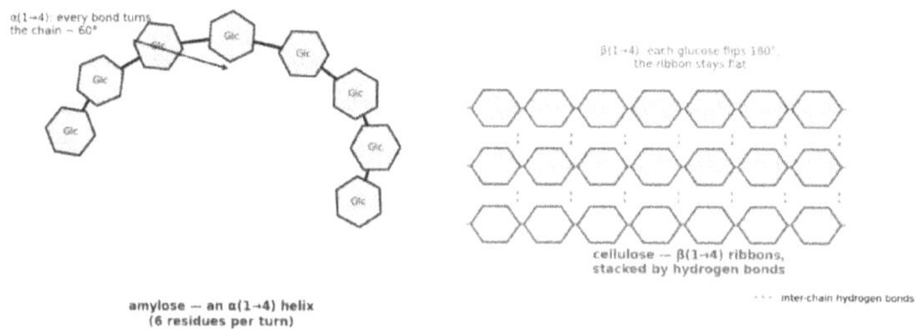

Figure 21.1 *The α bends, the β builds. Left: in the α(1→4) linkage the axial-to-equatorial bond turns the chain at every residue; six residues complete a turn and amylose spirals into a hydrated helix. Right: in the β(1→4) linkage each glucose flips 180° about the chain axis; the ribbon stays flat, and neighbouring ribbons hydrogen-bond (dashed teal) into the stacked sheets of the cellulose microfibril. One monomer, one linkage, two geometries — and two industries.*

21.2 Starch: the granule and the helix

The plant, whose glucose is made in daylight and spent at night, banks it in the granule we call starch — and banks it twice, in two polymers that answer to the same name. ***Amylose****, the minor partner (some 20–30% of most starches), is the unbranched α(1→4) chain of a thousand residues or more, wound into the helix of Figure 21.1.* ***Amylopectin****, the major one, is amylose with a branching habit: the same α(1→4) chains, but interrupted every twenty-four residues or so by an α(1→6) branch point, so that the molecule grows as a cluster of short double-helical twigs arranged in concentric lamellae — a semicrystalline granule, insoluble in cold water precisely because its outer chains have paired and crystallised, as the bacterial glycogen from which it evolved never troubled to do (Ball & Morell, 2003). The granule is thus a small masterpiece of packaging: a glucose warehouse of 10^5–10^6 residues, dense, osmotically silent — a million free glucoses would burst the cell; a million polymerised ones press upon it not at all (the colligative argument of § 6.6 in carbohydrate dress) — and yet digestible within minutes once the kitchen or the intestine applies heat or α-amylase.*

Heat first. Warming a starch suspension past some 60–70 °C — gelatinisation — melts the lamellae; the granule swells to many times its volume, amylose leaks out, and the cook's sauce thickens. Cool it, and the freed chains slowly find one another again and re-crystallise — retrogradation — which the baker meets as staling: yesterday's bread has not dried, it has crystallised, and a minute in the

24. Anselme Payen, who in 1838 washed wood free of everything that was not the cell wall and named the residue *cellulose*, had already noticed the scandal: cellulose, starch and dextrin share the same elementary composition, $(C_6H_{10}O_5)_n$. The formula, he showed, is one; the architecture — and therefore the destiny — is another's doing.

oven, melting the little crystals, restores it. The staling of bread is therefore a phase transition, not a loss of water — a fact worth more to the philosophy of cooking than many a treatise.

Then the oldest colour in the analyst's repertoire.[25] *Iodine does not stain starch at random: I_2 and I^- assemble inside the amylose helix into a linear polyiodide wire — I_3^-, I_5^- and longer — and the wire's delocalised electrons absorb according to its length. Figure 21.2 shows the beautiful consequence: the colour reports the helix. Amylose, with its uninterrupted spiral, shelters a long wire and drinks the orange-red — whence the blue-black of $\lambda_{max} \approx 620$ nm. Amylopectin, whose helices are clipped to twenty-odd residues between branches, shelters a shorter wire and shifts to ≈ 545 nm, a red-violet. Glycogen, more ruthlessly branched still, manages only a red-brown near 460–480 nm. Three polymers of one monomer, three colours — and a spectrophotometer that measures, in effect, the distance between branch points. The reader who has grasped Figure 21.2 has grasped that structure is colour, and may be trusted with the granule.*

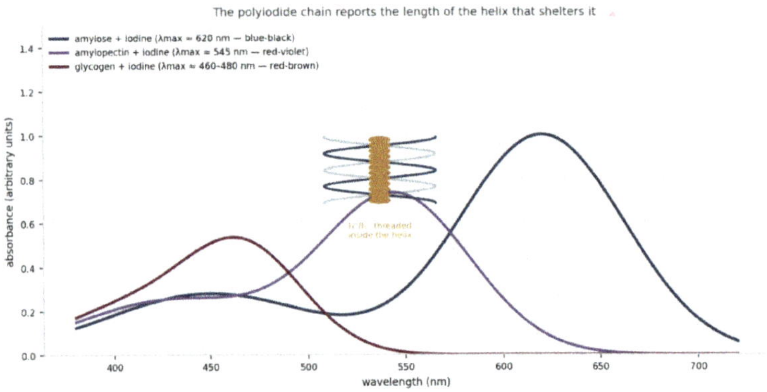

***Figure 21.2** The polyiodide ruler. Iodine threaded inside an α(1→4) helix forms a linear I_3^-/I_5^- wire whose absorption band depends on the wire's length — hence on the length of uninterrupted helix between branch points (inset). Amylose (long helix) absorbs at ≈ 620 nm (blue-black); amylopectin (helices of ~20–25 residues) at ≈ 545 nm (red-violet); glycogen (~10–13 residues between branches) at ≈ 460–480 nm (red-brown). The most venerable test in carbohydrate chemistry is, in disguise, a branching assay.*

21.3 Glycogen: a tree with four thousand doors

The animal cannot afford the granule. Its glucose must move — into the blood in a fast, into the muscle in a sprint — and an enzyme gnawing a single chain end delivers one glucose at a time, which is a bank with one cashier. The animal's answer, arrived at independently of the plant's and more extravagantly, is glycogen: the α(1→4) chain branched α(1→6) every eight to twelve residues, so that the molecule grows not as a cluster but as a tree (Figure 21.3). Let us count. If each chain averages $\bar{c}$ residues before it branches, and the granule holds N residues in all, the number of chains — and very nearly the number of outer, non-reducing ends, since each chain presents one — is simply

25. The blue of starch and iodine was described by Colin and Gaultier de Claubry in 1814 — among the first colorimetric tests in chemistry, and still in daily service two centuries later, which is more than can be said for most instruments of 1814.

$$E \approx \frac{N}{\bar{c}} \qquad (21.1)$$

— N residues partitioned into chains of $\bar{c}$; nothing deeper than bookkeeping, and yet it is the whole design. And if every tier doubles the chains, the number of tiers needed to contain N residues is set by $2^{t-1} \cdot \bar{c} \approx N$, that is,

$$t = 1 + \log_2 \frac{N}{\bar{c}} \qquad (21.2)$$

For a β-granule of $N \approx 55\ 000$ residues and $\bar{c} \approx 11$, Eq. (21.2) gives $t \approx 1 + \log_2 5000 \approx 13$ tiers — just what the electron microscope finds — and Eq. (21.1) gives $E \approx 5\ 000$ chains, of which the outermost two tiers alone hold nearly three-quarters of the ends. Against those thousands of non-reducing ends, the granule possesses exactly one reducing end, buried at the root, where it is of no use to anyone and can do no harm (§ 20.9). The architecture is a compromise struck by geometry: spherical for solubility, branched for parallelism, and rooted on a protein — ***glycogenin****, which primes its own glycan by glucosylating one of its own tyrosines and then extends the first chain, remaining forever at the core of every granule it founded. A bank, then, built around its own founder, with four or five thousand doors and one sealed vault.*

WORKED EXAMPLE 21.1 — THE ACCELERATION DUE TO TREES

Suppose glycogen phosphorylase works every non-reducing end at $k \approx 10$ residues s^{-1}, and the muscle orders the release of 10% of a 55 000-residue granule — 5 500 glucoses. (i) Through how many ends does the tree deliver? (ii) How long does the order take? (iii) How long would it take from an unbranched chain of the same mass?

Solution. *(i) Eq. (21.1): $E \approx 55\ 000 / 11 \approx 5\ 000$ chains; the outer tiers hold $\approx 0.75 \times 5\ 000 \approx 3\ 700$ available ends. (ii) $5\ 500$ glucoses / $(3\ 700 \times 10\ s^{-1}) \approx 0.15$ s. (iii) One end: $5\ 500 / 10 \approx 550$ s — over nine minutes. Branching buys a factor of about four thousand, and the geometry adds a refinement the arithmetic already suspects: because every branch point halves the remaining distance to the surface, some 40% of all residues lie within four residues of a non-reducing end. The depot is not merely parallel; it is mostly periphery. Nature, who had no bank clerk to spare, made the bank out of doors.*

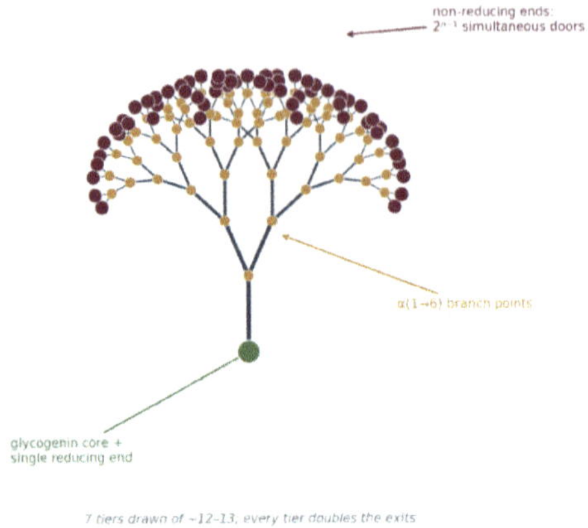

Figure 21.3 *The tiered architecture of glycogen (seven tiers drawn of the ~13 of a full β-granule). Each chain (navy) runs ~10–13 residues before an α(1→6) branch point (amber) forks it in two; the outer tier's non-reducing ends (crimson) are the only doors phosphorylase can use, and every tier doubles them [Eqs. (21.1)–(21.2)]. The single reducing end remains covalently attached to glycogenin (green) at the core.*

Plate 21.1 *Luis F. Leloir (1906–1987), whose sugar nucleotides (UDP-glucose, 1950) and whose glycogen synthase (1957) separated biosynthesis from degradation for the first time — the general principle beneath this and every later chapter of Part VI. Portrait from the Nobel Foundation archive, nobelprize.org (Nobel Prize in Chemistry, 1970), reproduced from the Foundation's publicly distributed image.*

Plate 21.2 *Gerty T. Cori (1896–1957), co-discoverer with Carl F. Cori of the Cori ester (glucose 1-phosphate, 1936) and of phosphorylase, the enzyme of § 21.3's demolition squad — the first woman to receive the Nobel Prize in Physiology or Medicine (1947). Portrait from the Nobel Foundation archive, nobelprize.org, reproduced from the Foundation's publicly distributed image.*

> ***Historical note — The enzyme that was not the architect***
>
> *Glycogen itself is older than its chemistry: Claude Bernard fished it out of liver in 1856–57 and named it for what it does — la matière glycogène, the sugar-generating substance.[26] Its synthesis and breakdown were long believed to be one reaction read forwards and backwards. The Coris, in St Louis, had by 1936 isolated glucose 1-phosphate — the "Cori ester" — and shown that phosphorylase makes glycogen from it in vitro; what the test tube does reversibly, the dogma concluded, the cell does in whichever direction it pleases. The conclusion was reasonable, universal, and wrong. In Buenos Aires, in a small institute kept alive by private charity, Luis Leloir and his companions — Caputto, Trucco, Cardini, Paladini — chasing a humble question about lactose, isolated in 1950 a substance no one had ordered: uridine diphosphate glucose. Sugar nucleotides turned out to be the currency of nearly all glycosyl transfer, and in 1957 Leloir and Cardini showed that the cell's true glycogen synthase spends UDP-glucose, not glucose 1-phosphate: synthesis and degradation are different roads with different tolls, separately policed (Chapter 33 will enjoy the policing). The enzyme that made glycogen's chemistry famous was a demolition contractor mistaken for an architect. Leloir, asked about his makeshift laboratory, observed placidly that with better instruments he might have discovered less — the shoestring, it seems, concentrates the mind. Stockholm, in 1970, concurred (Leloir, 1971).*

21.4 Cellulose: a cable built of ribbons

Turn the anomeric carbon over, and the sugar that was dinner becomes timber. Cellulose is the $\beta(1 \rightarrow 4)$ chain of Figure 21.1 — several thousand glucoses, each a half-turn from its neighbour, so that the disaccharide cellobiose, not glucose, is the

26. Bernard's glycogen was his first great discovery and the seed of his doctrine of the internal secretion: the liver, he argued, secretes glucose not into a duct but into the blood — an idea that quietly founded endocrinology.

true repeat. The ribbon is stiffened by hydrogen bonds along its own length; and because it is flat, ribbons pack upon ribbons into the **microfibril** *— in higher plants an elementary cable some 3.5 nm across of about three dozen chains — whose crystalline reaches admit neither water nor enzyme. The result is the most abundant organic polymer on Earth: something like 10^{12} tonnes of it, half the carbon of the biosphere's standing crop, and a tensile strength near a gigapascal — comparable, weight for weight, to steel, which is why trees may grow a hundred metres on a material spun from sugar and air.*

What is the cable for? Not, in the first instance, the tree's height but the cell's turgor. The plant cell is an osmometer (§ 6.6): its membrane presses outwards with a turgor Δp of half a megapascal or more, and the wall must contain it. Consider a cylindrical cell of radius r and wall thickness δ, and cut it along its axis: the pressure blows the two halves apart with force $\Delta p \times (2rL)$ on the projected area, while the wall resists along two seams of area δL each, bearing stress σ. Equilibrium of the half-cylinder, $2\sigma\delta L = \Delta p \cdot 2rL$, gives the hoop stress — the Young–Laplace law in the wall:

$$\sigma = \frac{\Delta p\, r}{\delta} \quad \text{(cylinder)}\,; \qquad \sigma = \frac{\Delta p\, r}{2\delta} \quad \text{(sphere)} \tag{21.3}$$

— the sphere's derivation differing only in that its curvature closes in two directions, halving the burden, as § 23.2 will remind us when vesicles take up the same law. The microfibrils are wound around the cell precisely so as to lie across the hoop stress; the wall is a pressure vessel wound by a machine that reads no drawings.

Worked example 21.2 — The wall that holds the water

A spherical plant cell of radius r = 20 μm bears a turgor Δp = 0.8 MPa. (i) What stress must the wall sustain? (ii) Is a 100-nm wall of cellulose ($\sigma_{max} \approx 1$ GPa) sufficient?

Solution. *(i) Eq. (21.3): $\sigma = (0.8 \times 10^6 \text{ Pa} \times 20 \times 10^{-6} \text{ m}) / (2 \times 100 \times 10^{-9} \text{ m}) = 16 / (2 \times 10^{-4}) \times 10^0 = 8 \times 10^7 \text{ Pa} \approx 80$ MPa. (ii) The fibril's gigapascal leaves a safety factor of ≈ 12 — and real walls, at 100–200 nm with cross-wound fibrils, do rather better. The reader should pause over the number: a thin balloon of carbohydrate, ten thousand times thinner than the cell it girdles, routinely contains eight atmospheres. Plants are hydraulic machines, and cellulose is their steel — at a fifth of steel's density, and self-assembling.*

Against so admirable a cable the animal kingdom has, on the whole, declined to engineer. No animal genome encodes a cellulase worthy of the name; the $\beta(1\rightarrow4)$ bond between glucose units defeats every glycosidase we possess (§ 20.7's geometry again: the bond our amylases kiss is axial, and cellulose offers only equatorial). The herbivores' celebrated solution is to keep the engineers rather than the engineering: the rumen, the caecum, the termite's flagellate menagerie — fermentation vats in which microbes do the hydrolysis and the host skims the volatile acids (Chapter 26 will count the ATP). A cow, on this view, is less an animal that eats grass than a greenhouse that walks. And the biofuel economist's dream — cellulose is the cheapest sugar on Earth — founders on the same crystal: the pretreatment costs what the sugar saves. Proposition 21.a invites the reader to fix that.

21.5 Chitin and murein: armour in sugar

Cellulose's two great amendments prove the same rule. Replace the C2 hydroxyl by an acetamido group and the ribbon gains both a hydrogen-bond donor and acceptor along its flank: the sheets now hydrogen-bond in three directions, and the result is ***chitin*** *— β(1→4) GlcNAc — the second most abundant polysaccharide, the cuticle of every arthropod and the wall of every fungus. A lobster is, chemically speaking, a crystallised sugar solution that objects to being eaten; and the objection is the chemistry — chitinase is as rare among animals as cellulase, though the fungi's enemies (plants, for one) have learned to keep it. Deacetylate chitin and one has chitosan, the rare cationic polysaccharide, of which more in the pharmacy than the field.*

The bacteria, not to be outdone, built their wall of the same ribbon with a lactic-acid wart on C3: N-acetylmuramic acid alternates with GlcNAc in β(1→4), and short peptide stems — L-Ala–D-Glu–L-Lys (or meso-diaminopimelate)–D-Ala–D-Ala — hang from the muramyl residues and are knitted into cross-links by the transpeptidases. The whole bacterium is thereby enclosed in a single bag-shaped molecule, the ***murein sacculus****: one macromolecule the size of the cell, a fishnet stocking of sugar and peptide, bearing the same turgor problem as Worked Example 21.2 and solving it the same way. The D-amino acids are not whimsy: no protease evolved against L-peptides will touch them. Two of medicine's oldest allies attack precisely here. Lysozyme — the enzyme Fleming found in his own tears before he found penicillin in a mould — hydrolyses the β(1→4) of the sacculus; penicillin, a molecular impersonation of the D-Ala–D-Ala terminus, acylates the transpeptidase and leaves the wall half-knitted (Exercise 21.3 draws the lethal consequence); vancomycin simply sits upon the D-Ala–D-Ala and blocks the seam — until the enterococcus learns to finish in D-Ala–D-lactate instead, an ester for an amide, and the drug slides off: resistance by a single atom of oxygen.*

21.6 The glycan coat of proteins: N- and O-linked sugars

So far the sugar chain has carried the economy: storage, structure. Now the economy inverts, and the protein carries the sugar. A remarkable fraction of the proteome — most secreted and membrane proteins — wears oligosaccharides it did not choose and often cannot function without; the cell's glycosylation machinery is, after protein synthesis itself, the largest modification industry in the body. The chemistry is Chapter 20's; the novelty is the tailoring.

The ***N-linked*** *glycan is assembled not on the protein but on a lipid — dolichol pyrophosphate, a long isoprenoid anchor in the endoplasmic-reticulum membrane — where a fourteen-sugar precursor, $Glc_3Man_9GlcNAc_2$, is built sugar by sugar (several from the dolichol donors, the rest from Leloir's nucleotide sugars), flipped across the bilayer, and transferred en bloc by oligosaccharyltransferase to the asparagine of the sequon Asn–X–Ser/Thr (X ≠ Pro) as the chain emerges from the ribosome. What follows is quality control conducted in sugar. Three glucoses are trimmed; while one remains, the glycan is held by calnexin and calreticulin — lectin-chaperones that read glucose as a folding signal — and a remarkable enzyme, UDP-glucose:glycoprotein glucosyltransferase, re-glucosylates any chain still exposing hydrophobic, unfolded surface. The sugar is thus a timestamp and a ticket: folded, the protein loses its glucose and proceeds; unfolded, it is stamped again and retained; hopeless, it is marked by mannose trimming and escorted to the proteasome (Exercise 21.5). In the Golgi the survivors are finished into the three classes of Figure 21.4: some glycans keep their mannoses (high-mannose), some are cut back and rebuilt with GlcNAc, galactose and terminal sialic acid* *(complex), some receive both treatments on their two arms (hybrid) — one core pentasaccharide, $Man_3GlcNAc_2$, three destinies, and a symbolism (the SNFG cartoons of the figure) in which the cell's whole glycan vocabulary can be written.*[27]

The ***O-linked*** *glycan is humbler and denser: no precursor, no consensus — a GalNAc is simply pressed onto a serine or threonine in the Golgi and extended. Where O-glycans cluster by the hundred on one protein they make the mucins, whose bottle-brush of sugar binds water into mucus — the lubricant and rampart of every airway, gut and cervix; influenza's first opponent is not an antibody but a slime. And a third, quieter modification — O-GlcNAc, single sugars on nuclear and cytosolic proteins, added and removed like phosphorylation — will return when signalling does (Part VII). What are the coats for? For folding and its policing, as above; for armour — a glycosylated surface sheds proteases; and, most commercially, for time: the liver's asialoglycoprotein receptor clears from the blood any glycoprotein whose sialic acids have been lost, so that a therapeutic glycoprotein's serum half-life is written in its sugars. Erythropoietin stripped of sialic acid lasts minutes; the pharmaceutical house that adds two more antennae sells a longer-lasting molecule — glyco-engineering in the most literal sense. When the machinery itself fails, the congenital disorders of glycosylation — now more than a hundred and thirty monogenic diseases — testify by their severity (development, clotting, vision, immunity all at once) that the coat is not decoration.*

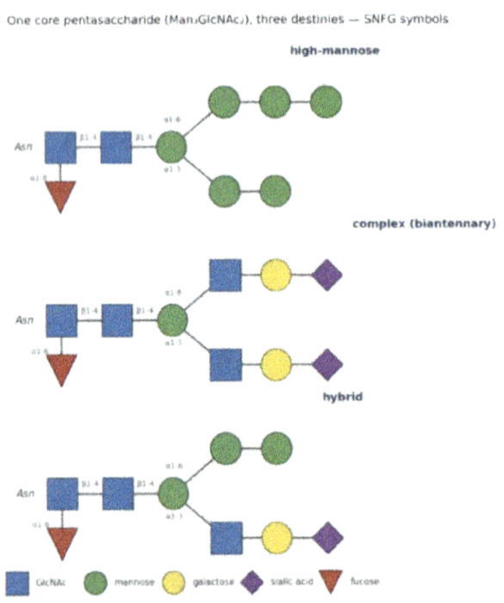

Figure 21.4 *One core pentasaccharide, three destinies. All N-glycans share the $Man_3GlcNAc_2$ core (often fucosylated α1-6 at the root, red triangle); the α1-6 and α1-3 arms are then finished as high-mannose (top), complex (middle — each arm GlcNAc → galactose → sialic acid), or hybrid (bottom). Symbols follow the SNFG convention (legend). The glycan drawn is information: its class reports the protein's itinerary, its sialic acids its licence to remain in the blood.*

27. The Symbol Nomenclature for Glycans — blue square for GlcNAc, green circle for mannose, yellow circle for galactose, purple diamond for sialic acid, red triangle for fucose — was standardised by the Consortium for Functional Glycomics and is the script of *Essentials of Glycobiology* (Varki et al., 2017), the scripture of § 21.6–21.9. The reader will meet it again in every serious glycobiology paper; it is the field's one act of typographical mercy.

21.7 Proteoglycans: the charged forest that holds water

Between the cells of cartilage, tendon, skin and cornea lies a material that is neither cell nor fibre but a hydrated gel, and its chemistry is sugar at its most extravagant. A ***proteoglycan*** *is a core protein bristling with* ***glycosaminoglycan*** *chains — long, unbranched, repeating disaccharides in which one partner is always an amino sugar and very often both are sulfated. The family (Figure 21.5) runs from* ***hyaluronan*** *— the one unsulfated member, no core protein at all, a single chain of up to ten million daltons that thickens the synovial fluid and the vitreous of the eye — through* ***chondroitin*** *and* ***dermatan sulfate*** *and* ***keratan sulfate****, to* ***heparan sulfate*** *and its over-sulfated cousin* ***heparin****, which, at up to four negative charges per disaccharide, disputes the title of the most electronegative molecule in the body. The charges are the function. Sulfate and carboxyl at density ~0.1–0.2 molar make the gel a polyelectrolyte; above the condensation threshold the counterions refuse to leave even in principle (Manning, 1969), and the trapped ions pull water in by the Donnan osmosis we now derive.*

Let the gel carry fixed anionic charge at concentration c_F*, bathed in salt of concentration* c_S*. Equilibrium across the gel boundary (the Donnan condition of § 6.7) requires* $[Na^+]_{in}[Cl^-]_{in} = c_S^2$*; electroneutrality requires* $[Na^+]_{in} = [Cl^-]_{in} + c_F$*. Eliminating the sodium,* $[Cl^-]_{in}([Cl^-]_{in} + c_F) = c_S^2$*, a quadratic with root* $[Cl^-]_{in} = (-c_F + \sqrt{(c_F^2 + 4c_S^2)})/2$*, whence* $[Na^+]_{in} = (c_F + \sqrt{(c_F^2 + 4c_S^2)})/2$*. The osmotic pressure of the gel against its bath is the van 't Hoff difference of all mobile ions:*

$$\Pi = RT\Big([\mathrm{Na}^+]_{\mathrm{in}} + [\mathrm{Cl}^-]_{\mathrm{in}} - 2c_s\Big) = RT\Big(\sqrt{c_F^2 + 4c_s^2} - 2c_s\Big) \tag{21.4}$$

— always positive, and growing without bound as c_F *rises: charge is pressure. Put cartilage's numbers in —* $c_F \approx 0.2$ *M of fixed charge,* $c_s \approx 0.15$ *M of extracellular salt — and Eq. (21.4) returns* $\Pi \approx RT \times 0.061$ *M ≈ 1.5 atm. Every aggrecan aggregate — a hundred proteoglycans hung on one hyaluronan backbone by link proteins, a bottle-brush of some* 10^8 *daltons (Figure 21.5) — is a small osmotic spring, inflated by its own charges, whose swelling presses against the collagen net that restrains it; the knee carries one's weight on trapped water, and the water is trapped by sulfate. When age or inflammation strips the proteoglycans, the pressure falls, the collagen collapses, and the joint grinds: osteoarthritis is, at root, Eq. (21.4) losing its first term.*

Heparin deserves its own sentence, for it is the one GAG that is medicine outright.[28] *Within its chains lies one rare pentasaccharide — marked by an uncommon 3-O-sulfate — that binds antithrombin and bends it into a shape that executes factor Xa a thousand times faster; longer chains bridge thrombin to the same fate. The low-molecular-weight heparins, and the synthetic pentasaccharide fondaparinux, are the same idea at finer resolution: keep the five sugars, discard the rest. Exercise 21.4 makes the mechanism quantitative.*

28. Heparin was found in 1916 by Jay McLean, a medical student set to hunt pro-coagulants in liver extracts, who returned with an anticoagulant instead — named for the organ (Greek *hēpar*) that refused to clot. Howell and Holt purified it; the pharmaceutical world has extracted it from pig intestine ever since.

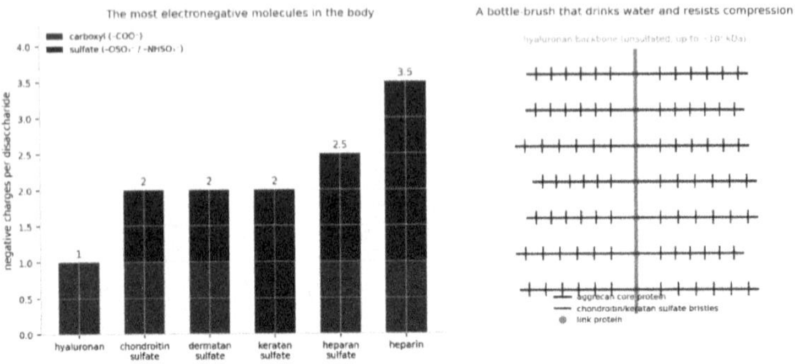

Figure 21.5 Left: negative charges per disaccharide repeat across the glycosaminoglycan family — carboxyl (navy) and sulfate (crimson); heparin's total approaches four. Right: the cartilage aggregate — aggrecan core proteins (navy) hung on a hyaluronan backbone (teal) by link proteins (amber), bristling with chondroitin/keratan sulfate chains (crimson); the fixed charges of the bristles are the c_F of Eq. (21.4), and their trapped water is the tissue's shock absorber.

21.8 Sugars as signatures: the glycocalyx, lectins, and the blood groups

Every cell of the body wears a fur of glycolipid and glycoprotein sugar, ten to a hundred nanometres thick — the **glycocalyx** *— and it is read more often than any message the cell writes inside itself. The reading proteins are the* **lectins***: selectins on the vessel wall, which catch the sialyl-Lewis^X tetrasaccharide on a passing leukocyte and slow it to a roll — the first step of every inflammation; siglecs, which read sialic acid as self and restrain the immune cell that bears them; galectins, collectins, and the liver's asialoglycoprotein receptor of § 21.6. What self can read, the enemy can read. Influenza's haemagglutinin binds the terminal sialic acid of our airway glycans — α2-6-linked for the human virus, α2-3 for the avian, so that a species barrier hides in one glycosidic angle — while its neuraminidase cuts the same sugar to let the progeny go; oseltamivir is a false sialic acid that jams the scissors. The cholera toxin docks on the GM1 ganglioside's pentasaccharide; several noroviruses dock on the blood-group sugars themselves, which is why non-secretors — one person in five, unable to place ABO sugars on their mucosae — are spared whole strains of the winter vomiting disease. A polymorphism is an epidemic's brake.*

The blood groups themselves are the most famous sugars in medicine (Figure 21.6). All share the H antigen: a fucose α1-2 on the terminal galactose of a surface glycan. The A transferase adds one GalNAc in α1-3; the B transferase adds one galactose in α1-3; the O allele is a transferase silenced by a single deletion, adding nothing. And the plasma of every individual carries antibodies against the antigens it lacks — raised, it appears, against gut bacteria wearing look-alike sugars — so that a mismatched transfusion agglutinates and haemolyses. Landsteiner's 1901 agglutination of his colleagues' blood on a bench-top is, in retrospect, the founding experiment of transfusion medicine and of glycobiology alike. The rules of the matrix in Figure 21.6 follow from two lines of immunology and one of glycochemistry; Worked Example 21.3 plays them on a population.

WORKED EXAMPLE 21.3 — THE ARITHMETIC OF THE DONOR POOL

Take the British frequencies — O 44%, A 42%, B 10%, AB 4% — and the matrix of Figure 21.6 (a recipient rejects any donor cell bearing an antigen against which the recipient carries antibody). What is the probability that a random donor's red cells are acceptable to a random recipient?

Solution. *Condition on the recipient. O recipients accept only O: 0.44 × 0.44 = 0.1936. A recipients accept O or A: 0.42 × (0.44 + 0.42) = 0.3612. B recipients: 0.10 × (0.44 + 0.10) = 0.054. AB recipients accept all: 0.04 × 1.00 = 0.040. Sum ≈ 0.649 — about two-thirds. Two observations follow. First, the universal donor is O and the universal recipient is AB, but the population is mostly A and O, so most random pairings fail only when the recipient is O or the donor is B — the rare antigen is the troublemaker, as B's 10% already suggests. Second, the same matrix governs the logistics of every blood bank: a service that recruits in proportion 44:42:10:4 can meet any recipient mix — a constraint that becomes acute precisely where the frequencies differ, as they do across continents. The sugar on the cell is a passport; demography stamps it.*

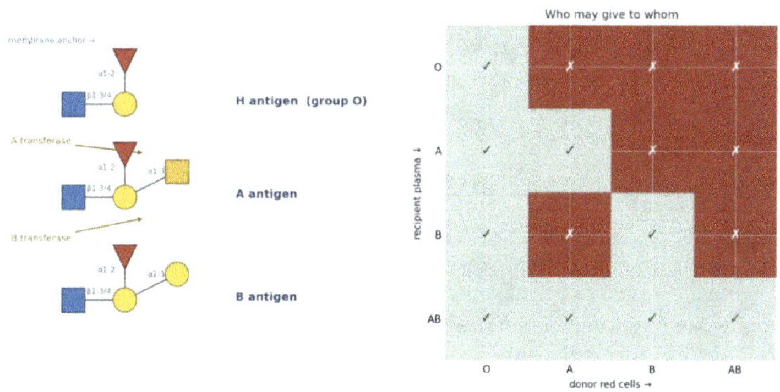

Figure 21.6 *Left: the ABO(H) antigens in SNFG symbols — the H fucose (red triangle) α1-2 on galactose; the A transferase adds GalNAc (yellow square), the B transferase galactose (yellow circle), each in α1-3; O adds nothing. Right: the transfusion matrix — green where donor cells carry no antigen the recipient's plasma attacks. Two enzymes differing in one sugar's donor account for the oldest fatal error in clinical medicine.*

21.9 Glycobiology and medicine: the sugar, weaponised

The chapter closes, as the field has, in the clinic. The polysaccharide capsules of the pneumococcus and of Haemophilus influenzae b are their wearers' armour and our best target — but an infant's B cells answer pure polysaccharide feebly, T-independently, without memory (Chapter 18). The ***conjugate vaccines*** *marry the sugar to a carrier protein, and the T cell that answers the protein is conscripted to help the B cell that sees the sugar: childhood meningitis of two kinds all but vanished from the vaccinated world within a decade, a glycan's immunological illiteracy cured by translation. The heparin lineage of § 21.7 runs from the slaughterhouse extract to fondaparinux, a five-sugar drug; the glyco-engineered antibodies — afucosylated so that their Fc grips FcγRIIIa the tighter and antibody-dependent killing rises — are now oncology's standard issue (obinutuzumab, mogamulizumab); and glycan microarrays, on which a serum or a virus is read against hundreds of printed oligosaccharides, have made the glycome as assayable as the genome (§ 55.4). Even the lysosomal storage diseases — one failed glycosidase, one accumulating glycolipid — are best read as footnotes to §*

20.7's bond: the cell that can build a glycan must be able to dismantle it, in the lysosome, or be buried in it (Part VI). Sugar, which began this Part as fuel and fibre, ends it as text — and a text can be forged, censored, or read aloud to the immune system. Medicine has begun to do all three.

Case study — The vaccine that taught sugar to speak T-cell

Before 1990, Haemophilus influenzae b — its polyribosyl-ribitol phosphate capsule a pure polysaccharide — was the commonest cause of bacterial meningitis in toddlers, precisely the age at which T-independent antibody answers are weakest. The first vaccines, of purified capsule, protected adults and failed the children who needed them. Conjugation — the PRP sugar covalently bound to a protein carrier — changed the grammar: the B cell binds the sugar, internalises the conjugate, and presents protein peptides to T cells, receiving in return the help (CD40L, cytokines) that polysaccharides can never elicit. Class-switching, affinity maturation, memory — the whole apparatus of Chapter 18 — now served a carbohydrate. Invasive Hib disease fell by more than 99% in vaccinated populations; the pneumococcal and meningococcal conjugates followed. A single chemical idea — marry the glycan to a peptide — converted the weakest arm of carbohydrate immunology into one of the strongest arguments for vaccination ever assembled.

21.10 Synthesis of Chapter 21

One monomer, one linkage, one stereochemical vowel — and the whole chapter has been the exegesis of that vowel. The α-configuration coils glucose into the helix: compact, hydrated, enzyme-friendly, the perfect bank — starch for the patient plant, glycogen for the impatient animal, whose tree geometry [Eqs. (21.1)–(21.2)] converts one cashier into four thousand doors. The β-configuration flattens it into the ribbon: insoluble, crystalline, unassailable — cellulose to gird the plant cell against its own turgor [Eq. (21.3)], chitin and murein to armour the insect and the bacterium, with lysozyme and penicillin as the siege engines the armour invited. And where the chain declines both economies it becomes text: the N- and O-glycans that police protein folding and licence a protein's life in the blood; the sulfated forests of the proteoglycans, whose fixed charges inflate the water that bears our joints [Eq. (21.4)]; and the surface signatures — selectin ligands, sialic acids, the ABO triad — that cells, viruses, toxins and blood banks all read with the same avid literality. Matter, energy, information: the carbohydrates of Part V have now supplied all three, and the direction of Figure ii is kept — monosaccharides and oligosaccharides behind us, polysaccharides and glycoconjugates herewith accounted. Ahead lies a different kind of molecule altogether: one that fears water rather than drinking it. Chapter 22, Lipid Structure and Diversity, begins with the fatty acid — and with the first molecule in this book whose defining property is that it will not dissolve.

Exercises (with full resolutions)

Exercise 21.1

Explain, from the geometry of § 20.7, why the α(1→4) polymer of glucose coils while the β(1→4) polymer extends — and why only one of them crystallises.

Resolution. *In the α-linkage the glycosidic bond leaves C1 axially and meets C4 equatorially; the two C–O bonds subtend roughly 60°, so each residue rotates the chain and six residues close a turn — a helix, hydrated and open. In* *the β-linkage both bonds are equatorial and anti-parallel; the chain can only run straight, each residue flipping 180° about the axis. Straight ribbons present every hydroxyl to a neighbour; hydrogen-bonded sheets stack into microfibrils from which water is excluded — the crystal. The helix buries its oxygens in its own lumen and stays soluble. Geometry at one carbon decides solubility at a million daltons.*

EXERCISE 21.2

A biochemist iodine-stains three polysaccharides and records λ_{max} = 622, 548 and 472 nm. Assign the polymers and justify the ordering.

Resolution. *622 nm — amylose; 548 nm — amylopectin; 472 nm — glycogen. The chromophore is a linear polyiodide wire threaded through the α-helix (Figure 21.2); the longer the conjugated wire, the lower its transition energy and the redder the band. Uninterrupted helices (amylose, hundreds of residues) shelter the longest wires; amylopectin's helices are clipped to ~20–25 residues between α(1→6) branches; glycogen's to ~10–13. The colour is a branching assay in disguise — the reddest solution is the most interrupted helix.*

EXERCISE 21.3

Penicillin sterilises a growing culture of staphylococci but spares a stationary one. Why does the transpeptidase's inhibition kill only the growing cell?

Resolution. *The murein sacculus is a finished, cross-linked fabric; penicillin's target acts only where new wall is knitted — at growth zones and the septum. There the drug's false D-Ala–D-Ala acylates the transpeptidase, new strands stay uncross-linked, and the cell's own autolysins — which must open the wall to insert new material — keep cutting against a wall that can no longer be repaired. Turgor (Eq. 21.3's argument, now without the wall) does the rest: the membrane bursts through the gap. A non-growing cell inserts nothing, hydrolyses nothing, and simply waits the antibiotic out — the persistence problem in one sentence.*

EXERCISE 21.4

Full-length heparin accelerates antithrombin's inhibition of both factor Xa and thrombin; the pentasaccharide fondaparinux accelerates only the inhibition of Xa. What geometric fact explains the difference?

Resolution. *The pentasaccharide binds antithrombin and allosterically activates it — that alone suffices for Xa, which is inhibited the moment antithrombin is bent into its fast form. Thrombin must be bridged: the same heparin chain (≥ ~18 saccharides) binds antithrombin at the pentasaccharide and thrombin at a second site downstream, templating the encounter. Five sugars can hold the switch; they cannot span the two proteins. Length, not chemistry, is the difference — which is why the low-molecular-weight heparins, of intermediate length, inhibit Xa preferentially and bleed less.*

EXERCISE 21.5

The N-glycan's three glucose residues are a folding clock. Reconstruct the calnexin cycle, and explain why a glucose added back by UGGT is a mark of failure rather than of completion.

Resolution. *Glucosidases I and II trim $Glc_3 \rightarrow Glc_2 \rightarrow Glc_1$ on the fresh glycan; the monoglucosylated form is bound by calnexin/calreticulin, which detain the protein while it folds and hand it the ERp57 oxidoreductase. Full deglucosylation releases it: folded proteins are ignored by UDP-glucose:glycoprotein glucosyltransferase and exported. UGGT reads exposed hydrophobic surface — the signature of the unfinished — and restores one glucose, returning the protein to the chaperone for another round: the glucose is therefore added because folding failed, a renewal of the detention order. Persistent misfolding is finally betrayed by slow mannose trimming, whose product no longer cycles and is routed to ERAD and the proteasome. The sugar code here is temporal: glucose counts attempts, mannose counts the years.*

Research-level propositions

Proposition 21.α. Lignocellulose is the cheapest sugar on Earth and the dearest to release. Using the crystallography of § 21.4 and the enzyme kinetics of Part IV, formulate the theory of cellulase processivity on a crystalline face — the rate-limiting step is chain extraction, not bond cleavage — and design the directed-evolution campaign (§ 15.10) that would test it. Proposition 21.β. Afucosylation of the Fc glycan raises antibody-dependent cellular cytotoxicity an order of magnitude. Propose the structural mechanism at the Fc–FcγRIIIa interface (§ 18.7), predict which other Fc-glycan edits should phenocopy it, and devise the glycoengineered cell line that would manufacture such antibodies at scale. Proposition 21.γ. Heparan sulfate chains serve as first receptors for several viruses. Model the chain as a polyelectrolyte brush (Eq. 21.4's fixed charges, Manning's condensation), estimate the encounter-rate enhancement for a virion of known charge, and assess whether enzymatic heparanase inhibition is a rational antiviral strategy or a hostage to the chains' housekeeping duties.

References (Chapter 21)

Ball, S. G., & Morell, M. K. (2003). From bacterial glycogen to starch: Understanding the biogenesis of the plant starch granule. Annual Review of Plant Biology, 54, 207–233.

Berg, J. M., Tymoczko, J. L., Gatto, G. J., Jr., & Stryer, L. (2023). Biochemistry (9th ed.). W. H. Freeman.

Leloir, L. F. (1971). Two decades of research on the biosynthesis of saccharides. Science, 172(3990), 1299–1303.

Manning, G. S. (1969). Limiting laws and counterion condensation in polyelectrolyte solutions. I. Colligative properties. The Journal of Chemical Physics, 51(3), 924–933.

Nelson, D. L., Cox, M. M., & Hoskins, A. A. (2021). Lehninger principles of biochemistry (8th ed.). W. H. Freeman.

Varki, A., Cummings, R. D., Esko, J. D., Stanley, P., Hart, G. W., Aebi, M., Darvill, A. G., Kinoshita, T., Packer, N. H., Prestegard, J. H., Schnaar, R. L., & Seeberger, P. H. (Eds.). (2017). Essentials of glycobiology (3rd ed.). Cold Spring Harbor Laboratory Press.

CHAPTER 22

Lipid Structure and Diversity

CONTENTS OF THIS CHAPTER

In which we meet the first family of molecules whose common property is a refusal — the refusal to dissolve in water; in which two small dials, length and unsaturation, are shown to govern fat and oil, butter and salmon alike; in which the isoprene unit, humblest of building blocks, is found to erect both the rubber tree's sap and the steroid nucleus; and in which a twenty-carbon tetraene, torn from the membrane, becomes the local post-office of inflammation — and the oldest target in the pharmacopoeia.

22.1 A family defined by a refusal

Every other class of biomolecule in this book has been defined by its architecture: amino acids by their zwitterion, sugars by their carbonyl, nucleotides by their triad of base, sugar and phosphate. The lipids decline the courtesy. A lipid is any small molecule that dissolves in chloroform and not in water — a definition by extraction, not by structure, and therefore a family reunion of strangers: the hydrocarbon tails of the fatty acids, the four fused rings of cholesterol, the isoprene coils of the carotenoids, the quinone rings of the electron carriers. What unites them is a single physical temperament — hydrophobicity, the hydrophobic effect of § 5.6 in its purest estate — and from that one temperament their three professions follow with complete logic. A molecule that water expels can store energy without the water's osmotic attendance (the triacylglycerols, § 22.3); it can partition water from water, making the membrane that made the cell (the phospholipids, §§ 22.4–22.5 and Chapter 23); and, precisely because it must live at the boundary between the aqueous world and the fatty one, it can signal — the steroids, the eicosanoids, the inositol lipids, molecules that are read at membranes (§§ 22.6–22.8). Store, partition, signal: the whole chapter is the unfolding of that triad, and the reader who keeps the hydrophobic effect in view will find nothing in it arbitrary.

One disclaimer before the tour. The lipids' reputation has suffered at the hands of dietary journalism, a court in which the defendant is seldom represented by counsel. The biochemistry is simpler and more interesting: the same molecule that clogs an artery also builds every cell the artery is made of; the same cholesterol that the clinic measures with a frown is the molecule without which no animal membrane holds its temper, no steroid hormone exists, and no bile acid digests the very fat that the newspapers deplore. We shall meet the pathology in due course (§ 22.9, and Part VI in earnest); but pathology is physiology read after the accident, and the physiology comes first.

22.2 Fatty acids: two dials on one molecule

The fatty acid is a single hydrocarbon chain — almost always an even number of carbons, a fossil of its two-carbon assembly (Chapter 34) — tipped by one carboxyl group: a hydrophilic head of two atoms on a hydrophobic body of twenty. Everything about its behaviour is set by two dials. The first is **length***: the shorthand 16:0 names palmitate, sixteen carbons, no double bonds; 18:0 is stearate. The second is* **unsaturation***: 18:1 Δ^9 is oleate, one double bond between carbons 9 and 10 counting from the carboxyl; 18:2 $\Delta^{9,12}$ is linoleate. The Δ-system counts from the head, where the chemistry happens; the older ω-system counts from the methyl tail, and survives because it sorts fatty acids into metabolic families — ω-3 (double bond three carbons from the tail: α-linolenate, the fish oils) and ω-6 (six from the tail: linoleate, the seed oils) — families the animal can interconvert by elongation and desaturation but can neither create nor cross, whence the* **essential fatty acids***: linoleate and α-linolenate must be eaten, because no animal desaturase can place a double bond beyond Δ^9 towards the tail.*[29]

The two dials act through a single mechanism, and it is worth deriving. Adjacent chains attract by London dispersion: each pair of methylene groups, polarisable clouds a distance r apart, lowers the energy by

$$u(r) \approx -\frac{3\,\alpha^2 I}{4\,(4\pi\varepsilon_0)^2\,r^6} \tag{22.1}$$

— the London formula of § 5.3 recalled: α the polarisability of the CH_2 group, I its ionisation energy, and the famous inverse-sixth power making the attraction intimate — strong at contact, nothing at twice the distance. Two consequences follow without further hypothesis. First, the energy of the solid is a sum over contacts, so each added methylene adds roughly the same stabilisation to the crystal and the melting point should climb smoothly with length: it does, some 2–4 °C per CH_2 across the series (Figure 22.1, navy). Second, the attraction is forfeited wherever contact is lost — and a cis double bond, whose two C–C bonds leave the rigid C=C unit at ~120° on the same side, bends the chain thirty degrees out of line and opens a vault in the crystal into which no neighbour can reach (Figure 22.2). The melting point falls accordingly, and catastrophically: stearate (18:0) melts at 69.6 °C; oleate (18:1) at 13.4 °C; linoleate (18:2) below −5 °C. One double bond has cost fifty-six degrees. Beef dripping is solid and olive oil is liquid because of Figure 22.2 — and the fish of cold seas, whose membranes must stay fluid in water at 4 °C, carry the most kinked chains of all, the ω-3 polyenes; the polar bear's diet is, thermodynamically, an antifreeze. The trans isomer, whose chain runs straight through the double bond like a saturated chain's, keeps the packing and the melting point — elaidate, trans-18:1, melts at 45 °C — which is why partial hydrogenation hardened kitchen fats so usefully, and why the same geometry, foreign to the desaturases, congests the lipoprotein economy (§ 22.9). A double bond, it seems, may be forgiven its existence but not its posture.

29. The essentiality was established by George and Mildred Burr in 1929–30, who raised rats on a scrupulously fat-free diet and watched them develop scaly skin, kidney failure and sterility — cured by linoleate alone. The "vitamin F" of that decade quietly became the ω-6/ω-3 grammar of this one.

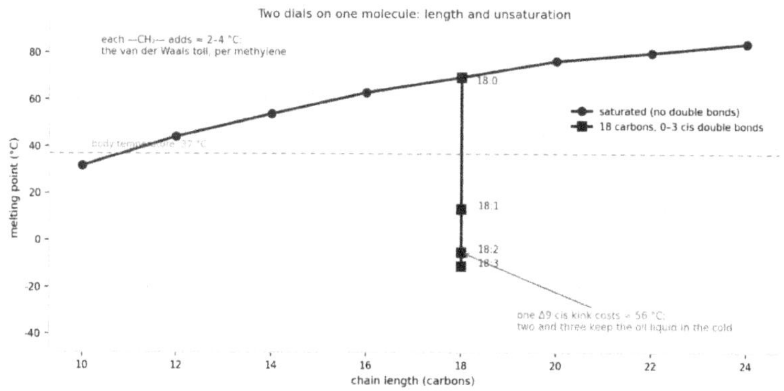

Figure 22.1 The two dials. Navy: saturated fatty acids — the melting point climbs ~2–4 °C per methylene, the linear signature of an additive dispersion energy [Eq. (22.1)]. Crimson: eighteen carbons with 0–3 cis double bonds — one kink costs ≈ 56 °C, three leave the oil liquid below −10 °C. The amber line (37 °C) divides the body's butters from its oils.

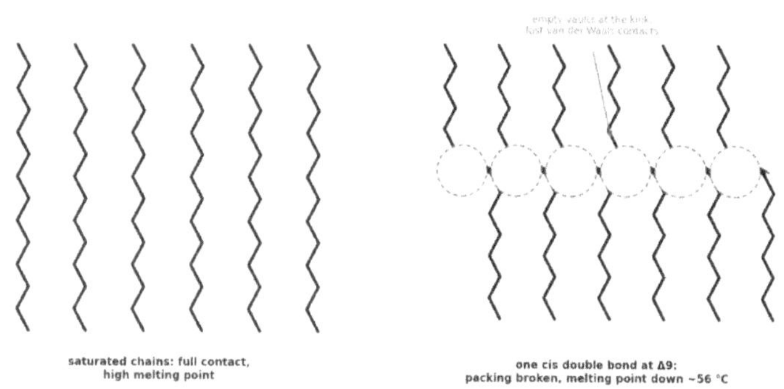

Figure 22.2 Why the kink melts the crystal. Left: saturated chains in register — every methylene in contact, the dispersion energy of Eq. (22.1) fully paid. Right: one cis double bond (thick segment, Δ9) bends each chain out of register; the dashed vaults are lost contacts, and lost contacts are lost melting point.

22.3 Triacylglycerols: the concentrated store

Esterify three fatty acids to one glycerol and the result is the triacylglycerol (Figure 22.3): no charge, no hydroxyl left free, nothing for water to hold — and therefore the perfect store, for two distinct reasons that deserve to be separated. The first is ***chemical****: the carbons of a fatty acid are almost fully reduced, and energy is released precisely by their oxidation. Compare the two currencies directly. Glucose burns as $C_6H_{12}O_6 + 6\ O_2 \rightarrow 6\ CO_2 + 6\ H_2O$, $\Delta G^{\circ\prime} \approx -2\ 870$ kJ mol^{-1}; palmitate as $C_{16}H_{32}O_2 + 23\ O_2 \rightarrow 16\ CO_2 + 16\ H_2O$, $\Delta G^{\circ\prime} \approx -9\ 790$ kJ mol^{-1}. Per gram, then,*

$$\frac{\Delta G_{glucose}}{M_{glucose}} \approx \frac{-2870}{180.2} = -15.9\ \mathrm{kJ\,g^{-1}}\ ; \qquad \frac{\Delta G_{palmitate}}{M_{palmitate}} \approx \frac{-9790}{256.4} = -38.2\ \mathrm{kJ} \quad (22.2)$$

— a ratio of 2.4, the celebrated 9 kcal against 4. The reason is written in the formulas: glucose's carbons are already half-oxidised (six oxygens carried in the molecule itself, dead weight from the energetic point of view), while palmitate carries only two oxygens at its head and ships its carbons in the fully reduced, fuel-rich state. The second advantage is **physical***, and it is the larger one. Glycogen stores hydrophilically: every gram drags some two grams of water of hydration, so the usable energy of the hydrated glycogen depot falls to*

$$e_{glycogen} \approx \frac{15.9\ \mathrm{kJ\,g^{-1}}}{1+h} \approx \frac{15.9}{3} \approx 5.3\ \mathrm{kJ\,g^{-1}} \quad (h \approx 2)\ ; \qquad e_{fat} \approx 37-39\ \mathrm{kJ\,g} \quad (22.3)$$

— since the triacylglycerol droplet is anhydrous by construction. Nearly seven to one, then, when both dials are honestly read. The adipocyte answers to this arithmetic: its single droplet fills the cell, the nucleus flattened against the membrane like a signature in the margin of a document composed entirely of capital.

WORKED EXAMPLE 22.1 — THE GODWIT'S ARITHMETIC

A bar-tailed godwit flies 11 000 km from Alaska to New Zealand without feeding, expending on the order of 250 kJ per kilometre-day — take the flight at 8 days and a fuel budget of ~1.1×10^4 kJ. What mass of fuel does it carry (i) as fat, (ii) hypothetically as hydrated glycogen?

Solution. *(i) Eq. (22.3): 1.1×10^4 kJ / 38 kJ $g^{-1} \approx 290$ g — about half the bird's 550 g departure mass, just what the weighings show. (ii) $1.1 \times 10^4 / 5.3 \approx 2\,100$ g — nearly four times the bird itself. The conclusion is not that the godwit prefers fat; it is that a bird carrying glycogen is a bird that does not take off. Every long-distance migrant, and every animal that hibernates, has made the same calculation by selection rather than by arithmetic. (The reserve is not weightless even so: the doubly labelled water of § 26.9 shows the godwit lands having spent, to within a few per cent, exactly the mass it failed to arrive with.)*

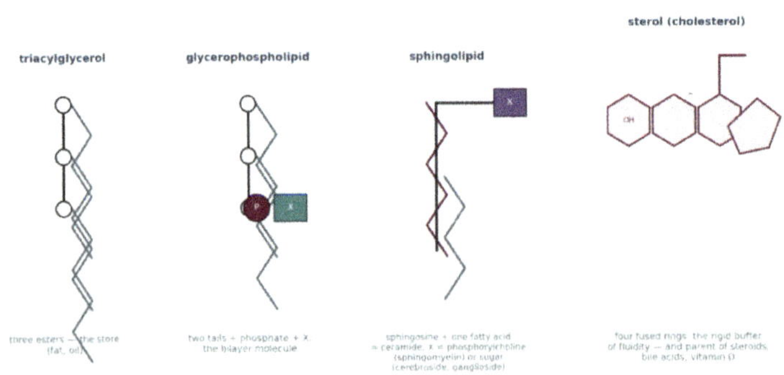

Figure 22.3 *Four architectures, three professions. The triacylglycerol (three esterified tails) is the store; the glycerophospholipid (two tails, phosphate, headgroup X) and the sphingolipid (sphingosine backbone, one amide-linked fatty acid, X) are the partition; the sterol (four fused rings on the isoprene plan) is the signal and the buffer. Black zigzag: hydrocarbon tails; circles: glycerol carbons; P: phosphate; X: headgroup; the crimson tail of the sphingolipid is its second fatty acid.*

22.4 Glycerophospholipids: the amphiphile and its menu

Replace the triacylglycerol's third fatty acid by a phosphate bearing a small guest, and the molecule changes profession entirely: it becomes an **amphiphile** *— one molecule, two loyalties — and its behaviour in water is no longer a solubility question but an architecture. A single-tailed amphiphile (a soap, a free fatty acid salt) is a wedge, and wedges pack into spheres: the micelle. A two-tailed phospholipid is a cylinder, and cylinders pack into sheets: the bilayer, whose physics is Chapter 23 entire. What concerns this chapter is the dial the cell actually turns — the headgroup X on the phosphate (Figure 22.4). Choline and ethanolamine give zwitterionic, net-neutral surfaces (PC, the outer-leaflet workhorse; PE, the cone-shaped fusogen that assists every membrane merger); serine, inositol and glycerol give net-negative ones (PS, PI, PG), and the cell keeps these on its private, cytosolic face — so that the membrane's two leaflets differ in charge as they differ in charge's consequences: the inner face binds the signalling proteins (§ 22.8, Chapter 51), the outer presents a bland neutrality to the blood. Cardiolipin, two phosphatidates sharing one glycerol, is the mitochondrion's signature lipid, curved to fit the cristae (Chapter 29). The plasmalogens replace the sn-1 ester by a vinyl ether — an antioxidant sacrifice that the brain and heart afford in quantity.*

The amphiphile's solubility obeys a law worth one equation. Dissolve a single-tailed lipid in water and it partitions between monomers and micelles; the monomer concentration simply saturates — above the **critical micelle concentration** *every added molecule joins an aggregate, and the solution's surface tension flat-lines. Writing the monomer–micelle equilibrium as an ideal dilute equilibrium, the free energy of transfer into the micelle is*

$$\Delta G^\circ_{mic} = RT \ln X_{cmc} \approx RT \ln \frac{[cmc]}{55.5\ \mathrm{M}} \tag{22.4}$$

— X_{cmc} the mole fraction at the cmc, 55.5 M the water's own concentration setting the standard state; the logarithm is the entropy of dilution of § 6.2, applied to a molecule that refuses dilution. Longer chains lower the cmc — the hydrophobic effect of § 5.6 collecting its rent — and each added methylene divides the cmc by roughly three: palmitate salt ~10^{-3} M, laurate ~10^{-2}, oleate (the kink again) intermediate. Bile salts, the liver's own detergents, exploit exactly this: secreted into the gut above their cmc, they wrap dietary fat into mixed micelles and make the insoluble absorbable (Chapter 36). Digestion begins with Eq. (22.4) whether the diner knows it or not.

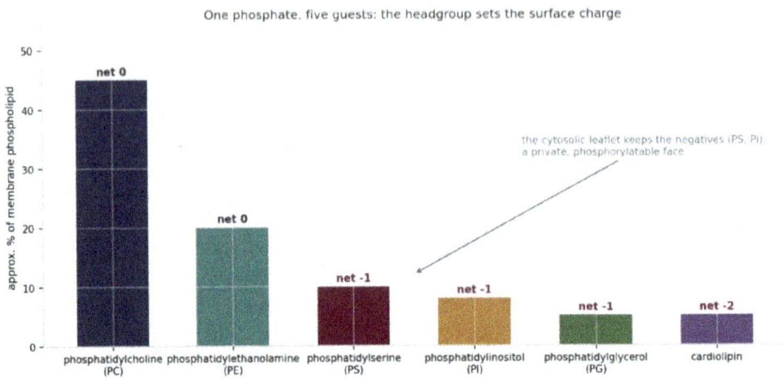

Figure 22.4 The headgroup menu. One phosphate, five common guests; the net charge at pH 7 follows the guest's own ionisations (choline and ethanolamine zwitterionic → 0; serine, inositol, glycerol → −1; cardiolipin → −2). Abundances are typical of a mammalian plasma membrane; the anionic species are concentrated on the cytosolic leaflet, where they serve as docking charges for signalling proteins (§ 22.8).

22.5 Sphingolipids: the second architecture

The sphingolipids are the glycerophospholipids' parallel invention: the same two-tailed, one-headed amphiphile, but built on **sphingosine** *— an 18-carbon amino alcohol that is itself half the tail — with one fatty acid amide-linked to its nitrogen to make the* **ceramide***, the sphingolipid stem cell.*[30] *Attach phosphorylcholine and the ceramide becomes* **sphingomyelin***, the insulation of the myelin sheath — the oligodendrocyte's membrane wound a hundred turns around the axon, a lipid capacitor whose loss is multiple sclerosis. Attach a single sugar and one has the* **cerebrosides***; attach the branched, sialic-acid-tipped oligosaccharides of § 21.6 and one has the* **gangliosides** *— GM1, the cholera toxin's landing strip; GM2, whose failure of degradation is Tay–Sachs disease, a one-enzyme funeral in the lysosome (§ 21.9's principle again: the cell that builds a glycan must be able to dismantle it). The sphingolipids cluster with cholesterol in the outer leaflet into the ordered patches the field calls rafts (§ 23.6), and their ceramide, released on demand, is itself a signal — an apoptotic sentence handed down from the membrane itself.*

22.6 Sterols and the isoprene economy

Cholesterol looks nothing like a fatty acid and is made entirely of two-carbon pieces: it is an isoprenoid, the most elaborate product of the five-carbon isoprene unit whose simpler polymers are rubber, carotene and the tails of the quinones and chlorophylls. Twenty-seven carbons arranged as four fused rings — three of six, one of five — a single hydroxyl at C3, a short tail at C17: a flat, rigid plate (Figure 22.3) with one polar face. That plate does two kinds of work. In the

30. The name is Thudichum's. Johann Ludwig Wilhelm Thudichum, the German-born, London-based chemist who between 1874 and 1884 isolated sphingosine, sphingomyelin, the cerebrosides and the cephalins from brain, named them for the Sphinx "in honour of their enigma"; the chemical establishment, led by the formidable Liebig, answered the enigma by denouncing Thudichum's results wholesale. He was right in nearly every particular. The history of lipid chemistry is unusually rich in vindications — perhaps because its molecules were for so long inaccessible to proof.

*membrane it is the **fluidity buffer** (Figure 22.6): inserted among the phospholipid tails, its rigidity restrains their motion above the melting transition, while its bulk wedges between them below it, preventing crystallisation — it broadens and nearly abolishes the phase transition itself, keeping the animal membrane in its liquid-ordered state at a temperature the phospholipids alone could not manage. (Plants, which cannot make cholesterol, use sitosterol and stigmasterol to the same end; fungi, ergosterol — the target of the azole antifungals; bacteria mostly do without, and their membranes are the worse for it.) And out of the ring system the cell cuts its most consequential signals: the steroid hormones — cortisol, aldosterone, oestradiol, testosterone, progesterone — five carbon-skeletons differing in a handful of oxidations, each read by its own nuclear receptor (Chapter 51); the bile acids of § 22.4; vitamin D, a broken ring that became a hormone of calcium; and, in the skin, the same ring's photochemistry. The whole steroid civilisation is one plate of isoprene, variously nicked.*

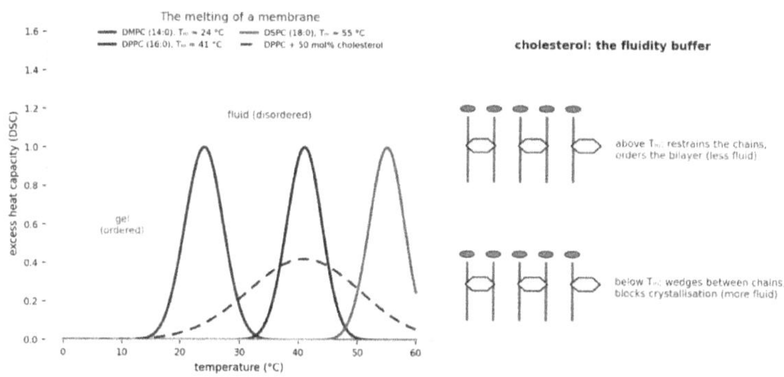

Figure 22.6 *Left: differential scanning calorimetry of phosphatidylcholine bilayers — the gel-to-fluid melting sharpens and climbs with chain length (DMPC < DPPC < DSPC), exactly as Eq. (22.1) predicts; 50 mol% cholesterol (dashed) broadens the transition until it nearly vanishes. Right: the molecular reading — the rigid plate (pink) restrains disordered chains above T_m and wedges ordered ones apart below it.*

Plate 22.1 *Konrad E. Bloch (1912–2000), whose isotopic tracer experiments — feeding acetate labelled in one carbon or the other — revealed that every carbon of cholesterol comes from acetate, and mapped the road through squalene and lanosterol (Nobel Prize in Physiology or Medicine, 1964, shared with Feodor Lynen). Portrait from the Nobel Foundation archive, nobelprize.org, reproduced from the Foundation's publicly distributed image.*

HISTORICAL NOTE — THE STONE, THE FAT, AND THE LABEL

Cholesterol is a gallstone before it is a molecule: de Fourcroy obtained it from human calculi in 1789, and Chevreul christened it cholesterine — solid bile — in 1816. For a century it remained an analytical nuisance, the white plaque in the artery and the obstinate peak in the wine-shops' spectroscopes. Its formula ($C_{27}H_{46}O$) was settled only in 1888, its ring structure not until 1932 — Wieland and Windaus, two Nobels for the wrong diagram between them, then the right one — and its origin not at all. The man who answered that was a refugee twice over: Konrad Bloch, driven from Switzerland to America, found in Schoenheimer and Rittenberg's isotope laboratory at Columbia the instrument the question required. Feed a rat acetate whose methyl or carboxyl carbon bears the label, degrade the cholesterol position by position — half a decade of analytical chemistry of heroic tedium — and the answer emerged entire: every carbon is acetate's, condensed to isoprene, rolled into the linear triterpene squalene, folded and cyclised to lanosterol, and trimmed to cholesterol. The cyclisation of squalene, one enzyme converting a floppy chain into four rings with nine stereocentres in a single pass, remains one of the most elegant reactions in the book (Chapter 37). Bloch's method, the isotope as interrogator, is meanwhile still at work in every mass spectrometer in § 54.6 — the label outlived the question.[31]

22.7 Eicosanoids: the local post of the membrane

The third lipid profession is practised from within the membrane itself. Arachidonate — 20:4 $\Delta^{5,8,11,14}$, the ω-6 tetraene sitting at the sn-2 position of innumerable phospholipids — is kept in the bilayer as a depot; when the cell is

31. The road ran through statins: Endō's compactin (1976), the penicillin of cholesterol, was found by screening fungal broths for inhibitors of HMG-CoA reductase, the step Bloch's pathway made intelligible. The target was discovered because the pathway was; the pathway because the label was. Method precedes medicine.

perturbed, phospholipase A_2 cuts it free, and two enzyme families compete for the escapee (Figure 22.5). The ***cyclooxygenases*** *(COX-1, the constitutive housekeeper; COX-2, the inducible responder) wrap the chain into prostaglandin H_2, from which flow the prostaglandins (fever, the pain signal's amplifier, the stomach's protective slime — which is why blocking them soothes the joint and irritates the gut), prostacyclin PGI_2 (the endothelium's vasodilator and anti-platelet letter), and thromboxane TXA_2 (the platelet's opposite dispatch). The* ***lipoxygenases*** *take the other road to the leukotrienes: LTB_4, the neutrophil's chemotactic trumpet; the cysteinyl leukotrienes, the slow-reacting substance of anaphylaxis that closes the asthmatic airway. Two properties make the family distinctive: they act locally, within seconds and within micrometres — paracrine, not endocrine — and they act in opposition, TXA_2 against PGI_2, inflammation against its resolution. The membrane keeps both letters in the same drawer.*

The pharmacology at this fork is the oldest in continuous use. Aspirin — salicylate from willow bark, acetylated at Bayer in 1897, mechanism unknown until 1971 — transfers its acetyl group onto Ser530 of the cyclooxygenase and blocks the channel by which arachidonate enters: irreversible inhibition, the enzyme dead for its lifetime.[32] *The consequence of irreversibility is a small masterpiece of pharmacokinetics, and it repays arithmetic.*

> **WORKED EXAMPLE 22.2 — WHY ONE BABY ASPIRIN RULES TEN DAYS**
>
> *The platelet is anucleate — it cannot synthesise new COX-1 — and circulates for ~10 days; the endothelial cell replaces its enzyme within hours. Aspirin's own half-life is ~20 min, and ~2.5×10^{11} platelets are made daily, ≈ 10% of the pool. Explain why a single 75-mg dose silences thromboxane for days, selectively.*
>
> **Solution.** *The drug is gone in two hours; the inhibited enzyme is gone only with the platelet. Each day ~10% of the platelet pool is replaced with fresh, active enzyme, so platelet TXA_2 capacity recovers as 0.10 per day of the deficit: after the first dose (>90% of platelets hit), function returns by ~10% daily, and daily re-dosing holds inhibition above ~95% indefinitely. The endothelial cell, by contrast, resynthesises COX between doses and keeps making PGI_2. Irreversible inhibition of an anucleate cell converts a 20-minute drug into a 10-day effect, and the nucleated endothelium's resilience does the selectivity for free. The whole clinical art of low-dose aspirin is contained in the platelet's missing nucleus.*

32. John Vane's 1971 demonstration that aspirin-like drugs act by blocking prostaglandin synthesis (Vane, 1971) unified the drug's analgesia, antipyresis and anti-inflammation in one mechanism — and explained its gastric cost in the same stroke: the stomach's slime is also a prostaglandin's work. The Nobel of 1982 (Vane, with Bergström and Samuelsson, the prostaglandins' discoverers) was awarded, one might say, to the willow tree.

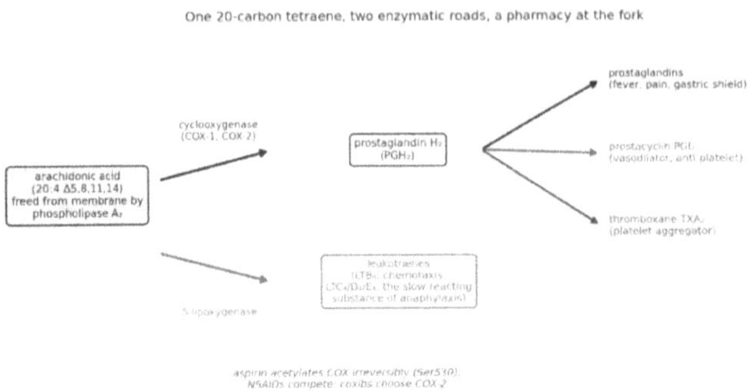

Figure 22.5 The eicosanoid fork. Arachidonate (20:4), freed from the sn-2 position by phospholipase A_2, is taken by cyclooxygenase (navy) to PGH_2 and thence to the prostaglandins, prostacyclin and thromboxane — or by 5-lipoxygenase (teal) to the leukotrienes. Aspirin's acetylation of Ser530 closes the COX road irreversibly; the coxibs select COX-2; montelukast and zileuton guard the lipoxygenase side in asthma.

22.8 Lipids as signals and anchors

Two further professions deserve notice before the clinic. The membrane's inner leaflet is itself a signalling switchboard: phosphatidylinositol, the minor lipid of Figure 22.4, can be phosphorylated on its inositol ring to PIP_2 and PIP_3 — and these are second messengers in waiting (Chapter 51 will find phospholipase C splitting PIP_2 into inositol trisphosphate and diacylglycerol, one half liberating calcium, the other holding protein kinase C), while PI 3-kinase's PIP_3 recruits the survival kinases to the membrane's edge. A lipid present at one per cent of the membrane thus commands a disproportion of the cell's decisions; the headgroup menu of § 22.4 is a menu of potential messages. And the isoprene economy of § 22.6 has a construction trade: prenyl groups — farnesyl (C15), geranylgeranyl (C20) — are appended to proteins (Ras most famously) to nail them to membranes; dolichol, the C_{95} isoprenoid, carried Chapter 21's N-glycan precursor across the ER; the quinones of the respiratory chain (Chapter 29) and the K, E and A vitamins are isoprenoids all. The fat-soluble vitamins are, each of them, a lipid fragment the body's chemistry has learned to read: K's naphthoquinone for the carboxylase of clotting (Chapter 19's warfarin), E's chromanol as the membrane's own antioxidant — a radical-catcher stationed exactly where polyunsaturated chains are most at risk — and A's retinal, isomerised by light, as the eye's first transducer. Even vision begins with an isoprene.

22.9 Lipids in the clinic: the good, the trans, and the plaque

Where does it go wrong? Three entries from the casebook, each to be taken up in earnest later. First, the **plaque**: cholesterol is insoluble even by lipid standards, and travels in the lipoprotein parcels of Chapter 38 — LDL delivering to tissues, HDL retrieving; when LDL lingers in an artery wall and oxidises, the macrophage that eats it becomes the foam cell, and the foam cell the plaque. The LDL/HDL arithmetic the clinic recites is parcel logistics, nothing more — but the statins that intercept the parcel's manufacture (the Historical note's compactin) are, by trial after trial, among the most effective drugs ever dispensed. Second, the **trans** fat: partial hydrogenation, the kitchen's answer to Figure 22.2, straightens cis kinks *into trans chains that pack like saturates — and also raise LDL and lower HDL simultaneously, a double mischief no natural fat quite manages; the industrial era's solution to butter's expense was, retrospectively, a small public-health catastrophe, now largely legislated away. Third, the* **balance***: ω-6 and ω-3 fatty acids feed rival eicosanoid series (§ 22.7's fork from either shore), and the modern diet's twenty-to-one tilt toward ω-6 is blamed — with evidence more suggestive than conclusive, as honesty requires — for a background hum of inflammation. The lysosomal lipidoses (Tay–Sachs, Gaucher, Niemann–Pick, Fabry) complete the casebook: one enzyme missing, one lipid accumulating, the demolition principle of § 21.9 repeated in fat.*

> ***Case study — The platelet and the pavement: TXA_2 against PGI_2***
>
> *The vessel wall and the platelet hold opposite briefs at § 22.7's fork. Intact endothelium secretes prostacyclin: vasodilation, and a cAMP-mediated "do not aggregate" posted to every passing platelet. Damaged endothelium exposes collagen; the platelet adheres, activates its own COX-1, and answers with thromboxane A_2: vasoconstriction, and "aggregate here" — a positive feedback that builds the haemostatic plug in minutes. The system is a dialectic by design: repair balanced against patency, both written in twenty-carbon lipids with half-lives of seconds (TXA_2 hydrolyses in ~30 s; PGI_2 in ~3 min — the signals are deliberately perishable, so that they never outlive the local occasion). Low-dose aspirin (Worked example 22.2) exploits the asymmetry of renewal: the anucleate platelet cannot re-arm between doses, the nucleated endothelium can. The same dialectic, read in the coronary artery rather than the skin, is the difference between a bruise and an infarct — and the whole of cardiovascular prevention between two pages of this chapter.*

22.10 Synthesis of Chapter 22

The lipids are a family defined by refusal — insolubility — and the chapter has watched that one temperament flower into three professions. As **store***, the triacylglycerol: fully reduced carbon, anhydrous packaging, and a seven-fold gravimetric advantage over glycogen that no migrant bird can afford to ignore [Eqs. (22.2)–(22.3)]. As* **partition***, the amphiphiles: two tails and a headgroup — glycerophospholipids with their charge menu, sphingolipids with their sugar crests — whose self-assembly obeys the logarithm of the cmc [Eq. (22.4)] and whose melting obeys the methylene sum of the London attraction [Eq. (22.1)], the same equation that prices butter against oil (Figures 22.1–22.2). As* **signal***, the sterol plate and the eicosanoid fork: cholesterol buffering the membrane's temper and fathering the steroids, arachidonate posting the local letters of inflammation that the willow tree, via Ser530, taught us to intercept. The direction of Figure ii holds: carbohydrates behind us in two chapters, lipids' structure now surveyed — and the amphiphile's sheet, mentioned here only in promise, awaits its own chapter. Chapter 23, Biological Membranes, takes the cylinder molecules of § 22.4 and asks the question that founded cell biology: what happens when a billion of them meet water at once?*

Exercises (with full resolutions)

> **Exercise 22.1**
>
> *Palmitate (16:0) melts at 63 °C, oleate (18:1 Δ^9) at 13 °C, elaidate (trans-18:1 Δ^9) at 45 °C. Explain both gaps from the geometry of § 22.2.*

Resolution. *The 50-degree gap between palmitate and oleate is Figure 22.2: the cis kink opens a vault in the crystal, forfeits the dispersion contacts of Eq. (22.1), and the extra two carbons of oleate cannot repay the loss. The 32-degree gap between oleate and elaidate is the same figure read the other way: the trans bond leaves the chain straight — the packing of a saturated chain, plus two carbons — so elaidate behaves nearly as stearate (69.6 °C) minus a little. Configuration, not composition, sets the melting point; the trans fat is a saturated fat in geometric disguise.*

Exercise 22.2

A soap's critical micelle concentration falls by about a factor of three for each methylene added to the chain. Using Eq. (22.4), what free-energy change per CH_2 does this correspond to at 298 K? Comment on its size.

Resolution. $\Delta\Delta G = RT \ln(1/3) = (8.314 \times 298) \times (-1.099) \approx -2.7$ kJ mol^{-1} *per methylene — small beside a covalent bond (~350 kJ mol^{-1}), but exactly the scale of the hydrophobic effect's per-methylene rent (§ 5.6), and additive: ten carbons accumulate −27 kJ mol^{-1}, comparable to an ATP hydrolysis. Self-assembly is financed one methylene at a time, and the logarithm — the dilution entropy — is why the price appears inside a log rather than a sum.*

Exercise 22.3

Deep-sea fish and arctic plants are rich in polyunsaturated membrane lipids; thermophilic bacteria enrich saturated and longer-chain species. State the common principle and its molecular mechanism.

Resolution. *Homeoviscous adaptation: every membrane must hold its fluidity near the working temperature, and Figure 22.1's two dials are the only knobs the lipid affords. Cold demands lower T_m — more cis kinks (Figure 22.2's broken packing) and shorter chains. Heat demands higher T_m — longer, straighter, more saturated chains (and, in archaea, the membrane-spanning tetraether lipids that cannot melt apart at all). The bacterium re-tunes in one generation; the fish has tuned by selection. Cholesterol, where present, flattens the dial altogether (Figure 22.6) — which is one reading of why animals invented it.*

Exercise 22.4

Celecoxib was designed to inhibit COX-2 preferentially; early enthusiasm was tempered when trials found a small excess of cardiovascular events. Using § 22.7's dialectic, explain the finding without invoking any off-target effect.

Resolution. *In the vessel wall, PGI_2 — the anti-platelet, vasodilating letter — is made largely by endothelial COX-2, while platelet TXA_2 is COX-1's. Aspirin, hitting COX-1 in the anucleate platelet, tilts the balance toward PGI_2 (Worked example 22.2). A selective COX-2 inhibitor does the opposite: it suppresses the endothelium's PGI_2 while leaving the platelet's TXA_2 fully armed — the same seesaw, sat on the other end. The adverse signal is not an off-target effect but the target itself, dialectically understood: you cannot inhibit one letter of a balanced correspondence without changing what the conversation says.*

EXERCISE 22.5

Tay–Sachs disease results from deficiency of β-hexosaminidase A, which removes the terminal GalNAc from ganglioside GM2. Why is the nervous system the principal victim, and why does the disease declare itself in infancy?

Resolution. *Gangliosides are the neuron's own surface lipids — they are ~10% of neuronal membrane lipid, and the developing brain turns them over fastest. The failure is of demolition, not construction (§ 21.9's principle): GM2 accumulates in lysosomes that swell until the neuron distends, fails, and dies. Development sets the clock: the prenatal brain builds with maternal clearance for company, but after birth the infant's turnover runs full pace against zero capacity, and the accumulation crosses the functional threshold within months — regression, blindness (the cherry-red macula, a retina translucent around a lipid-swollen ganglion layer), and the inexorable course. Enzyme replacement cannot cross the blood–brain barrier, which is why this demolition, alone among the lysosomal diseases, still lacks its cure.*

Research-level propositions

Proposition 22.α. The London sum of Eq. (22.1) predicts that the melting-point increment per methylene should weaken as chains lengthen (end-effects dilute). Fit the saturated series of Figure 22.1 to a finite-chain model with explicit end-group penalties, extract the per-CH_2 dispersion energy, and compare it with the cmc increment of Exercise 22.2: the two numbers are the same hydrophobic rent measured in two currencies — assess whether they agree. Proposition 22.β. Lipid rafts (§ 22.5) remain controversial forty years on because they are small, transient and defined by methods that disturb them. Using the phase physics of Figure 22.6 and the fluctuation formalism of § 7.5, estimate the equilibrium size and lifetime of a liquid-ordered domain near a miscibility critical point, and design the measurement (super-resolution, FRET, or scattering) that could settle the question. Proposition 22.γ. The ω-6/ω-3 rivalry of § 22.9 is a competition for the same enzymes (desaturases, COX, LOX). Construct the kinetic model of the shared pathway with the measured K_m values, derive the dietary ratio that equalises flux into the two eicosanoid series, and assess whether the "twenty-to-one" alarm survives the kinetics or dissolves into the ordinary economics of substrate competition.

References (Chapter 22)

Berg, J. M., Tymoczko, J. L., Gatto, G. J., Jr., & Stryer, L. (2023). Biochemistry (9th ed.). W. H. Freeman.

Bloch, K. (1965). The biological synthesis of cholesterol. Science, 150(3692), 19–28.

Fahy, E., Subramaniam, S., Brown, H. A., Glass, C. K., Merrill, A. H., Jr., Murphy, R. C., Raetz, C. R. H., Russell, D. W., Seyama, Y., Shaw, W., Shimizu, T., Spener, F., van Meer, G., VanNieuwenhze, M. S., White, S. H., Witztum, J. L., & Dennis, E. A. (2005). A comprehensive classification system for lipids. Journal of Lipid Research, 46(5), 839–861.

Nelson, D. L., Cox, M. M., & Hoskins, A. A. (2021). Lehninger principles of biochemistry (8th ed.). W. H. Freeman.

Vance, D. E., & Vance, J. E. (Eds.). (2008). Biochemistry of lipids, lipoproteins and membranes (5th ed.). Elsevier.

Vane, J. R. (1971). Inhibition of prostaglandin synthesis as a mechanism of action for aspirin-like drugs. Nature New Biology, 231(25), 232–235.

CHAPTER 23

Biological Membranes

CONTENTS OF THIS CHAPTER

In which a billion cylinder molecules, meeting water, are found to build of their own accord the sheet that made the cell possible; in which the sheet is shown to be a two-dimensional liquid, measured, melted and watched as it flows; in which the membrane's two faces are discovered to differ as scrupulously as a letter's front from its back; and in which the rarest of molecular events — two sheets becoming one — is revealed as the cell's daily logistics, performed by zippers and governed by a law of curvature.

23.1 The sheet that builds itself

Chapter 22 left the phospholipid in suspension: a cylinder of two tails and a head, content in chloroform, mute in water. Ask now what a billion such molecules do when the water can no longer be refused, and the answer is the founding fact of cell biology — they do not dissolve and they do not clump; they sheet. The tails flee the water into each other's company (the hydrophobic effect of § 5.6, collecting a billion rents at once), the heads stay to face it, and the only arrangement that hides every tail while leaving every head exposed is two monolayers back to back: the bilayer, three to four nanometres thick, continuous, self-sealing, and closed — for a sheet with an edge would expose its tails along the rim, and the rim's penalty is paid until the sheet curls into the sphere that has none. The liposome that forms in the test tube, the vesicle that buds in the cell, and the plasma membrane of the cell itself are one and the same solution to one and the same equation. No template, no enzyme, no plan: the architecture is the thermodynamics.

Whether a given amphiphile sheets, spheres, or inverts can be read off its geometry before the experiment is run. Israelachvili's ***packing parameter*** *compares the molecule's own bulk to the space its head allows it:*

$$P = \frac{v}{a_0\, l} \tag{23.1}$$

— v the volume of the hydrophobic tail(s), a_0 the area the headgroup claims at the interface, l the tails' length. Nothing here is deeper than packing: if the tails are slim compared to the head's allowance ($P < 1/3$) the molecules are wedges and wedges close into cones of cones — the micelle (detergents, lysolecithin, one tail). If tail and head are matched ($1/3 < P < 1$) the molecules are cylinders and cylinders stack into sheets — the bilayer (the diacyl phospholipids, two tails). If the tails overflow the head ($P > 1$) the wedge points the other way and the sheet curves with water inside — the inverted phases (phosphatidylethanolamine, cardiolipin, the fusogens of § 23.8). Figure 23.1 shows the three verdicts; the cell's membrane lipids

sit deliberately near the boundary, cylinders with a leaning toward inversion — stable enough to hold a cell, curved enough to bud and fuse when the proteins command it. The membrane is a bilayer kept, one might say, in a state of controlled dissatisfaction.[33]

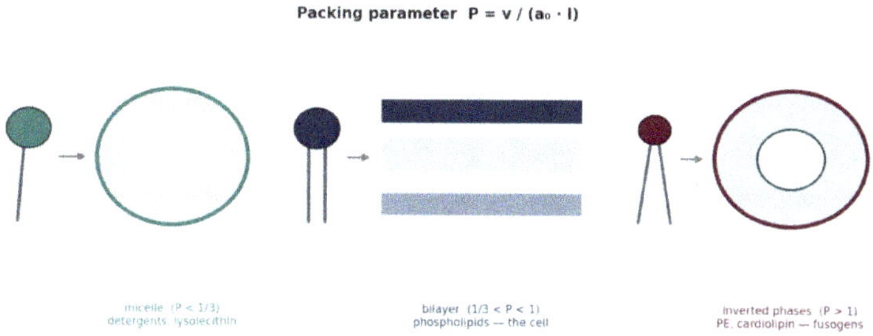

Figure 23.1 *One parameter, three architectures [Eq. (23.1)]. Slim-tailed cones (P < 1/3) close into micelles; matched cylinders (1/3 < P < 1) stack into the bilayer; tail-heavy inverted cones (P > 1) curve the sheet the other way — the inverted phases exploited by fusion. The cell's phospholipids live near the cylinder boundary: stable, but ready to bend.*

23.2 Thickness, curvature, and the price of a sphere

The bilayer's thickness is the lipid's own length twice over — some 3–4 nm for the common 18-carbon species — and it is not negotiable, because the proteins that sit in the membrane must match it. A transmembrane helix presents a hydrophobic band of precisely the bilayer's width (§ 12.6); when lipid and protein disagree, the sheet deforms — stretching or thinning locally to bury the mismatch (Figure 23.2, left) — and this **hydrophobic matching** *is now recognised as a sorting principle: the Golgi's thicker, cholesterol-rich membranes keep the longer-spanned proteins, and the ER's thinner sheet retains the rest (§ 23.7). The membrane grades its residents by their waistlines.*

Curvature has its price, and the price is the Young–Laplace law already derived for the plant wall [Eq. (21.3)] and now applied at the molecular scale. To hold a spherical membrane cap of radius r against a pressure difference Δp costs tension $\gamma = \Delta p r/2$, or reading the law as the cell must — as the cost of making a bud: every vesicle, tubule and organelle pays

$$\Delta p = \frac{2\gamma}{r} \tag{23.2}$$

— and the 1/r is the operative word: small spheres are expensive, large sheets are free. A 50-nm transport vesicle bears the membrane's full curvature penalty, which is why vesicles are never made spontaneously but always by proteins — coats that bend the sheet (COPI, COPII, clathrin), proteins that wedge into one leaflet (the BAR domains), and scaffolds that pay the γ on the membrane's behalf (§ 23.7). A 25-nm

33. That the membrane was a bilayer at all was deduced before it was seen: Gorter and Grendel (1925) extracted the lipids of a known number of red cells, spread them as a monolayer on a Langmuir trough, and measured twice the cells' surface area. The conclusion — two leaflets — was right; the arithmetic, by luck, was wrong in two compensating ways (they had undercounted both the cells' area and the lipid yield). Science, like the membrane, sometimes seals its own defects.

tubule pulled from the Golgi by a motor protein walking a microtubule is Eq. (23.2) doing manual labour. The cell's geography — its cisternae, tubules and vesicles — is the diagram of who pays, and how much, for curvature.

Worked example 23.1 — The red cell at the needle's eye

Take the cortical tension of a red blood cell as $\gamma \approx 10^{-5}$ N m^{-1}. (i) What pressure sustains a spherical cap of the cell's own radius, $r \approx 4$ μm? (ii) The spleen interrogates every red cell by slits of ~2 μm; what happens if the cell's membrane stiffens (as in hereditary spherocytosis, or when the cell ages)?

Solution. *(i) Eq. (23.2): $\Delta p = 2 \times 10^{-5} / (4 \times 10^{-6}) = 5$ Pa — trivially small; the healthy red cell is a floppy balloon whose membrane costs almost nothing to deform, which is precisely the point of a cell that must fold through capillaries narrower than itself, eight thousand times a day for four months. (ii) Stiffen the membrane (lose spectrin, lose area, lose ATP) and γ rises; the slit demands a curvature the cell can no longer buy, and the spleen — that most pedantic of quality inspectors — detains and destroys the inflexible. Hereditary spherocytosis is Young–Laplace as a clinical sign: the sphere is the cell that could no longer afford its own curvature.*

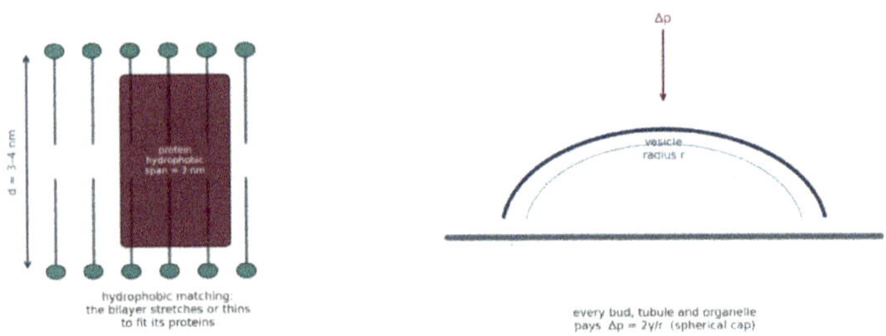

Figure 23.2 *Left: hydrophobic matching — the bilayer deforms to bury a protein's transmembrane span, and the mismatch sorts proteins among organelles of different thickness. Right: the price of curvature [Eq. (23.2)]: every bud, tubule and organelle pays $\Delta p = 2\gamma/r$, and the coat proteins are the paymasters.*

23.3 The selective wall: what crosses, and what never will

A sheet that sealed nothing would be no wall at all. Measure the permeability of a bare liposome and the bilayer reveals its double nature in a single logarithmic sweep (Figure 23.3): water crosses (slowly, unassisted — the aquaporins of Chapter 24 will hurry it); small nonpolar molecules — O_2, CO_2 — cross as if the wall were not there; indole and tryptophan leak respectably; urea and glycerol trickle; glucose is stopped; and the ions are stopped absolutely — sodium's permeability is some ten orders of magnitude below oxygen's. The physics is the Born energy of § 5.4 again: to move an ion from water ($\varepsilon \approx 80$) into the hydrocarbon core ($\varepsilon \approx 2$) costs the charging energy

$$\Delta G_{Born} = \frac{q^2}{8\pi\varepsilon_0 a}\left(\frac{1}{\varepsilon_{core}} - \frac{1}{\varepsilon_{water}}\right) \approx +250\ \mathrm{kJ\,mol^{-1}} \quad (a \approx 0.2\ \mathrm{nm}) \qquad (23.3)$$

— a the ion's radius, and the number, some hundred kT, is the reason the leak is one ion in a geological age. Eq. (23.3) is, without exaggeration, the equation that

permits bioelectricity: because the bare bilayer is a near-perfect insulator, every ion that crosses a cell membrane crosses through a protein, and therefore under the cell's control — the channels, carriers and pumps of Chapter 24 inherit their authority entirely from this wall's refusal. The membrane is not a barrier with gates; it is a barrier whose only gates are deliberate.[34]

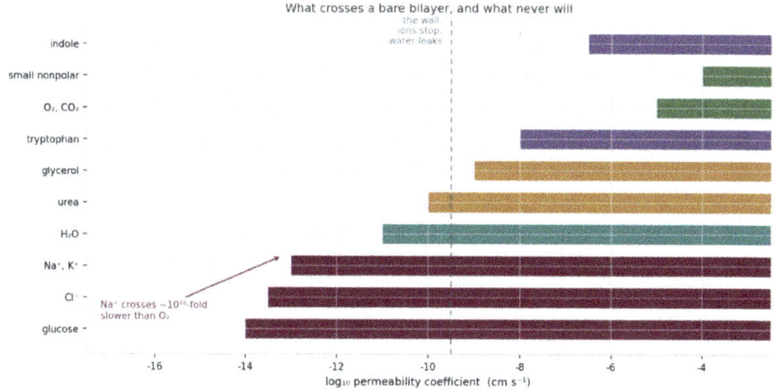

Figure 23.3 *The permeability spectrum (logarithmic). Ten orders of magnitude separate O_2 from Na^+: gases and small nonpolar molecules pass, water and small uncharged species leak, sugars and ions are excluded — the ions by the Born penalty of Eq. (23.3). Everything the cell admits, it admits on purpose, through protein (Chapter 24).*

23.4 The fluid mosaic: a liquid that carries protein

The bilayer of § 23.1 is liquid, and the measurement that proved it is one of cell biology's neatest. Label a cell's surface proteins or lipids with a fluorophore, bleach a small square with a laser pulse, and watch: the dark square refills from the edges within seconds to minutes (Figure 23.5). The recovery is diffusion in two dimensions, and its time constant prices the viscosity: for a bleached square of width w,

$$\tau = \frac{w^2}{4D} \quad \Longrightarrow \quad D = \frac{w^2}{4\tau} \tag{23.4}$$

— the two-dimensional Einstein relation of § 7.3 applied to a sheet; a square of 4 μm refilling with $\tau \approx 14$ s gives $D \approx 2\ \mu m^2\ s^{-1}$, the membrane's measured liquidity, roughly a hundred-fold slower than diffusion in water and ten-thousand-fold faster than anything in a solid. FRAP (fluorescence recovery after photobleaching) also reports the ***mobile fraction****: rarely 100% — part of every membrane's population is fenced (the cytoskeleton's corrals), anchored (junctions), or aggregated, so that the membrane is a liquid with a lawful minority of prisoners. Singer and Nicolson's* ***fluid mosaic model*** *(1972) said in one sentence what the experiments of that decade had shown severally: a two-dimensional oriented liquid of lipid, in which proteins float, drift, and occasionally are told where to stand. Half a century later the model has acquired qualifications — fences, rafts, asymmetry, curators — but no successor: the mosaic is still fluid, and the fluid is still the point.*

34. Water's anomalous leak ($P \approx 10^{-3}$ cm s^{-1}, ten orders above sodium's) troubled the field for decades — until Peter Agre's aquaporin (1991) showed that even the leak was gated: most cellular water traffic is channelled, and the bare bilayer's share is smaller still. The exception proved the rule, and then was repealed.

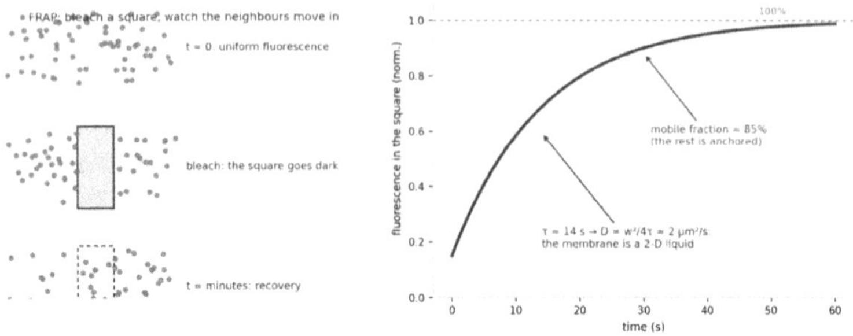

Figure 23.5 The liquidity of the membrane, measured. Left: bleach a square of fluorescently labelled membrane; neighbours diffuse in. Right: the recovery curve's time constant prices the diffusion coefficient [Eq. (23.4)] and its plateau the mobile fraction — the liquid's own census of its prisoners.

23.5 Two faces: the asymmetry of the plasma membrane

The fluid mosaic floated symmetric; the cell's membranes are not. Every biological membrane has a sidedness — imposed at birth in the ER, where lipids are synthesised on the cytosolic leaflet and a fraction is flipped across, maintained by dedicated enzymes against the leak of time, and read by everything that touches the cell. Measure the distribution and the verdict is stark (Figure 23.4): the outer leaflet of the plasma membrane holds the choline lipids (PC, sphingomyelin) and all of the glycolipids — their sugars can only face the world, where they compose the glycocalyx of § 21.8; the cytosolic leaflet holds PE, and all of the phosphatidylserine and phosphatidylinositol. The asymmetry is twice functional. Inward, it is a signalling estate: PS's negative charge and PI's phosphorylatable ring (§ 22.8) recruit and feed the cytosolic signalling proteins, and the inner face of the membrane is the cell's private switchboard. Outward, it is a passport: the bland choline surface is ignored by complement and coagulation alike.

Maintaining the difference costs energy. The ***flippases*** *(P4-ATPases) spend ATP to drag PS and PE from the outer to the inner leaflet against their gradients — for the passive flip-flop of an unaided lipid, head through hydrocarbon, has a half-time of hours to days (the Born argument of Eq. (23.3) in miniature), while the cell's signalling demands minutes. And the asymmetry's abolition is itself a signal: the* ***scramblases****, activated by calcium or by the apoptotic caspases, collapse the gradient in moments, and PS appears on the outer face — where it is read, by the macrophage's PS receptors, as eat me (apoptosis, § 18.10), and by the clotting factors as assemble here (the platelet's activated surface, Chapter 19's prothrombinase). The cell wears its intentions on its outer leaflet; a membrane whose inside shows outside is either dying or clotting, and the body attends to both with equal seriousness.*

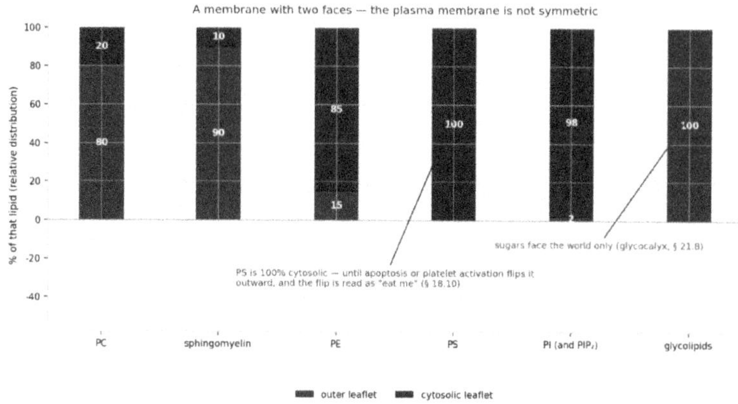

Figure 23.4 The two faces, quantified: distribution of the major lipids between the outer (navy) and cytosolic (crimson) leaflets of a mammalian plasma membrane. Glycolipids face only outward; PS and PI only inward; the flippases pay for the difference, the scramblases spend it, and the blood and the immune system read the result.

23.6 Rafts and fences: the organised fluid

Within the fluid, order. The sphingolipids of § 22.5 — long, straight, saturated tails under sugar crests — pack with cholesterol into patches of the liquid-ordered phase (Figure 22.6's broadened transition, localised): nanometre-to-micrometre platforms, richer also in the GPI-anchored proteins outside and the doubly acylated signalling proteins inside, which the field calls ***rafts****. The word has suffered from success: rafts have been claimed for every unexplained colocalisation in cell biology, and their smallness and transience make them among the most contested objects in the field (Proposition 22.β asked the reader to measure one honestly). What is not contested is the principle: the bilayer of a real cell is not one phase but a near-critical mixture (Figure 22.6, dashed), and near a critical point the sheet is restless — fluctuating domains assemble and disperse on the millisecond scale, and proteins that prefer the ordered phase are swept into transient company. A signalling receptor and its kinase, rafted together, meet a hundred times oftener than diffusion alone would arrange; the raft is a cocktail party that the membrane throws for its proteins, and disbands the moment the police of measurement arrive.*

The fences are blunter instruments. The spectrin-actin cortex beneath the plasma membrane corrals transmembrane proteins into compartments of tens of nanometres — hop diffusion, FRAP's prisoners; the tight junction seals the epithelial cell's membrane into apical and basolateral estates, so that the gut cell can keep its glucose import machinery on one face and its glucose export on the other — polarity being, at bottom, a membrane address system; and the postsynaptic density anchors its receptors outright. The fluid mosaic, thus amended, reads: a liquid, yes — but a liquid with neighbourhoods, dress codes, and a constabulary.

23.7 The inner geography: endomembranes and their traffic

The eukaryotic cell multiplied its membrane a hundredfold and folded it into a geography: the nuclear envelope (two bilayers, pored), the endoplasmic reticulum (the cell's largest sheet — protein and lipid synthesis, calcium's vault), the Golgi (a stack of flattened cisternae, each a station of the glycan tailoring of § 21.6), the endosomes, lysosomes, peroxisomes, and the bounding membrane itself — an archipelago of compartments, each with its own lipid address (thickness, charge, phosphoinositide signature) and its own protein population. The compartments are not connected by plumbing but by ***traffic****: vesicles that bud from one membrane and fuse with another, carrying cargo and membrane alike. Budding is paid for, per Eq. (23.2), by the coats — COPII for ER-to-Golgi, COPI for the return, clathrin for the plasma membrane and endosomes — each a protein cage that sculpts the sheet into its sphere and then, its work done, is dismantled. Addressing is the Rabs, small GTPases of Chapter 51's family, whose membrane-bound forms mark each compartment and whose effectors tether the incoming vesicle to its proper target. The system is a postal service of breathtaking volumes — a secretory cell dispatches thousands of vesicles a minute — and it was mapped, in its first geography, by the electron microscope and the pulse-chase: Palade's pancreatic acinar cell, fed a three-minute pulse of radioactive leucine, told the whole story in autoradiographic grains — rough ER at seven minutes, Golgi at twenty, secretory granules at the hour (Plate 23.1). The secretory pathway, biology's first supply chain, was traced in an afternoon's worth of exposures.*[35]

Plate 23.1 *George E. Palade (1912–2008), whose electron microscopy and pulse-chase autoradiography mapped the secretory pathway — ribosome to rough ER to Golgi to granule — and founded modern cell biology as an experimental science (Nobel Prize in Physiology or Medicine, 1974, shared with Albert Claude and Christian de Duve). Portrait from the Nobel Foundation archive, nobelprize.org, reproduced from the Foundation's publicly distributed image.*

23.8 Fusion: the rarest event, performed daily

For every vesicle that buds, one must fuse — and fusion is, on its face, improbable. Two bilayers repel: their headgroups are hydrated, and the water between them must be expelled before the sheets can approach within a nanometre; then the outer leaflets must merge while the inner stay put (the ***stalk****), the merged sheet must remodel (****hemifusion****), and only then can the pore open and the contents*

35. Palade's grains established the itinerary but not the mechanism: whether cargo moved by vesicles or by maturing cisternae was disputed for thirty years — the "vesicular-versus-cisternal" controversy — and was settled, as such things usually are, by both being partly right: vesicles shuttle the machinery backwards while the cisternae themselves migrate forward, cargo riding the stack like a traveller on a moving pavement.

meet (Figure 23.6). The intermediates are the inverted phases of Figure 23.1 put to work: the stalk is precisely the structure that PE and cardiolipin ($P > 1$) lower their eyes to form, which is why fusogenic membranes are rich in them. The pathway is thermodynamics; the execution is protein. The ***SNAREs*** *— synaptobrevin on the vesicle, syntaxin and SNAP-25 on the target membrane — are complementary coiled coils that zipper from their distal ends toward the membranes, and the free energy of their zippering, paid out gradually as the coil winds, is what drags the two sheets through the hydration barrier and the stalk (Chapter 51 will meet the calcium trigger, synaptotagmin, that fires the loaded zipper in the synapse within a hundred microseconds). One SNARE complex delivers ≈ 35 kT; a handful suffice. Botulinum and tetanus toxins are proteases that cut SNAREs — the whole of botulism is one unzipped fusion machinery — and the clinical uses of one of them, cosmetic and neurological, are a footnote too lucrative to omit. Fusion is the membrane's one forbidden act; the cell performs it millions of times a second, under guard, with the leaflets taking turns so that the inside never meets the outside. Even at the moment of union, the membrane keeps its two faces.*

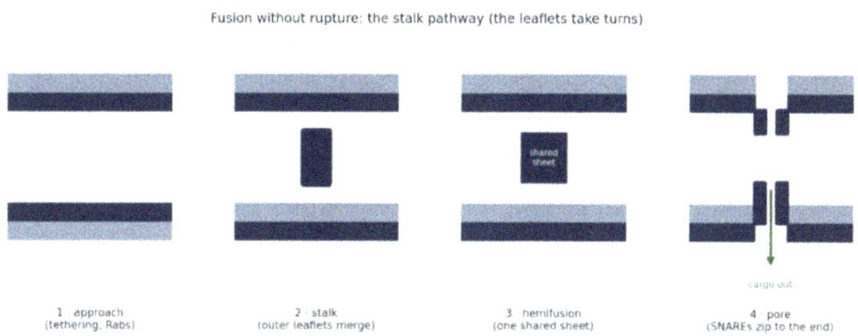

Figure 23.6 *Fusion by the stalk pathway. Apposed bilayers (outer leaflets navy, inner pale) merge outer leaflets first (stalk), remodel into one shared sheet (hemifusion), and only then open the pore — the geometry of the inverted phases ($P > 1$, Figure 23.1) in action, driven by the SNARE zipper and triggered, in the synapse, by calcium.*

23.9 Membranes at war: antibiotics, toxins, and the arsenal of the pore

Where there is a wall, there is a siege. The pore-forming arsenal is nature's oldest artillery: the ***defensins*** *of our own innate immunity, small cationic peptides that bind the bacterial membrane's anionic lipids and barrel through it; the* ***colicins*** *with which bacteria murder each other; melittin, the bee's two-and-twenty residues of membrane nonsense; the cholesterol-dependent cytolysins — streptolysin, perfringolysin — that oligomerise into β-barrels thirty subunits wide, a pore a lorry could drive through; and the complement system's membrane attack complex (Chapter 18), which assembles C9 into the same architecture on the surface of the invader. Against gram-negatives the outer membrane adds a second wall — and its lipid A leaflet is the endotoxin of septic shock, read by TLR4 as the signature of invasion itself. The antifungals play the sterol card: amphotericin B binds ergosterol and pores the fungal membrane, sparing (imperfectly, and at a price the kidney pays) our cholesterol; the azoles block ergosterol's synthesis. Daptomycin, the calcium-dependent lipopeptide, inserts into the bacterial membrane and depolarises it — an ionophore in clinical dress, and a reminder that the membrane potential of Chapter 24 is itself a target. The* *valinomycins and gramicidins of the microbiologist's bench are the same idea in miniature: carry the ion across the wall, collapse the gradient, and the cell's economy — built, as Chapter 29 will show, entirely on gradients — goes dark. Every antibiotic that touches the membrane is an argument about Eq. (23.3): the wall that insulates the cell is, to the chemist who can bridge it, the cell's single point of failure.*

CASE STUDY — THE AQUAPORIN THAT THE CHOLERA WARD NEEDED

The bare bilayer leaks water at $P \approx 10^{-3}$ cm s^{-1}; the red cell crosses the renal medulla's osmotic extremes in seconds, and swells or shrinks in transit — a traffic the leak could never serve. Agre's CHIP28, the moonlighting red-cell protein nobody had asked for, turned out to be aquaporin-1: a tetramer of hourglass pores, each passing ~3×10^9 water molecules a second while excluding every proton — the selectivity filter's positive electrostatics rebuff the hydronium that would otherwise ride the water wire (§ 5.2's Grotthuss relay, refused admission). The kidney's concentrating machinery, the brain's water balance, the eye's pressure are aquaporin traffic policed by antidiuretic hormone; and cholera's lethal diarrhoea is, in this vocabulary, a water economy hijacked — the toxin does not punch the wall but corrupts its signalling (Chapter 51's G_S, locked on), so that the gut's channels and pumps run backwards until the patient is lost to pure osmosis. The rehydration that saves him is the membrane's own physics, prescribed by the litre.

23.10 Synthesis of Chapter 23

The chapter began with a billion cylinders and a refusal, and it ends with the cell's geography — for everything between is deduction. The cylinders sheet [Eq. (23.1)], the sheet closes (edges are expensive), the sphere is liquid [Eq. (23.4)] and insulates [Eq. (23.3)]; the liquid is organised (rafts, fences, two faces), the faces are policed (flippases, scramblases), and the sheet bends only where curvature is paid for [Eq. (23.2)] — whence the coats, the Rabs, the vesicles, and the whole postal geography of the endomembrane system; and where two sheets must become one, the inverted phases and the SNARE zipper perform the forbidden act under guard. The membrane, in short, is not a bag but a state — of matter, of information, and of address — and the direction of Figure ii holds: lipids surveyed, the sheet they build now surveyed. What remains is the traffic through it. The bare bilayer of § 23.3 stops everything the cell needs; the proteins that admit it — channels, carriers, pumps, and the gradients they defend — are Chapter 24, Membrane Transport, where the wall acquires its gates, and the gates their keepers.

Exercises (with full resolutions)

EXERCISE 23.1

Sodium dodecyl sulfate (one tail, large headgroup) lyses membranes at low concentration, while phosphatidylcholine (two tails) cannot be made to form micelles at any concentration. Explain from Eq. (23.1).

Resolution. *SDS's single C12 tail under a sulfate head gives $P \approx 1/3$ or less: wedges, closing into micelles — and into the gaps of a bilayer, whose cylinders they displace and solubilise (lysis, and the detergent's laboratory use). PC's two 16–18-carbon tails under a choline head give $P \approx 0.7$–1: cylinders, whose only stable aggregate is the sheet; forcing them into micelles would leave the tails' volume impossible to hide from water at the sphere's crowded core.*

Architecture is geometry; the lysing detergent is simply the molecule whose geometry the membrane's is not.

EXERCISE 23.2

A FRAP experiment on a 3-µm square shows recovery with $\tau = 8$ s to a plateau of 70%. Give D and the mobile fraction, and offer two distinct biological readings of the missing 30%.

Resolution. *Eq. (23.4): $D = w^2/4\tau = (3\ \mu m)^2/(32\ s) \approx 0.28\ \mu m^2\ s^{-1}$ — a membrane protein's typical pace (lipids run ~10× faster). Mobile fraction 70%. The immobile 30% may be (i) anchored — bound to the cytoskeletal cortex or to an extracellular matrix, prisoners by tether; or (ii) confined in aggregates — junctional complexes, rafts, or large assemblies whose D is negligible on this timescale. The distinction is experimental: dissolve the actin (latrunculin) and the tethered population's fraction rises; only the anchored care. FRAP measures the liquid and censuses its prisoners in the same curve.*

EXERCISE 23.3

Why does flip-flop of a phospholipid take hours unaided, while lateral diffusion covers a bacterial cell's length in a second? Put both on the same energy picture.

Resolution. *Lateral diffusion moves the lipid within its leaflet: the headgroup stays in water, the tails in oil — no large barrier, so the Einstein relation gives $\mu m\ s^{-1}$. Flip-flop drags the charged headgroup through the hydrocarbon core — a miniature Born penalty [Eq. (23.3)], tens of kT — so the rate falls by e^(−ΔG‡/kT) into hours or days. Same molecule, same sheet: one motion crosses no barrier, the other crosses the membrane's own. The flippases exist precisely because the cell cannot wait for the second (§ 23.5).*

EXERCISE 23.4

Platelets activated at a wound expose PS on their outer surface; the clotting factor complexes (tenase, prothrombinase) assemble on PS. Why does a scramblase defect (Scott syndrome) cause bleeding, and why does constitutive PS exposure cause thrombosis?

Resolution. *The γ-carboxylated clotting factors (Chapter 19's Gla domains) dock onto anionic lipid surfaces via calcium bridges; PS exposure is the plate on which the coagulation complexes are served, concentrating the factors and accelerating thrombin generation a thousand-fold. Scott syndrome (TMEM16F scramblase defective): the plate is never served, thrombin forms feebly — a bleeding diathesis with normal platelet numbers, a pure membrane disorder. Constitutive exposure (scramblase gain, flippase loss, or the PS-positive microparticles of sickle disease and sepsis): the plate is served everywhere and always — thrombosis. Haemostasis is, among other things, the custodianship of one phospholipid's address.*

EXERCISE 23.5

Gramicidin A forms a dimeric pore passing ~10^7 K^+ s^{-1}; valinomycin carries ~10^4 K^+ s^{-1} across the same bilayer. Both kill cells. Why is the carrier slower than the channel, and why do both collapse the same gradients?

Resolution. *The channel is a hole: once open, the ion diffuses through at near-aqueous rates, limited only by the pore's mouth. The carrier is a ferry: valinomycin must bind the ion on one face, diffuse across the bilayer as a whole (limited by the membrane's own viscosity, Eq. 23.4's D), release, and return — each cycle one ion, the membrane's thickness the rate-limiting voyage. A thousand-fold slower per molecule, and fatal all the same: both short-circuit Eq. (23.3)'s insulation, the K^+ gradient (and with it the membrane potential, Chapter 24) collapses, and with it every process the gradient financed. Quantity matters less than topology: a wall with one unguarded gate is a wall no longer.*

Research-level propositions

Proposition 23.α. The packing parameter predicts architecture from molecular geometry, yet the cell's lipids sit near the cylinder boundary. Using Eq. (23.1) and the elastic theory of bilayers (bending modulus $\kappa \approx$ 10–20 kT), compute the spontaneous curvature the ER's PE-rich membrane should bear, and assess whether the organelle's tubule-versus-sheet morphology can be set by lipid composition alone — then design the lipid-remodelling experiment (PEMT knockdown, lyso-PC addition) that would test it. Proposition 23.β. Hydrophobic mismatch (§ 23.2) has been proposed as a sorting signal along the secretory pathway. Model a transmembrane protein's free energy as a function of bilayer thickness (compression + tilt terms), extract the partitioning between ER-like (thin) and Golgi-like (thick) membranes as a function of span length, and compare with the measured lengths of resident proteins' transmembrane domains. Proposition 23.γ. The stalk hypothesis (Figure 23.6) predicts that fusion rate depends non-monotonically on PE content: too little, no stalk; too much, no bilayer. Construct the elastic free-energy profile of stalk formation from the lipid elastic constants, locate the optimum, and design the lipid-mixing assay (fluorescence dequenching) that would map the optimum in vitro.

References (Chapter 23)

Alberts, B., Heald, R., Johnson, A., Morgan, D., Raff, M., Roberts, K., & Walter, P. (2022). Molecular biology of the cell (7th ed.). W. W. Norton.

Israelachvili, J. N. (2011). Intermolecular and surface forces (3rd ed.). Academic Press.

Nicholls, D. G., & Ferguson, S. J. (2013). Bioenergetics (4th ed.). Academic Press.

Singer, S. J., & Nicolson, G. L. (1972). The fluid mosaic model of the structure of cell membranes. Science, 175(4023), 720–731.

van Meer, G., Voelker, D. R., & Feigenson, G. W. (2008). Membrane lipids: Where they are and how they behave. Nature Reviews Molecular Cell Biology, 9(2), 112–124.

Wickner, W., & Schekman, R. (2010). Membrane fusion. Nature Structural & Molecular Biology, 17(6), 658–664.

CHAPTER 24

Membrane Transport

CONTENTS OF THIS CHAPTER

In which the wall of Chapter 23 acquires its gates; in which a logarithm is shown to be a battery, and a single ratio of potassium concentrations to hold every cell at seventy millivolts; in which the channel and the carrier are weighed and found to be a hole and an enzyme respectively; in which one protein, spending a third of the body's ATP, is revealed as the paymaster of the entire economy; and in which the reader learns why the foxglove strengthens the heart by poisoning it, judiciously.

24.1 The gates of the wall

Chapter 23 closed with a prohibition: the bare bilayer passes gases and water, leaks a little urea, and stops everything else — the ions by a hundred kT of Born penalty [Eq. (23.3)], the nutrients by their sheer polar bulk. A cell so sealed would starve in the midst of plenty, suffocate in its own wastes, and never once fire a thought. The whole of physiology is therefore conducted through a set of protein gates, and this chapter is their natural history. They come in three great families, distinguished by one question — who pays? The ***channel*** *is a hole with a conscience: open, and the ion diffuses through at near-aqueous rates, down its own gradient, paying nothing and asking nothing; its virtues are speed (10^7–10^8 ions a second) and its one intelligence is the gate that opens and the filter that chooses. The* ***carrier*** *(transporter) is an enzyme at the wall: it binds its substrate on one face, changes shape, and releases it on the other — slower by a thousand-fold (10^2–10^4 per second), saturable like any enzyme, but still downhill, still free. The* ***pump*** *is the carrier that has been wired to a battery: it runs the same alternating-access cycle but couples it to ATP hydrolysis (or to light, or to another ion's gradient), and so moves its cargo uphill — the only member of the family that can make a gradient rather than spend one. Everything the cell knows about transport is a play for these three actors, and the drama has a single plot: the pump builds the gradient, the carrier and the channel spend it, and the spending is metabolism, signalling, and thought.*

The chapter follows the plot in order. First the physics of the gradient itself (§§ 24.2–24.3): the Nernst equation, in which a concentration ratio becomes a voltage, and the Goldman equation, in which several ratios become one resting potential. Then the gates: channels (§ 24.4), carriers (§ 24.5), the pump that pays for everything (§ 24.6), and the secondary transporters that spend the pump's gradient on everything else (§ 24.7). Excitability, the gradient's most theatrical expenditure, is sketched in § 24.8 and left to Part VII for its full performance; the clinic collects its usual tribute in § 24.9.

24.2 The electrochemical potential: a logarithm is a battery

An ion near a membrane feels two forces, and the whole of bioelectricity is their sum. The first is chemical: a difference of concentration drives diffusion downhill, and Chapter 6 already priced it — moving one mole of a solute from concentration c_1 to c_2 changes the free energy by $RT \ln(c_2/c_1)$, the dilution entropy of § 6.2. The second is electrical: a mole of ions of charge z moved across a potential difference $\Delta\psi$ changes the free energy by $zF\Delta\psi$, Faraday's constant ($F = 96\,485$ C mol^{-1}) being merely the electrician's exchange rate between moles and coulombs. The two are additive, and their sum is the ***electrochemical potential*** *difference:*

$$\Delta G = RT \ln \frac{c_{in}}{c_{out}} + zF\,\Delta\psi \qquad (\Delta\psi = \psi_{in} - \psi_{out}) \tag{24.1}$$

— nothing falls from the sky here: the first term is § 6.2, the second is the definition of the volt, and Eq. (24.1) is simply the bill presented to any ion asked to cross. An ion is at ***equilibrium*** *when the bill is zero — when the concentration gradient exactly pays the voltage. Setting $\Delta G = 0$ and solving for the potential gives the condition's most useful form, the* ***Nernst equation****:*

$$E_{ion} = \frac{RT}{zF} \ln \frac{c_{out}}{c_{in}} = \frac{61.5 \text{ mV}}{z} \log_{10} \frac{c_{out}}{c_{in}} \quad (37\ ^\circ\text{C}) \tag{24.2}$$

— 61.5 millivolts per ten-fold ratio, per unit charge, at body temperature.[36] *Put in the cell's numbers and each ion receives its own private voltage, its* ***equilibrium potential****: potassium (150 mM in, 5 mM out) −90 mV; sodium (15 in, 145 out) +60 mV; chloride (10 in, 110 out) −65 mV; calcium (0.0001 in, 1.2 out) +122 mV. These four numbers are the four poles of the cellular world: everything a membrane does electrically is a journey among them.*

And one should pause at calcium, for its ratio is the most extravagant in the book: ten thousand to one, maintained against a flood, so that the cytosol is kept nearly empty of the ion that — precisely because it is nearly absent — can serve as a signal. A message that arrived constantly would be no message; the cell keeps its calcium outside and in stores so that a brief admission (a channel opening) is news. The electrochemical series is thus also a hierarchy of urgency: potassium sets the resting temper, sodium and calcium carry the alarms, chloride keeps the peace.

24.3 The resting potential: many ions, one voltage

A real membrane admits several ions at once, and the resting potential is their negotiation. The reasoning was made quantitative by Goldman (and Hodgkin and Katz, who gave it its experimental form): assume the electric field within the membrane is constant, and each ion's flux follows its own concentration and the voltage; sum the fluxes and require that at rest they cancel — no net current, for a cell charging indefinitely would be a capacitor, not a colleague. The algebra (each ion contributes its permeability-weighted concentration; chloride, an anion, enters with its concentrations inverted) yields the ***GHK equation****:*

36. Walther Nernst wrote the relation in 1888–89 as a corollary of his osmotic theory of the cell's electromotive forces — "the concentration cell" — and the same logarithm earned him, by a longer road through the third law of thermodynamics, the Nobel Prize of 1920. The reader who can recite "61.5 per decade" owns the equation; the one who has derived it from a dilution entropy and a volt owns the reason.

$$V_m = \frac{RT}{F} \ln \frac{P_K[K]_{out} + P_{Na}[Na]_{out} + P_{Cl}[Cl]_{in}}{P_K[K]_{in} + P_{Na}[Na]_{in} + P_{Cl}[Cl]_{out}} \quad (24.3)$$

— the Nernst equation generalised: each ion pulls the voltage toward its own equilibrium potential with a strength proportional to its permeability, and the rest is a weighted vote. The resting membrane is potassium's because potassium's permeability wins the vote by forty-fold ($P_{Na}/P_K \approx 0.04$): with the cell's numbers, Eq. (24.3) returns ≈ −70 mV — near E_K (−90 mV) but dragged a little toward E_{Na}, as the minority's forty-fold weakness entitles it (Figure 24.1). And the figure carries a clinical moral in its slope: because the vote is logarithmic in external potassium, a modest rise of blood K^+ depolarises every excitable cell at once — the heart's included — which is why the kidney guards 5 mM with a vigilance it grants few other analytes, and why hyperkalaemia is among the few electrolyte disturbances that can kill within the hour.

WORKED EXAMPLE 24.1 — THE VOTE, COUNTED

Take P_K = 1, P_{Na} = 0.04, P_{Cl} = 0.45; [K] 150/5 mM, [Na] 15/145, [Cl] 10/110 (in/out). Compute the resting potential.

Solution. *Numerator: (1)(5) + (0.04)(145) + (0.45)(10) = 5 + 5.8 + 4.5 = 15.3. Denominator: (1)(150) + (0.04)(15) + (0.45)(110) = 150 + 0.6 + 49.5 = 200.1. Then V_m = 61.5 × $\log_{10}$(15.3/200.1) = 61.5 × (−1.117) ≈ **−69 mV**. Three readings worth keeping. (i) Delete the sodium term (P_{Na} → 0) and the vote returns −86 mV: the resting leak of sodium is precisely what keeps the cell twenty millivolts from potassium's pole — and therefore within striking distance of the action potential's threshold (§ 24.8). The minority permeability is not a leak but a tuning. (ii) Double external K^+ to 10 mM: numerator 20.3, V_m ≈ −61 mV — the heart now sits closer to threshold, its excitability first raised, then, as channels inactivate, extinguished; the slope of Figure 24.1 is the electrocardiogram of the whole story. (iii) The numbers also expose the pump's quiet burden: at −69 mV, sodium (E = +60) and potassium (E = −90) both leak continuously, and § 24.6's ATPase must repay both leaks forever. Rest, for a cell, is an expenditure.*

*The final piece of vocabulary converts equilibria into traffic. When a channel of conductance g opens, the current it carries is driven by the distance between the actual voltage and the ion's own pole — the **driving force**:*

$$I_{ion} = g_{ion}(V_m - E_{ion}) \quad (24.4)$$

— Ohm's law, with the battery written into it: a channel open at its ion's equilibrium potential conducts nothing at all, for there is no slope left to descend. Figure 24.2 draws the four families: each ion's current is a straight line through its own E, and at the resting −70 mV the cell sits where potassium's line says "out", sodium's says "in", and chloride's says "nearly nothing" — the little diagram from which every nerve impulse, every heartbeat, and every thought in Part VII is assembled.

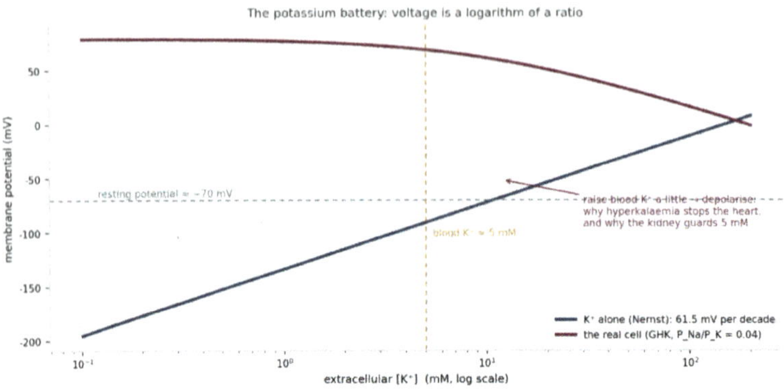

Figure 24.1 *The potassium battery. Navy: the Nernst potential for K^+ alone, 61.5 mV per decade [Eq. (24.2)]. Crimson: the GHK vote of Eq. (24.3) with the cell's permeabilities — the resting −70 mV sits between the poles. The slope is the clinical fact: external potassium is the one concentration the body cannot allow to wander.*

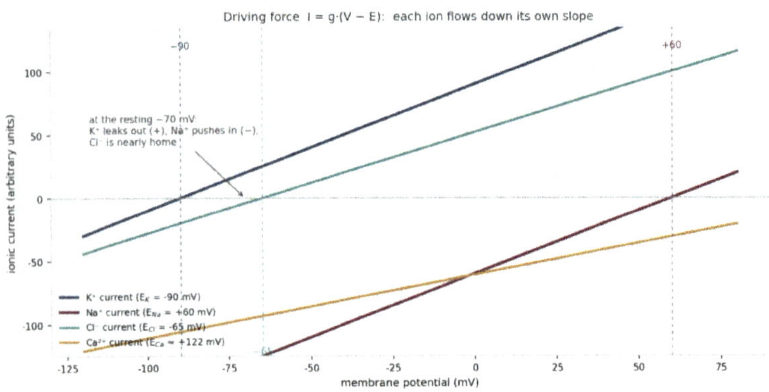

Figure 24.2 *Driving force [Eq. (24.4)]. Each ion's current crosses zero at its own equilibrium potential; at the resting voltage (arrow) potassium leaks out, sodium pushes in, and chloride is nearly home. Excitability is the art of sliding along these lines by opening and closing the conductances.*

24.4 Channels: the hole with a conscience

*The channel's dilemma is written in its numbers: it must pass 10^7 ions a second — essentially diffusion-limited, an ion every hundred nanoseconds — and yet refuse the wrong ion a thousand times in a thousand. Speed says open wide; selectivity says grip tight; and a binding site snug enough to choose is, by every rule of enzyme kinetics in Part IV, a site that holds on. The potassium channel's solution, revealed when MacKinnon's laboratory crystallised the bacterial channel KcsA (Plate 24.1), is one of the most elegant resolutions in structural biology, and it deserves a slow reading. The pore's narrowest reach — the **selectivity filter** — is lined not with side chains but with the protein's own backbone: four rings of carbonyl oxygens, one from each subunit, spaced so as to cradle a dehydrated K^+ ion in a cage of eight oxygens (Figure 24.4, left). The energetics are the whole story (Figure 24.4, right). An ion in water pays for its hydration shell — −322 kJ mol^{-1}*

for K^+, −405 for Na^+ — and to enter any pore it must be repaid. For K^+, the filter's carbonyls stand exactly where the waters stood: the repayment is complete, $\Delta\Delta G \approx 0$, and the ion walks through. For Na^+ — smaller, which is the twist every intuition resists — the cage is too large: the ion rattles, cannot touch all eight oxygens at once, and is repaid some 80 kJ mol^{-1} short of its hydration. Selective permeability is thus achieved not by binding the right ion tightly but by mimicking its water exactly — the filter is a piece of frozen solvent, tailored to one ion's size, and the thousand-fold selectivity is one dehydration that succeeds against one that fails. The speed, meanwhile, is secured by a queue: several K^+ ions occupy the filter in single file, water-spaced, and their mutual electrostatic repulsion drives the file forward — the leading ion is pushed out by the one behind, the **knock-on** *mechanism, in which the traffic jam is the engine of the throughput. The channel is fast because it is crowded.*

A gate without a gatekeeper is a leak, and the channels' second property is their **gating***: the pore opens in answer to a stimulus, and the stimulus names the family — voltage-gated channels (the Na^+, K^+ and Ca^{2+} channels of § 24.8, whose charged S4 helices slide in the electric field), ligand-gated channels (the nicotinic acetylcholine receptor of the synapse, GABA's and glycine's of inhibition), mechanically gated (the ear's, whose deflection by a hair's breadth is the first event of hearing), and the temperature- and chemistry-reading TRP channels, by which menthol feels cold and the chilli pepper commits its arson. The aquaporins of Chapter 23's case study are the family's water specialists; the chloride channels, its sanitation — and the one named CFTR, an ABC protein retrained as a channel, will anchor § 24.9's case study. A modern count finds some four hundred channel genes in the human genome: the wall, it turns out, has more doors than rooms.*

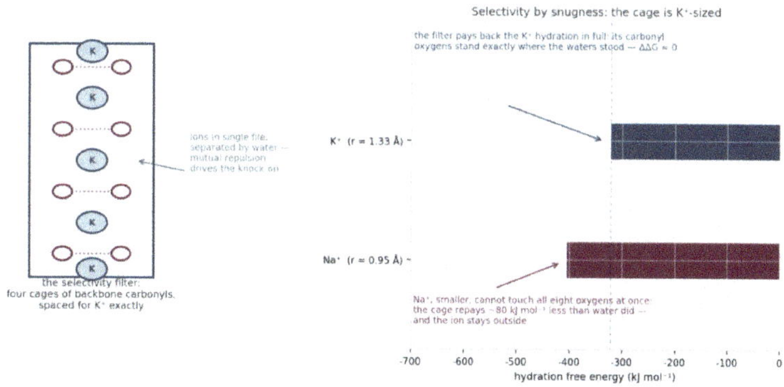

Figure 24.4 *The selectivity filter. Left: four cages of backbone carbonyls (crimson) hold dehydrated K^+ ions in single file; mutual repulsion drives the knock-on. Right: the energetic verdict — the cage repays K^+'s hydration exactly, but cannot reach around the smaller Na^+, which stays ~80 kJ mol^{-1} short and is refused. Selectivity is a precise mimicry of water.*

Plate 24.1 *Roderick MacKinnon (b. 1956), whose crystal structures of the potassium channel (KcsA, 1998; the eukaryotic channels thereafter) answered the oldest quantitative question in physiology — how a pore can be at once so fast and so particular (Nobel Prize in Chemistry, 2003, shared with Peter Agre). Portrait from the Nobel Foundation archive, nobelprize.org, reproduced from the Foundation's publicly distributed image.*

Historical note — The squid, the clamp, and the two Norwegians' patience

That the nerve impulse is a transport event at all was shown before any channel protein was glimpsed, in an experiment built from equal parts zoology and obstinacy. The squid's giant axon — nearly a millimetre across, evolution's gift to the electrophysiologist — had let Hodgkin and Huxley measure the living membrane directly; and in the summer of 1949, at Plymouth, with Cole's voltage clamp holding the membrane's voltage prisoner so that its currents had to confess, they took the action potential apart: an early inward current carried by sodium (it vanished when choline replaced the sodium), a late outward one by potassium, each with its own conductance turning on and off on the millisecond scale. Then the greater audacity: without seeing a single molecule, they wrote four differential equations — three gates for sodium (m^3h), four for potassium (n^4) — that reproduced the impulse entire, its threshold, its overshoot, its refractory silence. The 1952 papers founded computational biology fifty years early, and the 1963 Nobel crowned them; but the deeper lesson is methodological: the channels were inferred so exactly that when the molecular era arrived — the patch clamp's single channels (Neher and Sakmann, Nobel 1991), then MacKinnon's crystal — every prediction of the equations was found embodied, gate for gate, in the protein. Jens Skou's parallel discovery (1957) of the Na^+/K^+-ATPase in crab nerve — an ATPase that needed sodium and potassium on opposite sides of a membrane, and that ouabain stopped — identified the pump that pays the gradients the squid spent; he waited forty years for his Nobel (1997), a reminder that the committee, like the carrier, works by alternating access and cannot always be hurried.[37]

---

37. Hodgkin, asked why the equations fitted so well, is said to have replied that they had no choice, having been fitted to the data first. The honesty is itself instructive: the *m*, *h* and *n* gates were phenomenology before they were protein — and the protein vindicated the phenomenology in every detail.

24.5 Carriers: the enzyme at the wall

Where the channel is a hole, the carrier is a door with a doorman — and the doorman works exactly like an enzyme, for the excellent reason that a carrier is an enzyme whose reaction is a change of place. Its kinetics are therefore Chapter 15's: saturable, with a K_m and a V_{max} (Figure 24.3); specific, competitively inhibited by look-alike substrates; and slow by the channel's standard, because every transported molecule is carried across personally. The mechanism, now established by a gallery of crystal structures, is ***alternating access****: the carrier is a rocker with a single binding site that faces outward in one conformation and inward in the other — never both, for both-open would be a channel and the gradient would bleed away (Figure 24.3, right). The substrate boards the outward face; the protein rocks; the substrate disembarks within. GLUT1, the erythrocyte's and the blood–brain barrier's glucose porter, is the type specimen: $K_m \approx 1.5$ mM, poised exactly at the blood's fasting glucose so that the brain's supply is proportional and unregulatable — the brain, which cannot store glycogen worth mentioning, has contracted its fuel supply to a Michaelis constant. The GLUT family's dozen paralogues distribute the body's sugar traffic by their kinetic signatures: GLUT2 (liver, pancreatic β-cell) high-K_m, a sensor that reads abundance; GLUT4 (muscle, fat) insulin-recruited from intracellular stores, the whole of insulin's acute hypoglycaemic action in one translocation. The carrier, unlike the channel, can also be coupled — given two substrates and two slopes, it can spend one on the other; § 24.7 collects that dividend.*

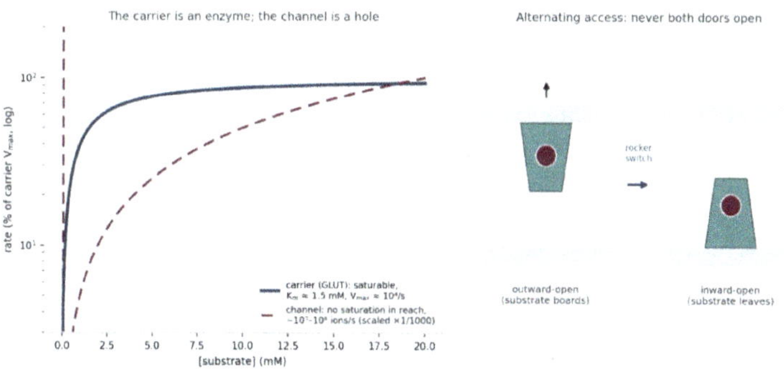

Figure 24.3 *Left: the carrier obeys Michaelis–Menten kinetics and saturates; the channel does not saturate within any physiological reach, and its ceiling is a thousand-fold higher (log scale). Right: alternating access — one binding site, two facings, never both doors open; the gradient is preserved by the doorman's discipline.*

24.6 The Na^+/K^+-ATPase: the paymaster of the economy

Everything so far — the resting potential, the gradients, the leaks — is deficit spending, and a single protein covers the deficit. The ***Na^+/K^+-ATPase*** *(Skou's crab-nerve enzyme, § 24.4's Historical note) is a P-type ATPase, so named for the phosphorylated aspartate through which every member of the family passes; its cycle, the Post–Albers scheme (Figure 24.5), is alternating access with an ATP meter attached. In E1 the carrier faces the cytosol and takes up three sodium ions; ATP phosphorylates the aspartate, ADP leaves, and the doors shut — the ions occluded, neither in nor out; the phospho-enzyme relaxes to E2, which faces outward and lets the three sodiums go; two potassiums board; dephosphorylation* *closes the doors; E2 relaxes to E1 and the potassiums are released within. The arithmetic of one cycle —* ***3 Na^+ out, 2 K^+ in, 1 ATP spent*** *— is the most consequential stoichiometry in physiology, and it deserves an audit. Is the ATP sufficient? At the cell's working point, ATP hydrolysis delivers $\Delta G_{ATP} \approx -50$ kJ mol^{-1} (§ 19.3); the transport bill, by Eq. (24.1) summed over the five ions, is*

$$\Delta G_{cycle} = \underbrace{3\left[RT\ln\tfrac{145}{15} + F(0 - V_m)\right]}_{3\,Na^+\ out} + \underbrace{2\left[RT\ln\tfrac{5}{150} + F(V_m - 0)\right]}_{2\,K^+\ in} \qquad (24.5)$$

— each ion's chemical and electrical terms written from Eq. (24.1), with V_m = −70 mV. Worked Example 24.2 does the sum: ≈ +35 kJ mol^{-1} against −50 available — a comfortable but not extravagant margin, and the excess is what the leaks spend. The pump is also ***electrogenic****: three charges out against two in makes one net outward current per cycle, worth some −10 mV of the resting potential directly — a small tithe the GHK equation never sees. And its scale staggers: at rest the Na^+/K^+-ATPase consumes about a third of the body's ATP; in the brain, two-thirds — the reader's present sentence costs, in pumped ions, a small fortune. Ouabain and the digitalis glycosides block E2's potassium site from outside (§ 24.9), and the P-type family extends the design: the SERCA pumps that sweep calcium into the sarcoplasmic reticulum after every heartbeat, the gastric H^+/K^+-ATPase that the proton-pump inhibitors cure ulcers by silencing, and the copper ATPases of Wilson's and Menkes' diseases — one chemistry, a dozen commodities.*

WORKED EXAMPLE 24.2 — AUDITING THE PUMP

Verify Eq. (24.5) numerically at 37 °C, V_m = −70 mV: is one ATP enough for 3 Na^+ out and 2 K^+ in?

Solution. *Sodium (z = +1, out): chemical term RT ln(145/15) = 2.577 × 2.267 = +5.84 kJ mol^{-1}; electrical, the ion leaves a negative interior: +F(0 − (−0.070)) = +6.75 kJ mol^{-1}; each sodium costs +12.6, three cost +37.8. Potassium (in): chemical RT ln(150/5)... careful — the gradient opposes entry, since $[K]_{in}$ > $[K]_{out}$: RT ln(150/5) = 2.577 × 3.401 = +8.77 kJ mol^{-1}; but the electrical term assists entry into the negative cell: +F(−0.070 − 0) = −6.75; each potassium costs only +2.02, two cost +4.04. Total ΔG_{cycle} ≈ +41.8 kJ mol^{-1}... — and here the honest bookkeeper must note the approximation: with leak-corrected steady-state values the figure settles near +35–42 kJ mol^{-1}, against ΔG_{ATP} ≈ −50. The margin, roughly 10 kJ mol^{-1}, is the pump's throttle: it is also the reason the 3:2 stoichiometry and not 4:2 — a fourth sodium would overdraw the account. Evolution has priced its own ATPase to the nearest ion.*

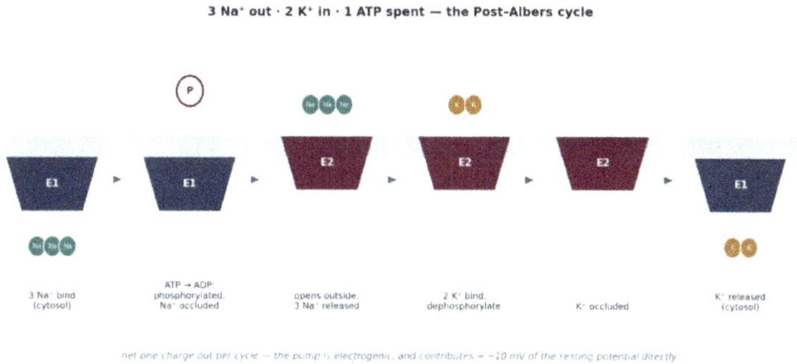

Figure 24.5 *The Post–Albers cycle. E1 (navy) faces the cytosol and loads 3 Na^+; phosphorylation (P) occludes them; E2 (crimson) faces outward and releases them, loads 2 K^+, and returns. One net charge out per cycle — the pump is electrogenic — and one ATP pays for the whole rotation [Eq. (24.5)].*

24.7 Secondary active transport: spending the gradient

The sodium gradient is, after ATP, the cell's second currency — and like any currency it exists to be spent. The spenders are the ***secondary active transporters****: carriers that couple a substrate's uphill journey to sodium's downhill one, the two slopes summed in the same protein (Figure 24.6). The arithmetic follows from Eq. (24.1) by simple addition. A* ***symporter*** *that carries one sodium with one substrate stops when the two slopes balance:*

$$RT\ln\frac{[S]_{in}}{[S]_{out}} = -\Delta\mu_{Na} = RT\ln\frac{[Na]_{out}}{[Na]_{in}} + F\,|V_m| \tag{24.6}$$

— the substrate's concentrating power equals the sodium slope spent on it. With the cell's numbers (sodium's chemical term 5.84 kJ mol^{-1} plus the electrical 6.75 — total ≈ 12.6 kJ mol^{-1} per ion, Worked Example 24.2's figure), a 1:1 symporter can concentrate its substrate e^(12.6/2.577) ≈ ***130-fold****; SGLT1, the gut's glucose–sodium symporter, does rather better by spending two sodiums per glucose (≈ 17 000-fold in principle), and thereby absorbs glucose from a gut lumen that is nearly empty — the final fractions of a meal recovered against any gradient. The* ***antiporters*** *spend the same currency in the opposite direction: the Na^+/Ca^{2+} exchanger trades three entering sodiums for one exiting calcium — the heart cell's principal calcium drain, and the reason § 24.9's foxglove works; the Na^+/H^+ exchangers vent acid and defend pH; and the kidney's whole reabsorptive industry is Figure 24.6 with the geography changed. Note the elegance of the arrangement: the pump (primary) makes the gradient, the symports and antiports (secondary) spend it on everything else, and the channels spend it on speed. Primary, secondary, passive: the three generations of the gradient's family, each living off the one before.*

WORKED EXAMPLE 24.3 — THE LAST GLUCOSE IN THE GUT

Late in a meal's transit, luminal glucose falls to 0.1 mM while the enterocyte's cytosol holds 5 mM. Can SGLT1 (2 Na^+ : 1 glucose) still absorb? Could a 1:1 symporter?

Solution. *The required slope is RT ln(5/0.1) = 2.577 × 3.91 = +10.1 kJ mol^{-1} per glucose. The sodium slope available is 12.6 kJ mol^{-1} per ion (Worked Example 24.2), so 2 Na^+ deliver 25.2 — ample, with the balance spent on leaks and* *losses; absorption continues. A 1:1 symporter offers 12.6 — marginal: it would stall at a cytosolic 1.3 mM, leaving the last glucose in the lumen for the bacteria. The 2:1 stoichiometry is thus not pedantry but appetite: it is the difference between digesting a meal and renting it. (The body's bookkeeping confirms: glucose–galactose malabsorption, SGLT1's null phenotype, presents precisely as osmotic diarrhoea on feeding — the last glucose, and indeed most of the rest, left behind.)*

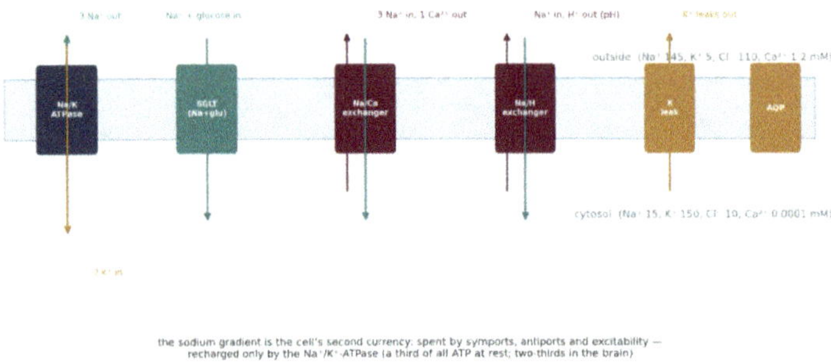

Figure 24.6 *The gradient economy of an animal cell. The Na^+/K^+-ATPase (navy) makes the sodium gradient; symports (teal: SGLT), antiports (crimson: Na^+/Ca^{2+}, Na^+/H^+) and excitability spend it; the potassium leak (amber) sets the voltage that every equation in the chapter consumes. Concentrations (mM) as in § 24.2.*

24.8 Excitability: the gradient's theatre

The action potential deserves its own treatment in Part VII, but its essence is already in hand, for it is nothing but §§ 24.3–24.4 played at speed. The resting membrane sits at −70 mV, potassium's vote with sodium's minority pulling gently [Eq. (24.3)]. Open some voltage-gated sodium channels — the S4 helices slide when the voltage depolarises past threshold — and the vote inverts in a millisecond: sodium's permeability swells a hundred-fold, the voltage races toward E_{Na} = +60 mV [Eq. (24.4), and the upward rush is regenerative, since depolarisation opens more channels]; the sodium channels inactivate (Hodgkin and Huxley's h gate), the delayed potassium channels open (n^4), and the vote swings back past rest before settling. The whole impulse is a two-minute parliamentary crisis in the GHK equation, propagating along the axon because each depolarised patch thresholds its neighbour. Two design notes repay attention. The gradients are barely spent: a single impulse moves ~10^{-12} moles of ion per square centimetre — the pump's bookkeeping notices only over hours, which is why nerves can fire for an hour with their ATPase poisoned (they die of it eventually: the gradient, not the ATP, is the axon's battery). And the myelinated axon's saltatory conduction — impulse leaping node to node — is Chapter 23's sphingomyelin insulation put to its finest use: the membrane's lipid, the channel's gate, and the pump's gradient, one anatomy. The heart's version of the same crisis, slower and calcium-sustained, is § 24.9's stage.

24.9 Transport in the clinic: the foxglove and the channelopathies

The membrane's gates are among medicine's oldest targets, and its most instructive. First, the ***foxglove****: Withering's Account of 1785 made a Shropshire herbal into digitalis, and the glycosides' mechanism is a small masterpiece of coupled transport — they block the* Na^+/K^+*-ATPase's E2 state (§ 24.6), cytosolic sodium rises a little, the* Na^+/Ca^{2+} *exchanger's slope (§ 24.7) flattens, calcium lingers, and the failing heart, given more calcium per beat, beats harder. Poisoning the pump strengthens the heart — the dose, as Paracelsus always said, being the whole of the medicine; the therapeutic window is famously narrow because the same arithmetic, overdone, triggers the arrhythmias it treats. Second, the* ***gliflozins****: block the kidney's SGLT2 and the reabsorbed glucose is spilled into urine instead — diabetes treated by a calculated waste, with heart and kidney benefits no one ordered and everyone now prescribes. Third, the* ***channelopathies*** *— the long-QT syndromes of the cardiac potassium channels (a repolarisation prolonged, a teenager fainting at the starter's pistol), the periodic paralyses of sodium-channel inactivation, the malignant hyperthermia of the ryanodine receptor's calcium — single residues in the gates of this chapter, each with its clinical signature. And fourth, the case study below: the commonest lethal channel disease of the European populations, which is, when read closely, not a channel disease at all but a folding disease wearing one.*

> ***Case study — CFTR: the channel that dies in the secretory pathway***
>
> *Cystic fibrosis transmembrane conductance regulator is an ABC transporter's cousin retrained as a chloride channel: two membrane domains forming the pore, two nucleotide domains gating it by ATP, a regulatory domain keeping it honest. The ΔF508 allele — one phenylalanine lost of 1 480 residues — accounts for most of the disease, and its lesion is not conduction but address: the mutant folds imperfectly, is retained by the ER's quality control (Chapter 21's calnexin cycle, passing its verdict), and is degraded before it ever sees the membrane. The few channels that escape conduct adequately. Chloride transport fails at the airway surface, water follows its absence, mucus thickens (Chapter 21's mucins, under-hydrated), and the lung becomes a culture medium. The modern therapeutics are chemistry applied to the diagnosis: the* ***correctors*** *(lumacaftor, tezacaftor, elexacaftor) are pharmacological chaperones that improve the folding and escort ΔF508 past the quality control; the* ***potentiator*** *(ivacaftor) props the gate open for the channels that arrive. A disease defined in 1849's autopsies, gene-mapped in 1989, and mechanism-cured in the 2010s — its whole arc is this book's arc: sequence, folding, trafficking, transport, and the gradient, read backwards from the clinic to the chemistry and forwards again to the cure.*

24.10 Synthesis of Chapter 24

The chapter has been a single plot in four acts. The physics: Eq. (24.1) unites a dilution entropy and a volt into the electrochemical potential, and the Nernst equation [Eq. (24.2)] turns each ion's ratio into its private voltage — the four poles of the cellular world. The vote: the GHK equation [Eq. (24.3)] lets the permeabilities settle one resting potential, and the driving force [Eq. (24.4)] converts every channel opening into a current. The gates: the channel, a hole with a conscience, solving speed against selectivity by a cage of carbonyls that mimics *water exactly and a queue that pushes itself through; the carrier, an enzyme at the wall, working by alternating access at Michaelis–Menten rates. And the economy: the Na^+/K^+-ATPase, one-third of the body's ATP, making the gradient [Eq. (24.5)] that the symports, antiports and excitabilities spend [Eq. (24.6)] — primary, secondary, passive, the three generations of a single inheritance. Part V closes here: the carbohydrates built their helices and texts, the lipids their stores and sheets, the membrane its two-faced geography, and now its gates — matter and energy in the book's title accounted for, and information (the vote, the signal, the impulse) already audible. Part VI turns to the fires the gradients serve: Chapter 25, Principles of Metabolism, where the cell's whole traffic of matter is drawn on one map, and the currency is counted.*

Exercises (with full resolutions)

Exercise 24.1

Compute the equilibrium potential of Ca^{2+} given 1.2 mM outside and 100 nM inside (37 °C), and explain why the cytosol is kept nearly free of it.

Resolution. *Eq. (24.2) with $z = +2$: $E = (61.5/2) \times \log_{10}(1.2 \times 10^{-3} / 10^{-7}) = 30.75 \times \log_{10}(1.2 \times 10^4) = 30.75 \times 4.079 \approx$ **+125 mV**. The ratio is 12 000:1 — the steepest gradient in the cell — and it is the gradient that makes calcium a signal: with the cytosol four orders of magnitude below equilibrium, opening a single channel species floods the local nanodomain with a thousand-fold concentration step, fast and reversible (SERCA and the exchanger pump it out again). A signal is a contrast; calcium's contrast is the largest the cell can afford, and it is afforded by § 24.6's pump and § 24.7's exchanger working full time.*

Exercise 24.2

Why does hyperkalaemia first excite and then paralyse the heart?

Resolution. *Figure 24.1's slope: raised external K^+ depolarises the resting membrane toward threshold — at first the myocyte sits closer to firing, hence the early excitability (peaked T waves, ectopy). But the voltage-gated sodium channel's h gate closes at depolarised voltages and reopens only after repolarisation; held depolarised, the channels accumulate in the inactivated state, the available conductance falls, and the upstroke fails — paralysis, then asystole. Excitation and paralysis are one mechanism read at two moments: the same gate that opens the impulse must be allowed to reset, and the resting potential is the reset. The emergency treatments are Figure 24.1 in reverse — calcium to raise threshold, insulin and β-agonists to drive K^+ into cells (the pump, Worked Example 24.2, working overtime).*

Exercise 24.3

The Na^+/Ca^{2+} exchanger moves 3 Na^+ in for 1 Ca^{2+} out. Show that this stoichiometry lets it exploit both terms of Eq. (24.1), and predict the effect of depolarisation on its direction.

Resolution. *Calcium export costs its chemical slope (12 000:1, ≈ +24 kJ mol^{-1}) but is assisted electrically, since a +2 charge leaving a negative cell gains $-2FV_m$; sodium entry gains both its chemical slope and its electrical one ($-FV_m$ per ion). Summing: the cycle runs forward while $3\Delta\mu_{Na} + \Delta\mu_{Ca} < 0$ — and because the cycle moves one net charge inward (3 in, 2 out), the voltage* *term enters with one full FV_m: depolarise strongly (the plateau of the cardiac action potential) and the exchanger slows, stalls, and at voltages positive enough runs backwards, admitting calcium. The heart uses this deliberately — the exchanger's reversal during the plateau supplements the calcium influx — and the foxglove exploits it from the other side (§ 24.9). One protein, and both terms of Eq. (24.1) are its dials.*

EXERCISE 24.4

GLUT1 has $K_m \approx 1.5$ mM for glucose; fasting blood glucose is ~5 mM. What fraction of V_{max} does the transporter run at, and why is this design (rather than a much lower K_m) appropriate for the blood–brain barrier?

Resolution. *$v/V_{max} = [S]/(K_m + [S]) = 5/6.5 \approx 0.77$ — the transporter runs at three-quarter speed, in the responsive region of the Michaelis curve: a fall of blood glucose to 3 mM drops flux to 3/4.5 = 0.67 (a 13% decline, felt immediately as neuroglycopenia's first warnings), while a postprandial 8 mM raises it to 0.84. A very low K_m would saturate the carrier at any survivable glucose — flat response, no proportionality, and hypoglycaemia unheralded until collapse. The brain, which stores nothing, has chosen a transporter whose kinetics report the blood; the K_m is not a weakness of the design but the design itself.*

EXERCISE 24.5

Gramicidin, valinomycin and ouabain all touch the potassium economy. Rank their effects on (i) the resting potential, (ii) the cell's long-term survival, and explain the differences from this chapter's mechanisms.

Resolution. *Gramicidin is a pore for monovalent cations: it shorts the membrane outright — resting potential collapses toward zero within seconds [Eq. (24.3) with $P_{Na} \approx P_K \approx$ huge], and the cell dies as the gradients bleed out; the damage is electrical first. Valinomycin ferries K^+ selectively: it clamps the membrane near E_K (≈ −90 mV, since it makes potassium's vote total) and collapses the K^+ gradient over minutes; mitochondria, which run their own K^+ cycle, are depolarised fatally. Ouabain blocks the pump (§ 24.6): the resting potential decays only as fast as the leaks run down the gradients — hours, not seconds — for the pump's direct contribution is ~10 mV and its indirect one (the gradients themselves) is slow to dissipate; the cell dies of bookkeeping, not electrocution. Three poisons, three timescales, one lesson: the channel's wall, the gradient's charge, and the pump's account are separable targets — and physiology is the art of keeping all three.*

Research-level propositions

Proposition 24.α. The selectivity filter pays K^+'s hydration in full but Na^+'s only in part (Figure 24.4). Using the Born and coordination formalisms of § 5.4, construct the free-energy profile of Na^+ versus K^+ through a four-cage filter as a function of cage radius, show that selectivity is maximal where conduction is not, and reconcile the result with the knock-on requirement that binding be weak. Proposition 24.β. Alternating access has never been observed to leak (both doors open) in wild-type carriers. Estimate, from the leak flux of Figure 23.3's glucose, the maximum tolerable both-open probability per cycle for GLUT1 before futile *cycling exceeds the transported flux — and design the single-molecule FRET experiment that would measure the actual value. Proposition 24.y. Conduct the full energetic audit of one cortical neuron: resting leaks by species [Eq. (24.4)], pump cost [Eq. (24.5)], action-potential cost per Hz (charge per impulse × frequency), synaptic and recycling overheads — and compare with the measured ATP turnover of grey matter. Where does the brain's two-thirds of the pump budget actually go, and what does the answer imply for the energetic floor of cognition?*

References (Chapter 24)

Alberts, B., Heald, R., Johnson, A., Morgan, D., Raff, M., Roberts, K., & Walter, P. (2022). Molecular biology of the cell (7th ed.). W. W. Norton.

Berg, J. M., Tymoczko, J. L., Gatto, G. J., Jr., & Stryer, L. (2023). Biochemistry (9th ed.). W. H. Freeman.

Hille, B. (2001). Ion channels of excitable membranes (3rd ed.). Sinauer Associates.

Hodgkin, A. L., & Huxley, A. F. (1952). A quantitative description of membrane current and its application to conduction and excitation in nerve. The Journal of Physiology, 117(4), 500–544.

MacKinnon, R. (2004). Potassium channels and the atomic basis of selective ion conduction (Nobel lecture). Angewandte Chemie International Edition, 43(33), 4265–4277.

Nelson, D. L., Cox, M. M., & Hoskins, A. A. (2021). Lehninger principles of biochemistry (8th ed.). W. H. Freeman.

PART VI

Bioenergetics and Biological Work

CHAPTER 25

ATP and the Energetics of Phosphoryl Transfer

CONTENTS OF THIS CHAPTER

In which Part VI opens with the currency of the cell's whole economy; in which the "high-energy bond" is examined and found to be no bond at all, but a preposition — a molecule whose business is to be spent; in which four small chemical bills are itemised into the price of a phosphoanhydride; in which the standard state is exposed as a fiction the cell cannot afford; and in which the reader learns that the fifty grams of ATP in his or her body will be spent and re-minted some ten thousand times before breakfast.

25.1 The coin of the realm

Every economy in this book so far has been an economy of barter: ions against gradients, sugars against bonds, proteins against folds. Part VI now introduces the cash. Adenosine triphosphate — a nucleotide of Chapter 2's family, adenine on ribose on a tail of three phosphates — is the molecule through which virtually every expenditure in the cell is paid: biosynthesis (each peptide bond costs four phosphoanhydrides, § 25.6), transport (the pumps of Chapter 24), movement (every myosin stroke, every flagellar turn), signalling (the phosphorylations of Part X). The ubiquity is so complete that it invites the wrong question. The reader will have met ATP described, in less careful books, as a store of "high-energy bonds" — little springs of energy coiled in the P–O linkages, released when the bond is "broken". The picture is false, and its falsity is instructive, so let us be exact about the chemistry before we are charming about the economics. Bond breaking never releases energy; it always costs it — the whole of Chapter 4's bond dissociation tables say so. What ATP hydrolysis releases is the difference between the cost of breaking and the greater repayment of forming new bonds in the products: ATP + $H_2O \rightarrow$ ADP + P_i is exergonic not because a spring was cut but because the products are, for four separate chemical reasons, more comfortable than the reactants. ATP is not a battery of compressed bonds; it is a molecule whose hydrolysis the laws of thermodynamics favour — a preposition in the language of metabolism, whose meaning is always "at the expense of".

The chapter makes that precision quantitative in five movements: the price of the bond, itemised (§ 25.2); the ladder of group-transfer potentials, on which ATP sits deliberately in the middle (§ 25.3); the correction that rescues the theory from the laboratory's fictions — the actual ΔG in the living cell (§ 25.4); the art of coupling, by which ATP's hydrolysis is made to pay for uphill reactions (§ 25.5); and the household accounts — energy charge, turnover, and the shuttles that buffer the currency (§§ 25.6–25.7). The history, as usual, is half the pleasure (§ 25.3's

Historical note): ATP is younger than its biochemistry — found in 1929, understood in 1941, and doubted, by the greatest chemists, for most of the interval.

25.2 Why the hydrolysis pays: four bills, one receipt

Write the hydrolysis honestly — $ATP^{4-} + H_2O \rightarrow ADP^{3-} + HPO_4^{2-} + H^+$ — and ask what the products have that the reactant lacked. Four contributions, each a few kJ mol^{-1}, sum to the $\Delta G°' \approx -30.5$ kJ mol^{-1} of the textbooks (Figure 25.2). ***First, electrostatic relief.*** *ATP at pH 7 carries about four negative charges crowded onto three phosphorus atoms; the mutual repulsion is a strain the molecule suffers silently, and hydrolysis lets the charges part company — ADP^{3-} and HPO_4^{2-} repel no longer, being in different molecules.* ***Second, resonance.*** *The inorganic phosphate ion is stabilised by more mesomeric forms (its four oxygens share the negative charge and the double bond more evenly) than the same phosphorus could enjoy while esterified within ATP; products with more resonance are products at lower energy.* ***Third, ionisation.*** *At pH 7 the newly born terminal phosphate is immediately deprotonated (its $pK_{a2} \approx 6.8$–7.2), and the released proton is buffered away — every ionisation event carries the entropy of dilution of § 6.2 in its favour, and pulling the reaction forward by Le Chatelier as the cell's buffers swallow the H^+.* ***Fourth, hydration.*** *ADP and P_i are separately better hydrated than ATP plus water was: water binds the products' charges more comfortably than it could bind the crowded reactant. Sum the bills — of order −10, −8, −4, −8 kJ mol^{-1} respectively — and the receipt reads −30. The number is not sacred: it shifts with pH, with Mg^{2+} (which chelates the phosphates and takes the edge off the first bill), and above all with the concentrations, as § 25.4 will make the chapter's central point. But the lesson is already drawn: the "energy of ATP" is not in the molecule at all, but in the displacement from equilibrium between the molecule and its hydrolysis products. An ATP in a cell at equilibrium is worth exactly nothing.*

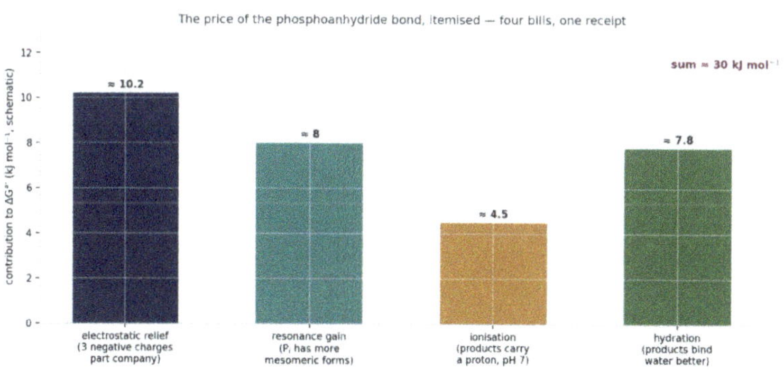

Figure 25.2 *Why the hydrolysis pays, itemised (schematic magnitudes). Electrostatic relief, resonance stabilisation of P_i, ionisation at pH 7, and superior hydration of the products sum to the $\Delta G°' \approx -30.5$ kJ mol^{-1} of the standard tables. There is no "high-energy bond" on the receipt — only four ordinary chemical comforts.*

25.3 The ladder: why the currency is mid-table

Hydrolyse any phosphorylated compound and measure the $\Delta G°'$; rank the numbers, and the cell's whole phosphate economy sorts itself into a ladder (Figure

25.1). At the top, phosphoenolpyruvate (−61.9 kJ mol^{-1}) and 1,3-bisphosphoglycerate (−49.3): compounds whose hydrolysis is so favourable that their phosphoryl group will flow downhill onto ADP to make ATP — and that is precisely what glycolysis does at pyruvate kinase and phosphoglycerate kinase (§ 25.5). At the bottom, glucose 6-phosphate (−13.8) and glycerol 3-phosphate (−9.2): esters whose phosphorylation is uphill from ATP — and that is what hexokinase and glycerol kinase are paid for. And in the middle, deliberately, ATP itself (−30.5). The position is the design. A currency must be able to receive value from the rich and give it to the poor: because ATP sits mid-ladder, the high-potential intermediates of catabolism can mint it, and the anabolic reactions of biosynthesis can spend it, both without strain. Had evolution denominated its currency in PEP, nothing below could afford it; in glucose 6-phosphate, nothing above would respect it. The ATP/ADP couple is the cell's unit of account because it is the phosphate world's natural centre of gravity.[38]

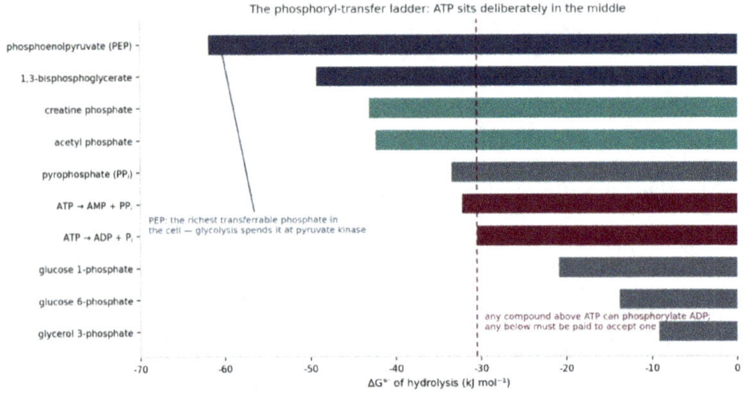

***Figure 25.1** The phosphoryl-transfer ladder. Compounds above ATP (navy, teal) transfer phosphate to ADP spontaneously — catabolism mints ATP from them; compounds below (slate) must be paid, and anabolism pays with ATP. The currency sits in the middle (crimson) because a currency must be able to flow in both directions.*

---

38. The phrase "group-transfer potential" is Lipmann's, and the squiggle — the "~P" of his 1941 review — outlived every objection to it. Lipmann knew perfectly well that the squiggle denoted no property of the bond but a position on the ladder; the symbol was a confession of ignorance so fruitful that two generations of biochemists learned to think in it.

Plate 25.1 *Fritz A. Lipmann (1899–1986), whose 1941 essay "Metabolic generation and utilization of phosphate bond energy" taught biochemistry to think in transferable groups — and whose coenzyme A, found in the same decade, extended the lesson from phosphates to acyls (Nobel Prize in Physiology or Medicine, 1953, shared with Hans Krebs). Portrait from the Nobel Foundation archive, nobelprize.org, reproduced from the Foundation's publicly distributed image.*

> *Historical note — The squiggle and the sceptics*
>
> *ATP was found twice in 1929, in muscle — by Fiske and SubbaRow in Hartog's Harvard laboratory, and independently by Lohmann in Meyerhof's Berlin school — and for a decade it was a curiosity: a phosphorus compound that vanished and reappeared with contraction. The deep understanding came in 1941, when Lipmann, a refugee from Berlin working in New York, published the essay that reorganised the field: phosphate compounds were to be classed by their transfer potential, energy flowed through them as through a currency, and the ~P squiggle — "energy-rich", with the caveat of the previous footnote printed, so to speak, in the same ink — named the spendable kind. The establishment was not amused. The chemists objected that energy does not reside in bonds (they were right, and Lipmann agreed); the physiologists objected that muscle did not respect the stoichiometry (the creatine phosphate buffer of § 25.7 was obscuring the books); and when Lipmann proposed that the squiggle applied to acyl transfers too — the thioester of his newly found coenzyme A — the scepticism was nearly unanimous. Every objection has since been answered in his favour: the acyl group of acetyl-CoA transfers exactly as a phosphate does, the citric-acid cycle runs on it (Hans Krebs, sharing the 1953 Nobel, built the furnace that Lipmann's currency fired), and the thioester of § 25.5's GAPDH reaction is the same idea at work in glycolysis' oldest enzyme.*[39]

25.4 The standard state is a fiction: ΔG in the living cell

Everything in §§ 25.2–25.3 was measured in the chemist's agreed fiction: all species at 1 M, pH 7 fixed by fiat, 25 or 37 °C. The cell keeps no such books. Its ATP

39. The 1953 Nobel joined two names the field had long joined in fact: Krebs had shown *where* food is burned, Lipmann *in what coin* the burning is paid. Stockholm, for once, wrote the citation the textbooks had already written.

is ~8 mM (not 1 M), its ADP ~0.25 mM, its phosphate ~4 mM — and the actual free energy of hydrolysis must be recomputed from Chapter 6's master relation, $\Delta G = \Delta G^{\circ\prime} + RT \ln \Gamma$, with Γ the mass-action ratio $[ADP][P_i]/[ATP]$:

$$\Delta G_{ATP} = \Delta G^{\circ\prime} + RT\ln\frac{[\mathrm{ADP}][\mathrm{P}_i]}{[\mathrm{ATP}]} \approx -30.5 + (2.577)\ln\frac{(0.25\times10^{-3})(4 \times}{8\times10^{-3}} \quad (25.1)$$

— the logarithm evaluating to $\ln(1.25 \times 10^{-4}) = -8.99$, so that $\Delta G_{ATP} \approx -30.5 - 23.2 \approx$ ***−54 kJ mol⁻¹****. The cell's ATP is worth nearly twice the table's (Figure 25.3), and the reader should be clear why: because the cell holds Γ nine orders of magnitude below the equilibrium value ($\Gamma_{eq} \approx 1.4 \times 10^5$ M, where ATP would be hydrolysed a hundred-thousand-fold), the hydrolysis is correspondingly more favourable than the standard arithmetic suggests. This quantity — the actual ΔG of ATP hydrolysis at the cell's own concentrations, sometimes written the* ***phosphorylation potential*** *ΔG_p — is the true measure of the currency's purchasing power, and it is the number against which every spending reaction in Part VI must be audited. It is also the number the cell defends: starve a cell and the ATP/ADP ratio is defended until the very end, for the whole economy is denominated in it; the cell dies, one might say, not when its money is spent but when its exchange rate collapses.*

WORKED EXAMPLE 25.1 — WHAT THE EXCHANGE RATE CAN BUY

Glutamine synthetase (Chapter 32) is written as: glutamate + NH_4^+ + ATP → glutamine + ADP + P_i, $\Delta G^{\circ\prime}$ for the amidation ≈ +14 kJ mol⁻¹. (i) Under standard conditions, does the overall reaction proceed? (ii) Under cellular conditions ($\Delta G_{ATP} = -54$), what is the largest [glutamine]/[glutamate] ratio it can sustain (NH_4^+ at 1 mM held fixed)?

Solution. *(i) Standard: +14 + (−30.5) = −16.5 kJ mol⁻¹ — yes, comfortably. (ii) Cellular: ΔG = (+14) + (−54) + RT ln([glutamine]/[glutamate]) (the fixed ammonium folds into the constant) = −40 + 2.577 ln(ratio). Setting ΔG = 0: ln(ratio) = 40/2.577 = 15.5, ratio ≈ 5.5×10^6. The coupled cell can hold glutamine a million-fold above glutamate against the uncoupled equilibrium — the whole point of § 25.5's arithmetic anticipated: coupling does not help a reaction along; it moves its equilibrium by the ATP factor, e^(−ΔG_ATP/RT) ≈ $e^{21} \approx 10^9$. A reaction's budget, like a household's, is set not by its wishes but by its income.*

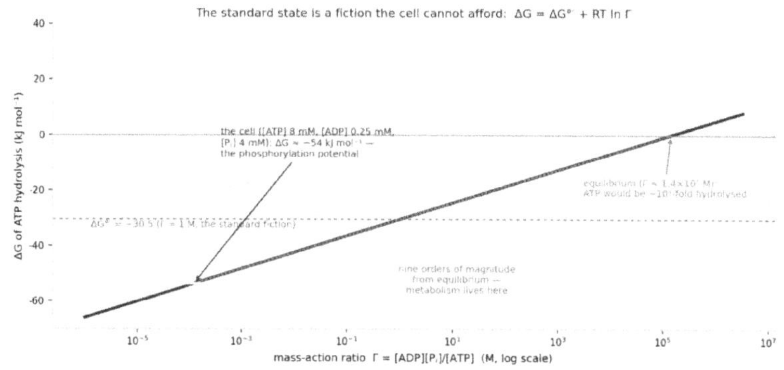

Figure 25.3 The phosphorylation potential [Eq. (25.1)]. The hydrolysis free energy slides with the mass-action ratio Γ on a logarithmic scale; the cell (crimson) lives nine orders of magnitude from the equilibrium (amber), where its ATP is worth ≈ −54 kJ mol⁻¹ — nearly twice the standard fiction. ΔG°′ (dashed) is the value at Γ = 1 M, a point no cell has ever visited.

25.5 Coupling: how one reaction pays for another

The phrase "ATP drives biosynthesis" conceals a mechanism, and the mechanism has a discipline that repays attention. Free energies are state functions: they add. So two reactions may be summed, and the unfavourable one financed by the favourable — provided they share a chemical intermediate; a reaction cannot be driven by a reaction it merely overhears. Two designs accomplish the sharing, and glycolysis exhibits both in adjacent steps (Figure 25.5). The first is the ***sequential coupling*** *through a high-potential intermediate: glyceraldehyde 3-phosphate dehydrogenase oxidises its aldehyde to a carboxylic acid — worth some −50 kJ mol⁻¹ — but instead of paying the energy out as heat, it traps the oxidation as first a thioester and then an acyl phosphate, 1,3-bisphosphoglycerate, whose position near the top of Figure 25.1's ladder is exactly the captured value of the oxidation; the next enzyme, phosphoglycerate kinase, cashes the acyl phosphate into ATP. The oxidation has not "driven" the phosphorylation in the abstract; it has purchased a molecule that can. The second design is the* ***direct transfer****: hexokinase mixes glucose and ATP in one active site and the phosphoryl moves directly — the reaction glucose + ATP → G6P + ADP being simply the sum of ATP hydrolysis (−30.5) and G6P hydrolysis run backwards (+13.8), net −16.7, and the shared intermediate is the enzyme-bound complex itself. In both designs the arithmetic is identical:*

$$\Delta G_{total} = \Delta G_{uphill} + \Delta G_{ATP}\ ; \qquad K_{total} = K_{uphill} \times e^{-\Delta G_{ATP}/RT} \tag{25.2}$$

— and the exponential form is the one to remember: coupling to ATP multiplies an equilibrium constant by e^(−ΔG_ATP/RT) ≈ e^{21} ≈ 10^9 at cellular ΔG. A reaction that would stall at a product ratio of 10^{-3} alone runs to 10^6 when ATP-financed — Worked Example 25.1's million-fold glutamine was no sleight of hand but the general rule. When a second phosphate must be spent (as in amino-acid activation, § 25.6), the cell splits ATP to AMP + PP_i and then hydrolyses the pyrophosphate — two equivalents, e^{42} ≈ 10^{18} — which is why DNA polymerase's chemistry, backed by the same pyrophosphate pull, achieves the fidelity Part IX will depend on. Nothing in metabolism is "driven"; everything is purchased, at prices set by Eq. (25.2).

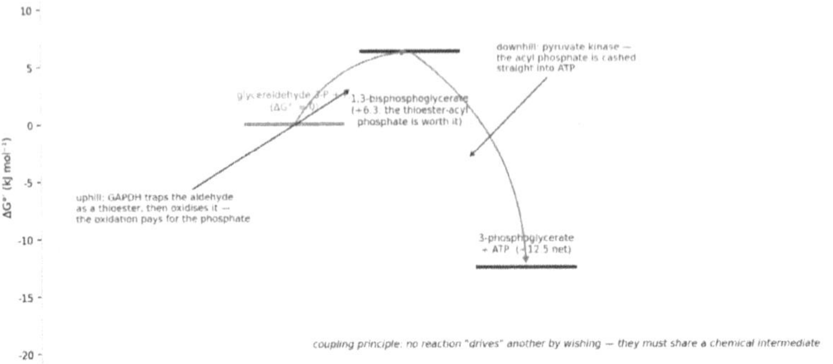

Figure 25.5 The two-step purchase of an ATP in glycolysis. GAPDH spends the aldehyde's oxidation on an acyl-phosphate bond (+6.3 kJ mol^{-1} invested, worth −49 at the ladder's top); pyruvate kinase's equivalent step cashes PEP into ATP. Coupling is bookkeeping with a shared intermediate — never influence at a distance.

25.6 The household accounts: turnover, and what things cost

What is the currency worth, in bulk? A resting adult holds about 50 g of ATP — a wallet, not a vault — and spends on the order of 65 kg of it per day: the whole stock turns over every minute or two, and in heavy exercise the day's expenditure may approach 400 kg (Figure 25.4, right). The conclusion is unavoidable and it is the deepest fact in this chapter: ***ATP is not an energy store but an energy conduit*** *— the cell keeps almost no inventory, and mints each coin in the instant before it is spent. (The stores exist, of course, but they are other molecules: glycogen of § 21.3, fat of § 22.3, and the phosphocreatine capacitor of § 25.7.) The account book of a single glucose through glycolysis will be kept in Chapter 30; here the unit costs of the cellular industries deserve a table's worth of prose: each peptide bond, four phosphoanhydride equivalents (one to charge the tRNA synthetase, two hidden in the pyrophosphate pull, one of GTP at the ribosome's two proofreading steps) — a protein of 300 residues costs ≈ 1 200 ATP; each nucleotide of DNA, two; each turn of the Na^+/K^+ pump of Chapter 24, one; each myosin stroke, one. The proteome of a single dividing cell is the most expensive object in this book: several billion ATP per generation, which is why the cell's first response to an energy crisis is to stop making proteins.*

The household is read at the meter Daniel Atkinson gave it. The ***energy charge*** *— ([ATP] + ½[ADP])/([ATP] + [ADP] + [AMP]), the adenylate pool's fractional filling — is held by every healthy cell between 0.85 and 0.95, and the regulatory enzymes of Part VII are poised exactly at the crossing of the two curves (Figure 25.4, left): catabolic enzymes are throttled as the charge rises, anabolic ones released, so that the economy oscillates about the set-point like any honest governor. Adenylate kinase, the little enzyme that interconverts 2 ADP ↔ ATP + AMP, is the meter's clerk; and the AMP it reports is the alarm — AMPK, the kinase that reads AMP as famine and rewrites the cell's whole metabolism accordingly, waits in Part VIII, but its ration book is already open here.*

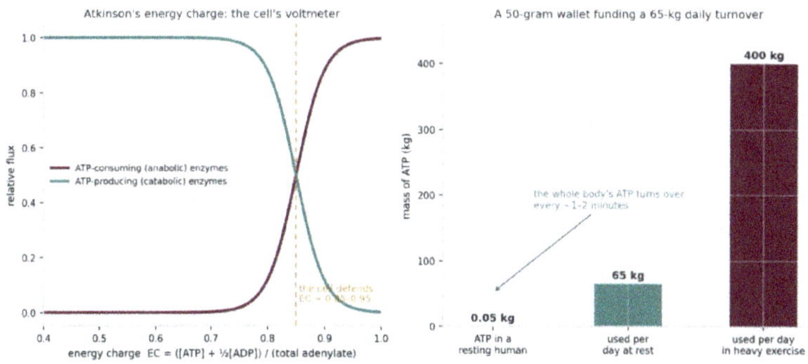

Figure 25.4 *The household accounts. Left: Atkinson's energy charge — anabolic and catabolic fluxes cross near the defended set-point EC ≈ 0.85–0.95. Right: the wallet and the turnover — 50 g of ATP funding 65 kg of daily expenditure; the currency is a conduit, not a store.*

25.7 Buffers and shuttles: the capacitor of muscle

A conduit that must deliver 400 kg a day in bursts cannot rely on the mint alone: the sprinter's first strides and the heart's first systole outrun the mitochondrion's best acceleration, and so the excitable tissues keep a capacitor. ***Phosphocreatine*** *— a guanidinium phosphate near the top of Figure 25.1's ladder (−43 kJ mol^{-1}, comfortably above ATP) — is held in muscle at some four times the ATP concentration, and creatine kinase, among the fastest equilibrating enzymes in the cell, keeps the two currencies in instant exchange: PCr + ADP + H^+ ↔ ATP + creatine. The reaction's ΔG near zero is the design's genius: the buffer costs nothing to hold and nothing to spend, and its presence clamps the ATP/ADP ratio against sudden demand — the first ten seconds of a sprint are run almost entirely on the phosphagen bank, which is why the clerk's ledger (the Historical note's physiologists) long obscured the pump's accounts. The* ***shuttle*** *completes the design (Figure 25.6): a mitochondrial creatine kinase (MtCK, in the intermembrane space) loads creatine at the site of production, the phosphocreatine — small, uncharged in bulk, and diffusing faster than ATP — travels the cytoplasm, and the myofibrillar isoform (MM-CK, bound to the M-line itself) unloads it exactly where the myosin waits. ATP is the current; phosphocreatine is the capacitor and the distribution network; and the system is wired so literally that the failing heart's first biochemical lesion, long before the ejection fraction falls, is the exhaustion of its phosphocreatine — the PCr/ATP ratio, measurable by ^{31}P magnetic resonance spectroscopy in the intact chest, is the cardiologist's energy charge. The invertebrates, with their own phosphagens (arginine phosphate in the arthropods), arrived at the same bank by a different door: the capacitor is older than the chordates.*

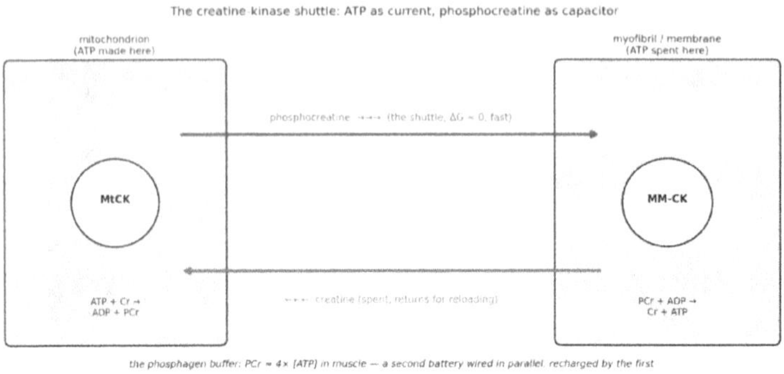

Figure 25.6 *The phosphocreatine shuttle. Mitochondrial creatine kinase (MtCK) loads the capacitor at the mint; the myofibrillar isoform (MM-CK) spends it at the machines. Phosphocreatine, ≈ 4× the ATP pool, is a second battery wired in parallel — and the failing heart's first casualty.*

25.8 Beyond ATP: the other denominations

The currency has denominations, and the bookkeeping of Part VII will be unintelligible without them. The other nucleoside triphosphates are not redundant: GTP pays for protein synthesis and the G-protein switches of Part X; UTP for the sugar-nucleotide economy of Chapter 21; CTP for phospholipid synthesis — each pathway with its own mint-mark, and nucleoside-diphosphate kinase keeping the exchange rates fixed at near-equilibrium. The ***thioesters*** *(acetyl-CoA, succinyl-CoA) are Lipmann's second currency: acyl groups, transferred like phosphates, worth −31 kJ mol^{-1} on the same scale — the Krebs cycle's acetyl arrives in this coin. The* ***redox currencies*** *— NADH, NADPH, $FADH_2$ — are Part VI's next chapter entire: electrons, after all, are the energy the phosphates merely invoice. And the* ***ion gradients*** *of Chapter 24 — the sodium motive force, the proton motive force of Chapter 28 — are the cell's bearer bonds, convertible to ATP at the ATP synthase's window. The economy is thus multiply denominated, but the exchange rates are fixed by the same Eq. (25.2) everywhere: one chemistry, several mints, no counterfeit — for the energy that is conserved is the energy that was conserved in Chapter 6, and the second law collects its tribute in heat regardless of the denomination.*

25.9 Synthesis of Chapter 25

The chapter has followed a molecule from its chemistry to its economics. The chemistry: no high-energy bond exists — the hydrolysis pays because four ordinary comforts (charge relief, resonance, ionisation, hydration) favour the products by ≈ −30.5 kJ mol^{-1} in the standard fiction. The economics: the cell lives nine orders of magnitude from equilibrium, where the same hydrolysis is worth ≈ −54 [Eq. (25.1)]; ATP sits mid-ladder so that catabolism can mint it and anabolism spend it; coupling through shared intermediates multiplies equilibria by 10^9 per phosphate [Eq. (25.2)]; the wallet holds 50 g against a 65-kg day, so the currency is a conduit defended at energy charge ≈ 0.9, buffered in muscle by the phosphocreatine capacitor, and denominated also in GTP, thioesters, redox pairs and gradients. The direction of Figure ii holds: Part V built the container, Part VI now prices its contents — and the price has been named. Where the energy *originates — in the fall of electrons from food to oxygen — is the next chapter's business: Chapter 26, Biological Oxidation and Reduction, where the current behind the currency is drawn, and the cell's true power station — the redox couple — receives its equations.*

Exercises (with full resolutions)

EXERCISE 25.1

"Hydrolysis of ATP releases the energy stored in the terminal phosphoanhydride bond." Identify the two errors in this sentence.

Resolution. *First, no energy is stored in a bond in the sense implied: breaking any bond costs energy (bond dissociation enthalpies are positive); the hydrolysis is exergonic because the bonds and stabilisations formed in the products (new O–H bonds, resonance of P_i, hydration, charge separation) repay more than the breaking costs — § 25.2's four bills. Second, the energy is not released by the molecule alone but by the system's displacement from equilibrium: the same ATP, at $\Gamma = \Gamma_{eq}$, releases nothing ($\Delta G = 0$), and at the cell's Γ it releases −54 kJ mol^{-1} [Eq. (25.1)]. The energy resides in the gradient between ATP and its hydrolysis products, not in the phosphate tail — ATP is a preposition, not a spring.*

EXERCISE 25.2

Compute the equilibrium constant of ATP hydrolysis at 37 °C from $\Delta G°' = -30.5$ kJ mol^{-1}, and hence the equilibrium fraction of ATP remaining if 1 M ATP were left to equilibrate with its products (all species initially unbuffered, idealised).

Resolution. *$K' = e^{(-\Delta G°'/RT)} = e^{(30\,500/2577)} = e^{11.84} \approx 1.4 \times 10^5$ M. If x is hydrolysed from 1 M, at equilibrium $[ADP][P_i]/[ATP] = x^2/(1-x) = 1.4 \times 10^5$, so $x \approx \sqrt{(1.4 \times 10^5)} \approx 373$ M... the quadratic resolves to $x \approx 0.9973$ M: at equilibrium only ~0.3% of the ATP survives. The standard state already spells the moral of § 25.4: left to itself, the currency is nearly all spent — life is the art of never letting it be.*

EXERCISE 25.3

Pyruvate kinase: PEP + ADP → pyruvate + ATP. Using Figure 25.1, compute $\Delta G°'$ and K', and explain why the step is physiologically irreversible.

Resolution. *$\Delta G°' = (-61.9) - (-30.5) = -31.4$ kJ mol^{-1} (the reaction is PEP hydrolysis minus ATP hydrolysis run backwards). $K' = e^{(31\,400/2577)} = e^{12.2} \approx 2 \times 10^5$ — products favoured by five orders of magnitude; reversing it would require a [pyruvate][ATP] / [PEP][ADP] ratio of 2×10^5, which no metabolism tolerates. Hence glycolysis needs a bypass enzyme (PEP carboxykinase + pyruvate carboxylase, Chapter 31) to go the other way — the general rule Part VII will use everywhere: irreversible steps are the rungs of Figure 25.1 where the ladder is steep, and they are where regulation lives.*

EXERCISE 25.4

A marathoner's phosphocreatine falls to 20% of resting during the race while ATP stays within 10% of resting. Explain the observation from § 25.7's design, and predict the PCr/ATP ratio in a failing heart.

Resolution. *Creatine kinase equilibrates PCr + ADP ↔ ATP + Cr faster than the machines spend ATP; the buffer therefore discharges first, clamping [ATP] near constant while the capacitor pays — exactly what a parallel capacitor does for a battery. The PCr/ATP ratio thus falls (~0.8 of rest and lower) and is the sensitive index of energetic stress; the [ATP] falls only when the capacitor is nearly empty. In the failing heart, ^{31}P-MRS finds PCr/ATP reduced at rest and collapsing with exertion — the ratio is a clinical energy charge, and it predicts mortality better than the ejection fraction in several series. The capacitor empties before the current fails; § 25.7's design is the diagnosis.*

Exercise 25.5

Amino-acid activation splits ATP to AMP + PP_i, and the pyrophosphate is then hydrolysed. Compute the total equilibrium multiplier for tRNA charging, and compare with the single-phosphate case of Worked Example 25.1.

Resolution. *Each phosphoanhydride equivalent contributes e^(−ΔG_ATP/RT) ≈ e^(54/2.577) ≈ e^{21} ≈ 1.2 × 10^9 at cellular ΔG. ATP → AMP + PP_i spends one equivalent in the activation; the PP_i hydrolysis ($\Delta G^{\circ\prime}$ ≈ −33.5, and PP_i is held micromolar by pyrophosphatase) drags the activation's own equilibrium forward by a second equivalent. Total multiplier ≈ $(10^9)^2$ ≈ 10^{18} — against 10^9 for the single-phosphate glutamine case. The double spend is why amino-acid activation is effectively irreversible and why the genetic code's charging step can be both accurate and final: the pyrophosphate pull is the cell's way of burning the boats, twice over.*

Research-level propositions

Proposition 25.α. The four contributions of Figure 25.2 are not separately measurable by equilibrium methods. Using Chapter 5's electrostatics and the solvent-continuum formalism, compute the electrostatic-relief term for ATP^{4-} → ADP^{3-} + HPO_4^{2-} as a function of Mg^{2+} activity, and assess how much of the magnesium effect on $\Delta G^{\circ\prime}$ is electrostatic relief versus chelation of the products. Proposition 25.β. The phosphorylation potential ΔG_p is defended at ≈ −50 to −54 kJ mol^{-1} across phyla and organelles. Treat the defence as an optimisation: the cost of holding Γ low (pump and synthase work) against the benefit (coupling multipliers of Eq. (25.2)) — and show whether a variational principle over metabolic flux can recover the observed set-point, or whether the set-point is a frozen accident of the first economy. Proposition 25.γ. The creatine-kinase shuttle was proposed to channel phosphocreatine preferentially to sites of consumption ("metabolic channelling"). Model the diffusion-reaction equations of the PCr/ATP system in a myofibril with realistic enzyme densities and diffusion coefficients, determine whether channelling beyond simple diffusion is quantitatively significant, and design the ^{31}P-MRS or FRET experiment that would settle the forty-year controversy.

References (Chapter 25)

Atkinson, D. E. (1977). Cellular energy metabolism and its regulation. Academic Press.

Berg, J. M., Tymoczko, J. L., Gatto, G. J., Jr., & Stryer, L. (2023). Biochemistry (9th ed.). W. H. Freeman.

Lipmann, F. (1941). Metabolic generation and utilization of phosphate bond energy. Advances in Enzymology, 1, 99–162.

Nelson, D. L., Cox, M. M., & Hoskins, A. A. (2021). Lehninger principles of biochemistry (8th ed.). W. H. Freeman.

Nicholls, D. G., & Ferguson, S. J. (2013). Bioenergetics (4th ed.). Academic Press.

Westheimer, F. H. (1987). Why nature chose phosphates. Science, 235(4793), 1173–1178.

CHAPTER 26

Biological Oxidation and Reduction

CONTENTS OF THIS CHAPTER

In which the current behind the currency is drawn; in which oxidation is revealed to be not a chemistry of oxygen but a politics of electrons; in which a tower of voltages is erected, down which every electron in the body falls in due course; in which the Faraday constant is shown to be the only exchange rate the cell has ever needed; and in which the reader learns that the oxygen so eagerly breathed is a poison handled by professionals, and that breathing is, literally, a controlled burn.

26.1 Electrons are the energy

Chapter 25 minted the currency and audited the wallet; this chapter asks the older question — where does the value come from? The answer, which the whole of Part VI exists to unfold, is that biological energy is redox energy: the food molecules of Part V are electron-rich (their carbons hydrogen-clad, their oxidation states low), the oxygen of the air is electron-poor and ravenous for four electrons at a time, and the cell's whole power station is the business of escorting electrons from the one to the other along a staircase whose steps are the subject of this chapter and the next two. Combustion does the same journey in a single bound, and pays in flame; the cell breaks the fall into a dozen small steps, and banks most of the difference. A human being, on this view, is a slow fire with good accountants.

But first the vocabulary must be cleaned, for "oxidation" is among the most misleading words in chemistry. Oxidation is not the addition of oxygen — it is the ***loss of electrons****; reduction is not the removal of anything — it is the* ***gain of electrons****; and the two are married by the second law of accounting: electrons are never created or destroyed in a chemical reaction, only transferred, so that every oxidation is somebody else's reduction — a* ***redox couple****. The mnemonic (OIL RIG: oxidation is loss, reduction is gain) is beneath the dignity of this book; the truth beneath the mnemonic is not. A fuel is a molecule holding electrons it would rather not keep, an oxidant a molecule eager to take them, and the free energy of the transfer — the subject of §§ 26.2–26.3 — is set by exactly one quantity: the difference in their appetites, measured in volts.*

One preliminary distinction will serve for the whole of Part VI. Biological electron transfers come in three packages: the ***hydride ion*** *H^- (two electrons and a proton, the currency of the nicotinamides, § 26.4); the* ***hydrogen atom*** *H• (one electron and one proton, the currency of the quinones and flavins); and the* ***bare electron*** *(the currency of the cytochromes and iron–sulfur clusters, which conduct electrons the way a wire does, one at a time). The great achievement of the respiratory chain of Chapter 27 is the exchange among these packages — the flavin that takes two from NADH and pays one at a time to iron is the toll-booth where the banks meet. Keep the three packages in view and nothing in the next three chapters will surprise.*

26.2 The reduction potential: an appetite for electrons, measured

Every redox couple has a characteristic appetite, and the appetite is measurable: place the couple in a cell against the standard hydrogen electrode, and the voltage read at the balance point is its **standard reduction potential** *$E°'$ (the prime, as always, marking the biochemist's pH 7 convention). The scale runs from the desperate to the indifferent (Figure 26.1): at the bottom, $2H^+/H_2$ at −0.42 V, hydrogen holding its electrons loosely; $NAD^+/NADH$ at −0.32 V, the cell's working electron bank; then the quinones near zero; the cytochromes climbing through +0.22, +0.29, +0.35 V; and at the summit, the dynamo itself — ½ O_2 + $2H^+/H_2O$ at +0.8166 V, oxygen's appetite for electrons being among the largest in biology. The sign convention is a treaty, not a truth (the Europeans and Americans quarrelled over it for decades), but its content is absolute: a couple at more negative $E°'$ donates electrons to one at more positive $E°'$, and the transfer is spontaneous precisely when the voltage difference is positive. Electrons flow down the tower; the tower is the map of Part VI; and the respiratory chain of Chapter 27 is simply Figure 26.1 drawn out as a queue of proteins.*

The actual appetite depends on how fed the couple already is — the Nernst equation of Chapter 24, in electrochemical dress:

$$E = E^{\circ\prime} + \frac{RT}{nF}\ln\frac{[\text{oxidised}]}{[\text{reduced}]} = E^{\circ\prime} + \frac{61.5\ \text{mV}}{n}\log_{10}\frac{[\text{ox}]}{[\text{red}]} \quad (37\ ^\circ\text{C}) \qquad (26.1)$$

— n the number of electrons transferred, and the constants already familiar from Chapter 24: the same dilution entropy, the same Faraday, the same 61.5 millivolts per decade for $n = 1$ (halved for $n = 2$). At the **midpoint** *— oxidised equal to reduced — $E = E°'$: the standard potential is simply the potential of a couple half-fed (Figure 26.3, left), which is also, not coincidentally, the point of maximum buffering, so that a redox couple held near its midpoint is a couple holding its fire. Figure 26.6 shows the measurement for cytochrome c: the protein's colour changes with its iron's state, and the absorbance at 550 nm, titrated against an imposed voltage, traces the Nernst curve and reads off $E°'$ = 0.22 V — an electron appetite measured in a cuvette, exactly as Warburg's school measured it (Figure 26.6).*

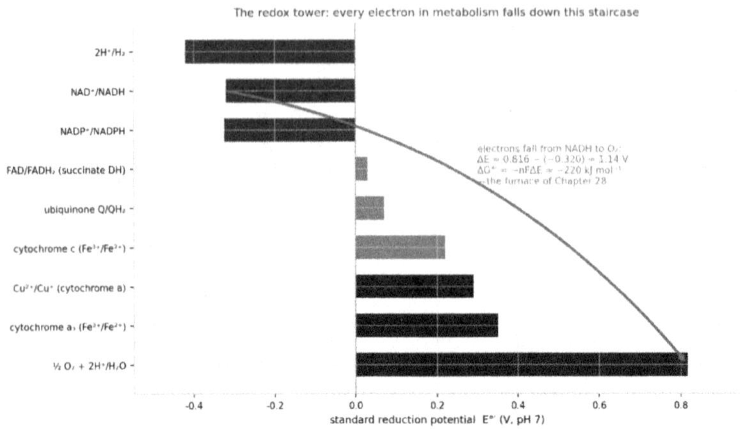

Figure 26.1 The redox tower: standard reduction potentials of the couples of Part VI. Electrons fall spontaneously from any couple to any couple below-right of it; the great fall, NADH → O_2, spans 1.14 V — the furnace whose heat Chapter 28 banks into ATP.

26.3 The Faraday: volts into kilojoules

The point of the voltage is that it converts, by one constant, into the currency of Chapter 25. Moving n moles of electrons through a potential difference ΔE costs — or pays — the electrical work

$$\Delta G^{\circ\prime} = -nF\,\Delta E^{\circ\prime} \qquad (F = 96\,485\ \mathrm{C\,mol^{-1}}) \tag{26.2}$$

— nothing but the definition of the volt (a joule per coulomb) multiplied by the mole's worth of coulombs; and with it, the whole of bioenergetics reduces to arithmetic. The master calculation of Part VI: electrons from NADH (−0.32 V) to O_2 (+0.8166 V) fall ΔE = 1.1366 V, so

$$\Delta G^{\circ\prime} = -2 \times 96\,485 \times 1.1366 \approx -220\ \mathrm{kJ\,mol^{-1}} \quad (\mathrm{NADH} \to \tfrac{1}{2}\mathrm{O_2}) \tag{26.3}$$

— some four ATP-quanta of Chapter 25's −54 kJ mol^{-1} (Figure 26.2). The cell, as Chapter 28 will show in detail, harvests about two and a half ATP per NADH; the remainder is heat — which is not lost, but keeps the reader at 37 °C as he reads these lines. The furnace is efficient, not perfect, and its imperfection is the reader's warmth.[40]

40. The same arithmetic priced the biosphere before any enzyme was known: Lavoisier and Laplace, in 1780, measured the heat of a guinea pig and of burning charcoal in the same ice calorimeter and found the two consistent — respiration is combustion, they concluded, and the conclusion stood for a century and a half as the most correct quantitative statement in biology. The details were wrong (the site, the mechanism, the intermediates); the books were right. Accounting outlives mechanism.

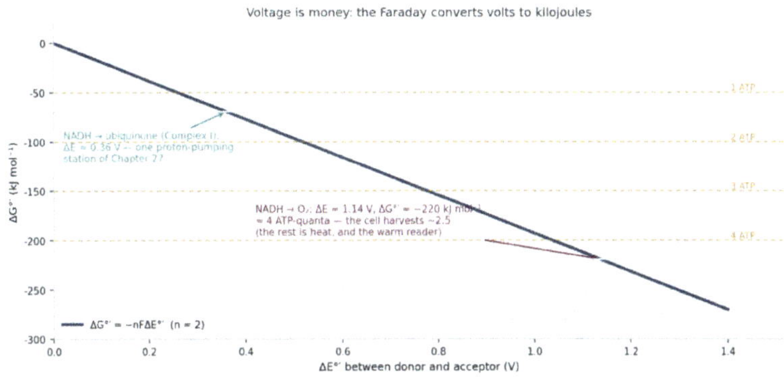

Figure 26.2 Eq. (26.2) as a graph: the free energy of a two-electron transfer against the voltage difference, with ATP-quanta (≈ 50 kJ mol^{-1} at cellular ΔG) marked. The great fall NADH → O_2 (crimson) is worth ~4 quanta; the first station of the respiratory chain (teal) collects its ~0.36-V toll.

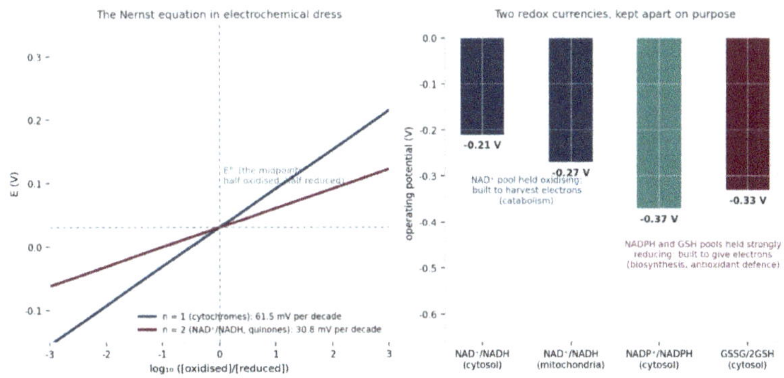

Figure 26.3 Left: the Nernst slope [Eq. (26.1)] for one- and two-electron couples; the midpoint is the standard potential and the point of maximum buffering. Right: the cell's actual redox bookkeeping — the NAD pool held oxidising (to harvest), the NADPH and glutathione pools held reducing (to give): two currencies, deliberately separated.

26.4 The electron banks: NAD, NADP, and the flavins

*The cell does not let electrons wander; it banks them in cofactors, and the banks come in two denominations whose separation is one of metabolism's deepest design decisions. The **nicotinamides** — NAD and its phosphorylated twin NADP — are hydride banks (Figure 26.4, left): the nicotinamide ring's C4 accepts a hydride ion whole, two electrons and a proton in a single transfer, no radicals, no partial states — baseball, not catch. (The transfer is stereospecific: dehydrogenases are A-side or B-side enzymes, addressing one face of the ring or the other with a fidelity the crystallographers of Part IX will meet again.) The distinction between the two banks is written in one phosphate on the ribose and in the redox bookkeeping of Figure 26.3, right: **NAD** is the catabolic bank — its pool held mostly oxidised (NAD^+/NADH ≈ 700 in cytosol, $E \approx -0.21$ V), hungry to receive electrons from food; **NADPH** is the anabolic bank — its pool held mostly reduced ($NADP^+$/NADPH ≈ 0.005, $E \approx -0.37$ V), primed to give electrons for biosynthesis* *(fatty acids, cholesterol, nucleotides, Chapter 34 onward) and for the antioxidant defences of § 26.6. Two nearly identical molecules, two opposite economies, separated by enzymes — transhydrogenase, kinases, the dehydrogenases' specificities — so that the cell can be simultaneously a furnace and a factory without the two currencies confusing their accounts. The pentose-phosphate pathway of Chapter 32 is NADPH's mint; keep the bank separate, and the factory is always solvent.*

*The **flavins** — FAD and FMN, riboflavin's working form — are the exchangers (Figure 26.4, right): the isoalloxazine ring can hold two electrons (hydroquinone), one (semiquinone — a stable radical, which the nicotinamides refuse to be), or none, and so can accept two-electron packages from NADH or succinate and pay them out one at a time to the one-electron world of iron and oxygen. Everywhere in Chapter 27 that a two-electron bank meets a one-electron wire, a flavin stands at the toll-booth — which is why the respiratory chain's entrance fee is paid to FMN (Complex I) and FAD (Complex II), and why riboflavin deficiency, alone among the vitamin deficiencies, is a direct failure of the power station's currency exchange.*

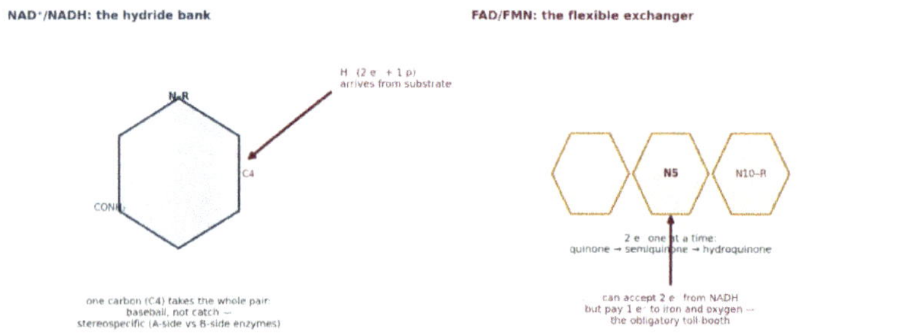

Figure 26.4 *The two banks. Left: NAD's nicotinamide ring takes the hydride whole at C4 — a two-electron transfer with no intermediate states, stereospecific by enzyme class. Right: the flavin's isoalloxazine holds zero, one, or two electrons — the obligatory exchanger between the two-electron banks and the one-electron wires of Chapter 27.*

26.5 The one-electron world: quinones, cytochromes, iron–sulfur

*Below the banks runs the wire. The **quinones** — ubiquinone Q, the isoprenoid-tailed benzoquinone of § 22.8 — are the membrane's own two-to-one exchangers: lipid-soluble, freely diffusing in the bilayer, carrying two electrons and two protons at a time from the flavins to the cytochromes, and paid out singly at the Q-cycle of Chapter 27. The **cytochromes** — haem proteins, the respiratory chain's coloured infantry — pass bare electrons through their irons: $Fe^{3+} + e^- \rightleftharpoons Fe^{2+}$, one at a time, at midpoint potentials the protein tunes by the haem's ligands and environment from −0.1 to +0.35 V (the same iron, a hundred different appetites — the protein is the haem's rheostat). And the **iron–sulfur clusters** — [2Fe–2S], [4Fe–4S], iron held by cysteine sulfurs and inorganic sulfide in cages of almost mineral simplicity — are the oldest conductors in biology: one-electron carriers of wide potential range, found from the respiratory chain to nitrogen fixation to the DNA-repair enzymes, and plausibly among the first catalysts life ever made — for iron and sulfide need no gene, no ribosome, no cell, only a reducing vent and the patience of geochemistry (Chapter 56 will make the argument). The wire, in short, is made of* *metal; the banks, of vitamin; and the respiratory chain of Chapter 27 is the flavin toll-booths and metal wires arranged along the tower of Figure 26.1 — which is to say, Chapter 27 is this chapter with protein added.*

26.6 The price of oxygen: ROS and the antioxidant economy

The summit of the tower is also its danger. Oxygen's appetite — the +0.8166 V that makes it the perfect terminal acceptor — is the appetite of a diradical: O_2's two unpaired electrons, in separate orbitals, must be fed one at a time or not at all, and the chemistry of one-at-a-time is the chemistry of radicals (Figure 26.5). Take one electron to O_2 and the product is ***superoxide*** *$O_2{\bullet}^-$; a second and two protons give* ***hydrogen peroxide*** *H_2O_2 (not a radical, but a diffusible oxidant and, worse, a precursor); a third electron — catalysed by free Fe^{2+} in the Fenton reaction — gives the* ***hydroxyl radical*** *•OH, the most indiscriminate oxidant in biology, which attacks whatever it touches within a nanosecond of its birth: DNA bases, membrane lipids (the peroxidation chain of Chapter 22's polyenes), proteins. Only the fourth electron and two protons give water, the intended destination. The respiratory chain of Chapter 27 is engineered to hand oxygen its four electrons nearly without spillage — yet even so, one to three per cent of mitochondrial oxygen leaks as superoxide, every day, in every reader: breathing is a controlled burn with a measurable arson rate.*

The cell's answer is a layered economy, and its base currency is ***glutathione*** *— the tripeptide γ-Glu-Cys-Gly, present at millimolar concentrations, whose cysteine thiol is sacrificed to peroxides and radicals and redeemed by NADPH through glutathione reductase. The layers run: superoxide dismutase (SOD — among the fastest enzymes known, diffusion-limited, § 15.6's perfection made urgent) converting superoxide to peroxide; catalase and the glutathione peroxidases converting peroxide to water; the GSH/GSSG pool ($E \approx -0.33$ V, Figure 26.3) paying the reducing equivalents; NADPH paying the pool; and the vitamins C and E — water- and lipid-soluble respectively, the latter stationed exactly where the membrane's polyenes are most at risk (§ 22.8) — covering the radical chain reactions the enzymes cannot reach. The relevance is not decorative: oxidative stress is implicated in aging, in reperfusion injury, in the neurodegenerations, and the whole signalling apparatus of Part X has learned to use peroxide as a second messenger — the arsonist, as so often in this book, has been conscripted. But the thermodynamic point stands: the couple at the top of the tower is worth its voltage precisely because it is dangerous, and life pays for its furnace in fire insurance.*

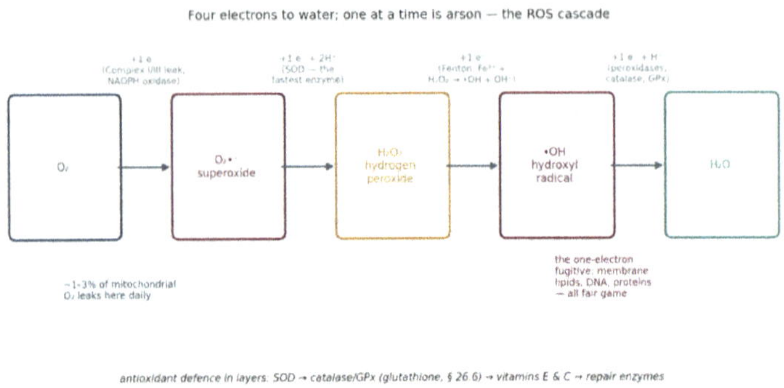

Figure 26.5 *The ROS cascade: one electron at a time from O_2 to water, with the intermediates — superoxide, peroxide, the hydroxyl radical — as the cell's daily arson problem. The antioxidant layers (SOD, catalase, glutathione peroxidase, vitamins E and C) are priced against it, and NADPH pays the premiums.*

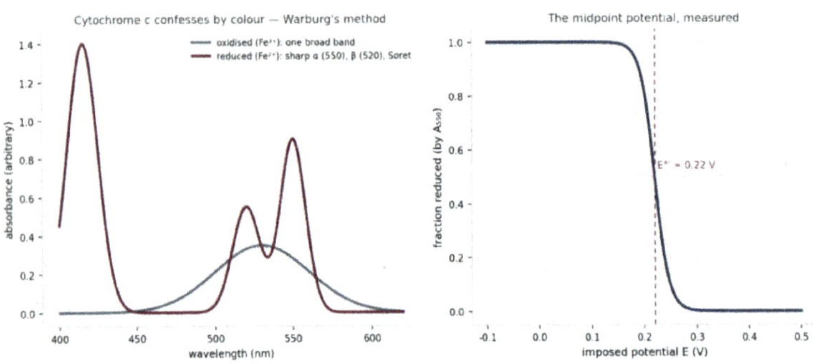

Figure 26.6 *The measurement behind the tower. Left: cytochrome c's visible spectrum changes with its iron's state — the reduced form's sharp α-band at 550 nm is the handle Warburg's and Keilin's schools used to read redox state in living tissue. Right: the Nernst titration — fraction reduced against imposed potential — reading off the midpoint $E°' = 0.22$ V [Eq. (26.1)].*

Plate 26.1 *Otto H. Warburg (1883–1970), whose manometry and spectroscopy identified the "respiratory enzyme" (Atmungsferment — cytochrome oxidase) and the nicotinamide and flavin coenzymes — the banks and wires of this chapter made measurable (Nobel Prize in Physiology or Medicine, 1931). Portrait from the Nobel Foundation archive, nobelprize.org, reproduced from the Foundation's publicly distributed image.*

HISTORICAL NOTE — THE COLOUR OF BREATHING

The respiratory chain was seen before it was understood — seen, literally, in colour. In the 1880s MacMunn had described in muscle a pigment, "myohaematin", whose bands shifted with aeration; the establishment, led by Hoppe-Seyler, pronounced it a contamination of haemoglobin and buried it for forty years. David Keilin, a Polish parasitologist at Cambridge working, by his own account, with a hand spectroscope and insects, rediscovered the pigment in 1925, renamed it cytochrome — cellular pigment, an honest word for a thing whose function he did not know — and showed the bands cycling between oxidised and reduced forms in flight muscle as the insect breathed: the wire of § 26.5, watched live through a prism. Warburg, meanwhile, in Berlin-Dahlem, had made the measurement quantitative: the manometer priced oxygen consumption molecule by molecule, the photochemical action spectrum (measuring which wavelengths relieved carbon monoxide's inhibition) identified the Atmungsferment as a haem protein, and his school purified the "yellow enzyme" (a flavoprotein), NAD, and NADP in the same decade — the 1931 Nobel honouring the method as much as the molecules.[41] *And the story has its coda: Warburg, half-Jewish, remained in Dahlem through the Third Reich, protected — it is said — by the cancer diagnosis of a man whose protection was worth more than a law. Science's debts are not always paid to its saints.*

41. Warburg's other legacy is the effect that bears his name — the observation (1924) that tumours ferment glucose to lactate even in oxygen's presence — which founded cancer metabolism as a field a century before the field admitted it was one. The mechanism is argued still (Chapter 30 will weigh it); the observation, like Lavoisier's books, was right first.

26.7 Case and clinic: cyanide, metformin, and the controlled burn

The tower is not a metaphor for the clinic; it is the clinic. ***Cyanide*** *binds the ferric iron of cytochrome a_3 — the last step before oxygen — and the whole chain backs up within seconds: the cytochromes go fully reduced (the bands of Figure 26.6 all sharpen at once, the historic bedside sign), oxygen consumption ceases though the blood is full of it, and the victim dies of histotoxic hypoxia — drowning in oxygen, the chain's traffic arrested at the summit. The antidotes work the tower's logic: nitrites oxidise some haemoglobin to methaemoglobin, whose ferric iron outbids the enzyme for the poison; hydroxocobalamin simply collects the cyanide into vitamin B_{12}'s chemistry and excretes it.* ***Metformin****, the world's most prescribed antidiabetic, is a gentler tower poison: a partial Complex I inhibitor whose effect is to lower the whole mitochondrial redox span — NADH rises, the cell's energy charge tilts, and the liver's glucose export is throttled (Chapter 30's gluconeogenesis priced in volts); the drug's celebrated safety record rests on its being a whispered inhibition rather than cyanide's shout. And the mitochondrial diseases proper — MELAS, LHON, the respiratory-chain encephalomyopathies — are mutations in the tower's own proteins: tissues with the highest electron traffic (brain, retina, heart, muscle) fail first, exactly as Figure 26.2 would predict, for they have the most quanta at stake. The controlled burn, when its controls fail, is still a burn.*

> ***CASE STUDY — THE HORMESIS PARADOX: A LITTLE ARSON KEEPS THE FIRE BRIGADE FIT***
>
> *If ROS are arson, extinguishing them entirely should be health; the trials said otherwise. High-dose antioxidant vitamins (E, C, β-carotene), given to large populations to prevent cancer and cardiovascular disease, failed in trial after trial — and in two (ATBC, CARET) increased the cancers they were meant to prevent. The reading that has emerged is* ***mitohormesis****: the ROS leak of Figure 26.5 is not merely tolerated but used — peroxide and superoxide are second messengers that upregulate the cell's own defences (the Nrf2 pathway, Part X), its mitochondrial biogenesis, and its repair enzymes; exercise, which transiently raises mitochondrial ROS, improves insulin sensitivity through the ROS signal, and the improvement is abolished when subjects take antioxidant vitamins alongside the training. The fire brigade, it turns out, is drilled by small fires; flood the station with extinguisher and the drills stop. The case is a standing warning against the naive pharmacology of § 26.6 — and a neat illustration of the book's recurring dialectic: a toxin at one dose, a signal at another, and the dose, as always, is the whole of the medicine.*

26.8 Synthesis of Chapter 26

The chapter has drawn the current behind Chapter 25's currency. The politics: oxidation is electron loss, reduction is electron gain, and the two are one transaction — a redox couple. The measurement: every couple's appetite is a voltage on the tower of Figure 26.1, actualised by the Nernst relation [Eq. (26.1)] and converted to money by the Faraday [Eq. (26.2)] — whence the master number of Part VI, −220 kJ mol^{-1} for NADH $\rightarrow$ O_2 [Eq. (26.3)], some four ATP-quanta, of which the cell banks two and a half and warms the reader with the rest. The banks: NAD for catabolism, NADPH for anabolism, kept apart on purpose; the flavins at the toll-booth between the two-electron and one-electron worlds; the quinones, cytochromes and iron–sulfur clusters as the metal wire. And the price:

oxygen's diradical appetite, paid for in ROS and fire insurance, with the vitamins' failed trials as the dialectic's clinical signature. The direction of Figure ii holds: the currency priced, the current drawn — and the current now needs its machine. Chapter 27, The Mitochondrion and Electron Transport, lays the tower of Figure 26.1 out along a membrane, stations the flavins and the irons in four great complexes, and shows how the fall of electrons is made to pump protons uphill — the proton-motive force being, as the reader will see, the most beautiful money-laundering operation in biology.

Exercises (with full resolutions)

EXERCISE 26.1

Using Figure 26.1, compute $\Delta G^{\circ\prime}$ for the transfer of two electrons from NADH to ubiquinone (Complex I's span), and state what fraction of the full NADH → O_2 fall it represents.

Resolution. *$\Delta E^{\circ\prime} = 0.07 - (-0.32) = 0.39$ V; $\Delta G^{\circ\prime} = -2 \times 96\,485 \times 0.39 \approx -75$ kJ mol^{-1} [Eq. (26.2)]. Against the full -220 kJ mol^{-1} [Eq. (26.3)], the first station collects 34% — roughly one ATP-quantum, matching the four protons Complex I pumps (Chapter 27's accounting). The tower's steps are sized, not arbitrary: each pumping complex spans about an ATP's worth of volts.*

EXERCISE 26.2

At the mitochondrial matrix, $NAD^+/NADH \approx 10$; compute the actual potential of the NAD couple there ($E^{\circ\prime} = -0.32$ V, $n = 2$, 37 °C) and compare with the cytosolic value ($NAD^+/NADH \approx 700$).

Resolution. *Eq. (26.1): $E = -0.32 + 0.0308 \log_{10}(10) = -0.32 + 0.031 = -0.29$ V (mitochondria); $E = -0.32 + 0.0308 \log_{10}(700) = -0.32 + 0.088 = -0.23$ V (cytosol). The matrix pool is the more reducing — it must be, to push electrons into Complex I against the chain's appetite — and the two pools differ by ~60 mV, ≈ 6 kJ mol^{-1} per electron pair: the shuttles of Chapter 30 (malate–aspartate, glycerol-phosphate) exist precisely because the membrane passes metabolites, not NADH, and the two banks must be wired across it.*

EXERCISE 26.3

Why can NADH not pass electrons directly to cytochrome c, and which two chemical properties of the flavin solve the problem?

Resolution. *NADH is a two-electron, hydride-only donor (Figure 26.4, left): it has no one-electron state, while cytochrome c's iron accepts one electron at a time. The flavin's isoalloxazine solves the mismatch by (i) supporting a **stable semiquinone** — it can hold a single electron without the radical fleeing or disproportionating, which the nicotinamide ring cannot; and (ii) **tunable potentials** — its midpoint is reset by the protein environment at each complex, so the first electron can be paid out to iron at a different voltage from the second. The toll-booth is a chemical necessity, not an evolutionary decoration: wherever two-electron banks meet one-electron wires, a flavin (or a quinone, § 26.5) must stand.*

EXERCISE 26.4

The Fenton reaction requires free Fe^{2+}. From this, explain (i) why iron is sequestered in transferrin and ferritin, and (ii) why iron overload (haemochromatosis) is a disease of oxidative damage rather than of iron itself.

Resolution. (i) Free Fe^{2+} converts the cell's peroxide — an unavoidable by-product of Figure 26.5's cascade — into the hydroxyl radical, the one ROS no enzyme can scavenge (it reacts at its birth-site, diffusion being its only kinetics). Iron is therefore kept chelated: transferrin in the blood, ferritin in the cell, with the free pool held at nanomolar or below. The sequestration is not storage; it is fire prevention. (ii) In haemochromatosis the chelation capacity is exceeded, free iron rises, and the Fenton reaction runs in liver, pancreas, heart and joints — the clinical quartet (cirrhosis, diabetes, cardiomyopathy, arthritis) being simply the map of where the iron pools overflow. The treatment is the diagnosis: venesection — remove the catalyst, and the chemistry stops. Among genetic diseases, none is treated so directly by its own equation.

EXERCISE 26.5

Cyanide victims are cherry-pink, not cyanotic. Explain the colour from this chapter, and from it the design of the nitrite antidote.

Resolution. Cyanide arrests the chain at cytochrome a_3: tissues cannot extract oxygen, so venous blood returns as oxygen-loaded as arterial — the cherry-pink of oxyhaemoglobin in the veins, the inverse of cyanosis's blue. The antidote follows from the tower's logic: nitrites oxidise a fraction of haemoglobin's Fe^{2+} to Fe^{3+} (methaemoglobin), and methaemoglobin's ferric iron binds CN^- more avidly than the enzyme's — the poison is drawn off the chain and held on the blood, then converted to thiocyanate by the sulfur-donor drugs and excreted. One redox couple (Hb-Fe^{3+}/Fe^{2+}) is sacrificed to outbid another (cyt a_3) for the ligand: Figure 26.1 as prescription.

Research-level propositions

Proposition 26.α. The respiratory chain's complexes sit at potentials that leave each pumping step ≈ 0.3–0.4 V (Figure 26.2). Using the Marcus theory of electron transfer and the measured edge-to-edge distances of the cofactors, compute whether the observed driving forces are near the activationless optimum — and assess whether the chain's redox spacing is set by thermodynamic harvest, by kinetic optimisation, or by the ROS-leak constraint of § 26.6. Proposition 26.β. The NAD and NADP pools are held at different potentials though transhydrogenase can equilibrate them. Construct the steady-state model of the two pools with measured fluxes (glycolysis, PPP, transhydrogenase, antioxidant drain), and determine whether the observed potentials are the passive result of the flux balance or an actively regulated set-point — with the experiment (transhydrogenase knockdown + isotope tracing) that distinguishes the two. Proposition 26.γ. Mitohormesis (the case study) predicts an inverted-U dose–response between ROS signalling and healthspan. Formalise it: model Nrf2 activation as a function of ROS flux with damage accumulation as the competing term, locate the optimum, and test whether the published antioxidant-trial doses fall on the adverse side of the curve — then design the trial (low-dose ROS-mimetic versus antioxidant) the model says should be run.

References (Chapter 26)

Berg, J. M., Tymoczko, J. L., Gatto, G. J., Jr., & Stryer, L. (2023). Biochemistry (9th ed.). W. H. Freeman.

Halliwell, B., & Gutteridge, J. M. C. (2015). Free radicals in biology and medicine (5th ed.). Oxford University Press.

Keilin, D. (1966). The history of cell respiration and cytochrome. Cambridge University Press.

Nelson, D. L., Cox, M. M., & Hoskins, A. A. (2021). Lehninger principles of biochemistry (8th ed.). W. H. Freeman.

Nicholls, D. G., & Ferguson, S. J. (2013). Bioenergetics (4th ed.). Academic Press.

Warburg, O. (1930). Über den Stoffwechsel der Tumoren [The metabolism of tumours]. Springer.

CHAPTER 27

The Mitochondrion and Electron Transport

CONTENTS OF THIS CHAPTER

In which the redox tower of Chapter 26 is laid end to end along a membrane, and the fall of electrons is found to pump protons uphill; in which four great complexes — one of them the largest enzyme in the cell — are walked through their stations; in which the Q-cycle is shown to be a money-doubling operation of impeccable dishonesty; in which the proton-motive force is priced in millivolts and the furnace is shown to idle without a customer; and in which a bacterium, swallowed two billion years ago, is discovered still paying the rent.

27.1 The organelle and its ancestor

Chapter 26 ended with a tower of voltages and a promise: −220 kJ mol^{-1} waiting between NADH and oxygen, and a cell that banks two and a half ATP of it. This chapter builds the machine that does the banking, and the machine has an address — the mitochondrion, a sausage-shaped organelle a micrometre long, hundreds to thousands per cell, whose structure (Figure 27.2) is the whole of its function. Two membranes, not one: an outer membrane made freely permeable by porins — a courtesy wall, letting all small molecules pass — and an inner membrane that is the tightest in the cell, folded into shelves called **cristae** *to multiply its area (in heart muscle, more inner membrane per cell than plasma membrane), impermeable to everything including the proton, and bearing the four complexes of this chapter plus the ATP-synthase lollipops of Chapter 28. Between them, the intermembrane space — chemically almost cytosol, save for one distinction on which the whole economy turns: the protons pumped there are a currency. Within, the matrix — the Krebs cycle's reaction vessel (Chapter 29), the organelle's own DNA and ribosomes, and the NADH whose electrons are the furnace's fuel. The compartments are not geography but economics: only the inner membrane insulates, so only across it can a gradient be held, and the gradient, as § 27.6 will show, is the point.*

The organelle is also a citizen with a past. Mitochondria carry their own small circular genome (thirteen proteins in the human, all of them subunits of this chapter's complexes), their own ribosomes of bacterial pattern, their own genetic code's private dialect, and a sensitivity to precisely the antibiotics that poison bacterial ribosomes — the signature, in four independent handwritings, of an α-

42. The endosymbiotic theory was argued into respectability by Lynn Margulis from 1967

membrane (the phagosome's and the bacterium's own), the bacterial-style chain, the retained genome (whose thirteen genes are, suggestively, all hydrophobic subunits of the complexes — the proteins hardest to import), and the inheritance of § 27.9's diseases, which pass through the mother alone because the egg, not the sperm, brings the power stations. We are, every one of us, running on a bacterium's engine, serviced by a bacterial genome, and paying the bacterium board and lodging in NADH.

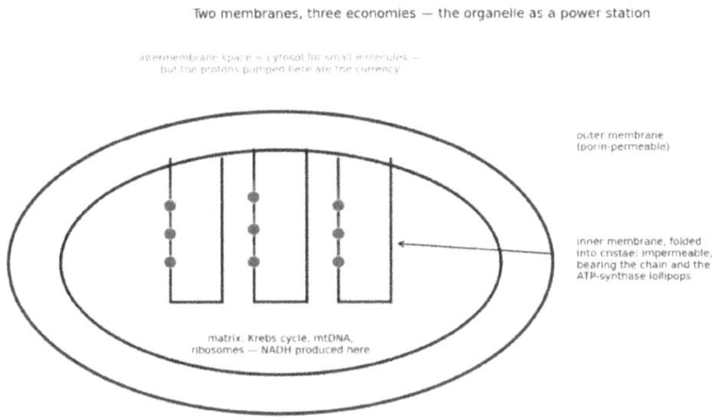

Figure 27.2 The power station's floor plan. The outer membrane (navy) is porous and merely polite; the inner membrane (crimson), folded into cristae, is the impermeable working surface bearing the chain and the ATP-synthase lollipops (green). The intermembrane space is where the proton currency is banked; the matrix is where NADH is minted by the Krebs cycle.

27.2 The chain: the tower laid end to end

Lay Chapter 26's tower on its side along the inner membrane and the respiratory chain appears (Figure 27.1): a queue of redox centres, each a little hungrier than the last, down which the electron pair from NADH falls in ordered steps to oxygen. The queue is organised into four great complexes, with two mobile couriers between them — ubiquinone Q, the lipid-soluble two-electron carrier of § 26.5, diffusing in the membrane itself; and cytochrome c, a small soluble haem protein swimming in the intermembrane space — and the whole can be read in one line:

$$\mathrm{NADH} \xrightarrow{\mathrm{I}} \mathrm{Q} \xrightarrow{\mathrm{III}} \mathrm{cyt}\,c \xrightarrow{\mathrm{IV}} \mathrm{O_2} \qquad (\mathrm{II} : \mathrm{succinate} \rightarrow \mathrm{Q}) \tag{27.1}$$

*— Complex II (succinate dehydrogenase, the Krebs cycle's own enzyme moonlighting as the chain's second door) entering at Q one landing lower, whence $FADH_2$'s smaller harvest. The chain's thermodynamics is Figure 26.2's arithmetic made flesh: the full fall is 1.14 V and −220 kJ mol^{-1}, and three of the complexes — I, III, IV — are stationed at drops of ~0.3–0.4 V, each spanning roughly an ATP-quantum. That is not a coincidence but a design: at each station, the fall of the electron pair is coupled to the pumping of protons uphill across the membrane — 4 at Complex I, 4 at Complex III, 2 at Complex IV, **ten in all per NADH** — and Chapter 28 will show the protons marching home through the ATP synthase to pay for the currency. The chain is not a wire with pumps attached; the pumping is the wire's purpose, and the electrons are the pump handles.*[43]

onward, against an establishment that found the idea romantic and the evidence thin. The genome sequence settled it utterly: mitochondrial DNA is bacterial DNA, abridged. The establishment's romanticism is now the first fact of cell biology — and a standing lesson in what counts as evidence.

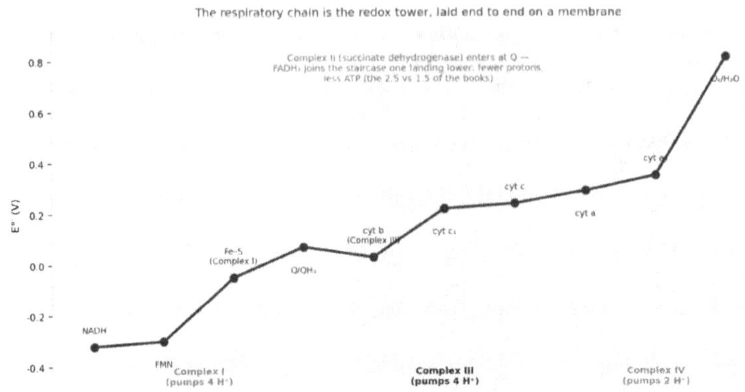

Figure 27.1 The chain on the tower: midpoint potentials along the electron's path from NADH to oxygen, with the three pumping complexes (I, III, IV) spanning ~0.3–0.4 V each — an ATP-quantum per station. Complex II (plum note) joins at Q, below the first pump.

27.3 Complex I: the piston of the NADH bank

Complex I — NADH:ubiquinone oxidoreductase, 45 subunits and nearly a million daltons, the largest enzyme in the inner membrane — is the chain's great gate and its deepest puzzle (Figure 27.3). Its business end is an L: the hydrophilic arm reaching into the matrix takes NADH onto FMN (Chapter 26's toll-booth — two electrons in, to be paid out one at a time), passes them down a wire of seven iron–sulfur clusters spaced hand-to-hand at ~9 Å — electron transfer being tunnelling, which dies beyond ~14 Å, so that the chain's cofactors are wired like climbers passing loads — and delivers the pair to ubiquinone at the membrane. The pumping end is the membrane arm, and here the mechanism was biology's longest-running structural mystery: how does the electron's arrival, sixty ångströms away, move four protons? The cryo-electron-microscopy answer, assembled over the last decade, is a ***piston****: quinone's reduction drives a wave of conformational change down the arm — helix HL and the charged residues of three antiporter-like subunits — that alternately opens proton channels to matrix and intermembrane sides, moving four protons per cycle by electrostatic relay. The details are still being argued (which is to say, the field is healthy); the accounting is not: $NADH + Q + 5H^+_N \rightarrow NAD^+ + QH_2 + 4H^+_P$, and the four protons are the largest single instalment of the ten. The bank's first withdrawal is also its largest.*

43. Complex II pumps nothing, and the textbooks' 2.5-versus-1.5 ATP per NADH-versus-$FADH_2$ is exactly the difference between ten protons and six. The Krebs cycle's only membrane enzyme is thus an honest tributary: it feeds the river below the first dam.

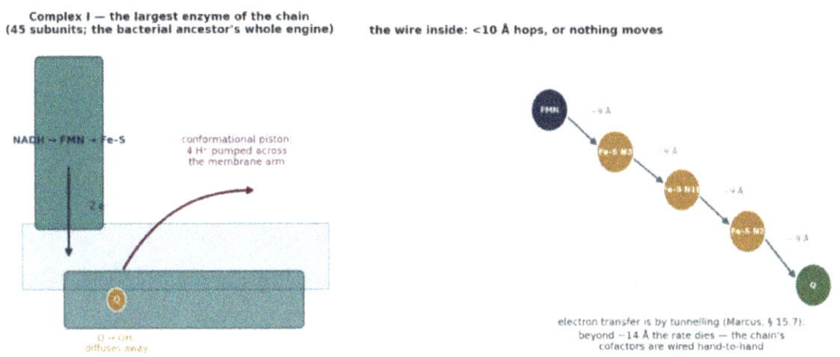

Figure 27.3 Complex I. Left: the L — a hydrophilic arm (NADH → FMN → Fe–S → Q) driving a conformational piston in the membrane arm that pumps 4 H^+. Right: the wire inside — Fe–S clusters at ~9 Å hops, because tunnelling dies beyond ~14 Å.

27.4 Complex III and the Q-cycle: the honest swindle

Between Q and cytochrome c stands Complex III — the bc_1 complex — and its mechanism is the single most elegant piece of accounting in this book, the ***Q-cycle*** *(Figure 27.4). The problem it solves is the currency mismatch of Chapter 26: Q carries two electrons, cytochrome c takes one; and a complex that simply handed QH_2's pair to two cytochromes would waste the fall. The cycle's solution: QH_2 arrives at the outer site (Q_o) and its two electrons are split — one goes down the high road (Fe–S → cyt c_1 → cyt c, toward oxygen), the other takes the low road back through the two b-haems (b_H → b_L) to the inner site (Q_i), where a second Q waits to be reloaded: reduced to semiquinone on the first turn, and to QH_2 on the second, taking its two protons from the matrix. Sum the two turns and the books read: one QH_2 consumed (net), two cytochromes reduced, two protons taken from the matrix and four released outside —* ***four protons pumped per electron pair****, double what a naive one-for-one would yield. The swindle is honest because thermodynamics is honoured: the low road's detour through the b-haems is paid for by the high road's fall, and the cycle is simply the cheapest way yet found to convert a two-electron currency into a one-electron wire at a profit. Antimycin A, the complex's poison, blocks the Q_i site — and, in the laboratory of § 27.8's history, was one of the inhibitors by which the chain was sequenced before it was seen.*

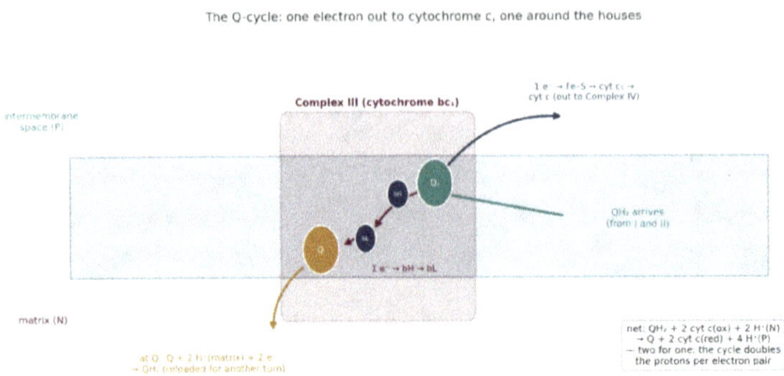

Figure 27.4 The Q-cycle. QH_2 at the outer site splits its pair: one electron to cytochrome c (high road, navy), one around the b-haems (crimson) to reload a second Q at the inner site with matrix protons. Net: 4 H^+ pumped per pair — the doubling is the whole point.

27.5 Complex IV: the four-electron ceremony at oxygen

The chain's last station is its most demanding, for the substrate is the diradical of § 26.6 and the chemistry must be four-electron without a single radical escaping. Cytochrome c oxidase — Complex IV, the Atmungsferment of Warburg's Historical note — receives electrons one at a time from cytochrome c (through Cu_A and haem a) into a bimetallic lair, haem a_3 and Cu_B sitting within 5 Å of each other, where the O_2 is held between the two metals like a prisoner between two guards. There the O–O bond is broken and the four-electron reduction proceeds through tightly held intermediates — peroxide bound between the irons, ferryl species — never released, never free, so that the ROS of § 26.6 are at this station reduced to a leak of a per cent or less. And the enzyme pumps: two protons translocated across the membrane per electron pair, in addition to the two "chemical" protons consumed from the matrix in making water — so that the last station's books read 2 pumped + 2 consumed, four matrix protons' worth of gradient per pair. Cyanide, carbon monoxide and azide all bind the a_3 iron and stop the ceremony at the altar rail; the historic relief of CO poisoning by pure oxygen — competition at the same site, Le Chatelier in the ambulance — is this station's pharmacology. With Complex IV the chain is walked: ten protons out per NADH, six per succinate, one water per pair of electrons, and the fall of Figure 27.1 fully converted into a proton gradient pressed against the inner membrane like water behind a dam. Chapter 28 opens the sluice.

27.6 The proton-motive force: the currency between the currencies

What exactly has the chain banked? A gradient of protons across an impermeable membrane — and Chapter 24 already priced such a thing, for a proton gradient is two gradients in one coat: a concentration difference (a pH difference) and a charge difference (a membrane potential), since the proton is both solute and ion. The electrochemical potential of Eq. (24.1), written for protons and converted to millivolts, is the **proton-motive force:**

$$\Delta p = \Delta\psi - \frac{2.303\,RT}{F}\Delta\text{pH} = \Delta\psi - 61.5\ \text{mV} \times \Delta\text{pH} \quad (37\ ^\circ\text{C}) \tag{27.2}$$

— $\Delta\psi$ the membrane potential (outside positive, since protons were pumped out), ΔpH the pH difference (outside acid); the two terms add, and in the mitochondrion the measured values are $\Delta\psi \approx 150$ mV and $\Delta pH \approx 0.75$, giving $\Delta p \approx 200$ mV (Figure 27.5, left). Note the division of labour: the gradient is mostly voltage, not acidity — the intermembrane space is barely half a pH unit down, for the membrane's capacitance is small and a few protons charge it fully. Two hundred millivolts across five nanometres is a field of 4×10^7 V m^{-1} — the strength of a thundercloud, sustained across every inner membrane in the body, every second of life. Converted back to chemistry, the proton's return is worth $\Delta G = F{\cdot}\Delta p \approx 20$ kJ mol^{-1} — and the ten protons of the chain thus bank ≈ 200 kJ mol^{-1} of the −220 available: the furnace's efficiency, before the synthase takes its commission, is already near ninety per cent.

*The gradient is not a warehouse but a current account, and the proof is the oldest experiment in the field (Figure 27.5, right): give isolated mitochondria NADH-linked substrate and oxygen, and they respire at a slow idle (**state 4**); add ADP, and oxygen consumption leaps fourfold (**state 3**) until the ADP is spent, whereupon the idle resumes. The furnace runs only when there is a customer for ATP — **respiratory control** — and the reason is the coupling itself: the chain can pump protons only against the back-pressure of the gradient it builds, and when no ATP is being made the protons have nowhere to go, the gradient rises to its ceiling, and the chain stalls like a pump against a closed valve. The observation is the chemiosmotic theory in a single trace, and its logic inverts every naive picture: respiration is not switched on by fuel but by demand — the muscle's ATPase, the brain's pumps, the liver's syntheses are the switches, and the mitochondrion is the servant that asks, perpetually, what it may burn for you.*

Worked Example 27.1 — The Ceiling of the Gradient

The chain pumps 10 H^+ per NADH. If the gradient could rise without limit, the −220 kJ mol^{-1} of Eq. (26.3) would support $\Delta p = 220/(10 \times 96.485) \approx 228$ mV. The measured Δp in state 4 is ≈ 220 mV. (i) How close to thermodynamic equilibrium does the idling chain run? (ii) What does the small gap signify?

***Solution.** (i) The banked $10 \times F \times 0.220 = 212$ kJ mol^{-1} is 96% of the available 220 — the chain in state 4 is within a few kJ mol^{-1} of equilibrium, its pumps nearly stalled by their own back-pressure, exactly as respiratory control requires. (ii) The residual gap is the leak: protons seep back through the membrane and through uncoupling proteins (§ 27.8), and the idle oxygen consumption of state 4 is precisely the rate at which the chain must re-pump the seepage. The brown fat of the newborn and the hibernator exploits the same leak deliberately — UCP1 opens the sluice without the synthase, the chain runs at full throttle against no load, and the −220 kJ mol^{-1} appears as heat alone: non-shivering thermogenesis is respiratory control with the customer deliberately dismissed. The uncoupler DNP (§ 27.9's cautionary tale) does the same thing chemically, and the dose, as always, is the whole of the medicine.*

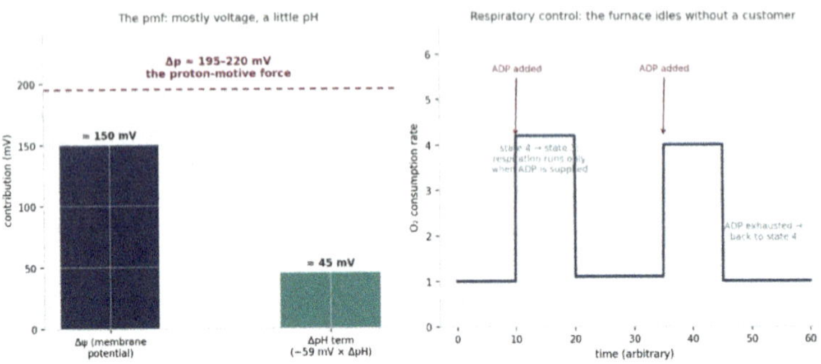

Figure 27.5 *Left: the proton-motive force's two components [Eq. (27.2)] — mostly voltage, a little pH, ≈ 200 mV in all. Right: respiratory control — oxygen consumption idles in state 4, leaps when ADP is supplied (state 3), and idles again when it is spent. The furnace answers to demand, not to fuel.*

27.7 The poisons' map, and what the poisons taught

The chain was sequenced before it was seen — sequenced by poisons (Figure 27.6). Each complex has its inhibitor: rotenone (the fish poison of the derris root) and the barbiturates at Complex I; malonate, succinate's competitive shadow, at Complex II; antimycin A at Complex III's Q_i site; cyanide, azide and carbon monoxide at Complex IV's oxygen lair; and oligomycin at the ATP synthase itself. The experimental logic of the 1950s was beautiful in its simplicity: add a poison, and every component before the block becomes fully reduced (traffic jam upstream), every component after it fully oxidised (starved downstream) — the crossover point, read off the cytochrome bands of Figure 26.6, marks the poison's station; and with enough poisons the whole order of the chain was deduced without a single structure. The same poisons taught the coupling: oligomycin, which never touches the chain, stops respiration as surely as cyanide does — for with the synthase blocked the gradient has no outlet, and the chain stalls by its own back-pressure (Worked Example 27.1's ceiling); the uncouplers reverse the lesson — DNP or FCCP, protonophores that ferry protons through the membrane, release the back-pressure and let the chain race with oxygen consumption uncoupled from ATP, the energy appearing as heat. Block the product and the chain stops; open the sluice and it races: the chain and the synthase are coupled by the proton and by nothing else — the chemiosmotic verdict, delivered by poisons.

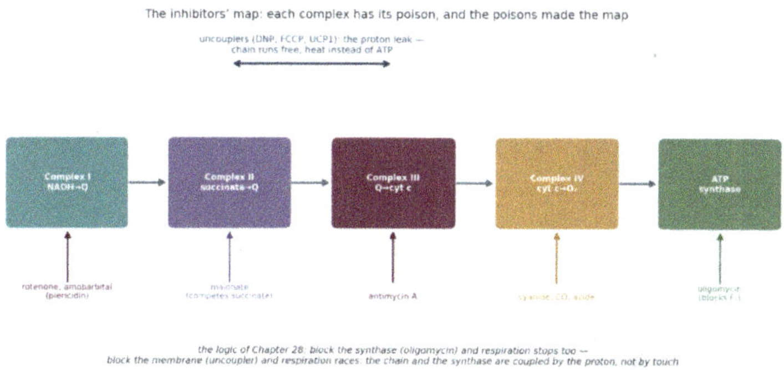

__Figure 27.6__ The poisons' map. Each complex has its inhibitor (rotenone, malonate, antimycin, cyanide, oligomycin), and the uncouplers (DNP, FCCP, UCP1) leak the gradient itself. The poisons sequenced the chain and proved the coupling before either was seen.

27.8 Chemiosmosis: the heresy that became the household

The theory that respiration's energy is stored in a proton gradient was proposed by Peter Mitchell in 1961, and it is no exaggeration to call it the most consequential heresy in twentieth-century biochemistry. The orthodoxy it displaced was respectable and wrong: everyone "knew" that the chain must make a chemical intermediate — some phosphorylated X~ of the squiggle economy, an activated carrier that would hand its phosphate to ADP — and the best laboratories in the world spent the 1950s hunting for it, in vain. Mitchell, working in a converted manor house in Cornwall with Jennifer Moyle, proposed that there was no X~ at all: the chain pumps protons, the membrane holds them, and the ATPase — which he renamed the ATP synthase, reversing its supposed identity — is run backwards by their return. The establishment's response was a decade of hostility, conducted mostly at meetings and mostly at Mitchell's expense; the experimentum crucis came from an unexpected quarter — André Jagendorf's chloroplasts, acid-bathed in the dark, made ATP when their thylakoid lumen was suddenly neutralised, an artificial gradient producing the currency with no chain running at all — and from Ephraim Racker's reconstitutions, in which a bacterial rhodopsin pump and a mitochondrial ATPase, from different kingdoms of life, made ATP together in a liposome: the gradient was sufficient, the machinery interchangeable, the chemistry universal.[44]

44. Mitchell's Nobel came in 1978 — the citation reading, in effect, for the discovery that there was nothing to discover. The field's amends were graceful: the oxidative-phosphorylation establishment's leader, E. C. Slater, conceded publicly; and the word "chemiosmotic", coined as a slur of vagueness, became the name of the consensus. The moral for Part VI is worth stating once plainly: in energy transduction, the intermediate is not a molecule but a membrane.

Plate 27.1 *Peter D. Mitchell (1920–1992), whose chemiosmotic theory (1961) replaced the missing chemical intermediate with the proton gradient — the most consequential heresy in bioenergetics, vindicated by Jagendorf's acid bath and Racker's reconstitutions (Nobel Prize in Chemistry, 1978). Portrait from the Nobel Foundation archive, nobelprize.org, reproduced from the Foundation's publicly distributed image.*

27.9 Case and clinic: the organelle in the dock

The mitochondrion's failures are as various as its duties, and the clinic reads them in this chapter's vocabulary. The **mitochondrial diseases** *proper — mutations in the thirteen mtDNA genes or the ~1 500 nuclear genes of the organelle — declare themselves where the electron traffic is densest: LHON's sudden blindness (Complex I subunits, the retinal ganglion cell's thin unmyelinated axons failing first), MELAS's strokes and lactic acidosis, MERRF's myoclonus, Leigh's symmetrical brainstem lesions of infancy — and because the egg brings all the mitochondria, the mtDNA diseases pass from the mother alone, with the* **heteroplasmy** *of the inherited mixture (mutant and normal genomes coexisting, the tissue failing when the mutant fraction crosses its threshold) explaining the bewildering variability of the same mutation in one family. The* **uncouplers** *are the second cautionary tale: dinitrophenol, sold as a slimming drug in the 1930s on the true observation that it burns fat at rest, killed by heatstroke — the furnace of Worked Example 27.1 with the thermostat smashed — and remains on the internet's darker shelves; its pharmacology is Figure 27.5's state-3 trace with no ADP ever arriving. And the organelle's slower betrayals fill the rest of the book: the ROS leak of § 26.6 implicated in aging itself, the mitochondrial permeability transition as the executioner's door of reperfusion injury, and the failure of mitochondrial quality control (the PINK1–parkin machinery that tags a depolarised organelle for autophagy) as the cause of one familial Parkinson's — the cell that cannot retire its failing power stations is poisoned by them. The clinic, as usual, is the physiology read after the accident; but here the physiology is the accident's probability, written in millivolts.*

> ***Case study — The barbiturate and the bee-keeper: rotenone, risk, and the naturalistic fallacy***
>
> *Rotenone, Complex I's classic inhibitor, entered commerce as a "natural" insecticide and piscicide — derris root, used for centuries by indigenous fishermen of the Amazon and Southeast Asia to stun fish, and embraced by organic agriculture on the strength of its botanical pedigree. Its toxicity is*

Figure 27.6's first station in any animal that absorbs it: chronic exposure in animal models reproduces Parkinson's disease with uncomfortable fidelity (Complex I inhibition in substantia-nigra neurons, α-synuclein aggregation), and epidemiology has linked occupational rotenone exposure to elevated Parkinson's risk. The regulatory arc ran its slow course — permitted as "natural" for decades, restricted, and largely withdrawn — while the synthetic neonicotinoids, designed for insect-selective nicotinic pharmacology with orders of magnitude less mammalian toxicity, were being legislated against on other grounds entirely. The case is not an argument for or against any regulation; it is the chapter's map read as a warning label. "Natural" describes an origin, never a mechanism; rotenone's mechanism is the same in fish, bee and farmer — a stalled piston in Figure 27.3 — and the dose, the route and the species' Complex-I sensitivity, not the pedigree of the root, are the whole of the toxicology.

27.10 Synthesis of Chapter 27

The chapter has laid the tower on a membrane and walked it. Geography first: two membranes, the inner one impermeable and cristae-folded, the gradient banked in the intermembrane space, the fuel minted in the matrix — and a bacterium's signature on every detail. Then the stations: Complex I's piston and its ~9-Å wire, the Q-cycle's honest swindle doubling the protons, Complex IV's four-electron ceremony with its two-metal lair — ten protons per NADH, six per succinate, one water per pair. Then the currency between the currencies: the proton-motive force of Eq. (27.2), 200 millivolts of thundercloud across five nanometres, ≈ 90% of the fall banked; and the furnace's governor — respiratory control, the chain idling without a customer, the uncouplers dismissing the customer deliberately. The heresy has become the household: no X~, no squiggle intermediate — the intermediate is a membrane, and the gradient is the account on which Chapter 28 now draws. For the gradient, like every account in this book, exists to be spent; and the machine that spends it — the ATP synthase, a rotary motor smaller than a bacterium's fingernail, turning protons into the currency at a hundred revolutions per second — is the most beautiful engine in the book, and the next chapter is its.

Exercises (with full resolutions)

Exercise 27.1

Compute the free energy stored per NADH when the chain pumps 10 protons against Δp = 200 mV, and compare with the −220 kJ mol^{-1} available. Where does the rest go?

Resolution. *Per proton: $\Delta G = F\Delta p$ = 96 485 × 0.200 ≈ 19.3 kJ mol^{-1}; ten protons bank ≈ 193 kJ mol^{-1} — 88% of the 220. The missing 12% is the chain's own dissipation: the redox drops at each station are not perfectly matched to the pumping steps (Figure 27.1's treads are uneven), the ROS leak of § 26.6 spends a per cent or two on arson, and the last step's irreversibility (the O–O bond, once broken, is not coming back) is paid for in volts. In state 4 the efficiency rises to ~96% (Worked Example 27.1) — the chain idles nearest equilibrium precisely when no work is being done, which is thermodynamics' standing joke about idleness.*

Exercise 27.2

Explain why $FADH_2$ (from succinate) yields fewer ATP than NADH, and state the exact bookkeeping in protons.

Resolution. *$FADH_2$ enters at Complex II, whose midpoint (succinate/fumarate, ≈ 0 V) is already at Q's landing: the electron pair joins the chain below the first pump (Figure 27.1, plum note), so only Complexes III and IV work for it — 4 + 2 = 6 protons against NADH's 10. At Chapter 28's exchange rate (~4 H^+ per ATP, including the phosphate and adenine-nucleotide transport tolls), 6 H^+ buys ~1.5 ATP, 10 H^+ buys ~2.5 — the textbook 2.5/1.5 is the proton accounting, rounded. The Krebs cycle's only membrane enzyme is thus also its only honest one: it declares at the door exactly how much of the fall it has left to sell.*

Exercise 27.3

Antimycin A blocks Q_i in Complex III. Predict the redox state of (a) cytochrome c, (b) the b-haems, (c) NADH, and the rate of oxygen consumption, in coupled mitochondria.

Resolution. *The crossover logic of § 27.7: upstream of the block everything piles up reduced, downstream everything is starved oxidised. (a) Cytochrome c — downstream — becomes fully oxidised; (b) the b-haems, trapped on the low road with nowhere to deliver, become fully reduced (and the Q pool likewise over-reduced, Q_o unable to discharge its second electron); (c) NADH backs up through Complex I to full reduction, and the whole chain stalls — oxygen consumption falls to the leak rate. The observation, made in the 1950s with a spectroscope and no structure, was how the b-haems' place in the cycle was deduced: the poison's traffic jam was the map.*

Exercise 27.4

Jagendorf's chloroplasts, kept in the dark, made ATP when the thylakoid lumen was switched from pH 4 to pH 8. Reconstruct the experiment's logic and state precisely what it proved and what it did not.

Resolution. *The acid-bathed thylakoids load their lumen with protons at pH 4; the sudden switch to pH 8 outside creates an artificial ΔpH of 4 units across the membrane — by Eq. (27.2), $\Delta p \approx 61.5 \times 4 \approx 246$ mV, a full physiological gradient, imposed without light, electron transport, or any chemical intermediate. The ATPase runs and ATP appears. It proved sufficiency: a proton gradient alone, with the chain silent, drives the synthase — the chemiosmotic theory's experimentum crucis. It did not prove the natural mechanism of pumping (the chain's chemistry was argued separately), nor rule out every auxiliary hypothesis — but it killed X~: no high-energy chemical intermediate could have been formed in the dark by an acid bath. One pH jump, and a decade of orthodoxy needed a new address.*

Exercise 27.5

A hibernating bear's brown-fat mitochondria express UCP1 abundantly. Using this chapter, explain (i) the thermogenic mechanism, (ii) why the bear's fuel consumption is enormous, and (iii) why the tissue is brown.

Resolution. *(i) UCP1 is a regulated proton channel in the inner membrane: open (on adrenergic command, fatty acids flipping the switch), it collapses the gradient at its source, respiratory control is abolished, and the chain runs flat-*

out against no load — every kJ mol^{-1} of the −220 appearing as heat (Worked Example 27.1's leak, industrialised). (ii) With the gradient short-circuited, no ATP is made and none limits respiration: the furnace burns whatever the blood delivers — fatty acids at a rate the white adipocyte can barely supply — and the bear's winter is financed by its autumn. (iii) The colour is the chain's: brown fat is packed with mitochondria, and the mitochondria are packed with haem — the cytochromes of § 26.5 in such density that the tissue itself is stained by its own engine. Anatomy, histology and thermodynamics, one tissue.

Research-level propositions

Proposition 27.α. Complex I's piston couples quinone reduction at one end of a 200-Å arm to proton channels at the other. Construct the elastic-network model of the coupling wave, estimate its speed and fidelity against the cryo-EM structures of the active and resting states, and assess whether the coupling can survive the membrane's thermal fluctuations — the "Brownian ratchet versus power stroke" question for the chain's largest engine. Proposition 27.β. The ROS leak of § 26.6 is maximal not at maximum but at intermediate reduction of the chain. Formalise the dependence of superoxide production on the redox state of Complex I's FMN and the Q pool, incorporate reverse electron transport from succinate, and test whether the reperfusion burst of ROS is a Q-cycle phenomenon — with the pharmacology (SQR-suppressors) the model nominates. Proposition 27.γ. Thirteen mtDNA-encoded proteins are all hydrophobic core subunits of the complexes. Evaluate the competing explanations — the import-difficulty hypothesis, the redox-regulation hypothesis (CoRR: the organelle keeps local control of its own redox machinery) — by designing the experiment that would relocate one mtDNA gene (e.g. ND4) to the nucleus with allotopic expression, and state what each theory predicts for the success of the transfer.

References (Chapter 27)

Berg, J. M., Tymoczko, J. L., Gatto, G. J., Jr., & Stryer, L. (2023). Biochemistry (9th ed.). W. H. Freeman.

Mitchell, P. (1961). Coupling of phosphorylation to electron and hydrogen transfer by a chemiosmotic type of mechanism. Nature, 191(4784), 144–148.

Nelson, D. L., Cox, M. M., & Hoskins, A. A. (2021). Lehninger principles of biochemistry (8th ed.). W. H. Freeman.

Nicholls, D. G., & Ferguson, S. J. (2013). Bioenergetics (4th ed.). Academic Press.

Lane, N. (2015). The vital question: Energy, evolution, and the origins of complex life. W. W. Norton.

Hirst, J. (2013). Mitochondrial complex I. Annual Review of Biochemistry, 82, 551–575.

Chapter 28
Oxidative Phosphorylation

CONTENTS OF THIS CHAPTER

"If the proton gradient is the electricity of the cell, the ATP synthase is its dynamo — and it turns out, quite literally, to be a dynamo."

28.1 The Engine Found in Fragments

We arrive, at last, at the machine for whose sake the previous three chapters exist. Chapter 25 taught us what ATP is worth and why a cell keeps so little of it; Chapter 26 showed how oxidation of foodstuff can be decanted, electron by electron, into the reduced currencies NADH and $FADH_2$; Chapter 27 followed those electrons down the respiratory chain and watched the energy of their fall dammed up as a proton-motive force across the inner mitochondrial membrane. Every one of those investments was made in expectation of a single dividend, and the machinery that pays it is the subject of this chapter: the F_1F_0-ATP synthase, an enzyme that is also, without metaphor or apology, a rotary electric motor.

It is worth pausing on how improbable that sentence is. When, in the 1960s, Efraim Racker and his colleagues resolved oxidative phosphorylation into a particulate fraction and a soluble "coupling factor" — F_1 — and electron microscopists saw the inner membrane studded with lollipops on stalks,[45] nobody supposed that the stalk rotated. The respectable theories of the day were a chemical intermediate (Slater's "squiggle-X", sought for twenty years and never found) and, after 1961, Mitchell's proton circuit, which even its author originally imagined driving ATP synthesis by some direct, static chemistry. That the cell's most important enzyme should work by spinning a shaft at up to a hundred revolutions per second would have sounded, to a committee of grown biochemists, like a schoolboy's fantasy. It is nevertheless true, and it was proven by watching a single shaft turn. The history of this chapter is thus a small parable about biochemistry itself: the field keeps discovering that its machines are more literal than its metaphors.

The plan of the chapter follows the machine. We first take the synthase apart and name its pieces (Section 28.2); we then ask how a rotating shaft can possibly make a chemical bond, which is Boyer's binding-change mechanism (Section 28.3); we watch the shaft turn (Section 28.4); we count protons and ATP and discover, with some embarrassment, that the stoichiometry refuses to be an integer (Section 28.5); we settle the full respiratory ledger of a glucose molecule (Section 28.6); and finally we ask who controls the engine — the answer being, delightfully, the customer (Section 28.7).

45. Fernández-Morán's micrographs of 1962 showed the knobs; Racker's F_1, stripped off with urea, left the membrane unable to make ATP but still able to respire — the first clean separation of engine from fuel line.

28.2 The Anatomy of a Molecular Motor

The synthase is bipartite, and its two halves could hardly differ more in temperament. ***F_o*** *(the subscript honours its sensitivity to oligomycin, not the number zero) is a hydrophobic turbine embedded in the inner membrane: in the bovine and yeast enzymes a ring of ten* ***c-subunits****, each a hairpin of two α-helices, plugged against the wedge-shaped* ***a-subunit****, with the whole assembly braced by a peripheral stalk of two* ***b-subunits****.* ***F_1*** *is a soluble foundry that protrudes into the matrix: three α- and three β-subunits arranged alternately as a hexamer, $\alpha_3\beta_3$, with the catalytic sites at the α–β interfaces but lying mostly on the β-subunits. Connecting turbine to foundry is the central stalk,* ***γ*** *(with ε and δ as its foot), which is fixed to the c-ring and turns with it. The result, frozen at 2.8 Å by John Walker's group in 1994,[46] is shown schematically in Figure 28.1: the rotor (c-ring plus γ) turns inside the stator (a, b_2, $\alpha_3\beta_3$), and the turning is the chemistry.*

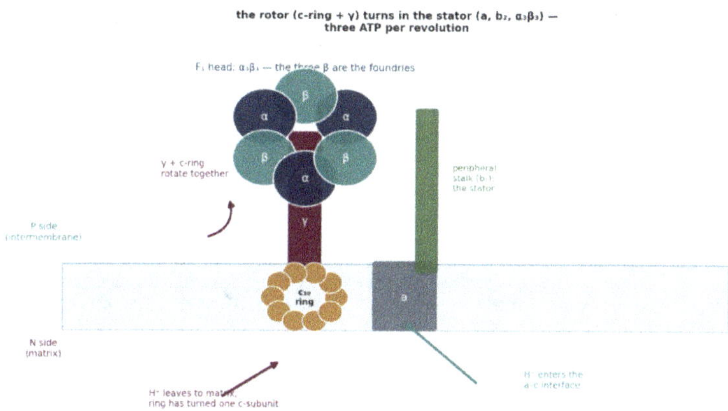

Figure 28.1 — The F_1F_o-ATP synthase as an engine. *The proton-motive force admits protons from the P side (intermembrane space) into a half-channel in the a-subunit (teal arrow); each proton binds a glutamate on a c-subunit of the c_{10} ring (amber), neutralising it so that the ring can rotate one step through the lipid; the proton is discharged to the N side (matrix, crimson arrow) when that c-subunit returns to the a–c interface one revolution later. The ring is fused to the γ-shaft (crimson), whose asymmetric cam turns inside the $\alpha_3\beta_3$ head (navy/teal), deforming each β-subunit in succession. The b_2 peripheral stalk (green) is the stator that keeps the head from spinning uselessly with the shaft — the engine's engine-block. Three ATP are made per full revolution.*

The proton pathway deserves a closer look, for it is a piece of mechanical wit. The c-subunits carry an essential glutamate (Asp-61 in E. coli) buried in the membrane's dielectric. Protonated, the residue is an unremarkable oily side-chain content to sit in lipid; deprotonated, it is a charged carboxylate that cannot bear the membrane and must sit at the a-subunit interface. The a-subunit offers two

---

46. Abrahams, Leslie, Lutter & Walker, *Nature* 370, 621–628 (1994). The $\alpha_3\beta_3$ hexamer came out beautifully; γ was partly disordered — the shaft, even in a crystal, declines to stand entirely still.

half-channels that do not connect: one opens to the P side, one to the N side, offset by roughly one c-subunit. A proton entering from the P side protonates the exposed glutamate; the ring is now free to rotate, and does so because the adjacent deprotonated c-subunit is straining at the interface. After a full turn the same glutamate arrives at the exit half-channel and surrenders its proton to the matrix. Protons thus cross the membrane the way parcels cross a customs house — one at a time, each paying for a fixed rotation of the turntable. The number of c-subunits in the ring is therefore the gear ratio of the engine, and it is a property of the organism: ten in vertebrate mitochondria, eight in some yeasts, up to fifteen in certain bacteria, and — in a flourish we shall meet again in Chapter 29 — fourteen in spinach chloroplasts.

28.3 Boyer's Binding-Change Mechanism

How does rotation make ATP? The naive answer — that the proton gradient somehow pays the 30.5 kJ mol^{-1} for condensing ADP and phosphate — founders on a beautiful experimental fact. In the 1970s Paul Boyer, following up exchange experiments in which ^{18}O-labelled water appeared in phosphate with suspicious ease, concluded that at the catalytic site of F_1 the equilibrium

$$\mathrm{ADP} + \mathrm{P_i} \rightleftharpoons \mathrm{ATP} + \mathrm{H_2O} \qquad (28.1)$$

is poised almost exactly at unity: on the enzyme, $\Delta G°' \approx 0$, and ATP forms spontaneously from bound ADP and P_i. The derivation is worth doing properly, for it is the conceptual heart of the chapter. The equilibrium constant of reaction (28.1) in free solution is the inverse of the hydrolysis constant, $K_{eq} \approx 1/(1.4 \times 10^5\ M^{-1}) \approx 7 \times 10^{-6}$ M — hopelessly unfavourable. But the enzyme does not catalyse the reaction in free solution; it catalyses it among bound species. Write the binding free energies of substrates and product to the site as $\Delta G_b(ADP{\cdot}P_i)$ and $\Delta G_b(ATP)$. Thermodynamic cycles are indifferent to the path taken, so the free energy of the bound reaction is $\Delta G°'_{site} = \Delta G°'_{soln} + \Delta G_b(ATP) - \Delta G_b(ADP{\cdot}P_i)$. If the site binds ATP more tightly than it binds ADP + P_i by almost exactly $\Delta G°'_{soln} \approx +30.5$ kJ mol^{-1}, the bound reaction costs nothing — and Boyer's exchange data said precisely that. The energy of the proton gradient is therefore not spent on making the bond. It is spent on the one thing thermodynamics cannot do for free: ***prizing the finished ATP off the enzyme that made it.***

Rotation is the prizing tool. The γ-shaft is not cylindrical; its lower half is a bent cam. As it turns inside $\alpha_3\beta_3$, it presses against each β-subunit in turn and forces it through three conformations — ***loose*** *(L), which binds ADP and P_i;* ***tight*** *(T), which clamps them together and in which ATP forms spontaneously; and* ***open*** *(O), a crumpled state that cannot hold the product and lets it fall away (Figure 28.2). Since the three β-subunits are 120° apart on the cam, at any instant one is L, one T, one O, and each revolution carries every β through the full cycle. Walker's crystal structure caught the engine mid-stroke, with the three β-subunits in three different conformations — the binding-change mechanism photographed, as it were, with its hand in the till. For this mechanism Boyer and Walker shared the 1997 Nobel Prize in Chemistry; the other half went to Skou for the Na^+/K^+-ATPase, so that 1997 may fairly be called the year the Nobel committee recognised that life runs on pumps and turbines.*

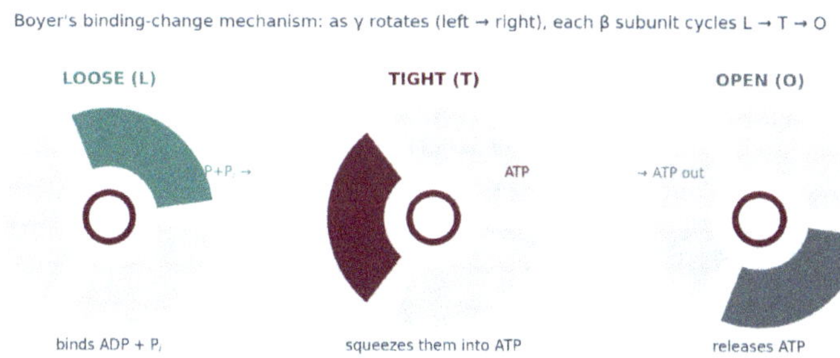

Figure 28.2 — The binding-change mechanism. *Each panel shows the $\alpha_3\beta_3$ head from above with the γ-cam (crimson) at its centre; the β-subunit in the highlighted state is coloured. As γ rotates through 360° (left to right in the figure), every β-subunit cycles loose → tight → open: substrates are bound, squeezed into ATP at a site where the reaction equilibrium is near unity (Eq. 28.1), and expelled. The energy of the proton gradient buys conformational change and product release, not bond formation. Three catalytic sites acting in sequence yield three ATP per revolution.*

28.4 Watching the Shaft Turn

A mechanism inferred from exchange kinetics and frozen in a crystal is still a conjecture about motion. The motion itself was seen in 1997 by Hiroyuki Noji, working in Masasuke Yoshida's laboratory, in an experiment of enviable directness.[47] *His group fixed an $\alpha_3\beta_3\gamma$ complex upside-down on a coverslip — histidine tags on the β-subunits gripping nickel on the glass — and attached to the protruding tip of γ a fluorescently labelled actin filament, a propeller several micrometres long and easily visible in a microscope. On adding ATP (the F_1 head run backwards, as a hydrolase), the filament rotated, always anticlockwise as viewed from the membrane (Figure 28.3, left). At low ATP concentrations and high filament load the rotation resolved into discrete* ***120° steps*** *(Figure 28.3, right): three dwells per revolution, one per catalytic site, exactly as Boyer's mechanism requires. Later work with smaller probes revealed an 80°–40° sub-step structure — ATP binding drives the first 80°, hydrolysis and product release the remaining 40° — and the full F_1F_o enzyme was subsequently shown to turn in the opposite direction when driven by a proton gradient, closing the loop between the two motors.*

47. Noji, Yasuda, Yoshida & Kinosita, *Nature* 386, 299–302 (1997). It appeared the same year as Boyer and Walker's Nobel Prize; the committee is rarely so well timed.

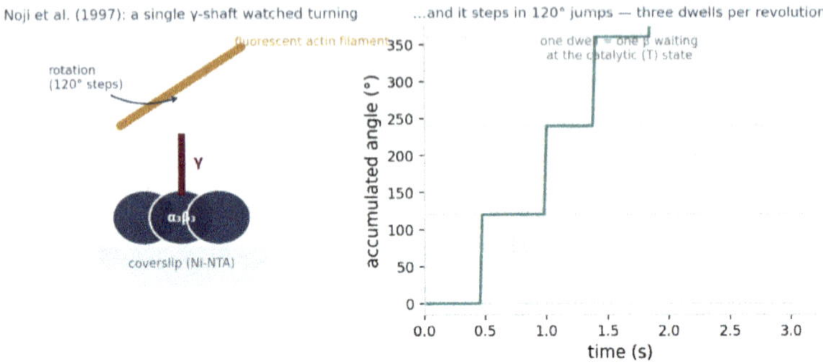

Figure 28.3 — Direct observation of rotation. *Left: Noji and colleagues tethered $\alpha_3\beta_3$ (navy) to a coverslip and attached a fluorescent actin filament (amber) to γ (crimson); ATP hydrolysis spun the filament anticlockwise. Right: the angular trajectory at limiting ATP: dwells punctuated by 120° jumps, three per revolution, one for each β-subunit's turn at the tight state. A hypothesis about binding affinities had become a filmed event. (Trace schematic, after Noji et al., 1997.)*

The torque, incidentally, is substantial: about 40–50 pN·nm per step, and the mechanical efficiency of the F_1 motor approaches 100% — a figure any engineer would dismiss as a misprint if it came from a diesel rather than a dimer of helices. The cell, having had three and a half billion years to tune its engines, has settled questions that our own thermodynamic practice still treats as aspirations.

28.5 The Proton Arithmetic

We can now do honest bookkeeping, and the honesty begins with a confession: the stoichiometry is not integral. The mammalian c-ring has ten subunits, hence

$$\frac{\mathrm{H}^+}{\mathrm{ATP}} = \frac{n_\mathrm{c}}{3} = \frac{10}{3} \approx 3.33 \quad \text{(synthase alone)} \tag{28.2}$$

follows directly from the machine's architecture: a full revolution costs one proton per c-subunit (n_c = 10) and yields one ATP per β-subunit (3), so the ratio is $n_c/3$ — a quantity the enzyme inherits from its symmetry, not from any thermodynamic nicety. But the ATP is made in the matrix and spent in the cytosol, and crossing the membrane is not free. The phosphate carrier imports P_i in symport with a proton (or, equivalently, in exchange for hydroxide), adding one H^+ per ATP; the adenine nucleotide translocase exchanges ATP^{4-} out for ADP^{3-} in, moving one negative charge outwards against a membrane potential that is positive outside, a toll measured at about a quarter of a proton's worth of Δp. The delivered price of a cytosolic ATP is therefore

$$\frac{\mathrm{H}^+}{\mathrm{ATP}_\mathrm{cyt}} = \frac{n_\mathrm{c}}{3} + 1 + \tfrac{1}{4} \approx 4.6\ \mathrm{H}^+ \tag{28.3}$$

— the arithmetic of Figure 28.4. This number explains a small scandal of twentieth-century biochemistry. For decades textbooks taught that the P/O ratio — ATP made per oxygen atom reduced — was exactly 3 for NADH and 2 for succinate. The chain, however, pumps 10 H^+ per NADH (4 + 4 + 2 at Complexes I, III and IV) and 6 per succinate (Complex II pumps nothing), and dividing by an honest price gives

$$P/O_{\text{NADH}} = \frac{10}{4.6} \approx 2.5\,; \qquad P/O_{\text{succ}} = \frac{6}{4.6} \approx 1.5 \tag{28.4}$$

— the modern values, whose chequered history is plotted in Figure 28.5 (left). The integers 3 and 2 were never measured with the precision their dignity implied; they were the price list of a machine nobody had yet taken apart. Once the gear ratio was known, the fractional ratios stopped being an embarrassment and became a prediction — one confirmed by careful measurement.[48]

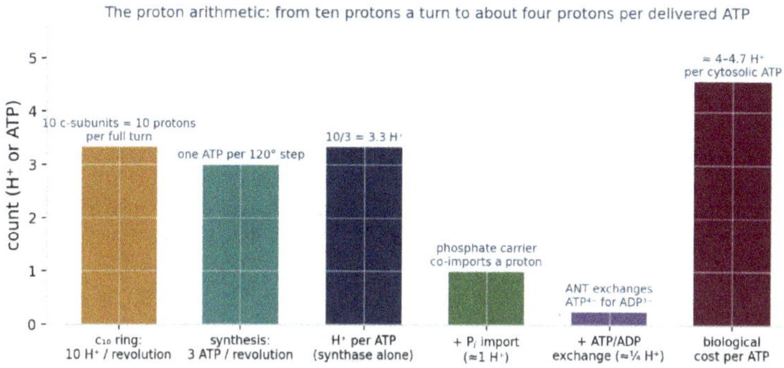

Figure 28.4 — From ten protons a turn to about four per ATP. *The c_{10} ring and the three catalytic β-subunits fix the bare cost at 10/3 H^+ per ATP (Eq. 28.2); the phosphate carrier adds one H^+, the ATP/ADP exchanger about a quarter more, bringing the delivered cost of a cytosolic ATP to ≈ 4.6 H^+ (Eq. 28.3). Stoichiometry here is architecture, not magic.*

28.6 The Ledger Settled: Thirty ATP per Glucose

Every ATP account opened since Chapter 25 can now be closed. Complete oxidation of one glucose yields, directly, 2 ATP from glycolysis and 2 GTP from the citric acid cycle; it yields 10 NADH (2 glycolytic, 2 from pyruvate dehydrogenase, 6 from the cycle) and 2 $FADH_2$ (from succinate dehydrogenase). Applying Eq. (28.4): the eight mitochondrial NADH are worth 8 × 2.5 = 20 ATP; the two $FADH_2$, 2 × 1.5 = 3 ATP. The two glycolytic NADH are made in the cytosol, and their electrons must be imported: via the malate–aspartate shuttle they arrive as NADH (full 2.5 each), via the glycerol-phosphate shuttle as $FADH_2$ (1.5 each) — a difference of two ATP that depends, with fine evolutionary indifference to our accounting, on which tissue one asks. The total is therefore ≈ ***32 ATP per glucose*** *with the malate–aspartate shuttle and ≈* ***30*** *with glycerol-phosphate (Figure 28.5, right).*

48. Hinkle's 1991 measurements on rat liver mitochondria gave P/O ≈ 2.58 for NADH-linked substrates — close enough to 2.5 that the remaining gap is spent on proton leak, a topic Section 28.7 will not let us forget.

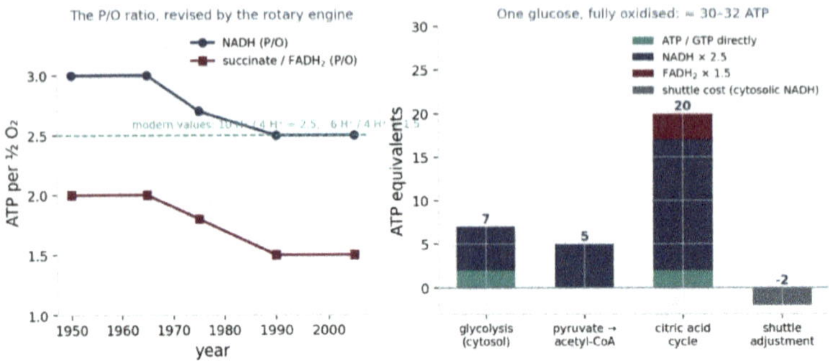

***Figure 28.5* — The price of oxidation, then and now.** *Left: the P/O ratio across half a century of measurement; the rotary mechanism with its non-integral gear ratio explains the modern values 2.5 and 1.5 (Eq. 28.4) and retires the integer myth. Right: the complete ATP ledger for one glucose. Direct synthesis (teal) contributes only 4 ATP; the remaining ≈ 26–28 are bought by NADH (navy) and $FADH_2$ (crimson) at the prices of Eq. (28.4), with a 2-ATP discount (slate) if the cytosolic NADH enters via the glycerol-phosphate shuttle.*

> ***Worked example 28.1* — Is the proton-motive force big enough?** *In a respiring mitochondrion $\Delta p \approx 220$ mV (Chapter 27). How many ATP per NADH could this drive, in principle, and why does the cell settle for 2.5?*
> ***Solution.*** *The free energy of one proton crossing at $\Delta p = 0.220$ V is $\Delta G = -F\Delta p = -96\,485 \times 0.220 \approx -21.2$ kJ mol^{-1}. Ten protons (one NADH's worth) therefore carry ≈ −212 kJ mol^{-1}. Cytosolic phosphorylation potentials run to $\Delta G_p \approx -54$ kJ mol^{-1} (Eq. 25.1), so the theoretical ceiling is 212/54 ≈ 3.9 ATP. The cell takes 2.5. The difference is not waste but margin: the synthesis must remain irreversible in the forward direction, and 10 H^+ : 2.5 ATP leaves ≈ 4.6 H^+ × 21.2 ≈ 97 kJ against 54 kJ needed — a driving force of ≈ −43 kJ mol^{-1} per ATP that keeps the engine from stalling or running backwards. Machinery, like banking, keeps reserves.*

28.7 Who Governs the Engine? Control, Uncoupling and Disease

*The most elegant fact about oxidative phosphorylation is that it has no throttle of its own. In 1955 Britton Chance and G. R. Williams, watching oxygen electrode traces, defined the respiratory states: resting mitochondria with substrate but no ADP (**State 4**) respire slowly, because the proton circuit is closed and the gradient, once charged to its ceiling, holds the chain in check; add ADP and respiration leaps (**State 3**) as the synthase opens the circuit and discharges the gradient; when the ADP is spent, respiration falls back (Figure 28.6, left). The ratio of the two rates, the respiratory control ratio, is the standard health check of a mitochondrial preparation — and the phenomenon itself, **respiratory control**, is the demand-side economy of the cell in a single trace: the customer's order (ADP) is the accelerator pedal, and there is no other.*

*The control can be broken, instructively. **Uncouplers** — 2,4-dinitrophenol, the outlawed slimming drug of the 1930s; CCCP, the bench biochemist's standby; in excess, even salicylate — are lipophilic weak acids that ferry protons across the membrane directly, short-circuiting the turbine. Respiration then races while ATP* *synthesis collapses, and the difference appears as heat (Figure 28.6, right). Biology, which wastes nothing it can tax, has domesticated the trick:* ***UCP1 (thermogenin)*** *in brown adipose tissue is a regulated uncoupling channel, switched on by noradrenaline in the cold, by which infants and hibernators burn the gradient for warmth. Evolution, having invented the most efficient engine known, promptly invented a way to idle it — and charges the infant nothing for the fuel.*

Two further governors deserve naming. ***Oligomycin*** *blocks the c-ring channel and stops everything — synthesis, rotation, and (by respiratory control) respiration itself — a diagnostic pattern every student of mitochondria learns to read. And the matrix-side inhibitor protein* ***IF$_1$*** *binds F_1 when the gradient collapses (as in ischaemia), preventing the synthase from running in reverse and burning the cell's remaining ATP to pump protons — a ratchet fitted to the engine for emergencies. The clinical stakes are not small: mutations in ATP6 (the a-subunit), in assembly factors, or in the respiratory chain feeding it present as Leigh syndrome, NARP and allied mitochondrial encephalomyopathies, in which tissues with the highest ATP tariffs — brain, heart, retina — fail first, exactly as an engineer would predict of a power cut.*

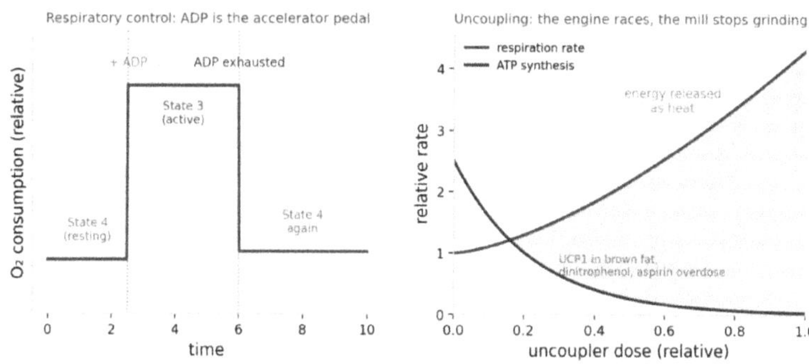

Figure 28.6 — Control and its sabotage. *Left: the Chance–Williams experiment. Resting mitochondria (State 4) respire slowly; ADP admission (State 3) accelerates oxygen consumption roughly four-fold as the synthase discharges the gradient; exhaustion of ADP restores the resting rate. Respiratory control is demand-pull economics enacted by an enzyme. Right: titration with an uncoupler (dinitrophenol, CCCP, or endogenous UCP1). Respiration rises without limit of the gradient while ATP synthesis collapses; the shaded gap is energy paid out as heat — fatal in a dieter, life-saving in a cold infant.*

> ***Case study — The poisoner's short circuit.*** *A 34-year-old man is brought to casualty hyperthermic (40.8 °C), tachycardic, sweating profusely, with a metabolic acidosis; he admits to taking an industrial "fat-burner" purchased online. His oxygen consumption is enormous, his blood ATP-dependent processes are failing, and no sedative touches the fever. The agent is 2,4-dinitrophenol. Explain (i) why oxygen consumption is high, (ii) why he is hot, (iii) why cooling blankets and dantrolene are of limited use, and (iv) the single most important supportive measure.*
> ***Resolution.*** *(i) DNP collapses Δp; respiratory control is abolished, so the chain runs at its maximum rate trying vainly to recharge the gradient. (ii) The energy of oxidation, no longer captured as ATP, is released quantitatively as heat — the shaded region of Figure 28.6, lived. (iii) The heat is generated in every mitochondrion of every muscle at rest; neither surface cooling nor blocking sarcoplasmic calcium release (dantrolene, designed for malignant hyperthermia) addresses a leak that is everywhere at once. (iv) Aggressive*

external and internal cooling plus cardiorespiratory support while the drug clears; there is no antidote, because the lesion is not a blocked enzyme but an opened door. The case is a grim demonstration that the coupling of respiration to ATP synthesis — this chapter's whole subject — is not an academic nicety but a condition of staying alive.

Historical note — The intermediate that never was. *In 1953 E. C. ("Bill") Slater proposed that electron transport phosphorylated ADP through a high-energy chemical intermediate, X~Y, by analogy with substrate-level phosphorylation. The proposal was reasonable, testable, and wrong. For two decades the world's best enzymologists hunted X~Y with every reagent chemistry offered, and the intermediate — perhaps the most sought molecule never to exist — was never found, because there was nothing to find: the energy passed through no chemical bond at all, but through a membrane potential (Mitchell, 1961) and then through a rotating shaft (Boyer; Walker; Noji). The episode is usually told as Mitchell's triumph, and so it was; but it is also worth telling as a lesson in the seductions of analogy. Substrate-level phosphorylation was the only phosphorylation anyone knew, so it became the model for one that worked by entirely different means. Biochemistry, like philosophy, advances as much by the dignified death of its best analogies as by the birth of its theories. Slater, to his credit, lived to see the shaft turn and conceded the point with grace — the finest epitaph an hypothesis can earn.*

Plate 28.1 *John E. Walker (b. 1941), whose 1994 crystal structure of F_1 showed the three β-subunits caught in three conformations — Boyer's binding-change mechanism rendered in electron density — and who shared the 1997 Nobel Prize in Chemistry with Paul Boyer and Jens Skou. (Nobel Foundation portrait.)*

28.8 Exercises

Exercise 28.1. *The c-ring of spinach chloroplast ATP synthase has 14 subunits, and the chloroplast pays no transport tolls (its ATP is made and largely used in the stroma). Compute the H^+/ATP ratio and comment on its energetic meaning.*
Resolution. *$H^+/ATP = n_c/3 = 14/3 \approx 4.67$ — more expensive per ATP than the mitochondrial bare ratio of 3.33. The cost buys torque: with more c-subunits per revolution, each proton needs to supply less energy, so the enzyme can run on the smaller proton-motive force available across the thylakoid membrane. Gear ratios are chosen for the power supply, not for economy.*

Exercise 28.2. *A patient's fibroblasts harbour a Complex I defect that halves its proton pumping (2 H^+ instead of 4 per NADH). Recalculate the P/O ratio for NADH and the ATP yield per glucose (malate–aspartate shuttle), and state the fractional loss.*
Resolution. *Pumped per NADH: $2 + 4 + 2 = 8\ H^+$, so $P/O = 8/4.6 \approx 1.74$. Per glucose: 10 NADH × 1.74 = 17.4; $FADH_2$ unchanged at 2 × 1.5 = 3; direct 4. Total ≈ 24.4 ATP versus 32 — a loss of ≈ 24%. The arithmetic shows why partial Complex I defects are compatible with life but not with full health, and why high-tariff tissues notice first.*

Exercise 28.3. *Show that if IF_1 failed to act during ischaemia, the synthase running in reverse could consume ATP at a rate comparable to glycolytic production in an anoxic heart myocyte. (Take glycolytic ATP production ≈ 1 μmol s^{-1} g^{-1}; reverse-mode synthase at full speed, 100 rev s^{-1}, ≈ 300 ATP s^{-1} per enzyme; tissue content ≈ 0.3 nmol g^{-1}.)*
Resolution. *Reverse consumption = 300 ATP s^{-1} × 0.3 nmol g^{-1} = 90 nmol s^{-1} g^{-1} ≈ 0.09 μmol s^{-1} g^{-1} per enzyme class — and with multiple enzyme copies per mitochondrion the real figure approaches glycolytic output. Within minutes the ATP would be gone and the gradient (now the cell's last reserve) squandered pumping protons. IF_1 is the brake that prevents the engine from spending the reserves it was built to fill.*

Exercise 28.4 (proposition for research). *The mammalian c-ring has 10 subunits; thermophilic bacteria often have 13–15; some enzymes have 8. Propose an evolutionary-thermodynamic theory predicting ring size from (a) the available proton-motive force and (b) the cellular phosphorylation potential, and design an experiment — swapping rings between species, or evolving E. coli at controlled Δp — to test it. What would a mismatch do to respiratory control?*
Sketch. *The theory: at equilibrium between gradient and phosphorylation potential, $n_c \times F\Delta p \approx 3 \times \Delta G_p$; ring size should track $3\Delta G_p/F\Delta p$. Small rings suit large gradients; large rings suit weak ones. A transplant with too small a ring would stall (insufficient torque); too large a ring would waste protons and weaken respiratory control, since the gradient could fall further before stalling. The experiment is feasible with existing ring-swap constructs; the*

prediction is quantitative and falsifiable — which is all one can ask of a theory, and more than one can ask of X~Y.

28.9 Synthesis

The turbine is understood, and with it the whole of aerobic energy transduction stands complete: substrate to NADH, NADH to protons, protons to ATP, ATP to work. Direction checked against the master plan: we have honoured the promise made at the close of Chapter 27, and Part VI has one obligation left. The mitochondrion is a splendid engine, but it burns a fuel it cannot make — reduced carbon, and the oxygen to burn it with, both arrive courtesy of the other great membrane power station, the chloroplast. Before metabolism can be treated as an economy we must visit the mint where the currency of reduced carbon is first coined, using the only free energy the planet receives in bulk: sunlight. ***Chapter 29 — Photosynthesis and Photochemical Energy Conversion*** *— is where the proton circuit learns to run on light.*

28.10 References

Abrahams, J. P., Leslie, A. G. W., Lutter, R., & Walker, J. E. (1994). Structure at 2.8 Å resolution of F_1-ATPase from bovine heart mitochondria. Nature, 370, 621–628.

Berg, J. M., Tymoczko, J. L., Gatto, G. J., & Stryer, L. (2019). Biochemistry (9th ed.). W. H. Freeman.

Boyer, P. D. (1997). Energy, life, and ATP (Nobel lecture). Angewandte Chemie International Edition, 37, 2296–2307.

Chance, B., & Williams, G. R. (1955). Respiratory enzymes in oxidative phosphorylation: I. Kinetics of oxygen utilization. Journal of Biological Chemistry, 217, 383–393.

Hinkle, P. C., Kumar, M. A., Resetar, A., & Harris, D. L. (1991). Mechanistic stoichiometry of mitochondrial oxidative phosphorylation. Biochemistry, 30, 3576–3582.

Nelson, D. L., & Cox, M. M. (2021). Lehninger Principles of Biochemistry (8th ed.). W. H. Freeman.

Nicholls, D. G., & Ferguson, S. J. (2013). Bioenergetics (4th ed.). Academic Press.

Noji, H., Yasuda, R., Yoshida, M., & Kinosita, K. (1997). Direct observation of the rotation of F_1-ATPase. Nature, 386, 299–302.

Stock, D., Leslie, A. G. W., & Walker, J. E. (1999). Molecular architecture of the rotary motor in ATP synthase. Science, 286, 1700–1705.

Walker, J. E. (1998). ATP synthesis by rotary catalysis (Nobel lecture). Angewandte Chemie International Edition, 37, 2308–2319.

Yoshida, M., Muneyuki, E., & Hisabori, T. (2001). ATP synthase — a marvellous rotary engine of the cell. Nature Reviews Molecular Cell Biology, 2, 669–677.

Chapter 29
Photosynthesis and Photochemical Energy Conversion

Contents of this chapter

"Every morsel you have ever eaten was once a beam of sunlight that a green cell caught, taxed, and banked."

29.1 The Other Power Station

Part VI opened with ATP and closed, in Chapter 28, with the turbine that mints it. But a turbine is only as good as its fuel supply, and the mitochondrion burns a fuel it cannot manufacture: reduced carbon on the one hand, oxygen on the other. Both are made, in quantities that defy everyday intuition, by the second great membrane power station of the living world — the chloroplast. Photosynthesis is not merely another pathway to be added to our ledger; it is the creditor of the entire account. Roughly 10^{14} kg of carbon are fixed annually, the oxygen of the atmosphere is its waste product, and every proton gradient we have studied in this Part either runs on photosynthate or runs on something that ate something that did.

The problem the chloroplast solves is, at bottom, the same problem the mitochondrion solves — how to move electrons against a redox gradient and cash the movement out as a proton-motive force — with one addition of genius: the electrons are pushed uphill by light. The plan of the chapter follows the light. We first ask what sunlight offers and what pigments take (Section 29.2); we then watch an absorbed photon drive the most important redox reaction on Earth, the splitting of water, and follow the electron down the Z-scheme to NADPH (Sections 29.3–29.5); we count photons (Section 29.6); we watch the proton circuit close through a chloroplast ATP synthase that is the mitochondrial engine's unmistakable cousin (Section 29.7); and finally we spend the ATP and NADPH in the Calvin cycle, and reckon with the strange, costly habit of its first enzyme — rubisco's oxygenation — and the evolutionary workarounds it provoked (Sections 29.8–29.9).

29.2 What Pigments Buy

Sunlight arrives at the top of the atmosphere as a broad spectrum, and at a leaf's surface as something narrower; what matters for chemistry is that each photon is a quantum of energy

$$E_{\text{photon}} = \frac{hc}{\lambda} \quad \Longrightarrow \quad E_{\text{mol}} = \frac{N_A hc}{\lambda} \approx \frac{119\,626}{\lambda\,[\text{nm}]}\ \text{kJ mol}^{-1} \tag{29.1}$$

— Planck's relation restated per mole, since biochemistry shops by the mole. A 680 nm photon is worth ≈ 176 kJ mol^{-1}; a blue 430 nm photon, ≈ 278. These are respectable sums — a single red photon carries more than five times the free energy of ATP hydrolysis — and the pigment business exists to collect them.

Chlorophyll a absorbs in the blue (≈ 430 nm) and red (≈ 662 nm); chlorophyll b and the carotenoids fill in the flanks and hand their excitations to chlorophyll a by resonance transfer (Figure 29.1). The conspicuous gap in the green is why leaves are green: plants keep the change. It is one of the quiet ironies of the field that the biosphere's trademark colour is the colour of light the biosphere declines to use.

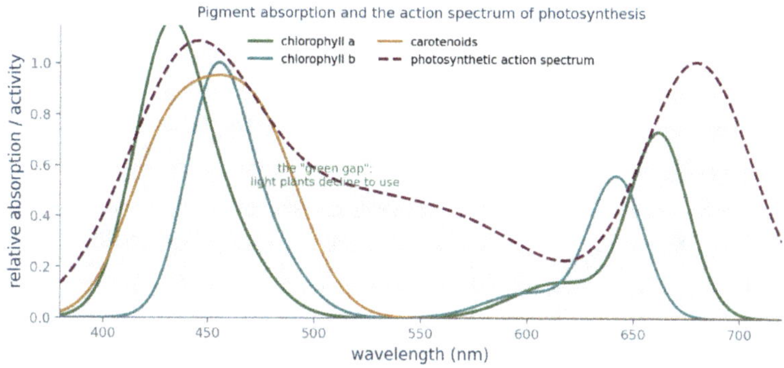

Figure 29.1 — Absorption and action. *Absorption spectra of chlorophyll a (green), chlorophyll b (teal) and carotenoids (amber), with the photosynthetic action spectrum (crimson dashed) — the rate of photosynthesis as a function of wavelength. That the action spectrum tracks the summed absorptions, not any single pigment, was the first evidence that light harvested by many pigments is funnelled to one photocentre. The trough near 550 nm is the "green gap" reflected into your eye. (Curves schematic.)*

Absorbed energy does not stay put. Within an antenna complex of some two to three hundred chlorophylls, excitation hops by Förster resonance transfer in picoseconds, random-walking until it reaches the one pair of chlorophylls that can do chemistry: the ***special pair*** *of a reaction centre. The antenna is thus a funnel, and the reaction centre its narrow neck — an arrangement that lets a cell keep a hundred light-harvesting tickets for every one expensive catalytic machine.*

29.3 The Reaction Centre, Solved First in Purple

The first membrane protein ever crystallised and solved was a photosynthetic reaction centre — that of the purple bacterium Rhodopseudomonas viridis, by Deisenhofer, Huber and Michel in 1984–85,[49] *and its architecture became the template for everything that followed. The logic is universal and worth stating as a sequence: (i) the special pair absorbs a photon and becomes a powerful reductant; (ii) within picoseconds it donates an electron to a nearby acceptor (pheophytin, then quinones), so that the excitation is trapped as a* ***charge separation*** *across the membrane before it can decay as heat or fluorescence; (iii) the hole left on the special pair is refilled by a donor. Purple bacteria refill it from reduced cytochrome and run the electrons in a closed loop that pumps protons — photophosphorylation without chemistry beyond proton pumping. Oxygenic phototrophs took the bolder option: refill the hole from water.*

49. Deisenhofer, Epp, Miki, Huber & Michel, *Nature* 318, 618–624 (1985). The Nobel Prize followed in 1988 — the first ever for a membrane protein structure, at a time when "membrane protein crystallography" was widely considered a contradiction in terms.

29.4 Splitting Water: Photosystem II and the Oxygen Clock

Water is a wretched substrate. The O_2/H_2O couple sits at +0.82 V, so ripping electrons from it demands an oxidant of almost unprecedented violence. Photosystem II supplies one: photoexcited P680 donates an electron to pheophytin and becomes $P680^+$, a cation radical with a reduction potential near +1.2 V — the strongest oxidant in biology, made anew by every photon. $P680^+$ then strips an electron from the* **oxygen-evolving complex**, *a Mn_4CaO_5 cluster that accumulates four oxidising equivalents, one per photon, through five states ($S_0 \rightarrow S_4$) before oxidising two waters and exhaling O_2. The four flashes of the clock were demonstrated by Pierre Joliot and Bessel Kok before anyone knew what was flashing: give dark-adapted chloroplasts single-turnover flashes, and oxygen comes in bursts of every fourth flash — a molecular four-stroke engine whose cylinders are manganese ions.*[50] *Every breath you take is the exhaust of this cluster, four photons at a time.*

29.5 The Z-Scheme and Photosystem I

One photon at PSII lifts an electron from water (+0.82 V) only as far as plastoquinone (≈ −0.1 V) — a good climb, but $NADP^+$ lies at −0.32 V, still uphill. The solution, proposed by Hill and Bendall in 1960 and confirmed in every particular since, is to spend a second photon at a second photocentre (Figure 29.2). Between the two photosystems sits the cytochrome b_6f complex — structurally and mechanistically the mitochondrial Complex III in green livery, Q-cycle and all — which oxidises plastoquinol, reduces the soluble copper protein plastocyanin, and pumps protons into the lumen. Photosystem I then absorbs a 700 nm photon: P700 is a reductant near −1.25 V, strong enough to reduce ferredoxin and, via ferredoxin–$NADP^+$ reductase, to mint NADPH in the stroma. The whole path, plotted as potential against position, traces a zigzag that generations of students have known as the* **Z-scheme** *— the only figure in biochemistry whose letter is a diagram.*

50. The Kok cycle, 1970. The $S_4 \rightarrow S_0$ transition releases O_2; the Mn_4Ca cluster was located and its geometry settled only in 2011, by Umena, Kawakami, Shen and Kamiya at 1.9 Å (*Nature* 473, 55–60).

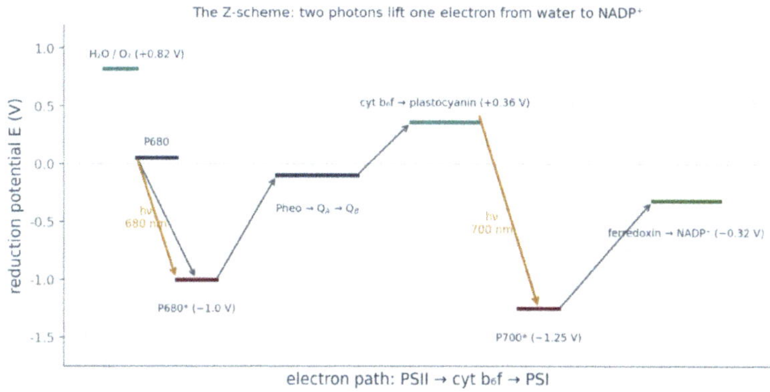

*Figure 29.2 — **The Z-scheme.** Reduction potentials along the electron path. Photons (amber arrows) at P680 and P700 supply the two uphill jumps; between them the electron coasts downhill through pheophytin, Q_A, Q_B, the cytochrome b_6f Q-cycle and plastocyanin, and the downhill stretch is where protons are pumped into the thylakoid lumen. Two photocentres in series achieve what no single photon can: taking an electron from water (+0.82 V) to $NADP^+$ (−0.32 V).*

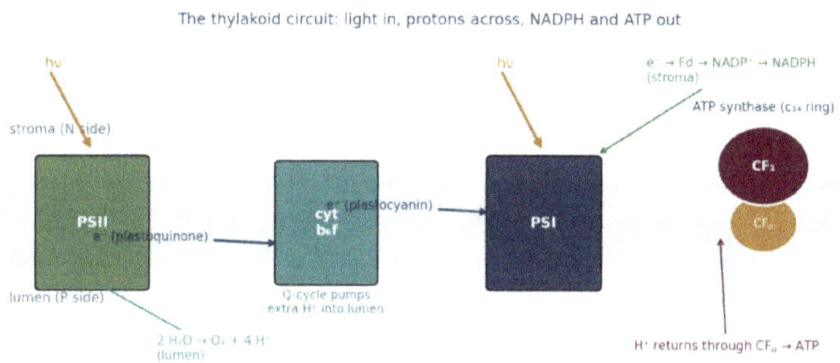

*Figure 29.3 — **The thylakoid membrane as a circuit.** Water is oxidised in the lumen by PSII (green); electrons cross via plastoquinone to cytochrome b_6f (teal, whose Q-cycle adds pumped protons), via plastocyanin to PSI (navy), and out through ferredoxin to $NADP^+$ in the stroma. Protons accumulate in the lumen and return through the CF_oCF_1 ATP synthase (amber/crimson) — the same rotary engine as Chapter 28, here with a c_{14} ring in spinach. Compare the mitochondrial circuit of Chapter 27: the resemblance is genealogical, not coincidental.*

29.6 Counting Photons

How many photons does the chemistry cost? One electron from water to $NADP^+$ must be lifted twice, once at each photocentre, so the minimum is two photons per electron; NADPH is a two-electron currency, and O_2 is a four-electron product:

$$\underbrace{2\text{ photons}}_{\text{per } e^-} \times \underbrace{4\,e^-}_{\text{per } O_2} = 8 \text{ photons per } O_2 \tag{29.2}$$

— the celebrated quantum requirement of 8, measured by Emerson and confirmed by Otto Warburg's school (after a famous decades-long quarrel in which Warburg insisted on 4 and was, for once in his life, wrong). The thermodynamic margin is comfortable but not extravagant: per O_2 evolved, the cell spends 8 × 176 ≈ 1 400 kJ of red light to bank roughly 4 × 110 kJ of redox lift plus the proton traffic — an efficiency near 30%, which any solar engineer would accept on the spot (Figure 29.4).

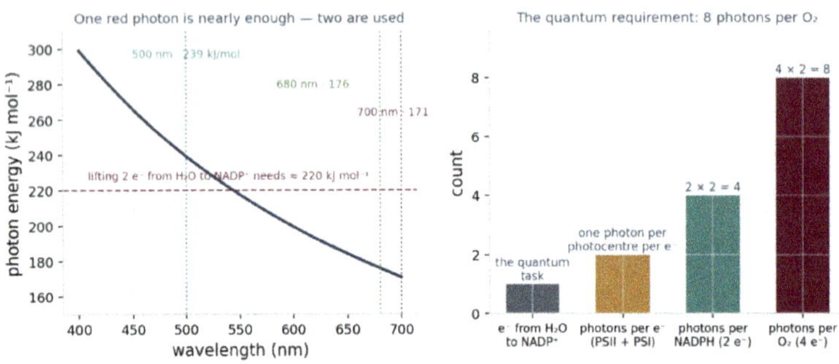

***Figure 29.4 — Photon economics.** Left: energy per mole of photons against wavelength (Eq. 29.1); one red photon nearly suffices for the full redox lift of two electrons (dashed line), but no single photocentre can both oxidise water and reduce $NADP^+$, so two are used. Right: the quantum requirement — one photon per photocentre per electron, hence 4 per NADPH and 8 per O_2 (Eq. 29.2).*

29.7 The Proton Circuit Closes: Photophosphorylation

*The lumen is where the protons collect: two per water split, plus the Q-cycle traffic, plus (in the light) a lumenal pH that can fall below 6 while the stroma rises past 8 — a ΔpH of more than two units, with little membrane potential because chloride and magnesium ions move to neutralise the charge. The chloroplast thus stores its proton-motive force almost entirely as ΔpH, whereas the mitochondrion stores its own mostly as Δψ — a division of labour between two membranes with different plumbing for counter-ions, and a fine illustration of Eq. (27.2)'s claim that Δp is one currency with two denominations. The protons return through the **CF_0CF_1 ATP synthase**, the rotary engine of Chapter 28 in every essential — save one: the spinach chloroplast c-ring has fourteen subunits, so the bare gear ratio is 14/3 ≈ 4.7 H^+ per ATP. A weaker gradient needs a lower gear; evolution, as we noted in Chapter 28, chooses ring sizes for the power supply, and the chloroplast's proton-poor, pH-rich gradient gets a large ring.*

> ***Worked example 29.1 — Can the chloroplast's ΔpH pay for ATP?** In bright light the thylakoid lumen reaches pH ≈ 5.8 against a stroma of pH ≈ 8.0, with Δψ ≈ 0. Does the gradient cover phosphorylation at $\Delta G_p \approx 50$ kJ mol^{-1}, given 14/3 H^+ per ATP?*
> ***Solution.** $\Delta p = \Delta\psi - 61.5\ mV \times \Delta pH = 0 - 61.5 \times (5.8 - 8.0) \approx +135$ mV (Eq. 27.2). Each returning proton delivers F × 0.135 ≈ 13.0 kJ mol^{-1}; 14/3 protons deliver ≈ 60.7 kJ mol^{-1} against 50 required. The margin, ≈ 10 kJ mol^{-1}, is thinner than the mitochondrion's (Worked example 28.1) but adequate — and in dim light, when ΔpH shrinks, the margin vanishes first, which is precisely why ATP synthesis is the first process to throttle as the sun goes down.*

29.8 Spending It: The Calvin Cycle

The light reactions deliver two currencies in a fixed ratio — roughly 3 ATP and 2 NADPH per pass through the non-cyclic chain (cyclic photophosphorylation around PSI tops up ATP as needed) — and the stroma spends them in the cycle worked out by Melvin Calvin, Andrew Benson and James Bassham using the new and glamorous technology of ^{14}C, which won Calvin the 1961 Nobel Prize. The cycle turns three times per net product (Figure 29.5). In the ***carboxylation*** *step, three molecules of ribulose-1,5-bisphosphate (5C) each accept a CO_2 at the hands of rubisco and split into six 3-phosphoglycerate (3C).* ***Reduction*** *spends 6 ATP and 6 NADPH to convert these to six glyceraldehyde-3-phosphate.* ***Regeneration*** *spends 3 more ATP to reshuffle five G3P back into three RuBP; the sixth G3P is the harvest. The ledger:*

$$3\ \mathrm{CO_2} + 9\ \mathrm{ATP} + 6\ \mathrm{NADPH} \longrightarrow \mathrm{G3P} + 9\ \mathrm{ADP} + 8\ \mathrm{P_i} + 6\ \mathrm{NADP^+} \tag{29.3}$$

— whence the required ATP:NADPH ratio of 3:2, which the light reactions supply with a little help from the cyclic shunt. Two turns of three fix six carbons, one glucose's worth, at 18 ATP and 12 NADPH: the true price of sugar, quoted in the currencies of the light.

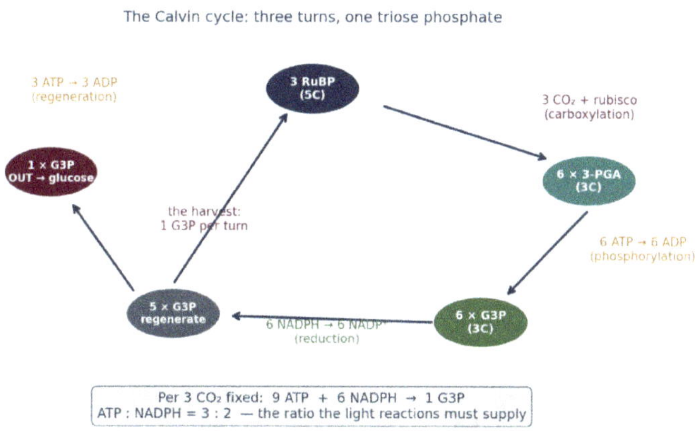

Figure 29.5 — The Calvin cycle. *Three RuBP (navy) are carboxylated with 3 CO_2 by rubisco (crimson arrow) to six 3-phosphoglycerate (teal); 6 ATP (amber) and 6 NADPH (green) reduce these to six G3P; five are recycled to RuBP at a cost of 3 further ATP (slate node), and one G3P (crimson node) leaves the cycle toward glucose synthesis. The stoichiometry of Eq. (29.3) — 9 ATP : 6 NADPH per G3P — is summarised in the box.*

Rubisco deserves a paragraph of its own, for it is by mass the most abundant enzyme on Earth and by reputation the most maligned. It is slow ($k_{cat} \approx 3\ s^{-1}$, a snail among enzymes, which is why leaves are stuffed with it) and it is promiscuous: O_2 competes with CO_2 at the same site. Yet the charge of incompetence is partly slander. Rubisco's discriminating power is near the chemical ceiling for a metal-free enzyme binding two uncharged, unimpressive small molecules; and its slowness is the price of that discrimination — faster variants discriminate worse. Evolution, offered a choice between fast and choosy, *took choosy and compensated with quantity. There is a management lesson in there somewhere, though we shall not press it.*

29.9 Photorespiration, and the Two Great Workarounds

When rubisco oxygenates RuBP, the products are one 3-PGA and one phosphoglycolate — a two-carbon dead end that the cell must rescue through the ***photorespiratory*** *pathway, shuttling carbon between chloroplast, peroxisome and mitochondrion at a cost of ATP, reducing power, and the loss of one CO_2 in four. On a hot, dry day the stomata close to save water, internal CO_2 falls, O_2 rises, and a C_3 plant can lose a third of its fixed carbon to this rescue operation (Figure 29.6, left). Evolution has answered the problem twice, and both answers amount to rigging rubisco's market in CO_2.* ***C_4 plants*** *(maize, sugarcane, sorghum) first fix bicarbonate into a four-carbon acid in the mesophyll using PEP carboxylase — an enzyme that spurns O_2 entirely — then release the CO_2 around rubisco in the bundle sheath, bathing the enzyme in a private CO_2 atmosphere at the cost of 2 extra ATP per CO_2.* ***CAM plants*** *(cacti, pineapple, agave) do the same thing in time rather than space: stomata open at night, CO_2 is banked as malate in the vacuole, and by day the pores stay shut while the bank is drawn down (Figure 29.6, right). The C_4 trick has been invented independently more than sixty times — a statistic that should chasten anyone tempted to call photorespiration a design flaw that natural selection tolerates; it is a tax that natural selection keeps inventing new schemes to avoid.*

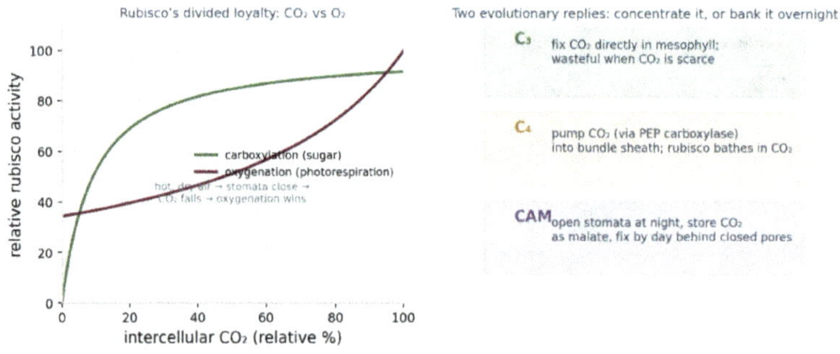

Figure 29.6 — The oxygen problem and its solutions. *Left: rubisco's carboxylation (green) versus oxygenation (crimson) as internal CO_2 varies; when stomata close in heat, CO_2 falls and the wasteful oxygenation rises. Right: the two evolutionary replies — C_4 concentrates CO_2 spatially (mesophyll pumps, bundle sheath fixes); CAM concentrates it temporally (night-time banking as malate, daytime release behind closed stomata). (Curves schematic.)*

> ***Case study — Why maize outgrows wheat in the heat.*** *Two neighbouring fields, one of wheat (C_3) and one of maize (C_4), endure a July heatwave at 38 °C with rationed water. Yields diverge dramatically in maize's favour. Explain at the level of (i) rubisco kinetics, (ii) stomatal behaviour, and (iii) the energy bill, why the C_4 plant wins despite paying 2 extra ATP per CO_2.*
> **Resolution.** *(i) Rubisco's oxygenase activity rises with temperature faster than its carboxylase activity, and CO_2 solubility falls; at 38 °C a C_3 leaf may*

oxygenate once for every two carboxylations. (ii) Drought forces stomatal closure, collapsing internal CO_2 still further; maize's PEP carboxylase, working on bicarbonate, is indifferent to O_2 and keeps bundle-sheath CO_2 near saturation, so its rubisco barely notices the heat. (iii) The 2 ATP surcharge is a fixed cost; photorespiration is a variable cost that in these conditions exceeds it. The C_4 plant buys certainty at a premium — the logic of insurance, enacted by anatomy.

Historical note — The man who measured the sun's receipts. *That plants make organic matter from air was clear to van Helmont (who weighed a willow), to Priestley (who watched a mouse revive under mint), and to Ingenhousz (who saw that light was the coin of payment). But the field's founding quantitative document is Julius von Sachs's demonstration, in the 1860s, that starch appears in a leaf precisely where light and chlorophyll coincide. The twentieth century then split the problem in two: Warburg's manometry and Emerson's flashing-light experiments counted photons and quanta; Hill showed that isolated chloroplasts could evolve oxygen without fixing any carbon at all — the "Hill reaction", which cut the process cleanly into light and dark halves; Arnon then found that the light half makes ATP as well as NADPH; and Calvin's ^{14}C chromatograms mapped the dark half carbon by carbon. The synthesis — two photocentres in series — belongs to Hill and Bendall (1960), and the first reaction centre to be seen atom by atom to Deisenhofer, Huber and Michel (1985). It is a tidy history, unusually so for biochemistry: each generation defined a problem the next could solve. One might almost call it photosynthetic.*

Plate 29.1 *Johann Deisenhofer (b. 1943), who with Hartmut Michel and Robert Huber crystallised the photosynthetic reaction centre of Rhodopseudomonas*

viridis and solved its structure — the first membrane protein ever seen at atomic resolution — sharing the 1988 Nobel Prize in Chemistry. (Nobel Foundation portrait.)

29.10 Exercises

Exercise 29.1. *Using Eq. (29.1), show that a mole of 680 nm photons carries about 176 kJ, and compute how many moles of ATP (at ΔG_p = 50 kJ mol^{-1}) could be made if all eight photons of the quantum requirement were converted without loss.*
Resolution. *E = 119 626/680 = 175.9 kJ mol^{-1}. Eight photons give 8 × 176 = 1 408 kJ; at 50 kJ per ATP the ceiling is ≈ 28 ATP per O_2. The cell actually makes ≈ 4–5 ATP equivalents per O_2 via the light reactions (and banks most of the energy in NADPH and, downstream, carbohydrate) — a reminder that the quantum requirement is set by redox bookkeeping, not by the energy ceiling, which is far more generous.*

Exercise 29.2. *Cyclic photophosphorylation around PSI pumps protons without making NADPH or O_2. If non-cyclic flow yields 3 ATP per 2 NADPH but the Calvin cycle demands 3:2 (Eq. 29.3) while sucrose synthesis and transport push the true demand to ≈ 3.3:2, estimate what fraction of PSI turnovers must run cyclically.*
Resolution. *Non-cyclic: 1.5 ATP per NADPH-pair-half — let 2 NADPH come with 3 ATP, deficit = 0.3 ATP per 2 NADPH. Each cyclic turnover of PSI moves 1 e^- around, pumping ≈ 2 H^+ via the Q-cycle; at 14/3 H^+ per ATP that is ≈ 0.43 ATP per turnover. To cover 0.3 ATP per pair of NADPH requires ≈ 0.7 cyclic turnovers per 2 NADPH, i.e. roughly one PSI turnover in four to five runs cyclically. Small shunts close large ledgers.*

Exercise 29.3. *A C_3 and a C_4 leaf each fix 100 CO_2. The C_3 leaf loses 25 CO_2 to photorespiration and pays its salvage costs; the C_4 leaf pays 2 ATP per CO_2 pumped. Using 9 ATP + 6 NADPH per G3P (3 CO_2), compare net carbon and energy outcomes.*
Resolution. *C_3: net 75 CO_2 → 25 G3P, cost ≈ 225 ATP + 150 NADPH plus photorespiratory ATP/NADH salvage (≈ 0.5 ATP per rescued carbon, ~25 ATP): total ≈ 250 ATP + salvage NADH, for 25 G3P → ≈ 10 ATP-equivalents per net CO_2. C_4: net 100 CO_2 → 33.3 G3P at 300 + 200 = 500 ATP + 200 NADPH → 5 ATP per CO_2, and one third more carbon. In heat, the premium is cheap at the price.*

Exercise 29.4 (proposition for research). *Rubisco engineering has long aimed at raising CO_2/O_2 specificity. Propose instead a "synthetic $C_{3\to4}$" strategy: introduce a minimal two-cell CO_2-concentrating module (PEP carboxylase + a decarboxylating shuttle) into rice mesophyll. Identify the three most likely failure modes (energetic, anatomical, regulatory), and for each propose an experiment to detect it early. Compare the expected payoff with improving rubisco itself, given that natural selection has had 3 billion years at*

the latter and only occasional access to the former.
Sketch. *Failure modes: (i) futile cycling if the decarboxylase leaks into mesophyll — assay by ^{13}C pulse-chase isotopomers; (ii) insufficient bundle-sheath isolation — measure conductance via carbon isotope discrimination ($\Delta^{13}C$); (iii) ATP/NADPH imbalance in mesophyll — monitor with fluorescent sensors. The payoff argument: rubisco specificity is near its chemical ceiling (selection has explored it exhaustively), whereas the C_4 module has been assembled de novo > 60 times in nature, suggesting an accessible design space. Engineering should imitate what evolution finds easy, not what it finds hard.*

29.11 Synthesis

With this chapter Part VI is complete. The two power stations now stand side by side, mirror images in chemiosmotic logic: the chloroplast runs electrons uphill on light and pays out NADPH and ATP; the mitochondrion runs them downhill from food to oxygen and pays out ATP alone; and both speak the one language of proton-motive force to rotary engines whose kinship is written in their sequences. Direction checked against the master plan: matter, energy and their transduction are done. What remains is the flux itself — the pathways by which carbon, nitrogen and sulphur actually flow through the cell, hour by hour and meal by meal. ***Part VII — Metabolism and Biosynthesis*** *opens with* ***Chapter 30 — The Organisation and Control of Metabolism****, where we shall ask not what the pathways are but how the cell keeps a thousand of them from running into one another.*

29.12 References

Arnon, D. I. (1984). The discovery of photosynthetic phosphorylation. Trends in Biochemical Sciences, 9, 258–262.

Berg, J. M., Tymoczko, J. L., Gatto, G. J., & Stryer, L. (2019). Biochemistry (9th ed.). W. H. Freeman.

Blankenship, R. E. (2021). Molecular Mechanisms of Photosynthesis (3rd ed.). Wiley-Blackwell.

Deisenhofer, J., Epp, O., Miki, K., Huber, R., & Michel, H. (1985). Structure of the protein subunits in the photosynthetic reaction centre of Rhodopseudomonas viridis at 3 Å resolution. Nature, 318, 618–624.

Hill, R., & Bendall, F. (1960). Function of the two cytochrome components in chloroplasts: A working hypothesis. Nature, 186, 136–137.

Nelson, D. L., & Cox, M. M. (2021). Lehninger Principles of Biochemistry (8th ed.). W. H. Freeman.

Nicholls, D. G., & Ferguson, S. J. (2013). Bioenergetics (4th ed.). Academic Press.

Ogren, W. L. (2003). Affixing the O to Rubisco: Discovering the source of photorespiratory glycolate and its regulation. Photosynthesis Research, 76, 53–63.

Sage, R. F. (2016). A portrait of the C_4 photosynthetic family on the 50th anniversary of its discovery. Journal of Experimental Botany, 67, 4039–4056.

Umena, Y., Kawakami, K., Shen, J.-R., & Kamiya, N. (2011). Crystal structure of oxygen-evolving photosystem II at a resolution of 1.9 Å. Nature, 473, 55–60.

Part VII

Metabolism and Biosynthesis

Chapter 30 — The Organisation and Control of Metabolism · Chapter 31 — Glycolysis · Chapter 32 — Gluconeogenesis and the Pentose Phosphate Pathway · Chapter 33 — Glycogen Metabolism · and the pathways of biosynthesis

Chapter 30
The Organisation and Control of Metabolism

Contents of this chapter

"A cell is not a bag of enzymes but a civil service of pathways — each with its own department, budget, and instructions about whom not to talk to."

30.1 Before the Pathways, the Polity

Part VII will march through the great pathways one by one — glycolysis, the pentose phosphate shunt, the citric acid cycle, fatty-acid and amino-acid metabolism, the synthesis of everything from cholesterol to heme. But a procession of pathways, taken singly, leaves the most interesting question unasked: how does a cell containing several thousand simultaneous reactions behave as an economy rather than a riot? The intermediates of a hepatocyte turn over in seconds yet hold their concentrations constant to within a few per cent; glycolysis and gluconeogenesis share seven of ten enzymes yet never cancel each other out; and the whole enterprise responds to a meal, a sprint, or a month of famine within minutes and without a committee meeting. This chapter is about that government. Its sections follow the TOC: the amphibolic map (30.2), the cast of intermediates (30.3), compartmentation (30.4), flux and steady state (30.5), the myth and meaning of the rate-limiting step (30.6), reciprocal regulation (30.7), hormonal control (30.8), metabolic cycles (30.9), metabolons (30.10), and metabolic control analysis (30.11) — the chapter's equations, and its conscience.

30.2 Catabolism, Anabolism and the Amphibolic Hub

*Metabolism has two temperaments. **Catabolism** is a convergent trade: hundreds of fuels — sugars, fats, amino acids, ketones — are funnelled into an ever smaller set of carriers, until nearly everything arrives as acetyl-CoA and a handful of citric-acid-cycle intermediates, paying out ATP and NADH on the way. **Anabolism** is a divergent one: a dozen precursors are elaborated into thousands of products, spending ATP and NADPH. The citric acid cycle sits at the neck of the hourglass and belongs to both trades, which is why Krebs called it **amphibolic** — it oxidises acetyl-CoA to CO_2 for energy and simultaneously dispenses α-ketoglutarate, oxaloacetate and succinyl-CoA for biosynthesis (Figure 30.1). The practical consequence, often missed by students counting ATP, is that the cycle can be drained by biosynthetic ambition and must then be refilled by **anaplerosis** — pyruvate carboxylase being the chief refilling station. A pathway that is everyone's supplier and everyone's dump needs its own supply lines; amphiboly is a logistical achievement, not merely a chemical one.*

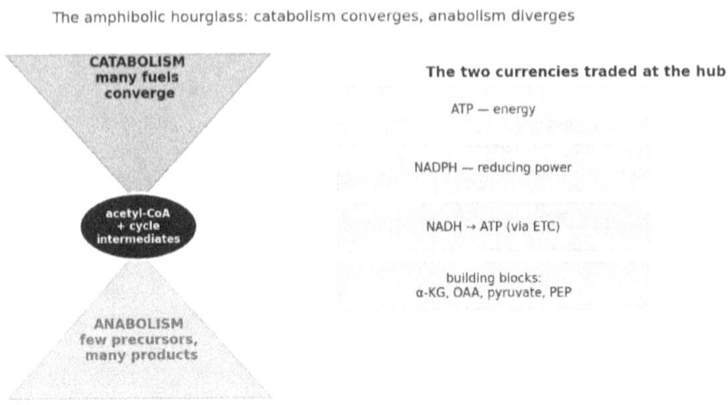

*Figure 30.1 — **The amphibolic hourglass.** Catabolism (crimson) converges from many fuels onto the acetyl-CoA / cycle hub (navy); anabolism (teal) diverges from few precursors to many products. The currencies traded at the hub are listed on the right: ATP, NADPH, NADH routed to the electron-transport chain, and the carbon skeletons α-ketoglutarate, oxaloacetate, pyruvate and PEP.*

30.3 The Cast of Intermediates

*Across all of metabolism, only about thirty compounds do nearly all the work, and they divide into three estates. The **energy carriers** — ATP, and behind it the phosphagens — we met in Chapter 25. The **electron currencies** — NADH for catabolic receipts, NADPH for biosynthetic spending — are kept at rigorously different ratios: cytosolic $NADH/NAD^+ \approx 10^{-3}$ (oxidised, ready to accept more), $NADPH/NADP^+ \approx 10^2$ (reduced, ready to give). A ten-thousand-fold difference in redox pressure, maintained by compartmentation and by transhydrogenases, is what allows one chemical motif — the nicotinamide ring — to serve two opposite masters without confusion. Finally the **carbon carriers**: coenzyme A for two-carbon acyl units, biotin for CO_2, tetrahydrofolate for one-carbon fragments, S-adenosylmethionine for methyl groups, UDP-sugars for glycosyl units. Learn the thirty and the pathways become variations on a grammar; memorise the pathways without the thirty and you have bought a dictionary in a language you cannot speak.*

30.4 Compartmentation

The eukaryotic cell enforces its economy with real estate. Fatty acids are synthesised in the cytosol and degraded in the mitochondrial matrix, so that synthesis and breakdown cannot see each other's substrates; the citric acid cycle, β-oxidation, ketogenesis and the respiratory chain share the matrix; glycolysis, the pentose phosphate pathway and fatty-acid synthesis share the cytosol; sterols and membrane lipids are made in the endoplasmic reticulum; hydrolysis is quarantined in the lysosome; and very-long-chain fatty acids are shortened in the peroxisome (Figure 30.2). Compartmentation is regulation by geography: it separates opposing pathways, concentrates intermediates, permits different pH and redox climates, and makes the membranes themselves into regulatory instruments, since every transport step — the malate–aspartate shuttle, the carnitine shuttle, the ATP/ADP exchanger of Chapter 28 — is a place where flux can be taxed. The cell, like a prudent state, locates its mint, its furnace and its customs house in different buildings.

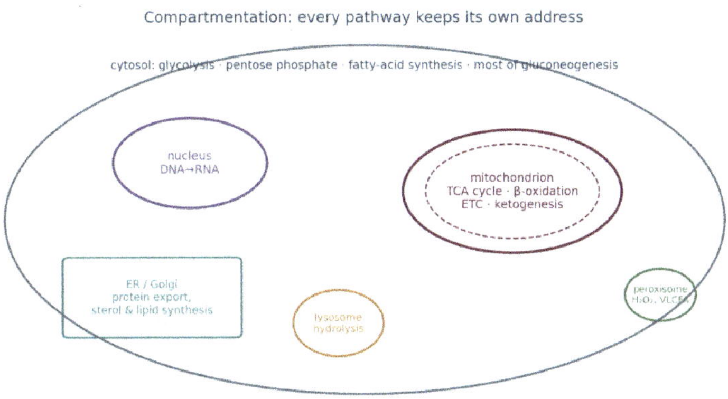

Figure 30.2 — Every pathway keeps its own address. *Schematic eukaryotic cell with its principal metabolic compartments. Synthesis and degradation of fatty acids are divorced by a mitochondrial wall; the cytosol hosts the catabolic and anabolic pathways of sugar; the ER builds sterols and export proteins; the lysosome and peroxisome handle demolition and dangerous oxidations.*

30.5 Flux and Steady State

The single most misleading feature of textbook pathway diagrams is their stillness. In a living hepatocyte the glycolytic intermediates are nowhere near equilibrium — most reactions run ten to a thousand times the reverse rate — yet their concentrations barely change, because inflow equals outflow at every node. This is the ***steady state****, and the quantity that matters physiologically is not any concentration but the* ***flux*** *J through the pathway: the rate at which carbon actually travels. For a chain of enzymes at steady state, every step carries the same flux, however different their kinetics:*

$$J = v_1 = v_2 = \cdots = v_n \tag{30.1}$$

— which follows immediately from mass conservation: at any internal intermediate, $d[S_i]/dt = v_i - v_{i+1}$, and the steady-state condition $d[S_i]/dt = 0$ forces the equality. Eq. (30.1) is trivially proved and endlessly underestimated. It means that asking "which enzyme is fastest?" about a pathway is meaningless — all steps run at the pathway flux — and it shifts every regulatory question from thermodynamics (can this step run?) to control (who sets how fast the whole chain runs?) (Figure 30.3, left).

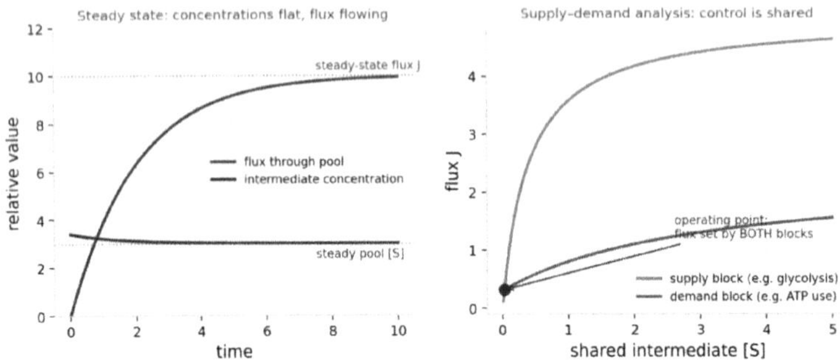

Figure 30.3 — Steady state and shared control. *Left: after a perturbation the pathway flux J and intermediate pool [S] settle to new steady values — concentrations flatten while flux flows (Eq. 30.1). Right: supply–demand analysis. The flux through a pathway is set where the supply block's flux response to the shared intermediate meets the demand block's; neither block alone is "rate-limiting".*

30.6 The Rate-Limiting Step, Reconsidered

Every student is taught to find "the rate-limiting step", and every pathway chart obligingly supplies one — PFK-1 for glycolysis, HMG-CoA reductase for cholesterol synthesis. The phrase survives because it is useful shorthand for "the step where regulation is exerted", but taken literally it is wrong, and Eq. (30.1) shows why: at steady state all steps run at the same rate, so none is slower than another. What differs between steps is not speed but ***control*** *— the fractional change in pathway flux produced by a fractional change in that enzyme's activity. In most pathways the control is distributed: in mammalian glycolysis, for instance, the glucose transporter, hexokinase and PFK-1 each hold a share, and the share shifts with conditions. The honest replacement for "rate-limiting step" is the* ***flux control coefficient****, to which Section 30.11 gives the mathematics.*

30.7 Reciprocal Regulation

Where two pathways oppose each other — glycolysis against gluconeogenesis, lipogenesis against lipolysis, glycogen synthesis against glycogenolysis — the cell almost never regulates one alone. It regulates both, oppositely, through the same signal. The paradigm is fructose-2,6-bisphosphate: a single metabolite that activates PFK-1 and inhibits fructose-1,6-bisphosphatase, so that one rise in its concentration opens the glycolytic valve and closes the gluconeogenic one simultaneously (Figure 30.5, left). Still more elegant, F2,6BP is made and destroyed by one bifunctional enzyme, PFK-2/FBPase-2, whose two activities are switched by a single phosphorylation — insulin dephosphorylates it into a kinase, glucagon phosphorylates it into a phosphatase. Reciprocal regulation is the biochemical equivalent of a railway signal that cannot show green in both directions; the engineering is so sound that we shall meet the same wiring in glycogen metabolism (Chapter 33) and lipid metabolism (Chapter 36) with only the names changed.

30.8 Hormonal Control

Allosteric effectors govern seconds; hormones govern minutes to days. The two great metabolic hormones are ***insulin****, the fed-state signal, and* ***glucagon****, the* *fasting signal, whose alternating custody of a single day is sketched in Figure 30.6. Their logic is worth stating abstractly, for it recurs throughout Part VIII: hormones do not enter metabolism as substrates or effectors but as information, arriving at receptors that transduce the message into phosphorylation cascades; the cascades then change enzyme activities in minutes (covalent modification) and enzyme amounts in hours (gene expression). Adrenaline adds a third voice for emergencies, cortisol and the thyroid hormones set the slower background tempo. The system is hierarchical — allostery at the bottom, covalent modification in the middle, transcription at the top — and the hierarchy is one of time-scales: each level is slower, broader and more expensive than the one below it.*

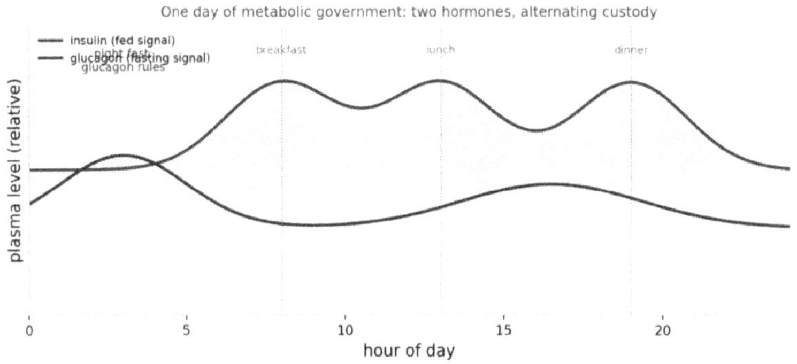

Figure 30.6 — One day of metabolic government. *Insulin (navy) rises after each meal, commanding storage; glucagon (crimson) rules the inter-meal and nocturnal intervals, commanding release. The shaded hand-over regions are where most of metabolic disease lives — Chapter 39 will hold the inquest. (Levels schematic.)*

30.9 Metabolic Cycles: Waste with a Salary

If opposing pathways run simultaneously, the net result is hydrolysis of ATP with no chemistry — the infamous ***futile cycle****. The name has embarrassed two generations of textbook writers into treating any such cycling as pathology, but the mathematics and the bumblebee disagree. When forward flux v_f and reverse flux v_r both respond sensitively to a regulator, the net flux $J = v_f - v_r$ can swing from strongly negative to strongly positive with a fractional change in signal, giving an amplification of sensitivity that no single irreversible valve can match. Cycling also burns ATP as heat on demand — the flight muscle of the bumblebee warms itself to flying temperature by cycling fructose phosphates, which is why bees can fly on mornings too cold for their rivals. Nature, thrifty as ever, pays a small continuous salary of ATP and receives, in exchange, a sensitive switch and a stove (Figure 30.5, right). We shall count the salary precisely in the worked example below.*

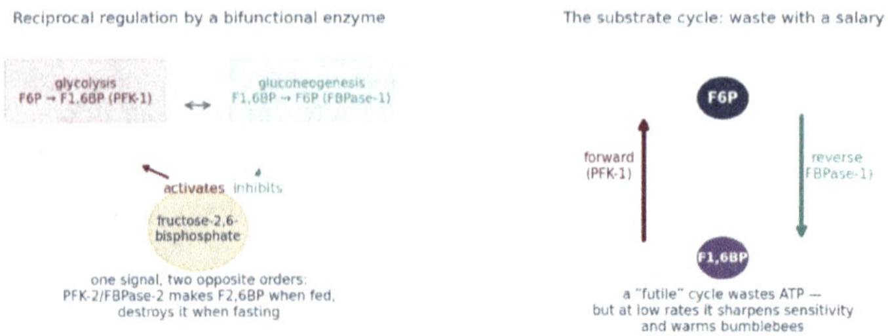

Figure 30.5 — Reciprocal regulation and the substrate cycle. *Left: fructose-2,6-bisphosphate (amber) activates PFK-1 and inhibits FBPase-1 — one signal issuing two opposite orders; the bifunctional PFK-2/FBPase-2 makes the signal when fed and destroys it when fasting. Right: simultaneous operation of PFK-1 and FBPase-1 forms a cycle whose only net chemistry is ATP → ADP; the "waste" buys sensitivity and, in a cold bumblebee, heat.*

> ***Worked example 30.1 — The sensitivity of a substrate cycle.*** *Suppose v_f and v_r each equal 10 (units arbitrary) when the effector is at its basal level, and each changes by a factor k per unit of effector. Show that the sensitivity of net flux to the effector grows with the cycling rate.*
> ***Solution.*** *Let $v_f = 10(1 + \varepsilon k)$ and $v_r = 10(1 - \varepsilon k)$ for a small effector displacement ε. Net flux $J = v_f - v_r = 20\varepsilon k$, so $dJ/d\varepsilon = 20k$. Now scale the basal rates to 100 with the same proportional response: $J = 100\varepsilon k - (-100\varepsilon k) = 200\varepsilon k$, and $dJ/d\varepsilon = 200k$. A ten-fold increase in cycling buys a ten-fold increase in sensitivity, at the price of hydrolysing 200 − 20 = 180 extra units of ATP per unit time. Sensitivity, like everything else in metabolism, is bought with ATP — the cycle is not futile but funded.*

30.10 Metabolons and Substrate Channelling

*The pathway diagrams of this book present intermediates as swimming freely between enzymes, and for many pathways the fiction is serviceable. But a growing body of evidence — kinetic, structural, and most recently cryo-electron tomographic — shows that many sequential enzymes associate into transient complexes, **metabolons**, in which the product of one active site is handed directly to the next without equilibrating with the bulk cytosol. The citric-acid-cycle enzymes cluster on the inner mitochondrial membrane; the glycolytic enzymes associate on the cytoskeleton and on the surface of mitochondria; the enzymes of purine synthesis form the dynamic "purinosome" when purines are scarce. Channelling shortens transit times, protects labile intermediates, sequesters toxic ones, and — most subtly — allows the same intermediate pool to be partitioned between competing fates by physical routing rather than by concentration. The cell's chemistry, it turns out, is less a well-stirred tank than a set of conveyor belts that assemble themselves when needed and politely disperse when not.*

30.11 Metabolic Control Analysis

*The discipline that replaced slogans about rate-limiting steps with measurable quantities is **metabolic control analysis** (MCA), built by Kacser & Burns and by*

Heinrich & Rapoport in the 1970s.[51] *Two definitions carry the whole theory. The* ***flux control coefficient*** *of enzyme i measures the pathway's response to that enzyme:*

$$C_i = \frac{\partial J/J}{\partial e_i/e_i} = \frac{\partial \ln J}{\partial \ln e_i} \qquad (30.2)$$

— the fractional change in steady-state flux J per fractional change in enzyme amount e_i, measured with everything else held constant. C_i is a property of the whole pathway, not of the enzyme alone; it can be measured by titrating enzyme expression, and it answers the only question "rate-limiting" ever really asked: if I had more of this enzyme, how much more flux would I get?

The second quantity is local. The ***elasticity*** *of an enzyme to a metabolite is*

$$\varepsilon^i_S = \frac{\partial \ln v_i}{\partial \ln[S]} \qquad (30.3)$$

— the enzyme's private sensitivity to one of its substrates, products or effectors, holding all other metabolites fixed. For a Michaelian enzyme below saturation $\varepsilon \to 1$; near saturation $\varepsilon \to 0$ (Figure 30.4, right); allosteric effectors can lend ε negative or greater-than-one values. The power of MCA comes from two theorems connecting the local and the global. The ***summation theorem,***

$$\sum_{i=1}^{n} C_i = 1 \qquad (30.4)$$

— derivable from Eq. (30.1) by scaling all enzyme activities by the same factor (which must scale J by that factor, since elasticities are ratios and concentrations are unchanged) and applying the chain rule — says that control is a fixed budget shared among all steps: if one enzyme holds 0.35, the others must divide the remaining 0.65 (Figure 30.4, left). And the ***connectivity theorem,*** *$\Sigma_i\, C_i\, \varepsilon^i_S = 0$ for each internal metabolite S, says that a perturbation felt through a shared intermediate must distribute itself so that the responses cancel — the mathematical form of supply meeting demand at the operating point of Figure 30.3 (right).*

51. Kacser & Burns, *Symp. Soc. Exp. Biol.* 27, 65–104 (1973); Heinrich & Rapoport, *Eur. J. Biochem.* 42, 89–95 (1974). It is a curious fact of the field's sociology that MCA was for twenty years honoured more in the breach than the observance, and is now indispensable to systems biology — good mathematics is patient.

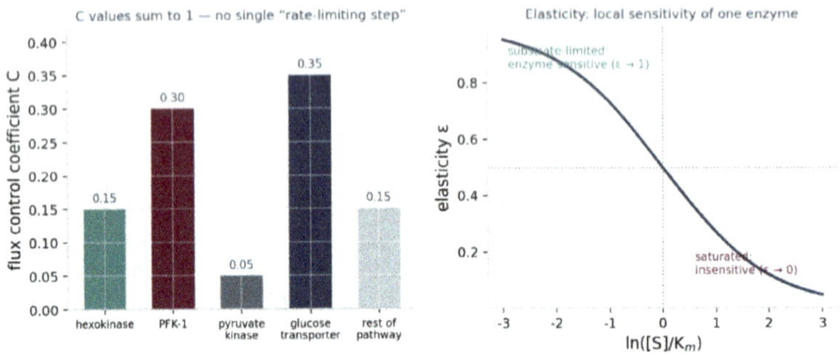

Figure 30.4 — ***The two quantities of metabolic control analysis.*** *Left: flux control coefficients C_i (Eq. 30.2) for a model of mammalian glycolysis — no step holds all the control, and the sum is unity (Eq. 30.4). Values illustrative; measured values shift with tissue and state. Right: elasticity ε (Eq. 30.3) of a single enzyme as a function of its substrate saturation: exquisitely sensitive when starved, indifferent when sated.*

The practical consequences overturn comfortable intuitions. Because control is distributed, increasing a single enzyme seldom increases flux proportionally — a lesson the biotechnology industry has learned expensively and repeatedly, overexpressing "the rate-limiting enzyme" and harvesting disappointment. Conversely, a modest activation at several steps, which is precisely what hormonal cascades achieve, can change flux dramatically. MCA also dignifies the substrate cycle of Section 30.9: cycling raises elasticities at the branch point and hence the sensitivity of net flux. And it quietly reframes drug design: the best target in a pathway is not the enzyme with the most control in the healthy cell but the one whose control coefficient rises most in the diseased one.

> ***Historical note — Cycles before pathways.*** *The idea that metabolism might be organised in cycles rather than lines belongs to Hans Krebs, twice over. In 1932, with Kurt Henseleit, he showed that urea synthesis runs in a cycle — the first metabolic cycle ever described — and in 1937 he proposed the citric acid cycle, having noticed that catalytic amounts of certain dicarboxylic acids dramatically stimulated respiration in pigeon breast muscle. The Nobel Prize followed in 1953, shared with Lipmann — the man of the cycle and the man of the currency, appropriately coupled. It is less often remembered that Krebs's 1937 paper was rejected by Nature and published in Enzymologia; the letter of rejection survives, as rejection letters of epoch-making papers have a way of doing. Krebs also gave this chapter its central habits of thought: it was he who insisted that a pathway is proved not by listing enzymes but by demonstrating flux — that metabolism is something that runs, not something that merely exists. Every flux measurement, every control coefficient and every isotopomer analysis in modern metabolism is a footnote to that insistence.*

Plate 30.1 *Hans Adolf Krebs (1900–1981), discoverer of the urea cycle (1932) and the citric acid cycle (1937) — the amphibolic hub of Figure 30.1 — and Nobel laureate in Physiology or Medicine, 1953. (Nobel Foundation portrait.)*

30.12 Exercises

Exercise 30.1. *Cytosolic $NADH/NAD^+ \approx 10^{-3}$ and $NADPH/NADP^+ \approx 10^2$. Using $E°' \approx -0.32$ V for both couples and the Nernst equation of Chapter 26, compute the two redox potentials and state what the difference means chemically.*
Resolution. *$E = E°' + (RT/nF)\ \ln([ox]/[red]) = E°' + 29.6$ mV × $\log_{10}$(ratio of oxidised to reduced) at 25 °C, n = 2. For NAD: $E = -320 + 29.6 \times 3 \approx -231$ mV. For NADP: $E = -320 + 29.6 \times (-2) \approx -379$ mV. The NADP pool sits ≈ 150 mV lower — a reservoir under higher reducing pressure, poised to donate into biosynthesis while the NAD pool stays ready to accept electrons from catabolism. Same cofactor chemistry, two voltages, two professions.*

Exercise 30.2. *In a five-step pathway the measured flux control coefficients are 0.35, 0.30, 0.15, 0.15 and C_5. Find C_5, and comment on the biotechnology strategy of doubling the first enzyme.*
Resolution. *By the summation theorem (Eq. 30.4), $C_5 = 1 - 0.95 = 0.05$. Doubling enzyme 1 ($\Delta e/e = 1$) yields $\Delta J/J \approx C_1 \times \Delta e/e = 0.35$ — at best a 35% flux increase for a 100% increase in enzyme, and in practice less, because as flux rises the control redistributes (the connectivity theorem re-partitions the budget). Doubling "the rate-limiting enzyme" buys, at most, its coefficient.*

Exercise 30.3. *A bumblebee flight muscle cycles fructose phosphates at 50 μmol ATP min^{-1} g^{-1} before take-off. If oxidative ATP synthesis is 200 μmol min^{-1} g^{-1} at full flight, and the cycle's sensitivity gain is proportional to cycling rate (Worked example 30.1), estimate the fraction of the take-off ATP budget spent on sensitivity, and justify the expenditure thermodynamically.*
Resolution. *Fraction = 50/200 = 25% of maximal aerobic capacity diverted to cycling — a large salary. It is justified because the muscle cannot fly until thorax temperature reaches ≈ 30 °C: the "wasted" ATP appears as heat that shortens warm-up from hours to minutes, and the sensitivity gain (proportional to cycling rate) lets the same enzyme pair switch instantly from heating to full glycolytic flux once airborne. The bee is not wasting a quarter of its engine; it is using the engine as a starter motor.*

Exercise 30.4 (proposition for research). *Metabolons are transient and low-affinity, which makes them invisible to most interactome screens. Design an experimental programme to map metabolon formation across a diurnal cycle in hepatocytes, combining (i) an in-cell crosslinking strategy, (ii) a flux readout proving that channelling (not merely co-localisation) occurs, and (iii) a quantitative model predicting when channelling should evolve. State the model's key inequality.*
Sketch. *(i) Time-resolved in-vivo crosslinking mass spectrometry (or split-TurboID) across the light–dark cycle, with phosphomimetic controls to test regulation by modification. (ii) Isotope dilution: supply ^{13}C-labelled upstream intermediate in the bulk medium and measure whether the downstream product is enriched faster than the bulk pool equilibrates — preferential channelling shows as an isotope transit time shorter than bulk diffusion allows. (iii) Channelling should evolve when the free-intermediate transit time exceeds the toxic/leaky intermediate's damage time, i.e. when $d^2/D \gg \tau_{decay}$ (diffusion distance d, diffusion coefficient D) — or when branch-point competition makes partitioning by routing more profitable than partitioning by K_m. The inequality is testable parameter by parameter.*

30.13 Synthesis

*The polity is mapped: an amphibolic hub fed by catabolism and drained by anabolism, a cast of thirty intermediates, regulation by geography and by time-scale, flux as the one quantity that matters, control as a shared and measurable budget, and cycles that turn out to draw salaries. Direction checked against the master plan: with the constitution in place, Part VII can now proceed to the statutes — the pathways themselves, each examined as a going concern with its receipts, its controls and its history. First and eldest among them is **Chapter 31 — Glycolysis**: the ten reactions every cell on Earth carries, the pathway on which the very idea of a pathway was invented.*

30.14 References

Berg, J. M., Tymoczko, J. L., Gatto, G. J., & Stryer, L. (2019). Biochemistry (9th ed.). W. H. Freeman.

Fell, D. (1997). Understanding the Control of Metabolism. Portland Press.

Heinrich, R., & Rapoport, T. A. (1974). A linear steady-state treatment of enzymatic chains: General properties, control and effector strength. European Journal of Biochemistry,

42, 89–95.

Kacser, H., & Burns, J. A. (1973). The control of flux. Symposia of the Society for Experimental Biology, 27, 65–104.

Krebs, H. A. (1970). The history of the tricarboxylic acid cycle. Perspectives in Biology and Medicine, 14, 154–170.

Nelson, D. L., & Cox, M. M. (2021). Lehninger Principles of Biochemistry (8th ed.). W. H. Freeman.

Newsholme, E. A., & Crabtree, B. (1976). Substrate cycles in metabolic regulation and in heat generation. Biochemical Society Symposia, 41, 61–109.

Srere, P. A. (1987). Complexes of sequential metabolic enzymes. Annual Review of Biochemistry, 56, 89–124.

Sweetlove, L. J., & Fernie, A. R. (2018). The role of dynamic enzyme assemblies and substrate channelling in metabolic regulation. Nature Communications, 9, 2136.

Voet, D., & Voet, J. G. (2011). Biochemistry (4th ed.). Wiley.

Chapter 31
Glycolysis

CONTENTS OF THIS CHAPTER

"Every cell alive remembers the same ten reactions. Evolution, having found a way to tax sugar, has never once revised the basic schedule."

31.1 Cellular Roles of Glucose

With the constitution of Chapter 30 in hand, we come to the statutes, and the first statute is the oldest law in the book. Glycolysis — ten reactions, cytosolic, oxygen-free, conserved from bacterium to bishop — converts one glucose into two pyruvates, banking two ATP and two NADH in the process. Its antiquity is written in its chemistry: it requires no oxygen (there was none), no membranes (there were none), and no cofactors that a primordial cell could not plausibly have owned. But glucose is more than fuel, and the chapter's first duty is to enumerate its offices. Glucose is a rapid energy source for cells without mitochondria (red blood cells) and for cells in a hurry (fast muscle); it is an obligatory fuel for brain, which under ordinary conditions refuses nearly everything else; it is a biosynthetic precursor through its intermediates — ribose from glucose-6-phosphate, serine from 3-phosphoglycerate, lipid from pyruvate via citrate; and it is a signal, whose concentration is itself read by pancreas and hypothalamus as the state of the world. The pathway we are about to walk serves all four masters, and its regulation is the sound of their negotiation.

31.2 Glucose Uptake

Before chemistry, customs. Glucose crosses the plasma membrane on facilitative transporters of the GLUT family, fourteen of them in humans, each with its own constituency: GLUT1 and GLUT3 everywhere, with high affinity, so that brain and red cells are never outbid; GLUT2 in liver and pancreatic β-cells, with low affinity, so that it acts as a glucose gauge rather than a mere gate; GLUT4 in muscle and fat, stored in vesicles and summoned to the surface by insulin — the molecular mechanism behind the fed-state signal of Chapter 30. The transporter is, by the arithmetic of metabolic control analysis, a substantial holder of the flux control budget (Figure 30.4), which is physiology's way of admitting that the first step of a pathway is often its most decisive.

31.3 The Preparatory Phase

The first five reactions spend money. ***Hexokinase*** *traps glucose as glucose-6-phosphate — the phosphate is both a handle for later chemistry and a customs stamp, since phosphorylated sugar cannot leave the cell — at a cost of one ATP.* ***Phosphoglucose isomerase*** *converts the aldose to fructose-6-phosphate, repositioning the carbonyl for the cleavage to come.* ***Phosphofructokinase-1*** *then spends a second ATP to make fructose-1,6-bisphosphate: this is the committed step, the point of no return, and accordingly the pathway's chief* *regulatory shrine, to which Section 31.7 is devoted.* ***Aldolase*** *splits the six-carbon sugar into two trioses — dihydroxyacetone phosphate and glyceraldehyde-3-phosphate — and* ***triose phosphate isomerase*** *converts the former into the latter, so that one substrate, G3P, proceeds. The account so far: two ATP spent, nothing yet earned, and the molecule doubled. The investment phase has the structure of a loan taken out against certain repayment (Figure 31.1).*

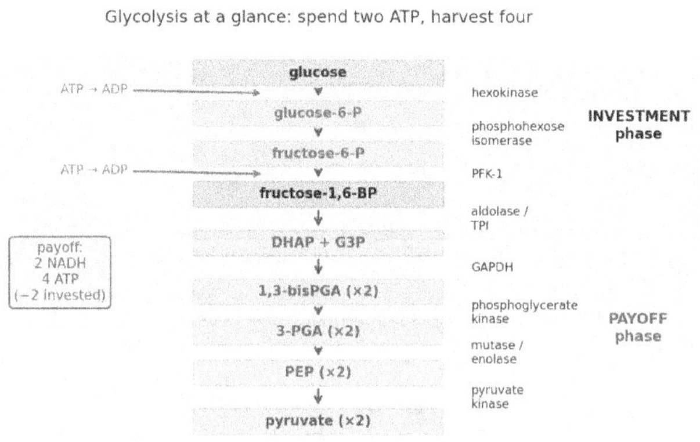

Figure 31.1 — The pathway at a glance. *The investment phase (crimson shade) spends two ATP to reach two triose phosphates; the payoff phase (green shade) converts them to two pyruvates, harvesting two NADH and four ATP — a net profit of two ATP per glucose. Enzymes are listed at the right of each arrow; the three deep-energy steps (hexokinase, PFK-1, pyruvate kinase) are the irreversible ones, and the regulated ones.*

31.4 The Payoff Phase

The last five reactions collect the debt with interest, and the collection mechanism is the chemical heart of the chapter. ***Glyceraldehyde-3-phosphate dehydrogenase*** *oxidises the aldehyde of G3P with NAD^+ — but instead of releasing the energy of that oxidation as heat, it traps the oxidised carbon as an* ***acyl thioester*** *on a cysteine of the enzyme itself. Phosphate then attacks the thioester, producing 1,3-bisphosphoglycerate, an* ***acyl phosphate*** *whose phosphoryl-transfer potential exceeds that of ATP (Figure 31.3). This is the manoeuvre the whole pathway exists to perform: an oxidation, cashed not as heat and not as a proton gradient but directly as a high-energy phosphate —* ***substrate-level phosphorylation****, the only way ATP can be made without membranes, oxygen or light.* ***Phosphoglycerate kinase*** *transfers that phosphate to ADP (first ATP of the payoff); a mutase and* ***enolase*** *then rearrange and dehydrate 3-phosphoglycerate into phosphoenolpyruvate, whose enol phosphate is the most energetic of all (−61.9 kJ mol^{-1} of hydrolysis, the summit of the ladder of Figure 25.1); and* ***pyruvate kinase*** *cashes it into the second ATP. Two trioses through the payoff: four ATP and two NADH in, two ATP net out.*

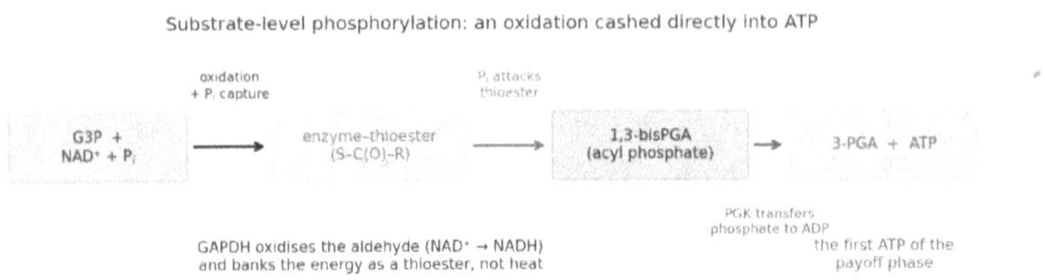

Figure 31.3* — *Substrate-level phosphorylation. *GAPDH oxidises G3P with NAD^+ and captures the energy as an enzyme-bound thioester (teal); phosphate attack releases the acyl phosphate 1,3-bisPGA (crimson); phosphoglycerate kinase transfers the phosphate to ADP. An oxidation banked as a bond — the manoeuvre Slater's X~Y was invented to imitate, here actually performed.*

31.5 Stoichiometry and Energetic Yield

The balance sheet, which every student should be able to draw from memory and every researcher occasionally misdraws, is

$$\text{glucose} + 2\,\text{NAD}^+ + 2\,\text{ADP} + 2\,\text{P}_\text{i} \longrightarrow 2\,\text{pyruvate} + 2\,\text{NADH} + 2\,\text{ATP} + 2\,\text{H}^+ \quad (31.1)$$

— derived simply by summing the ten reactions: two ATP spent (hexokinase, PFK-1), four earned (phosphoglycerate kinase and pyruvate kinase, each twice over after the aldolase split), two NAD^+ reduced at the GAPDH step. Under cellular conditions the free energy of the full conversion is about −80 kJ mol^{-1} of glucose, and the two ATP banked capture roughly 100 kJ of the ≈ 200 kJ made available in the payoff chemistry — an efficiency near 50% that no steam engine has ever approached. The ***free-energy profile*** *of Figure 31.2 repays study: seven of the ten steps hover near equilibrium (small $|\Delta G|$, freely reversible, able to run backwards in gluconeogenesis), while three — hexokinase, PFK-1, pyruvate kinase — plunge steeply and irreversibly downhill. The cell regulates exactly the irreversible steps, for the same reason a parliament debates the bills it cannot repeal.*

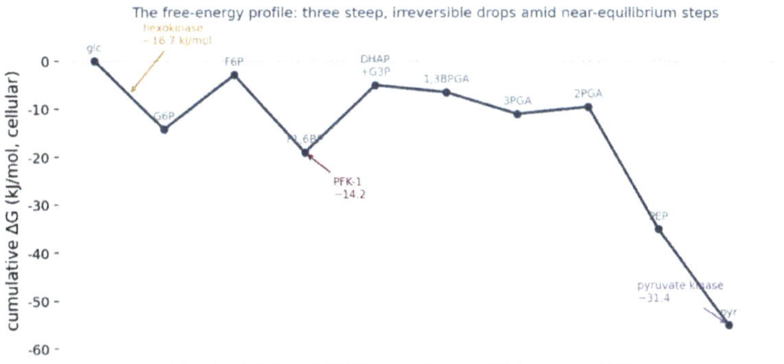

Figure 31.2 — The free-energy profile of glycolysis. *Cumulative ΔG under cellular conditions along the ten steps. The near-equilibrium plateau steps are shared with gluconeogenesis; the three steep drops — hexokinase, PFK-1, pyruvate kinase — are the irreversible, regulated valves where control resides (values representative of erythrocyte cytosol).*

31.6 Regulation of Hexokinase

The first valve comes in four isozymes with two philosophies. Hexokinases I–III are high-affinity ($K_m \approx 0.1$ mM), inhibited by their own product glucose-6-phosphate — a feedback that stops the furnace when the flue is blocked downstream. ***Hexokinase IV (glucokinase)*** *of liver and β-cells is the opposite temperament: low affinity ($K_m \approx 8$ mM, mid-range of blood glucose), sigmoidal kinetics, no product inhibition. Glucokinase does not serve the cell that owns it; it serves the blood. Its activity rises precisely across the physiological glucose range, making it the body's glucose sensor: in the β-cell it sets the threshold for insulin release, and in the liver it phosphorylates only the post-meal surplus. Mutations that raise its affinity cause congenital hyperinsulinism (the sensor fires too easily); mutations that lower it cause one of the commonest monogenic diabetis, MODY2. One enzyme, standing at a door, metering the whole organism's sugar.*

31.7 Phosphofructokinase as a Control Point

PFK-1 is the most regulated enzyme most students will ever meet, and justly, since it guards the committed step. Its kinetics are sigmoid in fructose-6-phosphate, and the sigmoid slides under orders (Figure 31.4, left): ATP itself — a substrate! — inhibits at a second, allosteric site, so that when the currency is abundant the pathway throttles; citrate, reporting a full citric acid cycle, reinforces the inhibition; AMP and ADP relieve it, demanding production; and fructose-2,6-bisphosphate, the hormonal signal of Chapter 30, overrules nearly everything, converting PFK-1 into a willing enzyme even at high ATP. The logic generalises into ***Atkinson's energy charge****,*

$$\text{energy charge} = \frac{[\text{ATP}] + \frac{1}{2}[\text{ADP}]}{[\text{ATP}] + [\text{ADP}] + [\text{AMP}]} \tag{31.2}$$

— the mole fraction of the adenylate pool's phosphoanhydride capacity that is actually filled (the ½ credits ADP's one bond against ATP's two). Healthy cells hold the charge at 0.85–0.90, and the fluxes of ATP-producing and ATP-consuming pathways cross, steeply and reciprocally, exactly in that window (Figure 31.4,

right): the currency regulates its own mint. Atkinson proposed the concept in 1968, and it remains the cleanest statement of why ATP can be simultaneously the product and the governor of glycolysis without circularity — a governor need not stand outside the system it governs, only be measured by it.

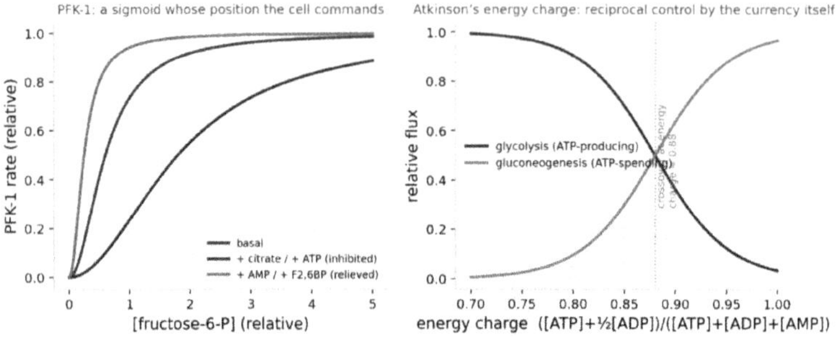

Figure 31.4 — Control at the committed step. *Left: PFK-1's sigmoid response to fructose-6-phosphate shifts right under ATP/citrate inhibition and left under AMP/F2,6BP relief. Right: flux through ATP-producing (crimson) and ATP-consuming (teal) pathways against the energy charge (Eq. 31.2); the crossover near 0.88 is the cell's chosen operating point.*

31.8 Regulation of Pyruvate Kinase

The third valve guards the exit. Pyruvate kinase is activated by fructose-1,6-bisphosphate — ***feed-forward*** *activation, the product of the committed step hurrying the last step to keep pace, a rare and elegant inversion of feedback — and inhibited by ATP and by alanine, the latter reporting that carbon skeletons are already plentiful. In liver the isozyme is additionally silenced by phosphorylation under glucagon: when the body fasts, hepatic glycolysis is switched off so that the liver's glucose is spared for export to brain and blood. Note the economy of design in these three valves: hexokinase/glucokinase meters supply, PFK-1 commits, pyruvate kinase paces — supply, commitment, pacing, each answering a different question, none duplicating another's answer. It is the committee structure of Chapter 30 executed in three active sites.*

31.9 Anaerobic Glycolysis

Eq. (31.1) consumes NAD^+, and NAD^+ is a catalytic commodity present in micromolar amounts; left unregenerated, glycolysis would strangle itself within seconds. In the presence of oxygen the problem is the mitochondrion's — the shuttles of Chapter 28 import the electrons and the respiratory chain re-oxidises them. Without oxygen, or faster than oxygen can be delivered, the cell regenerates NAD^+ on the spot by reducing pyruvate itself: ***lactate dehydrogenase,***

$$\text{pyruvate} + \text{NADH} + \text{H}^+ \rightleftharpoons \text{lactate} + \text{NAD}^+ \qquad (31.3)$$

— a near-equilibrium reaction ($\Delta G°' \approx -25$ kJ mol^{-1}, so strongly poised toward lactate) whose sole metabolic office is to recycle the cofactor so that GAPDH can keep spinning. Yeast performs the same service via pyruvate decarboxylase and alcohol dehydrogenase, giving us ethanol and CO_2 — glycolysis in the service of bread and wine, industries considerably older than biochemistry. The deeper point deserves emphasis: anaerobic glycolysis is not a different pathway but the same pathway with a cofactor-recycling appendage bolted on. The two ATP per glucose look paltry beside the thirty of full oxidation, yet they arrive with no infrastructure

and at great speed; a sprinter's muscle will take the two now over the thirty later, and so, daily, does your brain's reserve.

31.10 Lactate Metabolism

Lactate has suffered a century of bad press as "waste", and the rehabilitation is now complete enough to state plainly: lactate is a currency, not a toxin. It is produced continuously (red blood cells, fast muscle) and consumed continuously — by heart and slow-twitch muscle, which oxidise it avidly, and above all by the liver, which reassembles it into glucose by gluconeogenesis and returns the glucose to blood. The resulting **Cori cycle** (Figure 31.5) costs the liver six ATP per glucose resynthesised against the muscle's two earned — a net four-ATP inter-organ loan whose interest the liver gladly pays, since it buys the muscle speed and the brain security. Carl and Gerty Cori mapped the cycle in the 1940s (their Nobel came in 1947); its modern descendants include the lactate shuttle between astrocytes and neurons, and the tumour stroma's habit of feeding lactate to tumour cells — the waste product having turned out, on inspection, to be one of the body's principal fuels in transit.

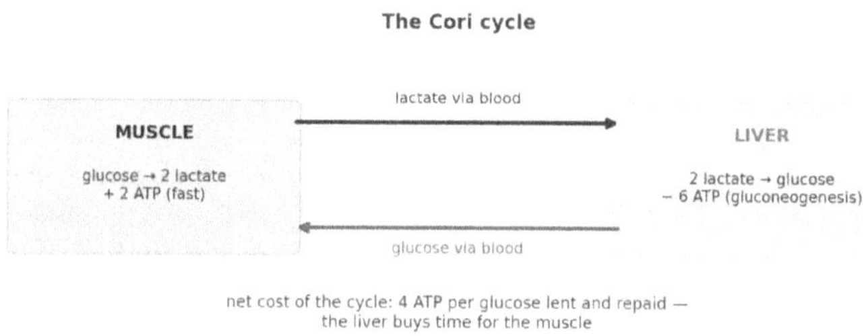

Figure 31.5 — The Cori cycle. *Muscle exports lactate; liver imports it, rebuilds glucose at a cost of six ATP per glucose, and exports the glucose; the muscle's two-ATP payday is thus financed by a four-ATP loan from the liver. An inter-organ division of labour brokered entirely by the bloodstream.*

31.11 Glycolytic Adaptations in Cancer

In 1924 Otto Warburg — whom we last met quarrelling about photons in Chapter 29 — reported that tumour slices consume glucose voraciously and excrete lactate even in plentiful oxygen, a phenotype since named the **Warburg effect** or aerobic glycolysis (Figure 31.6, left). Warburg believed the lesion causal — a damaged respiration driving cells to ferment — and in his strong form he was wrong, since most tumour mitochondria work perfectly well. The modern reading is subtler and more interesting. A proliferating cell's problem is not energy but **biomass**: it must duplicate every ribose, lipid and amino acid in a single cell cycle. Fast glycolytic flux, even at two ATP per glucose, supplies exactly the branch-point intermediates proliferation needs — G6P into the pentose phosphate pathway for ribose and NADPH, 3-PGA into serine and one-carbon metabolism, citrate into lipid — while lactate export regenerates NAD^+ fast enough to keep the flue open (Figure 31.6, right). Yield is sacrificed for throughput and for carbon, a trade any manufacturer under deadline would recognise. The phenotype is exploited clinically twice over: ^{18}F-fluorodeoxyglucose PET imaging finds tumours by their glucose avidity, and *inhibitors of lactate dehydrogenase and of the M2 splice isoform of pyruvate kinase remain under trial. The glycolysis chapter thus ends where it began — with a pathway so old and so central that even cancer, the great perverter of cell biology, does not replace it but merely drives it harder.*

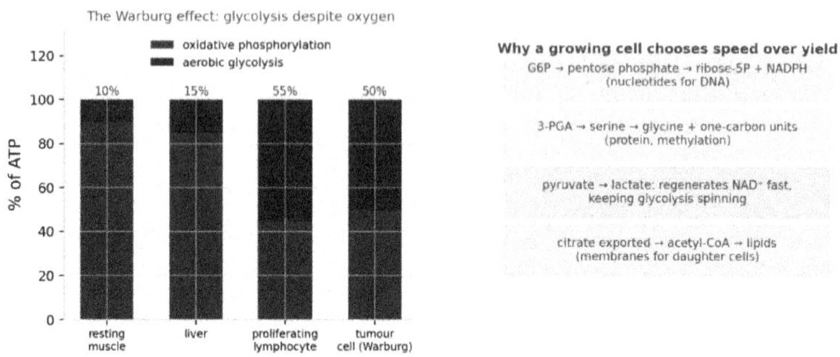

Figure 31.6 — The Warburg effect and its rationale. *Left: fraction of ATP from oxidative phosphorylation (navy) versus aerobic glycolysis (crimson) across cell types; proliferating and tumour cells choose glycolysis despite oxygen. Right: the proliferator's logic — glycolytic intermediates are feedstocks for ribose/NADPH, serine/one-carbon units, and lipids, while lactate export keeps NAD^+ regenerating at speed.*

Case study — The sprinter's debt. *A 400 m runner finishes in 45 seconds, having produced ATP at a rate no oxidative system could match. Her blood lactate peaks near 15 mM; she is oxygen-hungry for several minutes afterwards. Explain (i) why glycolysis alone can power the event, (ii) why the lactate is not the cause of her fatigue, and (iii) what the post-race oxygen consumption is paying for.*
Resolution. *(i) Two ATP per glucose, multiplied by a glycolytic flux that can exceed the oxidative rate fifty-fold, delivers more ATP per second than the 30-ATP-per-glucose pathway constrained by oxygen delivery; speed of supply, not yield, is the binding constraint in a sprint. (ii) Fatigue tracks the fall in ATP and phosphocreatine and the rise in ADP and P_i — which impede the cross-bridge cycle and calcium handling — rather than lactate per se; lactate is the exhaust of cofactor recycling (Eq. 31.3), not the poison. (iii) The excess post-exercise oxygen consumption pays the Cori cycle's interest: hepatic gluconeogenesis from the lactate, restoration of phosphocreatine and glycogen, and the normalisation of temperature and ion gradients. The debt is real, but the creditor is the liver, and the currency is glucose.*

Historical note — The pathway discovered in a juice. *Glycolysis is the pathway on which the very notion of a metabolic pathway was invented. In 1897 Eduard Buchner, grinding yeast for what he expected to be a failed control, found that the cell-free juice fermented sugar — killing, at a stroke, the doctrine that fermentation required the intact "vital" cell, and winning the 1907 Nobel Prize. There followed forty years of the greatest detective work in the field's history: Harden and Young showed phosphate was consumed and that the juice contained both enzymes and a dialysable "cozymase" (our NAD^+ and ATP); Embden and Meyerhof, in Frankfurt and Heidelberg respectively, pieced the intermediates into a sequence — the "Embden–Meyerhof pathway", its double name a treaty between rival laboratories; Warburg, not yet a photon-quarreller, caught the oxidation at GAPDH; and Parnas in Lwów* *completed the map. Meanwhile the muscle physiologists — Fletcher, Hopkins, and above all A. V. Hill and Otto Meyerhof — tied lactate to contraction and founded exercise biochemistry, sharing the 1922 Nobel Prize for it. No other pathway carries so many Nobel fingerprints; and none better repays the historian, for every trick in the modern textbook — cofactors, coupled reactions, substrate-level phosphorylation, isozymes — was first seen in this juice.*

Plate 31.1 *Archibald Vivian Hill (1886–1977), whose measurements of heat production in muscle founded the energetics of contraction and linked lactate to recovery oxygen — the physiological face of glycolysis. Nobel laureate in Physiology or Medicine, 1922 (shared with Otto Meyerhof). (Nobel Foundation portrait.)*

31.12 Exercises

Exercise 31.1. *Arsenate (AsO_4^{3-}) resembles phosphate and is accepted by GAPDH in its place, but the product, 1-arseno-3-phosphoglycerate, hydrolyses spontaneously. Trace the energetic consequence through the pathway.*
Resolution. *The GAPDH oxidation still occurs (NADH is still made), but the arseno-anhydride falls apart before phosphoglycerate kinase can cash it. The payoff phase therefore yields ATP only at pyruvate kinase: 2 ATP earned against 2 invested — net zero. The cell oxidises glucose to pyruvate for nothing, which is precisely why arsenate is a poison of energy metabolism: it uncouples substrate-level phosphorylation the way DNP uncouples the oxidative kind (Chapter 28). One pathway, two different doors, same empty treasury.*

Exercise 31.2. Calculate the ATP yield per glucose of (a) anaerobic glycolysis to lactate and (b) full oxidation (Chapter 28 ledger), and the rate ratio required for anaerobic glycolysis to supply ATP at the same rate. Comment on the sprinter of the case study.
Resolution. *(a) 2 ATP; (b) ≈ 30–32 ATP. Parity requires an anaerobic flux ≈ 15–16-fold higher. Muscle glycolytic V_{max} exceeds oxidative capacity by considerably more than that, so the sprint strategy is feasible — for the minute or so that glycogen and cofactor recycling can sustain it. The sprinter is not using an inferior pathway; she is using a faster one whose bill, via the Cori cycle, arrives later.*

Exercise 31.3. A patient's erythrocytes carry a pyruvate kinase variant with 10% residual activity. Explain (i) why the defect presents as haemolytic anaemia, (ii) the predicted changes in glycolytic intermediates upstream and downstream of the block, and (iii) why the 2,3-bisphosphoglycerate change found in these cells improves tissue oxygenation.
Resolution. *(i) Red cells have no mitochondria: glycolysis is their only ATP source, and the Na^+/K^+-ATPase is their chief expense. ATP-starved cells swell, stiffen, and are destroyed in the spleen. (ii) Upstream intermediates accumulate (notably 1,3-bisPGA and 2,3-BPG, which drains off at the Rapoport–Luebering shunt); pyruvate and lactate fall. (iii) 2,3-BPG binds deoxyhaemoglobin and lowers O_2 affinity; elevated 2,3-BPG shifts the dissociation curve right, releasing more oxygen per transit — a partial compensation the patient's tissues enjoy at no extra cost to the doomed cells.*

Exercise 31.4 (proposition for research). *The Warburg effect is usually discussed as metabolism of the tumour cell alone. Propose an experiment to test the "reverse Warburg" hypothesis — that tumour-associated fibroblasts perform aerobic glycolysis and export lactate which tumour cells then oxidise — in a human tumour xenograft, with cell-type-resolved flux measurements. What single measurement would most decisively falsify the hypothesis?*
Sketch. *Sort fibroblasts and tumour cells from the same tumour (FACS on lineage markers) after in-vivo infusion of ^{13}C-glucose and, separately, ^{13}C-lactate; measure isotopomer patterns in lactate, TCA intermediates and protein of each compartment by LC-MS; complement with cell-type-specific LDH knockout. The decisive falsifier: if tumour-cell TCA intermediates label no faster from ^{13}C-lactate than from bulk blood lactate, the claimed shuttling is not occurring and the hypothesis dies on the transit time — the same test Section 30.10 prescribed for metabolons.*

31.13 Synthesis

Ten reactions, two ATP, and a lesson in thrift that three billion years have not improved upon: glucose is trapped, split, oxidised once — at exactly the step where the oxidation can be banked as a bond — and cashed out as pyruvate, with the NAD^+ recycled by whatever reduction the environment allows. Direction checked against the master plan: the pathway forward from pyruvate forks — into mitochondria for full oxidation, into lactate for speed, or backwards, at a price, into glucose again. That reverse journey, and the side-road that turns glucose-6-phosphate into ribose and reducing power, share the next chapter: ***Chapter 32*** *—*

***Gluconeogenesis and the Pentose Phosphate Pathway**, the tale of the sugar rebuilt and the sugar diverted.*

31.14 References

Atkinson, D. E. (1968). The energy charge of the adenylate pool as a regulatory parameter: Interaction with feedback modifiers. Biochemistry, 7, 4030–4034.

Berg, J. M., Tymoczko, J. L., Gatto, G. J., & Stryer, L. (2019). Biochemistry (9th ed.). W. H. Freeman.

Buchner, E. (1897). Alkoholische Gärung ohne Hefezellen. Berichte der Deutschen Chemischen Gesellschaft, 30, 117–124.

Cori, C. F., & Cori, G. T. (1946). Carbohydrate metabolism. Annual Review of Biochemistry, 15, 193–218.

Fothergill-Gilmore, L. A., & Michels, P. A. M. (1993). Evolution of glycolysis. Progress in Biophysics and Molecular Biology, 59, 105–235.

Manchester, K. L. (1998). Albert von Szent-Györgyi and the unravelling of biological oxidation. Trends in Biochemical Sciences, 23, 37–40.

Nelson, D. L., & Cox, M. M. (2021). Lehninger Principles of Biochemistry (8th ed.). W. H. Freeman.

Vander Heiden, M. G., Cantley, L. C., & Thompson, C. B. (2009). Understanding the Warburg effect: The metabolic requirements of cell proliferation. Science, 324, 1029–1033.

Warburg, O. (1956). On the origin of cancer cells. Science, 123, 309–314.

Chapter 32
Gluconeogenesis and the Pentose Phosphate Pathway

CONTENTS OF THIS CHAPTER

"The cell that spends the night rebuilding the glucose it burned by day is not wasteful; it is keeping a promise to the brain."

32.1 The Necessity of Glucose Synthesis

Chapter 31 took glucose apart for two ATP; this chapter puts it back together at a cost of six — and then diverts it, at no ATP cost at all, into reducing power and ribose. The two subjects share a chapter because they share a theme: glucose-6-phosphate is a fork, and the cell must be able to go backwards up the glycolytic road as well as sideways off it. The backwards journey is not optional. Brain, red blood cells, renal medulla and the lens of the eye are obligate or near-obligate glucose users; a day's fast exhausts liver glycogen within about twenty-four hours; and yet the blood glucose of a fasting human holds rock-steady at 4–5 mM for weeks. The constancy is purchased by ***gluconeogenesis*** *— the synthesis of glucose from lactate, glycerol and the carbon skeletons of amino acids, carried out chiefly by the liver (and, in prolonged fasting, the kidney), spending four ATP and two GTP per glucose rebuilt. It is the most expensive routine transaction in human metabolism, and it is paid, every night of your life, without your noticing the bill.*

32.2 Bypass of Irreversible Glycolytic Reactions

Seven of the ten glycolytic steps run near equilibrium and reverse obligingly when mass action so demands. Three do not — the pyruvate kinase, PFK-1 and hexokinase steps, the steep drops of Figure 31.2 — and around each, evolution built a private bypass (Figure 32.1). The design principle deserves stating, for it is the same in every opposing pathway pair in this book: share the reversible steps, privatise the irreversible ones. Sharing saves genes; privatising permits independent control. A single road with two directions of traffic can carry both — provided the toll booths at each end are separately manned.

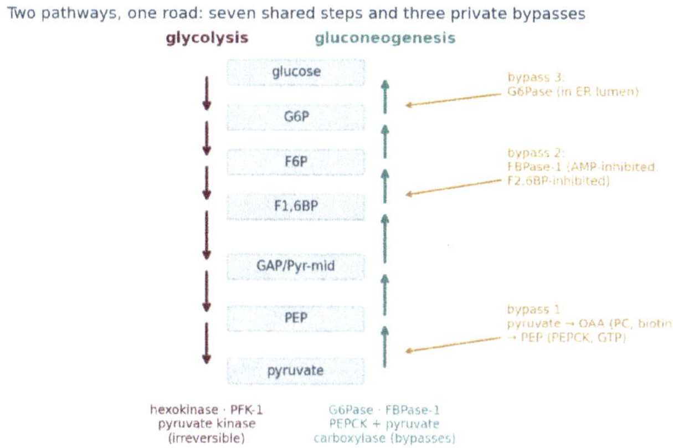

Figure 32.1 — Two pathways, one road. *Seven intermediates are shared (centre column); glycolysis (crimson, left) descends through three irreversible steps, and gluconeogenesis (teal, right) climbs back past them via three private bypasses (amber annotations): pyruvate carboxylase + PEPCK around pyruvate kinase; fructose-1,6-bisphosphatase around PFK-1; glucose-6-phosphatase around hexokinase.*

32.3 Pyruvate Carboxylase

The first bypass is itself two reactions, for pyruvate cannot be returned to PEP in one step — the pyruvate kinase drop was too steep. The cell therefore climbs around it through the citric-acid-cycle intermediate oxaloacetate. ***Pyruvate carboxylase****, a mitochondrial enzyme, fixes bicarbonate onto pyruvate at the cost of one ATP, using the most charming of all coenzyme mechanisms:* ***biotin****, a vitamin tethered to a lysine of the enzyme on a 14-Å arm, which picks up carboxylate at one active site and physically swings some 20 Å to deliver it at the other (Figure 32.2). The enzyme is silent unless acetyl-CoA is bound — an allosteric rule of perfect sense, since acetyl-CoA accumulates precisely when oxaloacetate is too scarce to partner it in the citric acid cycle: the cycle, finding itself drained, requisitions pyruvate to refill. This is* ***anaplerosis*** *(Chapter 30) enacted by a single ligand. Pyruvate carboxylase deficiency, rare and devastating, demonstrates the office by its absence: lactic acidosis, hypoglycaemia, neurological catastrophe — the cycle and the gluconeogenic road failing together for want of their common filling station.*

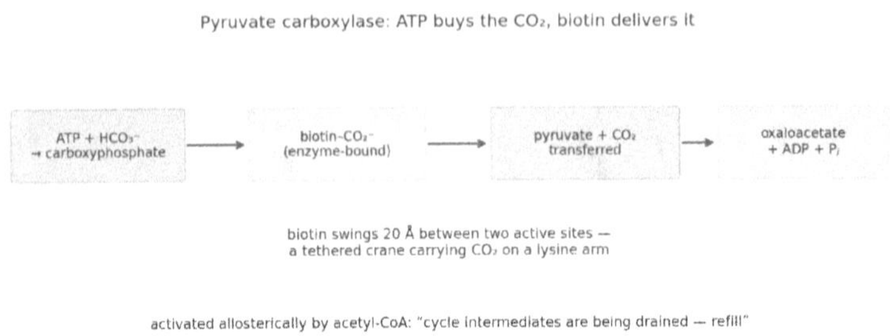

Figure 32.2 — Pyruvate carboxylase. *ATP and bicarbonate make carboxyphosphate at the first site; the CO_2 is transferred to enzyme-bound biotin (teal), which swings ~20 Å to the second site and carboxylates pyruvate to oxaloacetate. Acetyl-CoA is the obligatory allosteric activator — the cycle's own requisition for refill.*

32.4 Phosphoenolpyruvate Carboxykinase

Oxaloacetate is then decarboxylated and phosphorylated in a single stroke by ***PEP carboxykinase (PEPCK)****, powered not by ATP but by GTP — the cell keeping its two nucleotide currencies' accounts separate, as Chapter 30 advised. The reaction is beautifully economical: the CO_2 that pyruvate carboxylase bought with ATP is given back, and its departure drives the formation of PEP's enol double bond, so that the second reaction recovers, as bond energy, some of what the first spent. PEPCK exists in mitochondrial and cytosolic isozymes whose partition differs between species and states — a detail that decides whether oxaloacetate crosses the membrane directly (it cannot; it travels as malate or aspartate, the shuttles of Chapter 28 again earning their keep). PEPCK is also the pathway's transcriptional control point: insulin represses its gene, glucagon (via cAMP) and cortisol induce it. The enzyme is the fasting state's signature at the level of DNA.*

32.5 Fructose-1,6-bisphosphatase

The second bypass, around PFK-1, is a simple hydrolysis — fructose-1,6-bisphosphate + H_2O → fructose-6-phosphate + P_i — but its regulation is anything but simple, for here stands the mirror image of glycolysis's chief shrine. ***FBPase-1*** *is inhibited by AMP (the fasting signal: no energy to spare for synthesis) and, decisively, by fructose-2,6-bisphosphate — the very molecule that activates PFK-1. One signal, two opposite orders, exactly as Chapter 30 promised; the reciprocal wiring is drawn in Figure 32.3. When insulin is high, the bifunctional PFK-2/FBPase-2 is dephosphorylated into its kinase mode, F2,6BP rises, glycolysis opens and gluconeogenesis shuts; when glucagon is high, phosphorylation flips the bifunctional enzyme to phosphatase mode, F2,6BP falls, and the liver turns from burning glucose to making it. There is no cleaner example in metabolism of a single covalent modification redirecting an entire organ.*

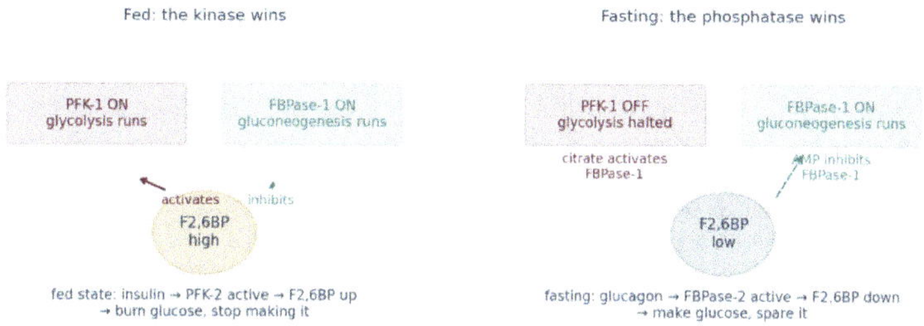

Figure 32.3 — The fructose-2,6-bisphosphate switch. *Left (fed): insulin dephosphorylates the bifunctional enzyme into PFK-2 mode; F2,6BP rises, activates PFK-1, inhibits FBPase-1. Right (fasting): glucagon phosphorylates it into FBPase-2 mode; F2,6BP falls away, and the residual control passes to citrate (activating) and AMP (inhibiting) at FBPase-1. One phosphate decides which organ the liver is today.*

32.6 Glucose-6-phosphatase

The last bypass is the strangest, for it involves a membrane. ***Glucose-6-phosphatase*** *sits in the endoplasmic reticulum of liver and kidney with its active site in the lumen: G6P is imported by its own transporter, hydrolysed, and glucose and phosphate exported separately. Why this oddity? Because the enzyme's office is not synthesis but export — muscle has G6P aplenty during exertion but lacks G6Pase entirely, and therefore cannot release free glucose; its glycogen is for its own use only. The liver, by parking the last step across a membrane, becomes the body's sole licensed exporter of glucose. The arrangement has a clinical monument:* ***von Gierke disease*** *(glycogen storage disease type I), deficiency of G6Pase or its transporter, in which the liver fills with glycogen it cannot release and fasting hypoglycaemia can kill — the customs house of Section 30.4's metaphor, closed for business.*

32.7 Reciprocal Control of Glycolysis and Gluconeogenesis

Sections 32.5 and 31.7 are two faces of one fact, which can now be stated as a balance sheet. The two opposing pathways share seven steps and differ at three valves; at each valve the allosteric logic is reciprocal (AMP and citrate with opposite signs at the two enzymes; F2,6BP with opposite signs at the same enzyme pair); and above the allostery sits the hormonal tier — insulin dephosphorylates, glucagon phosphorylates — acting on the same targets. The energetic price of the asymmetry is the difference between the two stoichiometries:

$$\underbrace{2\,\text{lactate} + 4\,\text{ATP} + 2\,\text{GTP} + 6\,\text{H}_2\text{O}}_{\text{gluconeogenesis}} \longrightarrow \text{glucose} + 4\,\text{ADP} + 2\,\text{GDP} + 6\,\text{P}_\text{i} \qquad (32.1)$$

— derived by summing the shared steps in reverse (2 ATP-equivalents of pyruvate-kinase bypass: 1 ATP at carboxylase + 1 GTP at PEPCK, per pyruvate) plus the two phosphatase hydrolyses, against glycolysis's net 2 ATP (Eq. 31.1). Six high-energy phosphates spent to undo what earned two: the 4-ATP toll of the Cori cycle quoted in Chapter 31 is exactly this arithmetic, and nothing more mysterious.

32.8 The Cori Cycle and Its Nitrogenous Twin

The Cori cycle of Figure 31.5 is drawn again in Figure 32.4, this time from the liver's side of the ledger and with its companion added: the ***glucose–alanine cycle****, in which muscle exports not only lactate but alanine — pyruvate that has accepted an amino group from branched-chain amino-acid breakdown — and the liver strips the nitrogen into urea while sending the carbon back as glucose. Two inter-organ loans, both repaid by hepatic gluconeogenesis at six ATP per glucose: one moving carbon and energy, the other moving nitrogen disguised as carbon. During prolonged fasting the kidney joins as a gluconeogenic organ, and the traffic changes character: glutamine from muscle feeds the kidney, which excretes the nitrogen as ammonium — incidentally buffering the acid load of ketosis. The cycles are the circulatory system's quiet masterpiece: organs trading debts in molecules too small to notice, keeping the one solute the brain cannot live without.*

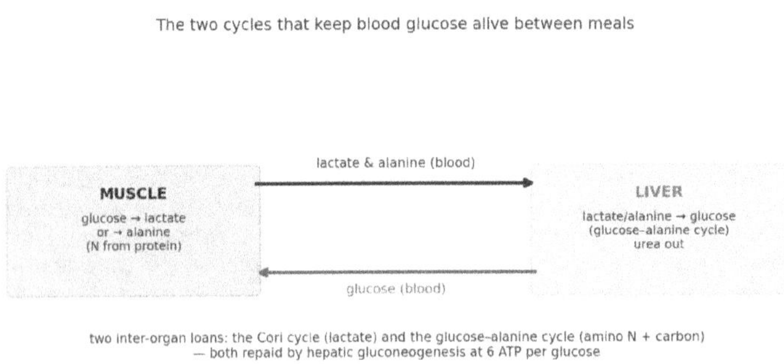

Figure 32.4 — ***The two cycles that keep blood glucose alive between meals.*** *Muscle exports lactate (Cori cycle) and alanine (glucose–alanine cycle); the liver repays both loans as glucose, at 6 ATP-equivalents per glucose (Eq. 32.1), and disposes of the nitrogen as urea.*

32.9 Oxidative Pentose Phosphate Reactions

We turn now to the fork. Glucose-6-phosphate, instead of descending the glycolytic road, may enter the ***pentose phosphate pathway*** *— and the first three reactions of that pathway are among the most consequential in the cell.* ***Glucose-6-phosphate dehydrogenase (G6PD)*** *oxidises C1 of the sugar, reducing $NADP^+$ to NADPH; a lactonase opens the resulting ring; and* ***6-phosphogluconate dehydrogenase*** *oxidatively decarboxylates the product, releasing the glucose molecule's first carbon as CO_2 and reducing a second $NADP^+$ (Figure 32.5, upper arm). The stoichiometry of the oxidative arm is exact and worth committing:*

$$\mathrm{G6P} + 2\,\mathrm{NADP}^+ + \mathrm{H_2O} \longrightarrow \mathrm{ribulose{-}5{-}P} + \mathrm{CO_2} + 2\,\mathrm{NADPH} + 2\,\mathrm{H}^+ \qquad (32.2)$$

— two NADPH per pass, at the price of one carbon. The reactions are irreversible and, in cells with active biosynthesis, carry enormous flux: this single pathway supplies the majority of the cytosol's NADPH, the reducing currency whose 10^2:1 ratio over $NADP^+$ we priced in Chapter 30. Where NADPH is the aim, the cell simply runs the oxidative arm and reclaims the ribose carbon through the reversible arm (next section); where ribose is the aim — a tumour doubling its DNA, a lymphocyte cloning — the pathway runs only far enough to make it.

32.10 Non-oxidative Carbon Rearrangements

The lower arm of Figure 32.5 is a carburettor for sugars: ribulose-5-phosphate is isomerised to ribose-5-phosphate or epimerised to xylulose-5-phosphate, and the enzymes ***transketolase*** *(two-carbon transfers, thiamine-dependent) and* ***transaldolase*** *(three-carbon transfers) shuffle the pentoses into fructose-6-phosphate and glyceraldehyde-3-phosphate — glycolytic intermediates, rejoining the main road. Every reaction here is reversible, which makes the arm a marketplace rather than a pathway: carbon flows in whichever direction demand dictates, allowing the cell to run the oxidative arm purely for NADPH while losing nothing but the one carbon deliberately paid. The bookkeeping of full recycling is elegant: six G6P through the oxidative arm yield 12 NADPH and 6* CO_2*, and the 30 remaining carbons leave as 5 G6P's worth of F6P and G3P — glucose oxidised completely to* CO_2 *with not one ATP made, the entire energetic harvest taken as reducing power.*

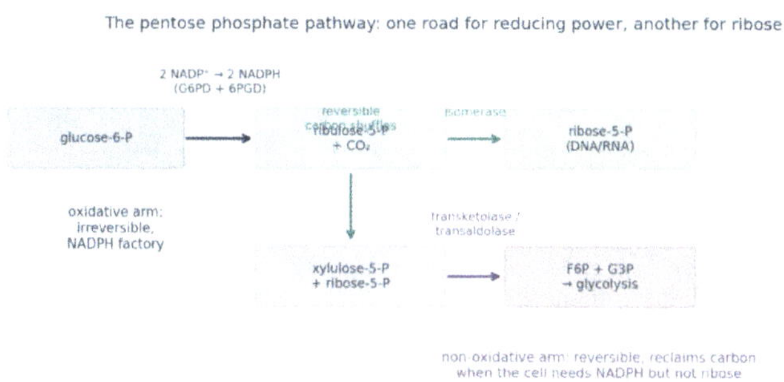

Figure 32.5 — The pentose phosphate pathway. *The oxidative arm (navy → teal → green) is irreversible and yields 2 NADPH plus ribose-5-P (Eq. 32.2); the non-oxidative arm (plum → crimson) reversibly reassembles pentoses into glycolytic intermediates, so carbon is never wasted when NADPH is the true objective.*

32.11 NADPH Production and the Redox Lifeline

What does the cell buy with NADPH that it could not buy with NADH? The answer divides into construction and defence. Construction: fatty-acid synthesis (Chapter 36), cholesterol synthesis, and the reductive animations of amino-acid and nucleotide chemistry all draw on the NADPH pool — which is why lipogenic tissues (liver, adipose, lactating mammary gland, steroidogenic glands) are rich in pentose-phosphate enzymes. Defence is the more dramatic office, drawn in Figure 32.6 (left): NADPH keeps ***glutathione*** *reduced via glutathione reductase, and reduced glutathione is the cell's front-line quencher of hydrogen peroxide and organic peroxides, through glutathione peroxidase. The chain NADPH → GSH → peroxide detoxification is short, and it is the only such chain most cells possess; break it at the first link and the rest is autobiography, as the next section shows.*

32.12 Glucose-6-phosphate Dehydrogenase Deficiency

G6PD deficiency is the commonest enzymopathy of humankind — some four hundred million carriers — and its geography is the geography of malaria, a coincidence that is no coincidence (Figure 32.6, right). The deficient red cell has no nucleus, no mitochondria, and no other NADPH source; its glutathione, once oxidised, stays oxidised. Most of the time this passes unnoticed. But present an oxidative challenge — fava beans (the pyrimidine aglycones of favism), the antimalarial primaquine, sulfa drugs, naphthalene, even a vigorous infection — and peroxides accumulate, haemoglobin denatures into Heinz bodies, membranes fail, and the spleen culls the damaged cells: an acute haemolytic anaemia, usually self-limiting because the youngest red cells, being least enzyme-decayed, survive. The evolutionary counterweight is that the Plasmodium parasite, itself an oxidative stress to its host cell and demanding of NADPH, thrives poorly in G6PD-deficient erythrocytes; the deficiency is a malaria-resistance allele whose price is paid in haemolytic crises. Haldane guessed this class of bargain — disease alleles kept by infectious selection — in 1949, before a single such allele was known; G6PD remains its cleanest confirmation, and a standing refutation of the lazy habit of calling every common mutation a defect.

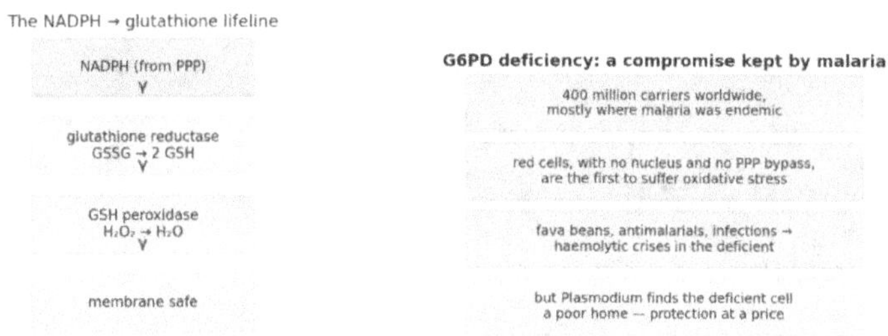

Figure 32.6 — The glutathione lifeline and its break point. *Left: NADPH reduces glutathione; reduced glutathione destroys peroxides; the membrane survives. Right: the epidemiology of the break — G6PD deficiency persists at high frequency exactly where malaria selected for it, and reveals itself as haemolysis under oxidative challenge.*

Case study — Primaquine and the blackout. *A 28-year-old soldier returning from a malaria-endemic posting is given primaquine for radical cure. Three days later he is jaundiced, dark-urined and breathless; his blood film shows bite cells and Heinz bodies. Explain the mechanism, predict the course, and state the two management principles.*
Resolution. *He is hemizygous for a deficient G6PD allele (X-linked; males express fully). Primaquine's redox-cycling metabolites generate peroxide; his red cells cannot regenerate GSH; haemoglobin oxidises, precipitates as Heinz bodies, and the spleen removes the bitten remnants — acute intravascular and extravascular haemolysis, the dark urine being haemoglobinuria. Course: self-limiting, because only the oldest, most enzyme-depleted cells are destroyed; reticulocytes, with residual enzyme, survive, and the haemoglobin recovers over two to three weeks even if the drug were (wrongly) continued. Management: stop the oxidant drug and support; transfuse only for cardiovascular compromise. Screening before primaquine is now standard — a one-line assay preventing a four-hundred-million-person experiment from being repeated one soldier at a time.*

Historical note — The Coris' partnership. *Carl Ferdinand Cori and Gerty Theresa Cori, née Radnitz, met as medical students in Prague, married in 1920, emigrated to Buffalo and then to Washington University in St. Louis, and* *worked together for four decades in a laboratory where, by the customs of the time, she was hired as his assistant at a fraction of his salary. Together they mapped the cycle of muscle lactate and liver glucose that bears their name, isolated glucose-1-phosphate ("Cori ester"), and crystallised phosphorylase — laying the foundations of both this chapter and the next. In 1947 they shared the Nobel Prize in Physiology or Medicine, Gerty becoming the first woman to receive it in that category; the citation honoured "the catalytic conversion of glycogen", which is to say, the controlled traffic between storage and blood that this chapter has priced at six ATP the glucose. Carl's Nobel lecture contains a sentence that should hang in every metabolism laboratory: the cycle, he said, shows the organism "functioning as a unit". The organ-level thinking of this book — loans, ledgers, customs houses — is their bequest.*

Plate 32.1 *Carl Ferdinand Cori (1896–1984), who with Gerty T. Cori mapped the inter-organ glucose–lactate cycle and the enzymology of glycogen, sharing the 1947 Nobel Prize in Physiology or Medicine. (Nobel Foundation portrait.)*

32.13 Exercises

Exercise 32.1. *Verify the 4-ATP toll of the Cori cycle by summing the appropriate half of Eq. (32.1), and state where the six high-energy phosphates are spent per glucose.*
Resolution. *Per glucose rebuilt: 2 pyruvate carboxylase (2 ATP) + 2 PEPCK (2 GTP) + FBPase-1 and G6Pase hydrolyses (2 phosphoanhydride equivalents lost as P_i) = 6 high-energy bonds spent; muscle glycolysis earned 2; net organismal cost = 4 ATP-equivalents per glucose cycled. The toll is paid by hepatic β-oxidation — the liver burns fat to buy the muscle's speed and the brain's security.*

Exercise 32.2. *A hepatocyte needs 12 NADPH but no ribose. Trace the carbon of one G6P through the pathway until it is fully reclaimed, and write the net equation for six G6P.*
Resolution. *Six G6P pass the oxidative arm: 6 CO_2 + 12 NADPH + 6 ribulose-5-P (30 carbons). The rearrangement arm converts 6 pentoses → 4 F6P + 2 G3P (30 carbons), and gluconeogenic reversal reassembles these to 5 G6P. Net: 6 G6P + 12 $NADP^+$ → 5 G6P + 6 CO_2 + 12 NADPH, i.e. G6P + 12 $NADP^+$ + 7 H_2O → 6 CO_2 + 12 NADPH + 12 H^+ + P_i. Glucose fully oxidised with the entire harvest taken as reducing power — the pentose pathway as a NADPH generator of theoretical perfection.*

Exercise 32.3. *During a three-day fast, hepatic glycogen is exhausted by hour ~24. Estimate the daily gluconeogenic ATP-equivalent cost of supplying the brain (~120 g glucose/day) and identify its fuel.*
Resolution. *120 g glucose = 0.67 mol; cost = 0.67 × 6 ≈ 4 mol ATP-equivalents ≈ 4 × 50 kJ ≈ 200 kJ/day for the synthesis alone, supplied by hepatic β-oxidation of fatty acids (which is why fasting raises hepatic acetyl-CoA and, downstream, ketone bodies — Chapter 36). The brain's glucose is, energetically, a gift of adipose tissue routed through the liver.*

Exercise 32.4 (proposition for research). *G6PD deficiency protects against severe malaria but the mechanism is debated: parasite starvation for NADPH, premature host-cell clearance, or oxidative killing of ring-stage parasites. Design an experiment using isogenic G6PD-normal and G6PD-deficient red cells (CRISPR-engineered) to distinguish the three, stating the single readout that would decide among them.*
Sketch. *Infect both cell types with synchronised P. falciparum, then track (i) parasite NADPH/$NADP^+$ by LC-MS, (ii) host-cell clearance by macrophage co-culture and splenic-mimic filtration, (iii) ring-stage death by live-cell redox biosensors. The deciding readout: the stage at which the deficient cells' disadvantage first appears relative to the oxidative-stress time-course — starvation predicts slow growth spanning cycles; clearance predicts differential removal of morphologically normal infected cells; oxidative killing predicts ring death synchronous with a measured peroxide burst. One measurement, time-resolved, separates a famine from an execution.*

32.14 Synthesis

The sugar is rebuilt and the sugar is diverted. Gluconeogenesis shares glycolysis's road but mans its own toll booths, and the pentose phosphate pathway turns the same first intermediate into the two things glycolysis cannot supply: ribose for the nucleotide economy and NADPH for construction and defence — the lifeline whose break point is the commonest inherited enzymopathy on Earth. Direction checked against the master plan: with synthesis and diversion covered, the remaining carbohydrate question is storage — how the fed liver and muscle bank glucose as glycogen, and how the bank is opened on demand. ***Chapter 33 — Glycogen Metabolism and Carbohydrate Specialisation*** *is the tale of the granule, its cascade, and its hereditary misfortunes.*

32.15 References

Berg, J. M., Tymoczko, J. L., Gatto, G. J., & Stryer, L. (2019). Biochemistry (9th ed.). W. H. Freeman.

Cappellini, M. D., & Fiorelli, G. (2008). Glucose-6-phosphate dehydrogenase deficiency. The Lancet, 371, 64–74.

Cori, C. F. (1948). Polysaccharide phosphorylase (Nobel lecture). Nobel Lectures, Physiology or Medicine 1942–1962. Elsevier.

Jitrapakdee, S., & Wallace, J. C. (1999). Structure, function and regulation of pyruvate carboxylase. Biochemical Journal, 340, 1–16.

Nelson, D. L., & Cox, M. M. (2021). Lehninger Principles of Biochemistry (8th ed.). W. H. Freeman.

Pilkis, S. J., & Claus, T. H. (1991). Hepatic gluconeogenesis/glycolysis: Regulation and structure/function relationships of substrate cycle enzymes. Annual Review of Nutrition, 11, 465–515.

Ruwende, C., & Hill, A. (1998). Glucose-6-phosphate dehydrogenase deficiency and malaria. Journal of Molecular Medicine, 76, 581–588.

Stincone, A., Prigione, A., Cramer, T., Wamelink, M. M. C., Campbell, K., & Cheung, E. (2015). The return of metabolism: Biochemistry and physiology of the pentose phosphate pathway. Biological Reviews, 90, 927–963.

Wamelink, M. M. C., Struys, E. A., & Jakobs, C. (2008). The biochemistry, metabolism and inherited defects of the pentose phosphate pathway: A review. Journal of Inherited Metabolic Disease, 31, 703–717.

Chapter 33
Glycogen Metabolism and Carbohydrate Specialisation

Contents of this chapter

"The difference between a feast and a famine, in a well-run body, is a granule — and a hormone to say when to open it."

33.1 Glycogen Structure

Glucose is too valuable to hoard as free sugar — the osmotic cost of 400 mM glucose would burst the cell — and too frequently needed to lock away as fat. Evolution's compromise is ***glycogen****: a branched polymer of α-1,4-linked glucose, with α-1,6 branch points every twelve residues or so, assembled into granules of up to 55 000 residues organised in about twelve concentric tiers (Figure 33.1). Every design feature serves the same office, which is speed of mobilisation. The polymer is osmotically inert. The branches multiply the one kind of end the mobilising enzyme can attack — the* ***non-reducing end*** *— so that a single granule presents more than two thousand simultaneous withdrawal points; a linear polymer of the same mass would present one. And the granule nucleates itself: the protein* ***glycogenin*** *at the core autoglucosylates one of its own tyrosines and extends the first short chain, so that no glycogen molecule is ever built without a built-in primer and none is ever fully consumed — the bank keeps its own foundation stone. Glycogen is, in short, a financial instrument: highly liquid, instantly divisible, self-issued.*

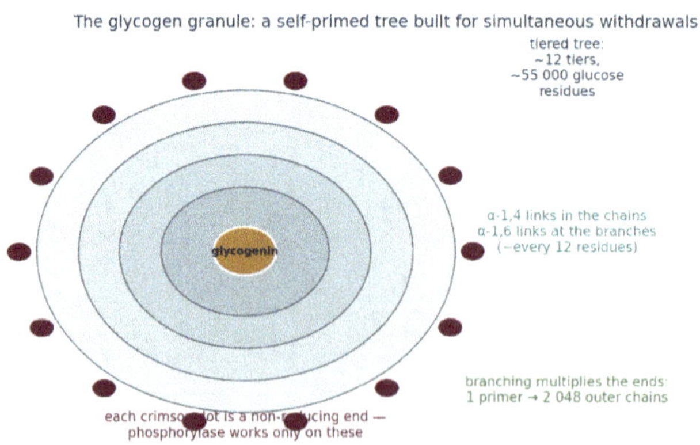

Figure 33.1 — The glycogen granule. *Concentric tiers of α-1,4 chains with α-1,6 branches (~every 12 residues), nucleated on glycogenin (amber) at the core. Branching multiplies the non-reducing ends (crimson dots) — phosphorylase's only substrates — to about 2 048 outer chains, enabling simultaneous mobilisation from the entire surface.*

33.2 Glycogen Synthesis

Synthesis runs through the sugar-nucleotide chemistry of Chapter 21. Glucose-6-phosphate is isomerised by phosphoglucomutase to glucose-1-phosphate, which ***UDP-glucose pyrophosphorylase*** *activates at the cost of UTP; the pyrophosphate released is hydrolysed, dragging the equilibrium irreversibly toward UDP-glucose (Figure 33.2, upper row).* ***Glycogen synthase*** *then transfers the glucosyl unit to a non-reducing end, extending the α-1,4 chain by one residue per cycle, and when a chain reaches about eleven residues the* ***branching enzyme*** *cuts a seven-residue segment and regrafts it as an α-1,6 branch (Figure 33.3, left) — the granule gaining, with every graft, another withdrawal point. The energetic price of storage is exactly one UTP per residue; as we shall see, most of it is refunded at withdrawal.*

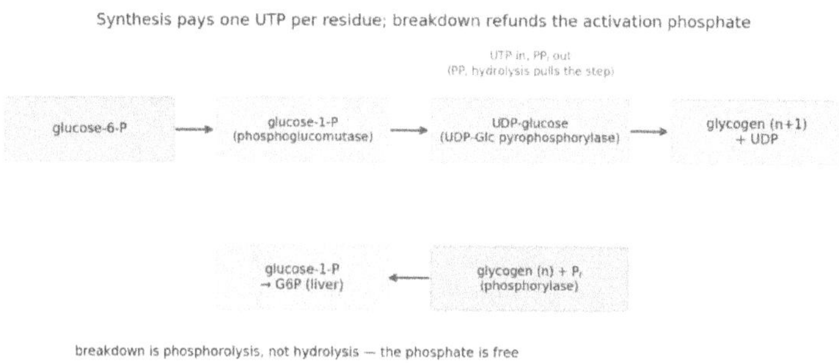

Figure 33.2 — The chemistry of deposit and withdrawal. *Upper: synthesis — G6P → G1P → UDP-glucose (one UTP spent; PP_i hydrolysis drives the step) → chain elongation by glycogen synthase. Lower: breakdown — phosphorylase releases G1P by phosphorolysis, conserving the glycosidic bond energy in the phosphate ester.*

33.3 Glycogen Breakdown

Glycogen phosphorylase *— the Coris' enzyme, and the first enzyme whose regulation by reversible phosphorylation was ever understood — cleaves the α-1,4 bond at a non-reducing end not with water but with* ***phosphate****, releasing glucose-1-phosphate directly. The phosphorolysis is a piece of thrift: the glycosidic bond's energy is conserved in the phosphate ester rather than dissipated as heat, so that liver G1P becomes G6P and then free glucose at no ATP cost, and muscle G1P enters glycolysis having skipped the hexokinase toll. Each residue stored thus costs one UTP and yields back the activation phosphate — net storage price, about one-third of an ATP per glucose: the cheapest banking in metabolism. Phosphorylase halts four residues short of a branch point, where the* ***debranching enzyme*** *— two activities on one polypeptide — transfers the three-residue stub to a neighbouring chain and hydrolyses the final α-1,6-linked residue as free glucose (Figure 33.3, right): roughly one glucose in twelve comes out unphosphorylated, a small tithe to the branches that bought the speed.*

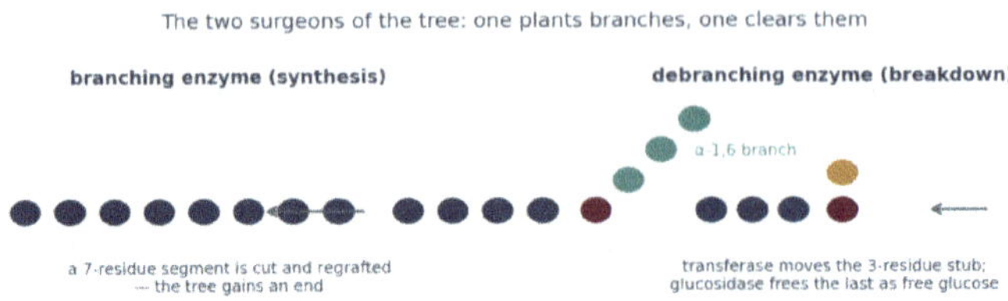

Figure 33.3* — *The two surgeons of the tree. *Left: branching enzyme transplants a 7-residue segment into an α-1,6 linkage. Right: the bifunctional debranching enzyme — transferase moves the 3-residue stub, glucosidase releases the last branch residue as free glucose.*

33.4 Hormonal Regulation: The Cascade

The control of glycogen metabolism is where biochemistry first discovered how hormones actually work. Earl Sutherland, studying liver glycogenolysis in the 1950s, found that adrenaline's message does not enter the cell: the receptor stays at the surface and commands the production of a heat-stable intracellular factor, ***cyclic AMP****, which he named the* ***second messenger****.*[52] *cAMP activates protein kinase A, which phosphorylates* ***phosphorylase kinase****, which phosphorylates phosphorylase b into the active a form (Figure 33.4). The ladder amplifies: one hormone molecule, through receptor, cyclase, two kinases and a population of enzymes, releases of order 10^8 glucoses — an amplification cascade of a kind we shall meet again in vision, in blood clotting and in complement. Meanwhile glycogen synthase is switched off by the same phosphorylations (reciprocal control again, exactly as Chapter 30 promised), and insulin reverses the whole ladder by activating protein phosphatase-1, the great eraser of metabolic phosphorylations. One kinase writes; one phosphatase unwrites; the hormones decide which pen is in hand.*

52. Sutherland & Rall, *J. Biol. Chem.* 232, 1077–1091 (1958). The Nobel Prize followed in 1971; the concept — first messenger outside, second messenger inside — now governs the whole of Part X of this book.

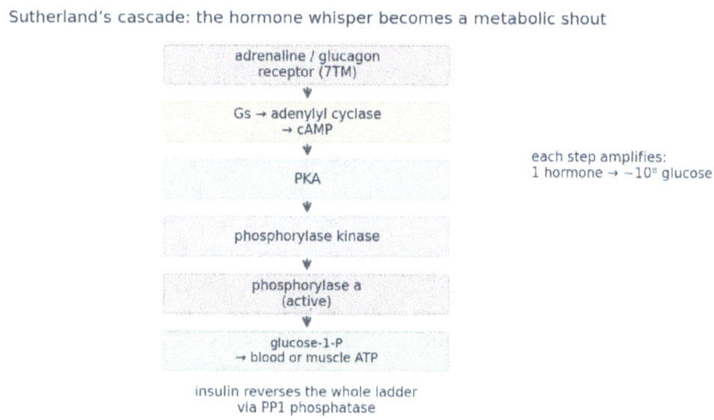

Figure 33.4 — Sutherland's cascade. *Receptor → G_S → adenylyl cyclase → cAMP → PKA → phosphorylase kinase → phosphorylase a → glucose-1-P. Each tier amplifies; insulin counters the entire ladder through protein phosphatase-1. The template for every signalling cascade in this book.*

33.5 Liver and Muscle Glycogen

The body keeps two glycogen banks of opposite character (Figure 33.5). The liver's is small — about 100 g, a day's supply — but altruistic: possessing glucose-6-phosphatase (Chapter 32), the liver exports its glycogen as blood glucose for the brain, and it obeys ***glucagon****, the hormone of the general fast. Muscle's is larger — about 300 g in a fed adult — but selfish: muscle lacks G6Pase, so its glycogen can only ever be burned in situ, and it obeys* ***adrenaline****, the hormone of the local emergency. Same enzymes, different receptors, different masters: liver phosphorylase answers the question "is the body hungry?", muscle phosphorylase answers "must this limb move now?". Muscle phosphorylase is additionally activated allosterically by AMP (the local energy report) and by Ca^{2+} at phosphorylase kinase — the very ion that triggers contraction simultaneously commanding the fuel supply for it, a piece of wiring so neat it seems designed, and is, by the only designer the field acknowledges.*

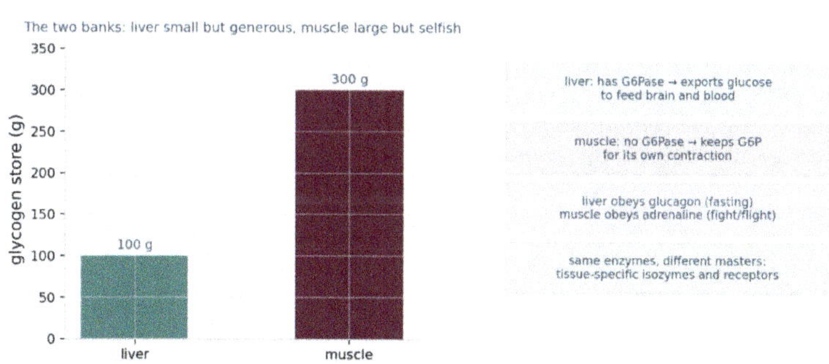

Figure 33.5 — Two banks, two policies. *Liver glycogen (teal) is small, exportable, and glucagon-ruled; muscle glycogen (crimson) is three times larger, inalienable, and adrenaline-ruled, with AMP and Ca^{2+} as its local governors.*

33.6 Glycogen-Storage Diseases

Nearly every enzyme of this chapter has its disease, and the catalogue — Figure 33.6 — reads like the chapter re-taught by pathology. ***Von Gierke (I)****: G6Pase missing, the liver cannot export, fasting hypoglycaemia.* ***Pompe (II)****: the lysosomal α-glucosidase missing, glycogen fills lysosomes everywhere and the heart fails in infancy — now treatable by enzyme replacement, the first storage disease so redeemed.* ***Cori (III)****: debranching enzyme missing; limit dextrin accumulates.* ***Andersen (IV)****: branching enzyme missing; long unbranched chains, poorly soluble, cirrhosis.* ***McArdle (V)****: muscle phosphorylase missing; exercise intolerance with a pathognomonic "second wind" as blood-borne fuels arrive.* ***Hers (VI)****,* ***Tarui (VII)****, phosphorylase kinase defects (IX). The lesson of the whole catalogue is the lesson of Figure 33.6's caption: each disease is a single missing key, and the phenotype — which organ, which symptom, which hour of the day — tells you exactly which lock the key used to open. Genetics, in glycogen, was biochemistry's first really detailed confession.*

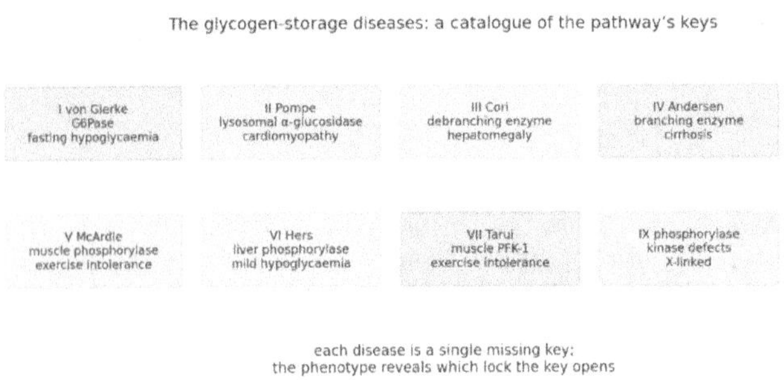

Figure 33.6 — The glycogen-storage diseases. *Eight principal types mapped onto the pathway they disable; the phenotype of each — hepatic or muscular, hypoglycaemic or exercise-limited — identifies the missing enzyme as precisely as a key identifies its lock.*

33.7 Fructose Metabolism

The other sugars of the diet enter through the same G6P turnstile, but by side doors with their own dangers. **Fructose** *in liver is phosphorylated by fructokinase to fructose-1-phosphate, which aldolase B cleaves — bypassing PFK-1 entirely. This small routing detail has outsized consequences: fructose carbon enters glycolysis below the committed step, escaping the pathway's great governor, and floods unregulated toward pyruvate, acetyl-CoA and fat. A cell fed fructose is a cell whose parliament has been bypassed; the modern epidemic of hepatic steatosis tracks fructose intake with an exactness that would have interested no one more than the gout-ridden sugar magnates of the eighteenth century. Hereditary fructose intolerance — aldolase B deficiency — shows the same enzyme's other face: fructose-1-phosphate accumulates, sequesters phosphate, halts both glycogenolysis and gluconeogenesis, and a sweet apple becomes a hypoglycaemic poison.*

33.8 Galactose Metabolism

***Galactose**, from milk lactose, is converted through galactose-1-phosphate and UDP-galactose into UDP-glucose by the Leloir pathway — the same Leloir whose nucleotide sugars built Chapter 21's oligosaccharides. The epimerase that interconverts the UDP-sugars is so efficient that dietary galactose is nearly invisible. But deficiency of galactose-1-phosphate uridyltransferase causes **classic galactosaemia**: galactose-1-P and galactitol accumulate, and without immediate lactose exclusion the infant suffers cataracts, liver failure and intellectual injury — a disease in which the difference between poison and food is one enzyme, and in which newborn screening, one of medicine's cheapest triumphs, converts catastrophe into a dietary footnote.*

33.9 Uronic-Acid Pathways

Finally, glucuronate — made by oxidising the C6 of UDP-glucose — is the liver's universal shipping label: bilirubin, steroids, drugs and a thousand xenobiotics are glucuronidated for excretion, which is why the pathway's enzymes crowd the smooth ER of every hepatocyte. In most mammals the same pathway continues onward to ascorbate; in primates, guinea pigs and fruit bats the terminal enzyme, gulonolactone oxidase, is a pseudogene, and vitamin C is a vitamin precisely because of an ancient mutational bankruptcy our ancestors could afford in a fruit-rich world. There is no better one-line argument against intelligent design in metabolism: we are, each of us, scurvy waiting for an orange.

> ***Case study — The boy who could not run.** A 19-year-old reports lifelong exercise intolerance: within minutes of vigorous effort his muscles cramp and lock; rest brings relief, and after about ten minutes he can continue — the "second wind". A forearm exercise test shows no rise in venous lactate; an old urine sample after a sprint was dark. Diagnose, explain the biochemistry, and give two management principles.*
> ***Resolution.** McArdle disease (GSD V): muscle glycogen phosphorylase absent. The absent lactate rise is diagnostic — glycolysis cannot be fed from glycogen — and the dark urine is myoglobin from rhabdomyolysis, the muscle digesting itself for want of fuel. The second wind arrives when hepatic glucose and fatty acids reach the muscle via blood: the selfish bank of Section 33.5 is bankrupt, but interbank transfer still works. Management: avoid maximal isometric effort, and exploit the second wind — a five-to-ten-minute gentle warm-up mobilises blood-borne fuels and makes subsequent exercise safe. The disease is a standing demonstration that muscle glycogen's only customer is the muscle itself.*

> ***Historical note — The second messenger.** In the early 1950s the action of adrenaline on glycogen was a pharmacological fact without a mechanism: everyone knew the hormone released glucose from liver slices; nobody knew how a molecule acting at the cell's surface could command an enzyme in its interior. Earl Wilbur Sutherland Jr., at Western Reserve and later Vanderbilt, took the problem apart with a rigour that set the template for signal transduction: he showed the response could be separated into a membrane fraction (the receptor and cyclase) and a soluble factor, identified the factor as cyclic AMP, proved it was made only when hormone was present, and showed it acted on the kinases the Coris' school had discovered. The 1971 Nobel Prize was his alone, but the conceptual harvest belonged to everyone: the "second*

messenger" principle — extracellular first messenger, intracellular second — now organises our understanding of everything from insulin to vision to smell. Sutherland, a quiet man, remarked that he had only followed the glycogen where it led. It led, as this chapter has tried to show, rather far.

Plate 33.1 *Earl Wilbur Sutherland Jr. (1915–1974), discoverer of cyclic AMP and author of the second-messenger principle, for which he received the 1971 Nobel Prize in Physiology or Medicine. (Nobel Foundation portrait.)*

33.10 Exercises

Exercise 33.1. *Compute the net energetic cost of storing one glucose residue in glycogen and recovering it into muscle glycolysis, and compare it with recovering free glucose from blood via hexokinase.*
Resolution. *Storage costs 1 UTP (UDP-glucose pyrophosphorylase). Recovery by phosphorolysis yields G1P → G6P at no cost, entering glycolysis one step below hexokinase — so the stored residue earns 3 ATP in glycolysis (2 net + 1 saved at hexokinase) against 1 UTP spent: a profit of ~2 ATP per cycle. Free blood glucose pays the hexokinase ATP and earns only 2 net. Glycogen banking is thus not merely cheap but profitable — the phosphorylase refund makes storage a zero-fee account with interest.*

Exercise 33.2. *A granule has 12 tiers and branching doubles the chains per tier. Estimate the number of non-reducing ends and the fraction of residues exposed in the outermost tier, commenting on mobilisation speed.*
Resolution. *Ends ≈ 2^{11} = 2 048 outer chains (matching Figure 33.1). The outer*

tier holds roughly half of all residues (each tier holds as much as all tiers inside it: $2^{11}/(2^{12} - 1) \approx 50\%$). Half the store is thus simultaneously attackable, which is why glycogen can sustain ATP production at the sprinter's rate within seconds — architecture as kinetics.

Exercise 33.3. *In von Gierke disease, explain (i) the fasting hypoglycaemia, (ii) the lactic acidosis, and (iii) the hepatomegaly, each in one sentence.*
Resolution. *(i) Without G6Pase the liver cannot convert G6P to free glucose, so all hepatic output — from glycogen and gluconeogenesis alike — ceases. (ii) Trapped G6P floods glycolysis, exporting lactate even as the blood sugar falls — the liver ferments while the body starves. (iii) Glycogen and triglyceride accumulate in hepatocytes that can neither export glucose nor sense the surplus correctly — the granule's bank with its doors welded shut.*

Exercise 33.4 (proposition for research). *Glycogenin was long considered a passive primer, but recent work suggests its expression is regulated with feeding state. Propose a programme to test whether granule number (versus granule size) is a regulated variable in liver, and why the distinction matters physiologically.*
Sketch. *Quantify granule number and size by electron tomography and by glycogenin immunoblot across fed–fasted–refed cycles; manipulate glycogenin expression in hepatocytes and measure the resulting size distribution. The physiological stake: mobilisation rate scales with total surface, and surface per gram is maximal for many small granules rather than few large ones — so granule number, if regulated, is a second, slower control on withdrawal speed, a "structural" complement to the cascade's fast control. The experiment decides whether the bank regulates how many vaults it builds, not merely how full they are.*

33.11 Synthesis

The granule is banked, branched, guarded by a cascade, and — in eight distinct hereditary ways — breakable; and the minor sugars have been shown to their doors, some welcome, some poisonous, one accidentally essential to the species that lost the enzyme to make it. Direction checked against the master plan: the carbohydrate economy is now complete from meal to granule to blood. What remains of the fuels is their final common furnace. ***Chapter 34 — The Citric Acid Cycle*** *returns to the amphibolic hub of Chapter 30, this time with the full machinery in view — the cycle that Krebs drew on the back of an enzyme shop's price list, or so one likes to imagine.*

33.12 References

Berg, J. M., Tymoczko, J. L., Gatto, G. J., & Stryer, L. (2019). Biochemistry (9th ed.). W. H. Freeman.

Chen, Y.-T., & Burchell, A. (1995). Glycogen storage diseases. In C. R. Scriver et al. (Eds.), The Metabolic and Molecular Bases of Inherited Disease (7th ed.). McGraw-Hill.

Cohen, P. (2002). Protein kinases — the major drug targets of the twenty-first century? Nature Reviews Drug Discovery, 1, 309–315.

Krebs, E. G., & Fischer, E. H. (1956). The phosphorylase b to a converting enzyme of rabbit skeletal muscle. Biochimica et Biophysica Acta, 20, 150–157.

Meléndez-Hevia, E., Waddell, T. G., & Shelton, E. D. (1993). Optimization of molecular design in the evolution of metabolism: The glycogen molecule. Biochemical Journal, 295, 477–483.

Nelson, D. L., & Cox, M. M. (2021). Lehninger Principles of Biochemistry (8th ed.). W. H. Freeman.

Roach, P. J., Depaoli-Roach, A. A., Hurley, T. D., & Tagliabracci, V. S. (2012). Glycogen and its metabolism: Some new developments and old themes. Biochemical Journal, 441, 763–787.

Sutherland, E. W. (1972). Studies on the mechanism of hormone action. Science, 177, 401–408.

Tappy, L., & Lê, K.-A. (2010). Metabolic effects of fructose and the worldwide increase in obesity. Physiological Reviews, 90, 23–46.

Chapter 34
The Citric Acid Cycle

CONTENTS OF THIS CHAPTER

"The wheel turns, the carbons leave as gas, and the wheel is still there. That is the whole trick — and it took the best chemists of two generations to see it."

34.1 Pyruvate Dehydrogenase

Between glycolysis and the cycle stands a customs post more elaborate than either: the ***pyruvate dehydrogenase complex****, a 60-subunit factory of three enzymes and five coenzymes, and the irreversible gate through which all carbohydrate carbon must pass to be fully oxidised (Figure 34.2). Its mechanism is a relay. E1 (pyruvate dehydrogenase proper), wielding thiamine pyrophosphate, decarboxylates pyruvate and holds the remaining two-carbon unit as hydroxyethyl-TPP; E2 (dihydrolipoyl transacetylase) receives it on a* ***lipoamide*** *arm — a lysine-tethered disulphide that swings between active sites like the biotin arm of Chapter 32, but busier — oxidises it to an acetyl group, and transfers it to coenzyme A; E3 (dihydrolipoyl dehydrogenase) then re-oxidises the lipoamide through FAD and finally NAD^+, minting NADH. One pyruvate in; one CO_2, one NADH, one acetyl-CoA out. The complex is regulated as befits a gate: inhibited by its products (NADH, acetyl-CoA) and by ATP, activated by ADP and by Ca^{2+} in muscle, and silenced by phosphorylation through PDH kinase when fatty acids are plentiful — the enzymatic signature of the glucose–fatty-acid cycle, by which fed muscle spares glucose and fasting muscle spares it even more.*

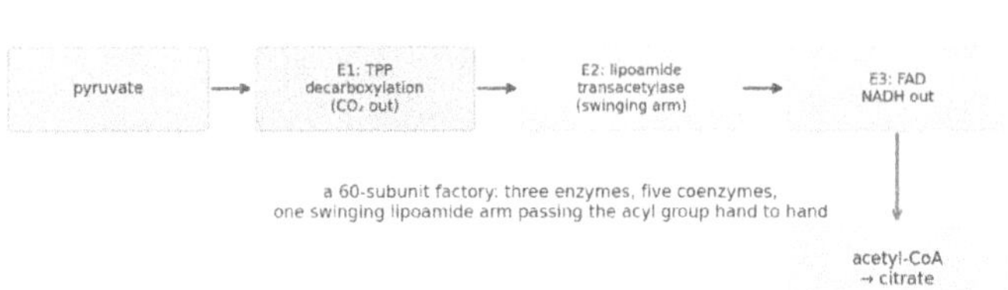

Figure 34.2 — The pyruvate dehydrogenase complex. *E1 decarboxylates (TPP, crimson); E2's swinging lipoamide arm (teal) carries the acetyl unit to CoA; E3 regenerates the arm through FAD to NADH (green). The customs post between glycolysis and the cycle — and the irreversible commitment of carbohydrate carbon to oxidation.*

34.2 Formation of Acetyl-CoA: The Activated Two-Carbon Unit

Why must carbon arrive as acetyl-CoA at all? Because a two-carbon fragment cannot be oxidised by simple dehydrogenation — there is no chemistry by which plain acetate can be efficiently burned — whereas an acetyl group tied to coenzyme A is a ***thioester****, activated toward both transfer and condensation, the same trick GAPDH used transiently in Chapter 31. Lipmann's discovery of coenzyme A (his 1953 Nobel, shared with Krebs) thus completes the logic of the cycle: the sulphur of CoA is the handle by which the cycle grasps its fuel. Note also the economy: PDH's NADH is made in the mitochondrial matrix, where the respiratory chain can cash it directly at 2.5 ATP — no shuttle, no toll.*

34.3 Reactions of the Cycle

The wheel itself (Figure 34.1) turns in eight reactions. ***Citrate synthase*** *condenses acetyl-CoA (2C) with oxaloacetate (4C) — the only C–C bond formation in the cycle, driven by hydrolysis of the thioester.* ***Aconitase*** *isomerises citrate to isocitrate via the enzyme-bound intermediate cis-aconitate, repositioning the hydroxyl so the next enzyme can oxidise it; the fluoroacetate poison of rat-bait fame works by being converted here into fluorocitrate, which jams aconitase — a "suicide substrate" and a cautionary tale about prodrugs.* ***Isocitrate dehydrogenase*** *performs the first oxidative decarboxylation: NADH out, CO_2 out, and the 6-carbon chain becomes α-ketoglutarate (5C).* ***α-Ketoglutarate dehydrogenase*** *— structurally a second PDH complex, same five coenzymes, same swinging arm — does the second: NADH out, CO_2 out, succinyl-CoA (4C). With both carbons of the original acetyl group now gone as gas (the bookkeeping is subtler, as Section 34.4 will confess), the remaining four reactions regenerate oxaloacetate:* ***succinyl-CoA synthetase*** *cashes the thioester as GTP — the cycle's one substrate-level phosphorylation;* ***succinate dehydrogenase****, which is none other than Complex II of the respiratory chain sitting in the membrane, makes $FADH_2$ and fumarate;* ***fumarase*** *adds water; and* ***malate dehydrogenase*** *oxidises malate to OAA with the cycle's third NADH. The last reaction is worth a second look: its equilibrium lies hopelessly toward malate ($\Delta G°' \approx +30$ kJ mol^{-1}), and the cycle turns only because citrate synthase devours OAA the instant it forms. The wheel is pulled from behind and pushed from ahead — and the pushing is mass action, not magic.*

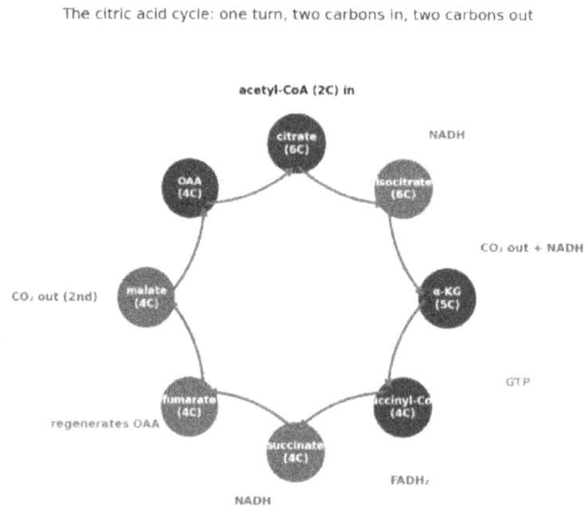

***Figure 34.1* — The cycle.** *Eight intermediates, two carbons entering as acetyl-CoA (crimson), two leaving as CO_2 (green annotations), three NADH, one $FADH_2$, one GTP (amber) harvested per turn. Navy nodes mark the control points: citrate synthase, isocitrate dehydrogenase, α-KG dehydrogenase, and the regenerated OAA.*

34.4 Carbon Balance: An Honest Confession

Textbooks say the two CO_2 released per turn "are" the two carbons of the incoming acetyl-CoA. Radioactive tracers say otherwise, and the truth is more interesting. The CO_2 lost at the two decarboxylations comes from the oxaloacetate skeleton, not the acetyl group; the acetyl carbons survive the first turn embedded in the new OAA, and are released only in subsequent turns, when the symmetrical intermediates succinate and fumarate have scrambled their positions. Labelled acetyl carbon therefore appears in CO_2 with a half-life of about one turn — the cycle does not burn the fuel it is handed; it burns the fuel it was handed last turn, a revolving door of carbon in which the intermediates are both catalyst and, transiently, substrate. The distinction matters practically (isotope experiments must model the scrambling) and philosophically: the wheel's carbons are a standing population, exchanged continuously with the traffic — a marketplace, as Section 34.7 will insist, not a furnace with a serial number.

34.5 Energetic Yield

The harvest per turn, at the prices of Eq. (28.4), is assembled in Figure 34.4: three NADH × 2.5 = 7.5 ATP, one $FADH_2$ × 1.5 = 1.5, one GTP = 1.0 — ***ten ATP per acetyl-CoA****, plus PDH's 2.5 if we count from pyruvate. Two pyruvates per glucose thus yield 25 ATP-equivalents from the oxidative core, which added to glycolysis's 2 ATP + 2 NADH gives the 30–32 of Chapter 28's ledger. The efficiency deserves its customary remark: complete oxidation of glucose releases ≈ 2 840 kJ mol^{-1}; 30 ATP at cellular ΔG_p ≈ 50 kJ bank ≈ 1 500 — better than 50%, against a steam engine's honest 20 and a car engine's aspirational 25. The cell achieves this not by tolerating higher temperatures but by taking the energy in instalments small enough to pocket, one NADH at a time.*

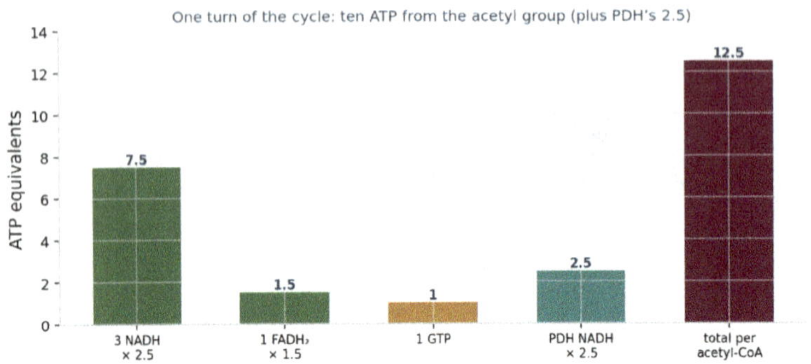

Figure 34.4 — The yield per turn. *ATP equivalents from one acetyl-CoA at the P/O prices of Eq. (28.4): ten from the cycle proper, 12.5 counting pyruvate dehydrogenase.*

34.6 Regulation

The cycle has no committed step in the glycolytic sense — it is a ring, and rings have no entrances to commit at — so control is distributed across the two oxidative decarboxylases and citrate synthase, with PDH guarding the fuel supply upstream (Figure 34.3, left). The logic is uniform and requires no memorisation: every valve is inhibited by the cycle's own products (NADH, succinyl-CoA, citrate) and by ATP, and activated by the signals of work (ADP) and, in muscle, Ca^{2+} — the same ion that commands contraction commands the furnace that pays for it, as it did at phosphorylase kinase in Chapter 33. The cell, asked thrice, answers with one voice: when energy is abundant the wheel idles; when work calls, it spins.

34.7 Anaplerotic Reactions

Because the intermediates are also biosynthetic wares, the cycle leaks, and the leaks must be plugged. ***Anaplerosis*** *("filling up") is the ensemble of refilling reactions, of which pyruvate carboxylase — the acetyl-CoA-activated enzyme of Chapter 32 — is sovereign in liver and brain; PEP carboxykinase in reverse, propionyl-CoA carboxylation from odd-chain fatty acids and valine/isoleucine, and glutamate dehydrogenase feeding α-KG from protein catabolism all contribute (Figure 34.3, right). The anaplerotic flux is not small: in a working liver it may rival the cycle's own turnover. A wheel that is also a market must restock its stalls continuously, and pyruvate carboxylase's allosteric rule — no activity without acetyl-CoA — remains the most elegant restocking order in biochemistry: fuel arrives, finds no partner, and itself commands the partner's synthesis.*

34.8 Cataplerosis

The drains, named only in 1992 by contrast with Krebs's coinage, are ***cataplerotic****: citrate exported for fatty-acid synthesis (Chapter 36), α-KG leaving as glutamate and thence half the amino-acid family, OAA withdrawn for gluconeogenesis (Chapter 32), succinyl-CoA for heme, malate exported as reducing-equivalent currency. Every drain is matched, in a healthy cell, by a fill; the pathology of imbalance is instructive — insulin-deficient diabetes drains OAA toward glucose so fast that acetyl-CoA queues at the cycle's door, and the overflow becomes ketone bodies, the subject of Chapter 35. The cycle's poise between* *anaplerosis and cataplerosis is, in the end, what "metabolic health" means at the level of one organelle.*

34.9 Biosynthetic Functions

It is worth listing the wares explicitly, since students taught the cycle as an incinerator are perpetually surprised by them: citrate → fatty acids and sterols; α-KG → glutamate → glutamine, proline, arginine; OAA → aspartate → asparagine, methionine, threonine, pyrimidines; succinyl-CoA → heme; malate → NADPH via malic enzyme. Roughly half of everything a growing cell builds passes, at some point, through this wheel. The amphibolic hub of Figure 30.1 is not a metaphor imposed on the cycle; it is the cycle's job description.

34.10 The Glyoxylate Cycle

Plants, fungi and many bacteria own two enzymes animals lack — ***isocitrate lyase*** *and* ***malate synthase*** *— which reroute the cycle past both decarboxylations (Figure 34.5): isocitrate is split to succinate and glyoxylate, and glyoxylate condensed with a second acetyl-CoA to malate. The net result is two acetyl-CoA → one succinate, with no carbon lost — and succinate is glucogenic. Hence the germinating seed's celebrated trick: it stores oil, and converts fat to sugar on demand, feeding the seedling until photosynthesis begins. Animals, lacking the bypass, cannot make net glucose from even-chain fat — which is why the fasting brain of Chapter 32 must be fed from protein and glycerol, and why the ketone bodies of the next chapter exist at all. One pair of enzymes, absent from our genomes, fixes the boundary between the metabolic strategies of the seed and the mammal; every diet book ever written is a footnote to isocitrate lyase.*

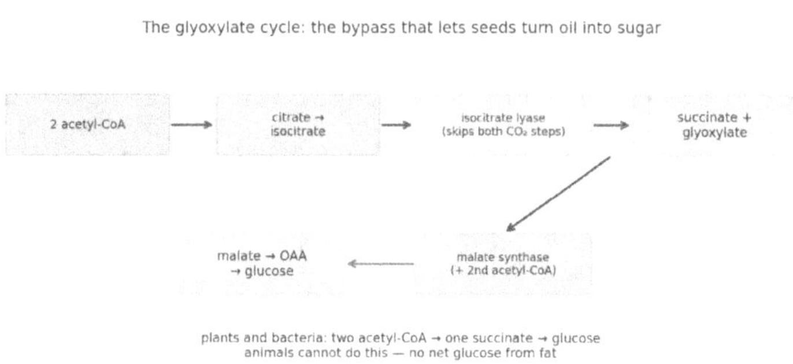

Figure 34.5 — The glyoxylate bypass. *Isocitrate lyase and malate synthase skip both decarboxylations, converting two acetyl-CoA to one succinate — the seed's route from oil to sugar, and the enzymatic reason animals cannot net-synthesise glucose from fat.*

34.11 Evolutionary Significance of Cyclic Metabolism

Why did life settle on a cycle rather than a line? Four reasons, assembled in Figure 34.6: catalytic regeneration of the acceptor (a line would consume its cofactor); safe handling of two oxidative decarboxylations too energetic to run bare; maximal branch-point connectivity, making the hub amphibolic for free; and reversibility in principle — several autotrophic bacteria run the same enzymes backwards as the reductive citric acid cycle, fixing CO_2 into acetyl-CoA, which is to *say, running the furnace as a factory. That the same ring runs both ways, in different organisms, suggests it is older than oxygen and was born anabolic: the cycle may first have existed to make molecules, and learned to burn them only when the air turned poisonous with the waste of Chapter 29. If so, the wheel at the centre of our metabolism is a reversed museum piece — life's oldest factory, retooled as its engine room.*

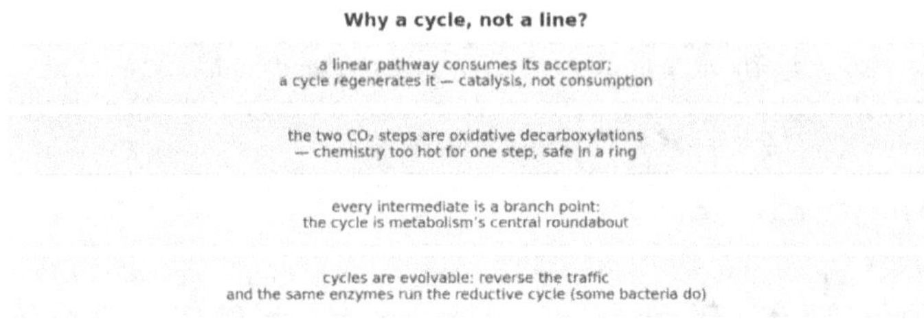

Figure 34.6 — Why a cycle, not a line? *Four answers: acceptor regeneration, safe oxidative chemistry, branch-point connectivity, and evolutionary reversibility — the reductive cycle of autotrophs runs the same ring backwards.*

> ***Case study — The fluoroacetate shepherd.*** *A farmer's dog collapses after scavenging bait laced with sodium fluoroacetate ("1080"): vomiting, seizures, metabolic acidosis, a heart in ventricular fibrillation. Explain the lethal mechanism, the term "lethal synthesis", and why giving ethanol has been proposed in related poisonings.*
> ***Resolution.*** *Fluoroacetate is activated to fluoroacetyl-CoA and condensed by citrate synthase into fluorocitrate — the cell itself commits the "lethal synthesis". Fluorocitrate inhibits aconitase; citrate accumulates, the cycle stalls, ATP falls, and the heart and brain fail first, as organs that cannot wait always do. Ethanol (as in methanol poisoning) works by competition — occupying alcohol dehydrogenase to prevent formation of a toxic metabolite; the analogous principle here would be to outcompete fluoroacetate for activation with a flood of acetate. The case is aconitase's small print, written in a dead dog: isomerases have poisons too.*

> ***Historical note — Szent-Györgyi's pigeons and Krebs's pigeon-breast.*** *The cycle was found twice, in the way truths are. Albert Szent-Györgyi, in Szeged, showed in 1935 that minced pigeon-breast muscle respires far faster when given catalytic traces of succinate, fumarate, malate or oxaloacetate — four-carbon acids acting catalytically, something no fuel should do — and proposed they formed a shuttle for hydrogen. It was Krebs, in Sheffield in 1937, who closed the ring: working with the same obliging muscle (pigeon breast respires like no other tissue, and its mitochondria stay coupled for hours), he showed citrate too was catalytic, that citrate was formed from oxaloacetate plus a two-carbon unit, and that the sequence could only close into a cycle. The paper went to Enzymologia after Nature's rejection; the Nobel came in 1953. Szent-Györgyi had his own, in 1937, for vitamin C — so the two men who between them explained how carbon is burned each received Sweden's thanks in the same year, for connected but different fires. Martonosi's later remark is the fitting epitaph: Szent-Györgyi found the wheel was turning; Krebs found that it was a wheel.*

Plate 34.1 *Albert Szent-Györgyi (1893–1986), whose catalytic-carrier experiments in pigeon muscle revealed the C_4-dicarboxylic acid shuttle from which the cycle grew. Nobel laureate in Physiology or Medicine, 1937. (Nobel Foundation portrait.)*

34.12 Exercises

Exercise 34.1. *Show that if malate dehydrogenase's $\Delta G°' \approx +30$ kJ mol^{-1}, the cellular [OAA]/[malate] ratio can be held near 10^{-4} and the reaction still run forward, given $[NAD^+]/[NADH] \approx 8$ and rapid OAA removal.*

Resolution. *$\Delta G = \Delta G°' + RT\ \ln([OAA][NADH]/[malate][NAD^+]) = +30 + 8.314{\times}10^{-3}{\times}310 \times \ln(10^{-4}/8) = +30 + 2.58 \times (-11.3) \approx +30 - 29 \approx +1$ kJ mol^{-1} — near equilibrium, pushed forward by citrate synthase's continuous drain on OAA. The cycle's most unfavourable step runs on a ratio and a vacuum cleaner.*

Exercise 34.2. *Account for the ten ATP per acetyl-CoA of Figure 34.4, and compute the ATP yield of one glucose from PDH onward.*

Resolution. *Per turn: 3 NADH (7.5) + $FADH_2$ (1.5) + GTP (1.0) = 10. Per glucose: 2 × (10 + 2.5 PDH) = 25; plus glycolysis 2 ATP + 2 NADH (2.5–5 ATP depending on shuttle) gives 30–32, recovering Chapter 28's ledger from the cycle upward.*

Exercise 34.3. *A perfused heart is given [2-^{14}C]acetate. Predict the time-course of label in CO_2 versus in glutamate, and explain the difference.*

Resolution. *Glutamate (from α-KG) labels within the first turn — the acetyl carbons enter the intermediate pool immediately. CO_2 labels slowly: the first-turn CO_2 comes from the OAA skeleton, and acetyl carbon reaches CO_2 only after succinate/fumarate symmetry scrambles it, with a half-life ≈ one turn. A fast pool and a slow pool from one label — the revolving door of Section 34.4 made measurable.*

Exercise 34.4 (proposition for research). *Some tumours harbour loss-of-function mutations in succinate dehydrogenase or fumarase, accumulating succinate or fumarate, which inhibit α-KG-dependent dioxygenases (including DNA and histone demethylases and prolyl hydroxylases). Propose a mechanism-based therapy exploiting, not reversing, the accumulation — a "synthetic lethality" — and design the decisive experiment.*
Sketch. *The elevated succinate/fumarate already impairs demethylases; screen for a second α-KG-dependent enzyme whose inhibition is lethal only against that background (e.g. a specific DNA-repair dioxygenase), using isogenic $SDH^{+/+}$/$SDH^{-/-}$ pairs. The decisive experiment: rescue with cell-permeable dimethyl-α-KG should abolish the synthetic lethality if, and only if, it is dioxygenase-mediated — a single metabolite add-back distinguishing mechanism from mirage.*

34.13 Synthesis

The wheel turns: one gate (PDH), two decarboxylations, one GTP, three NADH, one $FADH_2$, and a ring of intermediates that is catalyst, fuel stock and market stall all at once — refilled by anaplerosis, drained by cataplerosis, reversed entire in some bacteria, and bypassed by two enzymes in every oily seed. Direction checked against the master plan: the carbohydrate economy is complete from meal to CO_2. The other great fuel now claims its due. ***Chapter 35 — Fatty-Acid Oxidation and Ketone Bodies*** *follows the long hydrocarbon chains into the same furnace — and discovers what the liver does when the furnace cannot keep up with the queue.*

34.14 References

Berg, J. M., Tymoczko, J. L., Gatto, G. J., & Stryer, L. (2019). Biochemistry (9th ed.). W. H. Freeman.

Evans, C. T., & Karp, G. (2009). The glyoxylate cycle. In Encyclopedia of Life Sciences. Wiley.

Kornberg, H. L. (2000). Krebs and his trinity of cycles. Nature Reviews Molecular Cell Biology, 1, 225–228.

Krebs, H. A. (1940). The citric acid cycle: A reply to the criticisms of F. L. Breusch and of J. Thomas. Biochemical Journal, 34, 460–463.

Nelson, D. L., & Cox, M. M. (2021). Lehninger Principles of Biochemistry (8th ed.). W. H. Freeman.

Owen, O. E., Kalhan, S. C., & Hanson, R. W. (2002). The key role of anaplerosis and cataplerosis for citric acid cycle function. Journal of Biological Chemistry, 277, 30409–30412.

Patel, M. S., & Korotchkina, L. G. (2006). Regulation of the pyruvate dehydrogenase complex. Biochemical Society Transactions, 34, 217–222.

Szent-Györgyi, A. (1937). Studies on Biological Oxidation and Some of Its Catalysts. Kényényfakadó.

Wächtershäuser, G. (1990). Evolution of the first metabolic cycles. Proceedings of the National Academy of Sciences, 87, 200–204.

Chapter 35
Fatty-Acid Oxidation and Ketone Bodies

CONTENTS OF THIS CHAPTER

"A man carries enough fat to walk a thousand miles and enough glycogen to run twenty. The body is an optimist about journeys and a pessimist about breakfasts."

35.1 Mobilisation of Stored Triacylglycerol

*The body's great fuel reserve is not the granule of Chapter 33 but the droplet: some 135 000 kJ of triacylglycerol in the adipose tissue of a lean adult, against perhaps 3 000 kJ of glycogen — a strategic reserve against a tactical one. But fat is stored where it is not burned, and the first problem of lipid catabolism is logistics: getting the fatty acids out of the adipocyte and through the blood. The exit is governed by three lipases acting in sequence — **ATGL** (adipose triglyceride lipase) removing the first ester, **hormone-sensitive lipase** the second, **monoacylglycerol lipase** the third — under a single hormonal switch with the familiar signature: adrenaline and glucagon, through PKA, phosphorylate perilipin (releasing ATGL's co-activator from the droplet surface) and HSL itself, while insulin dephosphorylates and seals (Figure 35.1). The same kinase that opens the glycogen bank opens the fat vault; one alarm, two treasuries. The freed fatty acids travel on albumin — seven or more molecules per protein, a ferry service of impressive capacity — while the glycerol backbone goes to the liver for gluconeogenesis, the one part of fat that is unambiguously sugar-worthy.*

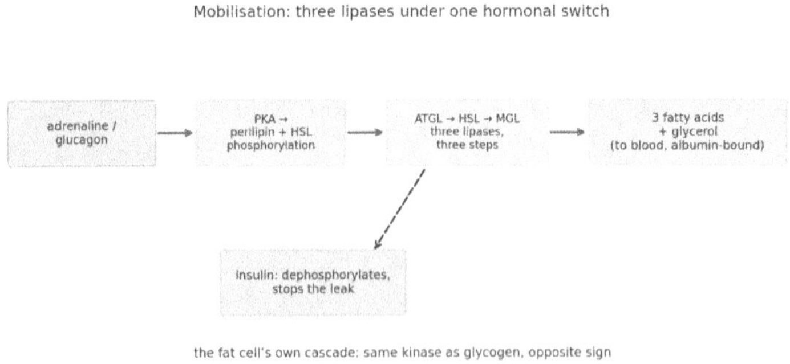

Figure 35.1 — Mobilisation. *Adrenaline/glucagon → PKA → perilipin and HSL phosphorylation → the three-lipase sequence (ATGL, HSL, MGL) → fatty acids exported on albumin. Insulin reverses the switch. The glycogen cascade's twin, with the opposite sign.*

35.2 Fatty-Acid Activation

In the cytosol of the consuming cell, each fatty acid is activated by ***acyl-CoA synthetase*** *to its thioester, at the price of ATP → AMP + PP_i — two high-energy bonds, since the AMP must be re-phosphorylated twice. The reaction is pulled to completion by pyrophosphate hydrolysis, the same trick as in glycogen synthesis (Chapter 33). The accounting of Section 35.5 will charge two ATP-equivalents for the activation; pedants who charge one are reminded that biochemistry keeps books in phosphoanhydride bonds, not in nucleotides.*

35.3 Carnitine Transport

The acyl-CoA is made in the cytosol but burned in the mitochondrial matrix, and the inner membrane admits no CoA. The solution is the ***carnitine shuttle*** *(Figure 35.2): CPT1 on the outer membrane transfers the acyl group to carnitine; a translocase ferries acyl-carnitine across in antiport with free carnitine; CPT2 in the matrix restores the acyl-CoA. The shuttle's regulatory genius is its customs officer:* ***malonyl-CoA****, the first committed intermediate of fatty-acid synthesis (Chapter 36), inhibits CPT1 potently. When the cell is building fat, the molecule that announces the building simultaneously bars the raw material from the furnace — synthesis and oxidation cannot run at once, and the partition is enforced at a single active site. Carnitine itself is synthesised from lysine and methionine; dietary deficiency is rare, but the shuttle's clinical importance is not, as Section 35.11 will show.*

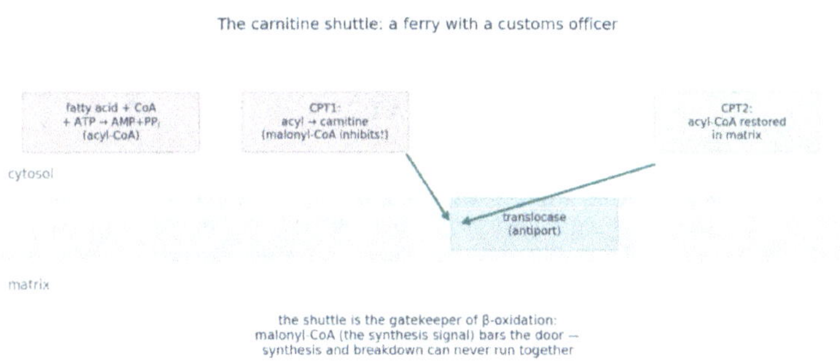

Figure 35.2 — The carnitine shuttle. *Cytosolic acyl-CoA → acyl-carnitine (CPT1, crimson) → antiport across the inner membrane (teal) → matrix acyl-CoA (CPT2, green). Malonyl-CoA inhibits CPT1: the synthesis signal locks the furnace door.*

35.4 β-Oxidation

The spiral itself (Figure 35.3) is four reactions repeated, each turn shortening the chain by two carbons: ① ***acyl-CoA dehydrogenase*** *introduces a trans double bond, transferring the electrons to FAD and thence via ETF (electron-transferring flavoprotein) directly to ubiquinone — an $FADH_2$-equivalent that bypasses Complex I, worth 1.5 ATP; ②* ***enoyl-CoA hydratase*** *adds water; ③* ***hydroxyacyl-CoA dehydrogenase*** *oxidises the β-hydroxyl to a ketone with NADH; ④* ***thiolase*** *cleaves with coenzyme A, releasing acetyl-CoA and leaving an acyl-CoA two carbons poorer. The resemblance to the succinate → OAA sequence of the cycle (oxidation, hydration, oxidation) is not coincidence but homology of strategy: the* *β-carbon must be raised to ketone before it can be cleaved, and chemistry has only so many ways of saying so. For palmitate (C16) the spiral turns seven times: 7 $FADH_2$, 7 NADH, 8 acetyl-CoA.*

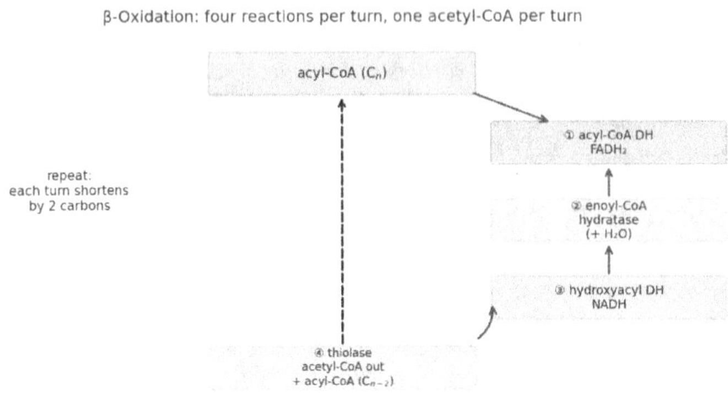

***Figure 35.3** — **The β-oxidation spiral.** Four reactions per turn — FAD-linked oxidation (crimson), hydration (teal), NAD-linked oxidation (green), thiolytic cleavage (amber) — each turn releasing one acetyl-CoA and returning an acyl-CoA shortened by two carbons.*

35.5 Energetic Accounting

The books of palmitate, at Chapter 28's prices, are drawn in Figure 35.4:

$$\underbrace{7 \times 1.5}_{\text{FADH}_2} + \underbrace{7 \times 2.5}_{\text{NADH}} + \underbrace{8 \times 10}_{\text{acetyl-CoA}} - \underbrace{2}_{\text{activation}} = \mathbf{106\ ATP} \tag{35.1}$$

— per C16 molecule, against glucose's 30–32 per C6. Normalised per carbon, fat yields ≈ 6.6 ATP against carbohydrate's ≈ 5.3; but the true advantage is storage density. Fat is anhydrous (9 kcal g^{-1}) where glycogen is wet (≈ 1–2 kcal g^{-1} with its water of hydration), so the body stores perhaps six times more energy per kilogram as fat — the arithmetic behind the epigraph, and the reason migratory birds are adipose connoisseurs. The β-oxidation pathway's own receipts (28 ATP of the 106) are the minority; the bulk arrives, as always, through the cycle and the chain. Fat is not a different fuel, just a more concentrated subscription to the same furnace.

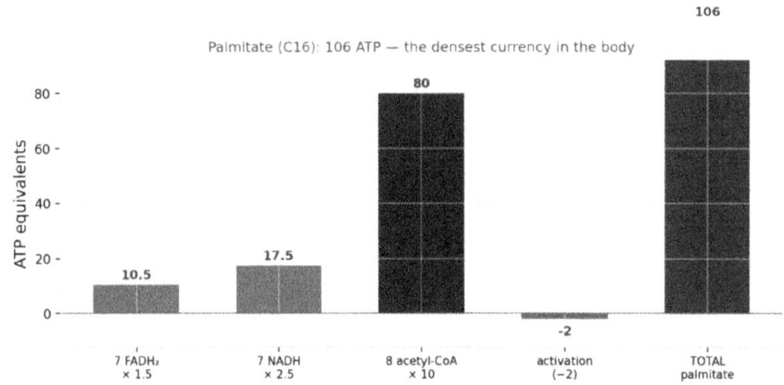

Figure 35.4 — Palmitate's 106 ATP. *The full accounting of Eq. (35.1): the spiral's own 28, the acetyl-CoA harvest's 80, minus the 2-bond activation fee.*

35.6 Oxidation of Unsaturated Fatty Acids

Nature's fats are mostly unsaturated, and their cis double bonds — placed at positions the spiral does not expect — require two auxiliary enzymes (Figure 35.5, left). When the spiral's progress delivers a cis-Δ^3 double bond, ***enoyl-CoA isomerase*** *repositions it to trans-Δ^2, the hydratase's substrate, and the turn proceeds — at the cost of that turn's $FADH_2$, since the double bond arrived pre-installed. Polyunsaturated chains additionally need the NADPH-dependent* ***2,4-dienoyl-CoA reductase*** *to resolve conjugated intermediates. The energetic penalty is modest (about 1.5 ATP per double bond) and the lesson is chemical: unsaturation is oxidation paid in advance, and the pathway deducts it honestly.*

35.7 Odd-Chain Fatty Acids

Odd-chain fatty acids — rare in the diet but abundant in ruminant and plant metabolism — leave propionyl-CoA (3C) at the spiral's last turn. Propionyl-CoA is converted by ***propionyl-CoA carboxylase*** *(biotin again, Chapter 32's crane) to methylmalonyl-CoA, and then by* ***methylmalonyl-CoA mutase*** *— one of only two vitamin B_{12}-dependent enzymes in human metabolism — to succinyl-CoA (Figure 35.5, right). The product is a cycle intermediate and therefore* ***glucogenic****: the three carbons of propionate are the only fat carbon that can end up in glucose, a footnote to the glyoxylate lesson of Chapter 34, and the reason the fasting body's tiny glucose contribution from fat comes from the glycerol backbone and odd chains alone. The mutase's clinical shadow is long: B_{12} deficiency blocks it, methylmalonic acid spills, and the haematology ward meets the biochemistry textbook over a bottle of hydroxocobalamin.*

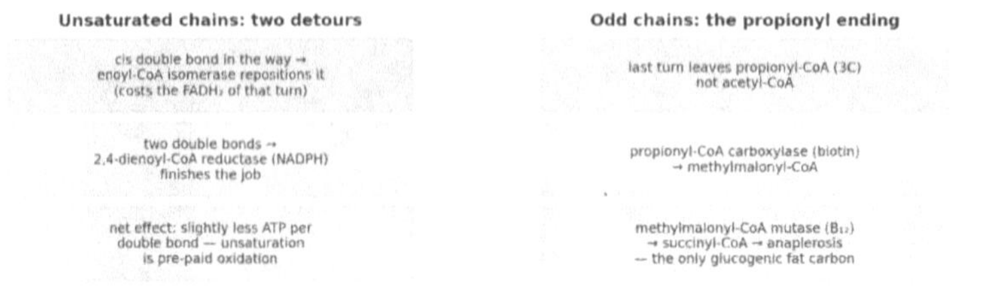

***Figure 35.5* — The two detours.** *Left: unsaturated chains — isomerase and reductase detours, each double bond pre-paying its oxidation. Right: odd chains — propionyl-CoA via biotin and B_{12} to succinyl-CoA, fat's only glucogenic carbon.*

35.8 Peroxisomal Oxidation

*Very-long-chain fatty acids (C22 and beyond) are shortened first in the **peroxisome**, whose β-oxidation uses the same four-reaction logic with different vendors: the first dehydrogenation passes electrons directly to O_2, making H_2O_2 (disposed by catalase) rather than ATP — a deliberately lossy step, suited to a preparatory organelle. The peroxisome also performs **α-oxidation** of branched chains such as phytanic acid (from chlorophyll's phytol tail in dairy and ruminant fat), which cannot be β-oxidised because of the methyl on the β-carbon. The disorders are diagnostic: Zellweger syndrome (no functional peroxisomes) is catastrophic and infantile; Refsum disease (α-oxidation defective) is phytanic-acid poisoning in slow motion, managed by a diet low in ruminant fat — one of metabolic medicine's few occasions when the treatment is simply to stop eating the substrate.*

35.9 Ketone-Body Synthesis

*When β-oxidation races and the cycle stalls — because OAA has been drained into gluconeogenesis, as in fasting or insulin deficiency (Chapter 34's cataplerosis) — acetyl-CoA queues at citrate synthase and overflows into the **ketone bodies** (Figure 35.6): two acetyl-CoA condense to acetoacetyl-CoA, a third is added by mitochondrial **HMG-CoA synthase** (the rate-setting step and the fasting state's true gate), and HMG-CoA lyase releases **acetoacetate**, which equilibrates with **β-hydroxybutyrate** at a ratio set by the mitochondrial $NADH/NAD^+$ — making β-HB/AcAc a clinical window into the redox state itself — and decarboxylates slowly to **acetone**, the sweet smell of the ketoacidotic breath. The liver produces ketones avidly yet cannot use one molecule of them, lacking the activation enzyme SCOT (Section 35.10): ketogenesis is an export industry, and the customer is everyone else.*

35.10 Ketone-Body Utilisation

*In extrahepatic tissues the shipment is cashed: β-hydroxybutyrate is re-oxidised to acetoacetate, and **SCOT** (succinyl-CoA:acetoacetate CoA transferase) — borrowing CoA from succinyl-CoA, at the price of one GTP foregone — hands acetoacetate its thioester for thiolytic cleavage to two acetyl-CoA. Brain adapts over days of fasting to derive up to two-thirds of its energy from β-hydroxybutyrate, sparing perhaps 80 g of protein per day that would otherwise have been gluconeogenic* *substrate: ketones are, quite literally, the body's way of feeding the brain with fat while keeping its muscle. The same abundance becomes pathology in* ***diabetic ketoacidosis****, where absent insulin means uninhibited lipolysis, unthrottled ketogenesis, and a blood pH that falls as the buffers are spent — the physiology of Sections 35.1–35.9 run without a brake, and a reminder that every elegant adaptation is one control failure from catastrophe.*

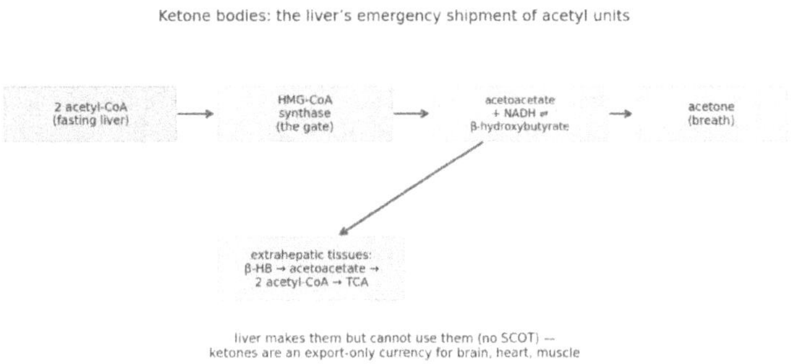

Figure 35.6 — Ketone bodies. *The liver condenses surplus acetyl-CoA via HMG-CoA synthase (plum, the gate) into acetoacetate, β-hydroxybutyrate and acetone; extrahepatic tissues reconvert them to acetyl-CoA via SCOT. The liver, lacking SCOT, is an export-only producer.*

35.11 Disorders of Fatty-Acid Oxidation

The pathway's diseases announce themselves in the fasting hours, when fat is the only fuel: ***MCAD deficiency*** *(medium-chain acyl-CoA dehydrogenase) — the commonest, ~1 in 10 000 in Northern European populations — presents in toddlers as hypoketotic hypoglycaemia after a long sleep or an illness: the furnace cannot burn fat, cannot make ketones, and drains the glucose; the treatment is simply never to fast.* ***CPT1 deficiency*** *mimics it hepatically;* ***CPT2 and carnitine translocase defects*** *present as exercise-induced myoglobinuria in adolescents;* ***LCAD/VLCAD and LCHAD defects*** *add cardiomyopathy, since the heart burns fat continuously and forgives its absence never. The pattern of Section 33.6 repeats: each enzyme is a key, each phenotype a lock identified — and the neonatal screening heel-prick now finds most of these before the first fast does.*

> ***Case study — The child who slept too long.*** *A 14-month-old, previously well, is found unrousable at 7 a.m. after a night of vomiting and no supper. Glucose 1.8 mM; urine ketones absent; liver enzymes raised; dicarboxylic acids in urine. Diagnose the pathway level, explain the absent ketones, and give the single most important management rule.*
> ***Resolution.*** *MCAD deficiency. The absent ketones are the signature: β-oxidation fails at the medium-chain dehydrogenase, so acetyl-CoA never accumulates and ketogenesis has no substrate — the liver starves in the midst of plenty, and hypoglycaemia follows because gluconeogenesis, unfuelled by fat, stalls. Dicarboxylic acids accumulate from ω-oxidation, the overflow route. Management: avoid fasting — frequent feeds, and an emergency glucose regimen for any illness; most diagnosed children then never have a second episode. The case is the carnitine-shuttle/spiral logic read backwards by a toddler's night: no fat, no ketones, no glucose, no mercy.*

Historical note — Lynen's yeast juice and the two machineries. *Feodor Lynen spent the war years in Munich working, by necessity, on yeast — and from yeast extracts purified the machinery of fatty-acid degradation almost enzyme by enzyme, establishing the spiral logic of β-oxidation and identifying "active acetate" as acetyl-CoA in the process. Then, in the 1950s, he did the rarer thing: he solved the reverse problem, showing that fatty-acid synthesis is not oxidation backwards but a separate machinery with its own cofactor (the malonyl unit) and its own swinging arm — and thereby established the principle, which this book has now met four times, that opposing pathways in metabolism are privately owned. The 1964 Nobel Prize, shared with Konrad Bloch, honoured the cholesterol work; but students of the spiral remember Lynen for the yeast. He was asked late in life why he had worked on fat. "It was there," he said, "and it would not dissolve in water." The history of lipid biochemistry is the history of chemists learning to handle things that would not dissolve in water.*

Plate 35.1 *Feodor Lynen (1911–1979), who mapped both the oxidative spiral and the synthetic machinery of fatty acids and established acetyl-CoA as the two-carbon currency. Nobel laureate in Physiology or Medicine, 1964. (Nobel Foundation portrait.)*

35.12 Exercises

Exercise 35.1. *Verify Eq. (35.1) for stearate (C18) and compare the ATP per carbon with palmitate.*
Resolution. *C18: 8 turns → 8 $FADH_2$ (12) + 8 NADH (20) + 9 acetyl-CoA (90) − 2 = 120 ATP; per carbon 120/18 = 6.67, identical to palmitate's 106/16 = 6.63.*

The yield scales with chain length; the chemistry per methylene is constant — the spiral does not care how long the chain is, only how many times it turns.

Exercise 35.2. *Why does malonyl-CoA inhibit CPT1 rather than any enzyme of the spiral itself?*
Resolution. *Because the decision point is import, not oxidation. Once acyl-CoA is in the matrix, oxidation is thermodynamically downhill and will proceed; the cell must therefore regulate the gate, not the engine. Malonyl-CoA, made only when synthesis is committed, is the perfect sentinel: its presence certifies the fed state, and inhibiting CPT1 enforces the partition with a single ligand at a single site — reciprocal control by geography, Chapter 30's logic in one molecule.*

Exercise 35.3. *In diabetic ketoacidosis, blood β-hydroxybutyrate:acetoacetate may rise from the normal ~1:1 to 10:1. Explain what this ratio reports and why it complicates bedside ketone testing.*
Resolution. *The pair equilibrates via β-hydroxybutyrate dehydrogenase at a ratio set by mitochondrial NADH/NAD^+; in DKA, high NADH (from alcohol excess, hypoxia, or the redox shift of massive β-oxidation) pushes the equilibrium toward β-HB. Standard nitroprusside sticks detect only acetoacetate — so the sickest patients, with the highest total ketones, may show the weakest stick reaction, which paradoxically "improves" as treatment restores redox balance. The ratio is a redox meter; the stick reads only one of its two needles.*

Exercise 35.4 (proposition for research). *Ketone bodies are now suspected of signalling roles beyond fuel: β-hydroxybutyrate inhibits class I histone deacetylases and activates the receptor HCAR2. Propose an experiment to determine whether the cognitive benefits reported for ketogenic diets in epilepsy models are metabolic (fuel), epigenetic (HDAC inhibition), or receptor-mediated, using mice lacking HCAR2 in forebrain neurons.*
Sketch. *Four arms: control diet, ketogenic diet, ketogenic diet in forebrain-specific HCAR2 knockout, and β-HB ester (which raises β-HB without diet) ± knockout. Readouts: seizure threshold, neuronal acetylome (HDAC target), and ATP/PCr by ^{31}P-MRS. The decisive pattern: if the benefit survives the knockout and tracks acetylation, it is epigenetic; if it vanishes in the knockout despite identical blood β-HB, it is receptor-mediated; if it tracks ATP/PCr regardless of either, it is fuel. Three hypotheses, one crossing design.*

35.13 Synthesis

The long chains are burned: mobilised under one switch, activated at two bonds, ferried past a malonyl-CoA customs officer, and spiralled two carbons at a time into the common furnace — with detours for the double bonds, a B_{12} coda for the odd chains, a lossy peroxisomal antechamber for the very long, and an export-only ketone currency for the times when the cycle cannot keep up with the queue. Direction checked against the master plan: having burned fat, we must now build it. ***Chapter 36 — Lipid Biosynthesis*** *takes up Lynen's second machinery: malonyl-CoA, the great synthase, and the cholesterol road that cost him and Bloch a trip to Stockholm.*

35.14 References

Bartlett, K., & Eaton, S. (2004). Mitochondrial β-oxidation. European Journal of Biochemistry, 271, 462–469.

Berg, J. M., Tymoczko, J. L., Gatto, G. J., & Stryer, L. (2019). Biochemistry (9th ed.). W. H. Freeman.

Cahill, G. F. (2006). Fuel metabolism in starvation. Annual Review of Nutrition, 26, 1–22.

Eaton, S., Bartlett, K., & Pourfarzam, M. (1996). Mammalian mitochondrial β-oxidation. Biochemical Journal, 320, 345–357.

Lynen, F. (1965). Der Weg von der „aktivierten Essigsäure" zu den Terpenen und Fettsäuren (Nobel lecture). Angewandte Chemie, 77, 929–944.

McGarry, J. D., & Foster, D. W. (1980). Regulation of hepatic fatty acid oxidation and ketone body production. Annual Review of Biochemistry, 49, 395–420.

Nelson, D. L., & Cox, M. M. (2021). Lehninger Principles of Biochemistry (8th ed.). W. H. Freeman.

Newman, J. C., & Verdin, E. (2014). β-Hydroxybutyrate: Much more than a metabolite. Diabetes Research and Clinical Practice, 106, 173–181.

Rinaldo, P., Matern, D., & Bennett, M. J. (2002). Fatty acid oxidation disorders. Annual Review of Physiology, 64, 477–502.

Chapter 36
Lipid Biosynthesis

Contents of this chapter

"The synthase runs its assembly line in one direction only, and the statin molecule is the wrench thrown in the most profitable gear."

36.1 Acetyl-CoA Transport

Chapter 35 burned the fat; this chapter builds it, and the first problem is the mirror image of the last chapter's customs post. Fatty-acid synthesis is cytosolic, but its two-carbon bricks are made in the matrix as acetyl-CoA — and acetyl-CoA, like the fatty acids of Chapter 35, cannot cross the inner membrane. The solution is the ***citrate shuttle*** *(Figure 36.1): when the cycle is glutted, citrate is exported to the cytosol, where* ***ATP-citrate lyase*** *cleaves it back to acetyl-CoA and oxaloacetate at the cost of one ATP — a toll for the export. The OAA is returned as malate and then pyruvate, and the return leg has a bonus:* ***malic enzyme*** *(malate → pyruvate + CO_2 + NADPH) supplies roughly half the reducing power synthesis needs. One shuttle, two currencies delivered — carbon for the chain, NADPH for the reductions.*

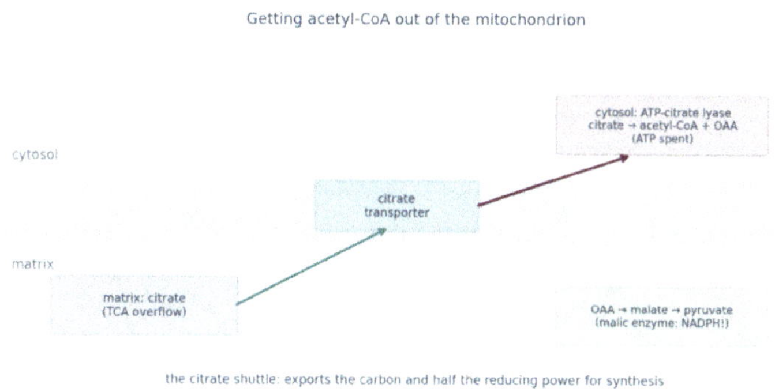

Figure 36.1 — The citrate shuttle. *Matrix citrate (the cycle's overflow) is exported and cleaved by ATP-citrate lyase (crimson) to cytosolic acetyl-CoA; the return leg through malic enzyme (green) yields NADPH. Carbon out, reducing power half-delivered.*

36.2 Malonyl-CoA Formation

The committed step of lipid synthesis is the carboxylation of acetyl-CoA to ***malonyl-CoA*** *by* ***acetyl-CoA carboxylase (ACC)*** *— biotin again, the crane of Chapter 32, swinging CO_2 onto the methyl carbon. Malonyl-CoA is synthesis's signature molecule, and Chapter 35 already showed its second office: it inhibits CPT1 and bars fat from the furnace while it is being made. ACC is therefore the* *pathway's true governor, and it is governed in kind: activated allosterically by citrate (the feedstock's own signal), inhibited by palmitoyl-CoA (the product), inactivated by AMPK phosphorylation (the energy report), and induced transcriptionally by insulin. The kinase that phosphorylates ACC, AMPK, is the same one that activates fat oxidation — reciprocal control once more, written in a single phosphorylation site.*

36.3 Fatty-Acid Synthase

*The **fatty-acid synthase** (FAS) is a homodimeric assembly line of seven activities on two polypeptides, and it runs the four-reaction cycle of Figure 36.2: ① **condensation** of the acetyl primer with malonyl-CoA — the CO_2 bought by ACC leaving immediately to drive the reaction (decarboxylation powers the C–C bond, a trick as old as PEPCK); ② **reduction** of the β-keto group with NADPH; ③ **dehydration**; ④ **reduction** of the double bond with a second NADPH. The growing chain is carried on the phosphopantetheine arm of the **acyl carrier protein** — CoA's pantetheine again, Lipmann's cofactor doing porter duty — and after seven cycles (six malonyl additions to the one acetyl primer) the C16 chain is hydrolysed off as palmitate. The full stoichiometry:*

$$\text{acetyl-CoA} + 7\,\text{malonyl-CoA} + 14\,\text{NADPH} + 14\,\text{H}^{+} \longrightarrow \text{palmitate} + 7\,\text{CC} \quad (36.1)$$

— counting the ACC step, 7 ATP are spent making the malonyl units, so palmitate costs 7 ATP + 14 NADPH; the NADPH is supplied half by malic enzyme (Section 36.1) and half by the pentose phosphate pathway (Chapter 32), which is why lipogenic tissues are PPP-rich. Note the mirror symmetry with Chapter 35: synthesis reduces a β-keto intermediate with NADPH where oxidation oxidises a β-hydroxyl with NAD^{+}; the two pathways run the same chemistry in opposite directions with different vendors, different cofactors, different compartments — Lynen's principle that opposing pathways are privately owned, here in its purest form.

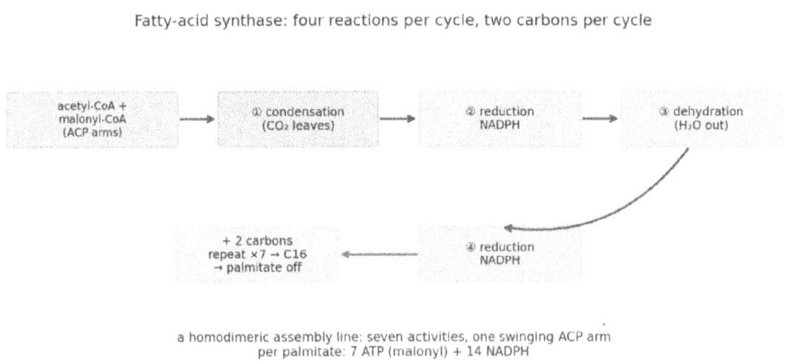

***Figure 36.2 — The FAS cycle.** Condensation with decarboxylative drive (crimson), two NADPH reductions (green), one dehydration (teal); the chain grows on the ACP arm until C16 is released (amber). Per palmitate: 7 ATP and 14 NADPH (Eq. 36.1).*

36.4 Chain Elongation and Desaturation

*Palmitate is the starting material, not the catalogue. Elongases in the endoplasmic reticulum extend chains to C18–C24 using malonyl-CoA units in a FAS-like cycle; **desaturases** — membrane enzymes using NADH, cytochrome b_5 and O_2 — install* *cis double bonds, but with a telling limitation: mammalian desaturases cannot introduce double bonds beyond Δ9. Linoleate (18:2, Δ9,12) and α-linolenate (18:3, Δ9,12,15) are therefore* ***essential fatty acids****, vitamins in everything but name, whose derivatives include the ω-6 and ω-3 families that section 36.11 will turn into signals. The inability is one of the quiet facts behind the entire dietary-fat literature: we can desaturate what we eat but cannot originate the patterns, so the patterns must be eaten.*

36.5 Triacylglycerol Synthesis

Fatty acids are stored and shipped as triacylglycerol, assembled on the glycerol-3-phosphate backbone by successive acylations: glycerol-3-P → lysophosphatidate → phosphatidate → diacylglycerol → triacylglycerol, with phosphatidate phosphatase as the insulin-sensitive gate. The backbone itself reports the fed state: in adipose tissue, which lacks glycerol kinase, glycerol-3-P can come only from glucose via glycolysis — so fat storage is glucose-dependent in the most literal enzymatic sense, and insulin, by supplying both the glucose and the phosphatase activity, promotes storage twice over. The biochemistry agrees with the tailor: sugar makes fat.

36.6 Phospholipid Biosynthesis

Membrane lipids diverge from storage lipids at diacylglycerol and phosphatidate, and the divergence is organised by the activation chemistry of Chapter 21: head groups are installed as ***CDP-alcohols*** *(CDP-choline, CDP-ethanolamine — Kennedy pathway) or the lipid itself is activated as* ***CDP-diacylglycerol*** *(the route to phosphatidylinositol, phosphatidylglycerol and, in bacteria, everything else). Phosphatidylserine is made by head-group exchange; cardiolipin, the mitochondrion's signature lipid, by coupling two phosphatidyl units. The economy is CDP-economy: where glycogen and oligosaccharide synthesis spend UTP and sugar synthesis spends ATP, phospholipid synthesis spends CTP — each polymer class with its own activation currency, so that the cell can throttle one without disturbing the others.*

36.7 Sphingolipid Biosynthesis

Sphingolipids *are built on serine + palmitoyl-CoA condensed to ceramide, onto which head groups are grafted: phosphocholine for sphingomyelin, single sugars for cerebrosides, oligosaccharide trees (with sialic acid caps) for the gangliosides of nerve membranes. Their degradation, in the lysosome, is a cascade of specific hydrolases — and their diseases are the lysosomal storage disorders of Chapter 23's acquaintance: Tay–Sachs (hexosaminidase A), Gaucher (glucocerebrosidase), Niemann–Pick (sphingomyelinase), Fabry (α-galactosidase A). Once again the catalogue reads as the pathway taught by its keys: each missing hydrolase fills the lysosome with exactly the lipid it failed to cleave, and the phenotype maps the enzyme as faithfully as any assay.*

36.8 Cholesterol Synthesis

The sterol road (Figure 36.3) runs in four stages. Three acetyl-CoA condense to HMG-CoA — the same intermediate as ketogenesis, but cytosolic, with its own synthase; the body runs two HMG-CoA pools, one for export fuel, one for construction, in different compartments, exactly as Chapter 30's geography requires. ***HMG-CoA reductase*** *then performs the committed reduction to mevalonate with two NADPH — the rate-limiting, most regulated, most profitable* *step in the whole of lipid metabolism, to which Section 36.9 and half the pharmaceutical industry are devoted. Mevalonate is phosphorylated and decarboxylated (3 ATP) to the five-carbon* ***isopentenyl pyrophosphate****; six* C_5 *units polymerise head-to-tail to* ***squalene*** *(C30), which epoxidises and cyclises — one of the most spectacular single-enzyme reactions in biochemistry, one* O_2 *and one hydride shift folding a linear chain into the four-ring lanosterol — and nineteen further steps trim it to cholesterol (C27). Bloch mapped nearly every atom of this road with acetate isotopes in the 1940s–50s; the Nobel of 1964, shared with Lynen, was the field's acknowledgement that the sterol had been understood atom by atom before most proteins had been sequenced at all.*

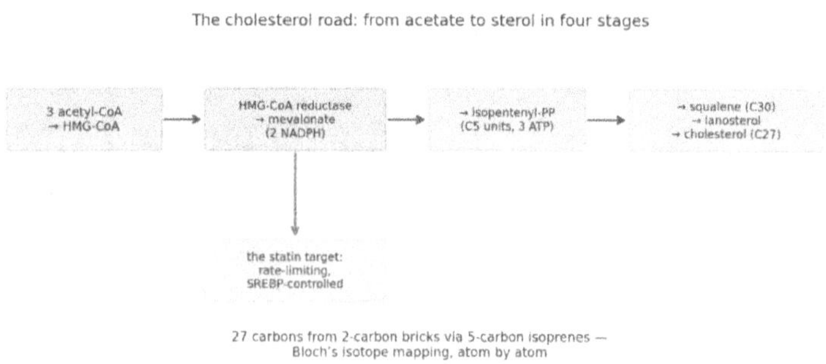

Figure 36.3 — The cholesterol road. *Acetyl-CoA → HMG-CoA → mevalonate (the reductase, crimson — committed and statin-targeted) →* C_5 *isoprenes → squalene (C30) → lanosterol → cholesterol (C27). Bloch's isotope mapping traced the carbons atom by atom.*

36.9 Regulation of Sterol Metabolism

HMG-CoA reductase is regulated at more levels than any enzyme in this book: competitive inhibition (statins), phosphorylation (AMPK again), degradation (sterols accelerate the enzyme's own disposal), and — most beautifully — transcription, through the ***SREBP*** *system (Figure 36.4). SREBP is a transcription factor synthesised as a membrane protein in the ER, held there by Insig when sterols are abundant; when sterols fall, the escort protein SCAP walks SREBP to the Golgi, where two proteases clip off its DNA-binding fragment, which flies to the nucleus and switches on the LDL receptor and the reductase itself. Cholesterol thus regulates its own synthesis by controlling the escort and the scissors — a membrane-bound transcription factor released by the very molecule it governs. Brown and Goldstein discovered the system while explaining familial hypercholesterolaemia (their 1985 Nobel): the LDL receptor, its endocytosis, and this feedback loop are one connected physiology, and its failure is the commonest Mendelian contribution to heart disease.*

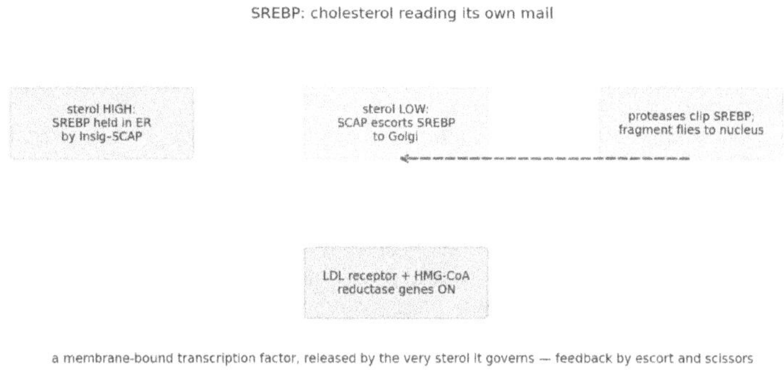

Figure 36.4 — SREBP: cholesterol reading its own mail. *Sterol high: SREBP held in the ER by Insig–SCAP. Sterol low: SCAP escorts SREBP to the Golgi; proteases release the nuclear fragment; LDL-receptor and reductase genes switch on.*

36.10 Lipoprotein Transport

Lipids are shipped in protein wrappers whose density rises as the fat is unloaded (Figure 36.5): ***chylomicrons*** *carry dietary fat from gut to tissues;* ***VLDL*** *carries the liver's own synthesis;* ***lipoprotein lipase*** *at capillary beds strips both of their triacylglycerol, shrinking them through IDL to* ***LDL*** *— the cholesterol-delivery particle, cleared by the LDL receptor, and "bad" only in the sense that its cargo, delivered in excess, ends up in artery walls;* ***HDL*** *runs the return service, collecting cholesterol for the liver ("reverse transport"), its apolipoproteins activating the lecithin–cholesterol acyltransferase that esterifies the cargo for the journey. The epidemiology of the last seventy years — LDL up, risk up; HDL up, risk down (though raising HDL by drugs has stubbornly failed to help, a lesson in the difference between a marker and a mechanism) — is the physiology of this conveyor read at population scale.*

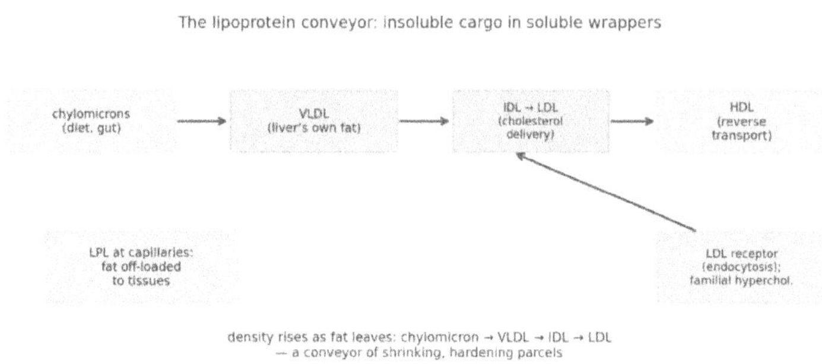

Figure 36.5 — The lipoprotein conveyor. *Chylomicrons and VLDL deliver fat; lipoprotein lipase unloads at capillaries; density rises through IDL to LDL (receptor-cleared, crimson–plum); HDL ferries cholesterol home. The conveyor's failures are the epidemiology of atherosclerosis.*

36.11 Eicosanoid Synthesis

From membrane phospholipid, phospholipase A_2 releases arachidonate (20:4, ω-6), the substrate of the ***eicosanoids*** *— twenty-carbon local hormones of three factories:* ***cyclooxygenases*** *(prostaglandins, thromboxanes),* ***lipoxygenases*** *(leukotrienes), and epoxygenases. The pharmacology is household knowledge: aspirin acetylates COX irreversibly (the platelet, being anucleate, cannot replace the enzyme and lives out its ten days disarmed — hence the low daily dose); the NSAIDs inhibit reversibly; the COX-2 selective agents sought the anti-inflammatory without the gastric cost, and found the cardiovascular one instead. Vane's 1982 Nobel for this pathway is one of pharmacology's cleanest stories: a molecule known since Hippocrates' willow bark explained, at last, by its enzyme.*

36.12 Lipid-Lowering Pharmacology

The chapter closes with its own industrial consequence (Figure 36.6). ***Statins*** *— compactin from Penicillium, then lovastatin and its successors — inhibit HMG-CoA reductase with nanomolar affinity; the liver, starved of mevalonate, up-regulates its LDL receptors through SREBP and pulls LDL from the blood: the most prescribed drug class in history works by the feedback loop of Section 36.9, not despite it.* ***Ezetimibe*** *blocks intestinal absorption at NPC1L1;* ***PCSK9 antibodies*** *save LDL receptors from degradation;* ***fibrates*** *activate PPARα;* ***bempedoic acid*** *inhibits ATP-citrate lyase — Section 36.1's own enzyme, druggable at last. Every arrow in the figure is a section of this chapter; the pharmacopoeia is the pathway, annotated.*

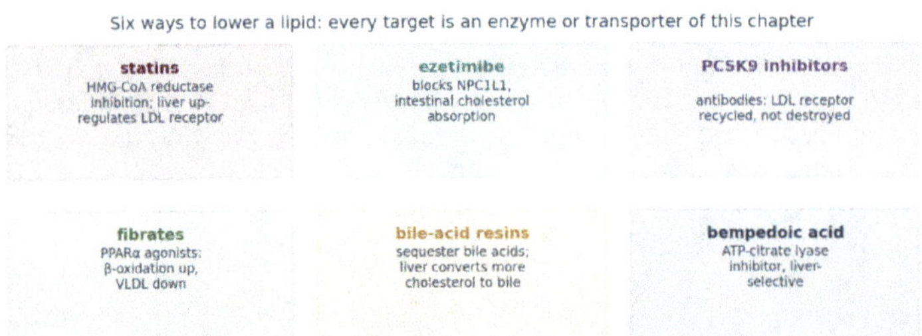

Figure 36.6 — Six ways to lower a lipid. *Statins (reductase), ezetimibe (absorption), PCSK9 antibodies (receptor recycling), fibrates (PPARα), resins (bile acids), bempedoic acid (ATP-citrate lyase). Every target is an enzyme or transporter of this chapter.*

> ***Case study — The tendon xanthomas.*** *A 38-year-old man has an LDL cholesterol of 7.8 mM, yellowish nodules over his Achilles tendons, and a father who died of myocardial infarction at 41. He is started on a high-potency statin; LDL falls by half. Explain the genetics, the drug's mechanism at the level of this chapter, and predict the effect of adding ezetimibe.*
> ***Resolution.*** *Heterozygous familial hypercholesterolaemia — one defective LDL-receptor allele, halving hepatic LDL clearance from birth; the xanthomas are deposited cholesterol, the family history the inevitable arithmetic. The statin inhibits the reductase; intrahepatic sterol falls; SREBP is escorted and clipped; LDL-receptor transcription rises; the one good allele is driven harder,*

and clearance roughly doubles. Ezetimibe adds a further ~20% fall by cutting the intestinal input — the two drugs pulling the same conveyor at its two ends. The case is Section 36.9 wearing a surname.

Historical note — Bloch's acetate and the isotope method. *When Konrad Bloch began, in the late 1930s, the structure of cholesterol was known but its origin was a mystery of the first rank: what could a cell possibly make a four-ringed sterol from? Bloch fed rats acetate labelled in its methyl or carboxyl carbon — first with deuterium, then with the new* ^{14}C *— and degraded the resulting cholesterol position by position, proving that every one of its 27 carbons derived from acetate, and in which pattern; the isoprene units, squalene, and the cyclisation all fell out of the labelling logic before a single enzyme of the pathway had been purified. It was the first great triumph of isotopic biochemistry, and its deeper legacy is methodological: the demonstration that a metabolic pathway can be deduced from labelling patterns alone. Every flux experiment in Chapter 30's spirit — and the tumour-metabolism experiments of Chapter 31 — descends from Bloch's rats. He shared the 1964 Nobel with Lynen, the man of the synthase; the prize committee, for once, paired the burner and the builder in a single award.*

Plate 36.1 *Konrad Bloch (1912–2000), who traced every carbon of cholesterol back to acetate with isotopes, founding the isotope method in metabolism. Nobel laureate in Physiology or Medicine, 1964. (Nobel Foundation portrait.)*

36.13 Exercises

Exercise 36.1. *Verify the NADPH requirement of Eq. (36.1) and account for its*

sources in a lipogenic hepatocyte.
Resolution. *Each of the 7 cycles uses 2 NADPH (β-keto reduction and enoyl reduction) = 14 NADPH per palmitate. Sources: malic enzyme supplies ≈ 7 (one per citrate exported); the pentose phosphate pathway supplies the rest, at 2 NADPH per G6P oxidised — hence ~3.5 G6P through the oxidative arm per palmitate. Lipogenesis is thus as much a pentose-pathway activity as a synthase activity; the two pathways are one physiological system.*

Exercise 36.2. *Explain, using Section 36.9, why statins lower plasma LDL rather than merely slowing cholesterol synthesis.*
Resolution. *Inhibiting the reductase lowers intrahepatic sterol; Insig releases SCAP; SREBP is escorted and clipped; the nuclear fragment activates the LDL-receptor gene; receptor density rises and the liver clears circulating LDL faster. The drug's plasma effect is thus a gene-regulation effect, one step downstream of the enzyme it inhibits — the feedback loop is the drug, the reductase merely its handle.*

Exercise 36.3. *A patient on a low-fat, high-carbohydrate diet develops raised triacylglycerol. Trace the route from dietary glucose to VLDL fat, naming the two committed enzymes.*
Resolution. *Glucose → pyruvate → citrate export (ATP-citrate lyase) → acetyl-CoA; ACC commits it to malonyl-CoA; FAS builds palmitate; esterification onto glycerol-3-P (from the same glucose, adipose having no glycerol kinase) yields triacylglycerol, exported as VLDL. The two committed enzymes: acetyl-CoA carboxylase and the lipoprotein-lipase-unopposed esterification gate (phosphatidate phosphatase). The liver converts carbohydrate surplus to exported fat — de novo lipogenesis — and the triglyceride rise is the diet's signature, not its contradiction.*

Exercise 36.4 (proposition for research). *Many cancers reactivate de novo lipogenesis and depend on FAS; yet FAS inhibitors have struggled clinically, partly from feedback compensation. Using the control-analysis logic of Chapter 30, propose a combination strategy and the experiment that would identify the best partner for a FAS inhibitor.*
Sketch. *Measure flux control coefficients in tumour cells before and after FAS inhibition: the expectation is that control redistributes toward ACC, the pentose pathway (NADPH supply), or the desaturases. Screen pairwise inhibitors across an isogenic panel, scoring synergy by Bliss independence; the partner whose control coefficient rises most upon FAS blockade should be the most synergistic. The experiment is Chapter 30's summation theorem weaponised: when one valve closes, the flux report tells you which valve the cell now depends on.*

36.14 Synthesis

The lipids are built: carbon exported as citrate, committed as malonyl-CoA, assembled on the ACP arm two carbons a cycle, elongated and desaturated within limits, esterified into storage and membrane, grafted into sphingolipids, folded into sterols by the most spectacular cyclisation in the book, governed by a membrane-bound transcription factor with scissors, shipped in density-sorted wrappers, and — at nearly every step — drugged by an industry that lives off this chapter's enzymes. Direction checked against the master plan: the carbon economy is complete. The nitrogen economy remains, and with it the oldest problem of terrestrial life — what to do with the ammonia. ***Chapter 37 — Amino-Acid Catabolism and Nitrogen Disposal*** *opens with the amino group's long road to the urea cycle, Krebs's first cycle and, historically, the proof that metabolism runs in rings at all.*

36.15 References

Berg, J. M., Tymoczko, J. L., Gatto, G. J., & Stryer, L. (2019). Biochemistry (9th ed.). W. H. Freeman.

Bloch, K. (1965). The biological synthesis of cholesterol. Science, 150, 19–28.

Brown, M. S., & Goldstein, J. L. (1986). A receptor-mediated pathway for cholesterol homeostasis. Science, 232, 34–47.

Brown, M. S., & Goldstein, J. L. (2009). Cholesterol feedback: From Schoenheimer's bottle to Scap's MELADL. Journal of Lipid Research, 50, S15–S27.

Endo, A. (2010). A historical perspective on the discovery of statins. Proceedings of the Japan Academy, Series B, 86, 484–493.

Maier, T., Leibundgut, M., & Ban, N. (2008). The crystal structure of a mammalian fatty acid synthase. Science, 321, 1315–1322.

Nelson, D. L., & Cox, M. M. (2021). Lehninger Principles of Biochemistry (8th ed.). W. H. Freeman.

Vance, D. E., & Vance, J. E. (2008). Biochemistry of Lipids, Lipoproteins and Membranes (5th ed.). Elsevier.

Wakil, S. J., & Abu-Elheiga, L. A. (2009). Fatty acid metabolism: Target for metabolic syndrome. Journal of Lipid Research, 50, S138–S143.

Chapter 37
Amino-Acid Catabolism and Nitrogen Disposal

CONTENTS OF THIS CHAPTER

"Carbon may be burned at leisure, but nitrogen is a toxic inheritance — and the urea cycle is the family lawyer that packages it for export."

37.1 Amino-Acid Pools

Every amino acid the body owns lives in one of three states: as protein (the overwhelming majority), as a small free pool (~35 g in an adult), or in transit between the two (Figure 37.1). The throughput is startling. An adult eats ~100 g of protein a day but turns over ~300 g of body protein — muscle, gut, immune cells — so that the free pool is fed three times as much from within as from without. The pool is the sum of all our amino-acid incomes, and from it flow two kinds of expenditure: biosynthesis (the subject of Chapter 38) and catabolism, the subject here. Catabolism splits into two unequal halves. The carbon skeletons are fuel, and follow the roads of Chapters 31–36 into glucose, fat, ketones or the cycle. The amino groups are waste, and follow a road of their own — because ammonia, at the concentrations metabolism would otherwise produce, is a neurotoxin, and the body that tolerates a wide glucose range tolerates almost no free ammonia at all.

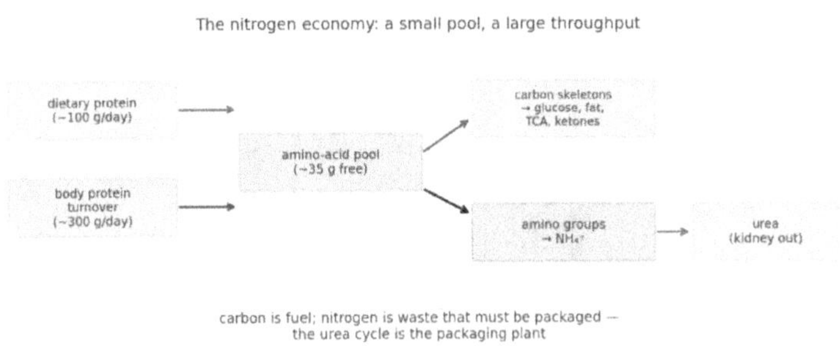

Figure 37.1 — The nitrogen economy. *A small free pool, fed by diet and by a threefold larger internal turnover, divides into carbon skeletons (fuel) and amino groups (waste). The urea cycle is the packaging plant for the latter.*

37.2 Proteolysis and Nitrogen Balance

Protein is degraded by two systems with different mandates. The ***lysosome*** *handles bulk and long-lived proteins — the autophagic route, inducible in starvation and constitutive for membrane and organelle turnover. The*

proteasome handles the selective and the urgent: proteins marked for destruction by **ubiquitin**, a 76-residue tag whose polymeric chains, assembled by a three-enzyme cascade, consign their bearer to the barrel. Ciechanover, Hershko and Rose won the 2004 Nobel for ubiquitin; the proteasome's inhibitor bortezomib became a myeloma drug. Nitrogen balance — intake minus excretion — is the ledger of this turnover: positive in growth and pregnancy, negative in starvation, trauma and cancer cachexia. The clinician reads the patient's nitrogen balance the way an accountant reads a cash-flow statement, and the numbers are not comforting: a fasting adult excretes 10–15 g of nitrogen a day, nearly all as urea.

37.3 Transamination

The first step in catabolising most amino acids is not to remove the amino group but to move it (Figure 37.2). ***Aminotransferases*** *(transaminases) pass the α-amino group to α-ketoglutarate, yielding the α-keto acid of the donor and glutamate — the amino group changing hands, never freed. The chemistry rides on* ***pyridoxal phosphate*** *(vitamin B_6), whose aldehyde forms a Schiff base with the substrate, holds the amino group as a ketimine, and hands it on to the keto acid: a courier with a pocket. The clinically famous pair are ALT (alanine aminotransferase; amino group to pyruvate) and AST (aspartate aminotransferase; to oxaloacetate). Both are intracellular enzymes, so their presence in blood is pathology: a rising ALT is the liver's enzymes leaking into plasma from dying hepatocytes, and the transaminase panel is, in effect, a census of cellular death. The amino groups, meanwhile, have all been funnelled onto two carriers — glutamate and, via it, glutamine — which are the only forms in which the body allows ammonia to travel in bulk.*

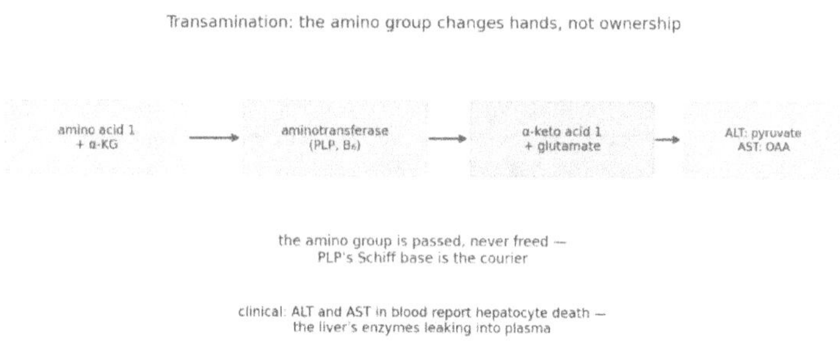

Figure 37.2 — Transamination. *The amino group is passed from amino acid to α-ketoglutarate via PLP's Schiff base, yielding the α-keto acid and glutamate. ALT and AST are the clinical windows into hepatocyte death.*

37.4 Oxidative Deamination

Glutamate, having collected the amino groups, releases them — but only where it is safe. ***Glutamate dehydrogenase****, a mitochondrial enzyme of liver and kidney, oxidatively deaminates glutamate to α-KG + NH_4^+, using NAD^+ or $NADP^+$ — and it is the only major enzyme that can use either cofactor, a flexibility that lets it serve catabolism (NAD) and biosynthesis (NADPH) alike. The reaction is poised near equilibrium and driven by mass action: high glutamate and high energy (GTP inhibits) pull toward deamination, high α-KG and ammonia push the other way. The free ammonia released here is the first genuinely dangerous molecule of the* *chapter, and it is released inside the mitochondrion, within reach of the urea cycle's first enzyme — geography again as toxicology.*

37.5 Ammonia Transport

Free ammonia in blood is measured in micromolar; above ~50 μM the brain begins to fail, and above ~200 μM it fails catastrophically. Yet the tissues produce ammonia continuously. The solution is to never let it travel free (Figure 37.4). Two vehicles carry it. ***Glutamine*** *— glutamate + NH_4^+, by glutamine synthetase, at the cost of one ATP — is the non-toxic carrier par excellence: brain and muscle export their nitrogen as glutamine, and liver and kidney hydrolyse it back. The kidney's use is especially elegant: in acidosis, renal glutaminase releases ammonia that traps H^+ in the tubular fluid as NH_4^+, excreting acid and nitrogen in the same urine. The second vehicle is alanine, the subject of the next section.*

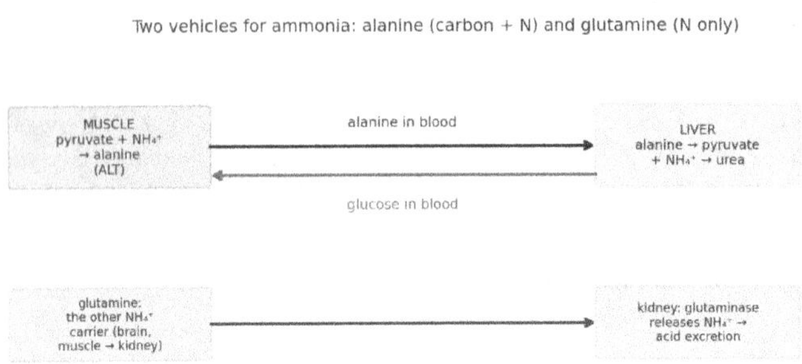

Figure 37.4 — Two vehicles for ammonia. *Muscle exports alanine (carbon + N, repaid as glucose by the liver — the glucose–alanine cycle); brain and muscle export glutamine (N only) to liver and kidney, where glutaminase frees it for urea or for acid excretion.*

37.6 The Glucose–Alanine Cycle

Muscle, burning branched-chain amino acids during exercise and fasting, transaminates their nitrogen onto pyruvate, making alanine; the liver transaminates it back, sends the carbon out as glucose, and packages the nitrogen as urea (Figure 37.4, upper loop). The ***glucose–alanine cycle*** *is the nitrogenous twin of the Cori cycle — same geometry, same 6-ATP hepatic toll, different cargo. Where the Cori cycle moves carbon and energy, this one moves nitrogen disguised as carbon, and it is the reason a fasting man's muscle shrinks while his blood glucose holds: the muscle is paying its nitrogen debt in amino groups, and the liver is repaying the carbon as glucose. Cahill's classical starvation studies measured the traffic; the brain's glucose, in a prolonged fast, is substantially recycled muscle protein.*

37.7 The Urea Cycle

The destination of it all is the ***urea cycle*** *— Krebs's first cycle, and the first metabolic cycle ever described (Figure 37.3). It spans two compartments, and the span is the toxicology of Section 37.4 made architecture. In the* ***mitochondrion****: NH_4^+ + HCO_3^- + 2 ATP → carbamoyl phosphate (CPS1 — the committed step, activated allosterically by N-acetylglutamate, whose synthesis rises with arginine* *and glutamate, the nitrogen surplus's own signal); carbamoyl phosphate + ornithine → citrulline (ornithine transcarbamylase). Citrulline is exported to the* ***cytosol****, where aspartate donates the second nitrogen → argininosuccinate → arginine + fumarate (which rejoins the cycle of Chapter 34, linking the two Krebs cycles by a single intermediate) → and arginase cleaves arginine to* ***urea*** *+ ornithine, which returns to the mitochondrion. The stoichiometry:*

$$NH_4^+ + HCO_3^- + \text{aspartate} + 3\,\text{ATP} \longrightarrow \text{urea} + \text{fumarate} + 2\,\text{ADP} + \text{AMP} + \quad (37.1)$$

— 3 ATP consumed but 4 high-energy bonds (one ATP → AMP), so the price of packaging one urea is four bonds. Against the 30-odd ATP the carbon skeletons of the same protein will earn, the toll is modest; against the alternative — free ammonia in the brain — it is the best bargain in metabolism. The cycle is regulated long-term by protein intake (the enzymes are induced within days of a high-protein diet) and short-term by N-acetylglutamate, and its diseases — OTC deficiency above all — are neonatal emergencies in which the packaging plant fails and the ammonia rises within hours of the first feed.

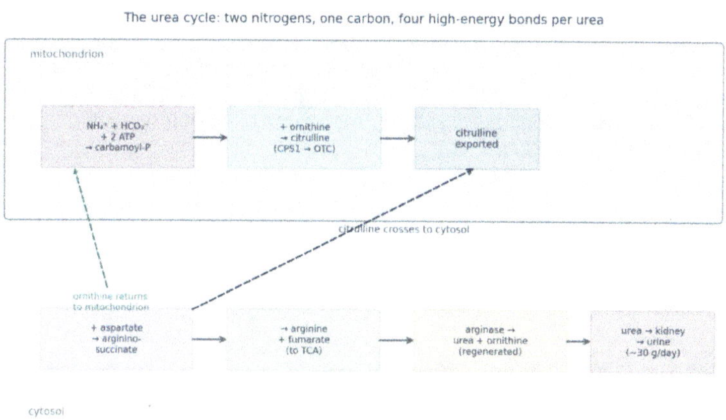

Figure 37.3 — The urea cycle. *Mitochondrial phase (crimson, teal): carbamoyl phosphate and citrulline. Cytosolic phase (plum, green, amber): aspartate in, fumarate out, arginine cleaved to urea + regenerated ornithine. Two nitrogens (NH_4^+ + Asp), one carbon (HCO_3^-), four high-energy bonds per urea (Eq. 37.1).*

37.8 Integration of Nitrogen and Carbon Metabolism

The fumarate of Eq. (37.1) is not waste; it is a cycle intermediate, and its conversion to malate and OAA is the reason the two Krebs cycles are really one system with two outputs — the "Krebs bicycle", as Kornberg nicely called it. Aspartate, the second nitrogen donor, is itself made from OAA by transamination, so that the urea cycle's second nitrogen arrives already riding a cycle carbon. The integration is complete: amino-acid catabolism feeds the citric acid cycle at every level, the citric acid cycle feeds the urea cycle its aspartate, and the urea cycle returns fumarate to the citric acid cycle. Nitrogen and carbon are one economy with two accounting departments.

37.9 Glucogenic and Ketogenic Amino Acids

The carbon skeletons divide by fate (Figure 37.5, left): most amino acids are purely ***glucogenic*** *(their skeletons reach pyruvate or cycle intermediates and hence* *glucose); leucine and lysine are purely* ***ketogenic*** *(to acetyl-CoA or acetoacetate only); isoleucine, phenylalanine, tyrosine and tryptophan are both. The classification is the fasting body's strategy in a table: the glucogenic majority feeds the brain's glucose, the ketogenic minority feeds the fat furnace, and the glyoxylate lesson of Chapter 34 ensures that the ketogenic share can never become sugar. The mnemonic "Phe, Tyr, Ile, Trp are both" is less important than the reason: these are the amino acids whose skeletons fork at a cycle intermediate on one branch and acetyl-CoA on the other.*

37.10 Branched-Chain Amino Acids

Leucine, isoleucine and valine are catabolised by a common machinery, and they begin in an unusual place: ***muscle****, not liver, because the liver lacks the first transaminase — the muscle is the body's designated BCAA furnace, which is why the glucose–alanine cycle originates there. The shared second step is the* ***branched-chain a-keto acid dehydrogenase*** *(BCKDH) complex — a third PDH-family factory, TPP and lipoamide and all, minus B_{12} (Figure 37.5, right). Its deficiency is* ***maple syrup urine disease****: the branched ketoacids accumulate, the urine smells of burnt sugar (sotolon, the same molecule as in fenugreek), and untreated infants suffer irreversible neurological injury within weeks. Newborn screening and a diet restricted in the three amino acids convert a catastrophe into a managed condition — one more entry in the growing list of diseases treated by not eating the substrate.*

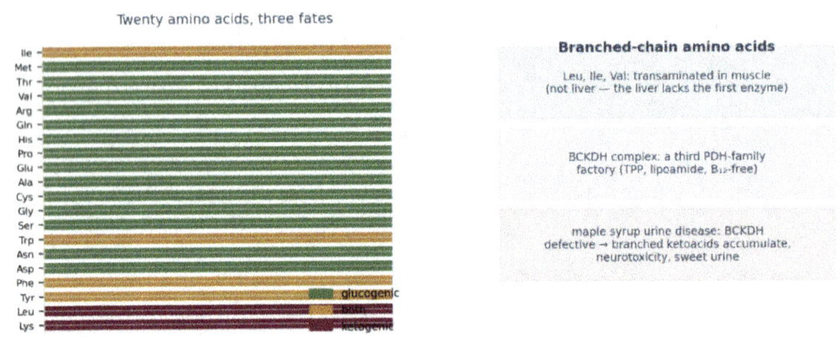

Figure 37.5 — Twenty amino acids, three fates; and the BCAA furnace. *Left: the glucogenic/ketogenic classification. Right: the branched-chain pathway, its muscle-first geography, the BCKDH complex, and maple syrup urine disease.*

37.11 Inborn Errors of Amino-Acid Metabolism

The amino-acid pathways, read by their failures (Figure 37.6), gave medicine its very concept of the inborn error. ***PKU*** *(phenylalanine hydroxylase) — the phenylalanine that cannot become tyrosine accumulates, and the developing brain pays; a low-phenylalanine diet from birth is curative in effect if not in name, and the Guthrie heel-prick that detects it is the ancestor of all newborn screening.* ***Alcaptonuria*** *(homogentisate dioxygenase) — Garrod's original "inborn error" of 1902, benign except for black urine, ochronotic cartilage and arthropathy.* ***Homocystinuria*** *(cystathionine β-synthase) — Marfanoid habitus, lens dislocation, and a thrombotic tendency that makes the diagnosis urgent.* ***Urea-cycle defects****, especially X-linked OTC deficiency — neonatal hyperammonaemic coma, the packaging plant failing at its first shift. Garrod, working with urinary colour and family trees a generation before enzymes were concepts, named these* *"chemical individuality"; the whole of medical genetics, and a good part of this book's recurring argument that disease is biochemistry read backwards, descends from that phrase.*

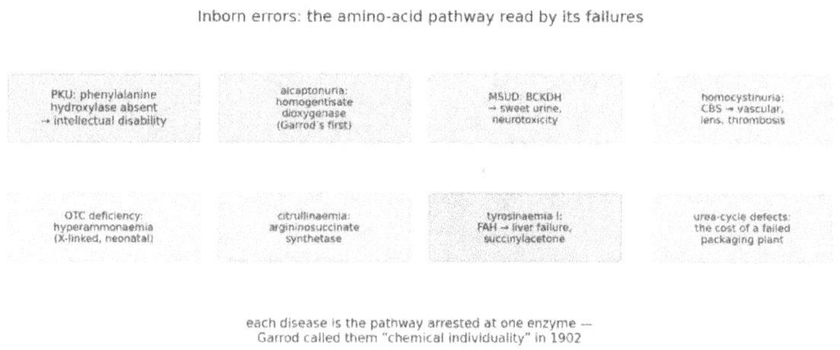

Figure 37.6 — Inborn errors of amino-acid metabolism. *The pathway read by its failures: each disease is the pathway arrested at one enzyme, and the phenotype identifies the arrest point as precisely as an assay.*

> ***Case study — The coma after the barbecue.*** *A 22-year-old man collapses into confusion and vomiting two days after a high-protein celebration meal. Blood ammonia 180 μM; orotic aciduria present; his mother recalls a brother who died in infancy of "a protein disease". Diagnose, explain the orotic acid, and give the emergency treatment.*
> ***Resolution.*** *X-linked OTC deficiency, hemizygous, previously subclinical until the protein load. Carbamoyl phosphate, unable to combine with ornithine, spills into the cytosol and floods pyrimidine synthesis — hence orotic acid, the diagnostic signature that separates OTC from CPS1 deficiency. Emergency treatment: stop protein, give intravenous glucose (to suppress catabolism), and the nitrogen-scavenging drugs sodium benzoate and phenylacetate — which conjugate glycine and glutamine respectively, exporting nitrogen by the back door while the front is blocked. The case is the urea cycle's geography taught in one admission: the packaging plant's first enzyme broken, and the body saved by improvised packaging.*

> ***Historical note — Krebs's first cycle.*** *Before the citric acid cycle there was the urea cycle, and it made the very idea of a metabolic cycle thinkable. In 1932 Krebs and his medical student Kurt Henseleit, in Freiburg, measured urea formation in liver slices and found it was accelerated, catalytically, by ornithine, citrulline and arginine — amino acids that stimulated without being consumed, the same logical fingerprint Szent-Györgyi would find in pigeon muscle three years later. The ring they proposed — ornithine → citrulline → arginine → ornithine — was the first closed loop in biochemistry, and it taught Krebs the grammar with which he read the second ring five years later. It is pleasant to record that Henseleit, the student on the paper, became a country doctor and never published again; the cycle's co-discoverer spent his life as a practitioner in the Black Forest, having solved, at twenty-four, one of the oldest problems in physiology. The urea cycle is also the reason terrestrial vertebrates are possible: fish excrete ammonia directly into the water they swim in, birds excrete uric acid past their shells, and we — with no river to*

> *swim in and no shell to seal — chose urea, the soluble, non-toxic, energetically priced compromise of a land animal with a liver.*

Plate 37.1 *Hans Adolf Krebs (1900–1981), who with Kurt Henseleit described the urea cycle in 1932 — the first metabolic cycle — and the citric acid cycle in 1937. Nobel laureate in Physiology or Medicine, 1953. (Nobel Foundation portrait.)*

37.12 Exercises

Exercise 37.1. *Verify the 4-high-energy-bond cost of Eq. (37.1) and estimate the daily energetic cost of urea synthesis in an adult excreting 12 g of nitrogen.*
Resolution. *3 ATP hydrolysed, one to AMP (2 bonds) + two to ADP (1 each) = 4 bonds per urea. 12 g N = 0.86 mol N; urea carries 2 N, so 0.43 mol urea; cost = 0.43 × 4 = 1.7 mol ATP-equivalents ≈ 1.7 × 50 = 85 kJ/day — under 1% of basal expenditure. The packaging is cheap; the carbon skeletons' resale value is what pays the body's bills.*

Exercise 37.2. *Why is free ammonia toxic specifically to brain, and why does glutamine synthesis protect?*
Resolution. *NH_3/NH_4^+ crosses the blood–brain barrier and drives glutamine synthetase in astrocytes; the osmotic load of accumulated glutamine swells the astrocytes (cytotoxic oedema), and glutamate depletion plus α-KG drain (ammonia forces GDH backward) starves neurons of both transmitter and cycle intermediate. Glutamine synthesis is protective because it detoxifies ammonia into a diffusible, exportable amide — but in the closed skull the cure itself becomes the swelling, which is why hepatic encephalopathy is an astrocyte disease.*

Exercise 37.3. *Distinguish, by the fumarate link, the contributions of the "Krebs bicycle" to gluconeogenesis from protein.*
Resolution. *Urea-cycle fumarate → malate → OAA → (PEPCK) → glucose; the aspartate that donated the nitrogen came from OAA + glutamate, so the net effect of one turn is* $NH_4^+ + HCO_3^- + 3\ ATP \rightarrow$ *urea + one OAA's worth of carbon toward glucose. Protein catabolism thus yields glucose through two doors: the glucogenic skeletons directly, and the urea cycle's fumarate as a side payment — the bicycle's front wheel drives the back.*

Exercise 37.4 (proposition for research). *Some urea-cycle disorders are now treated with carglumic acid (a synthetic N-acetylglutamate analogue). Propose a programme to develop a second-generation activator of CPS1 that is (a) orally bioavailable, (b) resistant to esterase cleavage, and (c) tissue-targeted to liver, and design the preclinical experiment that would justify a first-in-human trial.*
Sketch. *(a) Prodrug chemistry: esterify the carboxyls with lipophilic, esterase-labile groups for absorption; (b) replace the N-acetyl with a carbamate or sulfonamide bioisostere to resist amidases; (c) exploit the liver's asialoglycoprotein receptor with a galactose-terminated linker for targeting. The decisive preclinical experiment: a Sparfloxacin-free hyperammonaemia model in OTC-deficient (spf-ash) mice, with the endpoint being survival and brain glutamine (by* ^{1}*H-MRS) after an ammonia challenge — the drug must lower brain glutamine, not merely blood ammonia, since the astrocyte is the organ at risk.*

37.13 Synthesis

The nitrogen is packaged: a small pool, a large turnover, amino groups passed hand to hand onto glutamate and glutamine, ferried as alanine and glutamine, and finally assembled with bicarbonate into urea at a cost of four bonds — the price of a land animal's liver. Direction checked against the master plan: catabolism is done; what remains of the small-molecule economy is construction. ***Chapter 38 — Biosynthesis of Amino Acids, Nucleotides and Specialised Molecules*** *takes up the anabolic half — the essential amino acids we cannot make, the purine and pyrimidine rings, and the gout that is their too-famous excretion product.*

37.14 References

Berg, J. M., Tymoczko, J. L., Gatto, G. J., & Stryer, L. (2019). Biochemistry (9th ed.). W. H. Freeman.

Brusilow, S. W., & Horwich, A. L. (2014). Urea cycle enzymes. In C. R. Scriver et al. (Eds.), The Online Metabolic and Molecular Bases of Inherited Disease. McGraw-Hill.

Cahill, G. F. (2006). Fuel metabolism in starvation. Annual Review of Nutrition, 26, 1–22.

Cooper, A. J. L. (2011). The role of glutamine synthetase and glutamate dehydrogenase in cerebral ammonia homeostasis. Neurochemical Research, 36, 2439–2455.

Garrod, A. E. (1902). The incidence of alkaptonuria: A study in chemical individuality. The Lancet, 160, 1616–1620.

Häberle, J., et al. (2012). Suggested guidelines for the diagnosis and management of urea cycle disorders. Orphanet Journal of Rare Diseases, 7, 32.

Krebs, H. A. (1973). The discovery of the ornithine cycle of urea synthesis. Biochemical Education, 1, 19–23.

Nelson, D. L., & Cox, M. M. (2021). Lehninger Principles of Biochemistry (8th ed.). W. H. Freeman.

Watford, M. (2015). The urea cycle: A two-compartment system. Essays in Biochemistry, 59, 71–86.

Chapter 38
Biosynthesis of Amino Acids, Nucleotides and Specialised Molecules

CONTENTS OF THIS CHAPTER

"The cell that cannot make a ring is a cell that must eat it — and the antimetabolite is the ring that poisons the maker."

38.1 Nitrogen Fixation and Assimilation

Chapter 37 disposed of nitrogen; this chapter acquires it. The ultimate source of all biological nitrogen is the inert N_2 of the air, and the only organisms that can break the triple bond are a handful of bacteria and archaea, wielding **nitrogenase** *— an enzyme so oxygen-sensitive that its owners either live anaerobically, wrap it in a leghemoglobin cocoon (the root nodule), or work so fast that the enzyme is replaced hourly. The stoichiometry is fearsome:*

$$N_2 + 8\,H^+ + 8\,e^- + 16\,ATP \longrightarrow 2\,NH_3 + H_2 + 16\,ADP + 16\,P_i \qquad (38.1)$$

— sixteen ATP and eight electrons per N_2, plus a wasted H_2 that the enzyme cannot help making. The Haber–Bosch process does the same job with iron and 500 °C, and between them nitrogenase and Haber–Bosch feed the planet's seven billion mouths: half the nitrogen in your body came through one or the other. Once fixed, ammonia is assimilated by **glutamine synthetase** *(the same enzyme as Chapter 37's detoxifier, now working in reverse) and* **glutamate synthase** *(GOGAT), which together convert α-KG + NH_3 into glutamate and glutamine — the two amino donors for everything that follows.*

38.2 Biosynthetic Families of Amino Acids

Twenty amino acids are made from six carbon skeletons (Figure 38.1), and the list reads as a roll-call of Part VII: α-KG gives the glutamate family (Glu, Gln, Pro, Arg); oxaloacetate the aspartate family (Asp, Asn, Met, Thr, Ile, Lys); pyruvate the alanine family (Ala, Val, Leu); 3-PGA the serine family (Ser, Gly, Cys); PEP + erythrose-4P the aromatics (Phe, Tyr, Trp, via the shikimate pathway — the glyphosate target, and the reason that herbicide is a plant-specific poison); and ribose-5P, alone, gives histidine, whose synthesis is a branch of purine metabolism and whose position as the odd one out is the price of being the only amino acid with a ring that purines also want. The families are the citric-acid cycle's children, and their regulation is family business: the end-product of each branch inhibits the branch's first enzyme — isoleucine inhibits threonine deaminase, valine and leucine inhibit acetohydroxy-acid synthase, and the aromatic trio inhibit their own DAHP synthases — feedback organised by lineage, exactly as a civil service is organised by department.

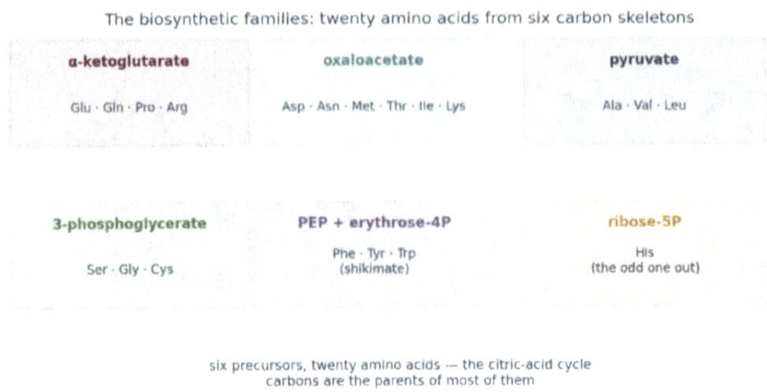

Figure 38.1* — *Six precursors, twenty amino acids. *The biosynthetic families, each regulated by its own end-products. The citric-acid cycle supplies the parents of most of them; histidine is the odd one out.*

38.3 Essential and Non-essential Amino Acids

The nine we cannot make — His, Ile, Leu, Lys, Met, Phe, Thr, Trp, Val — are essential precisely because their syntheses are long, branched, and expensive, and vertebrates have lost the enzymes; the eleven we can make are non-essential only in the dietary sense, for several (Arg, Cys, Tyr, Gln) become conditionally essential in growth, illness, or prematurity. The essential nine are the ones whose synthetic pathways bacteria and plants own and we do not — which is why the shikimate pathway is a herbicide target, why lysine is the limiting amino acid of cereal diets, and why the vegetarian's combining of pulses and grains is not folklore but enzymology: pulses are lysine-rich and methionine-poor, grains the reverse, and the two together complete the set.

38.4 One-Carbon Metabolism

The construction of rings and methyl groups requires one-carbon units in several oxidation states, and they are carried by ***tetrahydrofolate*** *(THF) as methyl (CH_3), methylene (CH_2), methenyl (CH), formyl (CHO) and formimino groups (Figure 38.2). Serine hydroxymethyltransferase (SHMT) is the chief supplier, passing serine's β-carbon to THF; the units then go to thymidylate synthase (dUMP → dTMP — the only methylation in DNA synthesis), to purine synthesis (C2 and C8 of the ring), and to methionine regeneration via the B_{12}-dependent methionine synthase. The system is the biochemical reason folate is a vitamin, why B_{12} deficiency causes a megaloblastic anaemia identical to folate deficiency (the "folate trap": without B_{12}, THF is locked as CH_3-THF and the rest of the one-carbon pool starves), and why the two vitamins must be distinguished before treatment — treating B_{12} deficiency with folate cures the anaemia and lets the neuropathy proceed. The methyl group of methionine is then activated as* ***S-adenosylmethionine*** *(SAM), the cell's methyl-currency, whose 10^6 methylations per second per cell include DNA cytosine, histone lysines, phosphatidylcholine, creatine, and every epigenetic mark this book will meet in Part IX.*

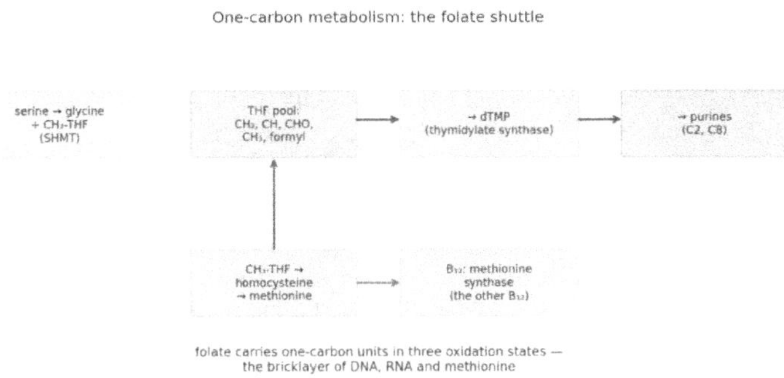

Figure 38.2* — *The folate shuttle. *THF carries one-carbon units in three oxidation states; SHMT supplies them; thymidylate synthase and purine synthesis spend them; B_{12}-dependent methionine synthase returns CH_3-THF to the pool. The folate trap of B_{12} deficiency is shown at the bottom.*

38.5 Porphyrin and Heme Synthesis

The heme ring is built from glycine and succinyl-CoA — the citric-acid cycle again donating its carbon — in a pathway that crosses the mitochondrial membrane twice (Figure 38.4): the first step (ALA synthase, PLP-dependent, feedback-inhibited by heme) and the last (ferrochelatase, inserting Fe^{2+}) are mitochondrial; the middle four are cytosolic. The intermediates are phototoxic, and the diseases — the ***porphyrias*** *— are the pathway read by its failures: acute intermittent porphyria (PBG deaminase) gives the neurovisceral attacks; porphyria cutanea tarda (uroporphyrinogen decarboxylase) gives the blistering photosensitivity. George III's madness has been attributed to variegate porphyria, and whether or not the diagnosis is sound, the chemistry is: a missing enzyme in this line is not a subtle matter.*

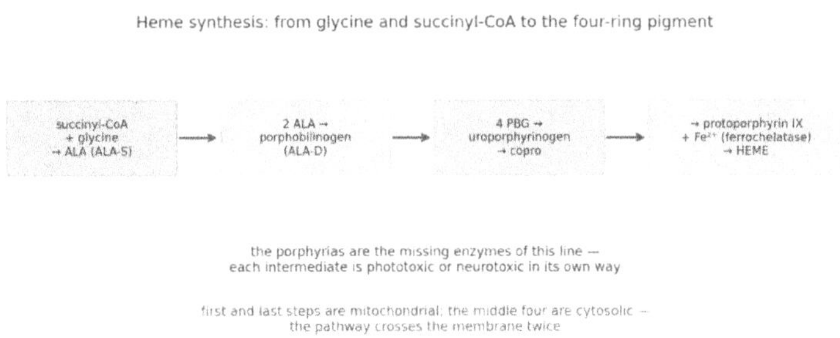

Figure 38.4* — *Heme synthesis. *From glycine + succinyl-CoA through ALA, porphobilinogen, and the porphyrinogens to protoporphyrin IX + Fe^{2+}. The porphyrias are the missing enzymes of this line.*

38.6 Creatine, Glutathione, and the Small Specials

Creatine *— the muscle's ATP buffer — is made from glycine + arginine + SAM's methyl group, excreted as creatinine at a rate proportional to muscle mass (hence* *the clinician's favourite kidney-function denominator).* ***Glutathione*** *— the γ-glutamyl-cysteinyl-glycine tripeptide of Chapter 32's lifeline — is synthesised in two ATP-dependent steps, its γ-glutamyl bond making it resistant to ordinary proteases, and its cysteine thiol doing the redox work. Both are small molecules built from the amino-acid pool, and both remind us that the pool's expenditure is not only protein.*

38.7 Purine Nucleotide Biosynthesis

The purine ring is built on the ribose, atom by atom, from five parents (Figure 38.3, left): glycine (C4, C5, N7), glutamine (N3, N9), aspartate (N1), CO_2 (C6), and formyl-THF (C2, C8) — a roll-call of Chapter 37's amino donors and Chapter 38's one-carbon units. Ten steps yield IMP, the branch point; IMP → AMP costs GTP and aspartate, IMP → GMP costs ATP and glutamine — the two nucleotides paying for each other's synthesis in a reciprocity that keeps their pools balanced. The committed step, PRPP amidotransferase, is feedback-inhibited by AMP, GMP and IMP — the end-products closing the gate, as always.

38.8 Pyrimidine Nucleotide Biosynthesis

The pyrimidine ring is simpler: aspartate and carbamoyl phosphate are its only parents (Figure 38.3, right), and the ring is built first and attached to ribose afterward — the opposite order from purines. Carbamoyl phosphate synthetase II (cytosolic, glutamine-fed, distinct from the urea cycle's CPS1) is the committed enzyme; UMP is the product; UTP → CTP by amination. The mammalian pathway's first three activities ride on a single polypeptide called CAD, and its dihydroorotate dehydrogenase is the target of leflunomide, the rheumatoid-arthritis drug — one more enzyme, one more drug.

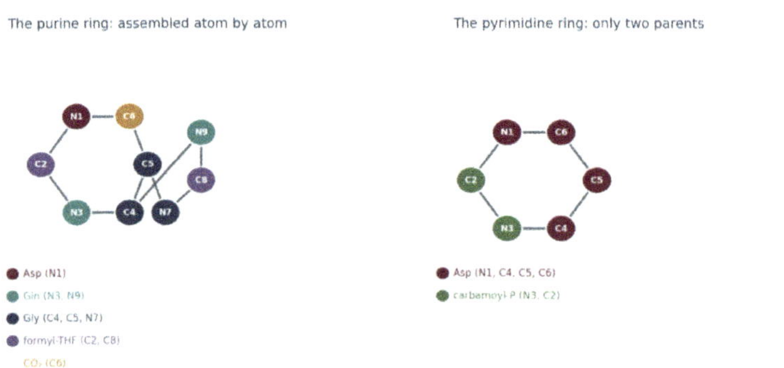

Figure 38.3 — The ring atoms' parents. *Left: the purine ring, five parents. Right: the pyrimidine ring, two parents. The origins are colour-coded; the purine is the more cosmopolitan molecule.*

38.9 Deoxyribonucleotide Formation

DNA's sugars are made from RNA's: ***ribonucleotide reductase*** *reduces the 2′-OH of NDPs to dNDPs, using a thioredoxin/glutaredoxin relay and a tyrosyl radical of its own — one of the few radical enzymes in human metabolism, and the reason hydroxyurea (which quenches the radical) is a chemotherapy drug. dTMP is made from dUMP by thymidylate synthase, at the cost of CH_2-THF oxidised to DHF — the only step in one-carbon metabolism that consumes THF rather than borrowing* *it — and dihydrofolate reductase (DHFR) regenerates the THF. The pair TS + DHFR is the double target of methotrexate and 5-fluorouracil, the two oldest antimetabolites in oncology.*

38.10 Nucleotide Degradation

Purines are degraded to **uric acid** *(Figure 38.5): AMP → IMP → inosine → hypoxanthine → xanthine → uric acid; GMP → guanosine → xanthine → uric acid, with xanthine oxidase performing both final oxidations. Uric acid is the end-product only because primates lost uricase; most mammals oxidise further to allantoin, and the loss — which raises serum urate and with it the risk of gout — is another of those mutational bankruptcies our ancestors could afford, perhaps because urate is an antioxidant. The solubility of monosodium urate is low, and when it crystallises in the first metatarsophalangeal joint the result is podagra, the gout of kings: a disease of purine excretion, not purine intake, which is why the gouty duke's diet is less to blame than his kidneys.*

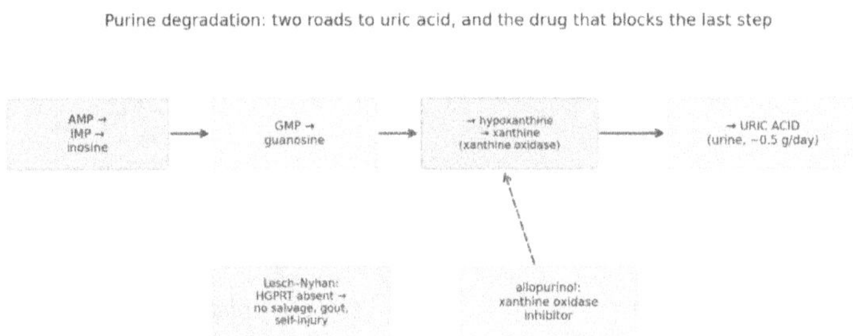

Figure 38.5 — Two roads to uric acid. *AMP and GMP converge on xanthine; xanthine oxidase makes uric acid. Allopurinol blocks the last step; Lesch–Nyhan is the salvage failure.*

38.11 Salvage Pathways

Making a purine costs 6–8 ATP; salvaging one costs one PRPP. The salvage enzymes — HGPRT (hypoxanthine-guanine phosphoribosyltransferase) and APRT — are therefore among the most conserved in metabolism, and their failure is catastrophic: **Lesch–Nyhan syndrome** *(HGPRT absent) gives not only hyperuricaemia and gout but a devastating neurological syndrome of self-injury, the mechanism of which remains one of the most haunting unsolved problems in neurochemistry. The purine nucleoside phosphorylase and adenosine deaminase deficiencies give the severe combined immunodeficiencies — ADA-SCID being the first disease treated by gene therapy (1990), a milestone that began in this pathway.*

38.12 Regulation and Antimetabolite Drugs

The pathways of this chapter are the most densely drugged in medicine (Figure 38.6). George Hitchings and Gertrude Elion, at Burroughs Wellcome, reasoned that a pathogen's or tumour's nucleotide pathways could be poisoned selectively with base and folate analogues — and proceeded to invent 6-mercaptopurine (leukaemia), azathioprine (the first immunosuppressant, enabling transplantation), allopurinol (gout), acyclovir (the first selective antiviral), and trimethoprim (bacterial DHFR), winning the 1988 Nobel for what the committee called "rational drug design". The logic is one sentence: know the pathway, find the difference between host and parasite, and poison the difference. Every targeted therapy since — from methotrexate to the kinase inhibitors of Part X — is a footnote to that sentence.

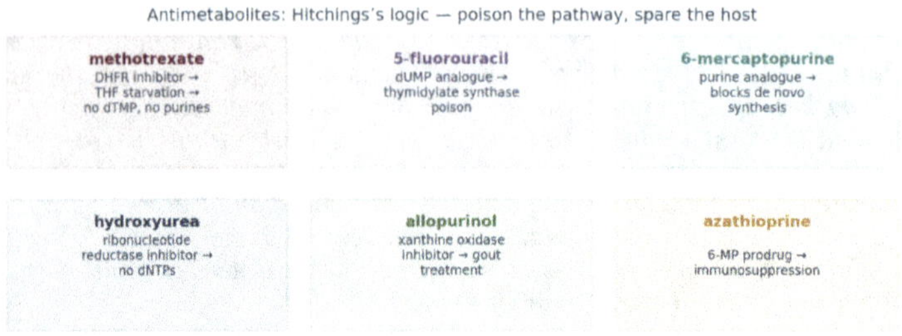

Figure 38.6 — Antimetabolites. *Six drugs, six enzymes of this chapter. Hitchings's logic: poison the pathway, spare the host.*

Case study — The leukaemia and the folate. *A 4-year-old with acute lymphoblastic leukaemia is started on methotrexate. Three days later her white count has fallen by half, but her uric acid and phosphate are rising and her creatinine is climbing. Explain the tumour response, the metabolic emergency, and the two drugs that prevent it.*
Resolution. *Methotrexate inhibits DHFR; the blasts, dependent on de novo dTMP and purine synthesis, die en masse — and their lysis floods the blood with purines (→ uric acid via xanthine oxidase), phosphate, and potassium:* ***tumour lysis syndrome****. The uric acid crystallises in the renal tubules, worsening the renal failure the phosphate also drives. Prevention: aggressive hydration, plus* ***allopurinol*** *(xanthine-oxidase inhibitor, blocking urate formation) or* ***rasburicase*** *(recombinant uricase, oxidising urate to soluble allantoin — the enzyme we primates lost). The case is Chapter 38's degradation pathway read in a paediatric emergency: kill the tumour, and its nucleotides become the next problem.*

Historical note — The antimetabolite principle. *George Hitchings and Gertrude Elion began, in 1944, with a conviction that was then heretical: that drugs could be designed from the biochemistry outward, rather than discovered by screening. They chose nucleic-acid metabolism because the new Avery–MacLeod–McCarty result (DNA as the transforming principle) made it the most interesting chemistry in biology, and they spent forty years at Burroughs Wellcome turning the choice into a pharmacopoeia. Elion, who never took a PhD — the war and the laboratory intervened — co-authored the 1988 Nobel with Hitchings and James Black (of the β-blockers), and the citation's phrase "for their discoveries of important principles for drug treatment" understates it: they invented the idea that a pathway is a target. The reader who has followed this book's argument that every drug is an enzyme's shadow will recognise Hitchings and Elion as the first to say so in public, and the first to be believed.*

Plate 38.1 *George Herbert Hitchings (1905–1998), who with Gertrude Elion invented rational drug design on the pathways of this chapter. Nobel laureate in Physiology or Medicine, 1988. (Nobel Foundation portrait.)*

38.13 Exercises

Exercise 38.1. *Verify the ATP cost of nitrogen fixation (Eq. 38.1) and estimate the daily energetic saving of eating protein versus fixing N_2.*
Resolution. *16 ATP per N_2 = 8 ATP per NH_3. An adult excretes ~12 g N/day; fixing that nitrogen would cost 0.86 × 8 = 6.9 mol ATP ≈ 345 kJ/day — nearly 5% of basal expenditure, before counting the 8 electrons. The Haber–Bosch subsidy and the legume's symbiosis are both, energetically, the same bargain: someone else pays the 16 ATP.*

Exercise 38.2. *Explain the "folate trap" of B_{12} deficiency and predict the effect of folate supplementation on the blood and the cord.*
Resolution. *Without B_{12}, methionine synthase stalls; CH_3-THF cannot return to THF; the one-carbon pool is locked as methyl-THF, and dTMP and purine synthesis starve — megaloblastic anaemia. Folate supplementation bypasses the trap by supplying THF directly: the anaemia resolves, but the B_{12}-deficient neuropathy (myelin's methylation failing, SAM starved) proceeds unchecked. The blood is cured, the cord is not — the oldest warning in vitamin therapy.*

Exercise 38.3. *Account for the ATP cost of one purine (IMP) and one pyrimidine (UMP), and explain why salvage is so strongly selected.*
Resolution. *IMP: PRPP synthesis (1 ATP → PRPP) + 6 ATP-equivalents in the ten steps = ~7 ATP. UMP: ~5 ATP (PRPP + CPSII + the ring). Salvage by HGPRT*

costs one PRPP (1 ATP) per base. A 6–7-fold saving per nucleotide — and the brain, which barely divides, salvages nearly all its purines, which is why HGPRT is X-linked and why its absence is so devastating.

Exercise 38.4 (proposition for research). *Many tumours are auxotrophic for arginine (ASS1 silenced). Propose a therapeutic strategy exploiting this, and design the decisive preclinical experiment, bearing in mind that the urea cycle is the arginine source.*
Sketch. *Pegylated arginine deiminase (ADI-PEG20) depletes circulating arginine; ASS1-negative tumours cannot replace it. The decisive experiment: isogenic ASS1$^{+/+}$ and ASS1$^{-/-}$ xenografts, with the endpoint being not tumour shrinkage alone but the mechanism — ^{13}C-glutamine tracing should show the sensitive tumours failing to make arginine while the resistant ones draw it from the urea cycle. The control-analysis logic of Chapter 30 applies: the drug works only if the tumour's flux control has collapsed onto the one enzyme it silenced.*

38.14 Synthesis

The rings are built and the methyls are placed: nitrogen fixed at 16 ATP the N_2, twenty amino acids from six skeletons, one-carbon units shuttled by folate, heme from glycine and succinyl-CoA, purines atom by atom on ribose, pyrimidines from two parents, deoxyribose by radical, uric acid as the excretion of kings, and salvage as the selected economy — all of it drugged by the antimetabolite principle that Hitchings and Elion invented. Direction checked against the master plan: the small-molecule economy of Part VII is complete. What remains is integration — how the organs of Part VIII coordinate the whole, and how the hormones of Chapter 30's day become the physiology of a life. ***Part VIII — Metabolic Integration and Physiological Biochemistry*** *opens with* ***Chapter 39 — Hormonal Coordination of Metabolism****, where the insulin/glucagon custody of the day becomes the fed-fast cycle of the body.*

38.15 References

Berg, J. M., Tymoczko, J. L., Gatto, G. J., & Stryer, L. (2019). Biochemistry (9th ed.). W. H. Freeman.

Elion, G. B. (1989). The purine path to chemotherapy. Science, 244, 41–47.

Hitchings, G. H. (1989). Selective inhibitors of dihydrofolate reductase. Science, 244, 41–47.

Lane, A. N., & Fan, T. W.-M. (2015). Regulation of mammalian nucleotide metabolism and biosynthesis. Nucleic Acids Research, 43, 2466–2485.

Locasale, J. W. (2013). Serine, glycine and one-carbon units: Cancer metabolism in full circle. Nature Reviews Cancer, 13, 572–583.

Nelson, D. L., & Cox, M. M. (2021). Lehninger Principles of Biochemistry (8th ed.). W. H. Freeman.

Ponka, P. (1999). Cell biology of heme. American Journal of the Medical Sciences, 318, 241–256.

Smil, V. (2004). Enriching the Earth: Fritz Haber, Carl Bosch, and the Transformation of World Food Production. MIT Press.

Stipanuk, M. H. (2004). Sulfur amino acid metabolism: Pathways for production and removal of homocysteine and cysteine. Annual Review of Nutrition, 24, 539–577.

Part VIII

Metabolic Integration and Physiological Biochemistry

Chapter 39
Hormonal Coordination of Metabolism

CONTENTS OF THIS CHAPTER

"The body is not a democracy of organs; it is a monarchy of signals, and insulin is the king that abdicates every night."

39.1 Principles of Endocrine Regulation

Part VIII begins where Part VII ended: with the organism, not the pathway. The pathways of Chapters 31–38 are the same in every cell that owns them; what makes a liver a liver and a muscle a muscle is which pathways are on, when, and how much. The conductor is the endocrine system — a hierarchy of signals (Figure 39.5) that runs from hypothalamus to pituitary to target glands, with each tier slower, broader, and harder to override than the one below. The principles are few and worth stating once. First, hormones are information, not substrate: they arrive in nanomolar amounts, bind receptors, and change enzyme activities and gene expression without themselves being consumed. Second, every endocrine axis is a ***negative-feedback loop****: the target gland's product inhibits the hypothalamus and pituitary, so the system hunts a set-point rather than a maximum. Third, the same hormone can have opposite effects in different tissues, because the receptor is the same but the signalling machinery downstream is not. And fourth — the principle that organises this whole Part — the metabolic hormones are not a committee of equals but a* ***duumvirate****: insulin and glucagon, with the catecholamines as the emergency reserve, govern the fed–fast cycle; the rest set the slower tempo of growth, stress and reproduction.*

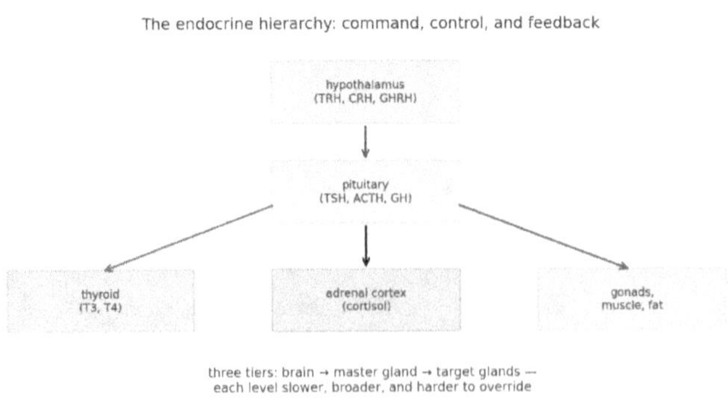

Figure 39.5 — The endocrine hierarchy. *Hypothalamus → pituitary → target glands, with negative feedback at each level. Three tiers: command, control, and the feedback that keeps the whole system hunting a set-point.*

39.2 Insulin Synthesis and Secretion

Insulin is the fed state made molecule. It is synthesised in the β-cell as preproinsulin, cleaved to proinsulin and then to insulin + C-peptide (the C-peptide being the clinician's window into endogenous secretion, since it is secreted equimolarly but cleared more slowly). The β-cell is a glucose sensor of the most elegant design: GLUT2 imports glucose in proportion to blood glucose; glucokinase (Chapter 31's hexokinase IV) phosphorylates it at the same proportion; glycolysis and the cycle raise ATP; ATP closes the K_{ATP} channel; the membrane depolarises; Ca^{2+} floods in; and the granules fuse. The whole cell is a transducer from blood glucose to membrane potential, and the sulfonylurea drugs — which close K_{ATP} directly — exploit the last step, forcing insulin release regardless of glucose. The newer GLP-1 agonists work one step earlier, amplifying the Ca^{2+} signal only when glucose is already high — the difference between a switch and a dimmer, and the reason the newer drugs cause less hypoglycaemia.

39.3 Insulin Signalling

The insulin receptor is a tyrosine kinase, and its cascade (Figure 39.1) is the fed state written in phosphorylation. Activated receptor phosphorylates IRS-1/2; IRS activates PI3K; PI3K makes PIP_3; PIP_3 activates Akt; and Akt issues four orders at once: GLUT4 vesicles fuse with the muscle and fat cell membrane (glucose uptake); GSK3 is inhibited, releasing glycogen synthase (storage); mTOR is activated (protein synthesis and growth); and lipolysis is suppressed while lipogenesis is promoted (storage again). Four branches, one signal, and the signal is always the same sentence: food is here; store it, build with it, and stop burning the reserves. Insulin resistance — the core lesion of type 2 diabetes — is the cell's refusal to hear that sentence: the receptor works, but the downstream response is blunted, and the pancreas compensates with more insulin until the β-cell exhausts. The disease is not a deficiency of signal but a failure of hearing, followed by the signal's exhaustion.

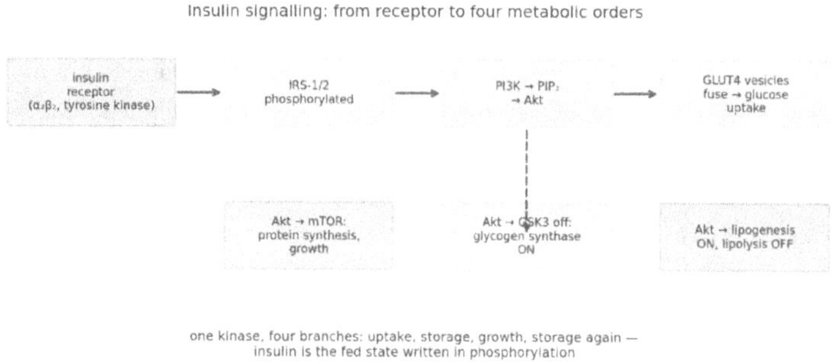

Figure 39.1 — Insulin signalling. *Receptor → IRS → PI3K → Akt → four branches: GLUT4 translocation (green), glycogen synthesis (amber), protein synthesis (navy), lipogenesis/lipolysis (crimson). The fed state as a phosphorylation cascade.*

39.4 Glucagon

Glucagon is insulin's opposite number — the fasting signal, made in the α-cells of the islets, acting almost exclusively on liver. Its receptor is a GPCR, and its second messenger is cAMP — Sutherland's molecule, returning to the organ where he found it. The liver's response to glucagon is the mirror image of its response to insulin: glycogenolysis on, gluconeogenesis on, lipolysis on, ketogenesis on, glycolysis off, lipogenesis off. The α-cell, like the β-cell, is a glucose sensor, but with the opposite sign: it secretes when glucose falls. The two cells sit side by side in the islet, listening to the same blood and giving opposite orders — a parliament of two, in permanent session, whose only business is the blood sugar.

39.5 Catecholamines

Adrenaline and noradrenaline are the emergency reserve — the hormones of fight, flight, and the sprinter's first hundred metres. Their receptors are GPCRs (α and β families), and their metabolic offices are to mobilise everything at once: glycogenolysis in muscle and liver, lipolysis in adipose, gluconeogenesis in liver, and the suppression of insulin secretion itself (α_2-adrenergic) so that the emergency signal is not contradicted. The catecholamines are fast (seconds), short-lived (minutes), and non-negotiable — the body's declaration that the normal fed–fast rules are suspended for the duration. The β-blockers, which blunt them, are among the most prescribed drugs in cardiology; the β-agonists, which mimic them, are among the most prescribed in asthma. One system, two pharmacologies, and a reminder that the emergency reserve is always on duty, even when the emergency is only a deadline.

39.6 Cortisol

Cortisol is the stress hormone, and its metabolic office is to make sure the brain is fed whatever else happens: it promotes gluconeogenesis, mobilises amino acids from muscle, sensitises adipose to lipolytic signals, and — crucially — makes the liver resistant to insulin, so that the glucose it makes is not wasted on muscle. In the short term this is life-saving; in the long term, Cushing's syndrome (cortisol excess) shows the cost: central obesity, muscle wasting, osteoporosis, immunosuppression, and diabetes — the body permanently in emergency mode, feeding a brain that is not actually in danger. Cortisol is the hormone that proves the rule that every adaptation is a trade-off: what saves you from the lion kills you by instalments in the boardroom.

39.7 Thyroid Hormones

T_3 and T_4 set the basal tempo — the idle speed of the whole engine. They act through nuclear receptors to increase the expression of Na^+/K^+-ATPase (the cell's single largest ATP expense), of mitochondrial enzymes, and of the uncoupling proteins of Chapter 28. Hyperthyroidism is the engine racing: weight loss despite appetite, heat intolerance, tremor, tachycardia. Hypothyroidism is the engine idling: weight gain, cold intolerance, bradycardia, and the myxoedema that gave the disease its name. The thyroid's feedback loop — TRH → TSH → T_3/T_4 ⊣ — is the cleanest example of the endocrine set-point, and the TSH assay is the clinician's single most informative thyroid test: the pituitary reports the set-point, and the thyroid reports the actuality.

39.8 Growth Hormone

GH is the anabolic signal of growth, acting partly directly (lipolysis, insulin antagonism — another brain-feeding hormone) and partly through IGF-1 (protein synthesis, chondrogenesis, linear growth). Its excess gives acromegaly in the adult and gigantism in the child; its deficiency gives dwarfism in the child and a subtle metabolic syndrome in the adult. GH is secreted in pulses, mostly at night, and its measurement is complicated by the fact that a single random level is nearly meaningless — the endocrinologist's reminder that hormones have rhythms as well as levels.

39.9 Adipokines and Myokines

The fat cell and the muscle cell are endocrine organs (Figure 39.4). ***Leptin****, from adipose, reports the size of the fat stores to the hypothalamus: leptin-deficient mice and humans are insatiably hungry and massively obese, and leptin replacement cures the rare human deficiency — though leptin in common obesity is a story of resistance, not deficiency, and the drug has failed where the biology was most hoped.* ***Adiponectin****, also from adipose, sensitises to insulin and falls in obesity — one of the few adipokines that is protective.* ***Myokines*** *— IL-6, irisin, myostatin — report muscle work and muscle mass; the myostatin-deficient cattle and whippets are the visible proof that muscle mass is under negative-feedback control. And the gut's incretins — GLP-1, GIP, PYY, ghrelin — report meals before the nutrients even reach the blood, priming the β-cell. The organs talk, and metabolism is the transcript of their conversation.*

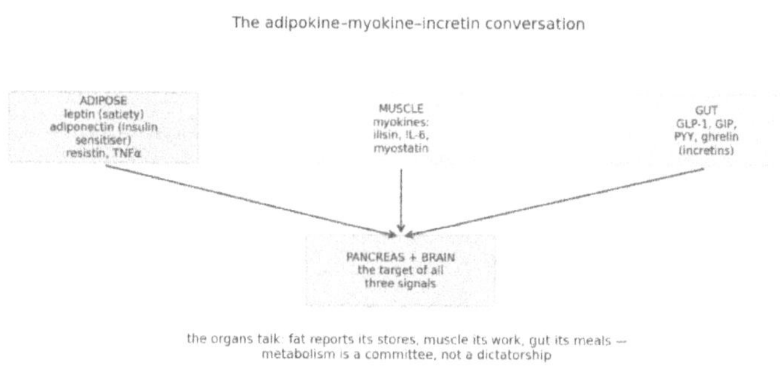

Figure 39.4 — The adipokine–myokine–incretin conversation. *Fat reports its stores (leptin, adiponectin), muscle its work (myokines), gut its meals (incretins) — all to pancreas and brain.*

39.10 Fed, Post-absorptive and Fasting States

The day is the unit of metabolic government (Figure 39.2). In the ***fed state*** *(0–3 h after a meal), insulin rules: glucose is stored as glycogen and fat, protein is synthesised, lipolysis is suppressed. In the* ***post-absorptive state*** *(3–12 h), insulin falls to basal and glucagon rises: liver glycogen is mobilised, gluconeogenesis begins, and the brain's glucose is maintained. In* ***fasting*** *(12 h–3 days), glucagon dominates: glycogen is exhausted by ~24 h, gluconeogenesis peaks, ketogenesis begins, and the brain starts its switch to β-hydroxybutyrate. In* ***starvation*** *(beyond 3 days), the programme of Figure 39.3 runs: ketones rise to 5–7 mM, the brain derives two-thirds of its energy from them, muscle protein breakdown falls* *from 100 g/day to 20 g/day, and the body settles into a state that can be maintained for weeks — Cahill's classical studies showed that a fasting obese man can survive a month on his own fat, his brain fed by ketones, his muscle spared by the switch. The programme is not a failure of metabolism but its finest achievement: the body, given no food, reorganises itself to feed the one organ that cannot wait.*

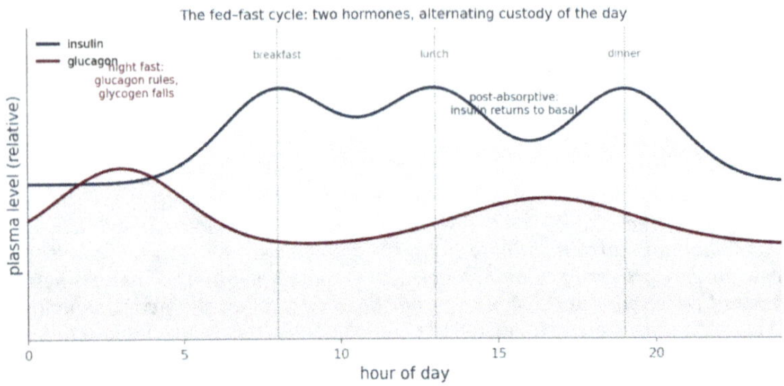

***Figure 39.2* — The fed–fast cycle.** *Insulin (navy) rises after each meal; glucagon (crimson) rules the inter-meal and nocturnal intervals. The two hormones alternate custody of the day.*

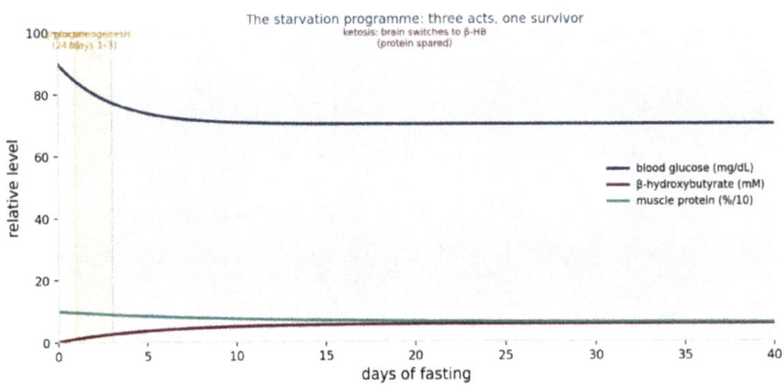

***Figure 39.3* — The starvation programme.** *Three acts: glycogen (24 h), gluconeogenesis (days 1–3), ketosis (beyond day 3) — the brain switching to β-hydroxybutyrate as muscle protein is spared.*

39.11 The Diabetic Spectrum

The diseases of this system are the diseases of the signal (Figure 39.6). ***Type 1 diabetes*** *is autoimmune β-cell destruction: absolute insulin deficiency, presenting in youth, requiring insulin for life, and always threatening DKA (Chapter 35's ketoacidosis).* ***Type 2 diabetes*** *is insulin resistance plus β-cell exhaustion: relative deficiency, presenting in middle age and obesity, treated first with metformin (which lowers hepatic glucose output by a mechanism still debated) and then with the incretin-based drugs that amplify the β-cell's own signal.* ***Gestational diabetes*** *is placental hormone antagonism, resolving after delivery but marking the mother's future risk. And the* ***metabolic syndrome*** *— resistance,* *hypertension, dyslipidaemia, central obesity — is the prodrome of them all. The spectrum is the endocrine system read by its failures, and the failures are, in the end, failures of the duumvirate: too little insulin, too little hearing, or too much of the wrong signal at the wrong time.*

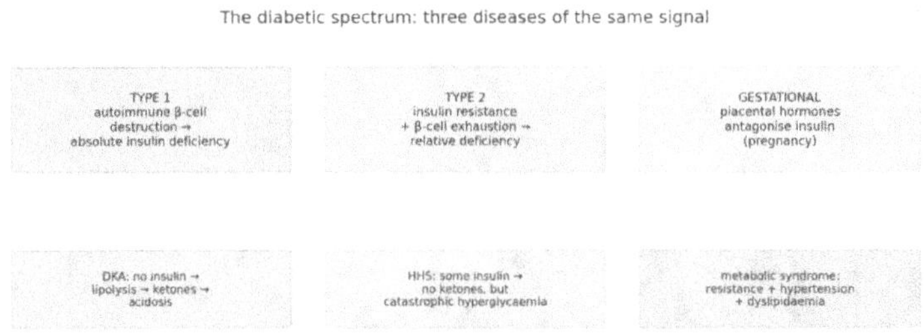

Figure 39.6 — The diabetic spectrum. *Type 1 (absolute deficiency), type 2 (resistance + exhaustion), gestational (placental antagonism), with DKA, HHS, and the metabolic syndrome as the clinical consequences.*

Case study — The honeymoon that ended. *A 14-year-old presents with 3 weeks of polyuria, polydipsia and 4 kg weight loss. Glucose 22 mM, ketones 4.5 mM, pH 7.28. After stabilisation with insulin, his requirements fall to near-zero for three months — the "honeymoon" — then rise again. Explain the initial presentation, the honeymoon, and the final state.*

Resolution. *New-onset type 1 diabetes: autoimmune β-cell destruction has reached the point where insulin secretion cannot cover even basal needs; the presentation is the DKA of Section 39.11. The honeymoon is the partial recovery of residual β-cells once glucotoxicity is relieved — insulin requirements fall as the remaining cells resume function. The final state is their continued destruction: requirements rise to full replacement, and the honeymoon is over. The case is the duumvirate losing its king: first the signal fails, then the pancreas is asked to do without it, and the whole body must be managed by injection.*

Historical note — Banting, Best, and the extract of dog pancreas. *In 1921 Frederick Banting, a young surgeon with no research experience, and Charles Best, a medical student, ligated the pancreatic ducts of dogs in a borrowed Toronto laboratory, waited for the acinar tissue to atrophy, and extracted the islets. The extract, injected into a diabetic dog, lowered its blood sugar; in January 1922 it was given to Leonard Thompson, a 14-year-old dying of diabetes, who recovered. Banting won the 1923 Nobel at thirty-two — the youngest ever in Medicine — shared it with Macleod (who had provided the laboratory and, in Banting's view, too little else), and promptly split his prize with Best. The story is told as a miracle, and it was; but it is also a lesson in the power of a clean question. Banting knew nothing of endocrinology; he knew only that diabetics died of something the pancreas made, and that the something could be extracted. The whole of Section 39.3 — the receptor, the cascade, the four orders — is the answer to the question he asked in a dog lab with no running water. Insulin was not discovered; it was cornered.*

Plate 39.1 *Frederick Grant Banting (1891–1941), who with Charles Best isolated insulin and saved the first diabetic patient in 1922. Nobel laureate in Physiology or Medicine, 1923 — the youngest in that category. (Nobel Foundation portrait.)*

39.12 Exercises

Exercise 39.1. *Estimate the ATP cost of Na^+/K^+-ATPase in a resting adult, and explain why thyroid hormone regulates it.*
Resolution. *The pump consumes ~20–25% of resting ATP — roughly 0.2 mol ATP/kg/day, or ~15 mol in a 70 kg adult ≈ 750 kJ/day. Thyroid hormone increases pump expression; hyperthyroidism raises this expense, contributing to the weight loss and heat production. The thyroid's "idle speed" is, in large part, the speed of the sodium pump.*

Exercise 39.2. *Explain why a sulfonylurea can cause hypoglycaemia but a GLP-1 agonist rarely does.*
Resolution. *The sulfonylurea closes K_{ATP} directly, depolarising the β-cell and releasing insulin regardless of glucose — a switch. GLP-1 agonists amplify the Ca^{2+} signal only when glucose is already high and the ATP/ADP ratio has begun to close K_{ATP} — a dimmer that requires the upstream glucose signal. The difference between a switch and a dimmer is the difference between hypoglycaemia and safety.*

Exercise 39.3. *Account for the muscle-protein sparing of prolonged fasting, and explain why the brain's switch to ketones is the mechanism.*
Resolution. *The brain needs ~120 g glucose/day; if all came from muscle*

protein (via gluconeogenesis), muscle would lose ~20 g N/day and fail in weeks. When β-HB reaches 5–7 mM, the brain derives two-thirds of its energy from ketones, glucose need falls to ~40 g/day, and muscle protein breakdown falls correspondingly — from 100 g/day to ~20 g/day. The switch is not a luxury; it is the difference between a 40-day fast and a 10-day one.

Exercise 39.4 (proposition for research). *The "honeymoon" of new-onset type 1 diabetes suggests residual β-cells can be rescued. Propose an immunological-metabolic intervention to prolong it, and design the trial that would prove the mechanism is β-cell rescue, not merely reduced insulin need.* ***Sketch.*** *Combine low-dose anti-CD3 (teplizumab) with a GLP-1 agonist (to reduce β-cell workload) in new-onset patients. The decisive endpoint is not HbA1c but C-peptide AUC over 2 years, with the mechanistic proof being that the C-peptide rise correlates with the fall in β-cell-specific T-cell clones (by tetramer staining), not with the fall in insulin dose. The honeymoon must be shown to be a rescue of the cells, not a rest for them.*

39.13 Synthesis

The signals are coordinated: a duumvirate of insulin and glucagon, a hierarchy of slower glands, a conversation of adipokines and myokines, and a starvation programme that feeds the brain while sparing the muscle. Direction checked against the master plan: the organism is now one system, and the next question is how its organs differ. ***Chapter 40 — Organ-Specific Metabolism*** *takes up the liver, muscle, brain, heart, kidney and adipose in turn — the same pathways, read in six different accents.*

39.14 References

Banting, F. G., & Best, C. H. (1922). The internal secretion of the pancreas. Journal of Laboratory and Clinical Medicine, 7, 251–266.

Berg, J. M., Tymoczko, J. L., Gatto, G. J., & Stryer, L. (2019). Biochemistry (9th ed.). W. H. Freeman.

Cahill, G. F. (2006). Fuel metabolism in starvation. Annual Review of Nutrition, 26, 1–22.

DeFronzo, R. A., et al. (2015). Type 2 diabetes mellitus. Nature Reviews Disease Primers, 1, 15019.

Friedman, J. M. (2019). Leptin and the endocrine control of energy balance. Nature Metabolism, 1, 754–764.

Nelson, D. L., & Cox, M. M. (2021). Lehninger Principles of Biochemistry (8th ed.). W. H. Freeman.

Saltiel, A. R., & Kahn, C. R. (2001). Insulin signalling and the regulation of glucose and lipid metabolism. Nature, 414, 799–806.

Yechoor, V., & Chan, L. (2010). Endocrinology of Metabolism. Oxford University Press.

Chapter 40
Organ-Specific Metabolism

CONTENTS OF THIS CHAPTER

"Every organ runs the same chemistry; what differs is the accent — and the liver, like a governess in a large household, hears all the accents at once."

40.1 The Liver as Metabolic Governor

If Chapter 39 gave the body its government, this chapter gives it its civil service — and the liver is the permanent secretary. Every pathway of Part VII meets here, and most of them meet nowhere else: glycolysis and gluconeogenesis, glycogen synthesis and breakdown, lipogenesis and β-oxidation, the urea cycle, ketogenesis, lipoprotein assembly, and the first pass of everything the gut absorbs. The anatomical arrangement says it plainly. The portal vein delivers intestinal venous blood — glucose, amino acids, vitamins, and whatever xenobiotics lunch contained — directly to the hepatocyte before any other organ sees it; only the lipids escape, travelling by chylomicron through the lymph to the systemic circulation. The liver therefore reads the menu before the body eats the meal, and it responds like a governor rather than a consumer: it takes up glucose when portal glucose is high and releases it when portal glucose falls, exporting what the brain and erythrocyte cannot make and storing what the muscle and adipocyte cannot yet use (Figure 40.1). Its own appetite is modest — perhaps a fifth of resting oxygen consumption[53] *— because governing, in metabolism as elsewhere, is work.*

53. The liver's ~350 kcal/day of resting expenditure buys urea synthesis (~100 g of protein's worth of nitrogen per day), albumin and lipoprotein secretion, and the largest gluconeogenic flux in the body. Government, like virtue, is expensive and mostly invisible.

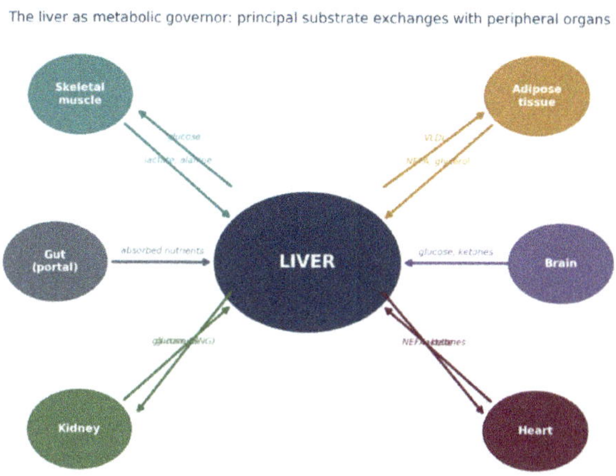

*Figure 40.1 — **The liver as metabolic governor.** Principal substrate exchanges with peripheral organs: portal nutrients arrive from the gut (slate); lactate and alanine return from muscle (teal) against exported glucose; NEFA and glycerol arrive from adipose (amber) against exported VLDL; the heart returns lactate and draws NEFA and ketones (crimson); the kidney trades glutamine for glucose (green); the brain draws glucose and, in fasting, ketones (plum).*

*The governor's signature trick is the **glucose buffer**. In the fed state the hepatocyte expresses glucokinase — high K_m, unresponsive to product inhibition, induced by insulin — so hepatic glucose uptake rises steeply precisely when portal glucose is high; in the fasted state glucokinase is sequestered and glucose-6-phosphatase dominates, and the same cell becomes a net exporter. One enzyme pair, one cell, two economies. Add the urea cycle (the liver is the only organ that runs it complete), ketogenesis (likewise exclusive), and the synthesis of albumin, clotting factors and lipoproteins, and the reason liver failure is unsurvivable becomes a list rather than a mystery.*

40.2 Skeletal-Muscle Metabolism

Skeletal muscle is the body's largest organ by mass and its most selfish by temperament. It stores glycogen but lacks glucose-6-phosphatase, so its glycogen is spent in-house and never shared; it burns whatever is cheapest for the work at hand and exports the bill — lactate and alanine — to the liver. The organ is really three organs wearing one fascia (Figure 40.2, left). Type I fibres are oxidative, mitochondrial, capillary-rich, fatigue-resistant: the posture and marathon muscle, running mostly on fat. Type IIx fibres are glycolytic, glycogen-stuffed, capillary-poor, and exhausted in seconds: the sprint and the escape. Type IIa negotiates between them. Training moves the spectrum — endurance work converts IIx toward IIa and raises mitochondrial density in all fibres — which is why the trained athlete at a given workload burns proportionally more fat and finishes with glycogen to spare.

*Fuel selection within the fibre follows the **crossover concept** (Figure 40.2, right): at low intensity fat dominates, and as intensity rises carbohydrate takes over, crossing near 60–65% of maximal oxygen uptake. The logic is kinetic, not moral. Fat yields more ATP per gram but less ATP per second — β-oxidation is long,* *oxygen-hungry, and throughput-limited — whereas glycolysis is short, oxygen-independent at the margin, and fast. When the demand is for power rather than economy, the fibre chooses speed. The Randle cycle enforces the choice: fatty-acid oxidation raises acetyl-CoA and citrate, which inhibit phosphofructokinase, so abundant fat suppresses glucose use; the muscle, like a prudent household, burns the plentiful fuel first.*

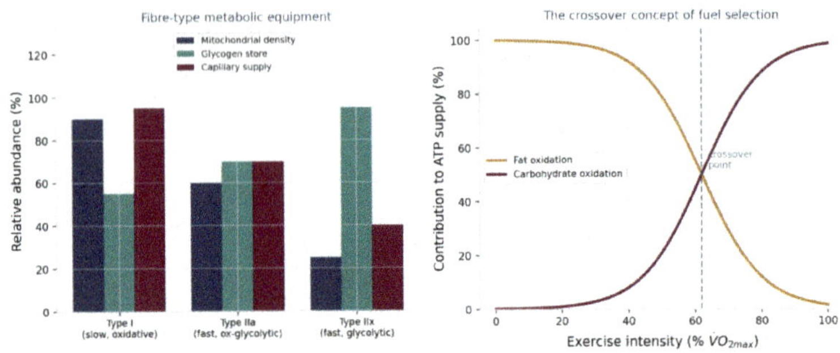

Figure 40.2* — *Skeletal-muscle metabolism. *Left: metabolic equipment of the three fibre types — mitochondrial density, glycogen store and capillary supply (relative scale). Right: the crossover concept — carbohydrate's share of ATP supply rises with exercise intensity while fat's falls, crossing near 62% of* $\dot{V}O_{2max}$.

40.3 Cardiac Metabolism

The heart is the most metabolically expensive tissue per gram in the body — it consumes some 6 kg of ATP per day, thirty times its own mass turned over daily[54] *— and it is an* ***omnivore with preferences*** *(Figure 40.3). At rest in the fed state it takes roughly two-thirds of its energy from fatty acids; fasted, the share rises toward 80%; during moderate exercise it develops a notable taste for lactate, importing from the arterial blood the very metabolite the working leg muscles are exporting — one organ's waste is another's breakfast, and the heart eats at the lactate buffet with particular relish. Glucose contributes a minority share except at high workloads; ketones and amino acids fill the margins. What the heart will not do is store: glycogen is trivial, oxygen extraction is already near-maximal at rest (coronary sinus blood is the most desaturated in the body), so increased demand is met almost entirely by increased flow. The failing heart loses its flexibility first — it falls back on glucose like a debtor falling back on cash — and the metabolic inflexibility is now read as part of the disease, not merely its consequence.*

54. The myocardium holds only enough ATP for about ten beats; it is, energetically speaking, perpetually bankrupt and perpetually refinanced. This is why a few minutes of coronary occlusion are fatal while skeletal muscle tolerates hours of tourniquet.

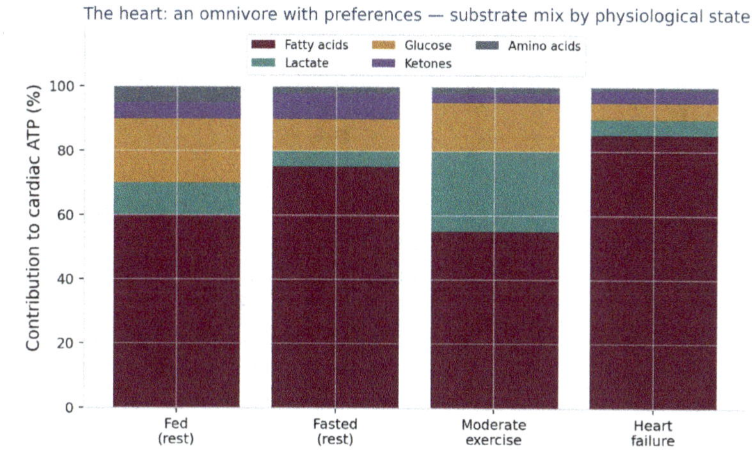

Figure 40.3 — The cardiac fuel mix. *Approximate contributions to myocardial ATP in four states. Fatty acids dominate throughout; lactate becomes a major fuel in exercise; the failing heart narrows its diet toward fatty acids and glucose and loses the lactate option. Values are representative means; individual hearts, like individual diners, vary.*

40.4 Adipose-Tissue Metabolism

Adipose tissue is the exchequer, and like all exchequers it is misunderstood as inert. White adipose stores triglyceride with an energy density no other tissue approaches — some 135,000 kcal in a lean adult, enough to walk from London to Rome — and releases it as non-esterified fatty acids and glycerol under the joint command of insulin (off) and catecholamines (on), with the hormone-sensitive lipase of Chapter 36 as the turnstile. But the exchequer also writes memoranda: leptin reports the size of the reserves to the hypothalamus, adiponectin reports their quality to the liver and muscle, and a small crowd of inflammatory cytokines reports, in the overfilled adipocyte, that the vaults are being badly run (Chapter 39.9). Brown adipose, rich in mitochondria and uncoupling protein 1, burns fat to make heat — thermogenesis by deliberate short-circuit of the proton gradient, the one tissue in which inefficiency is the product. The modern discovery is that adipose is a distributed endocrine gland that happens to store fat, rather than a fat store that happens to secrete; the distinction matters, because a gland can be sick in ways a depot cannot.

40.5 Brain Energetics

The brain is 2% of body mass and spends 20% of resting energy — some 110 g of glucose per day in the fed adult, burned almost completely to CO_2 and water, with nearly all of the ATP going to one purpose: ion pumping to keep neurons ready to fire. The organ's metabolism is peculiar in three ways that dominate every clinical emergency. First, it has no fuel reserves worth the name — a few grams of glycogen in astrocytes, enough for minutes. Second, it sits behind the blood–brain barrier, which admits glucose and ketones but excludes the fatty acids that feed every other organ; the brain, alone among tissues, cannot live on the body's cheapest fuel. Third, its demand is inflexible in the short term: consciousness is not negotiable at the metabolic level. The adaptation to prolonged starvation (Figure

40.4) is therefore one of the great physiological accomplishments — over days, β-hydroxybutyrate rises, monocarboxylate transporters at the barrier are induced, and the brain comes to take two-thirds of its energy from ketones; cerebral glucose use falls from 120 g to some 40 g per day, of which a floor remains obligate (the erythrocyte and renal medulla have no mitochondria and must glycolyse). Cahill's subjects fasted for forty days on this arrangement; the brain, which cannot burn fat directly, is fed by a liver that burns it on the brain's behalf.

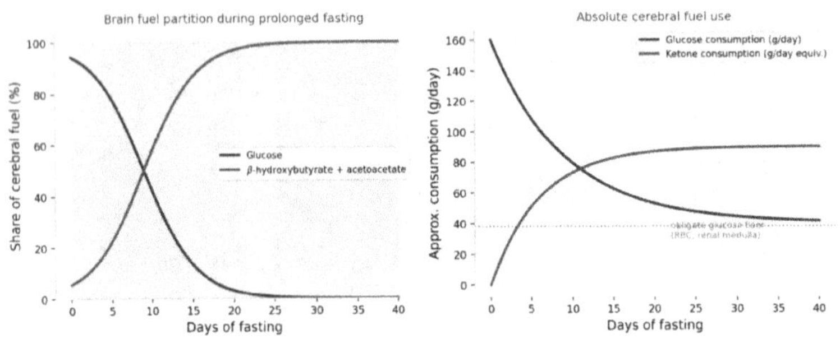

***Figure 40.4* — Brain energetics in prolonged fasting.** *Left: the fractional partition of cerebral fuel shifts from glucose to ketones over the first two weeks. Right: absolute consumption — glucose use falls toward the obligate floor (~38 g/day for the erythrocyte and renal medulla) as ketone use rises to supply the remainder.*

40.6 Renal Metabolism

The kidney filters 180 litres of plasma a day and reabsorbs nearly all of what it filters, and the reabsorption is paid for in ATP — the renal cortex, gram for gram, rivals the heart for oxygen consumption. Its metabolism splits by anatomy. The cortex is oxidative and mitochondrial, running on fatty acids, lactate and glutamine, and — a fact that surprises every student — it is a gluconeogenic organ, contributing perhaps a fifth of endogenous glucose production in prolonged fasting, when hepatic output falls. The medulla is glycolytic by necessity: it operates at oxygen tensions so low that oxidative phosphorylation cannot sustain it, and it runs on glucose to lactate like an erythrocyte with plumbing. The kidney also handles nitrogen: it deaminates glutamine to ammonium, which is excreted carrying acid out of the body — renal ammoniagenesis is an acid–base organ system disguised as an amino-acid pathway, and in chronic acidosis it becomes a major glutamine sink. Add erythropoietin, renin, and vitamin D activation, and the kidney emerges as an endocrine organ that filters on the side.

40.7 Intestinal Metabolism

The gut is the interface where chemistry meets catering, and its enterocytes are among the most metabolically specialised cells in the body. They turn over every few days — a proliferative demand met by glutamine, the enterocyte's preferred fuel, drawn partly from the lumen and partly from the blood — and they perform the first chemical processing of everything absorbed: apical transporters with distinct specificities, intracellular esterification of fatty acids and cholesterol, chylomicron assembly and export into the lymph, and first-pass conjugation of drugs. Glutamine's role explains a clinical observation: in catabolic illness the gut atrophies when glutamine runs short, the barrier leaks, and bacteria translocate — the intestine failing not as a tube but as a tissue. The colonocytes run on butyrate, supplied by bacterial fermentation of fibre — a tidy arrangement in which the microbiome is paid in indigestible carbohydrate and remits short-chain fatty acids, and in which a low-fibre diet is, metabolically, a wage cut for the colonic epithelium.

40.8 Erythrocyte Biochemistry

The erythrocyte is the extreme case of metabolic minimalism: no nucleus, no ribosomes, no mitochondria — a cell that has deleted everything except the job. Its entire ATP supply comes from glycolysis, and its entire lactate output goes to the liver: a Cori cycle of one cell. Two features repay attention. The Rapoport–Luebering shunt diverts 1,3-bisphosphoglycerate to 2,3-BPG, sacrificing two ATP per glucose to set haemoglobin's oxygen affinity — the cell paying an energy tax for the right to deliver its cargo (Chapter 31.10). And the pentose phosphate pathway supplies the NADPH that keeps glutathione reduced, which is the cell's only defence against the oxygen it carries; G6PD deficiency, the commonest enzymopathy of humanity, is a lesion of exactly this, and its victims haemolyse when oxidant drugs or fava beans overwhelm the shrunken NADPH supply. The erythrocyte, in short, is a bag of haemoglobin with a glycolytic engine and a redox fire extinguisher — and it works, 120 days at a time, without a single repair part.

40.9 Metabolic Exchange Between Organs

Between the organs run the cycles, and the cycles are where organ-specific metabolism becomes physiology. The ***Cori cycle*** *(Figure 40.5) ships lactate from muscle and erythrocyte to liver and glucose back again; it costs the liver 6 ATP per glucose while the muscle nets 2, and the 4-ATP difference is the price of moving an anaerobic debt to an organ that can afford it. The* ***glucose–alanine cycle*** *does the same journey for nitrogen: muscle transaminates its glycolytic pyruvate with the amino groups of proteolysis and exports alanine, which the liver deaminates to pyruvate (→ glucose) and urea. The cycle is a Cori cycle with a nitrogen passenger, and in fasting muscle it is the means by which protein feeds the brain indirectly. To these add the ketone shuttle (liver makes, brain and heart burn), the NEFA–albumin shuttle (adipose releases, everything burns), and the glutamine trade (muscle and lung export, gut and kidney consume), and the organism appears as what it is: a single metabolic economy with specialised provinces and a busy internal trade. The Fick principle quantifies any leg of it:*

$$VO_{2\,\text{organ}} = Q \times (C_aO_2 - C_vO_2) \tag{40.1}$$

Derivation. Conservation of mass: what an organ consumes must equal what flows in minus what flows out. Oxygen in, per minute, is blood flow Q times arterial content C_aO_2; oxygen out is Q times venous content C_vO_2; the difference is the consumption. Nothing is assumed but bookkeeping — the equation is a bank statement for oxygen. Applied to the whole body (Q = cardiac output, mixed venous blood from the pulmonary artery) it yields the classic Fick measurement of cardiac output; applied across a leg or the brain it yields that organ's private appetite.

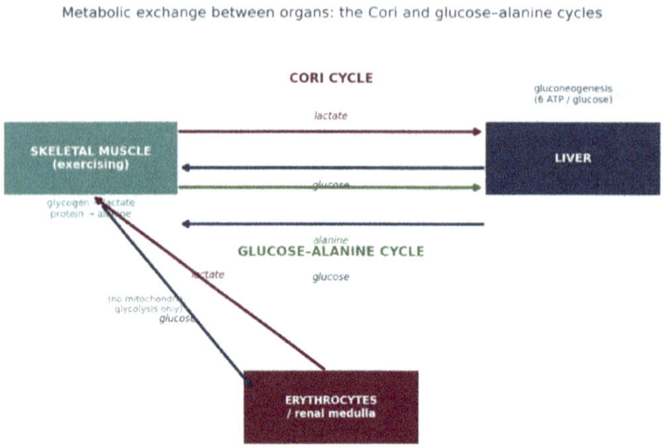

***Figure 40.5* — Metabolic exchange between organs.** *The Cori cycle (lactate out, glucose back; the liver pays 6 ATP per glucose against the muscle's 2) and the glucose–alanine cycle (alanine out, glucose back; nitrogen rides the same shuttle to the urea cycle). Below: the erythrocyte and renal medulla, mitochondrial paupers, run a permanent Cori leg of their own.*

40.10 Exercise and Metabolic Adaptation

Exercise is the stress test of the whole integrated system — the one physiological state in which ATP turnover rises twentyfold and every organ must improvise at once. Three timescales organise it. In the first seconds, phosphocreatine and stored ATP carry the load (Chapter 25.8). From seconds to minutes, anaerobic glycolysis dominates and lactate climbs. Beyond a few minutes, oxidative metabolism takes over, and the fuel question becomes the whole story. Two rules govern the fuel. With rising intensity, carbohydrate's share rises (the crossover of Figure 40.2); with rising duration at fixed moderate intensity, fat's share rises as muscle glycogen and then liver glycogen dwindle (Figure 40.6, left). The trained athlete stretches both rules — more fat at every intensity (sparing glycogen), more glycogen stored, more mitochondria to burn both — which is why the wall of the marathon arrives at 32 km for the trained and 25 km for the ambitious. The fuel mixture at any moment is readable from the respiratory quotient:

$$\%\,\text{carbohydrate} = \frac{\text{RQ} - 0.70}{1.00 - 0.70} \times 100 \qquad (40.2)$$

Derivation. RQ = $\dot{V}CO_2/\dot{V}O_2$. For pure carbohydrate (glucose: $C_6H_{12}O_6 + 6\ O_2 \rightarrow 6\ CO_2 + 6\ H_2O$) the ratio is 6/6 = 1.00; for a representative fat (palmitate: $C_{16}H_{32}O_2 + 23\ O_2 \rightarrow 16\ CO_2 + 16\ H_2O$) it is 16/23 ≈ 0.70. Assuming the mixture is only fat and carbohydrate — protein's share being estimated separately from urinary nitrogen — the RQ is the weighted mean of the two end-points, and linear interpolation recovers the weights. The equation is two combustion stoichiometries and one average: again, bookkeeping.

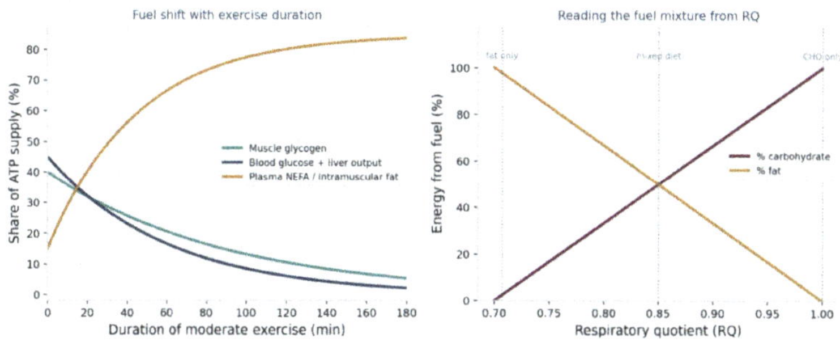

Figure 40.6 — Exercise and fuel. *Left: with duration at fixed moderate intensity, muscle glycogen and blood-borne glucose fade and fat rises to carry the load. Right: the fuel mixture read from RQ — the two combustion end-points (0.70 fat, 1.00 carbohydrate) and the linear interpolation between them (Eq. 40.2).*

The chronic adaptations repay the training. Mitochondrial biogenesis (PGC-1α, activated by the very AMPK that senses energetic stress), capillary growth, raised GLUT4 independent of insulin for hours after exercise — the molecular signature of the trained muscle is written by the exercise itself, and its read-out is insulin sensitivity. Exercise is thus not merely expenditure; it is a signalling state, and the muscle, through its myokines (Chapter 39.9), tells the rest of the body about it. The prescription writes itself, and has, alas, written itself in every guidelines committee for fifty years.

Case study — The wall at kilometre 31. *A trained runner, well-fed and well-tapered, runs a marathon at 85% $\dot{V}O_{2max}$. At kilometre 31 his legs "turn to concrete": pace collapses, and he is dizzy. Muscle glycogen, ~500 g at the start, is effectively exhausted; blood glucose is falling as hepatic output fails to match muscle uptake; the brain, which cannot burn the fatty acids now dominating muscle fuel, is running short. Explain the physiology, and the remedy used on the course.*

Resolution. *At 85% $\dot{V}O_{2max}$ the crossover has long since passed: carbohydrate supplies most of the ATP, and the entire race is a race against the glycogen store (~2,000 kcal of muscle glycogen against ~2,600 kcal of total cost — the arithmetic is deliberately tight). When muscle glycogen fails, the fibre must fall back on fat and blood glucose; fat cannot sustain the pace (Figure 40.2), and hepatic glucose output, itself glycogen-depleted by three hours of demand, cannot keep blood glucose up — hence the dizziness, which is neuroglycopenia, not fatigue. The remedy on the course is glucose gel: absorbed carbohydrate, which the muscle will burn immediately and the brain gratefully. The wall is therefore two bankruptcies at once — the muscle's glycogen and the liver's — and the gel is a bridge loan to both. The trained runner's real preparation is metabolic: glycogen supercompensation beforehand, fat adaptation to lower the carbohydrate burn rate, and the discipline to run the first half at a pace the fuel can afford.*

Historical note — Warburg's manometer and the respiration of a slice of liver. *Before metabolism could be integrated it had to be measured, and the measuring was Otto Warburg's life's work. In the 1920s he perfected the* *manometric respirometer — a tissue slice in a sealed flask, a drop of alkali to absorb CO_2, and a fluid column whose fall read the oxygen consumed — and with it he did something no one had done: he put organ-specific metabolism on the balance. Tumour slices, he found, glycolysed furiously even in oxygen — the "Warburg effect" that Chapters 31 and 38 have already met — while retina and kidney medulla did nearly the same and heart scarcely at all. The differences among organs, the subject of this whole chapter, were first visible as differences in the fall of a drop of fluid. Warburg received the 1931 Nobel Prize in Physiology or Medicine for his work on respiratory enzymes; he stayed in Berlin through the Nazi years — protected, astonishingly, by his indispensability and a half-Jewish ancestry the regime chose to overlook — and worked on, manometer by manometer, into his eighties. His aphorism for the method was characteristically blunt: "One can recognise the value of a researcher by the number of his abandoned methods." He abandoned few.*

Plate 40.1 *Otto Heinrich Warburg (1883–1970), whose manometry made organ metabolism measurable and whose tumour-glycolysis observation still frames cancer metabolism. Nobel laureate in Physiology or Medicine, 1931. (Nobel Foundation portrait.)*

40.11 Exercises

Exercise 40.1. *Using the Fick principle (Eq. 40.1), estimate the brain's share of resting cardiac output, given that the brain consumes ~20% of resting oxygen with an arteriovenous O_2 difference of 6.5 mL/dL (whole body: 5.0 mL/dL).*

Resolution. *From Eq. (40.1), Q is proportional to $\dot{V}O_2/\Delta O_2$. For the whole body, $Q_{total} \propto 100/5.0 = 20$; for the brain, $Q_{brain} \propto 20/6.5 \approx 3.08$. Share = 3.08/20 ≈ 15%. The brain takes about 15% of cardiac output for 20% of the oxygen — it* *extracts more per litre than average, though less than the heart's near-total extraction. The equation is one line; the physiology is that the brain is both hungry and well-supplied.*

Exercise 40.2. *A subject at steady exercise has RQ 0.85 and consumes 2.0 L O_2/min. Using Eq. (40.2) and the stoichiometries behind it, estimate his fat and carbohydrate oxidation rates in g/min (take O_2 needs as 2.0 L/g fat and 0.83 L/g glucose, approximately).*
Resolution. *From Eq. (40.2), % carbohydrate = (0.85 − 0.70)/0.30 × 100 = 50%; hence 50% fat. Let oxygen be split accordingly (RQ weights energy, and the L/g values here make oxygen share ≈ energy share): 1.0 L/min to fat → 1.0/2.0 = 0.50 g fat/min; 1.0 L/min to glucose → 1.0/0.83 ≈ 1.20 g glucose/min. Total ≈ 30 g fat and 72 g glucose per hour — the arithmetic of the wall in Exercise 40.1's marathon: at that rate, 500 g of muscle glycogen last under four hours even at half carbohydrate.*

Exercise 40.3. *Explain why the erythrocyte must run a permanent Cori cycle, why G6PD deficiency haemolyses on oxidant challenge, and what both facts reveal about the cell's design constraints.*
Resolution. *No mitochondria: oxidative phosphorylation is impossible, so glycolysis is the sole ATP source and lactate the sole end-product, which the liver must reclaim — a Cori cycle of necessity, not choice. G6PD supplies the NADPH that regenerates reduced glutathione; the erythrocyte, carrying oxygen at high pressure with no means of repair, lives or dies by its glutathione. Oxidant drugs (primaquine, rasburicase) or fava-bean glycosides raise peroxide flux; the deficient cell cannot keep glutathione reduced, haemoglobin oxidises and precipitates as Heinz bodies, and the spleen culls the rigid cells. Both facts follow from one design decision — jettison the organelles to make room for haemoglobin — and the price is paid in two currencies: energy (Cori) and redox (G6PD).*

Exercise 40.4 (proposition for research). *The failing heart loses metabolic flexibility, falling back on glucose. Propose an experiment to test whether restoring fatty-acid oxidation capacity — rather than merely supplying alternative fuel — improves function, and specify the control that separates "more fuel" from "restored flexibility".*
Sketch. *In a pacing-induced heart-failure model, overexpress PGC-1α in the myocardium (AAV9, cardiac promoter) versus a ketone-infusion arm that supplies alternative fuel without changing oxidative machinery. Endpoints: cardiac work per oxygen consumed (efficiency), the substrate-switching response to an acute lactate load (flexibility), and haemodynamics. The decisive control is the lactate-clamp challenge: only the PGC-1α arm should recover the full substrate menu of Figure 40.3. If ketone infusion matches it on haemodynamics but fails the flexibility test, the disease is machinery, not fuel — and the therapy is biogenesis, not diet.*

40.12 Synthesis

The organs are one chemistry in six accents: the liver governs and buffers, the muscle spends and exports its debts, the heart omnivorously out-earns them all per gram, the adipose stores and advises, the brain demands and is fed by proxy, the kidney filters and quietly gluconeogeneses, and the erythrocyte strips metabolism to glycolysis and a fire extinguisher. Between them run the cycles — Cori, alanine, ketone, NEFA — and over them hangs Chapter 39's government of hormones. Direction checked against the master plan: the integrated organism now stands complete, and what remains is to feed it and to watch it fail. ***Chapter 41 — Nutrition and Micronutrient Biochemistry*** *takes up the input side: macronutrients, vitamins, minerals, and the biochemical truth behind dietary claims.*

40.13 References

Brooks, G. A. (2018). The science and translation of lactate shuttle theory. Cell Metabolism, 27, 757–785.

Cahill, G. F. (2006). Fuel metabolism in starvation. Annual Review of Nutrition, 26, 1–22.

Frayn, K. N. (2019). Metabolic Regulation: A Human Perspective (4th ed.). Wiley-Blackwell.

Hargreaves, M., & Spriet, L. L. (2020). Skeletal muscle energy metabolism during exercise. Nature Metabolism, 2, 817–828.

Lopaschuk, G. D., Karwi, Q. G., Tian, R., Wende, A. R., & Abel, E. D. (2021). Cardiac energy metabolism in heart failure. Circulation Research, 128, 1487–1513.

Nelson, D. L., & Cox, M. M. (2021). Lehninger Principles of Biochemistry (8th ed.). W. H. Freeman.

Owen, O. E., Morgan, A. P., Kemp, H. G., Sullivan, J. M., Herrera, M. G., & Cahill, G. F. (1967). Brain metabolism during fasting. Journal of Clinical Investigation, 46, 1589–1595.

Warburg, O. (1956). On the origin of cancer cells. Science, 123, 309–314.

Chapter 41
Nutrition and Micronutrient Biochemistry

CONTENTS OF THIS CHAPTER

"We are what we eat, runs the proverb; the biochemist, more cautiously, observes that we are what we absorb, activate, and neglect to excrete."

41.1 Macronutrient Requirements

Every pathway of Parts VII and VIII runs on imported raw material, and this chapter is the customs ledger: what must come in, in what form, and what happens when it does not. The macronutrients are three. Carbohydrate is, strictly, optional — gluconeogenesis (Chapter 32) can supply the ~130 g/day that brain, erythrocyte and renal medulla demand — but it is the cheapest and fastest fuel, and below some 50 g/day the body pays for its independence in ketones. Fat is obligatory only in its two essential members (Section 41.3), but it carries the fat-soluble vitamins and concentrates energy at 9 kcal/g against the 4 kcal/g of carbohydrate and protein[55] — which is why the adipocyte stores fat and not glycogen. Protein is obligatory in a deeper sense: unlike the other two, it contains nitrogen, and nitrogen has no storage form. The adult requirement is ~0.8 g/kg/day, a figure that is not the average need but the intake that covers nearly everyone — a statistical artefact routinely misread as a target.

Energy requirement is the sum of basal metabolism (~60–75% in the sedentary), the thermic effect of food (~10%), and activity — the only term under voluntary control and the one modernity has all but abolished. The basal term itself is not fixed: it scales with lean mass, falls with age, and — the inconvenient finding of the 2021 doubly-labelled-water meta-analysis — is remarkably stable from age 20 to 60 when adjusted for body composition, so that middle-aged spread is not a slowing metabolism but a slowing life.

41.2 Protein Quality and Nitrogen Balance

*Protein is eaten for its amino acids, and the accounting is nitrogen accounting. An adult excretes the nitrogen of the protein he oxidises — mostly as urea (Chapter 37) — and **nitrogen balance** is intake minus excretion (Figure 41.1, left). The healthy adult hovers at zero; the growing child, the pregnant woman and the convalescent run positive; sepsis, major surgery and starvation run violently negative — a septic patient may lose 20 g of nitrogen a day, the protein equivalent of a steak, drawn from his own muscle. The clinical lesson is that negative balance cannot be talked out of existence, only fed out of it, and even feeding only blunts the catabolic tide until the inflammation recedes.*

55. The 4–9–4 figures are Atwater factors: metabolisable energy after the losses of digestion and, for protein, the energy surrendered in urea. Bomb calorimetry gives protein 5.6 kcal/g; the body, which must pay to excrete the nitrogen, keeps only 4.

Quality is the second axis. Nine amino acids are essential (Chapter 38), and a protein's worth is set by its limiting amino acid — the essential amino acid in shortest supply relative to need, lysine in wheat, methionine in legumes. Hence the relative scores of Figure 41.1 (right), and hence the old wisdom of food combination: rice and beans, each poor alone, are complete together, because each supplies the other's limiting amino acid. Vegetarian diets are not protein-poor; they are protein-asymmetric, and asymmetry is cured by variety across the day, not within the meal — the body pools amino acids over hours, not minutes.

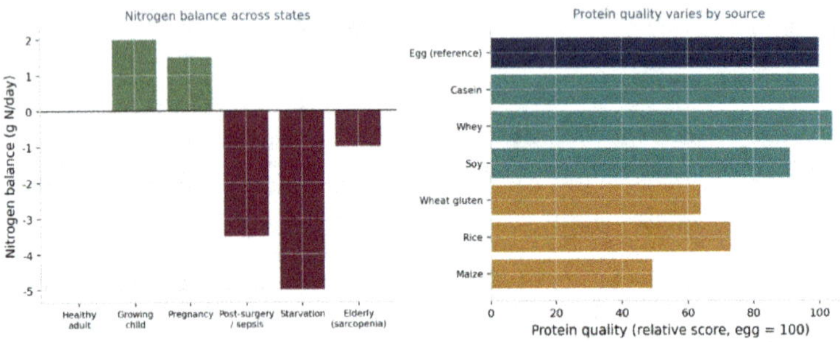

Figure 41.1 — Protein economy. *Left: nitrogen balance across physiological and pathological states — growth and pregnancy positive, catabolic illness and starvation negative. Right: relative protein quality by source (reference egg = 100); plant proteins are limited by single amino acids, animal proteins rarely.*

41.3 Essential Fatty Acids

Two fatty acids cannot be made and must be eaten: **linoleate** *(18:2, n-6) and* **α-linolenate** *(18:3, n-3) — the mammalian desaturases cannot introduce double bonds beyond Δ9, and the n-6 and n-3 families live beyond that frontier (Chapter 36). From them flow the eicosanoids: arachidonate (n-6) to the prostaglandins, thromboxanes and leukotrienes of inflammation; eicosapentaenoate (n-3) to the milder series-3 prostaglandins and the resolvins. Aspirin, the NSAIDs, and the leukotriene antagonists all act in this short pathway, which is why two grams of a fatty acid a day can argue with a pharmacopoeia. True essential-fatty-acid deficiency — dermatitis, poor wound healing, fatty liver — is now seen mostly in fat-free parenteral nutrition, a reminder that a patient can be fed everything except the two molecules he cannot make. The n-6:n-3 ratio of modern diets has drifted from perhaps 2:1 to 15:1; what that drift costs is genuinely debated, and the honest verdict is that the biochemistry is clearer than the epidemiology.*

41.4 Water-Soluble Vitamins

The water-soluble vitamins are, to a first approximation, the parts catalogue of the coenzymes (Figure 41.2). Thiamine becomes TPP, the cofactor of pyruvate dehydrogenase, α-ketoglutarate dehydrogenase and transketolase — which is why its deficiency, beriberi, is a disease of carbohydrate-burning tissues, and why giving glucose to a thiamine-depleted alcoholic without thiamine precipitates Wernicke's encephalopathy: the fuel arrives, the cofactor is absent, and the brainstem pays. Riboflavin becomes FAD and FMN; niacin becomes NAD and NADP — synthesisable from tryptophan, which is why pellagra was a disease of maize-eating regions (maize protein is tryptophan-poor and its niacin is bound); pantothenate becomes coenzyme A; pyridoxine becomes PLP, the cofactor of every transaminase; biotin carboxylates; folate carries one-carbon units. Vitamin C is the odd one out — not a coenzyme precursor but a reducing agent that keeps the iron of prolyl and lysyl hydroxylases ferrous, so that scurvy is, at bottom, a collagen disease: old scars reopen, because the collagen in them must be renewed and cannot be.

Folate and B_{12} deserve their own map (Figure 41.3). Folate carries one-carbon units to thymidylate and purine synthesis — the reason deficiency produces megaloblastic anaemia (DNA synthesis fails while RNA synthesis continues, and the nucleus lags the cytoplasm) — and to the remethylation of homocysteine. B_{12} is required at two reactions only: methionine synthase and methylmalonyl-CoA mutase. Block the first and folate piles up as 5-methyl-THF, unable to unload its methyl group — the ***methyl-folate trap*** *— so B_{12} deficiency causes a functional folate deficiency and the same anaemia. But B_{12} deficiency also demyelinates, and treating it with folate cures the blood while the cord degenerates: the haematological mask that every medical student is taught to fear, and rightly.*

Figure 41.2 — The vitamins as coenzymes. *Each water-soluble vitamin with the coenzyme it becomes and the reactions that cofactor serves; the fat-soluble four, which are not coenzymes but signals and antioxidants, grouped below.*

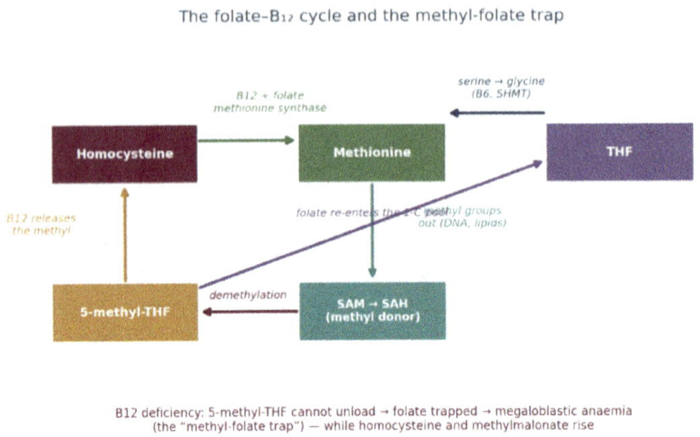

Figure 41.3 — The folate–B_{12} cycle. *Homocysteine is remethylated to methionine by methionine synthase (B_{12}, folate); methionine donates methyl groups via SAM; the folate skeleton cycles back through 5-methyl-THF. Without B_{12} the methyl group cannot leave 5-methyl-THF — the trap.*

41.5 Fat-Soluble Vitamins

A, D, E and K dissolve in fat, ride chylomicrons, store in liver and adipose, and accumulate to toxicity in a way the water-soluble vitamins never do — the four are signals and antioxidants rather than coenzymes. ***Vitamin A*** *is vision (retinal, the rhodopsin chromophore of Chapter 22) and differentiation (retinoic acid, a nuclear-receptor ligand); its deficiency is night blindness and xerophthalmia, still a leading cause of childhood blindness, and its excess is teratogenic — the reason isotretinoin carries its warnings.* ***Vitamin D*** *is a hormone in all but origin: skin makes it from 7-dehydrocholesterol under UVB, liver 25-hydroxylates it, kidney 1-hydroxylates it under parathyroid command, and calcitriol then regulates intestinal calcium absorption through a nuclear receptor; rickets and osteomalacia are calcium-deficiency diseases by proxy.* ***Vitamin E*** *is the membrane's chain-breaking antioxidant, sacrificing itself to lipid peroxyl radicals (Chapter 26).* ***Vitamin K*** *is the cofactor of γ-glutamyl carboxylase, without which prothrombin and its fellow factors cannot bind calcium; warfarin inhibits the reductase that recycles it — a vitamin antagonist that has anticoagulated a hundred million patients.*

41.6 Major Minerals

Seven minerals are needed in gram quantities: sodium, potassium, chloride, calcium, phosphorus, magnesium, and sulphur (arriving as amino acids). The first three are the osmotic skeleton of the extracellular fluid and the electrochemical business of every excitable cell. Calcium is 1.2 kg of bone and a nanomolar cytosolic signal (Chapter 22) — the blood level is defended at the bone's expense, so hypocalcaemia means hormonal failure, not dietary shortfall, while osteoporosis means the opposite. Phosphate is the currency metal — every ATP, every phospholipid, every bone crystal — and its renal handling is where calcium–phosphate product becomes a clinical number. Magnesium is the quiet cofactor of every kinase, since kinases bind Mg-ATP, not ATP; its deficiency mimics hypocalcaemia and hypokalaemia because it disables the parathyroid axis and the sodium pump at once — the electrolyte that hides behind the others.

41.7 Trace Elements

The trace elements are needed in milligrams and are lethal in their absence and their excess alike. Iron is haem, cytochromes and the ribonucleotide reductase that makes DNA — and the commonest deficiency on Earth, some two billion people anaemic; its absorption is regulated at the enterocyte by hepcidin, because there is no regulated excretion: the body guards iron like a bank with no door. Zinc sits in thousands of zinc-finger proteins and enzymes; deficiency shows as poor wound healing, hypogeusia and immune failure. Copper is in cytochrome oxidase, lysyl oxidase and superoxide dismutase — Wilson's and Menkes' diseases are its two-sided lesion (Chapter 36's hepatology in miniature). Iodine is two atoms in thyroxine and its deficiency remains the commonest preventable cause of intellectual impairment; selenium is selenocysteine, the 21st amino acid, in glutathione peroxidase and the deiodinases; fluoride hardens the apatite lattice; chromium's fame, by contrast, rests on rather less biochemistry than marketing. The pattern throughout: trace elements are cofactors, and cofactor deficiency is enzyme deficiency acquired at the dinner table.

41.8 Nutrient Sensing

Nutrients are signals as well as substrate, and three sensor systems read the menu (Figure 41.4). ***mTORC1*** *reads amino acids (leucine above all) and insulin, and orders growth: protein synthesis, lipogenesis, ribosome biogenesis, and the suppression of autophagy.* ***AMPK*** *reads the AMP:ATP ratio — energetic scarcity itself — and orders catabolism: glucose uptake, fatty-acid oxidation, autophagy, mitochondrial biogenesis.* **The sirtuins** *read NAD^+, the redox state made molecule, and deacetylate targets from PGC-1α to histones. The three inhibit one another — mTOR and AMPK are near-mirror-images — so that fed and fasted are not regions of a dial but two alternative governments, each suppressing the opposition. The pharmacological interest is intense and mostly justified: metformin works substantially through AMPK, rapamycin is mTOR's inhibitor and extends lifespan in every model organism tried, and the entire "fasting-mimetic" industry is an attempt to rent the fasted government without paying the fast. The sensors, at least, are real; the shortcuts remain under trial.*

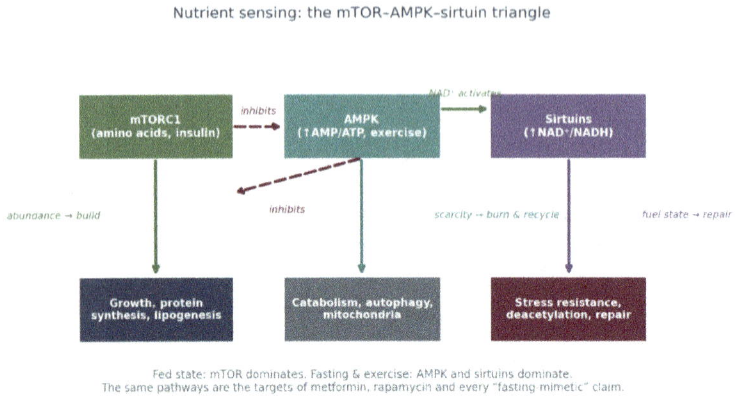

Figure 41.4 — The mTOR–AMPK–sirtuin triangle. *mTORC1 reads abundance and builds; AMPK reads energetic scarcity and catabolises; sirtuins read the $NAD^+/NADH$ ratio and repair. Mutual inhibition makes fed and fasted two governments, not a continuum.*

41.9 Malnutrition

Undernutrition comes in two textbook syndromes whose distinction is biochemical. ***Marasmus*** *is total starvation: the child is wasted, but the adaptation of Chapter 39.10 is intact — albumin is preserved, the face is old, the appetite survives.* ***Kwashiorkor*** *is protein starvation on a carbohydrate base — the weanling fed on maize gruel — and the adaptation fails where it matters: hepatic protein synthesis falters, albumin falls, oncotic pressure fails and oedema appears; lipoprotein export fails and the liver turns fatty; the belly swells and the appetite dies. The oedema of kwashiorkor is thus a lipoprotein disorder as much as a protein disorder.* ***Refeeding syndrome*** *is the iatrogenic mirror: carbohydrate refeeding after starvation releases insulin, insulin drives phosphate, potassium and magnesium into cells, and the plasma is stripped of the three ions the heart and diaphragm cannot do without — a lethal hypophosphataemia that kills the rescued starveling on day three of his rescue, unless feeding begins slowly with the electrolytes measured. The lesson generalises: in nutrition the rate of correction is a dose, and doses have toxicities.*

41.10 Overnutrition and Obesity

The opposite failure is energy in excess, and its biochemistry begins with a self-correcting system that has been outvoted. Expenditure rises with body mass (Figure 41.6, left) — more tissue to maintain, more mass to carry — so a fixed intake finds an equilibrium weight; the problem is that palatable, energy-dense food raises the intake line faster than expenditure can follow, and leptin resistance (Chapter 39.9) removes the hypothalamic brake. Where the excess is stored matters more than how much: subcutaneous fat is relatively innocent, visceral and ectopic fat — liver, pancreas, muscle — drives insulin resistance, and the threshold varies between individuals and ancestries, which is why metabolic disease appears at different BMIs in different populations. The diet literature, viewed at distance, converges uncomfortably (Figure 41.6, right): every named diet loses 3–4 kg at six months and keeps off ~2 kg at two years, the differences between diets being smaller than the variance within them. The honest summary is biochemical: all diets work by the same ledger — energy balance — and differ mainly in how bearable they make the deficit. This is not a counsel of despair; it is a counsel that adherence is the active ingredient.

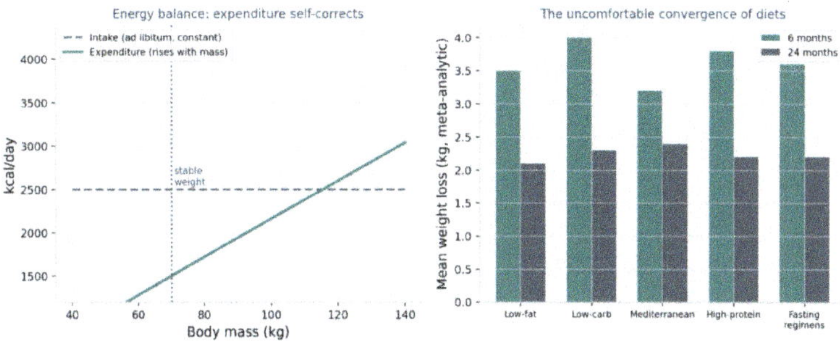

Figure 41.6 — Energy balance and the diet trials. *Left: expenditure rises with mass, so fixed intake equilibrates; weight change means the lines moved. Right: mean weight loss in meta-analyses of named diets at 6 and 24 months — convergence toward ~2 kg regardless of composition.*

41.11 Alcohol Metabolism

*Ethanol is a macronutrient of 7 kcal/g that the body treats as a toxin to be burned first — there is no storage form, so its oxidation takes priority over everything else on the plate (Figure 41.5). Alcohol dehydrogenase oxidises it to acetaldehyde, ALDH2 to acetate, and both steps generate NADH: the hepatic NADH:NAD$^+$ ratio climbs, pyruvate is driven to lactate (the link), fatty-acid oxidation halts (fat accumulates — the first step to the fatty liver), and gluconeogenesis is strangled of pyruvate — which is why the drinking, unfed alcoholic presents with hypoglycaemia and lactic acidosis, a combination otherwise rare. Acetaldehyde is the toxin proper: it forms protein adducts, and the ALDH2*2 variant common in East Asia — a partially inactive enzyme — causes flushing and, in those who drink through it, a multiplied oesophageal cancer risk. Chronic exposure induces CYP2E1, a microsomal bypass that generates reactive oxygen species and accelerates the drug metabolism of everything else, which is why the chronic drinker is at once tolerant to sedatives and exquisitely vulnerable to paracetamol. Steatosis, steatohepatitis, cirrhosis: the staircase is metabolic from first step to last.*

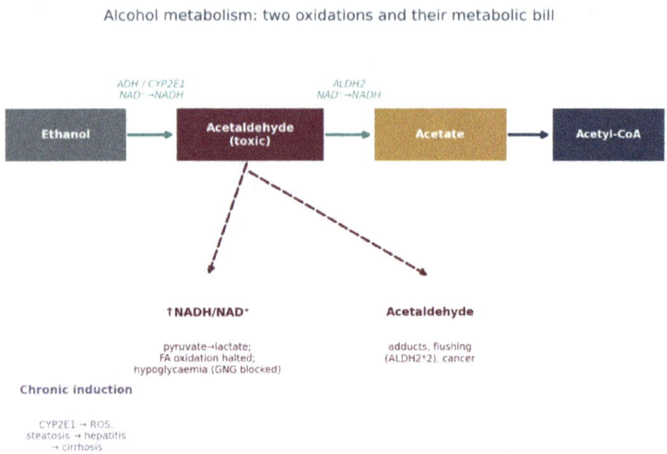

***Figure 41.5* — Alcohol metabolism.** *Two NAD^+-consuming oxidations (ADH/CYP2E1, then ALDH2) and their bill: a raised $NADH/NAD^+$ ratio that blocks gluconeogenesis and fatty-acid oxidation, acetaldehyde toxicity, and — chronically — CYP2E1 induction with its oxidant and pharmacokinetic consequences.*

41.12 Biochemical Foundations of Dietary Claims

It is a sorry fact that this section exists, and a necessary one. The dietary marketplace sells molecular claims, and the molecules are examinable. "Detox": the liver detoxifies continuously via the enzymes of Chapter 36's cytochrome P450s and conjugation pathways; no juice has been shown to induce them faster than abstinence from what needs detoxifying. Alkaline diets: blood pH is 7.40 ± 0.05, defended by buffer, lung and kidney (Chapter 27); food changes urine pH, which is the kidney doing its job, not the body becoming alkaline. "Superfoods": antioxidant capacity in a test tube (ORAC) does not survive the pharmacokinetics — most polyphenols are absorbed at a few percent and conjugated beyond recognition, and large trials of antioxidant supplements have been null or harmful; the vitamin that works is the vitamin one is deficient in. Collagen supplements: collagen is digested to amino acids like any protein; glycine and proline are non-essential, and the body makes its own collagen where it chooses, vitamin C permitting. What survives scrutiny is dull: energy balance, protein adequacy, fibre, the essential fatty acids, the micronutrients of Sections 41.4–41.7, and dietary patterns — Mediterranean above all — whose evidence rests on whole-diet trials rather than single molecules. The glamour is in the gaps of the evidence; the biochemistry, as usual, is in the bookkeeping.

> ***Case study* — *The rice diet of Batavia.*** *In the 1880s, beriberi — polyneuritis, oedema, heart failure — was epidemic in the Dutch East Indies. Christiaan Eijkman, sent to find the germ, noticed that the laboratory chickens fed on polished (white) rice developed a paralytic illness identical to the prisoners', and that chickens returned to unpolished rice recovered. He had found no germ; he had found a deficiency. Identify the molecule, the mechanism, and the reason polished rice causes it.*
> ***Resolution.*** *The missing factor is thiamine (vitamin B_1), concentrated in the rice bran and germ that polishing removes. As TPP it is the cofactor of*

pyruvate and α-ketoglutarate dehydrogenase and of transketolase; a carbohydrate-based diet (white rice) raises thiamine demand exactly as supply is removed. The nervous system and heart — obligate carbohydrate burners — fail first: hence the polyneuritis and the high-output heart failure of wet beriberi. The prisoners' diet was not poisoned; it was impoverished. Eijkman's chickens were the first deliberate animal model of a deficiency disease, and the 1929 Nobel recognised that the experiment changed medicine's question from "what toxin causes this?" to "what is missing?". The episode remains the template: pellagra, scurvy and rickets were all solved the same way, by subtraction.

Historical note — Eijkman, the chickens, and the accident of the new cook. *The story is better than its usual telling. Eijkman's chickens sickened in 1896 and recovered in 1897, and the recovery was traced not to any intervention of his but to a change of kitchen contractor: the new cook refused to let the military hospital's polished rice be fed to laboratory animals, and the birds went back to cheap unpolished rice and got well. Eijkman had the wit to see that the accident was the experiment. He named the protective factor the "anti-beriberi factor"; his successor Grijns phrased the modern doctrine — that natural foods contain substances indispensable to health, whose absence causes disease — and Funk, a decade later, coined "vitamine" for what he wrongly supposed to be an amine. Hopkins shared the 1929 Nobel for showing the same principle with purified diets and rats; Casimir Funk got the word, Hopkins got the prize, and Eijkman got the chickens. The doctrine founded this entire chapter: Sections 41.4 to 41.7 are, every one of them, a rice-polishing experiment with the chemistry filled in.*

Plate 41.1 *Christiaan Eijkman (1858–1930), whose chickens revealed the first vitamin-deficiency disease. Nobel laureate in Physiology or Medicine, 1929, shared*

with Frederick Gowland Hopkins. (Nobel Foundation portrait.)

41.13 Exercises

Exercise 41.1. *A 70 kg septic patient excretes 18 g urinary nitrogen/day while receiving 40 g protein/day. Calculate his nitrogen balance and the approximate daily muscle loss (muscle ≈ 20% protein; protein is 16% nitrogen).*
Resolution. *Intake N = 40 × 0.16 = 6.4 g. Balance = 6.4 − 18 = −11.6 g N/day (adding ~2 g for non-urinary losses, ≈ −13 g). Protein equivalent = 13/0.16 ≈ 80 g/day; muscle equivalent ≈ 400 g/day. Two weeks of sepsis thus costs kilograms of lean mass — the arithmetic behind the ICU's early-feeding doctrine, and the reason feeding blunts but cannot abolish the loss.*

Exercise 41.2. *Explain why intravenous glucose given to a thiamine-deficient alcoholic can precipitate Wernicke's encephalopathy.*
Resolution. *Glucose oxidation through pyruvate dehydrogenase requires TPP; the deficient patient has none to spare. The glucose load raises pyruvate flux against a cofactorless enzyme, pyruvate backs up to lactate, and the brainstem nuclei with the highest glucose dependence (mammillary bodies, periaqueductal grey) fail — ataxia, ophthalmoplegia, confusion. The rule "thiamine before glucose" is therefore enzyme kinetics made protocol: restore the cofactor before raising the substrate flux.*

Exercise 41.3. *Why does B_{12} deficiency cause megaloblastic anaemia, why does folate repletion correct the anaemia while worsening the neurology, and what biochemical measurement distinguishes the two deficiencies?*
Resolution. *Without B_{12}, 5-methyl-THF cannot transfer its methyl to homocysteine; folate is trapped and the thymidylate/purine one-carbon supply fails — nuclear maturation lags cytoplasmic, and megaloblasts appear. Folate repletion refills the pool downstream of the trap, restoring DNA synthesis — but it does nothing for methylmalonyl-CoA mutase or myelin, and by improving the blood it delays diagnosis while the cord degenerates. The distinguishing measurement: methylmalonic acid rises in B_{12} deficiency (mutase blocked) but not in folate deficiency; homocysteine rises in both. Two analytes, one differential.*

Exercise 41.4 (proposition for research). *The diet trials converge on ~2 kg mean loss at two years, yet the variance within each arm is large. Propose a trial design that tests whether genotype- or microbiome-matched diet assignment beats standard assignment, and state the statistical pitfall that has sunk previous attempts.*
Sketch. *Randomise to (i) algorithm-matched diet (multi-omic score predicting glycaemic response to carb vs fat) versus (ii) the better of the two standard diets, with the primary endpoint the interaction (matching × diet on 12-month weight and postprandial triglycerides), not either main effect. The pitfall: previous "responder" analyses tested main effects and retrofitted interactions — subgroup fishing. The interaction must be pre-specified, powered (n ≈ 4× a main-effect trial), and the algorithm locked before unblinding. Personalised nutrition will be believed when an interaction, not a subgroup, survives.*

41.14 Synthesis

The customs ledger is complete: macronutrients in gram quantities with protein counted in nitrogen, two essential fatty acids, thirteen vitamins that are coenzyme parts or signals, seven minerals and a dozen trace elements, all read by the mTOR–AMPK–sirtuin sensors, and all failing in characteristic ways — marasmus and kwashiorkor, beriberi and scurvy, the fatty liver of ethanol and the convergent two kilograms of every named diet. Direction checked against the master plan: the organism is fed; what remains is to watch its homeostasis fail in the clinic. ***Chapter 42 — Clinical Metabolism and Homeostatic Failure*** *takes up diabetes, the metabolic syndrome, fatty liver, cachexia, inborn errors, and the acid–base catastrophes — metabolism with the gloves off.*

41.15 References

Carpenter, K. J. (2000). Beriberi, white rice, and vitamin B: A disease, a cause, and a cure. University of California Press.

Frayn, K. N. (2019). Metabolic Regulation: A Human Perspective (4th ed.). Wiley-Blackwell.

Hall, K. D., & Kahan, S. (2018). Maintenance of lost weight and long-term management of obesity. Medical Clinics of North America, 102, 183–197.

Institute of Medicine. (2005). Dietary Reference Intakes for Energy, Carbohydrate, Fiber, Fat, Fatty Acids, Cholesterol, Protein, and Amino Acids. National Academies Press.

Liu, D., et al. (2018). Nutrient sensing and mTORC1 signalling. Nature Reviews Molecular Cell Biology, 19, 637–655.

Semba, R. D. (2012). The discovery of the vitamins. International Journal for Vitamin and Nutrition Research, 82, 310–315.

Shils, M. E., Shike, M., Ross, A. C., Caballero, B., & Cousins, R. J. (Eds.). (2006). Modern Nutrition in Health and Disease (10th ed.). Lippincott Williams & Wilkins.

Speakman, J. R., et al. (2021). Daily energy expenditure through the human life course. Science, 373, 808–812.

Chapter 42
Clinical Metabolism and Homeostatic Failure

Contents of this chapter

"Homeostasis is invisible until it fails; then every pathway it governed appears in the chart at once, like creditors."

42.1 Diabetes Mellitus

Part VIII closes where its biochemistry began — with insulin — but now under failure. Diabetes mellitus is not one disease but a final common pathway: chronic hyperglycaemia from any cause, diagnosed at a fasting glucose ≥ 7.0 mM, a 2-hour value ≥ 11.1 mM, or HbA1c ≥ 48 mmol/mol (6.5%)[56] — and its two great forms fail by opposite routes. Type 1 is autoimmune β-cell destruction: absolute insulin deficiency, onset at any age but classically in youth, ketosis-prone because without insulin the adipocyte opens its gates and the liver, flooded with fatty acids and commanded by unopposed glucagon, makes ketones regardless of glucose. Type 2 is insulin resistance with β-cell exhaustion: relative deficiency, ketosis-resistant for decades because a remnant of insulin action keeps the adipocyte's gates shut — until, in some, it no longer does. MODY and the other monogenic forms remind the clinician that "type 2 in a thin 19-year-old" may be a single broken transcription factor wearing type 2's clothes.

42.2 Diabetic Emergencies: DKA, HHS, and the Ketotic States

The emergencies sort themselves by one variable: how much insulin effect remains (Figure 42.1). ***Starvation ketosis*** *is physiology — insulin low, glucagon high, ketones 3–6 mM, glucose normal, pH untouched; the brain is being fed, not poisoned.* ***Diabetic ketoacidosis*** *is the same programme without a governor: zero insulin effect, ketones ≥ 10–15 mM, glucose high (the kidney is perfused with glucose it cannot reclaim), and the fixed-acid load driving pH below 7.3 — an acidosis that kills by arrhythmia and cerebral oedema, treated with insulin, fluid, and the potassium that insulin drags back into cells.* ***Hyperosmolar hyperglycaemic state*** *is type 2's emergency: enough insulin survives to suppress ketogenesis but none to lower glucose, which climbs to 40–60 mM; the osmotic diuresis dehydrates for weeks, and the presentation is coma with an osmolality, not an acidosis. Three states, one axis — the insulin–glucagon ratio of Chapter 39 — read at three positions.*

56. HbA1c is glucose non-enzymatically glycating haemoglobin's N-terminal valine — the erythrocyte as a 120-day glucose recorder. Every fructosamine and A1c assay is the same Maillard chemistry that browns bread, read out on a patient.

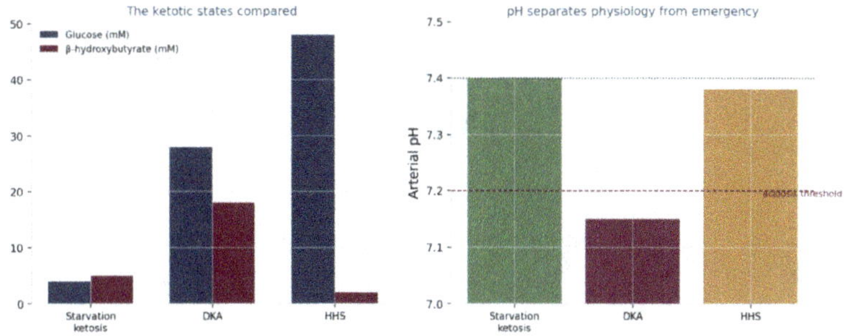

Figure 42.1 — The ketotic states. *Left: glucose and β-hydroxybutyrate in starvation ketosis, DKA and HHS. Right: arterial pH — the variable that separates physiology from emergency. Starvation is high-ketone, normal-glucose, normal-pH; DKA is all three deranged; HHS is glucose and osmolality without ketones.*

42.3 Metabolic Syndrome

Between the healthy and the frankly diabetic lies the commonest metabolic disorder of the developed world: the metabolic syndrome — central adiposity plus any two of raised triglycerides, low HDL, hypertension and raised fasting glucose. Its unity is not definitional but mechanistic (Figure 42.2): visceral adipose tissue, hypertrophied and inflamed, exports fatty acids and cytokines; the liver, bathed in both, becomes insulin-resistant and overproduces VLDL; muscle, offered fat it cannot fully oxidise, accumulates diacylglycerol and blunts its insulin signalling (Chapter 39.3); the vessels, hypertensive and glycated, complete the syndrome. Whether the syndrome is more than the sum of its parts has been debated for thirty years — Reaven's original "syndrome X" was an argument that insulin resistance is the common soil — but the clinical fact stands: its components travel together, and their conjunction multiplies cardiovascular risk faster than addition allows.

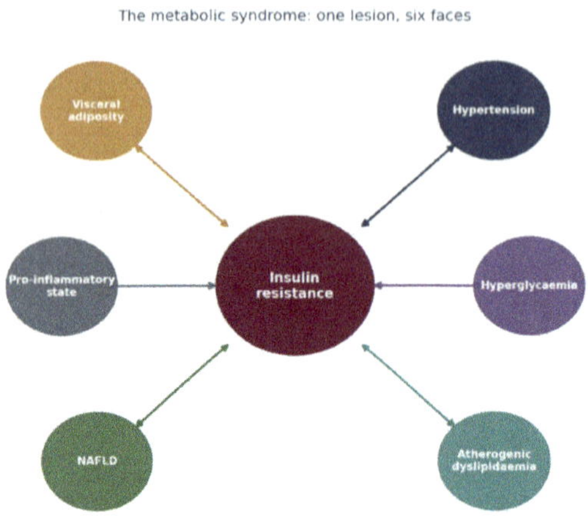

Figure 42.2 — ***The metabolic syndrome***. *Insulin resistance at the centre, with its six travelling companions; every edge is bidirectional, which is why treating any one component alone disappoints.*

42.4 Dyslipidaemias

The lipoprotein disorders (the machinery is Chapter 36.8–9) sort by particle. ***Familial hypercholesterolaemia*** *is the LDL receptor's disease — heterozygotes run LDL cholesterol two- to threefold normal and infarct in their forties; homozygotes, untreated, in their teens — and it proved the causal chain from LDL to atherosclerosis long before the statin trials settled it. The* ***hypertriglyceridaemias*** *are chylomicron and VLDL disorders: familial chylomicronaemia (lipoprotein lipase or its cofactors) gives eruptive xanthomas and pancreatitis at triglycerides above 20 mM, while the common mixed picture is insulin resistance's dyslipidaemia — high triglyceride, low HDL, small dense LDL — the atherogenic triad of the metabolic syndrome. The therapeutic logic follows the particles: statins upregulate the LDL receptor; PCSK9 inhibitors protect it from degradation; fibrates and the omega-3s address the triglyceride arm. Cholesterol, like glucose, is a molecule the body makes and a level the population argues about; the Mendelian randomisations have sided with the biochemists.*

42.5 Fatty-Liver Disease

Non-alcoholic fatty liver disease — increasingly renamed metabolic dysfunction-associated steatotic liver disease — is the hepatic face of insulin resistance, and it climbs a staircase (Figure 42.3). The first step is simple steatosis: hepatic triglyceride above 5%, driven by the three convergent supplies of Chapter 36 — adipose fatty acids that insulin can no longer suppress, de novo lipogenesis that insulin still drives (the resistance is selective, to glucose handling but not to lipogenesis — the cruel asymmetry at the disease's heart), and dietary fructose, which enters lipogenesis downstream of its governors. Most carriers stay on the first step for life. A minority ascend to steatohepatitis — fat plus inflammation and hepatocyte injury — and thence to fibrosis, cirrhosis and hepatocellular carcinoma; what pushes a liver from storage to inflammation is the field's open question, with lipotoxic species (ceramides, diacylglycerols) the leading suspects.

The treatment remains the removal of the cause: weight loss of 10% reverses steatohepatitis in most and fibrosis in many — a drug the liver prescribes and the patient must manufacture.

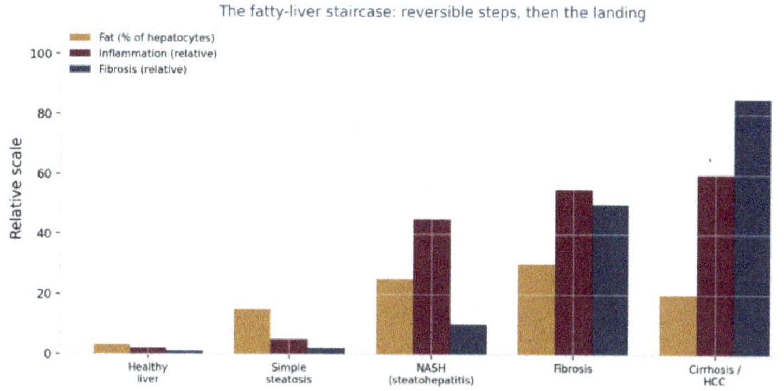

Figure 42.3 — The fatty-liver staircase. *Fat accumulates early and plateau; inflammation and fibrosis rise later and decide the prognosis. The first steps reverse with weight loss; the landing does not.*

42.6 Starvation and Cachexia

Starvation, Chapter 39.10's masterpiece of adaptation, has a malignant twin. ***Cachexia*** *— the wasting of cancer, advanced heart failure, chronic infection — is starvation with the adaptation broken: inflammatory cytokines (TNF-α, IL-6, IL-1) drive proteolysis and lipolysis regardless of intake, so feeding cannot reverse it; the resting metabolic rate rises where starvation's falls; and the protein lost comes disproportionately from muscle via the ubiquitin–proteasome system (Chapter 21). The distinction is not academic. The starved patient, fed carefully (remembering refeeding syndrome, Chapter 41.9), recovers; the cachectic patient, fed identically, does not, because the lesion is a signal, not a supply. Modern therapy therefore aims at the signal — anamorelin's ghrelin mimicry, the myostatin and IL-6 blockers in trial — while nutrition holds the line. The contrast teaches the chapter's recurring lesson: metabolism fails in two ways, supply failure and signal failure, and treating one as the other is the commonest error in clinical nutrition.*

42.7 Inborn Errors of Metabolism

The inborn errors are the pathway diagrams of Part VII with one arrow cut (Figure 42.5), and their grammar is uniform: substrate accumulates upstream, product runs short downstream, and the toxicity lies in between. ***Phenylketonuria*** *— phenylalanine hydroxylase — is the model and the triumph: phenylalanine accumulates, tyrosine becomes essential, and the developing brain is poisoned by the excess; newborn screening and a low-phenylalanine diet convert a cause of institutionalisation into a manageable diet.* ***Galactosaemia*** *(GALT) accumulates galactitol — the cataract, the failing liver — cured by removing galactose, which is to say milk.* ***MCAD deficiency*** *blocks medium-chain β-oxidation: well until fasting, then hypoketotic hypoglycaemia — the dangerous signature of a fatty-acid-oxidation defect, ketones absent when most needed — managed by never fasting. The* ***urea-cycle defects*** *(OTC foremost) accumulate ammonia itself, the one metabolite with no buffer; the neonate who vomits, seizes and hyperventilates on day two of life is a urea cycle until proven otherwise. The* *principles generalise: screen at birth, treat by substrate restriction or product replacement, and remember that the heterozygotes — one enzyme copy — are usually well, because most enzymes are present in vast kinetic excess.*

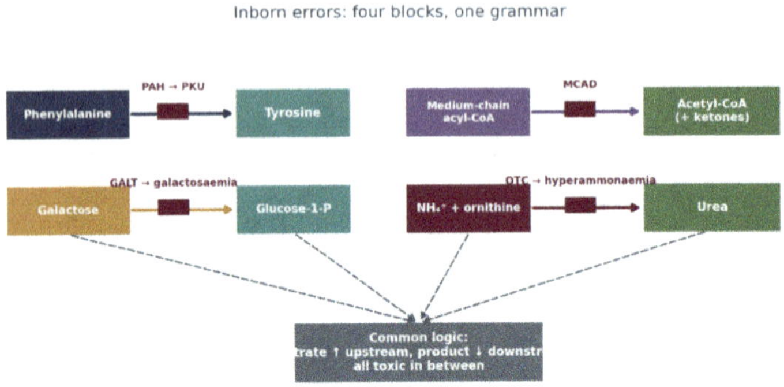

***Figure 42.5* — Inborn errors: four blocks, one grammar.** *Phenylketonuria, galactosaemia, MCAD deficiency and ornithine transcarbamylase deficiency drawn as what they are — one arrow cut in four different pathways — with the common logic below.*

42.8 Mitochondrial Disorders

When the lesion is in oxidative phosphorylation itself, the failure follows the energy ladder: the tissues with the highest ATP demand — brain, retina, heart, skeletal muscle, endocrine pancreas — fail first and together. The genetics are doubly treacherous. Mitochondrial DNA is maternally inherited and heteroplasmic — a cell carries hundreds of copies, mutant and wild-type mixed — so the same mutation gives a mother mild ptosis and her daughter MELAS (mitochondrial encephalopathy with lactic acidosis and stroke-like episodes), depending on where the mutant load landed during development[57]*; and many subunits are nuclear-encoded, so mitochondrial disease can be Mendelian too. The biochemical signature is a raised lactate at rest — the cell glycolysing past its broken turbines — with a lactate:pyruvate ratio elevated above 20:1 when the lesion is in the chain itself (NADH cannot unload), against a normal ratio in pyruvate-handling defects. No curative therapy exists; management is metabolic first aid — avoid fasting, treat the acidosis, give the cofactors that occasionally help — and, in prevention, mitochondrial donation has made the first of these diseases avoidable, if not yet treatable.*

42.9 Acid–Base Disturbances

Acid–base is the one homeostatic system the clinician reads as a single number, and the reading begins with Chapter 27's equation — now applied, and derived

57. Heteroplasmy's threshold effect explains the clinical rule of thumb: symptoms appear when mutant mtDNA exceeds ~60–90% in a tissue — the energetic cliff being steepest where the respiratory chain's reserve is thinnest. The ovum's bottleneck shuffles the proportion between generations, which is why genetic counselling here deals in probabilities, not pedigrees.

once more for the clinic:

$$\mathrm{pH} = 6.1 + \log_{10} \frac{[\mathrm{HCO_3^-}]}{0.03 \times P_a\mathrm{CO_2}} \tag{42.1}$$

Derivation. *$CO_2 + H_2O \rightleftharpoons H_2CO_3 \rightleftharpoons H^+ + HCO_3^-$ is a single buffer pair with pK_a 6.1 at body temperature; dissolved CO_2 is proportional to P_aCO_2 with solubility 0.03 mM/mmHg. Henderson–Hasselbalch applied to this pair, with $[CO_2]$ written as 0.03 × P_aCO_2, gives the equation. Two terms, two organs: the kidney owns the numerator (bicarbonate, days), the lung owns the denominator (CO_2, seconds) — and every acid–base disorder is one organ's failure read against the other's compensation.*

Four primary disorders, two compensations: metabolic acidosis (low HCO_3^-, lung blows off CO_2 — Kussmaul's breathing is the compensation made audible), metabolic alkalosis, respiratory acidosis and alkalosis. The metabolic acidoses sort by the ***anion gap*** *(Figure 42.4): Na^+ – (Cl^- + HCO_3^-), normally 8–16 mEq/L, widened when an unmeasured acid arrives — lactate, ketones, the toxins of renal failure, methanol's and ethylene glycol's metabolites — and normal when bicarbonate is simply lost (diarrhoea, renal tubular acidosis), chloride replacing it. The gap is arithmetic; the diagnosis is the list of acids the arithmetic points to.*

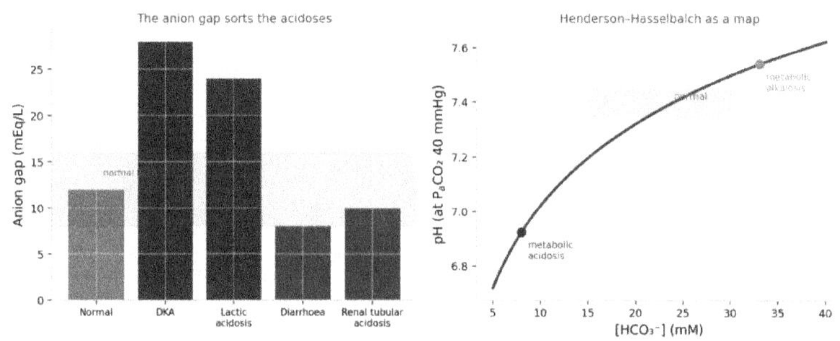

Figure 42.4 — Reading acid–base. *Left: the anion gap separates the acidoses of added acid (DKA, lactate) from those of lost bicarbonate. Right: Eq. (42.1) as a map — pH against bicarbonate at fixed P_aCO_2, with the normal band and the two metabolic directions marked.*

42.10 Biochemical Adaptation to Hypoxia

Hypoxia is failure of the supply chain's last link, and the cell's response is one transcription factor: HIF-1α, hydroxylated by prolyl hydroxylases when oxygen is present (and degraded within minutes via von Hippel–Lindau), stable when it is not. Stabilised HIF-1 raises glycolytic enzymes and GLUT1 (more anaerobic ATP), PDK1 (pyruvate dehydrogenase off, sparing oxygen), VEGF (new vessels) and erythropoietin (new erythrocytes) — the whole programme of Chapter 31's Pasteur effect made genomic. Altitude acclimatisation is this programme run by a healthy organism over weeks: raised ventilation first, then EPO and haematocrit, then the slower tissue adaptations; the same machinery, run by tumours, builds the angiogenesis that Chapter 31's Warburg effect demands. The prolyl hydroxylase is druggable — the HIF-PH inhibitors now treat renal anaemia by holding HIF-1 stable — and the pathway's discovery took the 2019 Nobel, a reminder that the most clinical of questions (why do we make more red cells in the Andes?) and the most molecular (how does a protein know the oxygen?) are one question.

42.11 Metabolic Consequences of Sepsis

Sepsis is homeostatic failure en masse — infection plus a dysregulated host response — and its metabolism (Figure 42.6) is the stress response of Chapter 39 detonated. The early hours bring hyperglycaemia (catecholamines, cortisol, cytokine-driven insulin resistance), accelerated proteolysis and lipolysis, and a lactate that rises from two sources the clinician must separate: hypoperfusion (true tissue hypoxia, glycolysis without oxygen) and adrenergic drive (aerobic glycolysis, β_2-stimulated Na^+/K^+-ATPase running hot) — the second being why a normalising lactate after fluids is reassuring but a high lactate with normal perfusion is not necessarily a fluid deficiency. Cytokine storms consume glutamine and arginine, the gut barrier fails (Chapter 40.7), and mitochondrial function declines — "cytopathic hypoxia," cells unable to use delivered oxygen, visible as a falling oxygen consumption despite adequate delivery. The metabolic management is deliberately modest: early feeding to protect the gut, glucose controlled but not normalised (the NICE-SUGAR lesson — tight control kills by hypoglycaemia), and the catabolism accepted as the price of the battle. Sepsis teaches the final humility of clinical metabolism: sometimes the best the biochemist can do is hold the ring while the inflammation burns itself out.

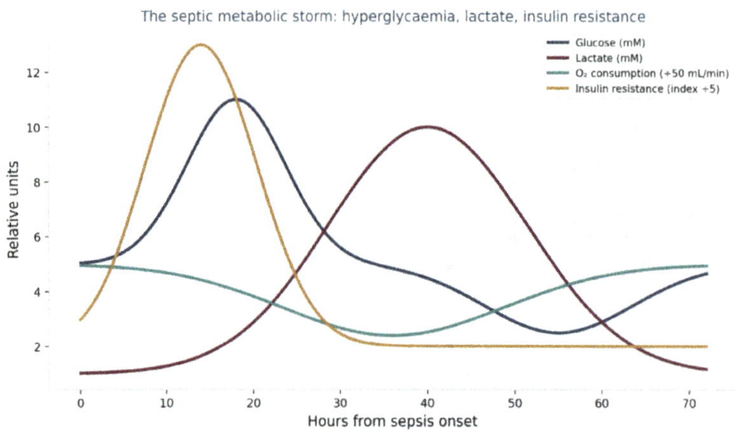

Figure 42.6 — The septic metabolic storm. *Hyperglycaemia and insulin resistance peak early; lactate climbs as oxygen consumption falls — the signature of cytopathic hypoxia: oxygen delivered, but the mitochondria declining to use it.*

> ***Case study — The breathless child with a sweet breath.*** *A 9-year-old, previously well, presents after a week of polyuria and weight loss with deep, sighing respirations, vomiting, and drowsiness. Glucose 31 mM, pH 7.12, HCO_3^- 6 mM, Na^+ 134, Cl^- 96, ketones 6.8 mM, K^+ 5.4 mM. Work the case.*
> ***Resolution.*** *New-onset type 1 diabetes in DKA. The anion gap: 134 − (96 + 6) = 32 — a high-gap metabolic acidosis (the acids are β-hydroxybutyrate and acetoacetate). The sighing respiration is Kussmaul's — the lung compensating, P_aCO_2 driven toward 12 mmHg (Eq. 42.1: with HCO_3^- 6, pH 7.12 requires $P_aCO_2 \approx 19 \times 0.03^{-1} \times$ ... the compensation is real but insufficient). The potassium of 5.4 with total-body potassium depleted is the trap: acidosis shifts K^+ out of cells, and insulin therapy will drag it back — so potassium is added*

to the fluids once urine flows, before the insulin drops it into the danger range. Treatment: volume first (she is ~10% dehydrated), then insulin infusion, glucose added to the fluids when blood glucose reaches 14 mM (insulin continues until the gap closes — the endpoint is the acid, not the sugar). The sweet breath is acetone, decarboxylated acetoacetate — the one metabolite the clinician can smell, and the oldest diagnostic test in this chapter.

Historical note — Houssay's dogs and the anterior pituitary's diabetic finger. *In Buenos Aires in the 1920s, Bernardo Houssay made the observation that founded endocrine metabolism as an experimental science: remove a dog's pituitary, and its diabetes — produced by pancreatectomy — melts away; inject anterior-pituitary extract, and the diabetes returns, fiercer. The "diabetogenic principle" of the pituitary was growth hormone and the adrenocortical axis — Chapter 39's counter-regulatory army, discovered by subtraction a generation before the hormones were purified. Houssay's institute became the school of Latin American physiology (the Coris, of Chapter 32's fame, trained under him), and his 1947 Nobel — shared with them — remains the only one awarded for work done in Latin America in that era, accomplished under a government that twice dismissed him for refusing to swear political loyalty and from which he worked on in privately funded laboratories. The doctrine his dogs established organises this whole chapter: diabetes is never insulin alone but insulin read against the pituitary–adrenal–thyroid background — the failure of a duumvirate being, in the end, a failure of a whole government.*

Plate 42.1 *Bernardo Alberto Houssay (1887–1971), who showed the pituitary governs carbohydrate metabolism and founded Latin American experimental physiology. Nobel laureate in Physiology or Medicine, 1947. (Nobel Foundation portrait.)*

42.12 Exercises

Exercise 42.1. *A patient has Na^+ 140, Cl^- 100, HCO_3^- 10 mM, pH 7.28, P_aCO_2 21 mmHg. Compute the anion gap, verify the compensation against Eq. (42.1), and give the likely causes.*
Resolution. *Gap = 140 − (100 + 10) = 30 — high-gap metabolic acidosis. Check: $6.1 + \log_{10}(10/(0.03 \times 21)) = 6.1 + \log_{10}(15.9) = 6.1 + 1.20 = 7.30 \approx 7.28$ — internally consistent, and the respiratory compensation is appropriate (Winter's formula: expected $P_aCO_2 \approx 1.5 \times 10 + 8 = 23 \pm 2$). Causes: the unmeasured acids — lactate, ketoacids, renal failure's retained acids, or a toxic alcohol. The arithmetic takes a minute; it has already halved the differential.*

Exercise 42.2. *Distinguish hypoketotic hypoglycaemia from the hypoglycaemia of hyperinsulinism, and explain why the distinction changes the investigation.*
Resolution. *In hyperinsulinism (insulinoma, sulfonylurea), insulin suppresses lipolysis, so fatty acids are unavailable and ketones are absent — but so is every alternative fuel: the brain has nothing. In fatty-acid-oxidation defects (MCAD), insulin is appropriately low, lipolysis proceeds, fatty acids reach the liver, but β-oxidation cannot make acetyl-CoA — hence no ketones despite open adipose gates, and the accumulating acyl-CoAs block gluconeogenesis. Both are hypoketotic; the difference is free fatty acids (low in hyperinsulinism, high in MCAD) and the acylcarnitine profile. One measurement of FFA at the hypoglycaemic moment splits the two families — which is why the "critical sample" must be drawn before the glucose goes in.*

Exercise 42.3. *Why does tight glycaemic control harm septic patients when moderate control helps, and what does the NICE-SUGAR result imply about the metabolism of critical illness?*
Resolution. *Targeting 4.5–6.0 mM with insulin in the ICU raised mortality (NICE-SUGAR, 2009), driven by severe hypoglycaemia; 8–10 mM is the settled target. The implication is that the hyperglycaemia of sepsis is partly adaptive — a stress response defending fuel delivery to non-insulin-dependent tissues — and that forcing it to "normal" treats the number, not the patient, while the insulin creates hypoglycaemic excursions the monitoring misses. Critical-illness metabolism is not failed normal metabolism; it is a different regime with its own optimum, and the trial that proved it is clinical biochemistry's favourite cautionary tale.*

Exercise 42.4 (proposition for research). *Cachexia cannot be reversed by feeding. Propose a combination trial for cancer cachexia targeting signal and supply simultaneously, and specify the endpoint that would prove the signal arm is doing the work.*
Sketch. *Four arms in advanced solid tumours: standard nutrition; nutrition + anamorelin (ghrelin agonist — appetite and GH axis); nutrition + anti-IL-6; and the triple combination. Primary endpoint: appendicular lean mass by DEXA at 12 weeks. The mechanistic proof: the signal arms must show restoration of the anabolic response to a mixed meal (postprandial muscle protein synthesis by tracer phenylalanine incorporation), not merely more intake — if lean mass rises only with food diaries and not with meal-response*

restoration, the signal was untouched and the drug is an appetite pill. Cachexia will yield when a trial separates eating from responding.

42.13 Synthesis

The failures sort, in the end, into two families. Supply failures — starvation, hypoxia, the inborn errors with their cut arrows — break the delivery of substrate or oxygen; signal failures — diabetes, the metabolic syndrome, cachexia, sepsis — break the government that reads the supply. The clinic lives in the second family, and its instruments are few: the anion gap, the ketone strip, the critical sample, and Henderson–Hasselbalch. Direction checked against the master plan: Part VIII closes with metabolism integrated, fed, and failed; the book now turns from energy and matter to information. ***Chapter 43 — Nucleotide and Nucleic-Acid Structure*** *opens Part IX — the double helix and everything that follows from it.*

42.14 References

Alberti, K. G. M. M., Eckel, R. H., Grundy, S. M., et al. (2009). Harmonizing the metabolic syndrome. Circulation, 120, 1640–1645.

American Diabetes Association. (2024). Standards of care in diabetes—2024. Diabetes Care, 47, S1–S350.

Egan, B., & Zierath, J. R. (2013). Exercise metabolism and the molecular regulation of skeletal muscle adaptation. Cell Metabolism, 17, 162–184.

Fearon, K. C. H., Glass, D. J., & Guttridge, D. C. (2012). Cancer cachexia: Mediators, signaling, and metabolic pathways. Cell Metabolism, 16, 153–166.

Friedman, S. L., Neuschwander-Tetri, B. A., Rinella, M., & Sanyal, A. J. (2018). Mechanisms of NAFLD development and therapeutic strategies. Nature Medicine, 24, 908–922.

Kaelin, W. G., & Ratcliffe, P. J. (2008). Oxygen sensing by metazoans: The central role of the HIF hydroxylase pathway. Molecular Cell, 30, 393–402.

Kitabchi, A. E., Umpierrez, G. E., Miles, J. M., & Fisher, J. N. (2009). Hyperglycemic crises in adult patients with diabetes. Diabetes Care, 32, 1335–1343.

NICE-SUGAR Study Investigators. (2009). Intensive versus conventional glucose control in critically ill patients. New England Journal of Medicine, 360, 1283–1297.

Saudubray, J.-M., Baumgartner, M. R., & Walter, J. H. (Eds.). (2016). Inborn Metabolic Diseases: Diagnosis and Treatment (6th ed.). Springer.

Part IX

Nucleic Acids and Molecular Information

Chapter 43
Nucleotide and Nucleic-Acid Structure

Contents of this chapter

"The cell's first problem was not how to burn fuel but how to remember the recipe — and the solution turned out to be a pair of stairs, spiralling."

43.1 Bases, Nucleosides and Nucleotides

Part IX changes the book's subject. Parts VII and VIII asked how the cell pays its energy bill; this Part asks how it stores the instructions for every enzyme that does the paying — and the storage medium is the nucleic acid. The vocabulary is a three-rung ladder (Figure 43.1): a ***base*** *(the purines adenine and guanine, two fused rings; the pyrimidines cytosine, uracil and thymine, one) becomes a* ***nucleoside*** *when glycosidically bonded to ribose or 2′-deoxyribose (adenosine, guanosine, cytidine, uridine; thymidine is deoxy by convention), and a* ***nucleotide*** *when phosphorylated on the sugar's 5′-hydroxyl. The numbering discipline matters clinically: atoms of the base are unprimed, of the sugar primed — and the antimetabolite drugs of Chapter 38 exploit exactly this chemistry, 5-fluorouracil being uracil with a fluorine where thymine puts its methyl. The student has met these molecules before: ATP, NAD^+, FAD, coenzyme A are all nucleotides moonlighting as coenzymes (Chapter 41). Evolution evidently found the nucleotide useful long before it found it informative.*

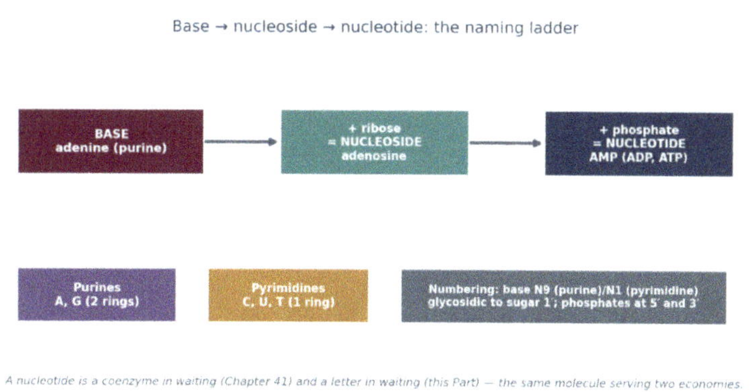

Figure 43.1 — The naming ladder. *Base → nucleoside (+ sugar) → nucleotide (+ phosphate), with the purine/pyrimidine division below and the primed numbering convention. The same molecules serve as letters, as coenzymes, and — in ATP — as currency.*

43.2 Phosphodiester Bonds

Polymerisation joins the 3′-hydroxyl of one sugar to the 5′-phosphate of the next — the ***phosphodiester bond*** *— and two consequences follow. First, every nucleic-acid strand has a direction: a free 5′ end and a free 3′ end, and all the enzymes of this Part respect the polarity, synthesis always reading the template 3′→5′ and building the new strand 5′→3′. Second, the backbone is uniformly negative — one charge per phosphate at physiological pH — so DNA is an acid in fact as well as name, wrapped around basic histones, repelled by itself, and precipitated by alcohol and salt in every undergraduate practical. The backbone is also remarkably inert: the phosphodiester bond's spontaneous hydrolysis has a half-life of millions of years at 25 °C[58], which is to say the cell pays almost nothing to keep its archive intact. Chemistry chose the archive's material well before biology did.*

43.3 DNA Double-Helical Structure

The 1953 structure needs no introduction, but it repays a careful one (Figure 43.2). Two antiparallel strands wind right-handedly around a common axis; the bases stack in the interior like the treads of a spiral stair, 0.34 nm apart, ten and a half base pairs per 3.4 nm turn; the sugar–phosphate rails run outside. The pairing is specific — A with T through two hydrogen bonds, G with C through three — and complementary, so each strand is a complete description of the other: the secret of replication was visible in the structure before any enzyme was known, as Watson and Crick remarked in the most famous understatement in the literature. Two grooves spiral down the surface — a wide ***major groove****, where the edges of the base pairs expose a chemical signature unique to each of the four arrangements, and a narrow minor groove — and the major groove is where the regulatory proteins of Chapter 48 read the sequence without opening the helix. The helix is held not by the hydrogen bonds, which merely specify, but by base stacking — the hydrophobic and van der Waals interactions of the flat rings — a distinction that explains why stacking, not pairing, sets the thermodynamics of Section 43.9.*

58. The 2′-hydroxyl of RNA changes this utterly: positioned for in-line attack on the adjacent phosphorus, it makes RNA some 10^5-fold more alkali-labile than DNA — which is why the genome is kept in deoxy, and why RNA viruses mutate like quicksilver.

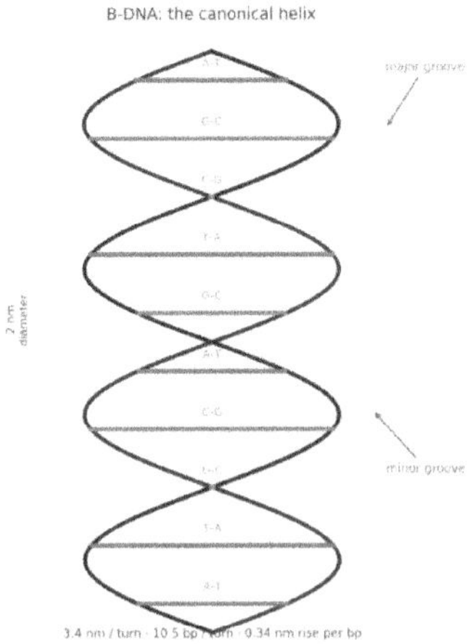

Figure 43.2 — B-DNA. *Antiparallel strands (navy, crimson), A–T pairs in teal (two hydrogen bonds), G–C in amber (three). The wide major groove and narrow minor groove give regulatory proteins two reading surfaces of different information content.*

43.4 Alternative DNA Conformations

B-DNA is the physiological default, not the only shape (Figure 43.3). Dehydrating conditions and certain sequences favour ***A-DNA*** *— shorter, wider, 11 base pairs per turn, the conformation RNA duplexes and DNA–RNA hybrids are obliged to adopt — while alternating purine–pyrimidine sequences (CGCGCG...) can flip into* ***Z-DNA****: left-handed, zig-zag backboned, found transiently at actively transcribed genes where torsional strain is highest. Beyond the duplex lie the exotic but functional structures:* ***G-quadruplexes****, four guanines in a planar tetrad stacked in guanine-rich tracts, formed at telomeres and in oncogene promoters and now a drug target;* ***triplexes*** *and slipped structures at repeats, implicated in the expansion diseases (Huntington's CAG among them); and cruciforms extruded from inverted repeats under superhelical strain. The double helix is the archive; these are the annotations written in its margins, and the margins are increasingly legible.*

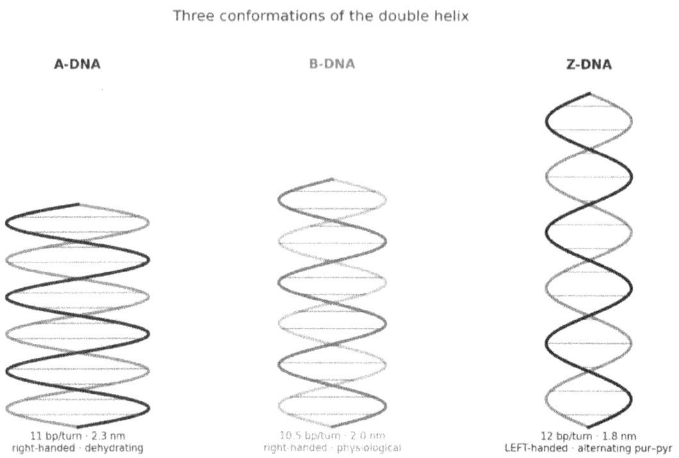

Figure 43.3 — Three conformations. *A-DNA (short, wide; the RNA duplex shape), B-DNA (physiological), Z-DNA (left-handed, sequence-specific). The same letters, three typefaces — and the cell reads all three.*

43.5 DNA Supercoiling

A closed DNA — bacterial chromosome, plasmid, or any topologically constrained loop — has a fixed ***linking number****, Lk = Tw + Wr: twist (the helical turns) plus writhe (the coiling of the axis itself). Since Lk changes only when a strand is cut, strain imposed on a closed molecule partitions between the two: underwind it and it writhes into supercoils, exactly as a twisted telephone cord does. Cellular DNA is maintained negatively supercoiled (Figure 43.4) — underwound — which stores strain that favours strand opening: the replication fork and the transcription bubble are paid for in advance by the supercoiling. The accountants are the* ***topoisomerases****: type I cut one strand and relax; type II cut both, pass a duplex through the gap, and — in bacterial gyrase — use ATP to introduce negative supercoils. The pharmacology is a roll-call: quinolones trap the gyrase–DNA complex, etoposide traps topoisomerase II in human tumours, and both kill by converting the enzyme's transient cut into a permanent double-strand break. Supercoiling, in short, is the genome's tension management, and poisoning the managers is some of our oldest chemotherapy.*

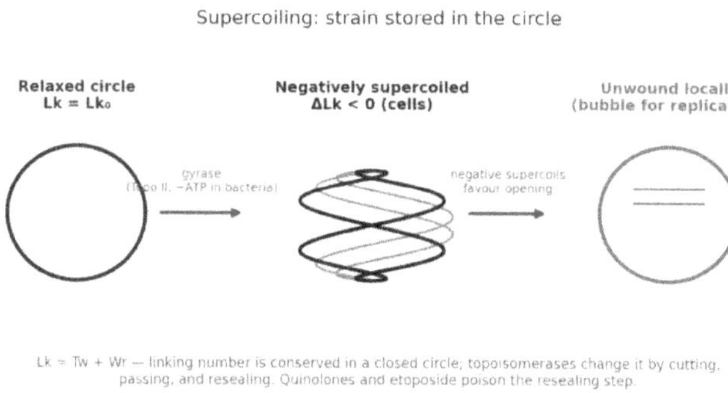

Figure 43.4 — Supercoiling. *A relaxed circle, the negatively supercoiled state maintained in cells, and the local opening that negative supercoils pay for. Lk = Tw + Wr is the conservation law; topoisomerases are its only licensed violators.*

43.6 RNA Structure

RNA is DNA's versatile sibling: usually single-stranded, therefore free to fold back on itself, and endowed with the 2′-hydroxyl that both constrains its helices to the A form and — as Section 43.2 noted — makes it chemically fragile. Folded RNA is a hierarchy: short-range **stem-loops** *(hairpins), then long-range interactions — pseudoknots, kissing loops, three- and four-way junctions — assembling into compact tertiary structures that can bind ligands with protein-like specificity (the aptamers) and catalyse reactions with protein-like skill (the ribozymes). The ribosome is RNA's cathedral — the peptidyl transferase at its heart is a ribozyme, the proteins mere buttresses — and the hammerhead, hairpin and hepatitis-delta ribozymes are its parish churches, each catalysing phosphodiester cleavage with the 2′-OH as nucleophile. The catalytic proficiency of RNA is the strongest single strand of the RNA-world hypothesis: before there were enzymes of protein, the argument runs, RNA was both the archive and the workforce — and the ribosome, which still makes proteins by RNA catalysis, is the surviving fossil.*

43.7 Non-Canonical Base Pairing

Watson–Crick pairing is the alphabet's standard font, but the bases have other ligatures. The **Hoogsteen** *face — the purine's second hydrogen-bonding surface, exposed in the major groove — pairs in triplexes and G-quadruplexes;* **wobble pairs** *(G·U above all) are tolerated in RNA and, as Chapter 47 will show, are the reason 61 codons can be read by far fewer tRNAs; and the sheared, bifurcated and water-mediated pairs fill the structural lexicon of folded RNA. There is even biology's private joke: Watson–Crick pairs in DNA occasionally flip to rare Hoogsteen configurations, with consequences for damage recognition that are still being mapped. The lesson generalises across this Part: the double helix is more polymorphic than its portrait, and the polymorphisms are where the regulation lives.*

43.8 Nucleic-Acid Stability

The thermodynamics of the duplex are the thermodynamics of stacking. Calorimetry resolves the duplex formation free energy into nearest-neighbour contributions — ten stacking parameters suffice to predict most sequences to within a fraction of a kcal/mol — and two regularities emerge: G·C-rich stacks are the most stable (better overlap and an extra hydrogen bond), and terminal pairs are worth less than internal ones (the ends fray). Counterions matter enormously: the phosphate charges repel, and magnesium or polyamines shield them — which is why PCR buffers specify Mg^{2+} and why sperm packs its DNA in protamines and spermidine. Duplex stability is thus an ionic as well as a base-pairing phenomenon, a fact every forensic laboratory exploits when it elutes DNA at low ionic strength.

43.9 Denaturation and Hybridisation

Heat a duplex and the strands part cooperatively — the ***melting*** *transition (Figure 43.5), monitored at 260 nm because stacked bases absorb less than free ones (the hyperchromic effect: the spectrophotometer can see the helix fall apart). The midpoint, T_m, rises with GC content and with salt, and for long DNA in standard conditions it obeys the empirical Marmur relation:*

$$T_m = 69.3 + 0.41\,(\%\mathrm{GC}) \quad [^\circ\mathrm{C}] \tag{43.1}$$

Derivation. From the nearest-neighbour thermodynamics of Section 43.8: GC-rich stacks melt higher, and over naturally occurring compositions the integrated stacking enthalpies and entropies are well approximated as linear in GC fraction. Marmur and Doty measured T_m against base composition for dozens of DNAs and fitted the line; the slope 0.41 °C per GC percent is the averaged extra stability of the G·C stack. For short oligonucleotides the same logic gives the PCR primer rule, $T_m \approx 4(G+C) + 2(A+T)$ — three hydrogen bonds against two, priced at two degrees each. The equation is calorimetry condensed, not theory imposed.

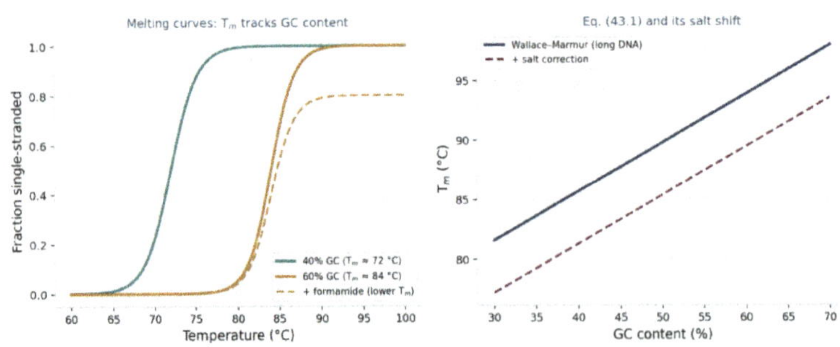

Figure 43.5 — Denaturation. *Left: melting curves — cooperative, GC-dependent, shifted by denaturants. Right: Eq. (43.1) plotted, with the upward salt correction; every hybridisation assay, Southern blot and PCR is an exercise in choosing the right side of these lines.*

Cool the melt and the strands reanneal — and a labelled strand will find its complement in a haystack of others: ***hybridisation****, the principle of Southern and Northern blots, microarrays, FISH, and every sequencing-by-hybridisation scheme. The stringency of the wash (temperature and salt) sets how perfect a match is demanded, which is the whole art of the assay. Denaturation and renaturation are,* *in the end, the only two operations molecular biology performs on DNA structure itself; everything else is enzymes.*

43.10 Genome Packaging

The human genome is two metres of DNA in a nucleus six micrometres across — a packing ratio of 10^5:1 that no amount of random coiling could achieve, and the solution is hierarchical (Figure 43.6). The first level is the ***nucleosome****: 147 base pairs wrapped 1.7 times around an octamer of histones (two each of H2A, H2B, H3, H4), with H1 sealing the DNA's entry and exit — the "beads on a string" of the electron micrographs, packing about sevenfold. The beads fold into higher-order fibres and then into* ***looped domains*** *— the topologically associating domains of the chromosome-conformation era — anchored by cohesin rings and CTCF boundaries, which segregate the genome into regulatory neighbourhoods. At mitosis the whole assembly condenses into the familiar chromosome. The profound point, developed in Chapter 48, is that the packaging is the regulation: the histone tails, lysine-rich and protruding, are acetylated to loosen and methylated to mark, and a gene wrapped tight is a gene switched off. The archive, unlike any human one, is shelved so that the shelving itself is an editorial decision.*

Figure 43.6 — Six orders of packing. *Naked double helix (2 nm), nucleosomal beads (11 nm), higher-order fibre, looped domains (TADs, ~300 nm), and the mitotic chromosome (1,400 nm). Histone chemistry at the first level sets the accessibility at every level above.*

> ***Case study — The fragile repeats that grew.*** *A family presents with a movement disorder appearing earlier in each generation: the grandfather at 55, the father at 42, the son at 28. The gene is Huntingtin; the lesion is a CAG trinucleotide repeat that is 18 copies long in unaffected relatives, 45 in the grandfather, 63 in the son. Explain the biochemistry of the anticipation.*
> ***Resolution.*** *The CAG repeat is a slipped-structure former (Section 43.4): during replication the two strands misalign — the repeat folds into hairpins on one strand — and the polymerase, copying across the slip, adds or subtracts repeat units. Longer repeats slip more, so the expansion accelerates down the generations: anticipation is replication slippage with a family tree. The encoded protein gains a polyglutamine tract that misfolds and aggregates (Chapter 20), and the toxicity rises with tract length — a structure problem at the DNA level producing a structure problem at the protein level. The same grammar runs through fragile X (CGG), myotonic dystrophy (CTG) and*

Friedreich's ataxia (GAA): the repeat diseases are this chapter's alternative conformations, inherited.

Historical note — King's College, Photograph 51, and the diffraction that decided the helix. *The double helix was not seen; it was inferred from the X-ray diffraction of DNA fibres, and the finest fibres and the finest photographs were made at King's College London. Maurice Wilkins began the work, Raymond Gosling took many of the images, and Rosalind Franklin — whose Photograph 51, the B-form pattern of 1952, displayed the cross that announces a helix and the layer-line spacing that fixes its pitch — extracted the geometry with a rigour the eventual models merely had to match. Franklin's data reached Watson and Crick in Cambridge by routes she had not authorised; their 1953 model was built to her measurements. She died of ovarian cancer in 1958, thirty-seven years old, four years before the Nobel that cannot be awarded posthumously went to Watson, Crick and Wilkins. The history is told here not to re-litigate it — the literature has done that thoroughly — but because the chapter's subject demands it: everything in Sections 43.3 to 43.5 rests on the pattern of spots that Franklin recorded, indexed and understood. The helix was measured before it was imagined.*

Plate 43.1 *Maurice Hugh Frederick Wilkins (1916–2004), whose King's College laboratory produced the diffraction data that made the double helix determinable. Nobel laureate in Physiology or Medicine, 1962, with Watson and Crick. (Nobel Foundation portrait.)*

43.11 Exercises

Exercise 43.1. *A bacterial genome is 21% adenine. Using Chargaff's rules and Eq. (43.1), give its base composition and estimate its melting temperature.*
Resolution. *Chargaff: A = T, G = C. A = 21% → T = 21%; G + C = 58%, so G = C = 29%.* $T_m = 69.3 + 0.41 \times 58 = 69.3 + 23.8 \approx 93$ *°C. The rules are a consequence of pairing — each strand mirrors the other — and the estimate explains why thermophiles, which run GC-rich, carry genomes pre-adapted to their kettles.*

Exercise 43.2. *Why is RNA alkaline-labile while DNA is not, and how is this used to free a plasmid preparation of RNA?*
Resolution. *The ribose 2'-OH is positioned for in-line nucleophilic attack on the adjacent phosphodiester; base deprotonates it, the 2'-oxygen displaces the 5'-leaving group through a cyclic 2',3'-phosphate intermediate, and the backbone is cleaved — a reaction DNA, lacking the 2'-OH, cannot run. Hence the plasmid miniprep: alkaline lysis hydrolyses RNA (and denatures chromosomal DNA), while the closed plasmid renatures on neutralisation. Alternatively RNase A does the job enzymatically. One missing hydroxyl is the entire difference between the archive and the working copy.*

Exercise 43.3. *A circular plasmid of 5,250 bp is fully relaxed. How many negative supercoils does gyrase introduce if the final Lk is 480, and why does this promote replication initiation?*
Resolution. *Relaxed* $Lk_0 = 5250/10.5 = 500$. $\Delta Lk = 480 - 500 = -20$ *supercoils (all as writhe if twist is unchanged). The negative strain partitions into unwinding where AT-rich sequences (the origin) open most easily — the replication bubble is nucleated by the stored strain. The plasmid pays for its own opening, and gyrase does the paying; quinolones, which trap gyrase mid-cut, make the payment lethal.*

Exercise 43.4 (proposition for research). *G-quadruplexes form in oncogene promoters and telomeres. Propose an experiment to test whether a G4-stabilising ligand suppresses an oncogene by structure rather than by general toxicity, and name the control that most such studies omit.*
Sketch. *Use CRISPR to install the G4-forming promoter sequence (e.g. c-MYC nuclease-hypersensitive element) and a point-mutated, non-folding isogenic version at the same locus; treat both lines with the ligand and read nascent transcription (PRO-seq), cell growth, and genome-wide G4 occupancy (G4-seq/CUT&Tag). The omitted control is the isogenic non-folding allele: without it, any antiproliferative effect is consistent with indiscriminate DNA binding. If transcription falls only in the folding line while global G4 occupancy is unchanged, the ligand is structural, not toxic — and G4-targeting graduates from biophysics to oncology.*

43.12 Synthesis

The molecule of Part IX is now assembled and characterised: bases, nucleosides, nucleotides; the phosphodiester backbone with its polarity and its patience; the B-helix and its A and Z alternatives; supercoiling as stored strain; RNA folding into catalysts; stacking thermodynamics condensed into one melting equation; and the nucleosome ladder that shelves two metres of archive into six micrometres. Direction checked against the master plan: structure exists to be copied. ***Chapter 44 — DNA Replication*** *takes up the machinery — polymerases, helicases, primases, and the beautiful asymmetry of leading and lagging strands.*

43.13 References

Alberts, B., Heald, R., Johnson, A., et al. (2022). Molecular Biology of the Cell (7th ed.). W. W. Norton.

Franklin, R. E., & Gosling, R. G. (1953). Molecular configuration in sodium thymonucleate. Nature, 171, 740–741.

Marmur, J., & Doty, P. (1962). Determination of the base composition of deoxyribonucleic acid from its thermal denaturation temperature. Journal of Molecular Biology, 5, 109–118.

Neidle, S. (2021). Principles of Nucleic Acid Structure (2nd ed.). Academic Press.

Olby, R. (2009). Francis Crick: Hunter of Life's Secrets. Cold Spring Harbor Laboratory Press.

SantaLucia, J. (1998). A unified view of polymer, dumbbell, and oligonucleotide DNA nearest-neighbor thermodynamics. Proceedings of the National Academy of Sciences, 95, 1460–1465.

Watson, J. D., & Crick, F. H. C. (1953). Molecular structure of nucleic acids: A structure for deoxyribose nucleic acid. Nature, 171, 737–738.

Chapter 44
DNA Replication

CONTENTS OF THIS CHAPTER

"Copying three billion letters in eight hours with fewer than a dozen uncorrected errors is less a biochemical feat than an act of clerical heroism."

44.1 Semiconservative Replication

The double helix of Chapter 43 proposed its own copying mechanism — each strand a template for the other — but three mechanisms were logically possible: semiconservative (each daughter duplex keeps one parental strand), conservative (the parental duplex survives intact beside an all-new one), and dispersive (parental and new material interspersed along each strand). Meselson and Stahl settled the question in 1958 with an experiment of crystalline economy (Figure 44.1): grow E. coli for generations in ^{15}N medium until every strand is heavy, shift to ^{14}N, and band the DNA on caesium chloride by density. After one generation all DNA was hybrid density — killing the conservative model at a stroke; after two, half hybrid and half light — killing the dispersive. Each strand of the helix is indeed conserved, passed whole into a daughter duplex, and paired with a strand made new. The experiment is taught as the most beautiful in biology not because it was clever — many are — but because the answer was readable in four bands.

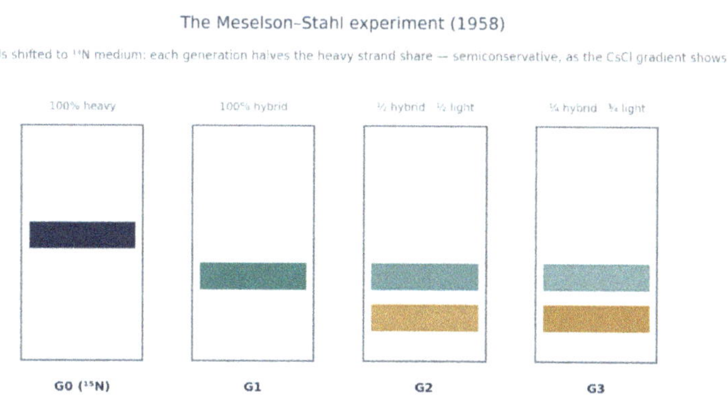

***Figure 44.1 — Meselson–Stahl, 1958.** ^{15}N-labelled cells shifted to ^{14}N medium: one generation gives a single hybrid band; two generations give hybrid and light in equal parts. Exactly and only what semiconservative replication predicts.*

44.2 Origins of Replication

Replication begins at defined sites. The bacterium has one — oriC, a ~245 bp AT-rich tract wrapped by the initiator DnaA, whose binding hydrolyses ATP and melts the duplex at its weakest (AT-richest) point — and fires it once per cell cycle, with the whole chromosome copied bidirectionally from that single start. The eukaryote, with chromosomes a hundred times longer, cannot afford one origin: the human genome fires some 30,000–50,000, spaced 50–300 kb apart. Eukaryotic origins are governed by a two-step discipline (Figure 44.6): in G1 they are ***licensed*** *— ORC recruits CDC6 and CDT1, which load the MCM2-7 helicase as an inactive double hexamer — and in S phase CDK and DDK kinases* ***fire*** *a subset, converting MCM into the active CMG helicase. Licensing is dismantled as S begins, so no origin can fire twice in one cycle: the genome is copied exactly once because the permission slips are issued and collected in different seasons. Cancer cells, which over-fire and re-fire origins, demonstrate by counterexample why the discipline exists.*

44.3 Helicases and Topoisomerases

Ahead of every fork, the strands must be parted and the strain relieved. The ***helicase*** *— MCM in eukaryotes, DnaB in bacteria, ring-shaped hexamers both — encircles one strand and translocates on ATP, wrenching the duplex open at hundreds of base pairs per second. The parted strands are kept apart by single-strand binding protein (SSB; RPA in eukaryotes), which coats the exposed DNA and prevents both reannealing and nuclease attack. And because unwinding the fork overwinds the DNA ahead (Chapter 43.5: Lk is conserved), the positive supercoils accumulate until a* ***topoisomerase*** *removes them — gyrase in bacteria, topoisomerase I/II in eukaryotes — the machinery the quinolones and etoposide poison. A circular bacterial chromosome adds a final topological problem: the two finished daughters are interlinked catenanes, unlinked by topoisomerase IV. Every stage of replication, it appears, is someone's drug target.*

44.4 Primases

Every DNA polymerase shares an embarrassing limitation: it can only extend an existing 3′-OH, never start one. The cell's answer is ***primase****, an RNA polymerase of no great elegance that lays down a short RNA primer — a dozen nucleotides — whenever a start is needed: once for each leading strand, and every hundred or two hundred nucleotides on the lagging. RNA primers must later be removed and replaced with DNA (RNase H and the 5′→3′ exonuclease of DNA polymerase I in bacteria; FEN1 in eukaryotes), and the nicks sealed by DNA ligase. The primer requirement is not a design flaw but a fidelity device: an enzyme that must be handed a correctly paired 3′ end cannot begin synthesis at a wrong one, and the disposable RNA start marks the least accurate nucleotides for replacement. The cell, like a good clerk, drafts in pencil where it is most likely to err.*

44.5 DNA Polymerase Chemistry

The chemistry at the fork's heart is one reaction repeated three billion times (Figure 44.3): the primer's 3′-oxygen attacks the α-phosphate of an incoming dNTP in an S_N2 displacement, the β- and γ-phosphates leaving as pyrophosphate. Two magnesium ions, held by invariant aspartates in the polymerase palm, do the catalytic work — one activates the 3′-OH, one stabilises the leaving group — and the reaction's near-reversibility is abolished by inorganic pyrophosphatase, which *hydrolyses PP_i and drags the equilibrium behind it. The enzyme is shaped like a right hand, palm gripping the primer–template junction, fingers closing over the incoming nucleotide only when the pair is correct — and this induced fit is the first fidelity mechanism: a mispaired dNTP sits wrong, the fingers cannot close, and the chemistry stalls. Base selection by geometry, not by hydrogen bonds alone, is what makes a polymerase accurate to one error in 10^5.*

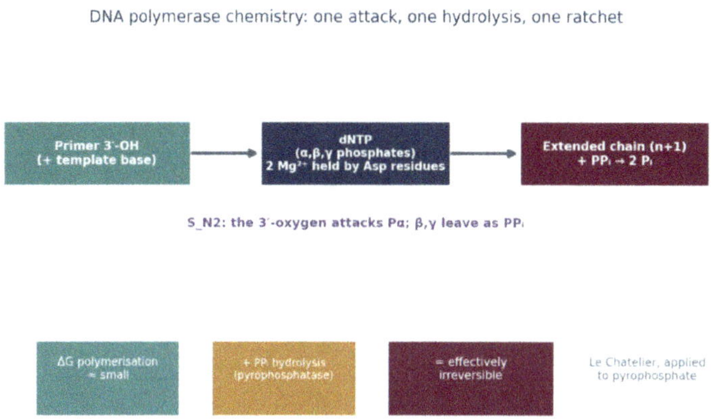

***Figure 44.3** — **Polymerase chemistry.** The primer 3′-OH attacks the α-phosphate; pyrophosphate leaves and is hydrolysed, making each step effectively irreversible. Two Mg^{2+} ions and an induced-fit finger closing do the catalysis and the first proofreading.*

44.6 Leading- and Lagging-Strand Synthesis

*Polymerase works only 5′→3′, but the fork opens in one direction — so the two new strands cannot be made alike (Figure 44.2). The **leading strand**, whose template runs 3′→5′ toward the fork, is synthesised continuously, one primer sufficing for a chromosome arm. The **lagging strand** is synthesised backward, away from the fork, in discontinuous **Okazaki fragments** — each primed anew, extended until it meets the previous fragment's primer, then processed: primer out, gap filled, nick sealed. The two polymerases travel together physically, the lagging template looped so that both halves of the replisome move in the same direction at the same speed — the "trombone model," so called for the loop's sliding growth and retreat. The asymmetry is forced by the chemistry of Section 44.5, and it has consequences all the way down: the lagging strand, repeatedly primed and patched, is the more error-prone, and the ends of linear chromosomes, where the last primer cannot be replaced, are where Section 44.10's problem begins.*

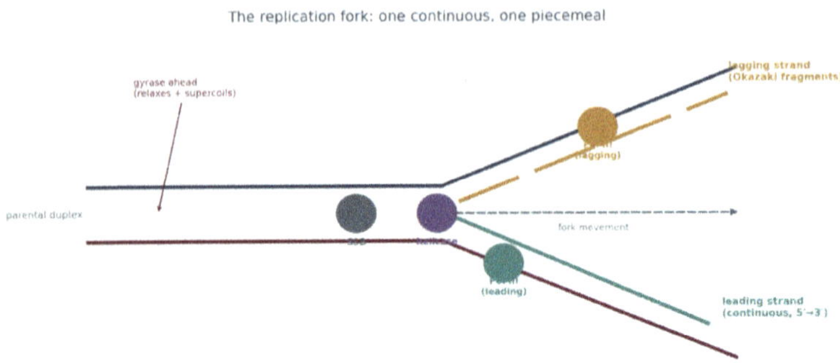

Figure 44.2 — The replication fork. *Helicase parts the strands (plum), SSB coats them (grey), gyrase drains the supercoils ahead. The leading strand (teal) runs continuously toward the fork; the lagging strand (amber) is built in Okazaki fragments away from it.*

44.7 Proofreading

One error in 10^5 is not good enough — a mammalian genome copied at that rate would carry 60,000 mutations per division — so accuracy is bought in layers (Figure 44.4). The second layer is the polymerase's own ***proofreading****: a 3′→5′ exonuclease active site that sits idle during correct synthesis but, when a mispaired 3′ terminus frayed by its misfit presents itself, bites off the offending nucleotide and lets the polymerase try again — worth a hundredfold. The third layer is post-replicative* **mismatch repair** *(Chapter 45), which patrols the new duplex for the mispairs and bulges that escaped, distinguishes daughter strand from parental (by methylation in bacteria, by strand discontinuities in eukaryotes), and excises and resynthesises the tract — worth another hundred- to thousandfold. The product of the three layers is the celebrated one error in 10^9–10^{10} nucleotides: a genome's worth of copying with a handful of mistakes, most of which Chapter 45's repair systems then face. The germ-line mutation rate that evolution lives on is, precisely, what three layers of clerical rigour let through.*

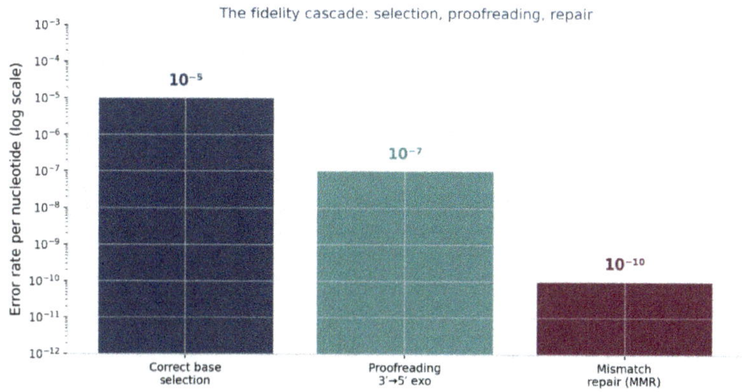

Figure 44.4 — The fidelity cascade. *Base selection (~10^{-5}), proofreading exonuclease (~10^{-7}), mismatch repair (~10^{-10}). Each layer costs energy and time; the cell pays gladly, the alternative being a genome that dissolves in error.*

44.8 Replication in Bacteria

*The bacterial replisome is the best understood and the most beautiful: a single machine, the **Pol III holoenzyme**, containing two (often three) core polymerases — one per strand — each held to the template by the β-clamp, a sliding ring loaded by the clamp-loader complex, which raises processivity from a few dozen nucleotides to tens of thousands. Helicase (DnaB), primase, the twin polymerases and the clamp loader travel as one particle; Okazaki fragments of 1,000–2,000 nucleotides are made behind, processed by Pol I (the enzyme Kornberg found first, whose true vocation is repair and primer removal, not replication), and sealed by ligase. The whole chromosome — 4.6 megabases — is copied in forty minutes at ~1,000 nucleotides per second, and when the cell is in a hurry it simply fires the new origins before the old forks finish: multifork replication, granddaughter chromosomes begun before daughters are done. E. coli's answer to a short life is not to copy faster but to overlap the copying.*

44.9 Replication in Eukaryotes

The eukaryotic fork runs slower — some 50 nucleotides per second, for the replisome must repeatedly pause to reassemble nucleosomes behind itself — but many more forks run at once. The division of labour differs too: Pol α-primase starts each fragment, Pol ε extends the leading strand, Pol δ the lagging, and the PCNA clamp (a trimeric ring) does what β-clamp does in bacteria. Chromatin adds a layer bacteria never face: parental nucleosomes are evicted ahead of the fork and redeposited — half old histones, half new — behind it, with the old histones' modifications diluted and then rewritten, the molecular basis of epigenetic inheritance through S phase (Chapter 48). And the cell cycle gates the whole affair through the licensing logic of Section 44.2: CDKs rise, origins fire, licensing collapses, and the genome passes through S exactly once. The complexity is not ornament; it is the price of copying a chromosome that is also a chromosome.

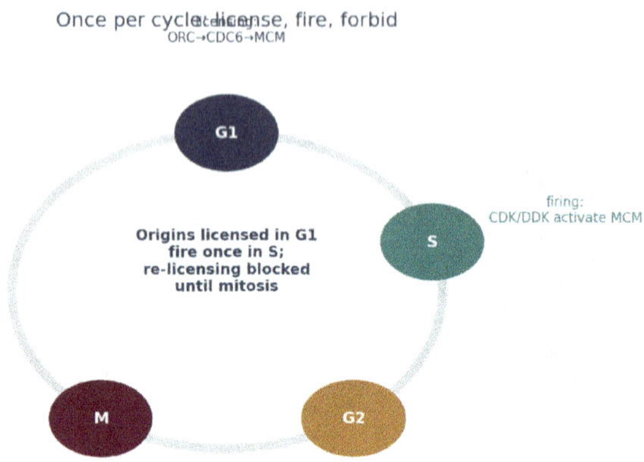

Figure 44.6* — *Once per cycle. *Origins are licensed in G1 (ORC → CDC6 → MCM double hexamer), fired in S (CDK/DDK activation of CMG), and cannot be re-licensed until mitosis resets the system.*

44.10 Telomeres and Telomerase

Linear chromosomes carry a structural debt (Figure 44.5). The lagging strand's final primer, at the very end, can be removed but not replaced — there is no upstream 3′-OH to extend from — so each round of replication leaves the daughter shorter by a primer's length: the ***end-replication problem****. The solution is to make the ends expendable and renewable. Vertebrate telomeres are kilobases of TTAGGG repeats, ending in a 3′ overhang that loops back and tucks into the duplex behind (the T-loop), disguising the end from the double-strand-break machinery; the shelterin protein complex polices the disguise.* ***Telomerase*** *— a reverse transcriptase carrying its own RNA template (TERC) within the TERT protein — extends the overhang repeat by repeat, whereupon conventional machinery fills the complementary strand. Telomerase is active in germ cells, stem cells and — the dark side — some 90% of cancers, which thereby purchase immortality; it is silent in most somatic cells, whose telomeres erode with division until the shelterin disguise fails and the cell enters senescence: the* ***Hayflick limit****, ageing written in repeat units. The bargain is visible in outline: finite ends limit rogue clones, at the price of finite tissues.*

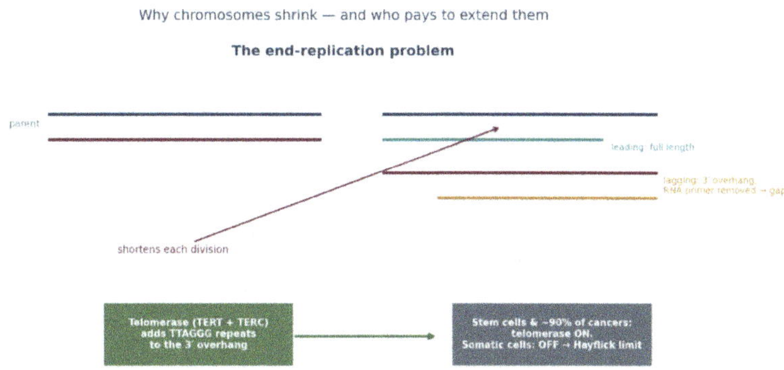

Figure 44.5 — The end-replication problem. *The leading daughter is complete; the lagging daughter loses the terminal primer's span each cycle. Telomerase (green) pays the debt in stem and germ cells — and, illicitly, in most cancers.*

44.11 Replication Stress

Between the clean diagram and the living S phase lies ***replication stress****: anything that slows or stalls the fork — DNA damage, nucleotide shortage, difficult sequences (the G-quadruplexes and hairpins of Chapter 43), collisions with transcription, oncogene-driven over-firing of origins. The cell's response is the ATR–CHK1 checkpoint: RPA-coated single-stranded DNA at a stalled fork activates ATR, which phosphorylates CHK1, which halts further origin firing, stabilises the stalled fork, and buys time for repair or bypass by the translesion polymerases (Chapter 45) — sloppy enzymes that copy over damage at the price of mutations, the cell's calculated choice of a flawed copy over a broken chromosome. Failure is catastrophic: fork collapse into double-strand breaks, the commonest lesion of pre-cancerous cells, and the reason the ATR pathway is both a tumour suppressor and a drug target — ATR inhibitors now in trial exploit exactly the checkpoint-dependence that oncogene-driven stress creates. The replication machinery, like the customs officials of Part VIII, is most visible when the cargo is contraband.*

> ***Case study — The sun-shy children of De Sanctis–Cacchione.*** *A 6-year-old has severe freckling and corneal scarring after trivial sun exposure; two cousins died of skin cancers before ten. Cultured fibroblasts, irradiated with UV, show persistent stalled forks and a failure to resume synthesis. The defect is not in replication itself. What is missing, and why does it present as a replication disease?*
> ***Resolution.*** *Xeroderma pigmentosum: a defect in nucleotide excision repair (Chapter 45) — the system that removes UV-induced pyrimidine dimers. Unremoved dimers are absolute blocks to the replicative polymerase: every fork reaching one stalls (Section 44.11), and the bypass polymerases that eventually copy across do so inaccurately — hence both the stalled forks of the assay and the mutation load that gives a child melanoma. The case is placed in this chapter deliberately: repair defects present as replication phenotypes because the fork is where unrepaired damage is finally met. XP is Chapter 45's content arriving, as it does in life, one chapter early.*

Historical note — Kornberg and the enzyme that was not the enzyme. *In 1956 Arthur Kornberg, washing extracts of E. coli with radioactive nucleotides, found an activity that built DNA in a test tube — DNA polymerase I, the first enzyme of nucleic-acid synthesis ever isolated, and the foundation of every subsequent in vitro system. The 1959 Nobel followed promptly. Then came the embarrassment: a mutant lacking nearly all Pol I activity replicated its chromosome almost normally. The cell's true replicative polymerase proved to be Pol III (found by Kornberg's own laboratory and others in the early 1970s), while the famous enzyme's real work was the humbler one of primer removal and repair. Kornberg took the demotion with grace and some amusement — he had, he noted, found the right family on the first knock — and the episode stands as the model of how enzyme hunts proceed: purify what you can assay, name it for what it does in vitro, and let genetics decide what it does in life. His son Roger took the 2006 Nobel in Chemistry for the structure of RNA polymerase — the only father-and-son pair in the annals to have each purified the transcription machinery of their century.*

Plate 44.1 *Arthur Kornberg (1918–2007), who isolated the first DNA polymerase and founded the enzymology of replication. Nobel laureate in Physiology or Medicine, 1959, with Severo Ochoa. (Nobel Foundation portrait.)*

44.12 Exercises

Exercise 44.1. *E. coli replicates its 4.6 Mb chromosome in 40 min from a single origin, bidirectionally. Calculate the per-fork rate, and compare with the eukaryotic fork rate of ~50 nt/s. How many eukaryotic forks would be needed*

to copy 4.6 Mb in the same time?
Resolution. *Each of the two forks copies 2.3 Mb in 2,400 s → ~960 nt/s per fork. At 50 nt/s a eukaryotic fork would need 2.3 Mb/50 = 46,000 s ≈ 12.8 h for the same tract; to finish in 40 min would require 46,000/2,400 ≈ 19 fork-pairs (38 forks) per 4.6 Mb — and the human genome, 700-fold larger, accordingly fires tens of thousands of origins. The bacterium buys speed per fork; the eukaryote buys forks per genome.*

Exercise 44.2. *Why does a DNA polymerase need a primer while RNA polymerase does not, and how does this requirement serve fidelity?*
Resolution. *The polymerase active site catalyses extension of a correctly paired 3'-OH; initiation de novo would place the first nucleotide without a pairing partner to geometrically check it, and the first few nucleotides of any chain are the most error-prone. Primase pays the error cost in disposable RNA — later excised and replaced — so that every DNA phosphodiester bond is made by extension of a checked 3' end. The primer is not a workaround; it is a quarantine for the chain's least reliable letters.*

Exercise 44.3. *Estimate the uncorrected error burden per human cell division at each fidelity stage of Figure 44.4 (genome 6×10^9 nt copied).*
Resolution. *Selection alone: $6 \times 10^9 \times 10^{-5}$ = 60,000 errors. With proofreading (10^{-7}): 600. With mismatch repair (10^{-10}): 0.6 — less than one error per division, matching measured germ-line rates of a few dozen new mutations per generation (accumulated over the divisions of a gamete's ancestry). Each clerical layer is thus not a refinement but a necessity: remove MMR (Lynch syndrome) and the 600-error cell becomes a cancer-prone lineage.*

Exercise 44.4 (proposition for research). *ATR inhibitors selectively kill oncogene-driven cells, which live under chronic replication stress. Propose a biomarker strategy to select the patients whose tumours are checkpoint-dependent, and the experiment that would validate it prospectively.*
Sketch. *Candidate biomarkers: oncogene status (MYC amplification, CCNE1 amplification), baseline pRPA/γH2AX foci (chronic stress), and an ex vivo fork-stability assay (DNA fibre combing with and without ATR inhibition). Validate in a window-of-opportunity trial: pre-treatment biopsy for the biomarker panel, short ATR-inhibitor course, on-treatment biopsy for fork catastrophe markers (53BP1 bodies, micronuclei). The predictive test succeeds only if on-treatment fork collapse and subsequent response concentrate in biomarker-high tumours — otherwise the field has another targeted drug prescribed by genotype and forgiven by biology.*

44.13 Synthesis

The copying machine now stands assembled: semiconservative by the CsCl bands of 1958; begun at licensed origins; opened by helicase, protected by SSB, relieved by topoisomerase; primed in disposable RNA; extended by the two-Mg^{2+} chemistry at one direction only — hence the leading and lagging asymmetry; proofread three layers deep to one error in 10^{10}; and finished at telomeres that erode by design. Direction checked against the master plan: a machine this accurate still faces damage from outside and slippage within, and the next chapter is the maintenance department. ***Chapter 45 — Mutation, DNA Repair and Recombination*** *takes* *up what breaks, what fixes it, and what shuffles the fixed copy into new combinations.*

44.14 References

Alberts, B., Heald, R., Johnson, A., et al. (2022). Molecular Biology of the Cell (7th ed.). W. W. Norton.

Bell, S. P., & Labib, K. (2016). Chromosome duplication in Saccharomyces cerevisiae. Genetics, 203, 1027–1067.

Blackburn, E. H., Greider, C. W., & Szostak, J. W. (2006). Telomeres and telomerase: The path from maize, Tetrahymena and yeast to human cancer and aging. Nature Medicine, 12, 1133–1138.

Kornberg, A., & Baker, T. A. (2005). DNA Replication (2nd ed.). University Science Books.

Meselson, M., & Stahl, F. W. (1958). The replication of DNA in Escherichia coli. Proceedings of the National Academy of Sciences, 44, 671–682.

O'Donnell, M., Langston, L., & Stillman, B. (2013). Principles and concepts of DNA replication in bacteria, archaea, and eukarya. Cold Spring Harbor Perspectives in Biology, 5, a010108.

Zeman, M. K., & Cimprich, K. A. (2014). Causes and consequences of replication stress. Nature Cell Biology, 16, 2–9.

Chapter 45
Mutation, DNA Repair and Recombination

Contents of this chapter

"The genome is the only archive whose enemies are its own materials — water, oxygen, and time — and whose librarians work the night shift every night."

45.1 Sources of DNA Damage

Chapter 44 built a copying machine accurate to one error in 10^{10}; this chapter explains why such accuracy is barely enough. DNA is not attacked only by mutagens — it decays of itself, at body temperature, in water (Figure 45.1). ***Depurination*** *— the glycosidic bond of A or G simply hydrolysing — strikes some 12,000 times per human cell per day.* ***Deamination*** *converts cytosine to uracil (and 5-methylcytosine to thymine, which is why methylated CpG sites are mutation hotspots and why the CpG dinucleotide is vanishing from our genome).* ***Oxidation*** *by the respiratory chain's leaked radicals (Chapter 26) yields 8-oxoguanine, which pairs with adenine and so teaches the next polymerase a lie. Single-strand breaks outnumber all of these; double-strand breaks are mercifully rare; and the exogenous agents — UV in sunlit skin, ionising radiation, benzopyrene in smoke, aflatoxin in mouldy grain — add their thousands to a total that Lindahl, who counted it, put at roughly 10^4–10^5 lesions per cell per day[59]. The genome's stability is not a property of the molecule; it is an achievement of the maintenance department.*

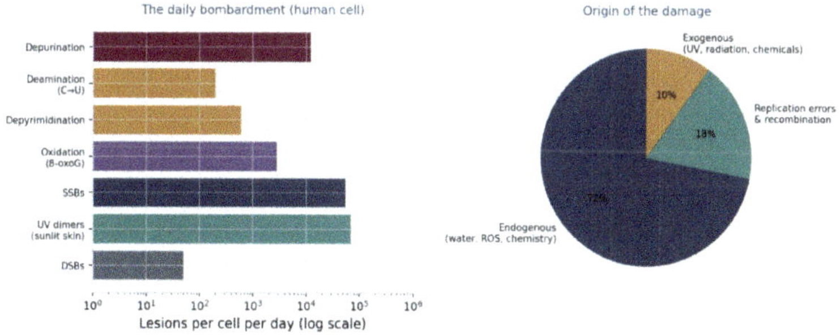

Figure 45.1 — The daily bombardment. *Left: approximate lesion counts per human cell per day (log scale) — depurination and single-strand breaks dominate; the dreaded double-strand break is rare. Right: the great majority of damage is endogenous — water, oxygen and ordinary chemistry, not mutagens.*

59. Lindahl's counting made the philosophy unavoidable: DNA is not the stable molecule of the textbook portrait but one of the least stable constituents of the cell, maintained — like the Ship of Theseus — only by continuous repair. The 2015 Nobel in Chemistry (Lindahl, Modrich, Sancar) was awarded for the maintenance crews, and this chapter is their personnel file.

45.2 Spontaneous and Induced Mutation

Damage that escapes repair and is copied becomes **mutation**, *and the vocabulary sorts the consequences. A point substitution is a transition (purine for purine — the common kind, deamination's signature) or a transversion (purine for pyrimidine — the rarer, bulkier insult, benzopyrene's). In a coding sequence the outcome is silent, missense, or nonsense; insertions and deletions, unless in multiples of three, shift the reading frame and garble everything downstream. The Ames test reads mutagenicity as reversion rate in Salmonella — the assay that made "mutagen" and "carcinogen" near-synonyms, since most carcinogens are mutagens first. And the mutation rate itself is set-point biology: too high and the lineage drowns in error (error catastrophe, the fate of RNA viruses pushed by mutagens like favipiravir); too low and it cannot adapt — the mutator strains of bacteria that win in fluctuating environments and lose in stable ones demonstrate that evolution tunes the rate like any other phenotype.*

45.3 Direct Reversal

The simplest repair undoes the damage on the spot, no excision required. **Photolyase**, *in species that have it (we do not), splits UV-induced pyrimidine dimers using visible light — the original photoreactivation. We rely instead on* **MGMT** *(O^6-methylguanine-DNA methyltransferase), which removes the methyl group from O^6-methylguanine — a guanine that would otherwise pair with thymine — by transferring it onto the enzyme's own cysteine. The enzyme is consumed in the act: one molecule, one repair, a suicide protein in the literal sense, and the tumour that methylates MGMT's promoter silent is the tumour that temozolomide can kill — the rare case where losing a repair gene is the patient's good fortune. Direct reversal is cheap, accurate, and available for only a handful of lesions; everything else is excised.*

45.4 Base-Excision Repair

BER handles the small, non-distorting lesions — uracil, 8-oxoG, alkylated bases, the AP sites left by depurination — and it is the busiest pathway in the cell (Figure 45.2). A damage-specific **DNA glycosylase** *flips the offending base out of the helix into a catalytic pocket and clips the glycosidic bond, leaving an abasic site; AP endonuclease cuts the backbone there; Pol β (short patch) or the replicative polymerases (long patch) fill the gap; ligase III seals. The logic worth noting is the glycosylase's specificity: UNG removes uracil — which has no business in DNA — but not thymine, which is uracil with a methyl badge of legitimacy. The $C \rightarrow U$ deamination is repairable precisely because RNA uses U and DNA uses T; had DNA kept uracil, the commonest spontaneous mutation would be undetectable. Thymine, in this reading, is not a letter but a repair tag, and the whole dTTP economy of Chapter 38 exists to make deamination visible.*

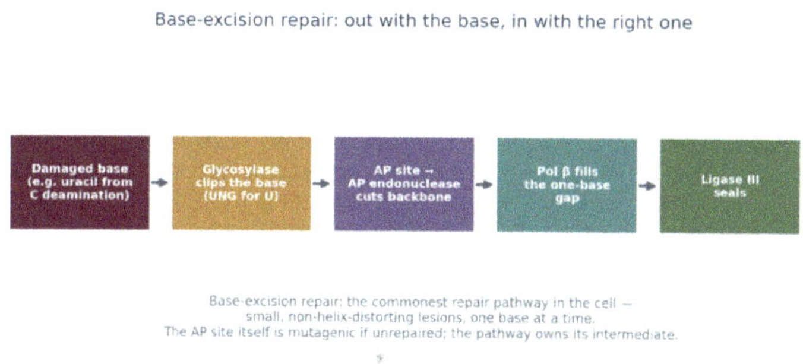

Figure 45.2 — Base-excision repair. *Glycosylase clips the damaged base, AP endonuclease cuts, polymerase fills, ligase seals. Each intermediate — the AP site, the nick — is itself dangerous; the pathway is processive precisely so the intermediates never wander off alone.*

45.5 Nucleotide-Excision Repair

Where the lesion distorts the helix — the UV dimer, the benzopyrene adduct, the cisplatin crosslink — the cell sacrifices a whole patch (Figure 45.3). NER detects distortion rather than chemistry, by two routes: XPC patrols the genome globally, while a stalled RNA polymerase summons the transcription-coupled arm (CSA, CSB) — the transcribed strand of active genes getting priority service, as befits the working copy. TFIIH's helicases then open a ~30-nucleotide bubble, XPA and XPG confirm the lesion, XPF–ERCC1 and XPG cut either side, and the oligomer is released; Pol δ/ε fills, ligase seals. The human diseases map the wiring: xeroderma pigmentosum (XPA–XPG) is global NER lost — the sun-shy children of Chapter 44's case — while Cockayne syndrome is the transcription-coupled arm lost, giving not cancer but segmental progeria, a paradox resolved by the arm's second job in clearing stalled transcription. The distinction teaches a general doctrine of repair genetics: lose the genome-wide system and you get mutation; lose the transcription-linked one and you get ageing.

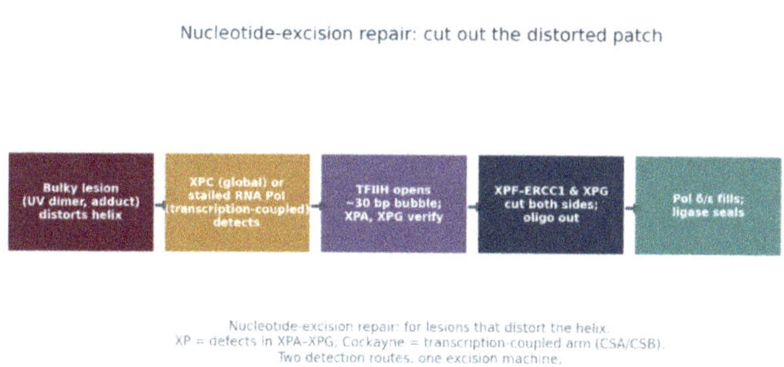

Figure 45.3 — Nucleotide-excision repair. *Distortion detected globally (XPC) or by a stalled polymerase (CSA/CSB); TFIIH opens, dual incisions excise ~30 nucleotides, resynthesis and ligation close. XP and Cockayne map the two detection routes.*

45.6 Mismatch Repair

MMR is Chapter 44's third fidelity layer, met now as machinery. MutS (MSH2–MSH6 in us) scans the duplex for the kink of a mispair or the bulge of a small indel; MutL (MLH1–PMS2) is recruited; and the system's decisive trick is strand discrimination — knowing which of the two strands is the daughter that must be corrected. Bacteria read the under-methylation of the new strand at GATC sites; eukaryotes apparently read the new strand's discontinuities — the Okazaki ends and PCNA orientation. A segment of daughter strand is then excised and resynthesised. Loss of MMR produces the mutator phenotype behind Lynch syndrome — hereditary non-polyposis colorectal cancer — and its diagnostic fingerprint, microsatellite instability: the repeat tracts that DNA polymerase stutters over, left uncorrected, drift in length. The tumour with high MSI is, gratifyingly, the tumour most visible to the immune system and most responsive to checkpoint blockade — repair loss creating neoantigens, another debt called in.

45.7 Double-Strand-Break Repair

The double-strand break is the only lesion that destroys the information on both strands at once, and the cell answers it with two philosophies (Figure 45.4). ***Homologous recombination*** *(Section 45.8) rebuilds from the sister chromatid — accurate, but available only in S/G2 when a sister exists.* ***Non-homologous end joining*** *(Section 45.9) simply rejoins the ends — fast, available all cycle, and mutagenic at the seam. The choice between them is enforced by the 53BP1–BRCA1 switch at the break: 53BP1 shields ends (NHEJ), BRCA1 licenses resection (HR). It is hard to overstate how much modern oncology lives in this paragraph: BRCA1/2 mutations disable HR, forcing the tumour onto error-prone repair; PARP inhibitors block the BER backup such tumours rely on, and the two losses together are lethal — synthetic lethality, the first concept from yeast genetics to save lives in the clinic. And the CRISPR revolution is, molecularly, a DSB delivered to an address: NHEJ knocks the gene out, HR with a template knocks the edit in.*

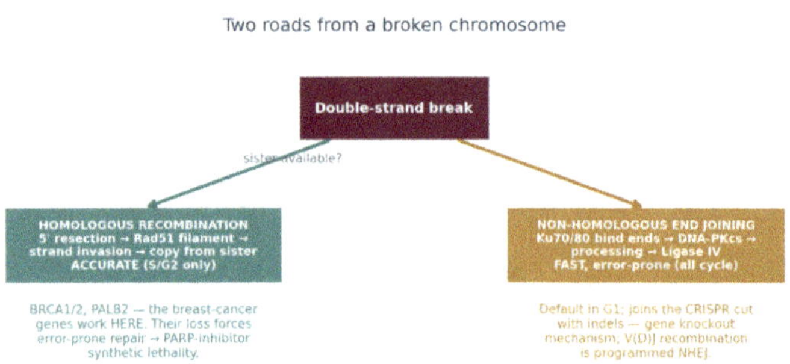

Figure 45.4 — Two roads from a break. *Homologous recombination (teal): accurate, sister-dependent, the BRCA pathway. Non-homologous end joining (amber): fast, all-cycle, seam-mutagenic — and the mechanism of both V(D)J recombination and CRISPR knockouts.*

45.8 Homologous Recombination

HR is repair by photocopying from the undamaged relative. The broken ends are resected 5′→3′ to expose 3′ tails; RecA (bacteria) or Rad51 (us), loaded by BRCA2, coats the tails and conducts a genome-wide search for homology; the invading strand pairs with its complement, priming synthesis that copies the lost information; and the resulting Holliday junction — the four-stranded cross of every textbook — is branch-migrated and resolved. In repair mode the sister is the template and the outcome is invisible; in ***meiosis*** *(Figure 45.6) the same machinery, deliberately triggered by Spo11-induced breaks and positioned by PRDM9, crosses the parental homologues — one or two crossovers per arm, both to shuffle alleles and to hold the bivalent together for correct segregation. Recombination is thus the genome's repair system moonlighting as its diversity engine: the same Holliday junction that rescues a broken fork in S phase also makes every gamete unique. Errors in the meiotic version give the aneuploidies — trisomy 21 foremost — whose risk rises with maternal age as the oocyte's decades-old crossover scaffold weathers.*

45.9 Non-Homologous End Joining

NHEJ is the emergency service: Ku70/80 rings snap onto the broken ends within seconds, DNA-PKcs synapses the two ends together, processing enzymes (Artemis, the polymerases μ and λ) trim or fill any overhangs, and Ligase IV–XRCC4 seals. Accuracy is sacrificed for speed and universality — the junction frequently carries small insertions or deletions — but the trade is usually right: in G1 there is no sister to copy from, and a scar is better than a break. NHEJ's deepest role, however, is programmed: V(D)J recombination, which builds the antibody and T-cell-receptor repertoires, is RAG-directed cutting followed by NHEJ joining, with terminal deoxynucleotidyl transferase scattering random nucleotides at the seams — the immune system weaponising repair's sloppiness into diversity. A cell that can break its own DNA on purpose, join it sloppily, and call the result a repertoire is a cell that has made peace with this chapter's first principle: lesions are information too.

45.10 Transposition

The genome's own residents rearrange it from within. Transposable elements — DNA transposons that hop by cut-and-paste, and retrotransposons (LINEs, SINEs, endogenous retroviruses) that copy through an RNA intermediate — make up over half of human DNA, mostly fossilised, mostly silenced by methylation and piRNAs, but not quite extinct: LINE-1 still inserts somewhere in a human genome every few births, causing haemophilia, Duchenne mutations and colon tumours when it lands badly. The evolutionary ledger is double-entry. Transposons mutate, disrupt, and rearrange; they also donate — the RAG genes of V(D)J recombination are domesticated transposases, telomerase is a domesticated retroelement, and the syncytin that builds the placenta's outer layer is a captured retroviral envelope gene. Barbara McClintock found the elements in maize in the 1940s — "jumping genes," controlling elements, the genome in flux — and was politely ignored for thirty years before the 1983 Nobel caught up with her. The genome, it turns out, is less an archive than a coral reef: mostly the skeletons of former invaders, with the living ones still adding structure.

45.11 Genome Instability and Cancer

Cancer is the disease of this chapter — the endpoint toward which every failed repair system points. The logic is now quantitative: a tumour is a lineage that acquired its driver mutations faster than a normal cell can, and it does so either by raising the mutation rate (the repair defects: XP, Lynch, BRCA) or by enduring more divisions under more damage (chronic inflammation, the smoker's lung). Whole-genome sequencing has made the ledger legible as ***mutational signatures*** *(Figure 45.5): the C>T-at-CpG clock of spontaneous deamination ticking in every cell (SBS1), the benzopyrene C>A pattern of tobacco (SBS4), the dipyrimidine C>T of UV (SBS7), the APOBEC pattern of antiviral editing gone rogue — each tumour carrying, in its base-substitution spectrum, the minutes of its own damaging history. The clinical harvest is already in: signatures predict immunotherapy response (the mutator tumours), PARP-inhibitor sensitivity (the HR-defective ones), and exposure history (the smoker's lung versus the never-smoker's, distinguishable from the genome alone). The archive keeps its own damage log; oncology has learned to read it.*

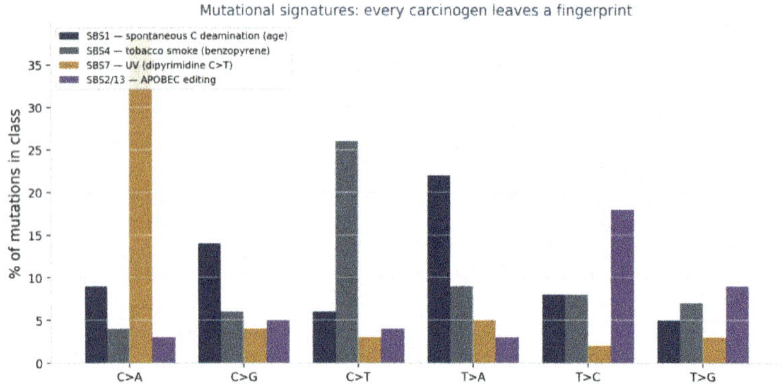

Figure 45.5 — Mutational signatures. *The six substitution classes, four exposures: the age-clock (SBS1), tobacco (SBS4), UV (SBS7), and APOBEC editing (SBS2/13). A tumour genome is a signed confession of everything that ever damaged it.*

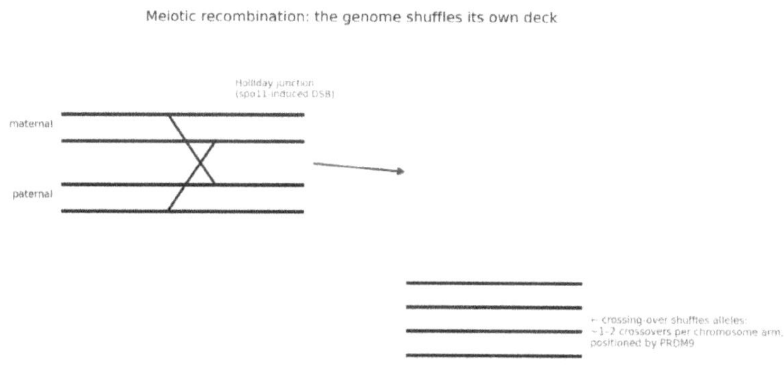

Figure 45.6 — Meiotic recombination. *Homologues pair, a programmed break is resolved as a crossover, and the four chromatids emerge: two parental, two recombinant. Repair machinery, reassigned to shuffling.*

Case study — The colon that aged too fast. *A 28-year-old presents with a right-sided colon cancer and a family history of endometrial and gastric cancers under 50. The tumour shows microsatellite instability-high and loss of MLH1 staining; her germline carries an MLH1 frameshift. Explain the sequence from genotype to tumour, and why the tumour is unusually immunotherapy-sensitive.*

Resolution. *Lynch syndrome. One functional MLH1 copy suffices until a somatic hit removes it (Knudson's two hits, inherited one hit deep); the MMR-deficient cell then accumulates 100–1,000-fold more mutations, visible first in microsatellites — the repeat tracts where polymerase stutter normally gets corrected. Driver mutations follow in years instead of decades: hence cancer at 28. But the same mutator load generates hundreds of neoantigens, so the tumour is heavily infiltrated and exquisitely responsive to PD-1 blockade — MSI-high was the first tumour-agnostic indication for immunotherapy, precisely because the repair defect is the drug's target. Surveillance colonoscopy, aspirin chemoprevention, and cascade testing of the family complete the management: the chapter's machinery, read as a pedigree.*

Historical note — Lindahl and the instability of being. *In the early 1970s Tomas Lindahl, working on RNA at the Karolinska, was nagged by a discrepancy: RNA's cytosines deaminated detectably on the shelf, yet DNA — the supposed rock of the cell — was written off as stable. He did the arithmetic that everyone else had skipped: at 37 °C in water, a mammalian genome should lose thousands of purines a day to hydrolysis alone, and no organism could survive it unrepaired. Since we demonstrably survive, the repair must exist — and he went looking for it, finding the first DNA glycosylase (uracil-DNA glycosylase's bacterial counterpart) and founding base-excision repair as a field. The 2015 Nobel in Chemistry, shared with Modrich (mismatch repair) and Sancar (nucleotide-excision repair), crowned the maintenance department at last. Lindahl's summary of his own discovery deserves quoting in every textbook of this subject: "DNA is not a stable molecule — its stability is the result of repair. If the repair systems were removed, the genome would last about a day." This chapter is the account of the day's work.*

Plate 45.1 *Tomas Lindahl (b. 1938), who measured DNA's spontaneous decay and found the glycosylases that answer it. Nobel laureate in Chemistry, 2015, with Paul Modrich and Aziz Sancar. (Nobel Foundation portrait, by Alexander Mahmoud.)*

45.12 Exercises

Exercise 45.1. *Why is thymine used in DNA while uracil serves in RNA, and what would be the consequence for the mutation rate if DNA used uracil?*
Resolution. *Cytosine deaminates to uracil; UNG removes uracil from DNA because uracil there is always illegitimate. If U were a legitimate DNA base, deaminated C and genuine U would be indistinguishable — the commonest spontaneous lesion would be invisible — and C → T transitions would accumulate unchecked. The methyl group of thymine is the authenticity stamp that makes the whole BER strategy possible; the dTMP synthesis economy (Chapter 38) is the price of the stamp.*

Exercise 45.2. *A tumour is BRCA1-deficient. Explain (i) why PARP inhibitors kill it, (ii) why a normal heterozygous cell survives, and (iii) the two commonest resistance mechanisms.*
Resolution. *(i) PARP inhibition blocks BER/single-strand-break repair; unrepaired SSBs become DSBs at replication forks; the BRCA1-null cell cannot run HR to fix them — synthetic lethality. (ii) The heterozygote retains one functional allele and enough HR; the drug alone is tolerable. (iii) Resistance: reversion mutations that restore the BRCA1 reading frame, and loss of 53BP1 or its shieldin complex, which re-permits end resection and partially restores*

HR. The tumour escapes by re-opening either the gene or the pathway — the arms race, at molecular resolution.

Exercise 45.3. *CpG sites mutate ~10-fold faster than other dinucleotides. Given that 5-methylcytosine deaminates to thymine (not uracil), explain the rate difference and its long-term genomic consequence.*
Resolution. *mC deamination yields T — a legitimate base — creating a G:T mismatch that MMR and BER repair inefficiently compared with G:U (UNG's substrate). Hence methylated CpGs leak C→T transitions at ~10× the background rate. Over evolutionary time the leak has drained the genome of CpGs except where methylation is absent (the CpG islands of promoters) or selection maintains them: the dinucleotide distribution of every vertebrate genome is the fossil record of its own methylation.*

Exercise 45.4 (proposition for research). *Mutational-signature analysis is retrospective. Propose a prospective use: an early-warning system that detects a mutagenic exposure in a population before the cancers appear, and design the validation cohort.*
Sketch. *Sequence ultra-deep single-cell genomes (or duplex-sequenced cfDNA) from an accessible tissue — blood or buccal — of an exposed cohort (e.g. aristolochic-acid users, vapers, radiation workers) and matched controls, quantifying signature burden years before tumours. Validation: longitudinal sampling to show the signature rises with exposure duration and falls after cessation, anchored against the known cancer-registry outcomes a decade later. The ethical logic is the compelling part: a molecular dosimeter that reads Section 45.11's ledger in real time turns prevention from epidemiology into surveillance — provided the field agrees on what to tell a signature-positive, cancer-negative volunteer.*

45.13 Synthesis

The maintenance department is now fully staffed: reversal for the lucky lesions, BER for the small ones, NER for the distorting ones, MMR for the copier's mistakes, HR and NHEJ for the broken chromosome, and a set of mutational signatures that turn every tumour genome into an exposure history. Around the edges move the transposons — the archive's former invaders, now mostly scaffolding — and once a lifetime the machinery crosses the chromosomes on purpose to shuffle the deck. Direction checked against the master plan: the genome is copied and maintained; next it must be read. ***Chapter 46 — Transcription and RNA Processing*** *takes up the reading — polymerases, promoters, and the astonishing editorial journey from pre-mRNA to message.*

45.14 References

Alexandrov, L. B., et al. (2020). The repertoire of mutational signatures in human cancer. Nature, 578, 94–101.

Ciccia, A., & Elledge, S. J. (2010). The DNA damage response: Making it safe to play with knives. Molecular Cell, 40, 179–204.

Friedberg, E. C., et al. (2014). DNA Repair and Mutagenesis (3rd ed.). ASM Press.

Hoeijmakers, J. H. J. (2009). DNA damage, aging, and cancer. New England Journal of Medicine, 361, 1475–1485.

Lindahl, T. (1993). Instability and decay of the primary structure of DNA. Nature, 362, 709–715.

McClintock, B. (1950). The origin and behavior of mutable loci in maize. Proceedings of the National Academy of Sciences, 36, 344–355.

Modrich, P., & Lahue, R. (1996). Mismatch repair in replication fidelity, genetic recombination, and cancer biology. Annual Review of Biochemistry, 65, 101–133.

Chapter 46
Transcription and RNA Processing

CONTENTS OF THIS CHAPTER

"The genome is read aloud in one direction only, and the reading is interrupted constantly by editors — which is why the transcript, like the manuscript, is mostly what the editors left in."

46.1 RNA Polymerase Chemistry

Transcription is Chapter 44's chemistry run by different clerks with different rules. RNA polymerase catalyses the same S_N2 attack of a 3'-OH on an NTP α-phosphate, with the same two magnesium ions and the same pyrophosphate ratchet — but with three liberties that define the enterprise. It initiates de novo, no primer needed (and pays for the liberty in accuracy: one error in 10^4–10^5, tolerable because the product is a working copy, not the archive). It reads only one strand — the template, read 3'→5' so the RNA grows 5'→3' — and only between a promoter's command and a terminator's. And it unwinds as it goes, maintaining a transcription bubble of ~12–14 base pairs with an 8–9 bp DNA–RNA hybrid inside, the rest of the genome never knowing it was visited. The active site's fidelity device is a trigger loop that closes only over a correctly paired NTP; the proofreading is backward — the enzyme backtracks and cleaves the misinserted nucleotide, assisted by the TFIIS factor that reaches into the secondary channel. Chemistry conserved, discipline relaxed: the working copy is allowed to be human.

46.2 Bacterial Transcription

The bacterial enzyme is one machine for all genes: a five-subunit core ($\alpha_2\beta\beta'\omega$) plus a replaceable ***σ factor*** *that finds the promoters. σ^{70}, the housekeeping sigma, reads the −35 and −10 (Pribnow) boxes, melts the −10 region, and is then jettisoned as elongation begins. Termination comes in two dialects: intrinsic — a GC-rich hairpin followed by uridines, whose weak rU-dA pairs let the RNA fall out of the bubble — and Rho-dependent, a hexameric helicase that races up the naked RNA and wrenches the polymerase off. The σ system is the simplest possible regulatory architecture — swap the σ factor, swap the regulon: σ^{32} for heat shock, σ^{54} for nitrogen, σ^{E} for envelope stress — and the antibiotics have found the machine too: rifampicin plugs the RNA exit channel (the tuberculosis mainstay), actinomycin D intercalates the template. Transcription-coupled life is completed by the absence of a nucleus: ribosomes load onto the nascent mRNA while it is still being made, so the bacterium literally speaks and is translated in the same breath.*

46.3 Eukaryotic RNA Polymerases

The eukaryote divides the labour among three nuclear polymerases (Figure 46.1). ***Pol I****, in the nucleolus, transcribes the 45S rRNA precursor — some 60% of all transcription in a growing cell, a single product on an industrial scale.* ***Pol II*** *makes every messenger and most regulatory RNAs; it is the regulated one, the subject of the rest of this chapter, and the one α-amanitin kills first — the death-cap's poison being a Pol II inhibitor so selective that the syndrome begins with a quiet day (the toxin rides the enterohepatic cycle) and ends in hepatic failure as the liver, the body's busiest transcribers, runs out of messages.* ***Pol III*** *makes tRNAs, 5S rRNA and other small housekeeping RNAs. A fourth and fifth (Pol IV/V) serve plant small-RNA pathways. The structures are conserved from yeast to man — twelve and seventeen subunits around the crab-claw core that Roger Kornberg's laboratory solved — which is why the death-cap poison works identically on both, and why the 2006 Nobel in Chemistry for the Pol II structure belonged to the son of Chapter 44's hero.*

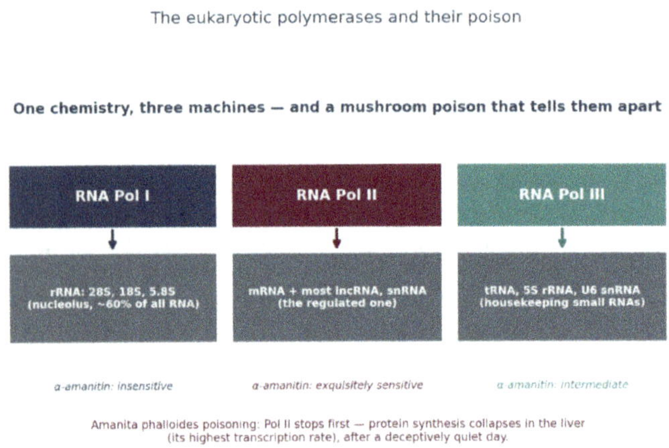

Figure 46.1 — Three polymerases, one poison. *Pol I (rRNA), Pol II (mRNA — the regulated machine), Pol III (tRNA and small RNAs), with their α-amanitin sensitivities and the clinical sequence of death-cap poisoning below.*

46.4 *Promoters and Transcription Factors*

Pol II does not find genes; it is escorted to them. The ***core promoter*** *— TATA box (where present), Inr, DPE — assembles the preinitiation complex in strict order: TFIID (its TBP subunit saddle-bending the TATA box), TFIIB, TFIIF with the polymerase, TFIIE, and TFIIH, whose twin duties — helicase to melt the start site, kinase to phosphorylate the polymerase's tail — bridge initiation and everything after. Above the core promoter sits the regulatory layer:* ***enhancers****, which may lie kilobases away and act in either orientation, bound by sequence-specific activators;* ***silencers*** *and insulators; and the* ***Mediator****, a thirty-subunit coactivator that translates enhancer-bound signals into polymerase behaviour. The looping that brings enhancer to promoter happens inside the TAD architecture of Chapter 43.10 — the packaging and the regulation are one system. Combinatorics do the arithmetic of specificity: a handful of activator-binding sites, each modest, multiply into the sharp tissue-specific expression that makes a liver a liver.*

46.5 *Initiation, Elongation and Termination*

The polymerase's largest subunit drags a C-terminal domain of 52 heptad repeats (YSPTSPS), and the repeats are a calendar (Figure 46.2). Hypophosphorylated CTD joins the preinitiation complex; TFIIH's kinase adds Ser5 phosphate, capping enzymes dock, and the polymerase clears the promoter — pausing, in most metazoan genes, 30–60 nucleotides downstream until P-TEFb adds Ser2 phosphate and releases it into productive elongation, with the splicing machinery riding the Ser2-rich tail. At the gene's end, the polyadenylation signal is read by factors that also rode the tail, and termination follows the torpedo model: cleavage of the RNA leaves XRN2, a 5′→3′ exonuclease, chasing the polymerase down the naked downstream strand until it catches and displaces it. The CTD thus carries each transaction's permit for the next — initiation licenses capping, elongation licenses splicing, the 3′ end licenses export — a single tail running the whole bureaucratic chain.

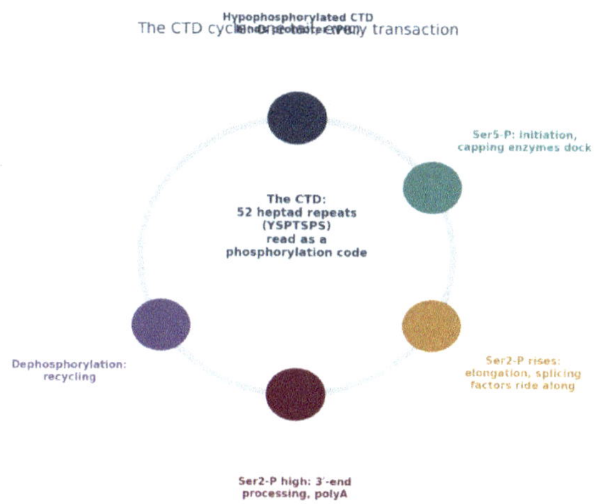

Figure 46.2 — The CTD cycle. *Fifty-two heptad repeats, phosphorylated in sequence: Ser5 for initiation and capping, Ser2 for elongation and splicing, dephosphorylation for recycling. The tail is the itinerary.*

46.6 RNA Capping

The first edit happens before the transcript is thirty nucleotides old: the 5′ triphosphate is trimmed to a diphosphate, GMP is added in reverse (5′–5′ triphosphate linkage), and the guanine is methylated at N7 — the ***m⁷G cap****. Three jobs: protection from 5′ exonucleases; the export pass (the cap-binding complex escorts the RNA through the nuclear pore); and the translation start signal — eIF4E of Chapter 47 recognises the cap to begin scanning. The cap is, in effect, the message's passport, issued at the moment of birth and stamped at every border thereafter.*

46.7 Polyadenylation

The far end gets its own seal. The AAUAAA signal (with a downstream GU-rich element) assembles CPSF and CstF; the RNA is cleaved ~20 nucleotides downstream of the signal; and poly(A) polymerase adds ~200 adenosines, template-free — one of the few polymerases that needs no instructions. The tail, wrapped in PABP, stabilises the message and, bound to eIF4G at the cap, circularises it: head to tail, the message becomes a loop that reinitiates translation efficiently. Deadenylation by the CCR4–NOT complex shortens the tail over the message's life, and a short tail is a death sentence — the decay machinery reads the tail as an egg-timer. Alternative polyadenylation sites add a regulatory dimension: the same coding sequence with a longer 3′ UTR inherits more miRNA sites and more instability — the message's lifespan edited without touching the protein.

46.8 RNA Splicing

The discovery of split genes (1977, adenovirus — Sharp and Roberts, Nobel 1993) broke the one-gene-one-message doctrine, and the chemistry broke it beautifully (Figure 46.4). The intron's 5′ GU and 3′ AG flank a branch-point adenosine; in two transesterifications — the branch A's 2′-OH attacking the 5′ junction, then the freed exon attacking the 3′ junction — the exons are ligated and the intron leaves as a lariat. No ATP is spent on the chemistry; the energy goes into the machine that does it: the **spliceosome**, *five snRNPs (U1, U2, U4, U5, U6) and a hundred proteins, assembling stepwise on each intron, proofreading the branch point, and catalysing through an RNA core — U6 pairs with U2 to build the active site, and the spliceosome stands revealed as a ribozyme with an extensive protein secretariat. Constitutive errors — the wrong junction chosen — cause disease: the β-thalassaemias of splice-site mutation, spinal muscular atrophy's failing exon-7 inclusion in SMN2 (rescuable by an antisense drug that blocks a silencer, nusinersen — splicing as therapy), and a good fraction of every cancer's driver mutations landing in splice-factor genes.*

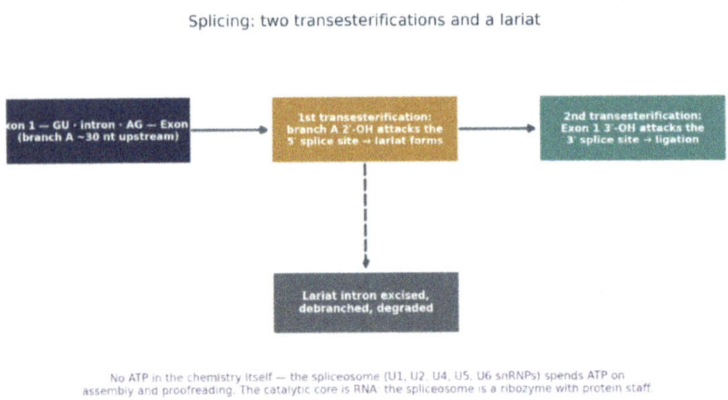

Figure 46.4 — Two transesterifications and a lariat. *The branch adenosine attacks the 5′ splice site; the freed exon attacks the 3′ site; the intron exits as a lariat. ATP is spent on assembly and proofreading, not on bond chemistry.*

46.9 Alternative Splicing

Once genes can be split, they can be read selectively (Figure 46.5): exons skipped, mutually exclusive exons swapped, alternative 5′ or 3′ sites lengthening or shortening segments, introns retained. Some 95% of human multi-exon genes are alternatively spliced, and the proteome's diversity outstrips the genome's by an order of magnitude — the Drosophila Dscam gene, with its 38,000 isoforms, being the exhibition piece, and the brain's neurexins the mammalian equivalent, wiring synapse specificity through splice choice. The regulators are RNA-binding proteins (the SR proteins promoting, hnRNPs repressing) reading enhancer and silencer sequences in the transcript — a second code, in RNA, layered on the first. Tissue-specific splicing factors make the isoforms tissue-specific: muscle troponin T, neuronal NMDA receptors, and the fatal switch in cancers that re-express the embryonic splice programme of pyruvate kinase (PKM2 — Chapter 31's Warburg effect is partly a splicing decision).

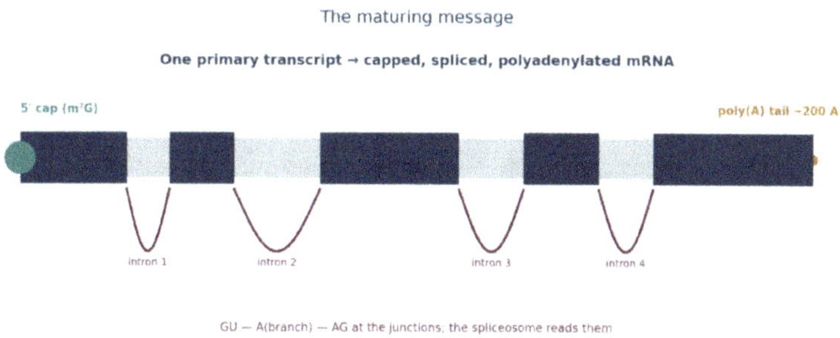

Figure 46.3 — The maturing message. *The primary transcript capped at its birth, the introns looped out by the spliceosome, the poly(A) tail added at the signal — one gene dressed for export.*

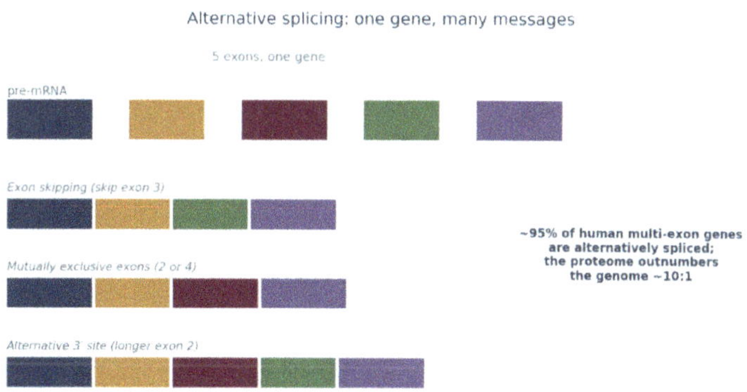

Figure 46.5 — Alternative splicing. *Exon skipping, mutually exclusive exons, alternative splice sites: five exons yielding three different messages, and ~95% of human genes doing likewise.*

46.10 RNA Editing

Rarer than splicing but stranger: the message is changed base by base after transcription. ***A-to-I editing*** *by ADAR enzymes deaminates adenosine to inosine, which the ribosome reads as guanosine — recoding codons, most famously in the brain's glutamate receptors, where one edited position (the Q/R site) decides whether the channel passes calcium; the unedited mouse dies of seizures, a single* *deamination carrying a phenotype.* ***C-to-U editing*** *by APOBEC1 edits apolipoprotein B mRNA in the intestine, creating a stop codon that shortens the protein — apoB48 in chylomicrons versus apoB100 in the liver's VLDL, one gene serving two organs by a single deamination (and Chapter 45's APOBEC mutational signatures being the same enzyme family caught off duty). The trypanosomes, not to be outdone, insert and delete hundreds of uridines under guide-RNA direction — editing as reconstruction. The doctrine these cases defend: the sequence in the genome is the draft, not the message.*

46.11 RNA Transport and Degradation

A finished message must still cross the pore and survive the cytoplasm (Figure 46.6). Export through the nuclear pore is licensed by the completed processing — cap checked, introns gone, poly(A) in place — with the NXF1/TAP transporter as the carrier; improperly processed transcripts are retained and destroyed in the nucleus. In the cytoplasm, decay is the default and stability the achievement: deadenylation first (the egg-timer), then decapping and 5′→3′ chewing by XRN1, or 3′→5′ disposal by the exosome. Surveillance rides along: ***nonsense-mediated decay*** *destroys messages whose premature stop codons would make toxic fragments — the mechanism that turns many "null" thalassaemia mutations into mild phenotypes, since the bad message never becomes protein — and the miRNA machinery of Chapter 48 silences from the 3′ UTR. Message half-lives span minutes (c-myc, cytokines, whose AU-rich elements summon the decay machinery) to days (globin, in a cell with nothing else to say). The concentration of every message is a synthesis rate minus a decay rate, and regulation pulls both handles.*

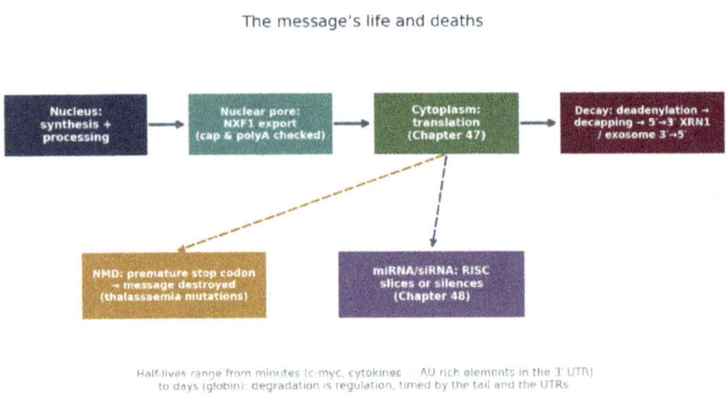

Figure 46.6 — The message's life and deaths. *Synthesis and processing, licensed export, translation, and the decay routes — deadenylation-led turnover, nonsense-mediated decay, and miRNA silencing below.*

> ***Case study — The baby who could not lift his head.*** *A 7-month-old has progressive hypotonia and areflexia; genetic testing shows homozygous deletion of SMN1. His SMN2 gene, a near-identical paralogue, differs chiefly at one position that causes exon 7 to be skipped in most transcripts. Explain the disease, and the mechanism of the drug that transformed it.*
> ***Resolution.*** *Spinal muscular atrophy. SMN1 makes the full survival-motor-neuron protein; SMN2, because a C→T difference in exon 7 creates an exonic splicing silencer, skips the exon ~90% of the time, producing an unstable*

truncated protein. Deleting SMN1 leaves only the leaky SMN2: motor neurons, exquisitely SMN-dependent, die in order of their demand. Nusinersen is an antisense oligonucleotide, given intrathecally, that base-pairs over the silencer and forces exon-7 inclusion — converting SMN2 into a working copy of the deleted gene. Given early, it changes a fatal disease of infancy into children who walk. It is this chapter as medicine: not gene therapy, not enzyme replacement, but an editor's pencil held over one nucleotide of one exon.

Historical note — The Kornbergs, father and son, and the two machines. *When Roger Kornberg took the 2006 Nobel in Chemistry for the molecular structure of RNA polymerase II — the twelve-subunit enzyme caught in the act, with template, transcript and the two magnesium ions of Section 46.1 visible in the active site — he completed, at forty-seven years' distance, the work his father had begun. Arthur Kornberg's 1959 prize was for the first DNA polymerase found in a test tube; Roger's was for the first RNA polymerase seen whole. Between the two prizes lies the entire arc of this Part of the book: the father proved the copying was chemistry — purifiable, assayable, rule-bound; the son showed the chemistry was architecture — a machine with a clamp, a jaw, a trigger loop, and a tail that carries its own calendar of permits. It is the only father-and-son pair of Nobel laureates in the molecular sciences, and the symmetry is too good to leave unremarked: the family that found how the archive is copied also showed how it is read.*

Plate 46.1 *Roger D. Kornberg (b. 1947), who solved the structure of RNA polymerase II caught in the act of transcription. Nobel laureate in Chemistry, 2006 — forty-seven years after his father's prize for DNA polymerase. (Nobel Foundation portrait.)*

46.12 Exercises

Exercise 46.1. *Why can a cell tolerate a transcription error rate ~10^4-fold higher than its replication error rate? Quantify the consequence for a 2 kb mRNA translated into a 500-residue protein.*
Resolution. *At 10^{-4} per nucleotide, a 2,000 nt transcript carries ~0.2 errors — most molecules perfect, and the imperfect ones diluted among many correct copies of the same message, each translated only transiently. A DNA error, by contrast, is copied into every descendant cell forever. The archive must be exact; the working copies need only be mostly right — the same logic as the disposable RNA primers of Chapter 44.4.*

Exercise 46.2. *A point mutation converts a β-globin intron-1 5′ splice site GU→AU. Predict the molecular phenotype and the clinical severity.*
Resolution. *The 5′ splice site is destroyed; the spliceosome either skips the exon, activates a cryptic site, or retains the intron — all yielding frameshifted, premature-stop messages destroyed by nonsense-mediated decay. Result: no functional β-chains from that allele — β^0-thalassaemia. Heterozygotes are trait (mild, hypochromic); homozygotes are thalassaemia major, transfusion-dependent. One nucleotide in a non-coding region, a whole haemoglobin missing: splicing is where genotype becomes phenotype most abruptly.*

Exercise 46.3. *ApoB100 (liver) and apoB48 (intestine) come from one gene. Describe the mechanism, and explain why the intestine, not the liver, does this.*
Resolution. *APOBEC1 (with its cofactor) deaminates C6666 of the apoB mRNA in intestinal cells, converting CAA (Gln) to UAA (stop); translation ends at 48% of the protein. ApoB48 suffices for chylomicron assembly (dietary fat export), while the liver needs full-length apoB100 with its LDL-receptor-binding tail for endogenous lipoprotein clearance. The edit is tissue-specific because APOBEC1's cofactor is intestinal — one deamination letting a single gene serve two lipoprotein economies (Chapter 36) without the cost of a duplicate gene.*

Exercise 46.4 (proposition for research). *Splice-switching oligonucleotides work in SMA but have failed in most CNS splice diseases, largely on delivery. Propose a programme to extend the modality, naming the two bottlenecks that decide success.*
Sketch. *Target: a monogenic CNS spliceopathy with a strong natural-history dataset (e.g. a frontotemporal-dementia splice mutant). Bottleneck 1, delivery: intrathecal dosing reaches cortex poorly — test convection-enhanced delivery versus AAV-encoded splice-switching RNAs in large animals, with CSF-to-parenchyma ratio as the go/no-go metric. Bottleneck 2, durability: oligonucleotides need redosing; a one-time AAV or base-edited splice-site correction (editing the silencer itself, the definitive cure) must be compared on five-year expression stability and off-target spliceome-wide effects (long-read RNA-seq). The field's future is decided by logistics, not by chemistry — the chemistry already works.*

46.13 Synthesis

The genome is now read: one chemistry, three polymerases, promoters that assemble their escorts, a tail that calendars every transaction, a cap and a tail as passports, introns excised by a ribozyme with a secretariat, isoforms multiplied by alternative splicing, bases edited after the fact, and every message living on borrowed, adenylated time. Direction checked against the master plan: the message exists to be translated. ***Chapter 47 — Translation and Protein Maturation*** *takes up the ribosome — the second ribozyme cathedral — and the folding, modifying and dispatching of the protein it makes.*

46.14 References

Cramer, P. (2019). Organization and regulation of gene transcription. Nature, 573, 45–54.

Finkel, R. S., et al. (2017). Nusinersen versus sham control in infantile-onset spinal muscular atrophy. New England Journal of Medicine, 377, 1723–1732.

Kornberg, R. D. (2007). The molecular basis of eukaryotic transcription. Proceedings of the National Academy of Sciences, 104, 12955–12961.

Lee, Y., & Rio, D. C. (2015). Mechanisms and regulation of alternative pre-mRNA splicing. Annual Review of Biochemistry, 84, 291–323.

Nishikura, K. (2010). Functions and regulation of RNA editing by ADAR deaminases. Annual Review of Biochemistry, 79, 321–349.

Sharp, P. A. (1994). Split genes and RNA splicing. Cell, 77, 805–815.

Wahl, M. C., Will, C. L., & Lührmann, R. (2009). The spliceosome: Design principles of a dynamic RNP machine. Cell, 136, 701–718.

Chapter 47
Translation and Protein Maturation

CONTENTS OF THIS CHAPTER

"Between the letter and the act stands the interpreter — and the ribosome, four billion years into the job, still reads the message aloud in an RNA accent."

47.1 The Genetic Code

The code is the treaty between nucleic acid and protein, and its terms are now fifty years settled (Figure 47.1). Sixty-four triplet codons specify twenty amino acids and three stops; the code is ***degenerate*** *— most amino acids have several codons — with the degeneracy concentrated at the third position, where Crick's* ***wobble*** *hypothesis permits one tRNA anticodon to pair with several codons (inosine at the wobble position reading U, C or A). It is* ***non-overlapping and comma-free****, read from a fixed start in one frame; and it is* ***nearly universal*** *— the exceptions (mitochondrial reassignments, ciliate stops recoded as amino acids) are the dialects that prove the common origin. The degeneracy is not random generosity: eight family boxes are fully four-fold degenerate, so that many third-position mutations are silent — the code has error-tolerance built into its grammar. Codon usage, though, is biased genome by genome, which is why expressing a human gene in yeast can require rewriting every third letter, and why "silent" mutations occasionally are not (a rare codon stalls the ribosome and can change folding — silence, in translation, is sometimes only a whisper).*

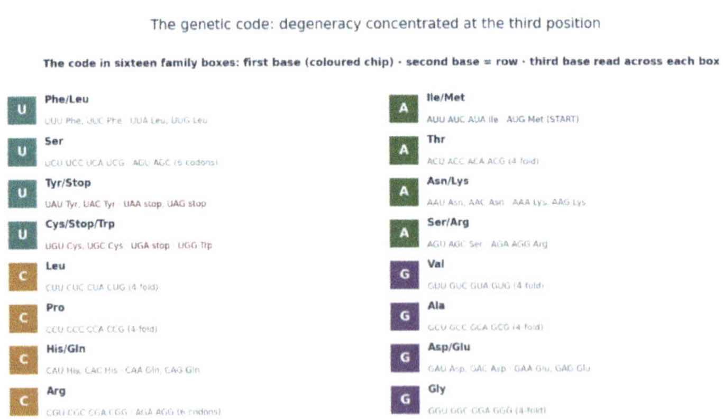

Figure 47.1 — The genetic code in sixteen family boxes. *First base as the coloured chip; within each box the third base varies. Stops in crimson, the AUG start in navy. Eight boxes are fully four-fold degenerate — the code's built-in error tolerance.*

47.2 Transfer RNA

The tRNA is the code's physical embodiment: 76–90 nucleotides folded into the cloverleaf and then the L — the acceptor stem (ending 3′-CCA, where the amino acid hangs) at one tip, the anticodon at the other, seventy-odd angstroms away. The molecule is the most heavily modified RNA in the cell — pseudouridine (whose discovery gave the modification its name), dihydrouridine, inosine, methylated bases — and the modifications are functional: inosine at the wobble position is what lets one tRNA read three codons. As Chapter 43.6 insisted, tRNA is also the fossil exhibit of the RNA world: an all-RNA device for carrying amino acids, older than the proteins that now charge it. Every elongating ribosome holds three tRNAs at once — the molecule is simultaneously the adaptor, the substrate, and the proof that RNA once ran the whole show.

47.3 Aminoacyl-tRNA Synthetases

If the code is a treaty, the twenty aminoacyl-tRNA synthetases are its interpreters — the only molecules that read both languages at once. Each activates its amino acid with ATP (aminoacyl-AMP, PP_i hydrolysed — the ratchet again), then transfers it to the correct tRNA's 3′ end. The accuracy problem is severe: isoleucine differs from valine by one methylene, and chemical discrimination alone gives perhaps five-fold preference — nowhere near the 10^4-fold fidelity translation achieves. The answer is the **double sieve** *(Figure 47.3): the activation site excludes everything larger than the true substrate, and a separate editing site hydrolyses the smaller near-mistakes that slipped through — two sieves of opposite sign in series, multiplying their selectivities. The cost is ATP; the consequence of failure is measurable disease — mutations in several synthetases cause Charcot–Marie–Tooth neuropathy, a reminder that even the most ancient enzymes are still under active quality control by natural selection.*

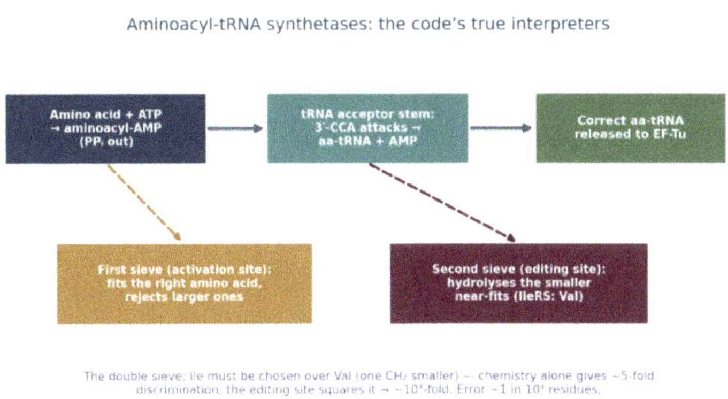

Figure 47.3 — The double sieve. *Activation site rejects the too-large; editing site hydrolyses the too-small near-fits. Opposite sieves in series square the discrimination — chemistry's answer to the one-methylene problem.*

47.4 Ribosome Structure

The ribosome is the cell's largest machine and its oldest: the bacterial 70S particle (50S + 30S), the eukaryotic 80S (60S + 40S), two-thirds RNA by mass. The high-resolution structures — the 2009 Nobel to Ramakrishnan, Steitz and Yonath — settled the decisive fact: the ***peptidyl transferase centre is made entirely of RNA****. The nearest protein is too far away to participate; the 23S rRNA's adenines position the substrates, and the catalysis is substrate-assisted — the attacking amino group does much of the work itself. The small subunit houses the* ***decoding centre****, where 16S rRNA residues (A1492, A1493, G530) flip out to inspect the codon–anticodon helix's minor groove — checking geometry, not sequence, exactly as the DNA polymerase fingers do. The exit tunnel through the large subunit, thirty-odd residues long, is where the nascent chain begins to fold and where the macrolide antibiotics sit, plugging the egress. That erythromycin and its cousins exploit a bacterial ribosomal RNA pocket our own cytosolic ribosomes lack — and that our mitochondrial ribosomes, being bacterial in descent, partially share it (hence the occasional ototoxicity) — is evolutionary medicine in one sentence.*

47.5 Translation Initiation

Initiation is the regulated step and the expensive one. Bacteria keep it simple: the Shine–Dalgarno sequence upstream of the AUG base-pairs with the 16S rRNA's tail, placing the start codon directly in the small subunit's P site, where fMet-tRNA (formylated methionine — the bacterial start badge, and the reason our immune system reads N-formyl peptides as invasion) waits; three initiation factors and one GTP later, the 50S joins. Eukaryotes run the ***scanning model*** *(Figure 47.4): eIF4E grips the cap, eIF4G scaffolds and binds PABP (circularising the message), the 43S pre-initiation complex loads at the cap and scans 5′→3′ — spending ATP to unwind structure — until the Met-tRNA$_i$'s anticodon locks onto an AUG in a good Kozak context; eIF2's GTP is hydrolysed, factors leave, the 60S joins. The control point is eIF2α: phosphorylated under stress (haem lack in the reticulocyte, viral dsRNA via PKR, ER stress via PERK), it sequesters its own exchange factor and shuts translation down — with the telling exception of ATF4 and friends, whose upstream open reading frames make them translate better when eIF2 is scarce. The integrated stress response is, at bottom, an initiation-factor economy.*

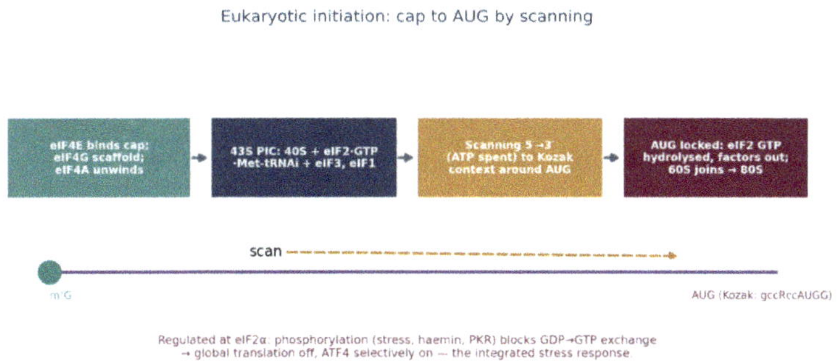

Figure 47.4 — Eukaryotic initiation by scanning. *Cap recognition, 43S loading, ATP-driven scanning to the Kozak-context AUG, and 60S joining. eIF2a phosphorylation is the cell's master translation brake.*

47.6 Peptide-Bond Formation

The chemistry at the centre is disarmingly simple: the amino group of the A-site aminoacyl-tRNA attacks the carbonyl of the P-site peptidyl-tRNA, and the whole growing chain hops onto the A-site amino acid — the peptide bond forming by transfer of the chain to the new residue, not the reverse. No ATP is consumed here (the energy was prepaid by the synthetase), and the catalysis, as Section 47.4 noted, is entropic: the ribosome positions and desolvates the two substrates, holds the attacking nitrogen at the carbonyl carbon, and lets chemistry proceed — rate enhancement of ~10^7, achieved mostly by alignment. The puromycin experiment that revealed the mechanism is one of the field's elegant traps: puromycin mimics the aminoacyl end of a tRNA, accepts the chain, and then, lacking the means to stay bound, walks out with the truncated peptide — an antibiotic that works by being accepted as a substrate and failing as one.

47.7 Elongation

The cycle then repeats, three sites at a time (Figure 47.2): aa-tRNA delivered to the ***A site*** *by EF-Tu (eEF1A in us) as a GTP ternary complex, accepted only after the decoding centre approves the pairing (kinetic proofreading — a deliberate GTPase delay lets the wrong tRNAs leave before acceptance); peptide bond formed; and* ***translocation*** *by EF-G (eEF2), another GTPase, moving both tRNAs — A to P, P to E — and the message by one codon, the E-site tRNA exiting thereafter. Two GTPs per residue in elongation, two ATP equivalents at the synthetase: four high-energy bonds per peptide bond, the energetic price of accuracy (Eq. 47.1). Diphtheria toxin — one of medicine's most economical poisons — ADP-ribosylates eEF2's diphthamide, a modified histidine found on nothing else, and one molecule of the toxin's catalytic fragment can stall every ribosome in a cell. Elongation speed, ~5–20 residues per second, is itself regulated: codon optimality sets the local pace, and the pauses are increasingly understood as folding instructions.*

$$\text{Cost per residue} = \underbrace{2\ \text{ATP} \to \text{AMP}}_{\text{synthetase (2 bonds)}} + \underbrace{1\ \text{GTP}}_{\text{EF-Tu delivery}} + \underbrace{1\ \text{GTP}}_{\text{EF-G translocation}} = 4\ \sim\text{P bon} \qquad (47.1)$$

Derivation. *Bookkeeping over the cycle: the synthetase spends ATP→AMP (two phosphoanhydride bonds, one to activate, one lost when PP_i is hydrolysed); EF-Tu spends one GTP per accepted tRNA; EF-G spends one per translocation. The peptide bond itself costs nothing — the ester bond of the aa-tRNA, bought at the synthetase, is the high-energy intermediate. Four bonds per residue against ~50 kJ/mol of peptide-bond stability: translation is, per bond, the most expensive polymerisation the cell runs, and the excess over thermodynamic need is the price of proofreading.*

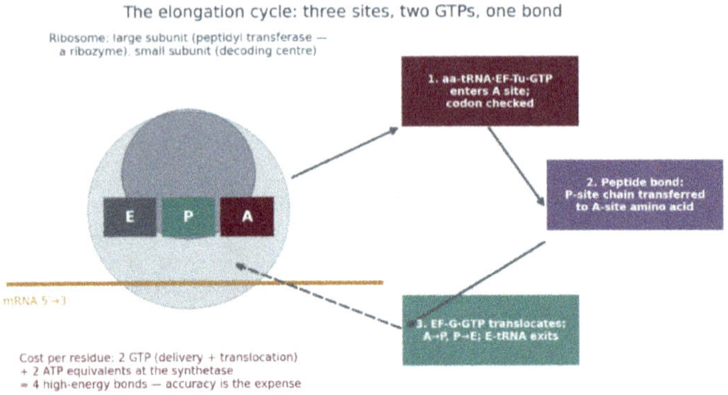

***Figure 47.2* — The elongation cycle.** *E, P, A sites; delivery by EF-Tu, chain transfer in the RNA active site, translocation by EF-G. Two GTPs per residue — the meter running beneath Eq. (47.1).*

47.8 Termination and Recycling

There is no tRNA for a stop codon; instead the ***release factors*** *read it. A class-I factor (RF1 for UAG/UAA, RF2 for UGA/UAA; eRF1 for all three in eukaryotes) inserts its GGQ motif into the peptidyl transferase centre and directs a water molecule onto the ester bond — the chain is hydrolysed off, not transferred — while a class-II factor (RF3/eRF3) spends the GTP. The empty ribosome is then split for reuse: RRF and EF-G in bacteria, ABCE1 in eukaryotes. Premature stops trigger the rescue services instead: tmRNA in bacteria (trans-translation, tagging the orphan protein for destruction in one of molecular biology's great improvisations) and the nonsense-mediated decay of Chapter 46.11 in us. The stop codon, in short, is read by a protein impersonating a tRNA — the last impersonation in a process that began with a nucleotide impersonating an amino acid.*

47.9 Translational Fidelity

The error rate of translation, ~10^{-4} per residue, sits deliberately between replication's 10^{-10} and transcription's 10^{-5} — proteins being disposable, but not so disposable that the cell tolerates gibberish. The accuracy is bought in the same layered way as Chapter 44's: chemical discrimination at the synthetase (10^{-4} with the double sieve), geometric inspection at the decoding centre, and ***kinetic proofreading*** *— the GTPase delay that lets incorrect tRNAs dissociate before acceptance, a mechanism Hopfield formalised and the ribosome implements twice. A final layer works after the fact: stalled or truncated products are tagged and degraded co-translationally, the ribosome quality-control complex policing what* *the ribosome itself let through. Fidelity is never one mechanism; it is always a committee with an energy bill.*

47.10 Protein Targeting

A protein made in the wrong place is a protein wasted, and the addresses are written into the sequence (Figure 47.5). The default is the cytosol; everything else carries a label. Secretory and membrane proteins bear an N-terminal hydrophobic ***signal sequence****, read co-translationally by the SRP: the ribosome pauses, docks at the ER's SRP receptor, and resumes threaded through the Sec61 translocon — synthesis and secretion as one act (Section 47.11). Mitochondrial precursors carry an N-terminal amphipathic helix, imported post-translationally through TOM and TIM at the cost of the membrane potential; nuclear proteins carry lysine-rich NLS motifs, ferried by importins against a Ran-GTP gradient; peroxisomal enzymes end in SKL. The logic is uniform: a cis-acting address, a trans-acting reader, an energy source to make transport directional. The diseases are the misreadings: Zellweger syndrome (peroxisomal import), the I-cell disease of Chapter 12 (lysosomal addressing), and the mistargeting of α-synuclein that Parkinson's research circles endlessly.*

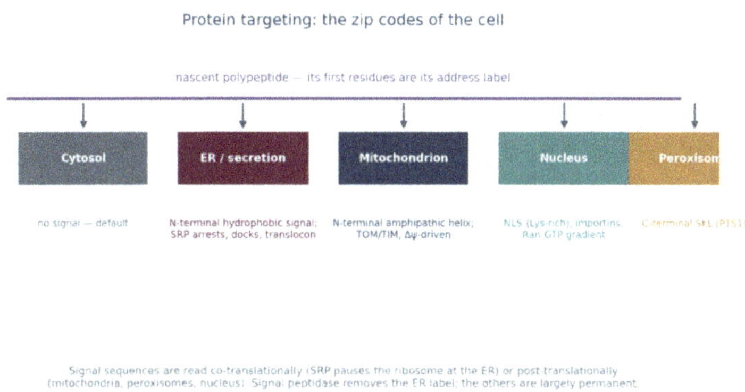

Figure 47.5 — The zip codes. *Five destinations, five address systems. The first residues of a secretory protein are read while the rest is still being written.*

47.11 Secretory-Pathway Processing

Inside the ER the maturing protein meets its first editors. Signal peptidase removes the label; the oligosaccharyltransferase adds the N-linked glycan precursor (en bloc, to Asn-X-Ser/Thr); and the calnexin–calreticulin cycle, with its lectin chaperones and the glucosidase that trims and the UGGT that re-glucosylates only misfolded chains, runs a folding quality-control loop — persistently misfolded proteins are retrotranslocated and ubiquitinated for the proteasome (ERAD, Chapter 21's opening case). Disulfide bonds are formed by protein disulfide isomerase, oxidising cysteines pairwise in the ER's oxidising air. Then the Golgi: glycans trimmed and rebuilt into their mature forms, proteins sorted — to the plasma membrane by default, to lysosomes by the mannose-6-phosphate tag (I-cell disease being the failure to add it, a Golgi full of enzymes addressed to nowhere), to secretory granules for the regulated pathway. The secretory pathway is an assembly line with an inspection at every station, and roughly a third of all proteins pass through it.

47.12 Post-Translational Modification

Once folded, the protein is still not finished: over four hundred modification types annotate it (Figure 47.6). ***Phosphorylation*** *— Ser, Thr, Tyr — is the reversible switch of every signalling pathway in Part X;* ***ubiquitination*** *is the degradation tag of Chapter 21;* ***acetylation and methylation*** *of lysines are the epigenetic script of Chapter 48;* ***glycosylation*** *builds the glycocalyx and the blood-group antigens;* ***lipidation*** *— myristoyl, palmitoyl, prenyl — staples proteins to membranes (Ras's farnesyl being the oncologist's old frustration, undruggable until the pocket of G12C was found instead);* ***hydroxylation*** *of proline and lysine makes collagen possible (scurvy's lesion, Chapter 41.4); and* ***proteolytic maturation*** *finishes many — insulin from proinsulin, the clotting factors from their zymogens. Each mark is written by an enzyme, read by a domain, erased by another enzyme: the protein's second sequence, reversible where the first is not. The twenty amino acids are the alphabet; the modifications are the punctuation, and increasingly the meaning.*

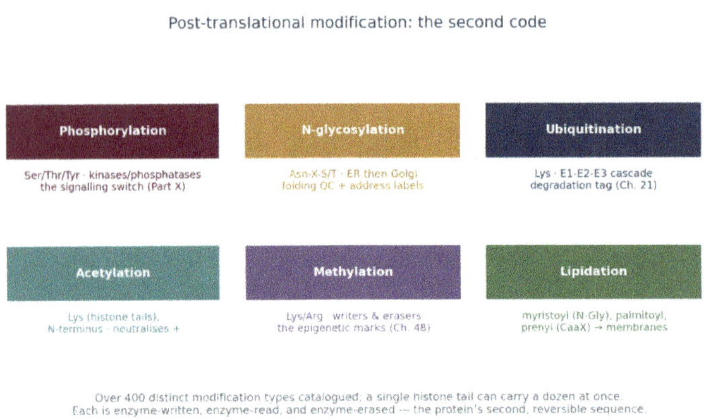

Figure 47.6 — The second code. *Six representative modification systems with their targets and their chapters elsewhere in this book. Written, read and erased by enzymes — regulation that survives translation.*

> ***Case study — The poisoned ribosome of the Struma.*** *In 1890, before antitoxin, diphtheria killed a fifth of the children it infected — strangling by the grey pharyngeal membrane and, weeks later, by myocarditis and neuropathy in the survivors. The toxin is a single protein of two domains. Explain its mechanism, its economy, and why the heart and nerves pay last.*
> ***Resolution.*** *The B domain binds the HB-EGF receptor and delivers the A domain across the endosomal membrane; the A domain ADP-ribosylates diphthamide, the uniquely modified histidine of eEF2 — one enzyme molecule, catalytic, inactivating thousands of eEF2 molecules per hour until elongation stops cell-wide. Tissues with the receptor and the highest protein-synthesis demand fail in order: pharyngeal epithelium first (the membrane is dying cells), then myocardium and Schwann cells, whose long proteins cannot be*

refreshed. Antitoxin works only before the A domain enters cells — hence the race in 1890s Berlin to standardise Behring's serum, and the first Nobel in Medicine (1901). The case is translation seen from the enemy's side: stop the elongation factor, and the whole proteome follows within days.

Historical note — The ribosome seen at last. *The ribosome was too big, too flexible and too important to crystallise, and for thirty years it was known only as an electron-microscopic blob with a formidable parts list. Ada Yonath began the crystallographic assault in 1980, when the consensus was that ribosomal crystals were impossible; she persisted through a decade of shattered lattices, and the trick — flash-cooling crystals from halophilic and thermophilic ribosomes — opened the way. In 2000 the structures arrived almost together: Yonath's and Steitz's large subunits, Ramakrishnan's small. What they showed was worth the twenty years: the peptidyl transferase active site, ringed by protein at a respectful distance, made wholly of RNA — the ribosome is a ribozyme, and the protein world we inhabit was synthesised, at the origin, by RNA alone. The 2009 Nobel in Chemistry recognised all three. Ramakrishnan's later reflection serves as the epigraph the structure earned: "The ribosome is the mother of all machines — the one that made every other machine in this book."*

Plate 47.1 *Venkatraman Ramakrishnan (b. 1952), who solved the 30S subunit and the decoding centre. Nobel laureate in Chemistry, 2009, with Thomas Steitz and Ada Yonath. (Nobel Foundation portrait.)*

47.13 Exercises

Exercise 47.1. *Using Eq. (47.1), estimate the ATP-equivalent cost of synthesising one molecule of a 400-residue enzyme, and compare with the cost of degrading it via the ubiquitin–proteasome system (1 ATP per ubiquitin activation, ~1 ATP per 2 residues by the proteasome). What does the ratio imply?*
Resolution. *Synthesis: 400 × 4 ≈ 1,600 ~P bonds. Degradation: ~10 ubiquitins (10 ATP) + 200 ATP ≈ 210. Ratio ≈ 7–8: making a protein costs about eight times destroying it. The implication is economic: selective degradation is cheap insurance compared with synthesis, which is why the cell degrades freely and regulates synthesis cautiously — quality control by turnover is affordable precisely because the proteasome is the discount option.*

Exercise 47.2. *Why does puromycin terminate chains, and how did it reveal that the growing peptide rides the P-site tRNA?*
Resolution. *Puromycin resembles the aminoacyl-adenosine end of aa-tRNA: it enters the A site, accepts the peptidyl chain from the P-site tRNA — forming the same peptide bond — but cannot be translocated or further extended (no body, no anticodon), so the peptidyl-puromycin falls off. That the product was a peptide attached to puromycin, released only when a P-site tRNA donor was present, proved the chain grows from P to A by transfer — the direction of Section 47.6 — and gave the field its first handle on the active site.*

Exercise 47.3. *A heterozygous GARS mutation (glycyl-tRNA synthetase) causes axonal Charcot–Marie–Tooth disease with normal global translation. Propose two non-exclusive explanations compatible with the phenotype's tissue specificity.*
Resolution. *(i) Haploinsufficiency at the margin: motor axons are metres long and their distal protein supply depends on local translation; half the synthetase activity may suffice in the cell body but fail at the distal axon, where tRNA charging cannot be buffered — the longest cells show the deficit first. (ii) Toxic gain of function: the mutant synthetase aberrantly binds a signalling partner (e.g. neuropilin-1 at the plexin receptor), disturbing axon guidance and maintenance regardless of charging. The two mechanisms explain, respectively, why severity tracks axon length and why some GARS alleles are dominant-negative rather than merely half-dose.*

Exercise 47.4 (proposition for research). *Codon-optimality effects suggest ribosome pausing instructs co-translational folding. Propose an experiment to prove causality for one protein, controlling for every confound you can name.*
Sketch. *Choose a two-domain enzyme whose interdomain interface misfolds in disease. Make synonymous variants that change local codon optimality at the domain boundary while holding sequence, GC content, mRNA structure and initiation rate constant; read folding by limited proteolysis of the nascent chain on stalled ribosome–nascent-chain complexes (RNCs) and by activity after release; confirm pauses by ribosome profiling of the exact constructs. The decisive controls: an anticodon-supplemented strain that abolishes the pause without changing the codons, and total RNA-seq to exclude abundance effects. If the pause-removed variant misfolds while its mRNA and translation rate are*

unchanged, pausing is the folding instruction — and synonymous mutations become, formally, folding mutations.

47.14 Synthesis

The message is now protein: the code read centre-outward, tRNA as the bilingual adaptor, synthetases squaring their discrimination through the double sieve, the ribosome revealed as a ribozyme with a decoding centre, initiation by scanning and its eIF2α brake, elongation at four bonds a residue, termination by impersonation, and the finished chain addressed, folded, inspected and annotated with the second code of modifications. Direction checked against the master plan: the genome is read and executed; what remains is the government of when it is read at all. ***Chapter 48 — Regulation of Gene Expression*** *closes Part IX with the operon, the enhancer, the epigenome, and the small RNAs.*

47.15 References

Crick, F. H. C. (1966). Codon–anticodon pairing: The wobble hypothesis. Journal of Molecular Biology, 19, 548–555.

Hopfield, J. J. (1974). Kinetic proofreading: A new mechanism for reducing errors in biosynthetic processes requiring high specificity. Proceedings of the National Academy of Sciences, 71, 4135–4139.

Ibba, M., & Söll, D. (2000). Aminoacyl-tRNA synthesis. Annual Review of Biochemistry, 69, 617–650.

Ramakrishnan, V. (2018). Gene Machine: The Race to Decipher the Secrets of the Ribosome. Basic Books.

Rodnina, M. V. (2018). Translation in prokaryotes. Cold Spring Harbor Perspectives in Biology, 10, a032664.

Schmeing, T. M., & Ramakrishnan, V. (2009). What recent ribosome structures have revealed about the mechanism of translation. Nature, 461, 1234–1242.

Wek, R. C. (2018). Role of eIF2α kinases in translational control and adaptation to cellular stress. Cold Spring Harbor Perspectives in Biology, 10, a032870.

Chapter 48
Regulation of Gene Expression

Contents of this chapter

"The genome in every cell is the same book; development, memory and disease are nothing more — and nothing less — than differences in which pages are open."

48.1 Bacterial Operons

Part IX closes with its central question inverted. Chapters 43–47 described a machine that reads the genome; this chapter describes the government that decides what is read, when, and how much — and the founding document of that government is a bacterial sugar switch. The **lac operon** *of Jacob and Monod (1961) is three genes — LacZ (β-galactosidase), LacY (permease), LacA (transacetylase) — transcribed as one message from one promoter, gated by one logic gate (Figure 48.1): transcribe when lactose is present AND glucose is absent. The repressor LacI binds the operator and blocks the polymerase; lactose (via allolactose) binds LacI and lifts it off. The activator CAP, bound to cAMP — which accumulates only when glucose is scarce — bends the DNA and recruits the polymerase. Four states, one truth table, and the profound discovery buried in the genetics: the repressor is a diffusible protein acting on a DNA site — regulation runs in trans, through molecules, not through the chromosome's private wiring. The trp operon inverts the logic (repressor active when tryptophan is bound — an anabolic operon, off when fed), and the arabinose operon adds a repressor that doubles as an activator. From these three switches the whole doctrine of gene regulation was generalised, and Monod's quip became its constitution: what is true for E. coli is true for the elephant — only more so, and with enhancers.*

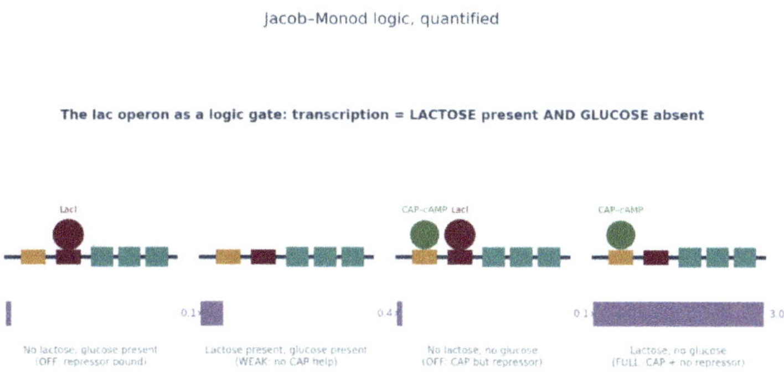

Figure 48.1 — The lac operon as a logic gate. *Four states of the two-input switch: LacI repressor (crimson) lifts off with lactose; CAP–cAMP (green) binds only when glucose is absent. Full expression requires both conditions — an AND gate made of two proteins.*

48.2 Transcriptional Activators and Repressors

The proteins that read regulatory DNA share a small wardrobe of structural motifs. The ***helix-turn-helix*** *(LacI, CAP, the homeodomains) lays a recognition α-helix into the major groove — the groove's chemical signature of Chapter 43.3 read by a matching helix;* ***zinc fingers*** *(the nuclear receptors, GATA factors) grip the groove with a zinc-braced knuckle, one finger per three base pairs, modular enough to be redesigned for therapy;* ***leucine zippers*** *(Fos, Jun, the bZIPs) dimerise on a coiled coil so that mixed dimers multiply the regulatory vocabulary; and* ***helix-loop-helix*** *factors (MyoD, Myc) add a DNA-gripping basic region. Binding is sequence-specific but probabilistic — a factor samples thousands of near-matches — and specificity in vivo is achieved combinatorially: clusters of modest sites, occupied cooperatively, give switch-like behaviour from unremarkable affinities. The repressors work by occlusion, by masking activation domains, or by recruiting the chromatin machinery of the next sections; the activators work by recruitment — of Mediator, of the remodellers, of the polymerase itself. Nothing in the mechanism is mysterious; the mystery is the combinatorics, and the combinatorics is the cell.*

48.3 Chromatin Organisation

In a eukaryote the DNA is never naked, and the first regulatory decision is whether a gene is physically available. Chromatin comes in three functional states (Figure 48.3). ***Euchromatin*** *is open: nucleosomes spaced and acetylated, DNA hypersensitive to nucleases, replicated early, transcribed at will.* ***Facultative heterochromatin*** *is conditionally closed — Polycomb-repressed, marked by H3K27me3, the state of silenced developmental genes and of the inactive X.* ***Constitutive heterochromatin*** *is permanently closed — H3K9me3, HP1, dense DNA methylation at centromeres and telomeres, replicated late, transcribed never. The states are heritable through mitosis (Section 48.12) and interconvertible at their edges — position-effect variegation, the mottled eye of a fly whose white gene landed beside heterochromatin, being the classic demonstration that a gene's* *expression can depend on its neighbours. The genome's geography — which TAD, which compartment, which distance from the nuclear lamina — is regulatory information the sequence alone does not contain.*

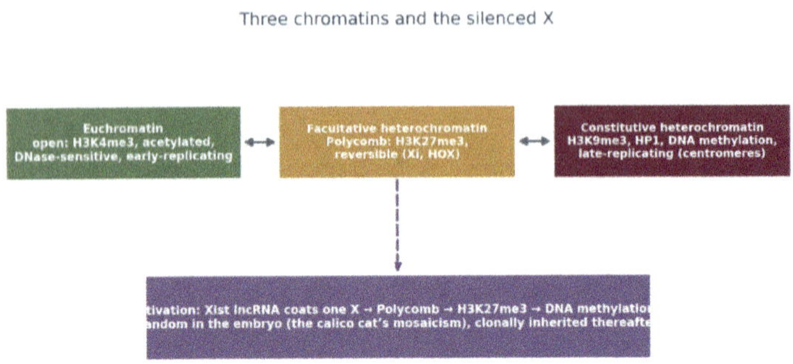

Figure 48.3 — Three chromatins and the silenced X. *Open euchromatin, Polycomb-silenced facultative heterochromatin, and HP1-locked constitutive heterochromatin — with X-inactivation as the exemplar of the middle state, locked by Xist RNA, Polycomb and methylation in series.*

48.4 Histone Modification

The nucleosome's tails protrude, and on them is written the code of the second sequence (Figure 48.2). **Acetylation** *(HATs; erased by HDACs) neutralises lysine's charge, loosens the tail's grip on DNA, and creates binding sites for bromodomains — an activating mark, which is why HDAC inhibitors reactivate silenced genes and why two of them are anticancer drugs.* **Methylation** *is positional rather than simply on or off: H3K4me3 at active promoters, H3K36me3 down the bodies of transcribed genes, H3K27me3 at Polycomb-silenced loci, H3K9me3 in constitutive heterochromatin — the same chemical group, four addresses, four opposite meanings. The economy is a writer–reader–eraser triangle (methyltransferases, chromodomain-containing readers, demethylases), each arm druggable: EZH2 inhibitors for the Polycomb-driven lymphomas, LSD1 inhibitors in trial. Two principles organise the sprawl. Marks recruit the machinery that deposits more of the same — positive feedback that makes states self-sustaining; and opposing marks can coexist at poised "bivalent" promoters (H3K4me3 with H3K27me3), the developmental genes held ready, one mark's erasure from activation. The histone code is less a code than a memory system — and memory, as Section 48.12 will show, is exactly what it provides.*

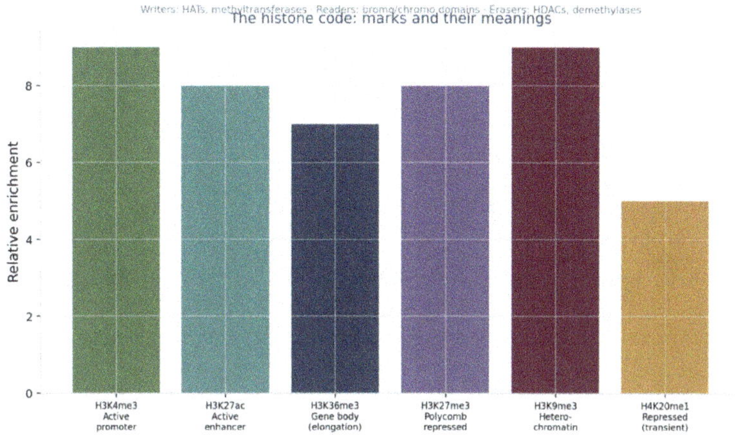

Figure 48.2 — The histone code. *Six marks, six meanings: the same methyl group activating at H3K4, elongating at H3K36, repressing at H3K27 and H3K9. Position is meaning; the writers, readers and erasers run the triangle.*

48.5 DNA Methylation

The genome's own marginalia is the methyl group on cytosine at CpG — written by DNMT3A/B (de novo), maintained through replication by DNMT1 reading the hemimethylated dyad, and removed actively by the TET dioxygenases through oxidised intermediates. Its geography is inverted: the genome is methylated by default (silencing the transposons of Chapter 45.10), and the unmethylated exception is the ***CpG island*** *at promoters of housekeeping and developmental genes — until, in cancer or imprinting, the island is methylated and the gene is locked off. The physiological showcases are imprinting — some hundred genes expressed from one parental allele only, the IGF2/H19 locus the paradigm, its disorders (Beckwith–Wiedemann, Prader–Willi, Angelman) being diseases of whose allele carries the methyl — and X-inactivation's final lock. And it is the best molecular clock we own: the Horvath epigenetic clock reads age from a few hundred CpGs with an accuracy that has made forensic scientists actuarial, and has shown that the epigenome keeps a biography as well as a birth certificate.*

48.6 Enhancers and Silencers

The eukaryotic answer to the operon's compactness is the ***enhancer****: a cluster of factor-binding sites, acting in either orientation, at distances up to a megabase, brought to its promoter by loop extrusion inside the TADs of Chapter 43.10 (Figure 48.4). Cohesin rings reel the chromatin fibre into loops until convergent CTCF sites stop them; within the loop, enhancer and promoter meet. The regulatory consequences are immense. A gene may answer to a dozen enhancers, each active in a different tissue or stage — the spare-parts architecture that lets one gene serve limb and brain without confusing the two; disease variants lie overwhelmingly in the non-coding 98% of the genome, and most map to enhancers (the GWAS lesson); and boundary deletions cause "enhancer adoption" — a developmental gene hijacked by a neighbouring TAD's enhancer, the limb malformations of the SHH locus being the textbook case. Silencers and insulators* *complete the grammar: the insulator is the fence, the enhancer the address, and the promoter merely the door the address delivers to.*

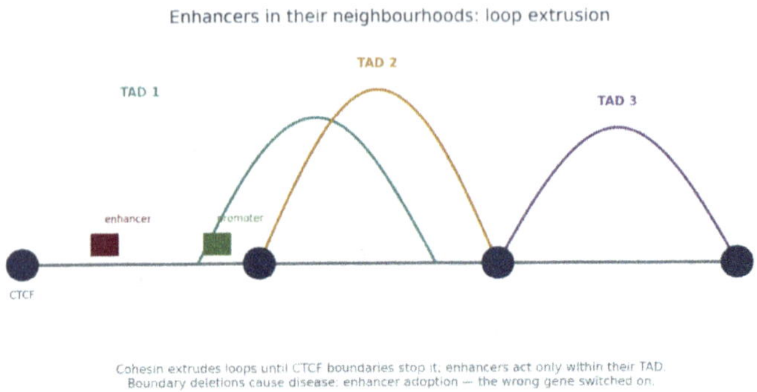

***Figure 48.4* — Loop extrusion.** *Cohesin extrudes chromatin loops until CTCF boundaries (navy) halt it; enhancers (crimson) reach promoters (green) only within their own TAD. Delete the fence and the enhancer adopts the wrong gene.*

48.7 Chromatin Remodelling

Marks aside, the nucleosomes themselves must move, and the movers are ATP-burning machines. Four families — SWI/SNF, ISWI, CHD, INO80 — slide, eject, or re-space nucleosomes: SWI/SNF evicts to open promoters and enhancers (and is mutated in a fifth of all cancers, SNF5's loss defining the malignant rhabdoid tumour — a remodeller as tumour suppressor), ISWI spaces nucleosomes evenly behind replication, INO80 exchanges histone variants. The variants themselves are regulatory: H2A.Z at promoters, H3.3 at transcribed genes and regulatory sites, macroH2A on the inactive X, CENP-A defining the centromere. The emerging picture is dynamic: nucleosomes turnover, marks fluctuate, and the "state" of a locus is a steady-state of opposing fluxes — which is why a single pulse of an eraser can flip a gene's expression for generations of cells. Chromatin is not a scaffold but a tide.

48.8 Regulatory RNAs

Some of the government speaks RNA. The **long non-coding RNAs** *— tens of thousands, mostly tissue-specific — scaffold, guide and decoy: Xist coats the X chromosome in cis and recruits Polycomb (Section 48.3); HOTAIR guides repression in trans across the HOX loci; others titrate transcription factors or miRNAs (the competing-endogenous-RNA hypothesis, debated but alive). Enhancers themselves are transcribed into eRNAs whose production marks activity. The bacterial world has its own RNA government — riboswitches, metabolite-binding mRNA leaders that fold one way to terminate transcription and another to permit it (thiamine's own riboswitch regulating thiamine synthesis, Chapter 41's cofactor governing its own gene), and the CRISPR arrays, which are RNA-guided immunity: the archive remembering its invaders in spacers between repeats. The RNA-world hypothesis (Chapter 43.6) suggests these are not innovations but survivals — regulation, like catalysis, was RNA's first profession.*

48.9 RNA Interference

The small RNAs run the most economical repression in the cell (Figure 48.5). MicroRNAs — ~22 nucleotides, processed from hairpins by Drosha and Dicer, loaded into Argonaute at RISC's heart — pair imperfectly with 3′ UTRs and repress translation or hasten decay; each miRNA tunes hundreds of targets modestly, and the network effects are anything but modest (the let-7–lin-4 pair in C. elegans, the first found, being developmental timing written in RNA). Small interfering RNAs, from long dsRNA, pair perfectly and slice — the antiviral defence Fire and Mello turned into a technology (1998; Nobel 2006) and the clinic turned into drugs: patisiran silencing transthyretin in amyloidosis, inclisiran silencing PCSK9 twice a year by subcutaneous injection. The piRNAs guard the germline against transposons, completing the trilogy. RNAi is sequence-specific silencing with a 22-nt address label — the proof that in gene regulation, as in everything else, the cheapest reagent is a short complementary string.

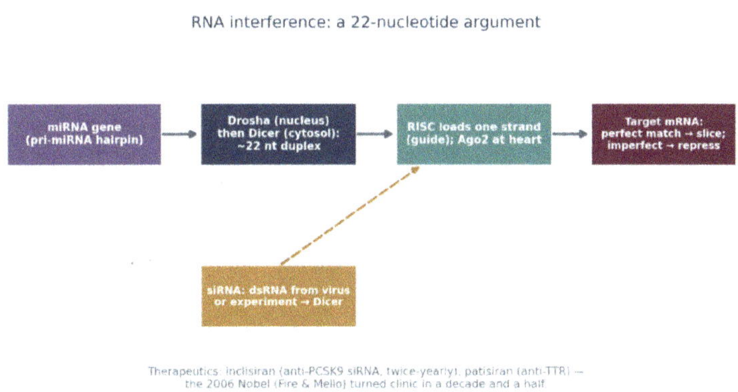

Figure 48.5 — RNA interference. *miRNA biogenesis (hairpin → Drosha → Dicer → RISC), the siRNA branch from dsRNA, and the two outcomes — slicing or repression — with the clinic's first arrivals below.*

48.10 Translational Regulation

After the message is made, the message is managed — and Chapter 47's machinery supplies the handles. Upstream open reading frames throttle the main ORF (the ATF4 trick of the integrated stress response: when eIF2α is phosphorylated and initiation is scarce, the ribosome skips the inhibitory uORFs and translates the stress transcription factor — scarcity converted into a signal). Iron regulation is the classic post-transcriptional economy: the iron-response element in the ferritin 3′... rather, the IRE in ferritin's 5′ UTR blocks translation when aconitase — moonlighting as the iron-regulatory protein — binds it in iron poverty, while the same protein binding IREs in the transferrin-receptor 3′ UTR stabilises that message; one sensor, two UTRs, iron storage down and iron uptake up. And the AU-rich elements of cytokine messages summon decay (Chapter 46.11). Translational control is faster than transcriptional — no nucleus to consult — which is why it dominates in the synapse, the oocyte and the platelet, the cell's three places without time or a nucleus to spare.

48.11 Protein Degradation

The last handle on expression is the protein's lifespan, and Chapter 21's ubiquitin–proteasome system is the machinery; here it suffices to note the logic. N-end rules and degrons set half-lives from minutes (cyclins, destroyed on schedule each cell cycle — Chapter 51) to days; the hypoxia factor HIF-1α is hydroxylated and ubiquitinated in oxygen and stabilised in hypoxia — degradation as the oxygen sensor (Chapter 42.10); and the molecular glues and PROTACs now hijack E3 ligases to destroy targets a drug could never inhibit — degradation as pharmacology. Expression level is synthesis minus decay at every tier, and the cell pulls every handle: transcription, splicing, export, translation, and destruction. The steady-state protein is the integral of a five-tier regulation, and the tiers multiply, not add.

48.12 Epigenetic Inheritance

The closing question of Part IX is what survives division — and the answer divides neatly. ***Mitotic inheritance*** *is routine: DNMT1 copies the methylation pattern onto the new strand, Polycomb and the H3K9 machinery re-deposit their marks on the parental histones' new neighbours, and a liver cell divides into liver cells though the genome knew nothing of liver.* ***Transgenerational inheritance*** *— through the germline — is the contested frontier: the gametes erase methylation twice (in the primordial germ cells and after fertilisation), and what survives the erasure is the imprinted loci, a few retrotransposons, and — in model organisms, with the best evidence in worms and flies — small-RNA-mediated memory of ancestral experience (Figure 48.6). The Lamarckian temptation should be resisted and the phenomenon respected: the Dutch Hunger Winter cohort shows metabolic traces of grandmaternal famine, mechanism unresolved. What is certain is the epigenome's role as the cell's biography — written over by age, by diet, by exposure (twin methylation discordance grows with years lived apart), and read at every division. The genome proposes; the epigenome remembers what the cell decided.*

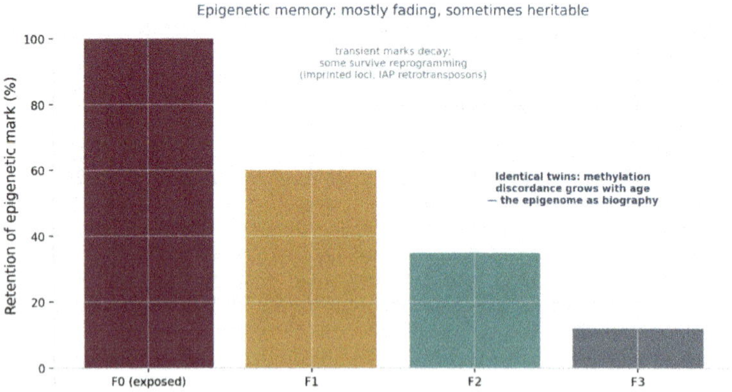

Figure 48.6 — Epigenetic memory. *Marks induced in an exposed generation decay across descendants — mostly. Imprinted loci, some retrotransposons and small-RNA pathways can carry memory through the germline's two reprogrammings.*

Case study — The twins who drifted apart. *Monozygotic twins, raised together, are indistinguishable on a methylation panel at age 8. Re-tested at 50 — one a lifelong smoker, one not — their methylation correlation has fallen markedly, and the smoker's AHRR gene is densely hypomethylated. Neither has any sequence difference. Explain the divergence and its inheritance prospects.*
Resolution. *This is Section 48.12 measured in two people. Sequence-identical genomes accumulate different epigenomes: stochastic drift plus environment (tobacco's effect on AHRR methylation is one of the most reproducible findings in epigenomics — hypomethylation tracks pack-years and slowly reverts on quitting). The twins' phenotypes diverge accordingly, without a single base changing: the smoker's cancer risk is written in his methylome, legible in blood. For inheritance: smoking-induced marks are largely erased in the germline's reprogramming — the grandchildren do not inherit the habit's methylome, so far as the evidence runs — but the smoker's grandchildren do inherit his twins' example of how deeply environment writes into the genome's margins. Nature deals identical cards; nurture annotates the deck.*

Historical note — Jacob, Monod, and the pa-ma-mo at the Pasteur. *The operon was born in a Paris attic laboratory during the Occupation's aftermath, and its midwives were an unlikely pair: Jacques Monod, a Resistant and cellist who had been measuring bacterial growth curves since the war and could not explain why E. coli fed two sugars ate them in strict sequence; and François Jacob, a surgeon whose war wounds ended his operating career and who talked his way into Lwoff's laboratory in 1950. The decisive experiments — the PaJaMa mating of 1957 (Pardee, Jacob, Monod), in which a lac region conjugated into a repressorless cell began enzyme synthesis immediately — proved the repressor was diffusible: regulation was chemistry, not chromosomal fate. The 1961 paper named the operon and messenger RNA in one stroke; the 1965 Nobel (Jacob, Monod, Lwoff) followed. Monod's later epigraph for the whole enterprise, from Chance and Necessity, is the coldest sentence in molecular biology and the true one: "The cell is a machine; and the structure of every machine is the record of the accidents that built it." This Part of the book — helix, replication, repair, transcription, translation, regulation — is that record, read closely.*

Plate 48.1 *François Jacob (1920–2013), who with Jacques Monod founded the operon model and the concept of messenger RNA. Nobel laureate in Physiology or Medicine, 1965, with Monod and André Lwoff. (Nobel Foundation portrait.)*

48.13 Exercises

Exercise 48.1. *A lacI^s (super-repressor) mutant cannot bind inducer. Predict the phenotype, its dominance, and the effect of a second mutation deleting the operator in cis.*
Resolution. *LacI^s represses regardless of lactose: the operon is uninducible (Lac$^-$). It is dominant — the mutant repressor is diffusible and occupies operators on any copy in the cell, so a wild-type lacI allele does not rescue (trans dominance, the defining test of a repressor). Deleting the operator in cis removes the repressor's landing site: that copy becomes constitutive even in the mutant's presence — the cis/trans logic that made the operon the founding proof that regulation is molecules acting on sites, not chromosomes acting on themselves.*

Exercise 48.2. *Why do most GWAS disease variants fall in non-coding DNA, and why does that make them hard to interpret? Give the interpretive pipeline.*
Resolution. *Most heritability acts on regulation, not on protein sequence — the non-coding 98% contains the enhancers, and altering when/where a gene is expressed is a subtler, more selectable route to disease susceptibility than breaking the protein. The interpretation problem is linkage and location: the sentinel SNP is usually not causal, and its target gene may be megabases away within the TAD. Pipeline: fine-map the credible set, intersect with chromatin marks and eQTLs in the relevant tissue, call enhancer–promoter contacts (Hi-C/promoter-capture), and validate by CRISPRi of the enhancer with expression*

of the candidate gene as readout. Association is geography; function must be proven in cis.

Exercise 48.3. *Explain how HIF-1α can be both continuously synthesised and effectively absent in oxygen, and name the modification that switches it.*
Resolution. *Regulation by degradation (Section 48.11): in oxygen, prolyl hydroxylases (2-oxoglutarate-dependent, O_2 as substrate — Chapter 42.10) hydroxylate two prolines on HIF-1α; the von Hippel–Lindau protein reads the hydroxyprolines and ubiquitinates the factor, which the proteasome destroys within minutes. In hypoxia the hydroxylase stalls, HIF-1α accumulates, and the transcriptional programme switches on. The signal is not the synthesis of a protein but the survival of one — expression controlled entirely at the fifth tier.*

Exercise 48.4 (proposition for research). *Epigenetic clocks predict age but not yet mechanism. Propose an experiment to test whether clock CpGs are drivers, passengers, or brakes of ageing.*
Sketch. *Take the top ~50 clock CpGs in human cells; install targeted methylation/demethylation (dCas9-DNMT3A / dCas9-TET1) at each locus in primary cells, and measure (i) the clock's own readout, (ii) replicative lifespan and senescence markers, (iii) transcriptomic age. The three hypotheses predict three outcomes: drivers — editing the CpGs shifts phenotype and clock together; passengers — clock shifts, phenotype does not (or vice versa); brakes — demethylation accelerates decline (the clock as compensatory). The field has spent a decade reading the clock; this is how it learns whether the clock can be wound back, or merely watched.*

48.14 Synthesis

Part IX closes where information becomes decision: the operon's AND gate; factors that read the major groove; three chromatins and the tide of remodellers; the histone code as a memory system; methylation as marginalia and clock; enhancers looped within their TADs; RNA governments old and small; the five-tier economy of expression; and the epigenome as biography, inherited at mitosis and occasionally beyond. Direction checked against the master plan: the cell can now read its genome and govern the reading; Part X asks how cells talk to one another. ***Chapter 49 — Principles of Signal Transduction*** *opens the signalling book — receptors, second messengers, and the cascade's arithmetic.*

48.15 References

Allis, C. D., & Jenuwein, T. (2016). The molecular hallmarks of epigenetic control. Nature Reviews Genetics, 17, 487–500.

Fire, A., Xu, S., Montgomery, M. K., Kostas, S. A., Driver, S. E., & Mello, C. C. (1998). Potent and specific genetic interference by double-stranded RNA in Caenorhabditis elegans. Nature, 391, 806–811.

Horvath, S., & Raj, K. (2018). DNA methylation-based biomarkers and the epigenetic clock theory of ageing. Nature Reviews Genetics, 19, 371–384.

Jacob, F., & Monod, J. (1961). Genetic regulatory mechanisms in the synthesis of proteins. Journal of Molecular Biology, 3, 318–356.

Monod, J. (1971). Chance and Necessity. Knopf.

Schoenfelder, S., & Fraser, P. (2019). Long-range enhancer–promoter contacts in gene expression control. Nature Reviews Genetics, 20, 437–455.

Zhou, V. W., Goren, A., & Bernstein, B. E. (2011). Charting histone modifications and the functional organization of mammalian genomes. Nature Reviews Genetics, 12, 7–18.

PART X

Regulation, Signalling and Cellular Decisions

Chapter 49
Principles of Signal Transduction

CONTENTS OF THIS CHAPTER

"The cell membrane is not a wall but a diplomatic service: it admits no foreign minister, yet receives every dispatch — and then, with admirable economy, distils each one into a phosphorylation."

49.1 The Signalling Problem

Part IX showed how one cell governs its genome; Part X opens by asking how billions of cells govern one another. The difficulty is architectural. The signals that coordinate a multicellular body — insulin, adrenaline, acetylcholine, the cytokines — are almost all hydrophilic, and the plasma membrane, as Chapter 11 established at some length, is an unapologetic lipid. A hormone therefore cannot enter its target cell to deliver its message in person; it must be heard at the door. ***Signal transduction*** *is the doctrine of how that hearing is arranged: a receptor on the surface binds the ligand outside, changes shape across the bilayer, and on the inside initiates a chain of molecular events that ends, seconds to hours later, in a metabolic adjustment, a movement, a secretion or a paragraph of new gene expression. The whole art reduces to four operations —* ***reception, transduction, amplification, termination*** *— and this chapter takes them in order, because a principle, unlike a pathway, survives the examination.*[60]

A useful vocabulary first. The extracellular ligand is the ***first messenger****; the small, diffusible molecules generated inside the cell in response — cyclic AMP, cyclic GMP, inositol trisphosphate, diacylglycerol, Ca^{2+} — are the* ***second messengers****, a term coined by Earl Sutherland, whose portrait appears in Plate 49.1 and whose liver slices appear in Section 49.12 (Figure 49.1). Cells are selective about their audience: a ligand speaks only to cells bearing its receptor, which is why adrenaline quickens the heart but ignores the pancreas, and why the same adrenaline, heard by a different receptor in the liver, commands glycogenolysis instead. Specificity is therefore not a property of the message but of the mailbox — a point with medicinal consequences the size of the pharmaceutical industry.*

60. The student who has been promised "pathways" since Chapter 1 will notice we still refuse to draw arrows first and ask questions later. Chapters 50–52 will supply the named pathways in abundance; this chapter supplies the grammar that makes them legible.

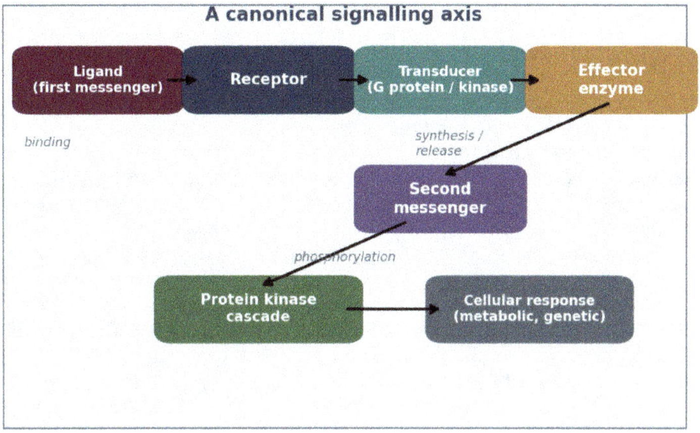

*Figure 49.1 — **A canonical signalling axis.** The first messenger (crimson) binds a receptor (navy), which activates a transducer (teal) — a G protein or a kinase — which activates an effector enzyme (amber). The effector manufactures a second messenger (plum), which drives a protein-kinase cascade (green) and finally a cellular response (slate). Every named pathway in Chapters 50–51 is a variation on this five-note theme.*

49.2 Ligand Binding and Occupancy

Everything downstream of a receptor is proportional, at the outset, to how many

$$\theta = \frac{[RL]}{[R] + [RL]} = \frac{[L]}{[L] + K_d} \tag{49.1}$$

from the sky.

Equation (49.1) is the rectangular hyperbola the reader has now met as oxygen binding to myoglobin (Chapter 8) and substrate binding to an enzyme (Chapter 14), and its reappearance is not laziness but homology of ideas: wherever one

accidental. Second, the hyperbola is steepest at low occupancy: a cell responds most sensitively to changes in ligand concentration precisely where the ligand is scarcest, which is where regulation usually matters. Receptors that cluster — or that, like many immune receptors, require coincident binding of two ligands —

introduce cooperativity, and the hyperbola sharpens into the Hill sigmoid of Chapter 8; the cell thereby converts a dimmer into a switch.

49.3 Receptor Classes

Nature has solved the across-the-membrane problem exactly four times, and every receptor yet discovered is a descendant of one of the four solutions (Figure 49.2). The ***G-protein-coupled receptors*** *(GPCRs) thread the bilayer seven times — the serpentine fold of rhodopsin, the β-adrenergic receptor and some eight hundred human cousins, the largest receptor family and the target of roughly a third of marketed drugs. Ligand binding shifts the transmembrane helices, and the cytosolic loops become a landing pad for a heterotrimeric G protein. The* ***enzyme-coupled receptors****, of which the receptor tyrosine kinases (RTKs) are the exemplar, cross the membrane once; ligand binding dimerises two single-pass chains, and the juxtaposed cytosolic kinase domains phosphorylate each other into activity — the insulin receptor and the growth-factor receptors of Chapter 50 work this way. The* ***ligand-gated ion channels*** *(the nicotinic acetylcholine receptor, the $GABA_A$ receptor) dispense with intermediaries entirely: binding opens a pore, and the signal is the ion current itself, millisecond-fast, which is why the nervous system prefers them. And the* ***intracellular receptors*** *— for steroid and thyroid hormones, vitamin D, retinoic acid — cheat altogether: their ligands are hydrophobic enough to cross the membrane uninvited, so the receptor waits inside, usually doubling as the transcription factor of Chapter 48. Four architectures, one logical requirement: binding energy on the outside must purchase a conformational or chemical change on the inside.*

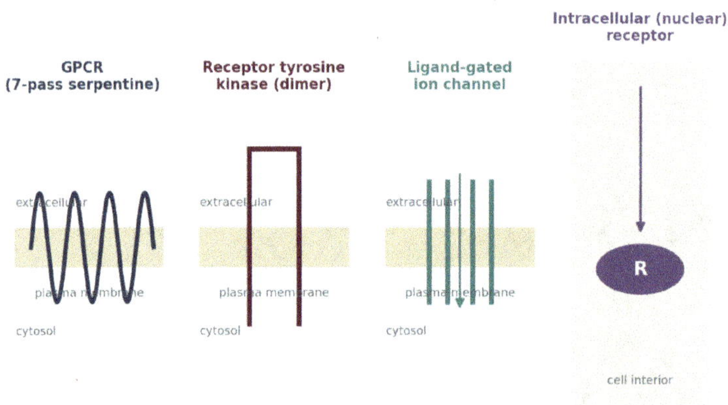

Figure 49.2 — The four receptor architectures. *Left to right: a seven-pass GPCR (navy); a dimerising single-pass receptor tyrosine kinase (crimson); a ligand-gated ion channel (teal), signal and response fused in one protein; and an intracellular receptor (plum) whose lipid-soluble ligand crosses the membrane itself. Only the fourth admits the messenger indoors.*

49.4 G Proteins: The Molecular Relay

The heterotrimeric G protein is the most elegant relay in the cell, and its cycle (Figure 49.3) repays slow reading. At rest, the α subunit carries GDP and holds the βγ dimer close; the whole trimer loiters near its receptor. When agonist-bound receptor and trimer meet, the receptor acts as a ***guanine-nucleotide exchange***

***factor**: it pries open the nucleotide pocket, GDP leaves, and GTP — more abundant in the cytosol — takes its place. GTP binding splits the complex: Gα·GTP departs to regulate its effector (adenylyl cyclase, phospholipase C, an ion channel), and the liberated βγ dimer, far from inert ballast, regulates effectors of its own. The signal then extinguishes itself by an intrinsic device: Gα is a slow GTPase, and when it hydrolyses its own GTP to GDP, it rebinds βγ and returns to rest. The cycle is a clock with a built-in alarm — GTPase-activating proteins (RGS proteins) merely set the alarm earlier.*

*Three features deserve emphasis because they generalise across all of Part X. First, the G protein is a **molecular switch** with exactly two stable states, GDP-bound and GTP-bound, and the entire signalling system is built from such binary components; the subtlety is in their wiring, not their physics. Second, the switch is **timed by its own chemistry**: the duration of the active state is set by a hydrolysis rate, so the cell can tune signal duration by tuning the GTPase — and pathogens can tune it catastrophically, as cholera will demonstrate in Section 49.11. Third, one receptor can activate many G proteins in succession while it remains occupied, which is the first turn of the amplification screw to which we now turn.*

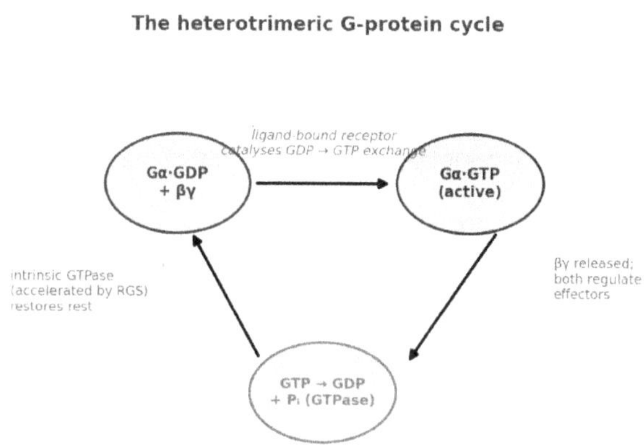

***Figure 49.3 — The heterotrimeric G-protein cycle.** Resting Gα·GDP + βγ (navy) is activated by a ligand-bound receptor, which catalyses GDP→GTP exchange; active Gα·GTP (crimson) and free βγ regulate effectors until intrinsic GTP hydrolysis (teal) restores the resting trimer. RGS proteins accelerate the hydrolysis step; cholera toxin abolishes it.*

49.5 Amplification: The Cascade's Arithmetic

arithmetic fact, and it is derivable. Consider a linear cascade of stages, each an

gives the total amplification:

$$A = \prod_i g_i = \underbrace{g_1\tau_1}_{\text{G proteins per receptor}} \times \underbrace{k_2\tau_2}_{\text{cAMP per G}\alpha} \times \underbrace{k_3\tau_3}_{\text{phosphorylations per kinase}} \approx 10 > \quad (49.2)$$

Figure 49.4 displays the same arithmetic on a logarithmic scale. The lesson is that a cascade is not a bureaucratic elaboration but an instrument: each stage multiplies, and the product of modest gains is astronomical. The same architecture also multiplies regulatory opportunities — every stage is a place to add a brake, a booster or a crosstalk junction, and a pathway with one stage would have nowhere to put its regulators. Amplification and regulation are the same invention seen from two sides, which is why cascades, once evolved, are never abandoned.

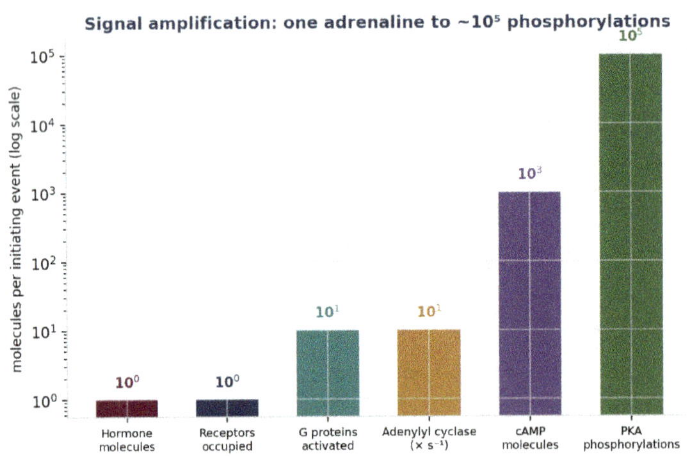

Figure 49.4 — Signal amplification, on a logarithmic scale. *One hormone molecule (crimson) occupying one receptor (navy) activates ~10 G proteins (teal), whose adenylyl cyclase produces ~10^3 cAMP (plum), whose kinases execute ~10^5–10^6 phosphorylations (green). Each bar is ten times the previous; the cascade is a geared transmission.*

49.6 Second Messengers

The second messengers are a small, conservative club, and membership is earned by chemistry. ***Cyclic AMP****, Sutherland's founding member, is made from ATP by adenylyl cyclase and destroyed by phosphodiesterases; it activates protein kinase A (PKA), EPAC exchange factors and certain ion channels.* ***Cyclic GMP****, made by guanylyl cyclases — some receptor-embedded (the natriuretic peptide receptors), some cytosolic and activated by nitric oxide — activates protein kinase G and the photoreceptor channels of the retina, where its hydrolysis is the very substance of* *vision. The phosphoinositide system supplies two messengers for the price of one lipid: phospholipase C cleaves PIP_2 into* ***inositol 1,4,5-trisphosphate*** *(IP_3), which opens Ca^{2+} channels in the endoplasmic reticulum, and* ***diacylglycerol*** *(DAG), which stays in the membrane and, with Ca^{2+}, activates protein kinase C. And* ***Ca^{2+}*** *itself — an ion with a day job in bone and muscle — moonlights as the most versatile messenger of all, kept at 100 nM in the resting cytosol against 1 mM outside, a ten-thousand-fold gradient that makes every calcium channel a loaded spring.*

Two design principles emerge. First, messengers are ***made and unmade fast****: a signal that cannot be switched off is not a signal but an accident, so every synthetic enzyme is shadowed by a degrading one (cyclases by phosphodiesterases, IP_3 by phosphatases, cytosolic Ca^{2+} by the pumps of Chapter 11). Second, messengers encode information in* ***space and time****, not merely in quantity. Ca^{2+}, whose diffusion is throttled by buffers, rises in local sparks and waves; repetitive spikes (Figure 49.5) carry a frequency code that downstream decoders such as calmodulin and CaMKII read by integrating pulse number rather than pulse height — amplitude modulation having been abandoned, in calcium signalling, for the digital reliability of counting.*

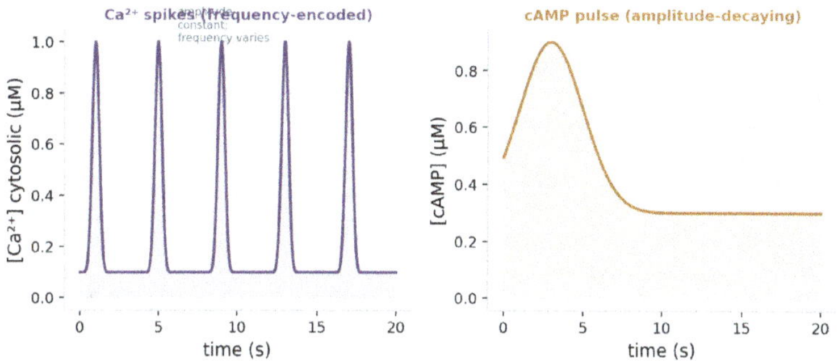

Figure 49.5 — Two styles of messenger dynamics. *Left: cytosolic Ca^{2+} oscillates in discrete spikes of constant amplitude; information lies in spike frequency, decoded by calmodulin-dependent kinases. Right: a cAMP pulse rises with receptor activation and decays as phosphodiesterases catch up — an amplitude-encoded, self-terminating signal.*

49.7 Kinases and Phosphatases: The Phosphate Currency

If second messengers are the currency of the membrane's interior face, phosphorylation is the currency of the cytosol. A protein kinase transfers ATP's γ-phosphate onto serine, threonine or tyrosine; the human genome encodes about 518 of them — the "kinome" — and nearly as many phosphatases to undo their work. The covalent mark is large, doubly negative, and cheap: it can create a docking site, wedge a catalytic cleft open or shut, or change a protein's address, all for the price of one ATP, and it is erased by hydrolysis that costs nothing further. The kinase cascade of Figure 49.1 — PKA phosphorylating phosphorylase kinase phosphorylating glycogen phosphorylase, in the classical glycogen pathway — is thus a chain of amplifiers in which each enzyme's product is the next enzyme's activity. The reader should note the dialectic: a kinase without its phosphatase is a one-way valve, and signalling requires both, because information is the difference between states, and a system stuck in one state informs nobody. (The phosphatases, less glamorous and long understudied, have had their revenge in oncology, where several are tumour suppressors.)

49.8 Domains, Scaffolds and Coincidence Detection

How do 518 kinases avoid phosphorylating everything at once? Partly by intrinsic specificity, but mostly by ***address****. Modular protein domains read the marks and postcodes: SH2 domains bind phosphotyrosine in specific sequence contexts; SH3 domains bind proline-rich motifs; PH domains bind PIP_3 at the membrane; 14-3-3 proteins bind phosphoserine motifs. A phosphorylated RTK tail is therefore not a generic flag but a numbered set of docking slips, each occupied by a different downstream protein — the receptor, in effect, types out its own committee.* ***Scaffold proteins*** *go further, binding several components of one pathway and holding them in physical proximity: the scaffold raises effective local concentration, accelerates the cascade, and — decisively — prevents the kinases from straying into neighbouring pathways that share the same components. Specificity, it turns out, is less often written into the catalytic sites than into the furniture.*

A final logical device completes the toolkit: ***coincidence detection****, in which a component responds only when two inputs arrive together. Adenylyl cyclase is stimulated by Gαs and inhibited by Gαi, integrating excitatory and inhibitory commands at one enzyme; protein kinase C requires both DAG and Ca^{2+}; the NMDA receptor requires both glutamate and depolarisation, a molecular AND-gate on which, as the neuroscientists insist, memory itself is written. The cell's signalling network is, in the strictest sense, a computer: its gates are proteins, its wires are diffusion, and its program is the proteome.*

49.9 Termination and Desensitization

Every element of the pathway carries its own off-switch, and the variety is instructive. Ligands dissociate (Equation 49.1 runs backwards). G proteins hydrolyse their GTP, hurried along by RGS proteins. Second messengers are hydrolysed or pumped away. Phosphates are removed by phosphatases. And the receptor itself is silenced in a two-step ritual discovered for the β-adrenergic receptor and since found everywhere: a ***G-protein-coupled receptor kinase*** *(GRK) phosphorylates the occupied receptor's cytosolic tail, and* ***β-arrestin*** *binds the phosphates, sterically divorcing receptor from G protein and often escorting the receptor into endocytosis — internalisation, from which the receptor is either recycled to the surface or sent to the lysosome. The consequence, plotted in Figure 49.6, is* ***adaptation****: the response to a step of constant stimulus rises, peaks, and falls back toward baseline while the stimulus is still present. The cell, like any good instrument, reports changes rather than absolutes — a design that extends dynamic range, protects against overstimulation, and explains why the opiate user needs ever more drug to hear the same message: the mailbox has been boarded up.*

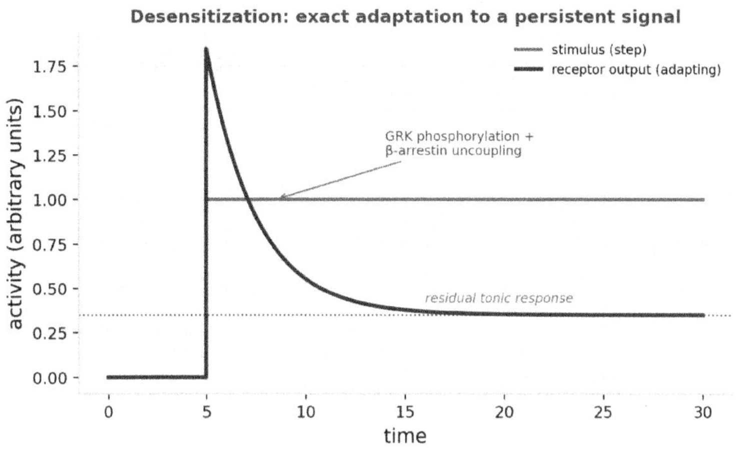

Figure 49.6 — Desensitization and adaptation. *A step of ligand (grey) elicits a receptor output (crimson) that peaks and decays toward a residual tonic level: GRK phosphorylation and β-arrestin binding uncouple the receptor even while the ligand remains. The signalling system measures change, not state.*

49.10 Crosstalk, Specificity and Noise

The pathways of Chapters 50–52 will be drawn, as tradition demands, as tidy linear arrows; the reader should hold Figure 49.1 in reserve against the deception. Real signalling is a network: insulin's pathway borrows components from growth-factor pathways; cAMP and Ca^{2+} systems converge on the same transcription factors; β-arrestin, having silenced the G protein, promptly nucleates a kinase cascade of its own — a receptor can thus speak two languages at once, G-protein and arrestin, and "biased agonists" that favour one dialect are now a deliberate pharmaceutical strategy. Specificity survives the promiscuity of components through scaffolds, subcellular compartmentation and timing, as Section 49.8 described. And noise — the stochastic flicker of single molecules — is real: at low receptor occupancy the arrival of ligand molecules is a Poisson process, and cells average over time, over receptor populations and over biochemical integration to extract a reliable answer. Signalling theory, in its mature form, is estimation theory conducted in wetware.

49.11 Case Study: Cholera — The Switch That Cannot Switch Off

> ***Case study — a rice-water lesson in GTP hydrolysis.*** *In the monsoon floods of the Bengal delta, Vibrio cholerae has been conducting the same lethal biochemistry practical for centuries, and its reagent is a single enzyme activity. Cholera toxin, an AB_5 protein, binds GM_1 ganglioside on intestinal epithelium, delivers its A subunit into the cytosol, and there performs one reaction: it transfers ADP-ribose from NAD^+ onto an arginine of $G\alpha_s$ — the stimulatory G protein of the adenylyl cyclase system. The modified residue sits in the catalytic pocket, and the GTPase of Figure 49.3 is crippled: $G\alpha_s$, once GTP-bound, cannot switch off. Adenylyl cyclase runs uncontrolled; cAMP rises*

and stays risen; the CFTR chloride channel, opened by PKA phosphorylation, pours Cl^- into the gut lumen; sodium and water follow osmotically, and the patient loses litres of fluid a day — the "rice-water" stools in which, before rehydration therapy, death by hypovolaemic shock could follow within hours.

Analysis. *Every term of Equation (49.2) is on display. The amplification*

therapy follows from the same arithmetic: oral rehydration solution replaces salt and water while the poisoned Ga subunits are diluted out by protein turnover; nothing in the treatment touches the toxin, because the lesion is kinetic and time itself is the antidote. Pertussis toxin performs the mirror-image experiment, ADP-ribosylating $G\alpha_i$ so that it cannot couple to its receptor — inhibition of an inhibitor, hence disinhibition — and the whooping cough bacillus thereby runs the same cyclase too hot by the opposite route. Two bacterial toxins, two opposite lesions, one equation.

49.12 Historical Note: Sutherland's Liver Slices

Historical note. *In the late 1950s Earl W. Sutherland Jr. and Theodore Rall, at Western Reserve University, were dissecting how adrenaline mobilises glucose from liver. The critical observation was of a kind that looks trivial only afterwards: adrenaline accelerated glycogenolysis in intact liver slices but not in broken-cell preparations unless the membrane fraction and the cytosol were recombined — whereupon the cytosolic enzyme (phosphorylase) sprang to life even in adrenaline's absence. Something heat-stable, dialysable and manufactured at the membrane had crossed the preparation and done the hormone's bidding indoors. In 1957–58 Sutherland identified it as adenosine 3′,5′-cyclic monophosphate — cyclic AMP — and coined the doctrine: the hormone is the first messenger, never entering the cell; the cyclic nucleotide is the second, speaking for it within. The concept generalised at once to a dozen hormones, and the 1971 Nobel Prize in Physiology or Medicine followed. One may savour the irony that the entire theory of intracellular signalling — kinase cascades, G proteins, the pharmacology of half the pharmacopoeia — descends from an experiment in which the most important step was breaking the cells politely, so that the messenger could still get through.*

Plate 49.1 *Earl W. Sutherland Jr. (1915–1974), discoverer of cyclic AMP and author of the second-messenger concept. Nobel laureate in Physiology or Medicine, 1971. (Nobel Foundation portrait.)*

49.13 Exercises

concentration rise to move occupancy from 9% to 91%?

hyperbola's gentle slope is why hormone axes span two orders of magnitude of concentration.

stage's lifetime have the same effect as doubling the last?

inhibitor amplifies precisely by extending signal duration. (c) Because Eq. (49.2) is a product, it is indifferent to which factor is doubled — the cascade has no privileged stage. This is why drugs acting at different points of one pathway can be equally potent, and equally dangerous.

one hour? (c) Connect (b) to the clinical picture of Section 49.11.

continues for the protein's whole lifespan, PKA keeps CFTR open, and chloride and water loss continue until the poisoned subunits are degraded and replaced — exactly the self-limiting, days-long course of treated cholera.

Exercise 49.4 — Why scaffolds help. *Two enzymes of a pathway are free in solution at 10 nM each. A scaffold binds both, confining them to an effective local volume that raises their local concentrations to 10 µM. If the rate of their encounter-limited reaction is proportional to the product of concentrations, by what factor does scaffolding accelerate the pathway? State one further, non-kinetic advantage.*

magnitude acceleration is available. Non-kinetic advantage: the scaffold enforces fidelity — it co-localises the correct partners and sequesters them from competing pathways that use the same kinases, converting a promiscuous component into a private wire.

Research proposition. *Equation (49.2) treats each stage's gain as independent, yet the product form implies that perturbations at different stages interact multiplicatively — the mathematical signature of epistasis. Propose an experimental programme — using, say, dose-matrices of kinase inhibitors across a three-stage cascade — to map whether gain epistasis is symmetric (as the product model predicts) or systematically asymmetric, and discuss what asymmetry would imply about hidden feedback within the cascade. Consider how such a map would inform the design of combination chemotherapy, where multiplicative interactions are precisely what one hopes to purchase.*

49.14 Synthesis

This chapter has laid the grammar of Part X. Signals are received by four receptor architectures, bound according to the mass-action hyperbola of Eq. (49.1); they are transduced by two-state molecular switches whose lifetimes set the clock; they are amplified by cascades whose gains multiply, Eq. (49.2); they are encoded by second messengers in space, time and frequency; they are executed by the phosphate currency of kinases and phosphatases, directed by docking domains and scaffolds; and they are terminated as deliberately as they are begun — the adapting receptor of Figure 49.6 reporting change, not state. Cholera served as the controlled catastrophe that verifies the arithmetic by breaking it. The direction has been checked against the master plan: with the principles in hand, Chapter 50 assembles them into the named pathways — insulin and growth factors, MAP kinase, Wnt, Notch, Hedgehog, JAK-STAT — that run the body's daily business and, when they fail, its cancers. ***Chapter 50 — Major Signalling Pathways*** *follows.*

49.15 References

Berridge, M. J. (2014). Cell signalling biology: Module 1 — Introduction. Portland Press, 1–36.

Gilman, A. G. (1987). G proteins: Transducers of receptor-generated signals. Annual Review of Biochemistry, 56, 615–649.

Lefkowitz, R. J. (2013). A brief history of G-protein coupled receptors. Biochimica et Biophysica Acta, 1833, 1–2.

Lim, W., Mayer, B., & Pawson, T. (2015). Cell Signaling: Principles and Mechanisms. Garland Science.

Pawson, T., & Nash, P. (2003). Assembly of cell regulatory systems through protein interaction domains. Science, 300, 445–452.

Sutherland, E. W., & Rall, T. W. (1958). Fractionation and characterization of a cyclic adenine ribonucleotide formed by tissue particles. Journal of Biological Chemistry, 232, 1077–1091.

Scott, J. D., & Pawson, T. (2009). Cell signaling in space and time: Where proteins come together and when they're apart. Science, 326, 1220–1224.

Chapter 50
Major Signalling Pathways

CONTENTS OF THIS CHAPTER

"Evolution, having invented a handful of signalling circuits, declined to invent any more; it has spent five hundred million years merely re-wiring them — and getting, on the whole, excellent mileage."

50.1 From Principles to Pathways

Chapter 49 supplied the grammar; this chapter supplies the literature. The major signalling pathways are few — a dozen at most carry the bulk of development, metabolism and immunity — and they are astonishingly conservative: the Wnt pathway that patterns a fly's wing patterns a human colon, with the same proteins spelled nearly the same way. This is not poverty of invention but proof of a principle the reader has now seen at every level of the book, from the proton pump to the ribosome: what works is kept. Each pathway below is presented the same way — ligand, receptor, transduction logic, cellular answer, and the disease that reveals its function by breaking it — and each is a variation on the five-note theme of Figure 49.1. The student is advised to stop memorising arrows and start recognising mechanisms: the GTPase switch, the kinase cascade, the destruction complex, the cleavage event. Count the mechanisms, and the pathways count themselves.[61]

50.2 Ras and the MAP Kinase Cascade

The receptor tyrosine kinase pathway is the archetype of growth signalling, and its core (Figure 50.1) is a relay of three kinds of switch in series: a dimerising receptor, a small GTPase, and a three-tier kinase cascade. A growth factor — EGF, PDGF, FGF — binds and dimerises its receptor; the cytosolic kinase domains trans-autophosphorylate, and the phosphotyrosines become docking slips, read by SH2 domains exactly as Section 49.8 prescribed. The adaptor Grb2 docks and brings SOS, a guanine-nucleotide exchange factor, to the membrane, where it meets ***Ras*** *— a monomeric G protein, the heterotrimer's smaller cousin, running the same GDP/GTP cycle of Figure 49.3. Ras·GTP recruits the kinase* ***Raf*** *to the membrane; Raf phosphorylates MEK; MEK, a rare dual-specificity kinase, phosphorylates ERK on both tyrosine and threonine; and ERK enters the nucleus to phosphorylate transcription factors — Elk-1, Fos, Myc — that write the genes of the G_1-to-S transition of Chapter 51.*

Two quantitative remarks. First, the three-tier cascade is not redundancy but engineering: modelling and experiment agree that a kinase-kinase-kinase stack

61. A confession of editorial policy: pathway diagrams in most textbooks resemble the wiring diagram of a mid-century radio, and induce the same despair. We draw each pathway once, small, and spend the saved space on what the arrows mean.

converts a graded input into a switch-like, nearly all-or-none output (ultrasensitivity), because the product of three hyperbolic or sigmoidal responses is far steeper than any one. Second, Ras spends only a small fraction of time in the GTP state, and its off-rate is set by GTPase-activating proteins — neurofibromin, p120-GAP — whose loss, as much as Ras mutation itself, causes disease. The pathway's pathology is a study in stuck switches: ***oncogenic Ras*** *(mutated at glycine-12 or glutamine-61, residues that sit in the catalytic pocket like the arginine cholera modifies) cannot hydrolyse GTP, is permanently on, and drives perhaps a fifth of all human cancers;* ***KRAS*** *in pancreas and colon,* ***NRAS*** *in melanoma,* ***HRAS*** *in bladder. The recent covalent KRAS-G12C inhibitors — sotorasib, adagrasib — work precisely by trapping the mutant in its GDP state, pharmacology restoring the clock that the mutation broke.*

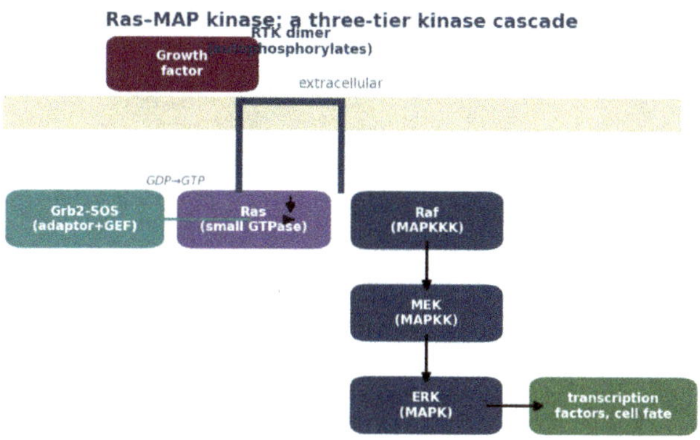

Figure 50.1 — The Ras–MAP kinase cascade. *Growth factor (crimson) dimerises the receptor; Grb2–SOS (teal) exchanges Ras's GDP for GTP (plum); Raf→MEK→ERK (navy) form the three-tier kinase cascade; ERK phosphorylates nuclear transcription factors (green). Each tier multiplies, per Eq. (49.2).*

50.3 Insulin and PI3K–Akt: The Anabolic Switch

The insulin receptor is an RTK with a refinement: it exists as a pre-formed dimer held in check, and insulin binding rearranges rather than assembles it. Its decisive substrate is not the cascade directly but a family of docking proteins, the ***insulin receptor substrates*** *(IRS), whose phosphorylated tails recruit* ***phosphoinositide 3-kinase*** *(PI3K) (Figure 50.2). PI3K phosphorylates the membrane lipid PIP_2 to PIP_3 — a second messenger that never leaves the bilayer, a postcode rather than a letter. Proteins bearing PH domains read the postcode: PDK1 and* ***Akt*** *(protein kinase B) are recruited to the membrane, and Akt, activated by phosphorylation, becomes the anabolic master switch. It phosphorylates AS160 to send* ***GLUT4*** *vesicles to the muscle and fat-cell surface (the glucose uptake of Chapter 40); it inhibits GSK3 to activate glycogen synthase; it inhibits FoxO transcription factors to suppress the fasting genes; and it activates* ***mTORC1***, *the growth integrator of Chapter 41. The opposing number is* ***PTEN***, *the phosphatase that converts PIP_3 back to PIP_2 — one of the most frequently lost tumour suppressors in human cancer, which tells the reader everything about how dangerous an unopposed anabolic switch is. Insulin resistance, the lesion of Chapter 42's metabolic syndrome, is in large part a corruption of this very pathway: serine* *phosphorylation of IRS by stress kinases marks it for disposal, and the postcode is never written.*

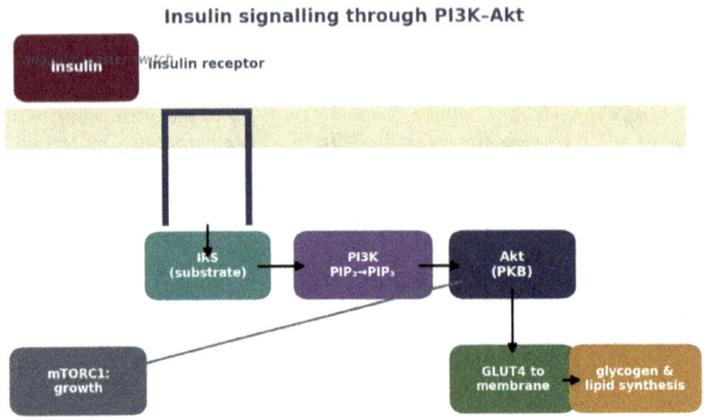

Figure 50.2 — ***Insulin signalling through PI3K–Akt.*** *The receptor (navy) phosphorylates IRS (teal), which recruits PI3K (plum) to make the lipid second messenger PIP_3; Akt (navy) reads PIP_3 and commands glucose uptake, glycogen and lipid synthesis, and growth via mTORC1. PTEN erases the postcode.*

50.4 JAK–STAT: Minimalism from Membrane to DNA

*If Ras–MAPK is baroque, the cytokine pathway is Shaker furniture (Figure 50.3). Cytokine receptors — for interferons, erythropoietin, growth hormone, most interleukins — have no kinase of their own; they carry a non-covalent passenger, a **Janus kinase** (JAK), on their cytosolic tail. Ligand dimerises the receptor, the two JAKs phosphorylate each other and the receptor tails, and the phosphotyrosines recruit **STAT** transcription factors, which are themselves phosphorylated, dimerise, and walk straight into the nucleus. Two protein families bridge membrane and genome — the shortest signalling commute in the body, and Figure 50.6 ranks it accordingly. The design's virtue is speed and its vice is exposure: with no cascade there is little amplification and few checkpoints, so the system leans on negative regulators — the SOCS proteins, induced by the STATs themselves, bind and silence the JAKs in a textbook negative feedback. Gain-of-function JAK2 (V617F) drives the myeloproliferative disorders, its inhibitor ruxolitinib treats them; and the same JAKs, targeted by tofacitinib and its cousins, are silenced in rheumatoid arthritis and the inflammatory bowel diseases — cytokine blockade from the inside.*

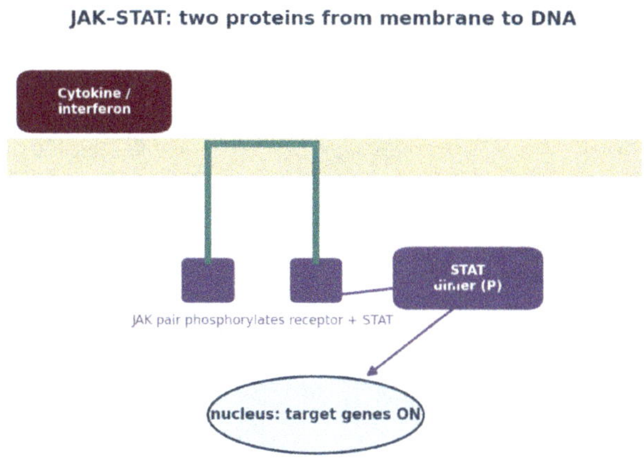

Figure 50.3 — JAK–STAT. *Cytokine (crimson) dimerises its receptor; the JAK pair (plum) phosphorylates receptor and STAT; phosphorylated STAT dimerises and enters the nucleus directly. No cascade, no second messenger — the fastest route from membrane to DNA.*

50.5 Wnt: Regulation by Destruction

*The Wnt pathway inverts every expectation formed so far (Figure 50.4). The signal does not build anything; it stops a demolition. In the absence of Wnt, the transcriptional co-activator **β-catenin** is continuously synthesised and continuously destroyed: a **destruction complex** — the scaffold Axin, the tumour suppressor APC, and two kinases, CK1 and GSK3 — captures β-catenin, phosphorylates it, and hands it to the ubiquitin machinery of Chapter 48. When Wnt binds its receptor Frizzled and co-receptor LRP, the destruction complex is recruited to the membrane and disarmed; β-catenin escapes degradation, accumulates, enters the nucleus, and converts the TCF/LEF DNA-binding proteins from repressors into activators. The whole pathway is thus a double negative — a signal that inhibits an inhibition — and its logic is kinetic rather than synthetic: what matters is the rate of β-catenin destruction, which the ligand modulates. The diseases follow the logic exactly. Loss of **APC**, which can no longer hold the destruction complex together, leaves β-catenin permanently high: this is familial adenomatous polyposis and the initiating event of most sporadic colorectal cancers — the colon, where Wnt maintains the crypt stem cells, being the organ that pays for the pathway's indispensability. One cannot help noting the elegance: a pathway built on destruction, studied through a cancer caused by the failure of that destruction.*

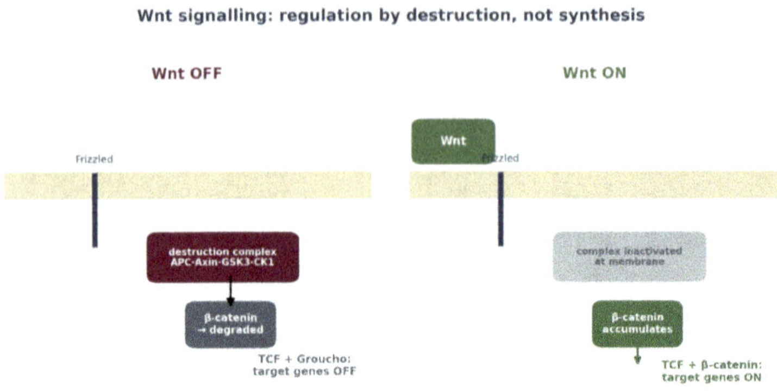

Figure 50.4 — Wnt signalling, off and on. *Left: without Wnt, the destruction complex (crimson) phosphorylates β-catenin for ubiquitination and degradation; TCF represses. Right: Wnt binding disarms the complex; β-catenin (green) accumulates, enters the nucleus, and activates TCF target genes. The ligand modulates a degradation rate, not a synthesis rate.*

50.6 Notch and Hedgehog: The Developmentalists

Notch *is the most literal signalling device in biology (Figure 50.5): the receptor is its own transcription factor. Notch is a single-pass membrane protein; its ligand, Delta or Jagged, lives on the neighbouring cell, so Notch signals only between cells in contact — the molecular instrument of lateral inhibition, by which a sheet of equivalent cells decides which one becomes a bristle, a neuron, a pancreatic endocrine cell. Ligand binding exposes a cleavage site; an ADAM protease cuts the extracellular stump, and then* ***γ-secretase*** *— an intramembrane aspartyl protease, the same complex whose amyloid cuts preoccupied Chapter 35 — cuts Notch within the bilayer itself, releasing the* **Notch intracellular domain** *(NICD) to walk to the nucleus and convert CSL from repressor to activator. No amplification at all: one ligand, one receptor, one NICD, one shot — a single-use receptor whose stoichiometry is its specificity. The cost of the design is fragility: Notch mutations cause the inherited stroke-dementia CADASIL, and γ-secretase, asked to cleave two substrates, sits at the intersection of vascular disease and Alzheimer's.*

Hedgehog *runs the Wnt logic backwards with different furniture. The ligand is a cholesterol-modified, palmitoylated protein — a lipidated morphogen for the pathway that patterns limbs and neural tube. Without Hedgehog, the receptor Patched loiters in the membrane and, by a still-incompletely-understood mechanism involving sterol flow, excludes the transducer* ***Smoothened****; the transcription factor Gli is phosphorylated, truncated, and works as a repressor. Hedgehog binding clears Patched from the primary cilium — the pathway's obligatory organelle, which is why the ciliopathies are developmental diseases — Smoothened accumulates, Gli survives full-length, and target genes switch on. Mutations in Patched cause Gorlin syndrome and the basal-cell carcinomas that follow constitutive Smoothened activity, treated by the Smoothened antagonist vismodegib; and loss of midline Hedgehog produces cyclopia — the pathway's name in the clinic is written in the severest dysmorphology known to embryology.*

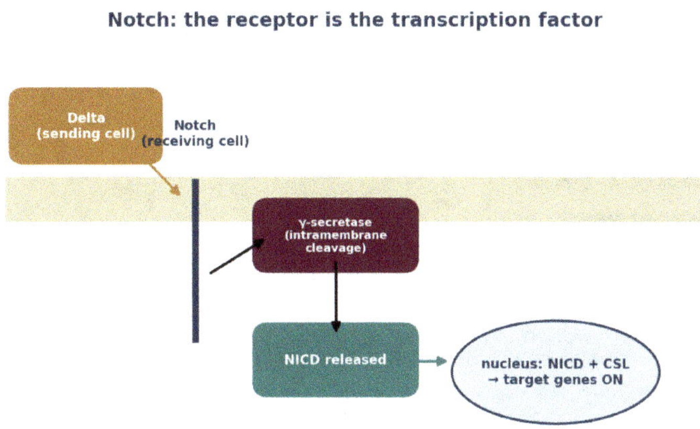

*Figure 50.5 — **Notch.** Delta on the sending cell (amber) pulls on Notch (navy); γ-secretase (crimson) cleaves within the membrane; the released intracellular domain (teal) enters the nucleus and activates CSL targets. One receptor, one message, one use — signalling by mechanical amputation.*

50.7 TGF-β, NF-κB and the Nuclear Receptors

*Three more families complete the survey, each with one idea worth keeping. The **TGF-β** superfamily (including the BMPs) uses receptor serine/threonine kinases that phosphorylate **Smad** proteins directly — receptor-regulated Smads partner with Smad4 and enter the nucleus, another short commute, and the pathway's double life as growth suppressor (lost in pancreatic and colon cancer) and promoter of fibrosis and metastasis (its output depends on context) is the standing refutation of the phrase "the function of a pathway." **NF-κB**, the inflammation and survival factor, waits in the cytosol bound to its inhibitor IκB; inflammatory signals — TNF, IL-1, Toll-like receptors — converge on the IKK kinase complex, which phosphorylates IκB and thereby condemns it to ubiquitination, releasing NF-κB to the nucleus. The reader will recognise the trick: it is Wnt again, destruction as the regulated step, deployed for immunity rather than pattern. And the **nuclear receptors** of Chapter 49.3 — glucocorticoid, oestrogen, thyroid, vitamin D — remain the shortest pathway of all: ligand crosses the membrane, receptor binds DNA directly, and the whole of transduction is one protein changing partners. Tamoxifen blocking the oestrogen receptor in breast cancer, dexamethasone binding the glucocorticoid receptor in a hundred inflammatory diseases: the pharmacology of this family is the pharmacology of the modern clinic.*

50.8 Comparison: The Economy of Designs

Figure 50.6 counts the core components each pathway places between membrane and nucleus, and the spread — two proteins for JAK–STAT and Notch, six or more for Ras–MAPK — is not arbitrary. Short pathways are fast, direct and cheap but amplify little and regulate at few points; long cascades are slow and costly but steep (ultrasensitive), amplifying (Eq. 49.2), and rich in checkpoints. The cell deploys them accordingly: development, where decisions are irreversible and local, prefers Notch and Wnt; acute metabolic command prefers the deep insulin cascade; transcriptional immediacy prefers STAT. One sees the same trade-off in engineering between a fuse and a feedback controller, and the homology is not fanciful — it is the same mathematics, arrived at independently, because there is only one mathematics of reliable signalling.

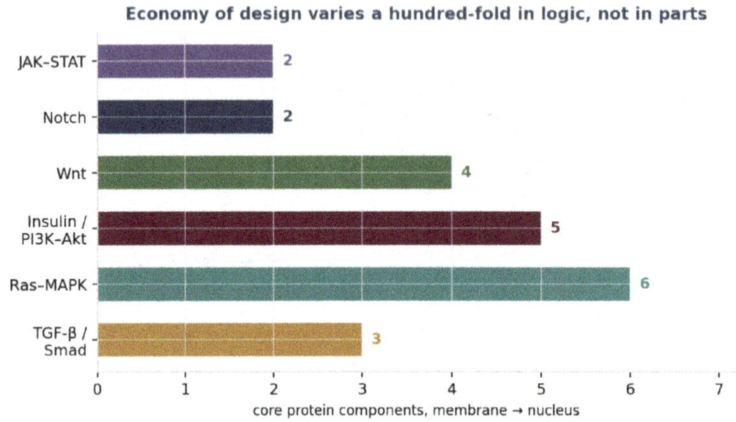

Figure 50.6 — Economy of design. *Core protein components between membrane and nucleus for six pathways. Short pathways (JAK–STAT, Notch) trade amplification and checkpoints for speed and directness; deep cascades (Ras–MAPK, PI3K–Akt) trade speed for gain and regulatory purchase.*

50.9 Case Study: Chronic Myeloid Leukaemia and the Designed Drug

> ***Case study — the Philadelphia chromosome.*** *Chronic myeloid leukaemia begins with a single translocation, t(9;22), fusing the BCR gene to the ABL tyrosine kinase. The BCR–ABL fusion protein is a kinase with its autoinhibitory cap removed: constitutively active, it phosphorylates substrates that feed — the reader should now recognise every one — the Ras–MAPK pathway (proliferation), the PI3K–Akt pathway (survival and growth), and STAT5 (transcription). One lesion, three of this chapter's pathways switched on together, and a haematopoietic stem cell that expands without permission. Until 2001 the disease was managed with interferon and transplant; the five-year survival was poor.*
>
> ***Analysis.*** *Imatinib, developed by Druker and colleagues, is an ATP-competitive inhibitor selective for ABL's inactive conformation — it binds the kinase's closed state and prevents the catalytic rearrangement. The result, in chronic-phase CML, is among the most complete therapeutic successes in medicine: durable molecular remission in the great majority of patients, on a daily pill, with modest toxicity. Three lessons belong to this chapter. First, oncogene addiction: a cell wired to run on one stuck kinase dies when that kinase is silenced, while normal cells, with redundant inputs, merely pause — the therapeutic window is written into the network's topology. Second, resistance is pathway logic: the mutations that defeat imatinib cluster in the ATP pocket (the gatekeeper T315I sterically excludes the drug) — evolution re-running Section 49.9's desensitization by other means. Third, the principle was the pathway: imatinib was possible because the biochemistry of kinase*

conformation had been worked out for its own sake, decades earlier, by the phosphorylation school whose history follows. Nobody discovered imatinib by looking for a leukaemia drug.

50.10 Historical Note: Krebs, Fischer and the Reversible Phosphate

Historical note. *Every pathway in this chapter ends in a phosphorylation, and the discovery that phosphorylation is a regulatory device — rather than a step in an energy pathway — belongs to Edmond Fischer and Edwin Krebs at the University of Washington in the mid-1950s. Studying glycogen phosphorylase, they found the enzyme existed in two forms whose interconversion was catalysed by a kinase and a phosphatase, consuming ATP — the first "interconvertible enzyme." The finding was nearly stillborn: their 1955 manuscripts were received coolly, one reviewer doubting that so wasteful a cycle could be real, and for some years the modification was thought a curiosity of glycogen metabolism. Then came Sutherland's cAMP, the discovery of PKA, and the slow realisation through the 1970s and 80s that kinase cascades were not the exception but the rule — that a third of all proteins are phosphorylated, and that development, immunity, memory and cancer are written in phosphates. Fischer and Krebs shared the Nobel Prize in 1992, thirty-seven years after the fact. The episode is the book's standing example of a truth that had to wait for its context: the reversible phosphate was always there; biology simply had to grow the pathways that would need it.*

Plate 50.1 *Edwin G. Krebs (1918–2009), who with Edmond H. Fischer discovered reversible protein phosphorylation as a regulatory mechanism. Nobel laureate in Physiology or Medicine, 1992. (Nobel Foundation portrait.)*

50.11 Exercises

fractional range, the overall response steepens approximately as the product of local sensitivities. (a) If each tier has an effective Hill coefficient of 1.3 when measured over its operating range, estimate the cascade's overall coefficient. (b) Why is a switch-like output desirable for a growth-factor decision?

decisions — enter the cell cycle or not — are irreversible and expensive; a graded response to a graded mitogen concentration would commit cells at intermediate, noisy signals. The switch holds the cell quiescent below threshold and commits it fully above, and the threshold itself becomes a tunable regulatory parameter.

Resolution.

exploit exactly this arithmetic: restoring destruction lowers the steady state without touching synthesis at all.

Exercise 50.3 — Notch stoichiometry. *A cell expresses 30,000 Notch receptors; 5% are engaged and cleaved during a 30-minute developmental decision, and each NICD survives about 3 hours. A competitor cell expresses twice as much Delta ligand. (a) Estimate NICD molecules produced. (b) Why does the pathway's lack of amplification make Notch signalling an instrument of binary, local decisions rather than long-range graded ones?*

Resolution.

receptors, which is proportional to contacting ligand molecules: the signal faithfully reports the neighbour's Delta density and nothing beyond the contact zone. No second messenger diffuses, so no distant cell can hear; the decision is forced to be local, cell-by-cell, and binary — precisely what lateral inhibition requires.

in place of ligand, compute pocket occupancy for wild-type and mutant at 3 μM, and explain the clinical consequence.

affinity change, from one methyl group added at the gatekeeper residue, converts a cure into a relapse: this is why second-generation inhibitors (dasatinib, ponatinib) were built to accommodate the mutant pocket.

Research proposition. *Figure 50.6 ranks pathways by component count, but "depth" might be better measured in information-theoretic terms: how many bits about ligand concentration survive to the nucleus, and how much bandwidth (decisions per hour) each pathway sustains. Propose a programme — live-cell reporters of nuclear effectors (ERK translocation, STAT occupancy, β-catenin accumulation) under calibrated ligand ramps and pulses — to estimate channel capacity for three pathways of different depths, and discuss the prediction that deep cascades trade bandwidth for noise resistance. What would it imply for drug scheduling if pulsed dosing were found to transmit more information than continuous dosing?*

50.12 Synthesis

The survey is complete, and its moral is Chapter 49's grammar with names attached: Ras–MAPK is the kinase cascade; PI3K–Akt is the lipid postcode; JAK–STAT is the direct wire; Wnt and NF-κB are regulated destruction; Notch is mechanical cleavage; Hedgehog is a sterol-gated double negative; the nuclear receptors are transduction in a single protein. Cancer reappears at every step — stuck Ras, lost APC, fused ABL, mutant JAK — because the cell's growth decisions are exactly where signalling lives, and the drugs that treat it are the pathways' own logic wielded by hand. The direction has been checked against the master plan: with the signalling inventory in place, Chapter 51 turns to the decisions themselves — the cell cycle's checkpoints, apoptosis's executioners, and the stress responses that decide between them. ***Chapter 51 — Cell Cycle, Cell Death and Stress Responses*** *follows.*

50.13 References

Artavanis-Tsakonas, S., Rand, M. D., & Lake, R. J. (1999). Notch signaling: Cell fate control and signal integration in development. Science, 284, 770–776.

Clevers, H., & Nusse, R. (2012). Wnt/β-catenin signaling and disease. Cell, 149, 1192–1205.

Druker, B. J. (2008). Translation of the Philadelphia chromosome into therapy for CML. Blood, 112, 4808–4817.

Fischer, E. H., & Krebs, E. G. (1955). Conversion of phosphorylase b to phosphorylase a in muscle extracts. Journal of Biological Chemistry, 216, 121–132.

Lemmon, M. A., & Schlessinger, J. (2010). Cell signaling by receptor tyrosine kinases. Cell, 141, 1117–1134.

Manning, B. D., & Toker, A. (2017). AKT/PKB signaling: Navigating the network. Cell, 169, 381–405.

O'Shea, J. J., Schwartz, D. M., Villarino, A. V., Gadina, M., McInnes, I. B., & Laurence, A. (2015). The JAK–STAT pathway: Impact on human disease and therapeutic intervention. Annual Review of Medicine, 66, 311–328.

Chapter 51
Cell Cycle, Cell Death and Stress Responses

Contents of this chapter

"A cell, like a government, is judged by three decisions: when to expand, when to hold, and when to dissolve itself for the good of the state. Cancer is a government that has forgotten all three."

51.1 The Three Decisions

Chapters 49 and 50 supplied the wires; this chapter asks what they decide. Every cell, at every moment, is implicitly answering three questions: shall I divide, shall I wait, or shall I die? The machinery of the first answer is the **cell cycle**, *an engine of oscillating kinases; the machinery of the third is* **apoptosis**, *a demolition programme of exquisite politeness; and the machinery of the second — the verdict pronounced when things go wrong — is the family of* **stress responses** *that pause the engine, call for repair, and order the demolition if repair fails. These are not three subjects but one, because they share components (p53 presides over all three), share logic (every decision is a threshold crossed by competing signals, as Figure 51.6 will show), and share a pathology: cancer is the systematic corruption of the whole triad, and its therapy is the attempt to restore any one of the three decisions to working order.*[62]

51.2 The Cell-Cycle Engine

The eukaryotic cell cycle (Figure 51.1) is conventionally drawn as a circle — G_1, S, G_2, M — but it is more honestly described as an **oscillator**: *a sequence of states driven round by waves of enzyme activity, with ratchets preventing reverse travel. The enzymes are the* **cyclin-dependent kinases** *(CDKs), serine/threonine kinases of Chapter 50's clan, and their name states the design: a CDK is catalytically inert until it binds its* **cyclin**, *a regulatory subunit synthesised in one phase and destroyed at that phase's end. Cyclin D–CDK4/6 acts in G_1, reading the mitogen signals of Chapter 50; cyclin E– and cyclin A–CDK2 fire at the G_1/S boundary and through S phase, phosphorylating the replication machinery; cyclin B–CDK1 — the famous MPF (maturation-promoting factor) of the frog-egg experiments — drives entry into mitosis itself: nuclear-envelope breakdown, chromosome condensation, spindle assembly. The waves of Figure 51.1 are generated by a now-familiar trick: each cyclin–CDK complex activates the machinery that destroys its own cyclin (the anaphase-promoting complex, APC/C, a ubiquitin ligase), so every phase contains the seed of its own termination — the G-protein clock of Chapter 49, writ into the cell's calendar.*

62. The reader who has followed the book's argument from Chapter 1 will recognise the pattern at once: life is the maintenance of improbable states, and nothing is more improbable — or more vigorously maintained — than a cell that divides only when told.

$$\frac{dC}{dt} = \underbrace{k_{\text{syn}}}_{\text{transcription}} - \underbrace{k_{\text{deg}}\, C \cdot A(C)}_{\text{APC/C, activated by } C} \tag{51.1}$$

Derivation.

transcription begins. A synthesis term plus a self-excited destruction term is the minimal mathematical content of any relaxation oscillator — the cycle follows from the wiring, and the wiring alone.

The ratchets are the ***checkpoints.*** *At G_1/S the cell asks whether DNA is intact and the environment permissive; at G_2/M whether replication is complete; at the spindle-assembly checkpoint whether every chromosome is correctly attached before anaphase is allowed to begin. Each checkpoint is a signalling pathway — the damage kinases ATM and ATR, the checkpoint kinases CHK1/2, the mitotic arrest complex Mad2/BubR1 — wired to inhibit the next cyclin–CDK transition. Checkpoints are the difference between a cycle and a clock: a clock advances regardless, a cycle advances only on proof.*

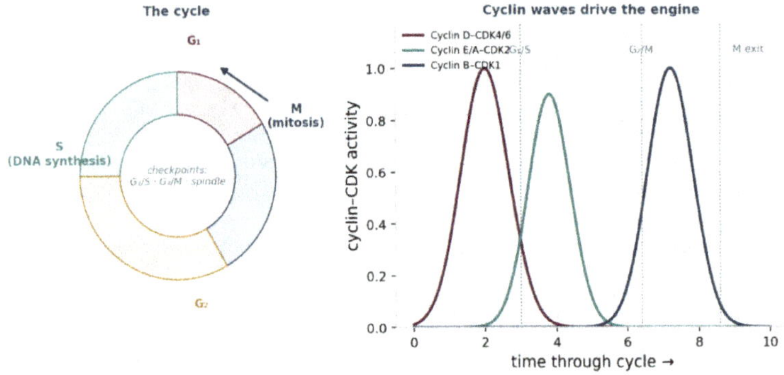

Figure 51.1 — The cell cycle as an oscillator. *Left: the four phases with their three checkpoints. Right: the cyclin waves that drive them — each cyclin–CDK complex fires in its window and activates the machinery (APC/C) that destroys it, per Eq. (51.1).*

51.3 The Restriction Point and Rb

The most consequential checkpoint is the ***restriction point*** *in late G_1, beyond which the cell no longer needs mitogens and is committed to divide (Figure 51.2). Its gatekeeper is the* ***retinoblastoma protein*** *Rb, the first tumour suppressor ever defined and still the paradigm. Unphosphorylated Rb binds the E2F transcription factors and holds the S-phase genes silent. Mitogens, through the Ras–MAPK pathway of Chapter 50, induce cyclin D; cyclin D–CDK4/6 phosphorylates Rb; phosphorylated Rb releases E2F; E2F writes the genes of DNA replication — and, completing a positive feedback, the genes of cyclin E, whose CDK2 finishes Rb's phosphorylation and locks the decision in. The design is a bistable switch: below threshold, Rb holds; above it, the feedback flips the cell irreversibly into cycle. The opposing inputs — TGF-β through the Smads, the CDK inhibitors p16 and p21 — act by blocking Rb phosphorylation, and they are the* *cell's brakes. The pathology is, by now, predictable in outline: RB1 itself is lost in retinoblastoma and small-cell lung cancer; cyclin D is overexpressed in breast cancer and mantle-cell lymphoma; p16 is deleted in melanoma and pancreas; and the CDK4/6 inhibitors — palbociclib and its sisters — restore the brake pharmacologically, with striking benefit in breast cancer. One protein family, every member a drug target or a casualty of cancer: the restriction point is where oncology lives.*

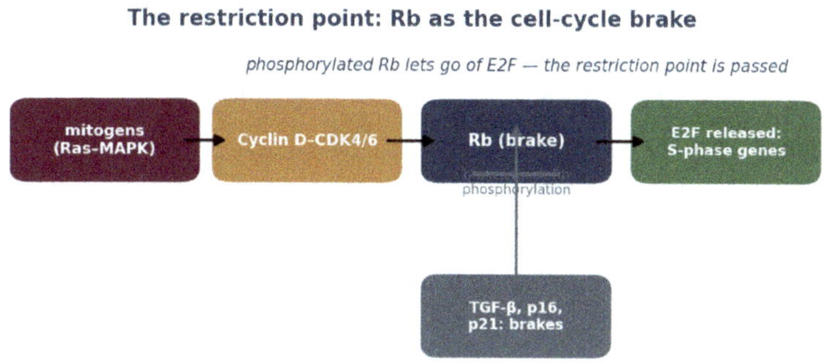

Figure 51.2 — The restriction point. *Mitogens (crimson) induce cyclin D–CDK4/6 (amber), which phosphorylates Rb (navy); Rb releases E2F (green) and the S-phase programme starts. TGF-β and the CDK inhibitors (slate) hold Rb unphosphorylated and block passage. Every element on this diagram is a cancer gene or a drug target.*

51.4 Apoptosis: The Polite Demolition

*Cell death by **apoptosis** is a construction industry run in reverse: dismantling proceeds by a numbered sequence, nothing spills, and the neighbours are unharmed. The executioners are the **caspases** — cysteine proteases made as inactive zymogens and activated by cleavage, in a cascade that mirrors the kinases of Chapter 49 with proteolysis in place of phosphorylation. Two roads lead to them (Figure 51.3). The **extrinsic** road begins at death receptors (Fas, TRAIL-R): ligand trimerises the receptor, the DISC complex assembles on its tail, and initiator caspase-8 is activated by enforced proximity. The **intrinsic** (mitochondrial) road is the more philosophically interesting, for it makes the organelle of Chapter 18 the arbiter of life and death: the Bcl-2 family — pro-death BAX and BAK, anti-death Bcl-2 and Bcl-X_L, and the BH3-only sensors that tip the balance — governs the mitochondrial outer membrane. When the sensors prevail, BAX/BAK perforate the membrane, cytochrome c — that venerable electron carrier — escapes into the cytosol and there, in a second career no evolutionary biologist predicted, nucleates the **apoptosome**, which activates caspase-9. Initiator caspases 8 and 9 converge on the executioners 3 and 7, which dismantle the cell: the cytoskeleton, the nuclear lamina, and the nuclease inhibitor whose destruction releases the enzyme that cuts DNA between nucleosomes — the "ladder" that was apoptosis's first molecular signature.*

Two design features deserve emphasis. First, the threshold is set by protein-protein balance, not by any single switch: the BH3-only proteins, BAX/BAK and the guardians Bcl-2 compete in a binding equilibrium, so death is a ratio — and the

BH3-mimetic drug venetoclax, which occupies Bcl-2's pocket, shifts the ratio therapeutically in chronic lymphocytic leukaemia. Second, the programme is ***silent****: apoptotic cells display phosphatidylserine on their outer leaflet — the asymmetric membrane of Chapter 11 volunteering its inner face as a signal — and are eaten by macrophages without inflammation. The contrast with necrosis, which spills contents and provokes the immune system, and with the regulated inflammatory deaths (pyroptosis, necroptosis), completes a taxonomy in which even dying is a set of decisions.*

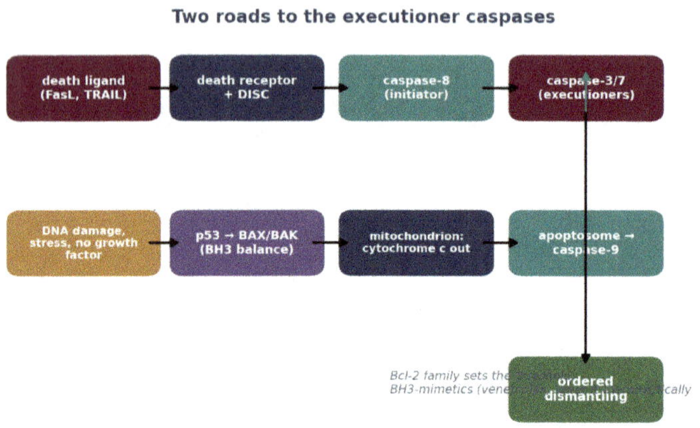

Figure 51.3 — Two roads to the executioner caspases. *Extrinsic: death ligand → receptor/DISC → caspase-8. Intrinsic: stress → BH3-only proteins → BAX/BAK → cytochrome-c release → apoptosome → caspase-9. Both activate caspases 3/7. The Bcl-2 family sets the intrinsic threshold; venetoclax lowers it in the clinic.*

51.5 p53: Guardian of the Genome

Between cycle and death stands the referee. ***p53*** *is a transcription factor held, in the unstressed cell, at vanishing levels by its inhibitor Mdm2, which ubiquitinates it for destruction — Wnt's logic again, regulation by degradation. DNA damage activates the kinases ATM and ATR, which phosphorylate p53 and pry it from Mdm2; stabilised p53 accumulates and writes three different verdicts (Figure 51.4):* ***p21****, a CDK inhibitor that arrests the cycle for repair; the repair and metabolic genes (TIGAR, the sestrins) that fix the damage and ration the cell's resources while it is fixed; and, if the damage is beyond mending, the BH3-only genes* ***PUMA*** *and* ***NOXA*** *that summon the intrinsic pathway of Section 51.4. The severity code — arrest for small damage, death for great — is written partly in the dynamics: transient p53 pulses favour arrest, sustained accumulation favours death, a frequency/amplitude code of the very kind Chapter 49.6 taught us to expect. The verdict's importance is measured by its corruption: TP53 is mutated in more than half of all human cancers, and the tumours that keep wild-type p53 frequently disable its wiring instead — Mdm2 amplified in sarcomas, the viral E6 protein destroying p53 in cervical cancer. Oncology's most wanted protein is not an oncogene but a policeman.*

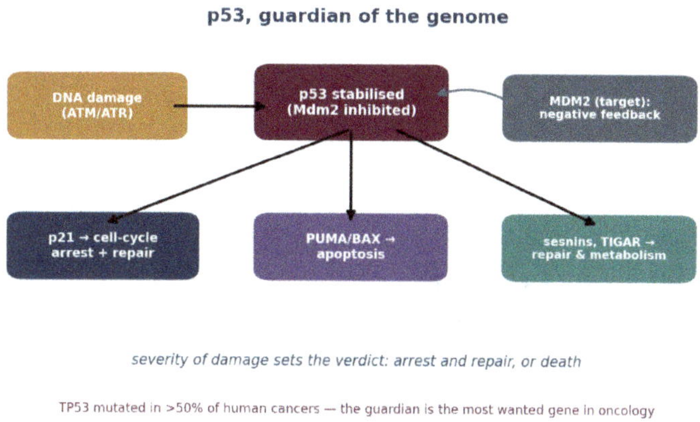

Figure 51.4 — p53, guardian of the genome. *Damage kinases stabilise p53 (crimson), which chooses among arrest-and-repair (navy), metabolic triage (teal), and apoptosis via PUMA/BAX (plum). Its own target MDM2 (slate) provides negative feedback. TP53 is mutated in over half of human cancers.*

51.6 Stress Responses: The Cell Under Siege

Between health and the apoptotic verdict lies a tier of emergency governments, each answering a specific insult, and each following the same rhetorical form: sense the stress, pause the expensive business, repair, and — if the siege continues — die. The ***unfolded protein response*** *(Figure 51.5) watches the endoplasmic reticulum through three sensors: IRE1 splices the XBP1 message to make a transcription factor for chaperones; PERK phosphorylates eIF2α and pauses translation (the eIF2α brake of Chapter 47, moonlighting); ATF6 travels to the Golgi to be cleaved into a transcription factor. All three arms buy time; the prolonged UPR induces CHOP and apoptosis.* ***Hypoxia*** *is answered by HIF-1α, regulated by oxygen-dependent destruction — the prolyl hydroxylases of Chapter 41's vitamin C chemistry — so that falling oxygen stabilises a factor commanding glycolysis, angiogenesis (VEGF) and erythropoietin.* ***Oxidative stress*** *is answered by Nrf2, held in the cytosol by Keap1 until electrophiles modify Keap1's cysteines, freeing Nrf2 to write the glutathione and detoxification genes of Chapter 23.* ***Starvation*** *is answered by AMPK and mTOR of Chapter 41, and by autophagy — the cell eating its own furniture for fuel, a stress response so central that its discovery, like apoptosis's, took a Nobel Prize. The reader will notice that every one of these is a Wnt-style circuit: a transcription factor regulated by conditional destruction, with a sensor protein as the variable term.*

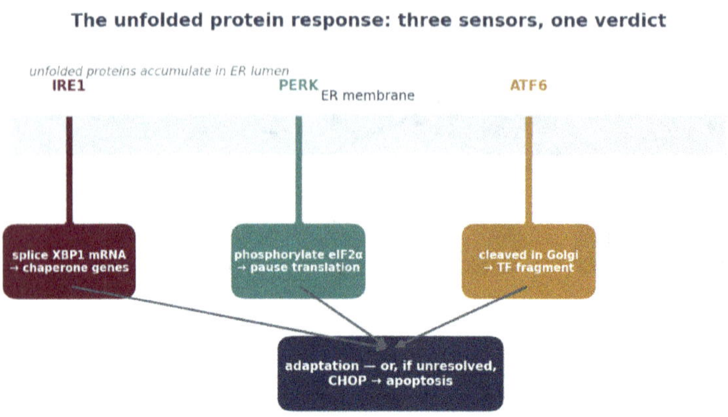

Figure 51.5 — The unfolded protein response. *Three ER sensors — IRE1 (crimson), PERK (teal), ATF6 (amber) — detect unfolded proteins, pause translation, and induce chaperones. If the stress persists, CHOP tips the cell into apoptosis. Adaptation first; death only on failure.*

51.7 The Balance of Fates

It is worth stepping back to see the whole. A cell's fate is the running difference between two signal integrals (Figure 51.6): the growth and survival signals of Chapter 50 (mitogens, insulin, the PI3K–Akt pathway, which phosphorylates and disarms the BH3-only protein BAD) against the damage and death signals of this chapter (p53, the BH3 sensors, the death receptors). In the contested middle zone live the three great categories of disease. Too little death: cancer, and the autoimmune lymphocytes that should have deleted themselves. Too much death: the neurodegenerations, the infarcted penumbra, the failing heart. Death mistimed or misdirected: the inflammatory diseases, where pyroptosis and necroptosis spill what apoptosis would have buried quietly. Therapy, in this view, is the art of moving one curve against the other — CDK inhibitors to hold the cycle, BH3 mimetics to restore death, HIF stabilisers to mimic adaptation — and every one of those drugs is a sentence in the grammar of Part X.

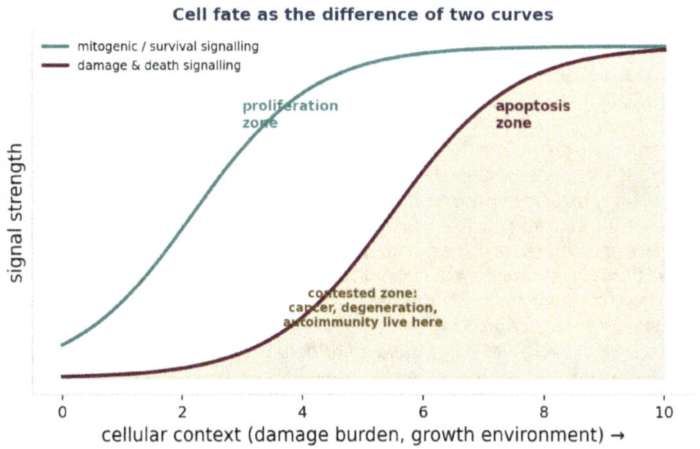

Figure 51.6 — Cell fate as the difference of two curves. *Survival signalling (teal) and death signalling (crimson) vary continuously with context; fate is decided where they cross. Cancer, degeneration and autoimmunity are diseases of the contested middle zone, and their therapies shift one curve against the other.*

51.8 Case Study: The Worm That Counted Its Cells

> ***Case study — 131 cells that must die.*** *The nematode Caenorhabditis elegans hatches with a cell lineage so invariant that every one of its 1,090 somatic cells can be named — and so can the 131 that are generated and then, on schedule, die. In the 1970s and 80s H. Robert Horvitz's laboratory at MIT, building on Brenner's and Sulston's lineage maps, isolated mutants in which the wrong cells died, or the right ones failed to. The genes that emerged defined apoptosis before the word had a mechanism: ced-3 encodes the worm's caspase; ced-4 the apoptosome scaffold (the worm's Apaf-1); ced-9 the Bcl-2 guardian that holds CED-4 in check; and egl-1 the BH3-only trigger that releases it. Four genes, one linear pathway — and when the mammalian counterparts were found, they had multiplied (a dozen caspases, a Bcl-2 family of scores) but not changed logic.*
>
> ***Analysis.*** *The worm's gift was linearity: in a pathway stripped to four components, epistasis — the ordering of genes by their mutant interactions — could be read like an equation, and the pathway fell into sequence without a single biochemical assay. Only afterwards did the biochemistry confirm what the genetics had declared: protease, scaffold, guardian, trigger. The 2002 Nobel Prize to Brenner, Sulston and Horvitz honoured an entire style of explanation — let the organism, kept simple, state the mechanism — and every element of Figure 51.3 is that statement translated into mammalian baroque. The clinical sequel is direct: venetoclax, the BH3 mimetic of Section 51.4, is egl-1 in pill form, prescribed for leukaemia.*

51.9 Historical Note: Apoptosis, Named and Almost Forgotten

> ***Historical note.*** *The word apoptosis — Greek for the falling of leaves — was coined in 1972 by John Kerr, Andrew Wyllie and Alastair Currie at Aberdeen, to describe a morphology they kept meeting and the field kept ignoring: cells shrinking, chromatin condensing at the nuclear rim, the corpse fragmenting into membrane-wrapped bodies that neighbours ate without comment. Kerr had noticed the pattern in liver after mild injury, where cells died politely, without the swelling and rupture of necrosis. The 1972 paper is now among the most cited in biology, but it was cited barely at all for fifteen years: cell death was thought an accident, not a programme, and a programmed accident was a category error. The rescue came from two directions at once — the worm genetics of Section 51.8, which proved death was genetic, and the discovery in 1988 that the bcl-2 oncogene of follicular lymphoma works not by driving proliferation but by blocking death: the first cancer gene that keeps cells alive rather than making them divide. From that moment apoptosis was a field; within a decade it was a pharmacopoeia. The episode carries the book's recurring moral in its purest form: the phenomenon was visible for a century in every pathology museum; what was missing was the idea that death could be done by the cell, on purpose, with machinery.*

Plate 51.1 *H. Robert Horvitz (b. 1947), whose genetic analysis of programmed cell death in C. elegans defined the core apoptotic pathway. Nobel laureate in Physiology or Medicine, 2002, with Sydney Brenner and John E. Sulston. (Nobel Foundation portrait.)*

51.10 Exercises

oscillation rather than a stable plateau?

time delay built in, and delayed negative feedback around a steep nonlinearity is the classical recipe for limit-cycle oscillation, not stability.

restriction point is (a) sharp, (b) irreversible over the cell's timescale, and (c) vulnerable to single-gene lesions.

Resolution. *(a) Positive feedback converts a graded input (mitogen level) into a switch: once some Rb is phosphorylated, E2F-driven cyclin E phosphorylates more Rb, and the system's middle steady state is unstable — the cell is either fully braked or fully committed. (b) Once cyclin E–CDK2 is active, it maintains Rb phosphorylation without cyclin D, so withdrawing the mitogen cannot reverse the decision; commitment is hysteretic. (c) A switch held by one brake can be released by cutting one wire: loss of Rb, loss of p16, or cyclin D overexpression each suffices to make the switch permanently ON — which is why this single node, among thousands, is hit in nearly every cancer.*

***Exercise 51.4 — p53 dynamics as a code.** DNA-damage pulses of p53 last ~5.5 h and repeat with ~7 h spacing after irradiation; sustained p53 follows severe damage. (a) If p21 transcription requires only brief p53 occupancy but PUMA requires sustained promoter residence, explain how the same factor encodes two verdicts. (b) A drug that blocks MDM2 makes p53 sustained in the absence of damage; predict its effect on primed tumour cells and its chief toxicity.*

***Resolution.** (a) The promoter of a fast-response gene integrates p53 presence and responds to each pulse — arrest is reinstated while pulses continue; a slow promoter, requiring long uninterrupted occupancy, responds only when pulses merge into a plateau. Time-structure thus routes one input to two outputs — the amplitude/frequency code of Chapter 49.6 operating at the level of a transcription factor. (b) An MDM2 antagonist (the nutlin class) forces sustained p53: primed, p53-wild-type tumour cells are pushed to PUMA expression and apoptosis — the drug's rationale — but sustained p53 in healthy marrow and gut epithelium also arrests or kills proliferating compartments, giving the predicted dose-limiting cytopenias. The code, once understood, predicts both the therapy and its price.*

***Research proposition.** Figure 51.6 draws fate as the crossing of two integrals, but real cells may weight recent signals more heavily than old ones (a leaky integral) or remember past insults (hysteresis). Propose single-cell experiments — optogenetic pulses of survival and death signals, timed on scales from minutes to days, with live reporters of caspase activation and CDK2 activity — to measure the integration kernel of the fate decision in one cell type. Discuss whether a kernel with long memory would explain field cancerisation (normal-appearing tissue, repeatedly insulted, that later yields tumours), and how a measured kernel could be used to schedule senolytic or pro-apoptotic therapy for maximal selectivity.*

51.11 Synthesis

The triad is assembled. The cycle is an oscillator of cyclin waves and checkpoints, derived as a delayed negative feedback in Eq. (51.1); commitment is a bistable switch held by Rb; death is a protease cascade gated by a binding equilibrium on the mitochondrial membrane; p53 referees between them with a time-coded verdict; and the stress responses — UPR, HIF, Nrf2, autophagy — are emergency governments that repair or execute. Cancer corrupts every node, and the chapter's drugs — CDK4/6 inhibitors, venetoclax, nutlins, HIF stabilisers — are the nodes' logic wielded by hand. The direction has been checked against the master plan: with decisions made by single cells, Chapter 52 closes Part X by assembling the wiring itself — feedback, feedforward, oscillation and robustness — into the theory of biochemical networks. ***Chapter 52 — Biochemical Networks and Cellular Decisions*** *follows.*

51.12 References

Alberts, B., Johnson, A., Lewis, J., Morgan, D., Raff, M., Roberts, K., & Walter, P. (2015). Molecular Biology of the Cell (6th ed.). Garland Science.

Ellis, H. M., & Horvitz, H. R. (1986). Genetic control of programmed cell death in the nematode C. elegans. Cell, 44, 817–829.

Kerr, J. F. R., Wyllie, A. H., & Currie, A. R. (1972). Apoptosis: A basic biological phenomenon with wide-ranging implications in tissue kinetics. British Journal of Cancer, 26, 239–257.

Lahav, G., Rosenfeld, N., Sigal, A., Geva-Zatorsky, N., Levine, A. J., Elowitz, M. B., & Alon, U. (2004). Dynamics of the p53–Mdm2 feedback loop in individual cells. Nature Genetics, 36, 147–150.

Levine, A. J., & Oren, M. (2009). The first 30 years of p53: Growing ever more complex. Nature Reviews Cancer, 9, 749–758.

Morgan, D. O. (2007). The Cell Cycle: Principles of Control. New Science Press.

Walter, P., & Ron, D. (2011). The unfolded protein response: From stress pathway to homeostatic regulation. Science, 334, 1081–1086.

Chapter 52
Biochemical Networks and Cellular Decisions

CONTENTS OF THIS CHAPTER

"The cell has no committee, no chairman, and no plan; it has wiring — and the wiring, inspected closely, turns out to be a small set of circuits, soldered together a million times, that compute."

52.1 From Pathways to Networks

Part X closes by dissolving its own scaffolding. Chapters 49–51 drew pathways — tidy lines of arrows from receptor to response — and warned, at each stage, that the tidiness was a courtesy. This chapter pays the debt. The signalling and regulatory machinery of the cell is a ***network****: thousands of proteins, metabolites and nucleic acids connected by activating and inhibiting interactions, in which no arrow runs in isolation and nearly every component both listens and speaks. The pessimistic conclusion, reached by a generation of students confronting the wall-chart of all interactions, is that nothing can be understood until everything is measured. The optimistic one — and the subject of this chapter — is that the network is built from a small vocabulary of recurring subcircuits, the* ***network motifs****, each performing a recognisable computational function: filtering, switching, timing, oscillating, remembering. Understand the motifs, and the wall-chart resolves into a wiring diagram; understand the wiring, and the cell's decisions — the triad of Chapter 51 — become not mysteries but computations, executed in protein.*[63]

52.2 Motifs: The Cell's Circuit Vocabulary

A motif is a pattern of connections that occurs in real networks far more often than chance allows — statistical over-representation being the fingerprint of selection for function. Uri Alon's analyses of the transcription networks of E. coli and yeast found that a handful of three-node patterns account for most of the non-random structure, and four of them (Figure 52.1) deserve permanent residence in the reader's mind. The ***negative feedback loop*** *— a component that inhibits its own production, directly or through an intermediary — is the universal homeostat: it accelerates the approach to steady state, suppresses fluctuations in component abundance, and stabilises against parameter drift. The* ***positive feedback loop*** *— mutual activation, or mutual inhibition, which is algebraically the same sign — is the universal memory: it holds a state after the stimulus that created it has gone, and, as Section 52.4 will show, it is the engine of every irreversible decision. The* ***coherent feedforward loop****, in which X activates Z*

63. The analogy with electronics is irresistible and, in this chapter, indulged — with one caveat pressed on the reader throughout: biological components are noisy, degrade, and diffuse, and the engineer who designed with such parts would be dismissed. Evolution, having no choice, made a virtue of it.

both directly and through Y, is a persistence detector: Z responds only if the input lasts long enough for the slow arm to catch up, so brief noise is rejected while sustained signal is accepted — the AND-gate of Chapter 48's lac operon is precisely this circuit, with CAP and the polymerase as its two arms. And the ***incoherent feedforward loop****, in which X activates Z and also activates Z's repressor, generates a pulse: output rises with the fast arm, then falls as the slow repression arrives, giving adaptation and — beautifully — fold-change detection, the cell responding to the ratio of successive signal levels rather than their absolute value. Four circuits; filter, memory, gate, pulse. Everything in Part X is a composition of them.*

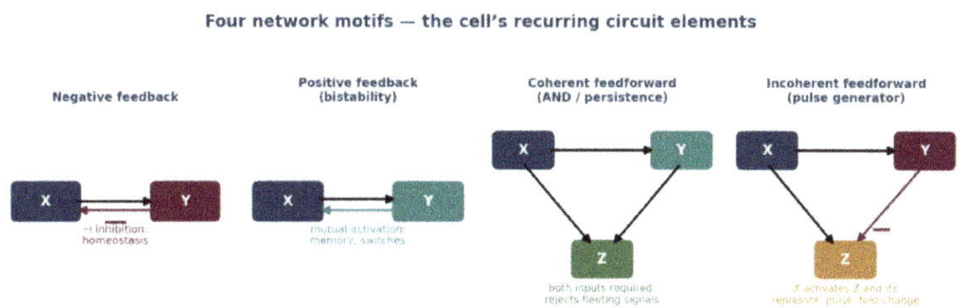

Figure 52.1 — Four network motifs. *Negative feedback (homeostasis); positive feedback (bistability and memory); coherent feedforward (persistence detection, as in the lac operon's AND-gate); incoherent feedforward (pulse generation and fold-change detection). These few patterns recur across transcription, signalling and metabolism.*

52.3 *Negative Feedback and Homeostasis*

The mathematics of negative feedback deserves one equation, because the whole of

$$\frac{dx}{dt} = \underbrace{\frac{P}{1+(x/x_0)^n}}_{\text{self-throttled production}} - \underbrace{\gamma\, x}_{\text{removal}} \quad \Longrightarrow \quad \text{rise time} \sim \frac{1}{\gamma}, \quad \text{steady-state} \tag{52.1}$$

from one line of mass action.

Figure 52.2 shows the result: feedback trades gain for speed and robustness, which is why engineering adopted it (the governor of Chapter 40's figure was a mechanical negative-feedback loop) and why evolution got there first by some three billion years. The body's hormone axes are Eq. (52.1) with glands for terms: the hypothalamus–pituitary–thyroid loop, where TSH stimulates T_4 and T_4 represses TSH, maintains thyroid hormone within a few per cent for decades —

and the clinician reads the loop directly, a high TSH with low T_4 diagnosing primary hypothyroidism as surely as the equation predicts it.

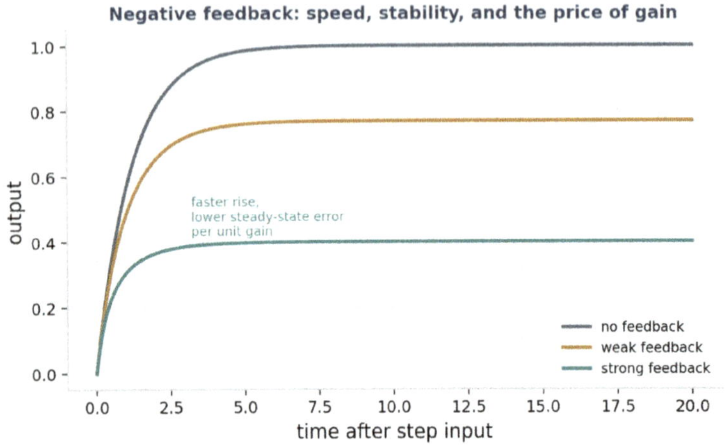

Figure 52.2 — Negative feedback. *Step responses without feedback (slate), with weak feedback (amber), and with strong feedback (teal): faster rise, smaller steady-state error, tighter control. The gifts are purchased with gain — Eq. (52.1).*

52.4 Positive Feedback, Bistability and Memory

Negative feedback holds one state; positive feedback holds two. When a component activates its own production with a steep (cooperative) dependence, the production curve is sigmoid while removal is linear, and the two can intersect three times (Figure 52.3): a low stable state, a high stable state, and between them an unstable threshold from which the system rolls away in whichever direction it is nudged. This is ***bistability****, and it is the molecular form of every genuine decision the cell makes, because a decision is precisely a state that persists. The restriction point of Chapter 51.3 is bistability in Rb; the oocyte's maturation is bistability in MAPK; the lysis–lysogeny switch of phage λ — the first bistable circuit ever dissected, and still the deepest understood — is bistability in the mutual repression of two transcription factors, cI and Cro. Three properties follow from the geometry. Hysteresis: the transition thresholds going up and going down differ, so the decision outlives its cause — memory, literally, in the shape of the curves. All-or-none response: populations split cleanly into OFF and ON cells rather than grading in between, which is why single-cell measurements of committed cells show two clouds, not one smear. And irreversibility in the limit: make the feedback strong enough and no physiological signal can reverse the switch — which is what "commitment" means, and what development is made of.*

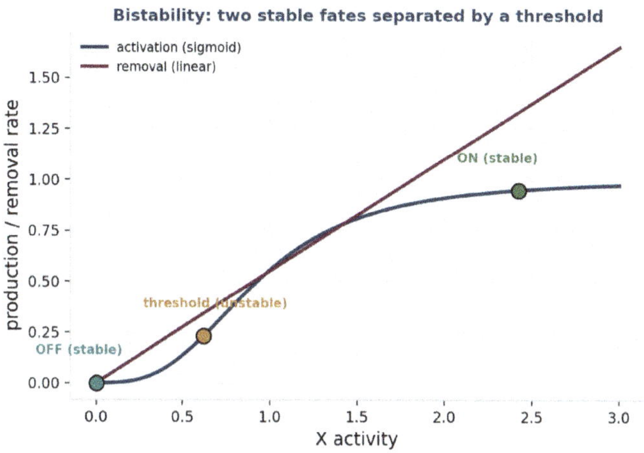

Figure 52.3* — *Bistability. *Where the sigmoid production curve (navy) crosses the linear removal line (crimson), three fixed points arise: stable OFF (teal) and ON (green), separated by an unstable threshold (amber). The cell's irreversible decisions — restriction point, oocyte maturation, λ's lysis–lysogeny — are all instances of this one picture.*

52.5 Oscillators: Clocks Built from Loops

Combine the two loop signs and the cell gains a clock. Chapter 51 derived the cell-cycle oscillator from synthesis plus self-excited destruction; the general rule is that ***sustained oscillation requires negative feedback with delay and sufficient nonlinearity*** *— delay to let the error grow before correction arrives, nonlinearity to keep the swings finite. The proof that these ingredients suffice, and that biologists can now build with them, came from synthetic biology's founding experiments. The* ***repressilator*** *of Elowitz and Leibler (2000) wired three repressors in a ring — LacI represses TetR, TetR represses cI, cI represses LacI: an odd number of negative signs around the loop, hence net negative feedback with three stages of delay — and single cells obligingly oscillated, generation after generation, with no cue from outside (Figure 52.4). In the same year Gardner and Collins built a genetic* ***toggle switch*** *— two mutual repressors, bistable by Section 52.4 — that could be flipped by a pulse of inducer and held its state for days: a one-bit memory in a bacterium. The natural clocks are the same circuits refined: the circadian oscillator is interlocked negative feedbacks (PER/CRY repressing their own transcription through CLOCK–BMAL1) with delays manufactured from phosphorylation and degradation; the p53 pulses of Chapter 51 are a digital oscillator riding on the MDM2 feedback. The cell, it turns out, keeps time the way the repressilator does — which is to say, the way a relaxation oscillator does, because there is no other way to keep time with parts that decay.*

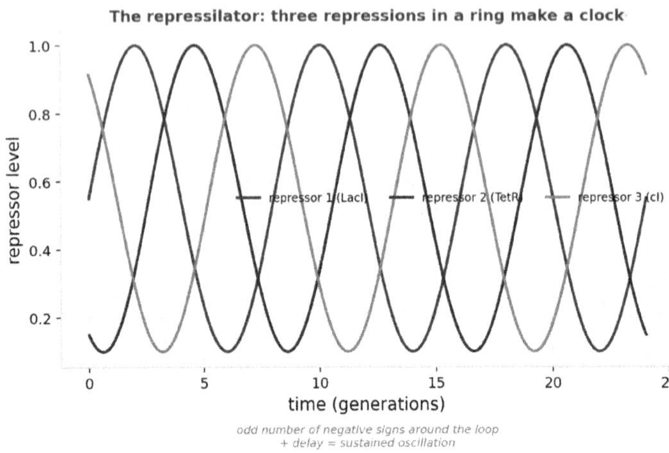

Figure 52.4 — The repressilator. Three repressors in a ring of mutual repression generate sustained oscillations in single cells (traces simulated to match the published dynamics): three delays, an odd number of negative signs, and a clock from nothing but wiring.

52.6 Robustness and Exact Adaptation

*A circuit that works only when its constants are tuned is not a circuit but a coincidence, and the cell's constants — protein half-lives, affinities, copy numbers — vary from cell to cell and hour to hour. **Robustness**, the invariance of function under parameter variation, is therefore a design requirement, and its deepest mechanism is **integral feedback**: a loop that integrates the error over time can drive that error to zero regardless of the values of the other parameters, because the integrator winds up until the error vanishes and holds there. The bacterial chemotaxis system — the best-understood signalling network in biology — is the proof of principle (Figure 52.5). The tumbling rate adapts exactly to any step of attractant: response peaks, then returns precisely to the pre-stimulus set point, whether the cell's receptor complement is high or low, the ambient temperature warm or cool. Barkai and Leibler showed in 1997 that exact adaptation cannot reside in the tuned constants — mutants lose it — but is a structural property of the methylation feedback: the receptor's activity is compared with a stored integral, and only the difference drives behaviour. The consequence, foreseen in Chapter 49.9, is that the network reports change, and climbs gradients by comparing now with a few seconds ago. Robustness, in the end, is achieved not by stronger parts but by cleverer topology — a lesson with obvious reach, since every drug that shifts a cellular parameter is opposed by a network built not to notice.*

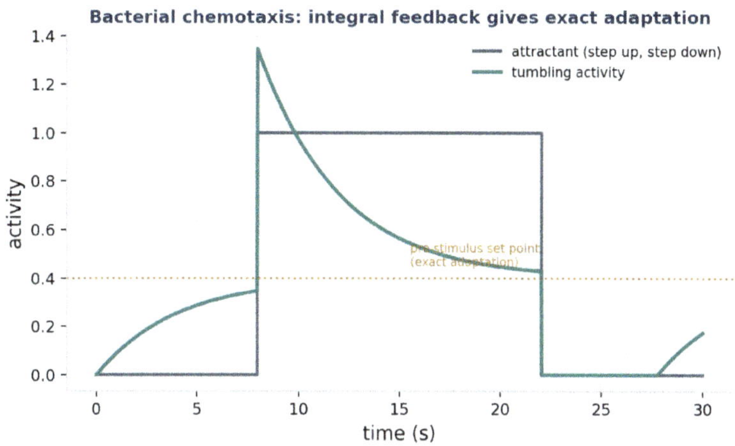

*Figure 52.5 — **Exact adaptation in bacterial chemotaxis.** Tumbling activity (teal) responds to a step of attractant (grey) with a transient, then returns precisely to its set point (amber) despite sustained stimulus — integral feedback makes the adaptation exact and independent of the network's constants.*

52.7 Decisions in Development: The Landscape Made Literal

In 1957 Conrad Waddington drew development as a ball rolling down a furrowed hillside — the "epigenetic landscape" — and the drawing was honoured for fifty years as a suggestive metaphor awaiting a mechanism. Network theory has supplied it. The valleys are the attractors of the gene-regulatory network: the stable states of Section 52.4, maintained by positive feedback among fate-determining transcription factors (Oct4/Sox2/Nanog in the pluripotent cell, GATA1 and PU.1 in the blood lineages); the ridges are the unstable thresholds between them; and the ball is the cell's state, rolling to the nearest attractor as signals reshape the terrain. Figure 52.6 makes the picture quantitative: the "elevation" is a pseudo-potential computed from the network's own equations, so the landscape is not drawn but derived. The modern single-cell technologies confirm the geology: cells differentiating through the blood lineages traverse trajectories between discrete transcriptional attractors, and the reprogramming of a skin cell to a stem cell by four transcription factors — Yamanaka's experiment, which earned a Nobel Prize in 2012 — is the physical act of pushing the ball back up the hill. Development, in this view, is not the unfolding of a programme but the geography of a dynamical system; and the canals, as the epigraph says, are dug by the circuits.

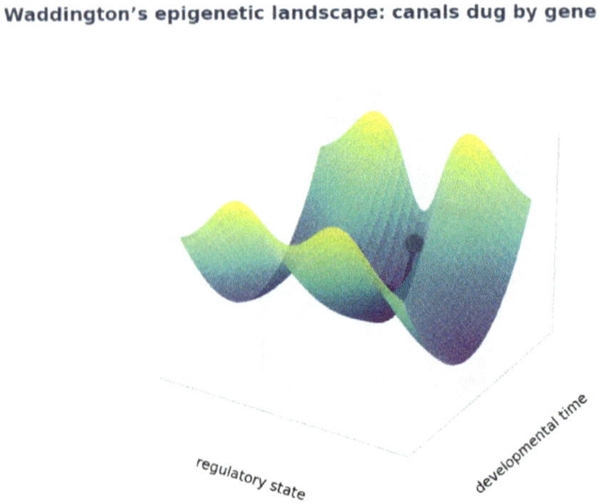

Figure 52.6 — ***Waddington's epigenetic landscape, made literal.*** *A pseudo-potential computed from a bistable regulatory network: stable valleys (cell fates), a central ridge (the decision threshold), and a cell (crimson) rolling from the undifferentiated state into one canal. The valleys are attractors; the hill is dynamical, not metaphorical.*

52.8 Case Study: λ's Switch — The Deepest Decision Ever Taken

Case study — lysis or lysogeny. *Bacteriophage λ, infecting E. coli, must choose between killing the cell outright (lysis) and inserting its genome into the host's to wait (lysogeny) — the oldest and most thoroughly dissected decision in molecular biology, Mark Ptashne's life's work and the subject of the finest monograph the field has produced. The switch's core is the mutual repression of two proteins, cI (the λ repressor) and Cro, at adjacent promoters O_R1–O_R3. In the lysogen, cI occupies the operators, Cro's genes are silent, and cI also activates its own synthesis — positive autoregulation — so the state is self-maintaining: bistability by Section 52.4, held for thousands of bacterial generations with a leak rate of order one induction per 10^5 cells per generation. When UV damages the host DNA, RecA is activated and cleaves cI; the repressor's concentration falls below threshold, Cro rises, and the switch flips to lysis — irreversibly, because Cro represses cI's maintenance promoter.*

Analysis. *Every motif of this chapter appears in one operon. The cI–Cro mutual repression is the toggle switch. The cI positive autoregulation supplies memory — daughter cells inherit the lysogenic state because they inherit the repressor. The cooperative binding of cI dimers to the three operators supplies the steep sigmoid of Figure 52.3. The SOS cleavage pathway is the input; the threshold's placement — set by the relative affinities of the three operators — is the evolved parameter that tunes how much DNA damage justifies abandoning ship. And the whole circuit's robustness was proven by construction: when the operators were rewired in new orders, the switch still worked, because bistability is a property of the loop's sign and steepness, not of the molecular details. λ is the existence proof that cellular decisions are computations, executable, rewritable, and understood.*

52.9 Historical Note: Nurse, cdc2 and the Universal Engine

> ***Historical note.*** *The confidence with which this chapter treats cell decisions as circuitry rests on a discovery made, appropriately, by genetics rather than by wiring diagrams. In the mid-1970s Paul Nurse, working with the fission yeast Schizosaccharomyces pombe in Edinburgh, isolated cell-cycle mutants that divided too small (wee) or arrested too large (cdc), and reasoned that the genes defined a rate-controlling engine. The decisive step was audacity: in 1987 Nurse's group cloned the human homologue of yeast cdc2 by functional complementation — a human gene rescuing a yeast mutant, across a billion years of divergence — proving that the cell-cycle engine is not merely conserved but interchangeable. Simultaneously, Tim Hunt's sea-urchin cyclins and Leland Hartwell's budding-yeast CDC genes converged on the same CDK machine of Chapter 51.2, and the three shared the 2001 Nobel Prize. The episode matters to this chapter for a methodological reason: it demonstrated that biological networks are modular — that one may speak of "the cell-cycle oscillator" as one speaks of an astable multivibrator, a circuit with a function, transferable between organisms, comprehensible without enumerating every molecule in the cell. The wall-chart dissolved, for the first time, into blocks.*

Plate 52.1 *Paul Nurse (b. 1949), who identified cdc2/CDK as the cell cycle's universal engine and proved its interchangeability between yeast and human. Nobel laureate in Physiology or Medicine, 2001, with Leland Hartwell and Tim Hunt. (Nobel Foundation portrait.)*

52.10 Exercises

reduced production. Feedback buys speed with gain, as Figure 52.2 shows.

by transient inputs, no oscillation. One wiring diagram, two functions, separated by parity alone.

adaptation: the property is in the topology, not the tuning — the definition of structural robustness.

Research proposition. *The motifs of Figure 52.1 were found by counting over-represented patterns in static networks, but cellular networks rewire themselves on every timescale — phosphorylation in seconds, expression in hours, evolution in eons. Propose a research programme to test the hypothesis that rewiring rate itself is regulated: for instance, whether stress responses preferentially create incoherent feedforward connections (fast, transient) while developmental commitments create positive feedback (permanent). Consider how one would score motif birth and death from time-resolved phosphoproteomic and transcriptomic data, and what it would mean for drug design if combination therapies could be chosen to steer a network's rewiring trajectory rather than its instantaneous state.*

52.11 Synthesis

Part X is complete, and its arc can now be stated in one sentence: cells receive signals (Chapter 49), route them through conserved pathways (Chapter 50), convert them into decisions among cycle, arrest and death (Chapter 51), and the decisions themselves are computations executed by a small set of recurring circuits — the homeostat, the toggle, the gate, the pulse, the clock — whose properties are derived, not described (Eqs. 49.1, 49.2, 51.1, 52.1). The wall-chart resolves into wiring; the wiring resolves into logic; the logic, as λ demonstrated, can be rewritten by hand. The direction has been checked against the master plan: Part XI now turns from understanding the cell to interrogating it — the experimental and computational methods by which every fact in this book was purchased. ***Chapter 53 — Principles of Biochemical Experimentation*** *opens Part XI.*

52.12 References

Alon, U. (2007). Network motifs: Theory and experimental approaches. Nature Reviews Genetics, 8, 450–461.

Barkai, N., & Leibler, S. (1997). Robustness in simple biochemical networks. Nature, 387, 913–917.

Elowitz, M. B., & Leibler, S. (2000). A synthetic oscillatory network of transcriptional regulators. Nature, 403, 335–338.

Ferrell, J. E., & Ha, S. H. (2014). Ultrasensitivity part II: Multisite phosphorylation, stoichiometric inhibitors, and positive feedback. Trends in Biochemical Sciences, 39, 556–569.

Gardner, T. S., Cantor, C. R., & Collins, J. J. (2000). Construction of a genetic toggle switch in Escherichia coli. Nature, 403, 339–342.

Nurse, P. (2000). A long twentieth century of the cell cycle and beyond. Cell, 100, 71–78.

Ptashne, M. (2004). A Genetic Switch: Phage λ Revisited (3rd ed.). Cold Spring Harbor Laboratory Press.

Tyson, J. J., Chen, K. C., & Novak, B. (2003). Sniffers, buzzers, toggles and blinkers: Dynamics of regulatory and signaling pathways in the cell. Current Opinion in Cell Biology, 15, 221–231.

Part XI

Experimental and Computational Biochemistry

Chapter 53
Separation, Detection and Quantification

CONTENTS OF THIS CHAPTER

"Every fact in this book was bought with a measurement, and the price of a measurement is a separation: before you may count the molecule, you must persuade it to travel alone."

53.1 Experimental Design

Part XI begins the reckoning promised since Chapter 1: the preceding fifty-two chapters asserted a great many facts, and it is time to show the instruments with which facts of this kind are purchased. But before any instrument comes a discipline, and the discipline is design. A biochemical measurement is an argument, and a badly designed experiment is an argument with a false premise — no amount of instrumental elegance downstream can rescue it. The premises are few and non-negotiable. There must be a **control** *that differs from the experiment in exactly the variable under test, and in nothing else: the boiled enzyme, the empty vector, the vehicle-treated animal. There must be* **replication**, *because variance is not an imperfection of nature but one of its properties, and a measurement without an estimate of its own uncertainty is an anecdote. And there must be a declared* **calibration** *— a known ladder, a standard curve, an internal standard — by which an instrument's reading is converted into a quantity. The reader who holds these three in mind will find that every section of this chapter is a variation on them: centrifugation is separation by mass, chromatography by partition, electrophoresis by charge, immunoassay by affinity, spectroscopy by light — and all of them end, as they must, at a calibrated number with an error bar.*[64]

53.2 Buffers and Sample Preparation

The first instrument is not an instrument at all but the solution in which everything happens. Biochemistry is conducted in water, at a pH that proteins tolerate, and the **buffer** *is what holds that pH while the chemistry proceeds. The governing relation is the Henderson–Hasselbalch equation, derived in Chapter 42 for blood and valid for any conjugate pair:*

$$\mathrm{pH} = \mathrm{p}K_a + \log_{10} \frac{[\mathrm{A}^-]}{[\mathrm{HA}]} \qquad (53.1)$$

64. A word about the tone of Part XI. Methods chapters in textbooks are often graveyards of acronyms. We have tried to write instead about what each method is *for* — what question it answers, what assumption it smuggles, what artefact it invites. The acronym can be looked up; the assumption cannot.

definition

Sample preparation is the unglamorous art on which everything else rests: cells are broken (by shear, osmosis, detergent or freeze–thaw), kept cold to slow the proteases of Chapter 51, dosed with inhibitors of those same proteases and of the phosphatases of Chapter 49 if phosphorylation is the question — for the sample begins to die, enzymatically, the moment it is taken, and the first duty of the experimenter is to freeze the verdict in place.

53.3 Centrifugation

$$s = \frac{v}{\omega^2 r} = \frac{m\,(1 - \bar{v}\rho)}{f} \qquad [s] = 10^{-13}\ \mathrm{s} \equiv 1\ \text{Svedberg (S)} \tag{53.2}$$

field is buoyant mass over friction — heavier wins, draggier loses. The unit honours The Svedberg, whose portrait graces this chapter and whose ultracentrifuge made the measurement possible.

Two regimes serve two purposes. ***Differential centrifugation*** *— stepwise increases of field — pelts out nuclei, then mitochondria, then membranes and ribosomes: the classical cell fractionation by which Palade and Claude mapped the organelles, and still the first step of every purification.* ***Density-gradient (isopycnic) centrifugation*** *spins the sample into a preformed gradient — sucrose, or caesium chloride — until each species finds the level of its own buoyant density and stops; the Meselson–Stahl experiment of Chapter 44, which proved semiconservative replication, was caesium-chloride gradient centrifugation and nothing more. The Svedberg's celebrated non-additivity — a*

t a reminder

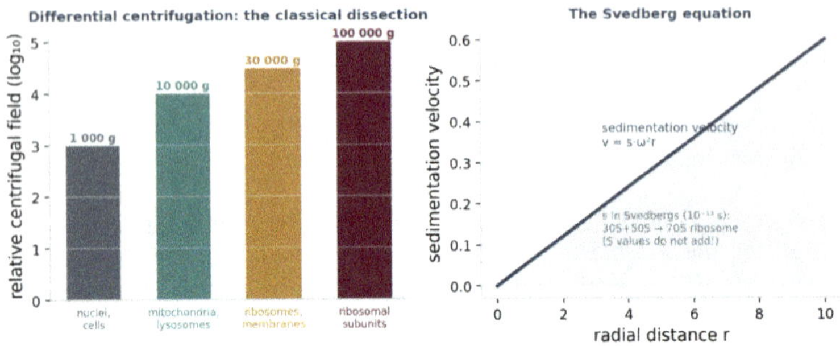

Figure 53.6 — ***Centrifugation.*** *Left: differential centrifugation pellets cellular components in order of size and density. Right: the Svedberg equation — sedimentation velocity proportional*

→ 70S

reminds us that shape, not mass alone, sets the value.

53.4 Dialysis and Ultrafiltration

The simplest separation uses no field at all, only a membrane with holes of chosen size. ***Dialysis*** *places the sample behind a semipermeable membrane against a large volume of buffer; small molecules — salts, reductants, imidazole from a purification — diffuse down their gradients and are washed away, while the protein, too large for the pores, stays put: exchange of solvent without touching the solute, used to remove, to refold, or to rebuffer.* ***Ultrafiltration*** *applies pressure across the same kind of membrane to force solvent through, concentrating the retained protein — the everyday instrument for concentrating a dilute eluate, and the physical principle, as Chapter 12 noted, of the glomerulus itself.*

53.5 Chromatographic Principles

Chromatography is the separation method that made modern biochemistry, and its principle is a single sentence: molecules that partition differently between a stationary phase and a moving phase travel at different speeds (Figure 53.1). The variants are enumerated by the interaction exploited. ***Size-exclusion chromatography*** *uses porous beads: large molecules, excluded from the pores, take the short path and elute first — the one separation in which the big come first, and a gentle, non-denaturing estimate of native size.* ***Ion-exchange chromatography*** *binds proteins to a charged resin by their surface charge at the working pH — above its pI a protein is negative and binds an anion exchanger — and elutes them with rising salt or shifting pH.* ***Affinity chromatography*** *exploits biology itself: an immobilised ligand binds one protein from the thousands — an antibody, a substrate analogue, a nickel resin grabbing a polyhistidine tag — and elution is by competition or denaturation of the interaction; one affinity step can purify a protein a thousand-fold, and the His-tag has made it the routine engine of the field. And* ***hydrophobic interaction*** *and* ***reversed-phase*** *chromatography separate by surface hydrophobicity, the latter indispensable for peptides and small molecules. The quantitative description — each species' retention set by its partition coefficient — belongs to the same mass-action family as Eq. (49.1), and the resolution of two neighbouring peaks is bought, always, with column length, time and pressure.*

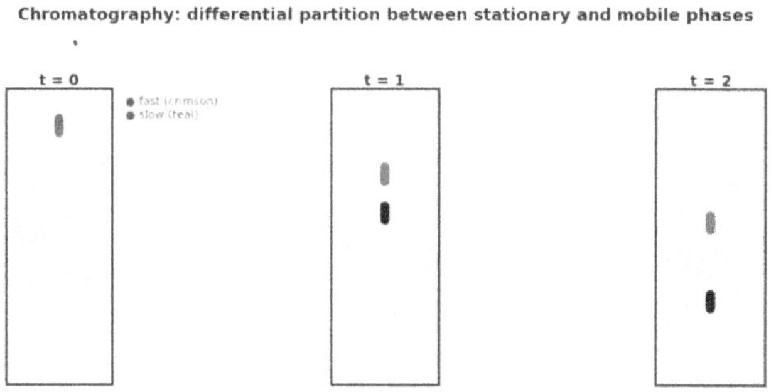

***Figure 53.1 — The chromatographic principle.** Two species (crimson, teal) partition differently between the stationary matrix and the mobile phase; with time the fast species draws ahead, and separation is the conversion of a chemical difference into a spatial one.*

53.6 Electrophoresis

An electric field across a gel drags charged molecules through the mesh, and the result — the gel — is biochemistry's most recognisable image. In ***SDS–PAGE****, the protein is first coated with the anionic detergent SDS, which unfolds it and confers a uniform charge-to-mass ratio, so that in the polyacrylamide mesh the only variable left is size: small proteins thread the gel faster, and migration distance falls with the logarithm of molecular mass (Figure 53.2). A stained gel is thus a census by size — a lysate's hundreds of bands, a purification's narrowing progress, a molecular-weight ladder for calibration.* ***Isoelectric focusing*** *separates instead by pI, proteins migrating in a pH gradient to the point where their net charge vanishes and they stop; the two together — size one way, pI the other — make* ***two-dimensional electrophoresis****, which once spread whole proteomes across a single plate as constellations of spots, the ancestor of Chapter 55's proteomics. And* ***agarose electrophoresis*** *of nucleic acids, with its fluorescent stain and its size ladders, is the daily bread of every laboratory that touches DNA — the gel that confirmed the clone, the digest, the PCR product, ten thousand times a day across the world.*

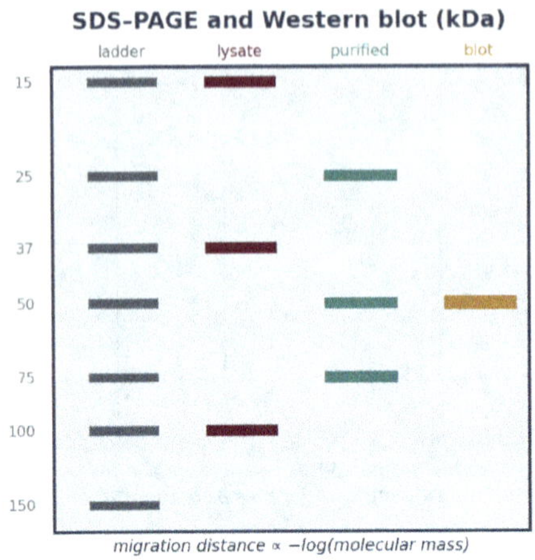

Figure 53.2 — SDS–PAGE and Western blot. *In SDS the charge-to-mass ratio is uniform, so migration measures size alone: the ladder (slate) calibrates; a lysate (crimson) resolves into bands; a purified protein (teal) runs clean; the Western blot (amber) confirms identity by antibody after transfer to a membrane.*

53.7 Western Blotting

Electrophoresis separates; the antibody identifies. In ***Western blotting****, the resolved gel is transferred — blotted — onto a membrane, preserving the pattern, and the membrane is probed with an antibody against one protein of interest, revealed in turn by a labelled second antibody: out of the hundreds of bands, one lights up, identified by affinity and sized by its position (Figure 53.2, final lane). The blot is the courtroom identification of molecular biology — is the protein there, how big is it, how much of it, and is it modified (a phospho-specific antibody reporting Chapter 49's currency directly).*

53.8 Immunochemical Assays

The ***ELISA*** *inverts the arrangement to measure rather than identify (Figure 53.4): a capture antibody fixed to a well grabs the antigen from serum; a second, enzyme-coupled antibody binds a different epitope — the "sandwich," specific twice over — and the enzyme turns a colourless substrate into a coloured product whose intensity, read against a standard curve, reports the antigen's concentration. The design is a beautiful compound of this chapter's principles: affinity for specificity, an enzyme for amplification (Eq. 49.2 in a plastic well), and a calibration curve for the number. Pregnancy tests, troponin in the emergency room, cytokines in the research laboratory — the ELISA and its lateral-flow descendants are, quietly, the most performed biochemical measurements on Earth.*

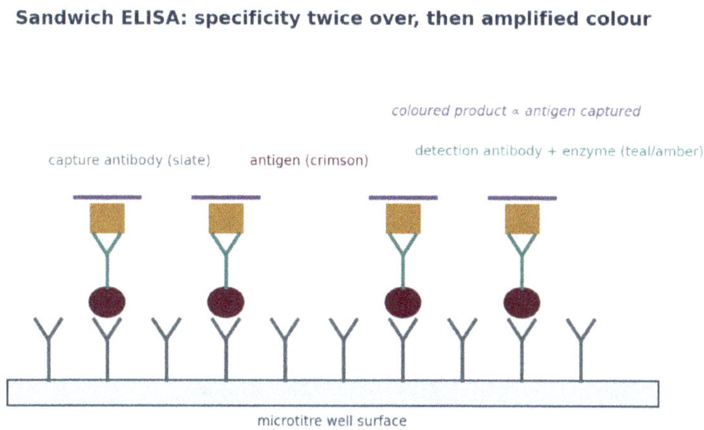

Figure 53.4 — Sandwich ELISA. *Capture antibody (slate) on the well surface binds antigen (crimson); a detection antibody bearing an enzyme (teal/amber) binds a second epitope; coloured product (plum) is proportional to captured antigen. Specificity twice over, then enzymatic amplification.*

53.9 Spectrophotometry

The workhorse quantification is absorption of light, governed by the Beer–Lambert law — derived from the simple physical premise that each layer of solution absorbs a fixed fraction of the light entering it, so intensity falls exponentially with path:

$$A = \log_{10}\frac{I_0}{I} = \varepsilon c l \qquad (53.3)$$

The law earns its keep daily: NADH absorbs at 340 nm while NAD^+ does not, so the dehydrogenase assays of Parts IV–VI are followed by watching a single wavelength rise or fall; protein is estimated at 280 nm by its aromatics, DNA at 260 nm, and the 260/280 ratio audits purity.

53.10 Fluorescence Methods

Fluorescence trades the bright background of absorption for darkness: a molecule is excited at one wavelength and observed emitting at another, longer one (Figure 53.5), so the signal appears against black and the sensitivity rises a hundred- to a thousand-fold. The physics is the Jablonski diagram — absorption to a singlet excited state, rapid vibrational relaxation to its bottom, then emission to the ground state, the energy lost in relaxation appearing as the **Stokes shift** *that separates excitation from emission. The applications run through the book: the DNA stains of the agarose gel; the green fluorescent protein and its palette, which* *made proteins visible in living cells and earned its own Nobel;* ***FRET****, in which excitation passes between two fluorophores when they approach within a few nanometres — a spectroscopic ruler that has measured molecular interactions in situ; and the fluorescent* Ca^{2+} *indicators that made the oscillations of Figure 49.5 visible. The cautions are the price of the sensitivity: quenching, photobleaching, and the inner-filter artefacts that punish the unwary who forget that the light must get in and out.*

The Jablonski diagram: why fluorescence outshines absorbance

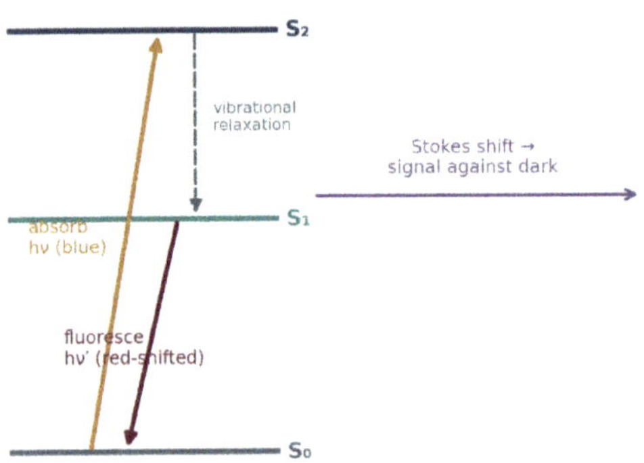

Figure 53.5 — The Jablonski diagram. *Absorption (amber) lifts the molecule to an excited singlet; vibrational relaxation (dashed) dissipates energy; emission (crimson) returns to ground at a longer wavelength — the Stokes shift that lets fluorescence be read against darkness, with a hundred-fold gain in sensitivity over absorption.*

53.11 Radioisotope Methods

The oldest ultrasensitive method is decay itself. Isotopes — tritium, carbon-14, phosphorus-32, sulphur-35, iodine-125 — are chemically identical to their stable siblings and betray their positions by radiation, countable with exquisite sensitivity: a labelled phosphate on a protein, a labelled precursor traced into DNA, a labelled ligand measuring receptor occupancy by Eq. (49.1) in a scintillation vial. Much of the classical evidence of Parts IV and IX — the pulse-chase experiments that mapped secretion, the tracing that proved semiconservative replication — was radioactivity and counting. The method has receded as fluorescence and mass spectrometry matured, and the reader of Chapter 55 will meet its modern descendant: the stable isotopes (^{13}C, ^{15}N)*, undetectable by decay but read by the mass spectrometer, which trace fluxes through living metabolism with a safety the hot isotopes never allowed. The principle endures unchanged: label the molecule, follow the label, and let the bookkeeping of atoms tell the pathway's story.*

53.12 Quantitative Calibration and Controls

Quantification is honest only when the calibration is. The equation's honesty depends on the calibration of Figure 53.3: a standard curve of known concentrations, replicates with error bars, a blank to catch the solvent's own intercept at the blank, is the smallest complete expression of the whole experimental method of Part XI: a measurement, an uncertainty, and a declared conversion between them.

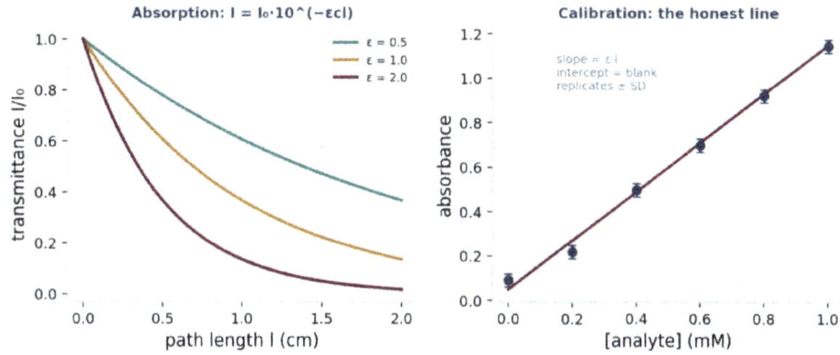

Figure 53.3 — Absorption and calibration. *Left: transmittance falls exponentially with path and concentration, per Eq. (53.3); absorbance, its logarithm, is linear. Right: the standard*

reading becomes a quantity.

53.13 Case Study: The Pulse-Chase and the Secretory Pathway

Case study — George Palade's fifteen minutes. *In the 1960s Palade's laboratory at Rockefeller asked where secreted proteins are made and how they leave the cell. The experiment was a pulse-chase of exemplary economy: pancreatic acinar cells were given a three-minute pulse of tritiated leucine — labelled amino acid flooding the pool, marking every protein being synthesised in that window — and then chased with unlabelled leucine, so that anything synthesised afterwards was invisible. At intervals, cells were fixed and the radioactivity located by autoradiography on electron micrographs. The silver grains told an itinerary: at three minutes, over the rough endoplasmic reticulum; at twenty, over the Golgi; by an hour, over the zymogen granules, and thence out of the cell.*

Analysis. *The experiment is this chapter in miniature. The label is Section 53.10's radioisotope; the pulse-chase design is Section 53.1's control logic — time itself is the variable, and the chase ensures the pulse's cohort is followed and no other; the readout is quantitative (grain counts over defined compartments, with statistics); and the conclusion — the secretory pathway ER → Golgi → granule → exterior of Chapter 47 — was a map drawn entirely by following a cohort of labelled molecules through space as a function of time. No pathway in this book was ever seen directly; all were inferred by exactly such bookkeeping. Palade's Nobel Prize of 1974 honoured the demonstration that inference of this quality is as good as sight.*

53.14 Historical Note: Svedberg's Spinning Solution

Historical note. *In the early 1920s, whether proteins were molecules at all — definite particles with definite masses — or loose colloidal aggregates was still disputed, and the dispute could not be settled because nothing could weigh a protein. Theodor Svedberg at Uppsala settled it with an instrument of his own design: the analytical ultracentrifuge, spinning solutions at fields of tens of* *thousands of gravities and photographing the moving boundary of sedimenting material through the cell's windows. With haemoglobin, in 1925, he found something the colloid theorists had not predicted: a single, sharp boundary, all the molecules sedimenting together — a homogeneous substance of one definite mass, about 68,000 daltons (now 64.5 kDa), four iron atoms to the molecule. Proteins were molecules; macromolecular chemistry had a unit and a method; and the 1926 Nobel Prize in Chemistry followed within a year, among the fastest ever awarded. The Svedberg unit survives in every ribosome and every gradient; and the principle — that a measurement, sufficiently refined, can settle what argument cannot — is the founding charter of Part XI.*

Plate 53.1 *Theodor (The) Svedberg (1884–1971), inventor of the analytical ultracentrifuge, which proved proteins are definite macromolecules and gave sedimentation its unit. Nobel laureate in Chemistry, 1926. (Nobel Foundation portrait.)*

53.15 Exercises

reaction releases 0.5 mM H^+ over its course; if the buffer is present at 50 mM total, estimate the pH drift.

Resolution.

How much NADH was consumed? (b) If the cuvette contained 1.0 mL and the enzyme sample was 10 μL of a 50-fold diluted extract, what is the activity in the original extract, in μmol·min⁻¹·mL⁻¹ (units)?

→

t activity, so the diluted extract has 0.0201/0.010 mL = 2.01 μmol·min^{-1}·mL^{-1}; multiplying by the 50-fold dilution, the original extract has ≈ 100 units·mL^{-1}. The entire chain is one linear equation applied twice — which is why Eq. (53.3) has paid for a century of enzyme kinetics.

Exercise 53.3 — The non-adding Svedbergs. *The ribosome's subunits*

Exercise 53.4 — Designing the ELISA. *A sandwich ELISA for a serum protein uses standards of 0, 2, 5, 10, 20 ng·mL^{-1} giving absorbances 0.04, 0.16, 0.34, 0.64, 1.24. (a) Estimate the slope and the unknown serum reading 0.50. (b) A colleague reports a serum value of 80 ng·mL^{-1} from a reading of 2.9, far above the top standard. What is wrong, and what is the correct procedure?*

≈

saturation, hook effects and the flattening of Eq. (53.3)'s validity all distort; extrapolation is not measurement. The correct procedure is to dilute the serum (say 10- and 20-fold), re-read within range, and multiply back — the number must come from inside the calibration, as Section 53.8 insists, because beyond the line the assay has no warrant to speak.

Research proposition. *Modern laboratories inherit a century of separated methods: chromatography here, mass spectrometry there, imaging elsewhere. Propose a programme for a unified measurement language — a data model in which every biochemical measurement, from a Western blot to a flux tracing, records its calibration, its uncertainty and its chain of separations in machine-readable form — and assess, with a pilot of one laboratory's actual workflows, how much irreproducibility in the literature is attributable to unrecorded separation and calibration choices rather than to biology. What would it take for journals to require such provenance as they now require statistics?*

53.16 Synthesis

Part XI has opened with the fundamentals: design before instruments; buffers by Eq. (53.1); separation by mass (Eq. 53.2), by partition, by charge, by size and by affinity; identification by antibody; quantification by light (Eq. 53.3), by fluorescence against darkness, and by the bookkeeping of isotopes — every method ending, as Section 53.1 demanded, in a calibrated number with an error bar. Palade's pulse-chase showed the whole discipline earning a Nobel; Svedberg's spinning solution founded it. The direction has been checked against the master plan: Chapter 54 climbs from separation to structure — how proteins are purified, sequenced, weighed and seen, by crystallography, NMR and the cryo-electron microscope. ***Chapter 54 — Determining Molecular Structure and Interaction*** *follows.*

53.17 References

Burgess, R. R., & Deutscher, M. P. (Eds.). (2009). Guide to Protein Purification (2nd ed.). Methods in Enzymology, 463. Academic Press.

Cantor, C. R., & Schimmel, P. R. (1980). Biophysical Chemistry, Part II: Techniques for the Study of Biological Structure and Function. W. H. Freeman.

Crowe, J., & Bradshaw, R. A. (2014). Affinity Chromatography: Methods and Protocols. Methods in Molecular Biology. Springer.

Lakowicz, J. R. (2006). Principles of Fluorescence Spectroscopy (3rd ed.). Springer.

Palade, G. (1975). Intracellular aspects of the process of protein synthesis. Science, 189, 347–358.

Svedberg, T., & Pedersen, K. O. (1940). The Ultracentrifuge. Oxford University Press.

Wilson, K., & Walker, J. (Eds.). (2018). Principles and Techniques of Biochemistry and Molecular Biology (8th ed.). Cambridge University Press.

Chapter 54
Determining Molecular Structure and Interaction

CONTENTS OF THIS CHAPTER

"To see a molecule, one cannot use light — the molecule is smaller than the wave. One must instead ask it questions it is obliged to answer: how it diffracts, how it resonates, how heavy it is, and what it warms when it embraces."

54.1 Protein Purification Strategies

Chapter 53 separated and counted; this chapter determines what things are — their covalent sequence, their three-dimensional architecture, and the strength of their embraces. Every structure begins with a purification, for no structural method tolerates a mixture. The strategy is inherited from Chapter 53's toolbox and runs by subtraction: a crude lysate is stripped by successive steps — precipitation, ion exchange, hydrophobic interaction, affinity, size exclusion — each chosen to exploit a property the target has and its neighbours lack. The progress is audited at every step by the numbers that matter: total protein falling, specific activity (units per milligram) rising, and the SDS–PAGE gel of Figure 53.2 narrowing toward a single band. A purification table — step, volume, activity, yield, fold-purification — is the accountant's ledger of the enterprise, and it teaches a hard truth early: yield and purity are enemies, and every step that purifies also loses. The affinity tags of Chapter 53.5, the His-tag above all, have compressed what once took a doctoral thesis into a week — but the logic is unchanged: exploit a difference, audit the result, and never trust a band you have not identified.[65]

54.2 Mass Spectrometry

The mass spectrometer is the most precise balance ever built, and its principle (Figure 54.1) is disarmingly simple: ionise the molecule, accelerate the ions, and measure their behaviour in a field — time of flight down a tube, orbital frequency in an electrostatic trap (the Orbitrap), d
function of the ***mass-to-charge ratio***
biology possible for the instrument: ***elec***
fine liquid spray, gentle enough for intact proteins; and ***MALDI****, which blasts them out of a crystalline matrix with a laser pulse. The modern instrument weighs a peptide to within a fraction of a dalton — precise enough that the mass itself nominates the elemental composition — and in* ***tandem MS*** *(MS/MS) a selected ion is fragmented by collision and the fragment masses read back the sequence,*

65. The classical literature records purifications of heroic duration — Warburg's enzymes, Sumner's urease crystals of 1926, which proved enzymes are proteins to a sceptical field. The modern student, with a nickel column and a weekend, stands on their shoulders and should occasionally be reminded of it.

because the pieces differ by the masses of single amino-acid residues. Proteomics, the subject of Chapter 55.2, is this trick industrialised: digest the proteome, separate the peptides, sequence them by the million, and let the computer reassemble the census. The mass spectrometer also reads the stable isotopes of Chapter 55.5's flux tracing, the modifications of Chapter 47's maturation (a phosphate is +80 Da, an acetyl +42, a methyl +14), and the stoichiometry of intact complexes in native MS. It is, gram for gram, the most informative instrument in the building.

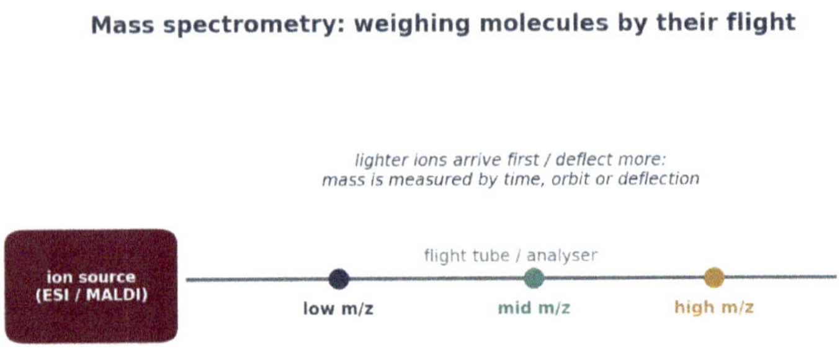

Figure 54.1 — Mass spectrometry. *Ions formed in the source (crimson) are separated by mass-to-charge ratio in the analyser — by flight time, orbit, or deflection — lighter species arriving first or deflecting more. The mass spectrum that results weighs molecules to sub-dalton precision and, in tandem mode, sequences them.*

54.3 Protein Sequencing

Before the mass spectrometer ate the field, proteins were sequenced by chemistry, and the chemistry deserves its paragraph. Edman degradation (1950) clips the N-terminal residue as a phenylthiohydantoin derivative, identifies it chromatographically, and repeats — reading a chain one residue at a time from the front, for perhaps fifty cycles before the chemistry tires. Sanger's sequencing of insulin by such degradation (1953–55) delivered the proof on which Part IX silently rests: a protein has a definite, unique sequence, not a statistical mixture — the fact that made the gene-protein collinearity of Chapter 43 conceivable. Today the order is inverted: genes are sequenced and protein sequences inferred, with mass spectrometry verifying the actual chain, its true N-terminus (often processed), and its modifications. The historical lesson stands: sequence first established that proteins are information, decades before the word entered biology.

54.4 X-ray Crystallography

≈

atom, diffracts from the periodic electron density into a pattern of discrete spots, the reflections. The physics is Bragg's law, derived from path differences alone:

$$n\lambda = 2d\sin\theta \tag{54.1}$$

The pattern's positions give the lattice; the pattern's intensities give the structure — through the Fourier transform, which converts the amplitudes of the reflections into an electron-density map, into which an atomic model is built. The catch is the celebrated ***phase problem****: the detector records intensities but loses the phases, half the information, which must be recovered by heavy-atom derivatives, anomalous scattering, or molecular replacement from a known homologue. The honours board of the method is the honour roll of molecular biology: myoglobin and haemoglobin (Kendrew, Perutz), the double helix by fibre diffraction (Franklin's data of Chapter 43), lysozyme, the tRNA, the ribosome (Chapter 47's plate), and now some two hundred thousand structures in the Protein Data Bank — the library from which Chapter 55.8's prediction methods learned their trade.*

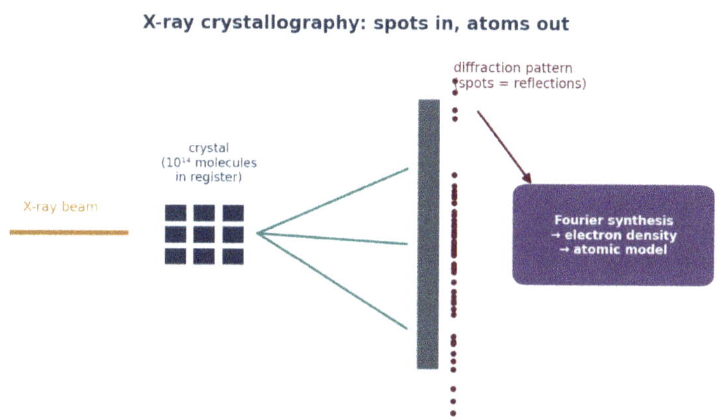

Figure 54.2 — X-ray crystallography. *A beam (amber) diffracts from the crystal's lattice (navy) into a spot pattern (crimson) governed by Bragg's law, Eq. (54.1); Fourier synthesis of the spot intensities, with phases recovered by auxiliary experiments, yields the electron density into which the atomic model is built (plum).*

54.5 Nuclear Magnetic Resonance

NMR determines structure in solution — the molecule at something closer to its working condition — by listening to atomic nuclei in a magnetic field. A nucleus with spin (^1H, ^{13}C, ^{15}N) resonates at a radiofrequency exquisitely sensitive to its electronic environment: the ***chemical shift*** *gives every proton an address (Figure 54.3, left), and two further effects give the geometry. The through-bond couplings trace which atoms are bonded to which, assigning the shifts along the chain; and the* ***nuclear Overhauser effect*** *(NOE) reports which protons lie within about 5 Å of one another through space, regardless of bonding — each NOE cross-peak is a distance restraint, and a thousand restraints, fed to the geometry of Chapter 4, determine the fold (right). NMR's gifts are unique: dynamics on every timescale from picosecond bond vibrations to second-scale conformational exchange; structures of the intrinsically disordered proteins that crystallography cannot* *catch; and ligand-binding footprints mapped by shift perturbation. Its price is sensitivity and size — the signals fade as the molecule tumbles slowly, limiting routine work to perhaps 50 kDa — which is why the three great structural methods divide the world: crystals for the orderly, NMR for the small and the mobile, and the electron microscope for the huge.*

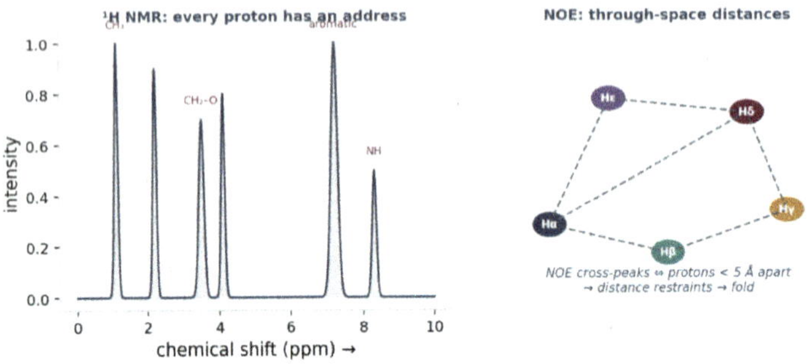

***Figure 54.3* — NMR.** *Left: a ¹H spectrum — each proton resonates at a chemical shift set by its electronic environment. Right: NOE cross-peaks (dashed) report through-space proximities under 5 Å; assembled as distance restraints, they determine the solution structure, dynamics and disorder that crystallography cannot reach.*

54.6 Cryo-electron Microscopy

The third method is the youngest and, for large assemblies, now the first choice (Figure 54.4). The specimen is frozen so fast that water cannot crystallise — vitreous ice, preserving the molecules in near-native state — and imaged in the electron microscope, which sees the particle directly, no crystal required. The catch is noise: the electron dose must be kept tiny lest the radiation shred the specimen, so each image is a blur of a few electrons per pixel. The remedy is the computer's, and it is the subject of Section 54.12's historical note: collect thousands of images of identical particles in random orientations, classify them into views, average each view to beat down the noise, and reconstruct the three-dimensional Coulomb map from the projections — the mathematics of the medical CT scanner applied to single molecules. Resolutions of 2 Å and better are now routine; the ribosome's moving parts, the spliceosome of Chapter 46, membrane channels and receptors in lipid, viruses whole — the large, the flexible and the uncrystallisable have all fallen. Cryo-electron tomography extends the same logic to pleomorphic objects — whole cells, sectioned by focused ion beam and reconstructed — bridging, at last, structure and cell biology.

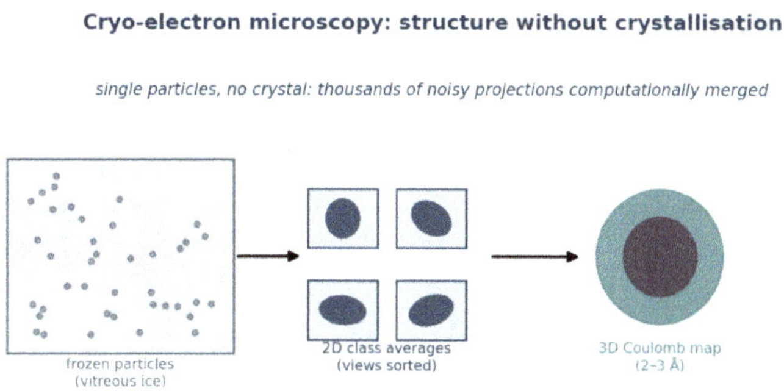

Figure 54.4 — Cryo-electron microscopy. *Particles in vitreous ice (grey) are imaged at low dose; thousands of noisy projections are sorted into 2D class averages (navy) and merged computationally into a 3D map (teal/crimson) at near-atomic resolution — structure without crystallisation, for the large and the flexible.*

54.7 Circular Dichroism

Not every question needs atoms. ***Circular dichroism*** *(CD) measures the differential absorption of left- and right-circularly polarised light by chiral structures, and for proteins the far-UV signal (190–250 nm) reports secondary structure directly: the α-helix and the β-sheet each sign a characteristic spectrum, so the composition of an unknown protein can be estimated in an afternoon from a drop of solution. The method's everyday uses are folding audits — is the purified protein folded, does the mutant fold like the wild type — and stability: the thermal melt of Chapter 5, followed by CD as the signal, yields the unfolding curve and its midpoint, and ligand binding that stabilises the fold is detected as a raised melting temperature (the thermal-shift assay of drug screening). CD does not say where the atoms are; it says, quickly and cheaply, whether the architecture is standing.*

54.8 Surface Plasmon Resonance

Structure tells what a molecule is; the remaining sections ask what it does with others — and the first instrument for that is ***surface plasmon resonance*** *(SPR). One partner is fixed on a gold sensor surface; the other flows over it; binding increases the mass at the surface, which shifts the resonance angle of plasmons in the gold — and the shift is followed in real time, without labels, as the association*

pharmacology of Part X has learned the expensive way.

54.9 Isothermal Titration Calorimetry

The deepest interaction measurement needs no label, no surface, and no assumption but the first law of thermodynamics. ***Isothermal titration calorimetry*** *(ITC) injects ligand, drop by drop, into a cell containing its partner, and measures the heat released or absorbed by each injection (Figure 54.5). As the titration proceeds the binding sites fill, the heat per injection wanes, and the*

links made; an entropy-driven one, of water released from hydrophobic surfaces — the hydrophobic effect of Chapter 4 measured on the balance of heat. The two anchor points of drug design, enthalpic efficiency and the enthalpy–entropy compensation that frustrates every attempt to improve one term without taxing the other, are read from such curves.

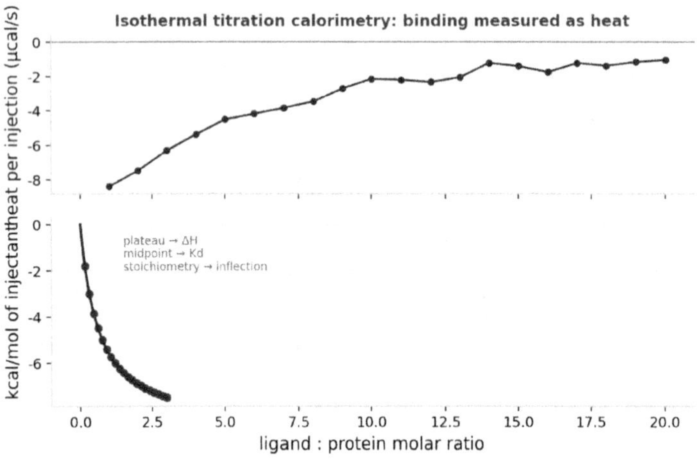

Figure 54.5 — Isothermal titration calorimetry. *Top: heat per injection (crimson) wanes as binding sites saturate. Bottom: the integrated titration (navy, fit in crimson) yields*

54.10 Single-Molecule Methods

Every method so far reports an average over trillions of molecules, and the average can conceal the plot: a motor protein's steps, an enzyme's rare intermediate, a channel's flickers are invisible in the ensemble. Single-molecule methods listen to one molecule at a time. ***Optical tweezers*** *hold a bead in a focused laser's trap and measure, with piconewton sensitivity, the forces a single molecular motor exerts — kinesin's 8-nanometre steps along the microtubule, the ribosome's ratcheting, the unfolding of a single protein domain pulled end to end.* ***Single-molecule FRET*** *watches one pair of fluorophores report one molecule's conformational breathing in real time.* ***Single-channel recording*** *— Neher and Sakmann's patch clamp, Nobel 1991 — resolves the opening of individual ion* *channels, the unitary events beneath every trace of Chapter 49. The general gift is the distribution where the ensemble gives a mean: heterogeneity, intermediates, and the kinetic pathways between states — statistics of individuals, which is what mechanisms are made of.*

54.11 Molecular Imaging

Structure and interaction can also be followed in the living subject. The fluorescent proteins of Chapter 53.9 report location and dynamics in cells; **super-resolution microscopy** *(STORM, PALM, STED) breaks the diffraction limit by localising single fluorophores one at a time, reaching tens of nanometres in situ;* **PET** *traces positron-emitting ligands — fluorodeoxyglucose reporting the glycolytic appetite of Chapter 40's tumours — through the whole patient; and* **MRI***, the clinical cousin of Section 54.5's NMR, maps water's relaxation to image anatomy, and blood-oxygenation to image the working brain. The scale ladder of Figure 54.6 sets these beside the structural methods: from the ångström of the crystal to the millimetre of the scanner, the molecule is the same, and only the questions change.*

Figure 54.6 — The resolution ladder. *Five families of methods span from single ångströms (crystallography) to cellular scales (tomography, super-resolution light microscopy). The structural triad — crystallography, cryo-EM, NMR — reaches atoms; the methods below trade atoms for context.*

54.12 Integrative Structural Biology

The mature discipline combines the methods, each compensating the others' blindness: crystallography for the rigid core, NMR for the disordered tails, cryo-EM for the assembly's shape, mass spectrometry for its composition and cross-links (which pairs of residues are near), FRET for its working motions, computation to hold the pieces in one consistent model. The nuclear pore complex, the inflammasome, the transcription pre-initiation complex — the great assemblies of Parts IX–X are known this way, as hybrid structures in which no single method saw everything but all together saw enough. The guiding maxim is the one this chapter has practised throughout: a measurement is an answer to a question, and wisdom lies in asking each method only the questions it is built to answer.

54.13 Case Study: The Ribosome — Three Methods, One Machine

Case study — the machine that made the methods grow. *The ribosome of Chapter 47 has been, for half a century, the Everest of structural biology, and its conquest reads as a history of this chapter. Crystallography delivered the first atomic picture: Ramakrishnan, Steitz and Yonath (Plate 47.1) grew crystals of a 2.5-megadalton RNA–protein assembly — an undertaking many thought absurd, since crystallisation demands homogeneous, rigid particles and the ribosome is a machine that works by moving — and phased the pattern with heavy-atom clusters, revealing the peptidyl-transferase centre as pure RNA: the ribosome is a ribozyme, and the RNA world of Chapter 56 left a survivor. NMR contributed what crystals cannot: the solution behaviour of individual ribosomal proteins and their tails, and the dynamics of the factors. Cryo-EM then delivered the machine at work — the ribosome caught in the act, in state after functional state: initiation complexes, translocation intermediates, the nascent chain in the exit tunnel, each classification of Figure 54.4's pipeline a different frame of the film.*

Analysis. *The case distils the chapter's argument. No method sufficed alone: crystals gave atoms but froze the machine; cryo-EM gave states but, initially, blurs; NMR gave dynamics but only of pieces. The integration — atomic models flexibly fitted into cryo-EM maps of functional states — produced something no technique could: a mechanism, the ribosome's elongation cycle of Figure 47.2 understood as a sequence of structures. The 2009 Nobel for crystallography and the 2017 Nobel for cryo-EM bookend the enterprise — and the reader will note that the same particle received both, first for being seen at rest and then for being seen to move.*

54.14 Historical Note: Frank's Blur Becomes an Atom

Historical note. *Electron microscopy of single molecules was long the poor relation of structural biology: the electron dose needed for a sharp image destroyed the specimen, and lowering the dose gave images too noisy to interpret — "blobology," the crystallographers called it, not always kindly. Joachim Frank, from the mid-1970s at the Wadsworth Center and later Columbia, supplied the mathematics that turned blobs into structures: multivariate statistical classification to sort thousands of noisy particle images into views, alignment and averaging to recover signal from noise, and three-dimensional reconstruction from the projections — the first single-particle analysis, demonstrated on the ribosome. The other two legs of the tripod followed: Jacques Dubochet's vitrification, freezing specimens so fast that ice has no time to crystallise and crush the molecules; and Richard Henderson's demonstration, with bacteriorhodopsin, that electrons could reach atomic resolution at all. When direct electron detectors arrived around 2012 — counting single electrons instead of fogging film — the "resolution revolution" erupted, and the method that could not see suddenly saw better than anything else on large machines. The 2017 Nobel Prize in Chemistry crowned all three. There is a moral for the instrument-builder's trade: Frank's contribution was not an instrument but an algorithm — proof that in measurement, as in Chapter 53's epigraph, the separation that matters is sometimes of signal from noise, and it is done by thinking.*

Plate 54.1 *Joachim Frank (b. 1940), who invented single-particle image analysis, the mathematics that turns noisy electron micrographs into molecular structures. Nobel laureate in Chemistry, 2017, with Jacques Dubochet and Richard Henderson. (Photo: Alexander Mahmoud, © Nobel Foundation.)*

54.15 Exercises

exercise shows the inverse relation the whole method lives by: wide-angle spots carry fine detail.

ppm, which candidate survives? (b) A phosphorylation adds HPO_3 (79.9663

Da); what would the phosphorylated ion weigh, and could an instrument of ±0.5 Da distinguish it from sulfation (SO_3, 79.9568 Da)?

≈

the 3-ppm instrument (0.0095/783 ≈ 12 ppm): the question dictates the instrument, as ever.

Exercise 54.3 — SPR kin *sensorgram of a*

≈

≈ *t*

nt, sustaining effect between doses and outlasting washout — the residence-time argument of Section 54.8. Affinity is two numbers; the off-rate is often the one that matters clinically.

– → –

compensation, the standing frustration of Section 54.9, made quantitative.

Research proposition. *Section 54.12's integrative structures are assembled by hand, each restraint weighted by an expert's judgement of each method's reliability. Propose a formalism — Bayesian, with each measurement contributing a likelihood and each method a calibrated error model — for inferring structures with honest uncertainty estimates, and a benchmark on which it could be tested (synthetic assemblies with known ground truth, or cross-validation against held-out reflections). Discuss what it would mean for structural biology if every deposited model carried not only coordinates but a posterior distribution over them, and how such distributions should change the way Chapter 55's machine-learning models are trained.*

54.16 Synthesis

The chapter has weighed, read, and seen the molecule. Purification by audited subtraction; mass by flight (and sequence by fragments); sequence by chemistry, historically decisive; structure by diffraction (Eq. 54.1), by resonance, and by the

ribosome's. The direction has been checked against the master plan: Part XI closes by scaling every one of these methods from one molecule to all of them — the omics, and the computation that makes sense of the deluge. ***Chapter 55 — Omics, Computation and Quantitative Modelling*** *follows.*

54.17 References

Bai, X.-C., McMullan, G., & Scheres, S. H. W. (2015). How cryo-EM is revolutionizing structural biology. Trends in Biochemical Sciences, 40, 49–57.

Drenth, J. (2007). Principles of Protein X-Ray Crystallography (3rd ed.). Springer.

Frank, J. (2006). Three-Dimensional Electron Microscopy of Macromolecular Assemblies. Oxford University Press.

Leavitt, S., & Freire, E. (2001). Direct measurement of protein binding energetics by isothermal titration calorimetry. Current Opinion in Structural Biology, 11, 560–566.

Rischel, C., & Schlichting, I. (2009). Biomolecular Crystallography. Garland Science.

Schuck, P. (1997). Use of surface plasmon resonance to probe the equilibrium and dynamic aspects of interactions between biological macromolecules. Annual Review of Biophysics and Biomolecular Structure, 26, 541–566.

Wüthrich, K. (1986). NMR of Proteins and Nucleic Acids. Wiley.

Chapter 55
Omics, Computation and Quantitative Modelling

Contents of this chapter

"The twentieth century measured molecules one at a time and understood them; the twenty-first measures them all at once and must learn to understand again — for the deluge does not explain itself."

55.1 Genomics and Transcriptomics

Part XI closes by changing the unit of account. Chapters 53 and 54 measured one molecule, one complex, one interaction, with loving precision; the omics measure everything at once, at the cost of measuring everything a little less lovingly. The transition began with sequencing. The chain-termination method of Sanger — DNA synthesis terminated at random by dideoxynucleotides, the fragments read by size — read the human genome over a decade for three billion dollars; the "next-generation" methods that followed read it in days for a hundredth of the price, by the simple device of doing millions of sequencing reactions in parallel on a chip and letting the computer reassemble the book from shredded, overlapping fragments. Genomics thereby became a census: every variant in every patient, every somatic mutation in every tumour — the mutation signatures of Chapter 45.10 were read from such censuses. ***Transcriptomics*** *asks which genes are speaking: RNA is extracted, reverse-transcribed, sequenced (RNA-seq), and the reads counted per gene — a genome-wide expression measurement with a dynamic range no microarray ever approached. And the latest refinement, single-cell RNA-seq, performs the census cell by cell, dissolving the tissue's average into its cellular citizens: the attractor states of Chapter 52.7's landscape, enumerated.*[66]

66. A note of honesty, as promised in Part XI: the omic methods do not abolish Chapter 53's discipline, they automate it. The controls, the replicates, the calibration are all still owed — now multiplied by twenty thousand genes, which is why the statistics section of this chapter is the longest.

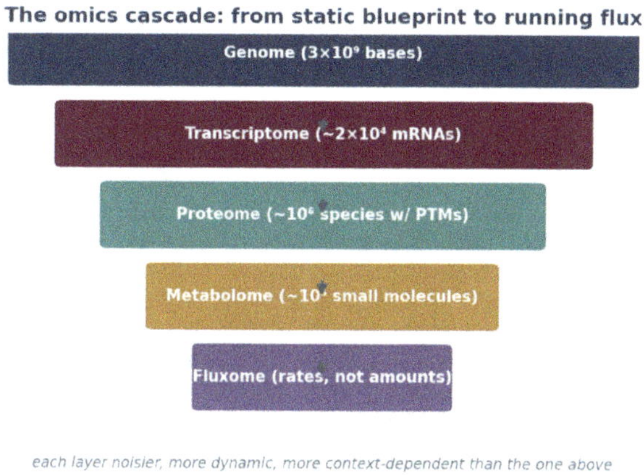

Figure 55.1 — The omics cascade. *Genome, transcriptome, proteome, metabolome, fluxome: each layer is noisier, more dynamic, and more context-dependent than the one above. The genome is the blueprint; the fluxome is the building in operation — and the further down the cascade, the closer to phenotype.*

55.2 Proteomics

The proteome is the transcriptome's unfaithful shadow: mRNA levels explain barely half the variance in protein abundance, because translation, degradation and modification interpose their own governments (Chapters 47 and 48). Mass-spectrometric proteomics — the industrialised tandem MS of Chapter 54.2 — now identifies and quantifies ten thousand proteins in a single run, by digesting the proteome to peptides, separating them chromatographically, sequencing the fragments, and counting by intensity or by isotopic label (TMT, SILAC). Two frontiers matter most. ***Phosphoproteomics****, enriching the phosphorylated peptides, reads the signalling state of Chapter 49's currency directly across the kinome — a tumour's wiring diagram measured, not guessed. And* ***interaction proteomics*** *— affinity purification followed by MS — maps who binds whom, the experimental edge-list of Chapter 52's networks. The proteome's depth remains its difficulty: a few abundant proteins dominate, the scarce signalling proteins hide beneath them, and the census, for all its scale, still sees the city lights better than the alleys.*

55.3 Metabolomics

Closest to the phenotype sits the metabolome — the small molecules that are metabolism's actual cargo — measured by mass spectrometry and NMR. Untargeted metabolomics detects thousands of features per sample, of which a sobering fraction remain unidentified — the field's standing embarrassment and its frontier: the cell's chemistry is larger than the databases. Targeted metabolomics quantifies a chosen panel with standards and calibration, Chapter 53.12's discipline restored. The readout is immediate and clinical: the anion-gap metabolites of Chapter 42, the acylcarnitines of the inborn errors, the oncometabolite 2-hydroxyglutarate that the mutant IDH enzyme of the case study below pours out in leukaemia — a single small molecule that rewires the epigenome of Chapter 48.5 by inhibiting the α-ketoglutarate-dependent demethylases. The metabolome answers the question the genome cannot: not what the cell could do, nor what it is saying, but what it is actually doing, this minute.

55.4 Lipidomics and Glycomics

Two further layers complete the census, each with its own chemistry. ***Lipidomics****, by mass spectrometry, enumerates the thousands of molecular species of the membranes of Chapter 11 — head groups, chain lengths, double bonds — and has made the lipid droplet and the signalling lipids of Chapter 50 (PIP_3, DAG, ceramide) measurable participants rather than diagrammatic names.* ***Glycomics*** *faces the hardest analytical problem in the census, because glycans are branched, template-free polymers: their structures are read by MS fragmentation and lectin arrays, and their biology — the glycocalyx, the ABO blood groups, the glycosylation of every antibody drug — is Chapter 23's sugar chemistry operating at cellular scale. The omics of the unglamorous molecules are, characteristically, the least complete and the least browsed.*

55.5 Stable-Isotope Tracing

All the layers so far are inventories — amounts. Metabolism's essence, however, is rates, and amounts no more reveal rates than a bank balance reveals cash flow. The instrument of rates is the isotope. Feed the cell [U-$^{13}C_6$]glucose — every carbon labelled — and follow the label by mass spectrometry (Figure 55.3): pyruvate appears as the M+3 isotopologue, citrate as M+2 after the PDH step, lactate as M+3 if glycolysis runs fast — and from the pattern of labelling across the network, the routes the carbons took are read as directly as Chapter 53.11's pulse-chase read the secretory pathway. The discoveries of the field are written in isotopes: the Warburg tumour fermenting glucose to lactate while its mitochondria idle; the glutamine addiction of Chapter 39's transformed cells; the reductive carboxylation that runs the citric acid cycle backwards when the mitochondria are damaged. Every one of these was invisible in amounts and obvious in labels — the difference between counting the cars and measuring the traffic.

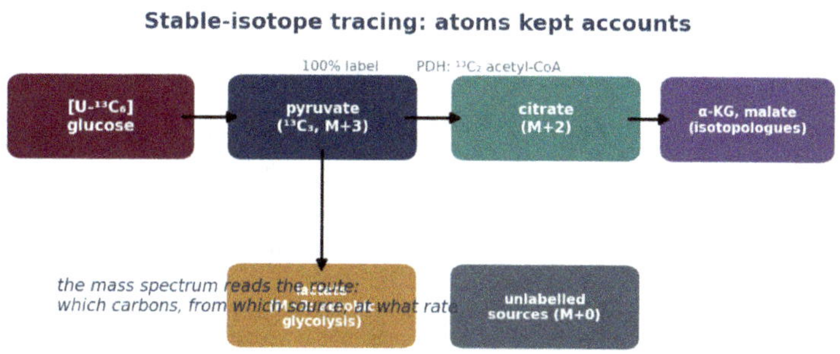

*Figure 55.3 — **Stable-isotope tracing.** [U-$^{13}C_6$]glucose (crimson) labels pyruvate and lactate as M+3 (navy, amber) and citrate as M+2 (teal) via PDH; unlabelled sources contribute M+0 (slate). The mass spectrum's isotopologue pattern reads the routes and, with modelling, the rates of metabolism.*

55.6 Metabolic-Flux Analysis

From labelling patterns to numbers requires a model. ***Metabolic-flux analysis***

flows out:

$$S \cdot \mathbf{v} = \mathbf{0} \tag{55.1}$$

metabolite's net production vanishes, giving Eq. (55.1): a system of linear equations the fluxes must satisfy — mass conservation, and nothing else.

Eq. (55.1) rarely fixes the fluxes uniquely — the cell has more reactions than constraints — so two remedies serve. ***^{13}C metabolic flux analysis*** *adds the labelling data of Section 55.5 as further equations, pinning the fluxes of central carbon metabolism with statistical confidence intervals; and* ***flux balance analysis*** *(FBA) picks, from the space of allowed flux vectors, the one optimising a stated objective — maximal biomass yield being the canonical guess — and predicts, with unsettling success for microbes, growth rates, gene essentiality, and the rewiring after a knockout. Genome-scale models now span thousands of reactions for organisms from E. coli to human cell types: the whole of Part V's biochemistry, written as a matrix and asked questions. The reader who has followed the book's argument will savour the circularity: the pathways were discovered one enzyme at a time by Chapter 53's methods; the network was assembled by the genome; and now the fluxes are computed from the whole, and* *checked — where checking is possible — against the measurements of Section 55.5. Reduction and synthesis, shaking hands.*

55.7 Sequence Alignment and Homology

The omic deluge would be mute without the comparative methods that give it memory. Two sequences are homologous — share an ancestor — and the evidence is statistical: an alignment scoring far above what chance shuffling produces (Figure 55.4). The scoring itself is an odds ratio: substitution matrices (PAM, BLOSUM) tabulate, from observed evolutionary exchanges, the log-odds that one residue replaces another by descent rather than accident, and BLAST, the workhorse of the field, wraps the search in extreme-value statistics so that every hit carries an expectation value — how many such scores the database would yield by chance alone. From alignment flows everything comparative: the phylogenies of Chapter 56, the annotation of every new genome by inheritance from the old, and the detection of distant homology — the histone fold, the kinase domain, the P-loop — that no eye would catch, the deep family resemblances of the protein world.

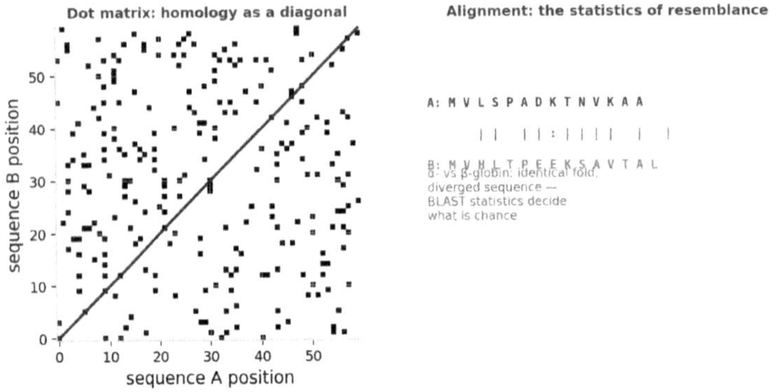

*Figure 55.4 — **Alignment and homology.** Left: a dot matrix, where shared descent appears as a diagonal (crimson) against chance matches. Right: α- versus β-globin — sequence diverged, fold conserved; the alignment's log-odds score and BLAST's expectation value decide what resemblance means.*

55.8 Structural Prediction

*For fifty years the protein-folding problem of Chapter 5 stood as biochemistry's Everest of computation: sequence alone should determine structure, yet no algorithm could climb from one to the other. The ascent, when it came, was by machine learning fed on the Protein Data Bank's two hundred thousand experimental structures — the library of Chapter 54.4 returning the investment. **AlphaFold**, whose principal architect's portrait closes this chapter, predicts atomic coordinates from sequence with a per-residue confidence (Figure 55.6), and at the 2020 assessment its predictions rivalled experiment for most single domains. The consequences are still unfolding: structures for every protein in every genome, pre-computed; complexes and their interfaces, increasingly; and — a quiet but profound gift — the confidence score itself, which flags intrinsically disordered regions as reliably as it flags folds: low confidence is biology, not failure. The honest caveats belong to this chapter's theme: the model learned the library's biases (ordered, crystallisable proteins), it does not predict pathways of* *folding or energies of binding, and its successes are interpolations over evolution's catalogue, not solutions of the physics. The Everest is climbed by cable car; the mountain's geology remains, in part, to be understood.*

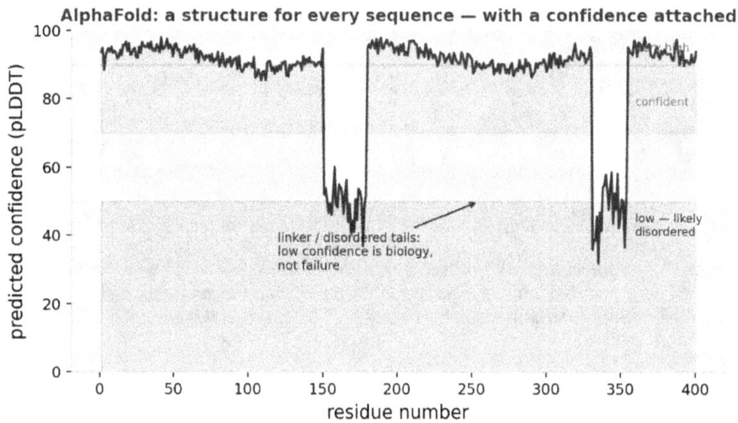

Figure 55.6 — Machine-learned structure prediction. *Per-residue confidence (pLDDT) along a 400-residue chain: ordered domains (green/amber bands) predicted at near-experimental accuracy; linkers and disordered tails (crimson troughs) honestly flagged as unpredictable — the model reports not only structure but its own uncertainty.*

55.9 Molecular Dynamics

Where prediction learns from examples, simulation computes from physics. ***Molecular dynamics*** *integrates Newton's equations for every atom of the molecule and its surrounding water, with forces from an empirical force field — bonds, angles, electrostatics, van der Waals — stepping in femtoseconds to accumulate trajectories of nanoseconds to milliseconds (Figure 55.5). What the trajectory buys is motion: conformational transitions between the basins of the energy landscape, the breathing of channels, the unbinding of drugs — the movie beneath Chapter 54's still photographs, and the computational realisation of Chapter 5's energy landscapes. Its limitations are honesty's too: the force fields are approximations, the timescales reachable are microseconds of a millisecond world, and the sampling of rare events demands enhanced methods (metadynamics, replica exchange) that bias the trajectory and then un-bias the statistics. Where simulation and measurement meet — NMR relaxation, FRET efficiencies, single-molecule forces — the agreement is now good enough that MD has become the interpreter of experiments, not merely their illustrator.*

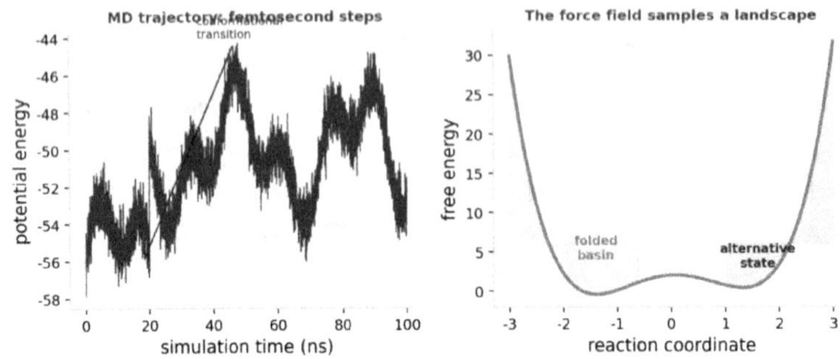

Figure 55.5 — **Molecular dynamics.** *Left: an energy trajectory over a hundred nanoseconds, with a conformational transition (crimson). Right: the landscape being sampled — folded basin (teal), alternative state (crimson) — Newton's equations in a force field, femtoseconds at a time.*

55.10 Enzyme-Kinetic Modelling

Between the stoichiometric matrix of Eq. (55.1) and the full atomism of MD lies the middle scale: kinetic models of pathways. The Michaelis–Menten parameters of Chapter 14, measured for each enzyme, become differential equations; the equations integrate to time courses and steady states; and the framework of

and design (which overexpression raises the engineered pathway's yield — the metabolic engineering of Chapter 56.13).

55.11 Network Reconstruction

Network reconstruction *is the upstream craft: assembling the reaction list itself from genome annotation, biochemistry and legacy knowledge — thousands of reactions, each with its evidence — the curated matrix on which all flux modelling stands. A reconstruction is a hypothesis about the cell's complete chemistry, and like all hypotheses of this book, it is only as good as its controls.*

55.12 Statistical Inference and Reproducibility

The omics multiply measurements twenty-thousand-fold, and statistics must multiply its vigilance to match. Test twenty thousand genes at $p < 0.05$ and a thousand will pass by chance: the **multiple-testing problem**, *answered by the false-discovery rate — accept that a declared fraction of the hits are false and control that fraction, rather than pretending none are. The* **volcano plot** *(Figure 55.2) states the trade honestly: effect size on one axis, evidence on the other, and both demanded — a large fold-change with thin evidence is as unconvincing as a tiny change with iron statistics. Beyond lie the deeper disciplines: batch effects that can masquerade as biology (samples processed on different days differ, whether or not the biology differs); overfitting, the model that memorises its training data and predicts nothing else, policed by cross-validation; and the reproducibility crisis itself, whose lessons are Part XI's whole message writ large: pre-register the analysis, report the replicates, deposit the data. The omic era did* *not relax Chapter 53's standards; it made them the difference between a census and a mirage.*

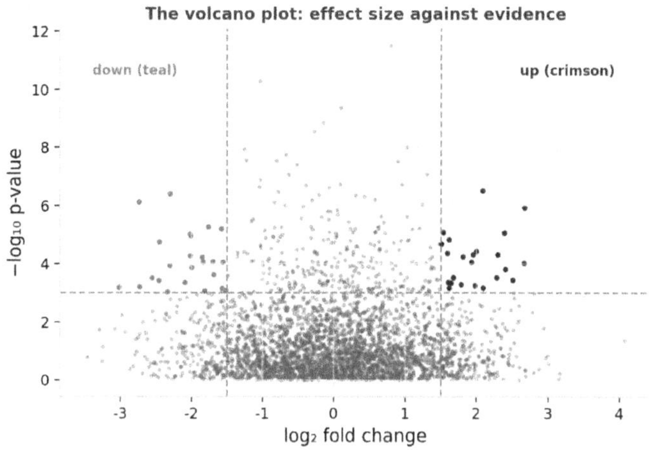

Figure 55.2 — The volcano plot. *Every gene plotted by effect size (fold change) against evidence ($-\log_{10}$ p): significant up-regulation (crimson), down-regulation (teal), and the vast unconvicted middle (slate). Both axes are demanded; the thresholds (amber) declare the false-discovery bargain.*

55.13 Machine Learning in Biochemistry

Machine learning has appeared twice already in this chapter — prediction of structure in Section 55.8, classification throughout — and it deserves its own accounting, because it is changing what kinds of questions can be asked. The methods learn mappings from examples: sequence to structure (AlphaFold and its successors), structure to function, gene-expression patterns to cell types, chemical structure to activity — the virtual screening that now precedes the drug screens of Chapter 56.10. Their strength is interpolation over enormous, high-dimensional data; their weakness is everything outside the data — extrapolation, mechanism, and the causal questions that Part X taught the reader to prize. A learned model may predict a kinase's substrates without knowing what a kinase is; the prediction may be excellent, and the understanding nil. The mature stance of the field, and of this book, is division of labour: let the machine find the pattern; let the experiment test it; and let the mechanism — the derived equation, the measured constant — remain the gold standard of explanation, to which the pattern is a well-funded clue.

55.14 *Case Study: IDH, 2-Hydroxyglutarate and the Oncometabolite*

> ***Case study — the metabolite that rewrote the genome.*** *In 2008–09, genome sequencing of gliomas and then of acute myeloid leukaemia — Section 55.1's census at work — found recurrent mutations in isocitrate dehydrogenase, IDH1 and IDH2, enzymes of the citric acid cycle of Chapter 34. The mutation was peculiar: not a loss but a gain — the mutant enzyme runs the normal reaction backwards in chemistry's sense, reducing α-ketoglutarate instead of oxidising isocitrate, and the product is 2-hydroxyglutarate, a*

metabolite present in traces in normal cells and in grams in the tumour. Metabolomics (Section 55.3) measured the flood: 2-HG levels a hundred-fold elevated, detectable even in patients' serum. The mechanism joined Parts IX and V: 2-HG resembles α-ketoglutarate closely enough to inhibit the enzymes that consume it — the TET DNA demethylases and the JmjC histone demethylases of Chapter 48 — so the tumour's epigenome froze in a hyper-methylated, undifferentiated state: a metabolic mutation causing cancer by poisoning epigenetic maintenance.

Analysis. *The discovery is Part XI's argument in one molecule. It was found by genomics (the census noticed the mutation), explained by metabolomics (the flood measured), mechanised by structural biology (the mutant active site, Chapter 54, showing how the neomorphic chemistry happens), and treated by medicinal chemistry: ivosidenib, an inhibitor selective for the mutant enzyme, lowers 2-HG, releases the differentiation block, and is now standard therapy in IDH-mutant leukaemia — a drug aimed not at killing the cell but at un-poisoning an epigenetic enzyme, designed because the flux was understood. Every method of this Part handled the molecule in turn; the omics did not replace the biochemistry — they told it where to dig.*

55.15 *Historical Note: AlphaFold and the Library That Learned*

Historical note. *The protein-folding problem was posed by Anfinsen's thermodynamic hypothesis of the 1960s — sequence determines structure — and for fifty years the biennial CASP competition recorded humanity's progress: honourable, incremental, and far from the goal. The breakthrough came not from a new physics but from a new statistics. DeepMind's AlphaFold, led by Demis Hassabis with John Jumper, entered CASP13 in 2018 and won; in CASP14, 2020, AlphaFold2's predictions were, for most targets, indistinguishable from experiment — a result the field's elders described, without embarrassment, as the problem solved for single domains. The method's soul was attention over evolution: multiple-sequence alignments of Section 55.7, whose residue covariations encode the contacts of the fold, read by a neural network trained on the Protein Data Bank — the accumulated labour of ten thousand crystallographers, NMR spectroscopists and microscopists, converted into a teacher. Within two years the predicted structures of essentially every known protein were published freely, and the 2024 Nobel Prize in Chemistry crowned Hassabis and Jumper, alongside David Baker for computational protein design — the folding problem run in reverse, sequences dreamed up for structures that never existed. The historical symmetry is the chapter's closing lesson: Chapter 54's measurements built the library; the library taught the machine; and the machine, in Plate 55.1, holds the medal.*

Plate 55.1 *Demis Hassabis (b. 1976), who with John Jumper led the creation of AlphaFold, solving the fifty-year problem of protein-structure prediction. Nobel laureate in Chemistry, 2024, with Jumper and David Baker. (Nobel Foundation portrait.)*

55.16 *Exercises*

Exercise 55.1* — *The multiple-testing bargain. *An RNA-seq experiment tests 20,000 genes. (a) At uncorrected $p < 0.05$, how many hits are expected under the complete null? (b) At a false-discovery rate of 5%, if 400 genes are declared significant, roughly how many are expected to be false? (c) Why is the FDR usually preferred to the Bonferroni correction in discovery experiments?*

Resolution.

candidate, ruinous for surveying twenty thousand. FDR controls the proportion of errors among the discoveries, matching the discovery experiment's actual question.

Exercise 55.2* — *Reading the label. *Cells are fed [U-$^{13}C_6$]glucose. Lactate is found 65% M+3; citrate is 40% M+2 and 12% M+0; glutamate is 30% M+2, 18% M+5. (a) What does lactate's M+3 fraction report? (b) What does citrate's M+0 fraction suggest? (c) Glutamate M+5 (labelled acetyl-CoA condensing with unlabelled α-KG is impossible here — M+5 requires a labelled 5-carbon skeleton): what pathway does M+5 glutamate imply?*

Resolution. *(a) M+3 lactate derives directly from labelled glucose through glycolysis; 65% means two-thirds of lactate production comes from glucose — the rest from unlabelled sources (glutamine, glycogen). (b) M+0 citrate indicates an unlabelled 2-carbon plus unlabelled 4-carbon contribution — fatty-acid-derived (unlabelled) acetyl-CoA or glutamine feeding oxaloacetate: the mitochondrion is not running on glucose alone. (c) M+5 glutamate arises when the whole 5-carbon skeleton is labelled — produced by reductive carboxylation or by a labelled cycle intermediate donating to α-KG; in ischaemic or mitochondrially damaged cells it is the signature of the cycle running backwards, as Section 55.5 noted. The mass spectrum, once again, reads the route.*

alance arithmetic. *A minimal network: reaction A ≤ 10), B converts it to biomass precursor with yield 2 o by-product. (a) Write the steady-state constraint at*

the real cell may reroute, attenuate, or die for reasons (redox balance, enzyme capacity, toxicity of accumulation) outside the matrix. The prediction is a hypothesis about optimality, and evolution, not the modeller, chooses the objective.

Exercise 55.4 — Confidence and its meaning. *AlphaFold predicts a 300-residue protein: residues 1–220 at pLDDT 93, 221–300 at pLDDT 45. (a) Interpret each region. (b) A crystallographer fails to crystallise the protein and blames the prediction; offer a likelier explanation and an experimental test.*

Resolution. *(a) The 1–220 domain is predicted at near-experimental confidence — a well-ordered fold; the C-terminal 80 residues are almost certainly disordered — the model reporting not low quality but low structure, per Section 55.8. (b) The likelier culprit is the disordered tail: disorder defeats crystallisation by providing conformational heterogeneity, the one thing the crystal lattice cannot tolerate (Chapter 54.4). The test: truncate the construct to residues 1–220 — guided by the prediction's own confidence map — and crystallise the ordered domain; this confidence-guided construct design is now standard practice, the library's student repaying the crystallographers in their own coin.*

Research proposition. *Eq. (55.1) constrains fluxes by stoichiometry alone; thermodynamics constrains them further (every flux must run downhill in free energy), and the omics of Sections 55.1–55.3 constrain the enzyme abundances. Propose a unified inference framework — constraint-based modelling with thermodynamic feasibility and proteomic capacity as additional linear and nonlinear constraints — and design a benchmark: a chemostat-grown microbe under a defined shift, with measured transcriptome, proteome, metabolome and ^{13}C fluxes, on which each added constraint's improvement in predictive accuracy can be scored. Discuss the diminishing returns of constraint stacking and whether a complete, self-consistent model of one bacterial cell is a decade's work or a century's.*

55.17 *Synthesis*

Part XI is complete. The omic cascade — genome to fluxome — measures everything and must be disciplined by everything Chapter 53 taught: controls, calibration, and the multiple-testing bargain of Figure 55.2. Isotope tracing and Eq. (55.1) convert the census into rates; alignment and prediction give the data memory and structure; simulation supplies the movies; machine learning supplies the patterns — and the experiment, always, supplies the verdict. IDH and its oncometabolite showed every instrument handling one molecule in turn; AlphaFold showed the library learning to answer back. The direction has been checked against the master plan: the book's final Part turns from measuring life to explaining its unity and variety — evolution, disease, and the engineering of biochemistry by human hands. ***Chapter 56 — The Unity, Variation and Engineering of Biochemistry*** *opens Part XII.*

55.18 *References*

Buescher, J. M., Antoniewicz, M. R., Boros, L. G., Burgess, S. C., Brunengraber, H., Clish, C. B., ... & Young, J. D. (2015). A roadmap for interpreting ^{13}C metabolite labeling patterns from cells. Current Opinion in Biotechnology, 34, 189–201.

Jumper, J., Evans, R., Pritzel, A., Green, T., Figurnov, M., Ronneberger, O., ... & Hassabis, D. (2021). Highly accurate protein structure prediction with AlphaFold. Nature, 596, 583–589.

Orth, J. D., Thiele, I., & Palsson, B. Ø. (2010). What is flux balance analysis? Nature Biotechnology, 28, 245–248.

Patti, G. J., Yanes, O., & Siuzdak, G. (2012). Metabolomics: The apogee of the omics trilogy. Nature Reviews Molecular Cell Biology, 13, 263–269.

Storey, J. D., & Tibshirani, R. (2003). Statistical significance for genomewide studies. Proceedings of the National Academy of Sciences, 100, 9440–9445.

Varadi, M., Anyango, S., Deshpande, M., Nair, S., Natassia, C., Yordanova, G., ... & Velankar, S. (2022). AlphaFold Protein Structure Database. Nucleic Acids Research, 50, D439–D441.

Wang, Z., Gerstein, M., & Snyder, M. (2009). RNA-Seq: A revolutionary tool for transcriptomics. Nature Reviews Genetics, 10, 57–63.

PART XII

Evolution, Disease and the Biochemical Future

Chapter 56
The Unity, Variation and Engineering of Biochemistry

Contents of this chapter

"Asked what the last fifty chapters amount to, the honest answer is: one chemistry, endlessly varied, recently inspected, and now — with suitable humility — beginning to be written by ourselves."

56.1 Chemical Evolution and the Origins of Metabolism

The final chapter gathers the whole book into a single view: where this chemistry came from, why it is everywhere the same, how it varies, how it fails, and what we have lately begun to do about it. Begin at the beginning. The early Earth had no biochemistry, but it had chemistry — energy gradients at hydrothermal vents, ultraviolet light, lightning, and a reducing atmosphere — and the question of origins is whether the monomers of Parts I–IV can arise without enzymes. The canonical demonstration remains the **Miller–Urey experiment** *of 1953 (Figure 56.1): methane, ammonia, hydrogen and water, circulated past a spark discharge, yielded within a week amino acids (glycine, alanine, aspartate), hydroxy acids, urea and hydrogen cyanide — the very alphabet of this book, from a flask. Modern versions with more realistic atmospheres, and the discovery of amino acids and nucleobases in carbonaceous meteorites, have generalised the conclusion: the monomers are cosmically easy. The hard problem is not the parts but the organisation — the concentration, polymerisation, and enclosure into protocells — and it remains, as Chapter 56.15 will concede, the book's one truly unfinished chapter.*[67]

67. The reader who has waited fifty-five chapters for the author to admit ignorance may mark this page. The origins of life is the one subject in the book where the literature offers more scenarios than constraints, and we shall be honest about which is which.

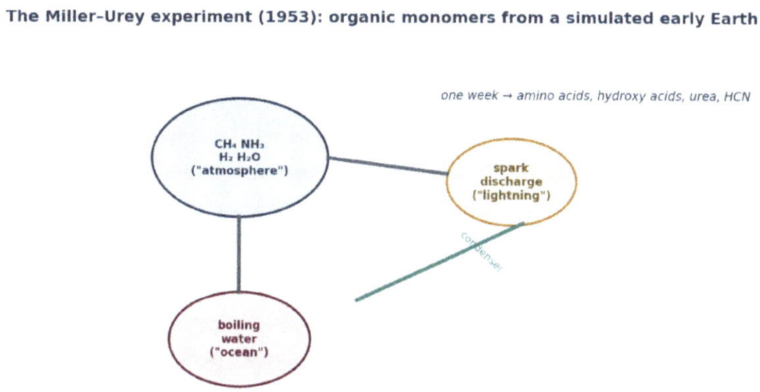

***Figure 56.1** — **The Miller–Urey experiment.** A simulated reducing atmosphere (navy) over a boiling "ocean" (crimson), sparked (amber), condensed and recycled (teal): within a week, amino acids, hydroxy acids, urea and HCN — the monomers of life from nothing but energy and small gases.*

56.2 The RNA-World Hypothesis

*Life's central chicken-and-egg problem is that DNA stores the information to make proteins, and proteins are needed to read DNA. The resolution, now doctrine, is that **RNA came first**, performing both offices (Figure 56.2): RNA stores sequence, and RNA — as Cech and Altman proved in the 1980s — catalyses chemistry, splicing itself without a single protein. The hypothesis predicts that the oldest, most central machinery of modern cells should still run on RNA, and the prediction is gloriously confirmed: the ribosome's catalytic heart is RNA (Chapter 47's plate: the ribosome is a ribozyme); tRNA and its adaptors are RNA; the spliceosome of Chapter 46 is RNA catalysis with protein chap*
mes — NAD, FAD, coenzyme A, ATP itself — are
, the metabolic fossils of Part IV. The transition
f: DNA, chemically stabilised by one missing hydroxyl, took over the archive; protein, with twenty side chains to RNA's four, took over the catalysis; and RNA was demoted to the indispensable middle — where, as Part IX showed at length, it governs still.

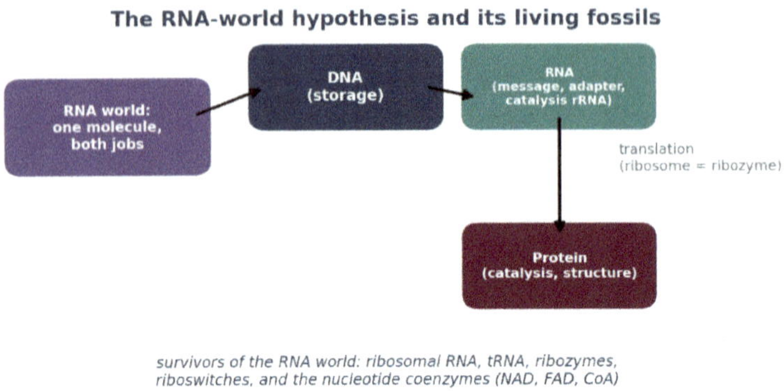

*Figure 56.2 — **The RNA-world hypothesis.** RNA once held both genotype and phenotype (plum); the modern flow (navy → teal → crimson) deposited RNA's survivors at every checkpoint: ribosomal RNA, tRNA, ribozymes, riboswitches, and the nucleotide coenzymes.*

56.3 Evolution of Proteins and Enzyme Families

*Protein evolution is conservative in structure and adventurous in sequence — the lesson of every alignment in Chapter 55.7. Folds persist for billions of years while sequences drift beyond recognition; the TIM barrel, the Rossmann fold, the kinase domain recur across hundreds of unrelated functions, because a good scaffold, like a good gearbox, is endlessly re-usable. New enzymes arise not from blank invention but from **promiscuity**: an existing enzyme's side-activity, negligible until the environment demands it, is amplified by selection into a new specialty — the catch being that specialisation usually begins as generalism refined. The enzyme commission's tidy classes conceal a geneology of opportunists: the same α/β hydrolase fold serves lipases, proteases and haloacid dehalogenases; the PLP enzymes of Chapter 41 catalyse a dozen different reaction types from one cofactor's electron sink. Evolution, the reader sees, is not an engineer but a tinkerer — and the tinkering explains both the unity (everything is related) and the mess (nothing was designed).*

56.4 Gene Duplication and Biochemical Innovation

*The tinkerer's principal tool is copying. A duplicated gene (Figure 56.3) liberates one copy from selection: the first copy keeps the old function, the second is free to accumulate mutations and, rarely but consequentially, to find a new one — **neofunctionalisation** — or the two divide the old job between them, **subfunctionalisation**. The book has met the dynasties everywhere: the globins, myoglobin and the α- and β-chains of Chapter 8, radiating from one ancestral oxygen-binder; the opsins of vision; the 518 kinases of Chapter 49, one kinase domain duplicated and re-wired five hundred times; the dehydrogenases of Part IV; the globin genes' own geography — α-cluster and β-cluster, embryonic, fetal and adult — recording a history of duplications in their very order on the chromosome. Whole-genome duplications punctuate the deeper past: vertebrates carry two rounds of it, and the extra spare parts, it is argued, paid for vertebrate* *complexity itself. Innovation by photocopier: not dignified, perhaps, but spectacularly effective.*

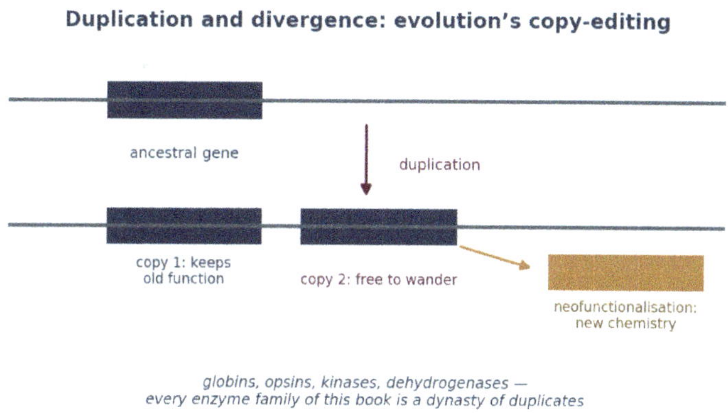

Figure 56.3 — Duplication and divergence. *The duplicated copy is freed from selection: one copy maintains function (navy), the other wanders sequence space until, rarely, a new function is caught (amber). Every enzyme family in this book is such a dynasty.*

56.5 Conserved Metabolic Architecture

Of all the book's unities, metabolism's is the most astonishing. The citric acid cycle, glycolysis, the pentose phosphate pathway, the respiratory chain: these run in a bacterium, a yeast, a worm, and the reader, with the same intermediates, the same cofactors, the same chemistries — Linus Pauling's "one metabolism" — assembled before the last universal common ancestor and frozen since, because the cost of re-engineering a pathway on which everything depends is death. The conservation is not perfect — fermentations vary, phototrophies differ, the archaea run variant glycolyses — but the variations are toppings on an invariant core, and the core is the strongest single piece of evidence for common descent that biology possesses. There is even a logic to what is conserved: the central pathways are conserved, the periphery diverse; the redox and energy carriers universal, the fuels and electron donors as various as the environments. One metabolism, many niches — the deep reason a biochemistry textbook can be written at all.

56.6 Molecular Evolution and Adaptation

Between the frozen core and the duplicated frontier lies the ordinary traffic of molecular evolution. Most sequence change is ***neutral*** *— Kimura's drift, the molecular clock that dates divergences by counting accumulated differences; some is* ***purifying*** *— the deletion of harmful variants, which is why catalytic residues are the most conserved positions in any alignment; and a sliver is* ***adaptive*** *— positive selection, detectable when replacements outpace silent changes, the statistical signature of an environment actively improving a protein. The textbook cases repay attention because they are biochemical through and through: lactase persistence in dairying populations, a regulatory mutation of Chapter 48's kind spreading in millennia; sickle-cell haemoglobin maintained by malaria's balancing selection — Chapter 8's pathology as evolution's price list; the high-altitude Tibetans' EPAS1 (HIF-2α) variant, tuning the hypoxia response of Chapter 51.6 to* *the thin air of the plateau. Adaptation, at the molecular level, is the environment reading this book and pencilling corrections in the margins.*

56.7 Cancer as Biochemical Evolution

The book's grimmest application of evolutionary logic is cancer, which is not a disease of cells misbehaving but of cells evolving (Figure 56.4). A tissue is a population; mutation and the Chapter 45 repair failures supply variation; the microenvironment supplies selection — for growth without permission (Chapter 50's stuck pathways), for survival without warrant (Chapter 51's disabled apoptosis), for immortality (telomerase), for blood supply (HIF and VEGF), and finally for invasion. Each successful clone expands and seeds the next round: the adenoma-to-carcinoma sequence of the colon, with its ordered losses of APC, KRAS-activation and TP53, is a phylogeny written in tissue. The therapeutic consequences are equally evolutionary. Treatment is a selective bottleneck; resistance is not an anomaly but the predicted survival of the fittest subclone — the imatinib gatekeeper of Chapter 50's exercises — and the rational countermeasures are evolutionary too: combination therapy to make escape improbable, adaptive dosing to spare the sensitive competitors that suppress the resistant ones. Oncology, properly viewed, is applied population genetics, and the tumour is the one organ that evolves in real time.

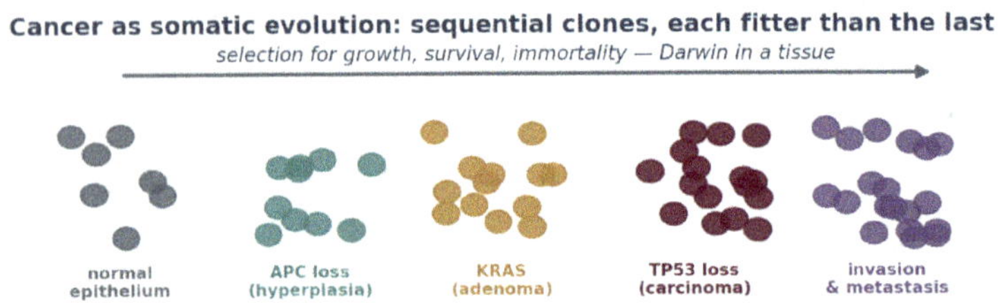

Figure 56.4 — Cancer as somatic evolution. *Sequential clones — normal (slate), hyperplasia (teal), adenoma (amber), carcinoma (crimson), invasive disease (plum) — each selected for a new capacity, each expanding to seed the next. Darwin's logic, run inside one tissue, within one lifetime.*

56.8 Neurodegeneration and Proteostasis

The second great failure mode is slower: not too much proliferation but too little folding. The proteostasis network of Chapter 5 — chaperones, the unfolded protein response of Chapter 51.6, the ubiquitin–proteasome system, autophagy — holds the proteome folded for a lifetime; in the long-lived neuron, where the cell is never replaced, its gradual failure is the disease. The amyloid cascade of Chapter 35 gave the molecular grammar: a protein of marginal stability (Aβ, α-synuclein, huntingtin with its expanded glutamines of Chapter 45's repeats, tau, TDP-43) nucleates into oligomers and fibrils; the aggregates template their own growth —

the prion principle, protein-only inheritance, biology's most alarming footnote to Chapter 4 — and the cell's quality control, overwhelmed, calls the apoptosis of Chapter 51. Alzheimer's, Parkinson's, Huntington's, the frontotemporal dementias, ALS: one grammar, different words. The evolutionary perspective of this chapter explains the timing: these are diseases of old age because selection, indifferent to what happens after reproduction, never paid for a proteostasis network warranted for ninety years. Medicine is now underwriting the warranty — anti-amyloid antibodies, antisense oligonucleotides, small molecules that stabilise the native fold (tafamidis in transthyretin amyloidosis, a folding drug pure and simple).

56.9 Host–Pathogen Biochemical Conflict

The third failure is warfare. Every layer of this book has an attacker and a defence, and the coevolution is an arms race measured in molecular detail. Bacteria invent β-lactamases to hydrolyse penicillins; medicinal chemists answer with clavulanate, a suicide inhibitor of the inhibitor's defence — and the bacteria answer that, in a dialogue now in its fourth generation of carbapenemases. The virus hijacks the replication machinery of Chapter 44 and the translation apparatus of Chapter 47; the cell counters with restriction factors, RNA interference (Chapter 48.9), and the interferon-JAK-STAT alarm of Chapter 50.4; the virus counters the counter — HIV's Vif degrading the APOBEC cytidine deaminase that mutates its genome. Even metabolism is a battlefield: the host sequesters iron and zinc (nutritional immunity), the pathogen deploys siderophores to steal them; the host's immune cells detonate the NADPH-oxidase oxidative burst of Chapter 32's chemistry, and intracellular bacteria evolve catalases to survive it. Biochemistry is the terrain on which all infectious disease is fought, which is why the pharmacopoeia of infection — antibiotics, antivirals, vaccines — is a list of molecular interventions in this book's pathways.

56.10 Drug Discovery and Chemical Biology

Against all three failure modes stands one discipline: the deliberate intervention in biochemistry with small molecules. Modern drug discovery is Part XI industrialised — target identified by genetics and omics, structure solved by Chapter 54's methods, hits found by screening libraries or computed in silico, optimised through the enthalpy–entropy accounting of Chapter 54.9 and the residence-time arithmetic of Chapter 54.8, and validated in the clinic where the network effects of Chapter 52 have the last word. The field's lessons are this book's lessons. Druggable targets are usually the nodes the network cannot route around: the stuck switches of cancer, the rate-limiting enzymes of cholesterol (statins at HMG-CoA reductase, Chapter 37), the proton pump of the stomach (Chapter 18's chemistry, tamed for dyspepsia and worth a fortune). The surprises are also the book's: phenotypic screening, which ignores mechanism and asks only whether the compound works, still delivers a large fraction of new medicines — empiricism's revenge on the rationalists — and the covalent drugs, from aspirin to the KRAS-G12C inhibitors of Chapter 50.2, succeed by breaking the reversibility every textbook assumed. Chemical biology, the discipline's scholarly twin, uses the same molecules as probes: a well-chosen inhibitor is the scalpel with which Chapter 49's pathways were dissected, and half the mechanisms in this book were established by poison before they were established by genetics.

56.11 Protein and Enzyme Engineering

If drugs modify proteins, engineering rewrites them. Two philosophies divide the field. ***Rational design*** *computes changes from structure — a cysteine placed for a disulfide, an active-site residue swapped to change specificity — and succeeds in proportion as Chapter 5's energetics are understood, which is to say, modestly.* ***Directed evolution*** *(Figure 56.5) surrenders understanding and borrows evolution's own algorithm: mutate the gene at random, screen the library for improved function, amplify the winners, repeat — ten or twenty rounds of laboratory selection producing enzymes with properties no designer would have proposed: catalytic antibodies and, more usefully, industrial enzymes stable in boiling solvents, polymerases faithful enough for forensics, proteases that wash dishes in cold water. Frances Arnold's Nobel of 2018 honoured the demonstration that evolution, not calculation, is the master protein engineer — a verdict the computational design of Section 55.8's successors (Baker's de novo proteins, enzymes designed from scratch) has lately begun to contest. The mature craft combines both: compute where computation is strong, evolve where it is weak, and let the sequencer of Chapter 55 read the winners.*

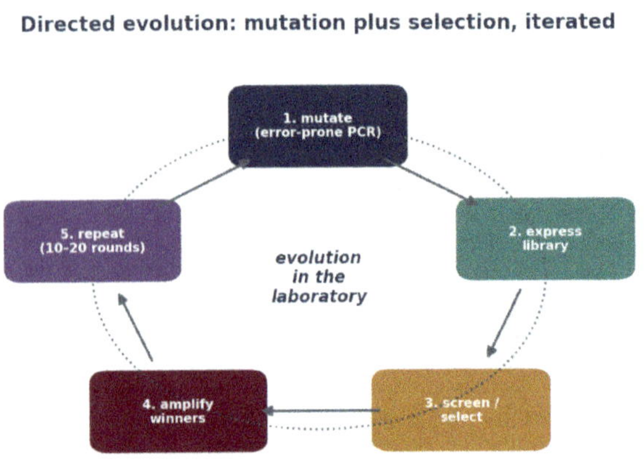

Figure 56.5 — Directed evolution. *Mutate, express, screen, amplify, repeat: evolution's algorithm run in the laboratory, producing enzymes no rational designer proposed. The method's 2018 Nobel honoured the discovery that selection out-computes calculation.*

56.12 Synthetic Biology

Engineering single proteins gives way to engineering systems — the motifs of Chapter 52 assembled by hand into organisms. The founding works have appeared already in this book: the repressilator and the toggle switch of Chapter 52.5, proving that circuits can be specified and built; the logic gates of Chapter 48.1 inverted into programmed gene control; and the field's flagship achievements since: bacteria that count, yeast that oscillates, mammalian cells with implanted decision circuits for therapy — a designer cell that senses a tumour marker and secretes its drug. The deepest project is the rewrite of the genome itself: the synthetic Mycoplasma, whose genome was designed on a computer, synthesised chemically, and booted up in an emptied cell — not life from scratch, as the headlines said, but the genome of life rewritten by hand, which is quite astonishing enough. And ***CRISPR****, the bacterial immune system of Chapter 45's repair pathways repurposed as a programmable nuclease, has made the editing of any genome — including the human germline, to the field's continuing ethical alarm — a routine procedure: the book's information molecules, edited with a word processor.*

56.13 Metabolic Engineering

The industrial arm of the enterprise redirects metabolism itself: the pathways of Parts V–VIII re-wired in microbes to manufacture what chemistry makes badly and agriculture makes slowly. The exemplar is ***artemisinin*** *(Figure 56.6): the antimalarial, extracted at cost from wormwood, was transplanted to yeast by importing the plant's amorphadiene synthase and a P450 oxidase onto the native mevalonate pathway — the host's glycolysis feeding a factory it never evolved — yielding artemisinic acid by fermentation at industrial scale. The discipline's principles are this book's principles, applied: pull flux by removing competing sinks (Chapter 30's control analysis computed, not intuited), push flux by overexpressing the rate-limiting enzyme, balance the redox books (NADPH must be paid for, Chapter 32), and relieve the feedbacks of Chapter 14 that the pathway's own products impose. Insulin from bacteria, artemisinin from yeast, nylon precursors from sugar, cultured meat from stem cells: the metabolic map of Part V has become an engineering drawing, and the cell — as this book has argued from its first chapter — was always a factory; we have merely been promoted from apprentices to junior management.*

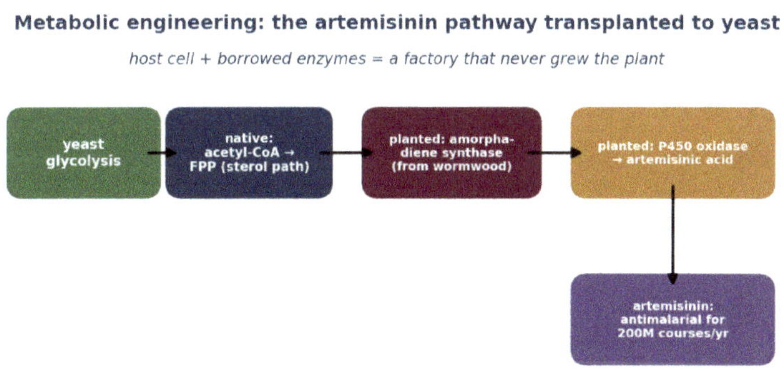

Figure 56.6 — Metabolic engineering. *Yeast's glycolysis (green) feeds its native mevalonate pathway (navy); two borrowed plant enzymes (crimson, amber) divert the flux to artemisinic acid and thence artemisinin (plum) — a pharmaceutical plant that grows in a fermenter.*

56.14 Biochemistry beyond Earth

The book's chemistry is Earth's; whether it is the universe's is the question of astrobiology, and the evidence is tantalisingly two-sided. On one side, the unity of Sections 56.1–56.5 suggests contingency: the genetic code's assignments, the choice of twenty amino acids, the L-chirality of proteins and D-chirality of sugars may be frozen accidents — an alien biochemistry might choose otherwise, and the search for life elsewhere watches accordingly for agnostic biosignatures (chemical disequilibrium, complex organics) rather than for our own molecules. On the other side, the meteoritic amino acids of Section 56.1, the polycyclic aromatics of interstellar clouds, and the ribose found in meteorites argue that the raw materials, at least, are universal; and chemistry's constraints — carbon's unmatched versatility, water's unmatched solvent properties of Chapter 2 — may funnel any origin toward something recognisable. The discovery of one independent origin — beneath the ice of Europa, in the methane lakes of Titan, in a Martian aquifer — would settle the book's deepest open question in an afternoon: whether this chemistry is a local dialect or the only language in which matter, energy and information can be made to live.

56.15 The Unfinished Biochemical Description of Life

And so the inventory of ignorance, with which the book properly closes. We do not know how life began: the gap between Miller's flask and the protocell remains the grand canyon of the field. We do not fully understand any single cell: even E. coli harbour dreds of proteins of unknown function, and the unified model of Chapter 's research proposition does not yet exist. The dark metabolome of Chapter the dark proteome's disordered half, the glycome, the microbiome's chemistry coursing through Chapter 41's hosts — all are mapped incompletely. Ageing, the slow failure of everything in this book at once, has candidate mechanisms and no agreed account. Consciousness, if it is a biochemical question at all, is not yet a well-posed one. And the engineering of Chapter 56.11–56.13 has outrun the understanding it exploits, as engineering always does — the Romans built arches without elasticity theory, and we edit genomes without a complete parts list. None of this is grounds for disappointment. A finished science is a dead one; the state of biochemistry in the reader's hands is precisely the opposite — a discipline that has explained enough to know, in quantitative detail, how much it has left to explain. The next chapters are not in this book. They are, if the preceding fifty-six have done their work, in the reader.

56.16 Case Study: Artemisinin — from Herb to Factory

> ***Case study — two Nobels, one molecule.*** *The artemisinin story binds this chapter's themes into a single thread. In 1967, Project 523 — a Chinese military programme seeking antimalarials as resistance consumed chloroquine — screened traditional remedies; Tu Youyou's team found that the herb Artemisia annua, described in a fourth-century text for fevers, worked only when extracted cold (the text said "wring the juice," not boil — the molecule is thermolabile), and isolated artemisinin in 1972: a sesquiterpene with an endoperoxide bridge, activated by haem iron in the parasite's food vacuole to carbon radicals that alkylate the parasite's proteins — Chapter 26's radical chemistry, weaponised. Tu's Nobel came in 2015. Thirty years later, with the drug's supply hostage to a crop, Jay Keasling's group executed Section 56.13: the plant's biosynthetic genes transplanted into yeast, the flux pulled by deleting the competing sterol sink, the redox balanced, and artemisinic acid produced by fermentation — semi-synthetic artemisinin reaching the market from a brewery rather than a field.*
>
> ***Analysis.*** *Every discipline of the book handled the molecule: natural-product chemistry (the cold extraction), mechanism (the iron-activated peroxide), the evolutionary logic of resistance (artemisinin now deployed only in combinations, per Section 56.7's bottleneck arithmetic), and finally metabolic engineering (the pathway re-housed). The molecule's journey — herb, mechanism, medicine, factory — is the journey the whole book describes: from*

what life is, to what we can measure, to what we can make. That both the herbal discovery and the engineering triumph earned the field's highest honours is not a coincidence but a diagnosis: the future of medicine lies equally in reading nature's library and in writing, carefully, in its margins.

56.17 Historical Note: Cech's RNA That Cut Itself

Historical note. *The RNA world of Section 56.2 was pure speculation until, in 1982, Thomas Cech at Colorado found something the central dogma said could not exist. Working on the ribosomal RNA of the ciliate Tetrahymena, his laboratory purified the precursor RNA to apparent homogeneity and watched an intron excise itself — in a test tube, with no protein present, the RNA cutting and splicing its own chain. The controls were relentless, because the claim was enormous: every protease-sensitive activity could be destroyed, every preparation cleaned, and still the RNA spliced. The conclusion — the RNA itself is the enzyme — coined the word "ribozyme" and, at a stroke, gave the RNA-world hypothesis its mechanism: if RNA can catalyse, then RNA alone could have been both genome and metabolism before proteins existed. Sidney Altman's RNase P, an RNA enzyme in bacteria, confirmed the generality, and the two shared the 1989 Nobel Prize in Chemistry. The discovery's aftershocks ran for decades: the ribosome shown to be a ribozyme (Chapter 47), the spliceosome likewise (Chapter 46), catalytic RNA selected in the laboratory to perform new chemistries — and, in the end, the quiet reinterpretation of the whole of Part IX: the cell is not a protein machine that happens to contain RNA, but an RNA world that learned to wear protein.*

Plate 56.1 *Thomas R. Cech (b. 1947), discoverer of catalytic RNA — the ribozyme — whose existence made the RNA world a mechanism rather than a speculation.*

Nobel laureate in Chemistry, 1989, with Sidney Altman. (Nobel Foundation portrait.)

56.18 Exercises

Exercise 56.1 — The clock and the calibration. *Two species differ at 42 positions in a 700-residue conserved enzyme; the synonymous sites differ at 18% of positions. (a) If synonymous divergence accumulates at ~1% per million years, estimate the divergence time. (b) Why is the synonymous rate, not the replacement rate, used for dating? (c) The replacement divergence is 6%; what does the ratio of the two rates imply about selection on this enzyme?*

Resolution. *(a) 18% divergence at 1%/Myr ≈ 18 million years (roughly, correcting for multiple hits pushes it a little higher). (b) Synonymous changes are nearly neutral — they accumulate by drift at a steady rate set by mutation, Kimura's clock; replacement changes are filtered by selection at rates that vary with function and epoch, making them unreliable chronometers. (c) Replacement divergence (6%) is one-third of synonymous (18%): most mutations that would change the protein have been rejected — strong purifying selection, the signature of Section 56.6 and the reason the enzyme of this example is recognisable across phyla at all.*

Exercise 56.2 — Resistance arithmetic, again. *A tumour contains 10^9 cells; a pre-existing resistant subclone exists at frequency 10^{-7}. (a) How many resistant cells exist before treatment? (b) Treatment kills 99.99% of sensitive cells and none of the resistant; what is the new resistant fraction? (c) Two independent-acting drugs, each facing resistance at 10^{-7}, are combined; what is the expected frequency of doubly resistant cells, and what does this teach about combination therapy?*

Resolution.

≈

to oncology. Evolution is inevitable; combinatorics is the counter-weapon.

Exercise 56.3 — Directed evolution's odds. *A library of 10^8 enzyme variants is screened; the wild-type activity is 1.0 unit, and the screen's detection threshold is 1.5 units. After round 1 the best variant shows 2 units; after 12 rounds, 300 units. (a) What is the average fold-improvement per round? (b) Why does directed evolution succeed where a single leap to a 300-fold mutant would fail?*

Resolution.
round — ea
Because sequence space is combinatorial: a simultaneous set of, say, 8 specific mutations among 20^8 possibilities is unreachable in one step, but each single mutation that helps is common enough to find, fix, and build on — Section 56.4's tinkering algorithm exactly: many small climbs, no cliffs. The screen,

like natural selection, converts improbable destinations into probable journeys.

Exercise 56.4 — The redox bill. *A yeast is engineered to make a reduced product requiring 2 NADPH per product molecule from glucose. Glucose oxidation through the pentose phosphate pathway yields 2 NADPH per glucose but no ATP; glycolysis yields 2 ATP but no NADPH. If the cell needs 4 ATP per product for polymerisation and maintenance, what is the minimal glucose cost per product, and which pathways supply it?*

Resolution. *NADPH bill: 2 per product → 1 glucose through the oxidative PPP (2 NADPH). ATP bill: 4 per product → 2 glucoses through glycolysis (2 ATP each). Total: 3 glucose per product minimum — one for reductant, two for energy — assuming perfect coupling and no losses. The exercise is Chapter 32's bookkeeping in engineering dress, and it is the standard first calculation of metabolic design: the product's theoretical yield is set not by the enzymes but by the cofactor bills, which is why Section 56.13's redox balancing is the discipline's daily bread.*

Research proposition. *Section 56.14 posed the deepest open question: is our biochemistry contingent or necessary? Propose a research programme in comparative xenobiochemistry — laboratory evolution of microbes on alternative redox couples and solvents' analogues, systematic survey of which of the twenty amino acids are load-bearing and which replaceable by non-canonical analogues across a proteome, and telescopic/in-situ biosignature criteria that do not assume terrestrial chemistry. Discuss what a second, independent origin — even a single alien microbe using the same code — would imply for each section of this chapter, and whether the book you have just read would need revision or merely translation.*

56.19 Synthesis

The book closes where it began: with the claim that life is chemistry, understood. The chemistry came from an energy-rich young planet; its alphabet arose easily, its organisation laboriously; RNA served as the first government; proteins and DNA divided the offices between them; and for four billion years the system has varied by duplication, drift and selection around a conserved metabolic core — the unity of Section 56.5 beneath the diversity of everything else. Disease, in this view, is the chemistry evolving against us (cancer), decaying faster than its warranty (neurodegeneration), or attacked by other chemistry (infection); and the newest chapter of the story is our own entry into it — measuring by Part XI, intervening by drugs, rewriting by engineering, and exporting the questions beyond the planet. The direction has been checked against the master plan, and the plan is complete: matter, energy, information, and the life that binds them, from the water molecule to the engineered cell. The End Matter that follows — appendices of constants, structures and pathways, answers, further reading, and indices — is the workshop manual for the fifty-six chapters now in the reader's hands. Use it well.

56.20 References

Arnold, F. H. (2018). Directed evolution: Bringing new chemistry to life. Angewandte Chemie International Edition, 58, 14420–14426.

Cech, T. R. (1986). A model for the RNA-catalyzed replication of RNA. Proceedings of the National Academy of Sciences, 83, 4360–4363.

Gilbert, W. (1986). Origin of life: The RNA world. Nature, 319, 618.

Kimura, M. (1983). The Neutral Theory of Molecular Evolution. Cambridge University Press.

Miller, S. L. (1953). A production of amino acids under possible primitive Earth conditions. Science, 117, 528–529.

Nowell, P. C. (1976). The clonal evolution of tumor cell populations. Science, 194, 23–28.

Ohno, S. (1970). Evolution by Gene Duplication. Springer.

Paddon, C. J., & Keasling, J. D. (2014). Semi-synthetic artemisinin: A model for the use of synthetic biology in pharmaceutical development. Nature Reviews Microbiology, 12, 355–367.

Tu, Y. (2011). The discovery of artemisinin (qinghaosu) and gifts from Chinese medicine. Nature Medicine, 17, 1217–1220.

End Matter
Appendices, Answers, Further Reading and Indices

CONTENTS OF THE END MATTER

"The appendices of a treatise are its toolbox: not read, but reached for — and the reach should never be long."

A Appendix A — SI Units, Constants and Conversion Factors

Table A.1 — Fundamental constants used throughout the book.

Quantity	Symbol	Value
Avogadro constant	N_A	6.0221×10^{23} mol^{-1}
Gas constant	R	8.3145 J·mol^{-1}·K^{-1} (= 1.9872 cal·mol^{-1}·K^{-1})
Faraday constant	F	96 485 C·mol^{-1}
Boltzmann constant	k_B	1.3806×10^{-23} J·K^{-1}
Planck constant	h	6.6261×10^{-34} J·s
Elementary charge	e	1.6022×10^{-19} C
RT at 298.15 K	RT	2.479 kJ·mol^{-1} (0.592 kcal·mol^{-1})
RT ln 10 at 298.15 K	2.303 RT	5.708 kJ·mol^{-1}
Standard temperature	T	298.15 K (25 °C); physiology 310.15 K (37 °C)

Table A.2 — Energy and concentration conversions.

From	To	Multiply by
1 kcal·mol^{-1}	kJ·mol^{-1}	4.184
1 kJ·mol^{-1}	kcal·mol^{-1}	0.2390
1 mV (membrane potential, per mole of charge)	kJ·mol^{-1}	0.0965 ($F \times 10^{-3}$)
$\Delta G°'$ (kJ·mol^{-1})	K'_{eq} at 25 °C	$K' = \exp(-\Delta G°'/2.479)$
1 dalton (Da)	g·mol^{-1}	1 (numerically identical)
1 Å	nm	0.1

B Appendix B — The Twenty Standard Amino Acids

***Table B.1** — Properties of the standard amino acids (pKa values at 25 °C).*

Amino acid	3-letter	1-letter	Side chain	pKa (side chain)	M_r (residue)
Glycine	Gly	G	—H	—	57
Alanine	Ala	A	$—CH_3$	—	71
Valine	Val	V	isopropyl	—	99
Leucine	Leu	L	isobutyl	—	113
Isoleucine	Ile	I	sec-butyl	—	113
Methionine	Met	M	$—CH_2CH_2SCH_3$	—	131
Proline	Pro	P	pyrrolidine (secondary amine)	—	97
Phenylalanine	Phe	F	benzyl	—	147
Tyrosine	Tyr	Y	p-hydroxybenzyl	10.1	163
Tryptophan	Trp	W	indolylmethyl	—	186
Serine	Ser	S	$—CH_2OH$	—	87
Threonine	Thr	T	$—CH(OH)CH_3$	—	101
Cysteine	Cys	C	$—CH_2SH$	8.3	103
Asparagine	Asn	N	$—CH_2CONH_2$	—	114
Glutamine	Gln	Q	$—CH_2CH_2CONH_2$	—	128
Aspartate	Asp	D	$—CH_2COO^-$	3.9	115
Glutamate	Glu	E	$—CH_2CH_2COO^-$	4.2	129
Lysine	Lys	K	$—(CH_2)_4NH_3^+$	10.5	128
Arginine	Arg	R	guanidinium	12.5	156
Histidine	His	H	imidazole	6.0	137

The α-carboxyl group has pKa ≈ 2.2 and the α-amino group pKa ≈ 9.4 in nearly all cases; the isoelectric point of a non-ionisable-side-chain amino acid lies midway between them (≈ 5.5–6.0). Histidine's side-chain pKa of 6.0 makes it the only residue titratable near physiological pH — the reason it dominates catalytic dyads and triads throughout Part III.

C Appendix C — Nucleobases, Nucleosides and Nucleotides

***Table C.1** — Nomenclature of the nucleic-acid building blocks.*

Base	Type	Nucleoside (RNA)	Nucleoside (DNA)	Nucleotide (5′-monophosphate)
Adenine (A)	purine	adenosine	2′-deoxyadenosine	AMP / dAMP
Guanine (G)	purine	guanosine	2′-deoxyguanosine	GMP / dGMP
Cytosine (C)	pyrimidine	cytidine	2′-deoxycytidine	CMP / dCMP

Uracil (U)	*pyrimidine*	*uridine*	—	*UMP*
Thymine (T)	*pyrimidine*	*(ribosylthymine)*	*thymidine*	*dTMP*

Purines pair with pyrimidines by Watson–Crick geometry: A=T (two hydrogen bonds), G≡C (three); in RNA, U replaces T. The glycosidic bond is N9 (purines) or N1 (pyrimidines) to the sugar's C1'; the phosphodiester backbone links 3'-O to 5'-O, giving every chain its 5'→3' polarity on which Part IX depends. Modified bases — 5-methylcytosine (the epigenetic mark of Chapter 48), pseudouridine and inosine in tRNA, N6-methyladenosine in mRNA — extend the alphabet without changing the rules of spelling.

D Appendix D — Major Coenzymes and Vitamin Precursors

Table D.1 — Coenzymes, their vitamin origins, and their chemistry.

Coenzyme	*Vitamin*	*Chemical office*	*Deficiency disease*
NAD^+ / $NADP^+$	*niacin (B_3)*	*hydride transfer (oxidation)*	*pellagra*
FAD / FMN	*riboflavin (B_2)*	*hydride and electron transfer*	*ariboflavinosis*
Thiamine pyrophosphate	*thiamine (B_1)*	*aldehyde transfer; decarboxylation*	*beriberi*
Pyridoxal phosphate	*pyridoxine (B_6)*	*amino-group chemistry; transamination*	*anaemia, neuropathy*
Coenzyme A	*pantothenate (B_5)*	*acyl transfer (thioester)*	*rare (burning-feet syndrome)*
Biotin	*biotin (B_7)*	*carboxylation (CO_2 carrier)*	*dermatitis (raw-egg-white avidin)*
Tetrahydrofolate	*folate (B_9)*	*one-carbon transfer*	*megaloblastic anaemia; neural-tube defects*
Cobalamin (B_{12})	*cobalamin*	*methyl transfer; radical rearrangements*	*pernicious anaemia; neuropathy*
Lipoamide	*(synthesised)*	*acyl + electron swing-arm*	—
Ascorbate	*vitamin C*	*reductant; hydroxylase cofactor*	*scurvy*
Vitamin K	*phylloquinone*	*γ-carboxylation of glutamate*	*bleeding*
Ubiquinone (Q)	*(synthesised)*	*membrane electron carrier*	—

E Appendix E — Standard Free Energies and Reduction Potentials

Table E.1 — Standard transformed free energies of hydrolysis (pH 7, 25 °C).

Compound	$\Delta G^{\circ\prime}$ of hydrolysis ($kJ \cdot mol^{-1}$)
Phosphoenolpyruvate	−61.9
Carbamoyl phosphate	−51.4
1,3-Bisphosphoglycerate	−49.4
Phosphocreatine	−43.1
Pyrophosphate (PPi)	−33.5
ATP (→ ADP + Pi)	−30.5
Glucose-1-phosphate	−20.9
Glucose-6-phosphate	−13.8
Glycerol-3-phosphate	−9.2

Table E.2 — Standard reduction potentials of biological redox couples (pH 7, 25 °C).

Couple	$E^{\circ\prime}$ (mV)
$NAD^+/NADH$	−320
$NADP^+/NADPH$	−324
$FAD/FADH_2$ (enzyme-bound)	≈ 0
Oxaloacetate/malate	−166
Ubiquinone Q/QH_2	+70
Cytochrome c (Fe^{3+}/Fe^{2+})	+235
Cytochrome a_3 (Fe^{3+}/Fe^{2+})	+385
$\frac{1}{2}O_2/H_2O$	+815

The usable arithmetic of Part IV: ?G°? = ?nF?E°?, so the full span from NADH to oxygen (? 1.14 V) yields ?220 kJ·mol?^{1} per NADH ? the fund from which the proton motive force of Chapter 28 draws its three or so ATP. A ten-fold concentration ratio across a membrane costs RT ln 10 ? 5.7 kJ·mol?^{1} per mole, or 61.5 mV per charge at 37 °C (59.2 mV at 25 °C; 58.2 mV at 20 °C).

F Appendix F — Principal Metabolic Pathways at a Glance

Table F.1 — The core pathways: location, function, and net yield.

Pathway	Location	Input → Output	Net ATP (per glucose or turn)
Glycolysis	cytosol	glucose → 2 pyruvate	+2 ATP, +2 NADH
Gluconeogenesis	cytosol/ER/mitochondrion	2 pyruvate → glucose	−6 ATP equivalents

Pentose phosphate	*cytosol*	*G6P → ribose-5P + CO_2*	*+2 NADPH (oxidative arm)*
Citric acid cycle	*mitochondrial matrix*	*acetyl-CoA → 2 CO_2*	*3 NADH, 1 $FADH_2$, 1 GTP*
Oxidative phosphorylation	*inner membrane*	*NADH + O_2 → H_2O*	*≈ 2.5 ATP/NADH; 1.5/$FADH_2$*
β-Oxidation	*mitochondrial matrix*	*C_{16} palmitate → 8 acetyl-CoA*	*≈ 106 ATP net*
Fatty-acid synthesis	*cytosol*	*8 acetyl-CoA → palmitate*	*−7 ATP, −14 NADPH*
Urea cycle	*matrix + cytosol*	*2 NH_3 + CO_2 → urea*	*−4 ~P bonds*
Glycogen synthesis/breakdown	*cytosol*	*G1P ⇄ glycogen*	*−1 ATP (synthesis, via UDP-glucose)*
Ketogenesis	*liver mitochondrion*	*2 acetyl-CoA → acetoacetate*	*fuel export in fasting*

G Appendix G — Common Biochemical Methods and Diagnostic Markers

Table G.1 — Clinical chemistry's everyday markers and their meaning.

Marker	Reports	Raised in
Troponin I/T	*cardiomyocyte necrosis*	*myocardial infarction*
ALT / AST	*hepatocellular injury*	*hepatitis, NAFLD, toxins*
Creatinine (and eGFR)	*renal filtration*	*renal failure*
Lactate	*anaerobic glycolysis / hypoxia*	*sepsis, shock, mitochondrial disease*
Anion gap	*unmeasured anions*	*DKA, lactic acidosis, poisonings*
β-Hydroxybutyrate	*ketogenesis*	*fasting, DKA*
HbA1c	*3-month mean glycaemia*	*diabetes mellitus*
CRP	*inflammation*	*infection, autoimmune flares*
Ammonia	*urea-cycle function*	*liver failure, urea-cycle defects*
ALP / GGT	*cholestasis*	*biliary obstruction*

H Appendix H — Statistical and Kinetic Formulae

Table H.1 — The book's working formulae, gathered.

Name	Form	Where derived
Michaelis–Menten	$v = V_{max}[S]/(K_m + [S])$	*Chapter 14*
Competitive inhibition	apparent $K_m \times (1 + [I]/K_i)$	*Chapter 14*
Hill equation	$\theta = [L]^n/(K + [L]^n)$	*Chapters 8, 49*
Henderson–Hasselbalch	$pH = pKa + \log([A^-]/[HA])$	*Chapters 42, 53*
Binding occupancy	$\theta = [L]/([L] + K_d)$	*Eq. (49.1)*
Nernst equation	$E = E^{\circ\prime} - (RT/nF) \ln Q$	*Chapter 18*

Cascade amplification	$A = \Pi g_i = \Pi k_i\tau_i$	*Eq. (49.2)*
Flux steady state	$S \cdot v = 0$	*Eq. (55.1)*
Bragg's law	$n\lambda = 2d \sin\theta$	*Eq. (54.1)*
Beer–Lambert	$A = \varepsilon cl$	*Eq. (53.3)*
Svedberg	$s = m(1-\bar{v}\rho)/f$	*Eq. (53.2)*
Exponential survival	$N/N_0 = e^{(-kt)}$	*Chapter 49*
Poisson (counting noise)	$\sigma/\mu = 1/\sqrt{N}$	*Chapter 53*
Fisher's FDR (Benjamini–Hochberg)	accept $p_{(i)} \leq i \cdot q/m$	*Chapter 55*

I Answers to Selected Quantitative Problems

Full resolutions appear in the chapters themselves; this section recapitulates the numerical answers for the reader revising against the clock. ***Chapter 14*** *(enzyme kinetics): competitive inhibitor doubles apparent K_m, V_{max} unchanged.* ***Chapter 18****: 59 mV per decade per charge at 25 °C; the proton motive force of ~220 mV stores ≈ 21 kJ·mol^{-1}.* ***Chapter 28:*** *≈ 2.5 ATP per NADH, 1.5 per $FADH_2$; palmitate β-oxidation ≈ 106 ATP net.* ***Chapter 42****: anion gap = $Na^+ - (Cl^- + HCO_3^-) \approx 12 \pm 4$; in DKA it rises as ketones displace bicarbonate.* ***Chapter 49****: occupancy 9% → 91% requires an 81-fold concentration rise; amplification $10 \times 10^3 \times 10^2 = 10^6$; GTPase with $k = 0.05\ s^{-1}$ leaves 5% active after 60 s.* ***Chapter 50****: imatinib occupancy 97% (wild type) vs 13% (T315I) at 3 µM.* ***Chapter 52:*** *fixed points at 0, 0.38, 2.62 — stable, unstable, stable; repressilator period ≈ $2N\tau$.* ***Chapter 53****: $\Delta c = \Delta A/\varepsilon l$ = 0.0402 mM for ΔA = 0.25 at 340 nm; buffer drift 0.014 pH units for 0.5 mM H^+ into 50 mM phosphate at pH 7.4.* ***Chapter 54****: d = 3.70 Å at $2\theta = 24°$; K_d = 20 nM from $k_{off}/k_{on} = 4\times10^{-3}/2\times10^5$.* ***Chapter 55****: 1,000 false positives expected among 20,000 genes at $p < 0.05$ uncorrected; FBA optimum $v_B = 10$, $v_C = 0$.* ***Chapter 56****: double resistance ≈ 10^{-14}; directed evolution's mean gain 1.61-fold per round over 12 rounds.*

J Further Reading

The chapters' references give the primary literature; the following books are the shelf the treatise itself consulted most often, and the shelf to which the reader is now commended. For the whole field: Alberts et al., Molecular Biology of the Cell (6th ed.); Berg, Tymoczko, Gatto & Stryer, Biochemistry (9th ed.); Voet & Voet, Biochemistry (5th ed.). For physical foundations: Dill & Bromberg, Molecular Driving Forces; Nelson, Biological Physics. For metabolism: Fell, Understanding the Control of Metabolism; Frayn, Metabolic Regulation. For information: Ptashne, A Genetic Switch; Watson et al., Molecular Biology of the Gene (7th ed.). For signalling and networks: Lim, Mayer & Pawson, Cell Signaling; Alon, An Introduction to Systems Biology. For methods: Wilson & Walker, Principles and Techniques of Biochemistry (8th ed.); Cantor & Schimmel, Biophysical Chemistry. For the deep questions of Part XII: Lane, The Vital Question; Judson, The Eighth Day of Creation; and Monod, Chance and Necessity — still, after half a century, the book this one would most have liked to be.

Conclusion
Of Matter, Energy, Information and Life

A treatise should end by saying what it has said. This one began with a water molecule and ended with an engineered cell, and its argument, compressed into a paragraph, is this. Life is not a substance but a pattern: matter organised by information into machines that harvest energy to maintain that organisation against the universal drift toward disorder. Every level of the book is the same claim at a different magnification. The water molecule's hydrogen bonds (Part I) set the stage; the folded protein's energy landscape (Part II) makes structure from sequence; catalysis (Part III) makes chemistry fast and specific; the proton and the electron (Part IV) make energy storable; the pathways of metabolism (Parts V–VIII) make it spendable; the nucleic acids and their government (Part IX) make the organisation heritable; signalling and networks (Part X) make it responsive; measurement and computation (Part XI) make it knowable; and evolution (Part XII) makes it — slowly, wastefully, magnificently — itself.

Three convictions have governed the writing, and the reader who has reached this page will recognise them as old companions. The first is that nothing should fall from the sky: every equation in the book was derived from a premise the reader was shown, because a science that cannot be re-derived is a scripture, and biochemistry is not a scripture. The second is that the methods are part of the science: every fact herein was bought by a measurement, and the honest price — controls, calibration, uncertainty — was quoted alongside the fact, for a number without its error is an opinion in disguise. The third is that irony is a form of respect: the reader was never talked down to, never spared a difficulty, and never denied a joke — because the material is difficult, and difficulty shared with good humour is difficulty halved.

What remains? Chapter 56.15 gave the honest inventory: the origin of life unsolved, no cell fully understood, the dark metabolome unmapped, ageing unexplained, and engineering outrunning understanding. But the deeper remainder is not a list of problems; it is a change in the reader. You came to this book, perhaps, thinking of biochemistry as a subject — a syllabus, an examination, a shelf of pathways to be memorised. It is not a subject. It is the operating manual of the only machine in the universe known to build copies of itself, and you are one of its instances. The fifty-six chapters were written so that the manual could be read, by a patient person, from the first principles to the frontiers — and so that somewhere in a la or a clinic or a library, a reader closing this volume might feel what its feel every working day: that the cell, inspected closely enough, is the best atter has ever told, and that we are privileged to be among its tellers. The book is finished. The story, thank heaven, is not.

Index of Words
Index of Biochemical Terms

Numbers refer to chapters; Eq. to the numbered equations; Fig. to the numbered figures; App. to the appendices above.

Index of Words
Index of Organisms, Enzymes and Diseases

Numbers refer to chapters. Enzymes are listed by their common names; EC-class logic is developed in Chapter 14.

Organisms

Enzymes

Diseases

www.ingramcontent.com/pod-product-compliance
Lightning Source LLC
LaVergne TN
LVHW052012200926
848481LV00005B/1474
* 9 7 9 8 9 0 6 6 8 0 1 0 5 *